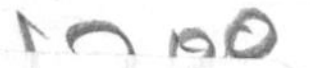

Mathematics at Work

FOURTH EDITION

Mathematics at Work

Practical Applications of Arithmetic, Algebra, Geometry, Trigonometry and Logarithms to the Step-by-Step Solutions of Mechanical Problems, with Formulas Commonly Used in Engineering Practice and a Concise Review of Basic Mathematical Principles.

by

Holbrook L. Horton

Edited by

Henry H. Ryffel
Edward E. Messal
Robert E. Green

INDUSTRIAL PRESS, INC.
200 Madison Avenue, New York, New York 10016

Library of Congress Cataloging-in-Publication Data

Horton, Holbrook Lynedon, 1907-
Mathematics at work : practical applications of arithmetic, algebra, geometry, trigonometry, and logarithms to the step-by-step solutions of mechanical problems, with formulas commonly used in engineering practice and a concise review of basic mathematical principles / by Holbrook Horton. – 4th ed. / edited by Henry H. Ryffel, Edward E. Messal, Robert E. Green.
688 p. 153x22.9 cm.
Includes index.
ISBN 0-8311-3083-0
1. Mathematics. 2. Mechanics, Applied Problems, exercises, etc.
I.Ryffel, Henry H. II, Messal, Edward E. III. Green, Robert E. (Robert Edward), 1932- . IV. Title.
QA39.2.H66 1999
620.1'001'51—dc21 99-14606
CIP

Industrial Press Inc.
200 Madison Avenue
New York, New York 10016-4078

Sponsoring Editor: John Carleo
Project Editor: Sheryl Levart

MATHEMATICS AT WORK—FOURTH EDITON

16 15 14 13 12 11 10 9 8 7 6 5 4 3 2

Some Preliminary Information About this Manual

The idea of a mechanical problem manual came into being as a result of the many letters of inquiry received by the editors of MACHINERY'S HANDBOOK, asking for assistance in solving specific problems which were basically mathematical in character and relating to mechanical work. These requests had to do with all types of problems. Some had proved difficult because of a forgotten or misunderstood mathematical rule or principle. Others were more complex and in some cases considerable ingenuity was required to find a solution.

It became evident that if a selected number of such problems were presented in book form to show in detail just how fundamental mathematical principles and methods of solution are applied, it would be of great help to those working in the shop and drafting room. Such a presentation would not merely "point the way" to solving each one in step-by-step fashion.

Selection of Problems. The problems which have been selected are, for the most part, those likely to be encountered in practical mechanical work. A few problems of a more theoretical nature have also been included to illustrate certain principles or methods which may be widely applied. Since there are an endless number and variety of mechanical problems, it was obviously impossible to include examples of every type. Hence, many mechanical problems may be encountered which are not similar to any presented in this manual. However, it is believed that the basis of classification will, in many cases, permit finding a solution, even when the problem itself is quite different from any in the manual.

Classification of Problems. The classification of problems which has been used is a mathematical one; that is, problems illustrating some common mathematical principle or method have been grouped together. A glance at the Table of Contents will show the subjects under which the problems have been grouped for convenient reference. When looking up information about solving a given problem, the mathematical principle which forms the basis of the problem or the method of solution which seems to be called for should be used as the reference key. This arrangement makes it possible to apply the manual to a much broader range of problems than might be apparent at first.

Method of Presentation. Each problem has been presented in such a way as to show: (1) what the problem is about; (2) how to analyze it and develop a method of attack and solution; (3) what formula is required, if a formula is applicable; (4) how this formula is derived or, if no formula is given, what the step-by-step procedure is for mathematically solving the problem; and (5) how a typical example is worked out.

Thus, if the reader has an exactly similar problem and is interested only in the solution, and not in how the solution is obtained, he can take the formula given, insert his own particular data into it and solve his problem directly. If, however, he is interested in how the formula was derived, he can follow its derivation through in a step-by-step fashion. Possibly, he may be in doubt as to just how the problem should be analyzed. After reading the analysis and having the method of attack clearly in mind, he may wish to go ahead from that point "on his own," in deriving a formula and applying it to the solution of his problem. The formula and derivation which are given in the text may then be used for a comparison of his method of derivation and a check on its accuracy.

An attempt has been made to present each problem in such a manner as to duplicate, as far as possible, personal instruction. In order to accomplish this, the analysis, derivation and complete solution are carried through step-by-step. Questions likely to arise have been answered, where practicable, and doubtful points clarified by further explanations and examples.

Fundamentals of Mathematical Subjects. As a general rule, the solutions of many problems are clarified when the fundamentals on which they were based are made available for quick and easy reference. With this in mind, a digest of certain important elements in arithmetic, algebra, geometry, trigonometry and logarithms is given in the first five chapters; but since this is essentially a practical working manual, no attempt has been made to provide an exhaustive treatment of any of these subjects. Enough has been presented, however, to make these five chapters, alone, a valuable source of reference.

Aids in Computation. Methods of computation, which have been the source of frequent inquiries, and the chapters relating to them should be of general interest to all who are engaged in the solving of practical problems by mathematics. In many instances, for example, the solution of a given problem can be worked out satisfactorily by applying an approximate formula with considerable less work involved than if an exact formula were employed. Chapter 16 explains approximate formulas and tells why and how they are used. Chapter 20 tells how to find a ratio which approximates a given ratio to any degree of accuracy required.

Some problems cannot be solved readily by exact mathematical formulas. In such cases a practical approach is to work out a trial-and-error solution. Chapter 22 explains how this is done and gives examples of the procedure.

Empirical Formulas. As many users of this book realize, empirical formulas are intended to be practical even though they may not strictly conform to any mathematical theory or derivation. In certain cases, workers in shops or drafting rooms have claimed that a particular mathematical formula, which turned out to be an empirical one, was either incorrect or could not be derived mathematically. Chapter 24 gives the justification for such "working formulas" and explains how they are developed and applied.

Handling of Errors. In interchangeable manufacture, as is well known, it is recognized that attempts to achieve perfection in the measurement of machine parts would be impracticable; consequently certain tolerances or allowable errors are specified, and then means are provided for keeping such tolerances within the prescribed limits. Similarly, in practically all mathematical calculations some approximate values are allowed because perfection either is impossible or does not serve any practical purpose. Chapter 23 points out salient facts about allowable errors and the systematic handling of them in practical computations.

Questions and Answers. The chapter on refresher questions in mathematics, mechanics, and strength of materials is based on inquiries about troublesome definitions and relationships in these fields. A review of this chapter should be of considerable value in clarifying one's ideas about the fundamental points covered.

Tables of Working Data. In the last section of the book there are a number of tables selected to meet the practical needs of those who are working with mathematical problems. These include a table of mathematical signs and abbreviations; a table of mensuration formulas for all types of geometrical figures; a table of prime numbers; a table of fractions and their decimal equivalents; a table of natural logarithms; a table of trigonometric functions; and a table of weights and measures. Various other tables of useful data will be found throughout the book.

Taken together, the review of mathematical fundamentals, the comprehensive discussion of problems and their solutions, the explanation of special aids in computation, and the many tables of useful data were planned to make this manual an indispensable working tool for all who are engaged in the mathematical solution of mechanical problems.

Use of Calculators. In addition to using this manual to facilitate understanding of the methods of mathematics in industry, the reader may benefit from practice in the use of a handheld calculator to rework some of the problems for which detailed solutions are provided. This is particularly appropriate for those examples that use logarithms and trigonometric functions in the solution of a problem and for those problems that involve finding roots, reciprocals, and powers of numbers.

Contents

Introduction

Basic Mathematical Principles

Right- and Oblique-angled Triangle Problems

Problems Relating to Tapers and Angles

Problems Involving Arcs, Circles and Vees

Use of Approximate Formulas

Problems in Mechanics and Strength of Materials

Gear Ratio Problems

Methods and Formulas for Special Conditions

Miscellaneous Problems and Refresher Questions

Mathematical Tables

Index

1

Arithmetic—Some Useful Elements You May Have Forgotten

1. FACTORS AND FACTORING

Factors. The factors of a number are those numbers whose continued product is the number. Thus, 3, 3 and 2 are factors of 18 since $3 \times 3 \times 2 = 18$.

Factoring. Factoring is the process of finding the factors of a number. The following general principles will be of service:

(a) 2 is a factor of any number whose right-hand figure is an even number or zero. Thus, $28 = 2 \times 14$; $10 = 2 \times 5$.

(b) 3 is a factor of any number the sum of whose digits is divisible by 3. Thus, in the number 1869, $1 + 8 + 6 + 9 = 24$; $\frac{24}{3} = 8$; $1869 = 3 \times 623$.

(c) 4 is a factor of any number when the number expressed by the two right-hand digits of the given number is divisible by 4. Thus, in the number 1844, $\frac{44}{4} = 11$ and $1844 = 4 \times 461$.

(d) 5 is a factor of any number whose right-hand digit is 0 or 5. Thus, $10 = 5 \times 2$; $15 = 5 \times 3$.

(e) 6 is a factor of any even number which is divisible by 3. Thus, $\frac{360}{3} = 120$ and $360 = 6 \times 60$.

(f) 8 is a factor of any number when the number expressed by the three right-hand digits of the given number is divisible by 8. Thus, in the number 1640, $\frac{640}{8} = 80$ and $1640 = 8 \times 205$.

(g) 9 is a factor of any number when the sum of the digits of that number is divisible by 9. Thus, the sum of the digits in the number 1899 is 27, $\frac{27}{9} = 3$, and $1899 = 9 \times 211$.

Prime Number. A number which has no factors is a prime number, as 1, 3, 5, etc. *Note:* Every prime number but 2 and 5 has 1, 3, 7, or 9 for its unit or right-hand figure.

Prime Factors. The prime factors of a number are those prime numbers whose continued product is the number. Thus, the prime factors of 28 are 2, 2, and 7.

How to Use Table of Prime Numbers and Factors. A table of prime numbers and prime factors of numbers that are not prime up to 9600 is given in Chapter 27. This table can be used to find all the prime factors of a given number as follows:

1. Look up the number in the table. If the number is not itself prime, as indicated by the letter P, a single prime factor will be given.
2. Divide the number by this prime factor.
3. Take the resulting quotient and referring to the table, determine if it is prime.
4. If not, divide this quotient by the prime factor given for it.
5. Check to see if this second quotient is a prime number.
6. If not, divide it by the prime factor given.
7. Continue in this way until the resulting quotient is found to be a prime number.

The prime factors taken successively from the table with the last quotient, which is also prime, constitute the prime factors of the original number.

Example: Find the prime factors of 2067. Referring to the table, 3 is a prime factor of 2067. $2067 \div 3 = 689$. Referring to the table, 13 is a prime factor of 689. $\frac{689}{13} = 53$. Referring to the table, 53 is a prime number. Hence the prime factors of 2067 are 3, 13, and 53.

Numbers Prime to Each Other. Two or more numbers are said to be prime to each other when they have no common factor except 1. Numbers which are not themselves prime may be prime to each other. Thus, 22, 13, and 9 are prime to each other but 22 and 9 are not prime numbers.

Cancellation is the striking out of equal factors from the numerator and denominator of a fraction. This operation does not change the value of the fraction but aids in carrying out the division indicated.

Example: Multiply 819 by 323 and divide this product by the product of 741 by 119. From the table of prime factors in Chapter 27 it will be found that:

$$819 = 3 \times 3 \times 7 \times 13 \qquad 741 = 3 \times 13 \times 19$$

$$323 = 17 \times 19 \qquad 119 = 7 \times 17$$

Hence $$\frac{819 \times 323}{741 \times 119} = \frac{\cancel{3} \times 3 \times \cancel{7} \times \cancel{13} \times \cancel{17} \times \cancel{19}}{\cancel{3} \times \cancel{13} \times \cancel{19} \times \cancel{7} \times \cancel{17}} = 3$$

2. DIVISORS AND MULTIPLES

Greatest Common Divisor or Highest Common Factor. The greatest common divisor or highest common factor of two or more numbers is the *largest* number which will divide each of them without a remainder. It is found by taking the product of the prime factors common to all of the numbers. *Note:* If any prime factor occurs more than once in *each* of the numbers, it is set down as many times as it occurs in *each* number to find the product equal to the greatest common divisor.

Example: Find the greatest common divisor of 24, 48 and 60.

The prime factors of these three numbers are as follows:

$$24 = 2 \times 2 \times 3 \times 2$$

$$48 = 2 \times 2 \times 3 \times 2 \times 2$$

$$60 = 2 \times 2 \times 3 \times 5$$

Note that the factor 2 occurs twice and the factor 3 once in each of these numbers. Hence the greatest common divisor is:

$$2 \times 2 \times 3 = 12$$

Where only two numbers are involved, the greatest common divisor can be found by dividing the greater number by the lesser. If there is a remainder, the previous divisor is divided by it. If there is again a remainder, the second divisor is divided by it. This process of dividing the last divisor by the last remainder is continued until the remainder is zero. The final divisor will be the greatest common divisor.

One of the uses of the greatest common divisor is to reduce a common fraction to its lowest terms as is explained under 3. FRACTIONS.

Example: Find the greatest common divisor of 1316 and 91.

```
              14                       2                   6
Step 1  91)1316        Step 2  42)91        Step 3  7)42
           91                      84
           406                      7
           364
            42
```

Therefore 7 is the greatest common divisor.

Least Common Multiple. The least common multiple of two or more numbers is the *smallest* number that is divisible by each of the given numbers. It is found by taking the product of the different prime factors of the numbers, each factor being taken the greatest number of times that it occurs in *any* of the given numbers.

Example: Find the least common multiple of 24, 48 and 60. The prime factors of these three numbers are as follows:

$$24 = 2 \times 2 \times 3 \times 2$$

$$48 = 2 \times 2 \times 3 \times 2 \times 2$$

$$60 = 2 \times 2 \times 3 \times 5$$

Note that the greatest number of times that factor 2 occurs in any of the numbers is four; the greatest number of times that factor 3 appears is once and the greatest number of times that factor 5 appears is once. Hence the least common multiple is

$$2 \times 2 \times 2 \times 2 \times 3 \times 5 = 240$$

Another method of finding the least common multiple of several numbers is to take some prime number which is a factor of two or more of them. This factor is used as a divisor for those numbers which are evenly divided by it. The quotients so obtained and the undivided numbers are again set down and another (or the same) prime number which is a factor of two or more of them is taken as a divisor. This process is repeated with the quotients and the undivided numbers until no two of them can be factored by a prime number. The remaining numbers are then said to be prime to each other. The least common multiple is the product of all of the divisors and the remaining numbers.

Example: Find the least common multiple of 13, 30, 36 and 54.

Step 1. Divide by the prime factor 2.

13 (not divisible) $\frac{30}{2} = 15$ $\frac{36}{2} = 18$ $\frac{54}{2} = 27$

Step 2. Divide by the prime factor 3

13 (not divisible) $\frac{15}{3} = 5$ $\frac{18}{3} = 6$ $\frac{27}{3} = 9$

Step 3. Divide by the prime factor 3

13 5 $\frac{6}{3} = 2$ $\frac{9}{3} = 3$

Step 4. Multiply the successive divisors and the remaining numbers

$$2 \times 3 \times 3 \times 13 \times 5 \times 2 \times 3 = 7020$$

which is the least common multiple.

A more compact arrangement of the successive steps can be set down as follows:

2	13,	30,	36,	54
3	13,	15,	18,	27
3	13,	5,	6,	9
	13,	5,	2,	3

Least common multiple $= 2 \times 3 \times 3 \times 13 \times 5 \times 2 \times 3 = 7020$

3. FRACTIONS

A Common Fraction, sometimes called a *vulgar fraction*, is simply an expression of division. Thus, $\frac{3}{4}$ means that 3 is to be divided by 4. The result of this division is the equivalent decimal fraction, 0.75.

To Reduce a Common Fraction to Its Lowest Terms, divide the numerator and denominator by their greatest common divisor. Thus, for the fraction $\frac{42}{826}$, the greatest common divisor is 14 and the fraction becomes $\frac{3}{59}$.

To Reduce a Common Fraction to Higher Terms, multiply both numerator and denominator by any given number. The value of the fraction remains unchanged. Thus

$$\frac{4}{5} \times \frac{3}{3} = \frac{12}{15}; \qquad \frac{8}{9} \times \frac{7}{7} = \frac{56}{63}.$$

A Proper Fraction is one in which the numerator is less than the denominator. Thus, $\frac{4}{23}$ is a proper fraction.

An Improper Fraction is one in which the numerator is greater than the denominator. Thus, $\frac{42}{9}$ is an improper fraction.

A Complex Fraction is one having a fraction or a mixed number in the numerator or denominator.

A Mixed Number is a number consisting of a whole number and a fraction.

To Reduce a Mixed Number to an Improper Fraction, multiply the whole number by the denominator of the fraction, add the numerator to this product, and, under the sum, write the denominator. Thus, the mixed number $13\frac{5}{8}$ becomes $\frac{13 \times 8 + 5}{8} = \frac{109}{8}$. The reduction of an improper fraction to a mixed number is accomplished by actual division of the numerator by the denominator. Thus

$$\frac{109}{8} = 109 \div 8 = 13\tfrac{5}{8}.$$

Addition and Subtraction of Fractions. Only like quantities can be added or subtracted. Hence, for these operations the fractions must be reduced to a common denominator, which is preferably the least common multiple of all of the denominators. Thus, to add the fractions $\frac{3}{4}$, $\frac{4}{5}$, and $\frac{5}{8}$, they are reduced to the least common denominator which is 40. Then, $\frac{3}{4} = \frac{30}{40}$, $\frac{4}{5} = \frac{32}{40}$, $\frac{5}{8} = \frac{25}{40}$, the sum of which is $\frac{87}{40}$. Again, to subtract $\frac{5}{10}$ from $\frac{6}{7}$, the denominators are reduced to 70, thus

$$\frac{6}{7} - \frac{5}{10} = \frac{60}{70} - \frac{35}{70} = \frac{25}{70} = \frac{5}{14}.$$

Multiplication and Division of Fractions. To multiply fractions, find the product of the numerators for a new numerator and that of the denominators for a new denominator. To divide fractions, invert the divisor and multiply as above. Thus,

$$\frac{1}{2} \times \frac{3}{4} \times \frac{5}{8} = \frac{1 \times 3 \times 5}{2 \times 4 \times 8} = \frac{15}{64}$$

$$\frac{2}{3} \div \frac{5}{7} = \frac{2}{3} \times \frac{7}{5} = \frac{14}{15}$$

$$\frac{15}{19} \times \frac{3}{11} \div \frac{4}{5} = \frac{15}{19} \times \frac{3}{11} \times \frac{5}{4} = \frac{225}{836}$$

4. DECIMALS

The principles and operations of common fractions are fully applicable to decimal fractions, since a decimal is equivalent to a common fraction whose denominator is 10, 100, or 1000, etc. If these are to be applied, the decimal fraction is first converted to its common fractional equivalent.

A table of fractions and their decimal equivalents is given in Chapter 27.

Addition and Subtraction. These operations are performed as with whole numbers, except that the decimal fractions are so placed that the decimal point in one is below that in the other. Thus,

$$\begin{array}{r} 0.8375 \\ +\ \underline{0.691} \\ 1.5285 \end{array} \qquad\qquad \begin{array}{r} 0.8375 \\ -\ \underline{0.691} \\ 0.1465 \end{array}$$

Multiplication and Division are performed as if the numbers were whole numbers. Then, in multiplication, as many decimal places are pointed off in the product as there are in both factors together, and, in division, the number of decimal places in the divisor is subtracted from those in the dividend to find the number of decimal places in the quotient, enough ciphers having been added to the dividend to make its decimal places equal to or greater than those in the divisor. Thus, $3.53 \times 0.6 = 2.118$ and $3.5 \div 0.05 = 3.50 \div 0.05 = 70$ or $3.500 \div 0.05 = 70.0$.

Reduction to Decimals. To reduce a common fraction to a decimal, annex a decimal point and one or more ciphers to the numerator and divide by the denominator. Point off as in decimal division. Thus, $\frac{3}{8} = 3.000 \div 8 = 0.375$; $1\frac{3}{8} = \frac{11}{8} = 11.000 \div 8 = 1.375$.

Reduction from Decimals. A decimal is equivalent to a common fraction whose denominator is 10 or multiples of 10. To reduce a decimal to a common fraction in its simplest form, write in the denominator and reduce the resultant fraction to its lowest terms. Thus, $0.125 = \frac{125}{1000} = \frac{1}{8}$.

Circulating Decimals. In circulating or recurring decimals, the same figure or group of figures is repeated indefinitely. Thus, $\frac{2}{3} = 0.6666$, etc., and $\frac{3}{11} = 0.272727$, etc. For compactness, the group in a circulating decimal may be written but once and a dot is placed over the first and last figure. Thus, $\frac{1}{3} = 0.\dot{3}$ and $\frac{7}{37} = 0.\dot{1}8\dot{9}$. To multiply or divide one circulating decimal by another, first reduce them both to common fractions.

To Reduce a Circulating Decimal to a Common Fraction, use for the numerator the figures of the entire decimal, minus those figures in it which do not repeat, and, for the denominator, as many nines as there are repeating figures, followed by as many ciphers annexed as there are decimal places before the repeating figures. Thus, in the decimal 0.76354, the circulating part of the decimal is (354), hence

$$0.76354 = \frac{76354 - 76}{99900} = \frac{76278}{99900} = \frac{12713}{16650}$$

5. COMPOUND NUMBERS

Compound Number. A compound number is a number consisting of various denominations, or more than one unit. Thus, 9 pounds 6 ounces; 14 yards 2 feet 4 inches.

To Change a Compound Number from a Higher to a Lower Denomination. Multiply the given number by the number of units of the lower denomination required to make one unit of the higher denomination. Thus,

$$1.89 \text{ feet} = 1.89 \times 12 \text{ inches} = 22.68 \text{ inches}$$
$$2.35 \text{ square feet} = 2.35 \times 144 \text{ square inches} = 338.4 \text{ square inches}$$

To Change a Compound Number from a Lower to a Higher Denomination. Divide the given number by the number of units of the lower denominations which are required to make one unit of the higher denomination. Thus, 150 square inches = 150 ÷ 144 square feet = 1 square foot 6 square inches = $1\frac{6}{144}$ square feet = 1.04 square feet.

A table of weights and measures, useful in the conversion of compound numbers to higher and lower denominations, will be found in Chapter 27.

6. RECIPROCALS

The Reciprocal of any number, as 2, is equal to one divided by that number, or 1/2. The reciprocal of a fraction, as 3/4, is the fraction inverted, since $1 \div \frac{3}{4} = 1 \times \frac{4}{3} = \frac{4}{3}$.

The use of reciprocals often saves labor in long division, as the quotient can be obtained by multiplying the dividend by the reciprocal of the divisor. Thus, the reciprocal of 244 is 0.0041 to four decimal places. To divide 13 by 244 or to reduce $\frac{13}{244}$ to a decimal, we have 13 × 0.0041 = 0.0533.

7. RATIO AND PROPORTION

Ratio is the relation of one quantity to another of the same kind, in respect to magnitude. This relation is the quotient obtained by dividing the first quantity by the second. The first quantity is called the *antecedent*, the second is the *consequent*, and the two form the *terms* of the ratio. Thus,

$$\text{ratio of 6 to 3} = 6 : 3 = \tfrac{6}{3} = 2$$
$$\text{ratio of 3 to 6} = 3 : 6 = \tfrac{3}{6} = \tfrac{1}{2}$$

When a ratio is stated as a fraction, the principles of the multiplication and division of fractions, as already given, apply to it.

A Reciprocal or Inverse Ratio is the reciprocal of the original ratio. Thus, the inverse ratio of 5 : 7 is 7 : 5.

In a **Compound Ratio,** each term is the product of the corresponding terms of two or more simple ratios. Thus,

$$\begin{array}{r} 8 : 2 = 4 \\ 9 : 3 = 3 \\ \underline{10 : 5 = 2} \\ 8 \times 9 \times 10 : 2 \times 3 \times 5 = 4 \times 3 \times 2 \\ 720 : 30 = 24 \end{array}$$

Proportion is an equality of ratios. Thus, 6 : 3 = 10 : 5, or 6 : 3 : : 10 : 5, or $\frac{6}{3} = \frac{10}{5}$.

The **Extremes** are the first and last terms of a proportion.

The **Means** are the second and third terms of a proportion.

The **Proportionals** are all four terms of a proportion.

Rules Governing Proportions. The product of the means equals the product of the extremes. Thus in the proportion 25 : 2 = 100 : 8, the product 25 × 8 equals the product 2 × 100.

If three terms in a proportion are known, the remaining term may be found by one of the following rules:

(1) The first term is equal to the product of the second and third terms divided by the fourth.

(2) The second term is equal to the product of the first and fourth terms divided by the third.

(3) The third term is equal to the product of the first and fourth terms divided by the second.

(4) The fourth term is equal to the product of the second and third terms divided by the first.

Examples: Let x be the term to be found, then:

(1) $x : 14 = 4 : 7$ $\qquad x = \dfrac{14 \times 4}{7} = 8$

(2) $4 : x = 6 : 17$ $\qquad x = \dfrac{4 \times 17}{6} = 11\dfrac{1}{3}$

(3) $\frac{1}{2} : 6 = x : \frac{3}{8}$ $\qquad x = \dfrac{\frac{1}{2} \times \frac{3}{8}}{6} = \dfrac{3}{96} = \dfrac{1}{32}$

(4) $5 : 11 = 19 : x$ $\qquad x = \dfrac{11 \times 19}{5} = 41\dfrac{4}{5}$

Mean Proportional. If the second and third terms in a proportion are the same, either is said to be the mean proportional between the first and the fourth terms. Thus, in the proportion $9 : 3 = 3 : 1$, 3 is the mean proportional between 9 and 1.

Finding the Mean Proportional between Two Numbers is accomplished by multiplying the two numbers together and extracting their square root. Thus, the mean proportional between 27 and 3 is found as follows: $27 \times 3 = 81$, $\sqrt{81} = 9$. Thus 9 is the required mean proportional.

Third Proportional. When the second and third terms of a proportion are the same, the fourth term is the third proportional to the first term and the second, or to the first term and the third. Thus, in the proportion $8 : 4 = 4 : 2$, 2 is the third proportional to 8 and 4.

Reciprocally Proportional. Two quantities are said to be reciprocally proportional when one is proportional to the reciprocal of the other. Thus, algebraically, in the proportion $x : 1 = 1 : y$, $xy = 1$; therefore, $x = \dfrac{1}{y}$ and $y = \dfrac{1}{x}$. Thus x and y are reciprocally proportional to each other, so that if x increases, y decreases proportionally, and vice versa. In other words, x increases proportionally with the reciprocal of y and y with the reciprocal of x.

8. POWERS

A Power of a number is the product obtained by multiplying the number by itself a given number of times.

Involution is the process of raising a number to a given power.

The Index or Exponent indicates the power of a given number or the number of times the given number is thus used as a factor. It is written as a figure placed to the right and above the number. Thus, $4 \times 4 = 16 = 4^2$, 16 is the second power of 4, and 2 is the exponent of that power.

To Multiply Two Powers of the Same Number, add their exponents, thus: $4^2 \times 4^3 = 4^{(2+3)} = 4^5 = 1024$.

To Divide One Power by Another, both of the same number, subtract the exponent of the divisor from that of the dividend, thus: $5^6 \div 5^2 = 5^{(6-2)} = 5^4 = 625$.

9. ROOTS

A Root is one of the equal factors into which a number can be resolved, exactly or approximately. Again, a root of a number is one of the equal factors whose continued product is the number. Thus, $4 \times 4 = 16$, and 4 is the square root of its own second power; it is also the cube root of 64, its third power, and so on. Hence, any given number is the root of its various powers.

Evolution is the process of thus resolving the number into these equal factors, the number of which is shown by the exponent of the root.

Roots are indicated by either the radical sign, as $\sqrt{2}$ = square root of 2, $\sqrt[3]{2}$ = cube root of 2, etc., or by fractional exponents, as $2^{1/2} = \sqrt{2}$ and $2^{2/3} = \sqrt[3]{2^2}$, in which the denominator is the exponent of the root and the numerator that of the power whose root is to be extracted. A negative sign before the exponent indicates the reciprocal. Thus,

$$2^{-1/2} = \frac{1}{2^{1/2}} = \frac{1}{\sqrt{2}} \quad \text{and} \quad \frac{1}{2^{-1/2}} = \frac{2^{1/2}}{1} = 2^{1/2} = \sqrt{2}.$$

Fractional Exponents often aid in simplifying complex expressions. Thus,

$$\sqrt[3]{2^9} = 2^{9/3} = 2^3 = 8$$

and $$4^{1/6} = 4^{1/2 \times 1/3} = \sqrt[3]{4^{1/2}} = \sqrt[3]{\sqrt{4}} = \sqrt[3]{2} = \text{also } 2^{1/3}$$

Square Root. A number may be considered as the square of either of two *equal* factors. If these equal factors are *entire*, the square root can be found *exactly;* if the equal factors are not entire, the root found is *approximate*, the degree of exactness depending on the number of decimal places. Thus, 64 is a *perfect square* whose exact square root is 8, and 50 is an *imperfect square* whose approximate square root is 7.0711.

To Extract the Root of a Fraction, extract the root of the numerator for a new numerator and that of the denominator for a new denominator. Thus,

$$\sqrt[3]{\frac{27}{64}} = \frac{\sqrt[3]{27}}{\sqrt[3]{64}} = \frac{3}{4}.$$

It is sometimes more convenient to reduce the fraction to a decimal first. Thus,

$$\sqrt[3]{\frac{27}{64}} = \sqrt[3]{0.4219} = 0.75.$$

Surd. A surd is a root which cannot be extracted exactly, i.e., it is an irrational number. Thus, $\sqrt{2}$ is a surd.

Approximate Roots. If a given number is not a perfect square, its square root cannot be found exactly. The simplest way of obtaining the square root to any required degree of accuracy, is to annex ciphers to the number and then extract the root. Thus, $\sqrt{5.0000} = \sqrt{5}$ to within 1/100. In general, to extract the nth root of a whole number to within $\frac{1}{p}$, multiply the number by p^n, extract the nth root of the product, and divide this root by p. The result will be the required approximate root. Thus, let the cube root of 5 to within $\frac{1}{100}$ be required. Here, $n = 3$, $p = 100$, and the required approximate root is $\frac{(\sqrt[3]{5 \times 100^3})}{100}$. The same principles apply to a mixed number, such as 3.1416, and to a decimal fraction. When common fractions are reduced to decimals, this method can be used.

10. POSITIVE AND NEGATIVE NUMBERS

Positive and Negative Numbers are numbers which have relative values.

The Relative Value of a number is its value with relation to some established reference value, usually 0. Thus, on a thermometer scale, all the degrees above the 0 point are *plus* or *positive* and all the degrees below the 0 point are *minus* or *negative.*

The relative value $+3$, then, is 3 units more than, or to the right of, or above, or in some arbitrary direction on a scale with relation to 0, and the relative value -3 is 3 units less than, or to the left of or below or in some arbitrary direction on a scale with relation to 0.

The Absolute Value of a number is its value regardless of its sign. Thus, $+3$ and -3 have the same absolute value of 3.

Note: Negative numbers should always be enclosed within parentheses when written in line with other numbers. Thus, $17 + (-13) - 3 \times (-0.76)$. In this particular example, 3 is a positive number, the minus sign before it indicating that it is to be subtracted from the number preceding it.

Rules for Adding and Subtracting Positive and Negative Numbers. (1) When adding two or more positive or two or more negative numbers, their sum is equal to the sum of their absolute values and has the same sign as that of the numbers being added.

Example:

$$2 + 8 + 10 = 20$$

$$(-3) + (-6) + (-4) = -13$$

(2) When adding a positive and a negative number, their sum is equal to the difference of their absolute values and has the same sign as the number having the larger absolute value, thus:

Example:

$$-6 + 2 = -4$$

$$9 + (-3) = 6$$

(3) When adding several positive and negative numbers, first add the positive and negative numbers separately and then add their respective sums. Follow Rule 2 for this last operation.

Example:

$$4 + (-6) + 8 + (-2) = 12 + (-8) = 4$$

(4) When subtracting one positive number from another positive number or one negative number from another negative number, the remainder is equal to the difference of their absolute values and has the same sign as the numbers being subtracted if the minuend has a larger absolute value than the subtrahend but the opposite sign if the minuend has a smaller absolute value than the subtrahend.

Examples:

$$8 - 10 = -2$$

$$(-6) - (-4) = -2$$

$$(-6) - (-9) = 3$$

(5) When subtracting a negative number from a positive number or vice versa, the remainder is equal to the sum of their absolute values and has the same sign as the minuend.

Examples:

$$8 - (-2) = 10$$

$$(-2) - 3 = -5$$

Multiplication and Division of Positive and Negative Numbers.
(6) When multiplying two positive or two negative numbers together, their product is equal to the product of their absolute values and is positive.

Examples:

$$18 \times 16 = 288$$

$$(-9) \times (-27) = 243$$

(7) When multiplying a positive and a negative number together, their product is equal to the product of their absolute values and is negative.

Example: $182 \times (-16) = -2912$

(8) When dividing a positive number by a positive number, or a negative number by a negative number, their quotient is equal to the quotient of their absolute values and is positive.

Example:

$$196 \div 28 = 7$$

$$(-1064) \div (-76) = 14$$

(9) When dividing a positive number by a negative number or vice versa, their quotient is equal to the quotient of their absolute values and is negative.

Example:

$$5190 \div (-346) = -15$$
$$-2698 \div 19 = -142$$

11. SPECIAL CASES OF MULTIPLICATION AND DIVISION

Meaning of Dividing a Number by Zero. If any number is divided by zero, the result is infinity, thus:

$$\frac{1}{0} = \infty \qquad \frac{26}{0} = \infty \qquad \frac{3798}{0} = \infty$$

The reason why this is so can be shown by taking some number and dividing it by successively smaller and smaller decimals, thus:

$$\frac{95}{0.1} = 950; \qquad \frac{95}{0.01} = 9500; \qquad \frac{95}{0.0001} = 950{,}000;$$

and

$$\frac{95}{0.0000001} = 950{,}000{,}000$$

It will be noted that as the divisor becomes smaller and smaller, the quotient becomes larger and larger. It can be inferred, therefore, that as the divisor becomes increasingly small and approaches zero, the quotient will become increasingly large and approach infinity. The end result is when the divisor is infinitely small or equal to zero, then the quotient will become infinitely large or equal to infinity.

Meaning of Multiplying a Number by Zero. If any number is multiplied by zero, the result is zero, thus:

$$4 \times 0 = 0; \qquad 2{,}567 \times 0 = 0; \qquad \text{and} \qquad 987{,}691 \times 0 = 0$$

The reason why this is so can be shown by taking some number and multiplying it by successively smaller and smaller numbers, thus:

$$457 \times 0.1 = 45.7; \qquad 457 \times 0.001 = 0.457;$$

$$457 \times 0.00001 = 0.00457; \quad \text{and} \quad 457 \times 0.00000001 = 0.00000457$$

It will be noted that as the multiplier becomes smaller and smaller, the product becomes smaller and smaller also. It can be inferred, therefore, that as the multiplier becomes increasingly small and approaches zero, the product will also become increasingly small and approach zero. The end result is when the multiplier is infinitely small or equal to zero, then the product will become infinitely small and equal to zero also.

Meaning of Factorials. The expression 6! is read *six factorial* and equals $1 \times 2 \times 3 \times 4 \times 5 \times 6$. In other words, the factorial sign ! means multiply the integer after which it appears by every integer which is less than that integer. The product of a factorial is called a *continued product.*

The factorial $N!$ is equal to 1 when $N=0$, by definition.

Proof:

Let $N=1$. Then $N! = N(N-1)! = 1 \times (1-1)! = 1 \times 0!$. Thus, for the identity $N! = N(N-1)!$ to be true, 0! must be equal to 1.

2

Algebra—The Shorthand of Mathematics

1. DEFINITIONS

An Algebraic Expression consists of letters, numbers and connecting signs which show the operations to be performed. Thus $7ax$ means 7 times a times x and $ax - x^2$ indicates that the square of x is to be subtracted from the product of a and x.

A Coefficient is the number, letter or expression which shows how many times the quantity which follows is to be taken additively. Thus, in $2x$, 2 is a coefficient of x and indicates $x + x$. Similarly, in $2ax^2$, $2a$ is the coefficient of x^2 and indicates that x^2 is to be taken $2a$ times.

An Exponent is a small number or letter written above and to the right of a quantity, which shows how many times the quantity is to be taken as a factor. Thus, in the term a^4, 4 is the exponent of a and indicates that a is to be taken as a factor four times or $a \times a \times a \times a$.

A Term is that part of an expression which either stands by itself or is separated from other terms by plus or minus signs. Thus, in $ab - bc$, ab and bc are separate terms; in $4a^2c + 3b^3c$, $4a^2c$ and $3b^3c$ are separate terms.

A Monomial is an expression consisting of but one term.

A Binomial is an expression consisting of but two terms.

A Trinomial is an expression consisting of three terms.

A Polynomial is any expression consisting of two or more terms.

Like Terms or **Similar Terms** are terms in which the elements other than the coefficients are alike. Thus, $8a^2b^2$, $16a^2b^2$, $9a^2b^2$ are like terms.

Unlike Terms or **Dissimilar Terms** are terms in which those elements other than the coefficients are unlike. $8a^2b^2$, $6ab^3$, $9ac^2$ are all unlike terms.

2. FUNDAMENTAL OPERATIONS

Addition. The Algebraic Sum of like terms having the same sign, either + or −, is their arithmetical sum. It is their *net* sum, when some of the terms are + and some −. Hence, if the terms are *like* and have the same sign, find their absolute sum and prefix this sign; but, if the signs are *unlike*, find the difference between the + and − terms and prefix the sign of the greater of the two. Thus, $a + 3a + 5a = 9a$ and $a + 3a - 6a = -2a$.

To add polynomials, place like terms, prefixed by their proper signs, in the same column and proceed as above. Thus,

$$\begin{array}{r} a^2 + 4ab + 4b^2 \\ a^2 - 4ab + 4b^2 \\ \hline 2a^2 \qquad\quad + 8b^2 \end{array}$$

Subtraction. To subtract one algebraic quantity or expression from another, change the signs of the subtrahend and proceed as in addition. Thus, to subtract $a^2 - 2ab + b^2$ from $a^2 + 2ab + b^2$,

$$\begin{array}{c} a^2 + 2ab + b^2 \\ -a^2 + 2ab - b^2 \\ \hline 4ab \end{array}$$

When brackets preceded by a minus sign are removed to simplify an expression, the signs of the terms enclosed by them must be changed. Thus, $a - (b - c) = a - b + c$.

Multiplication. *The Law of Signs* in multiplication is that if two terms are multiplied together, their product will have a plus sign when they have like signs and their product will have a minus sign when they have unlike signs. Thus,

$$a \times b = ab; \qquad (-a) \times (-b) = ab$$

$$(-a) \times b = -ab; \qquad a \times (-b) = -ab$$

The Law of Exponents in multiplication is that the exponent of the product of two or more powers of a given quantity equals the *sum* of the exponents of the several factors. Thus,

$$a^2b \times ab^3 = a^{(2+1)}b^{(1+3)} = a^3b^4$$

When multiplying two algebraic expressions together, all of the terms of the first expression are multiplied by the first term of the second expression. Then all of the terms of the first expression are multiplied by the second term of the second expression, and so on. The products of these multiplications of terms which constitute like terms are set down one under the other and added, due consideration being given to their respective signs. Thus,

$$\begin{array}{lrrrr}
\text{Multiply} & 2a^2 & -\ 3ab & +\ 4b^2 & \\
\text{by} & & ab & +\ b^2 & \\
\hline
 & 2a^3b & -\ 3a^2b^2 & +\ 4ab^3 & \\
 & & 2a^2b^2 & -\ 3ab^3 & +\ 4b^4 \\
\hline
\text{Product} & 2a^3b & -\ a^2b^2 & +\ ab^3 & +\ 4b^4
\end{array}$$

Division. *The Law of Signs* in division is that when one term is divided by another and their signs are alike, their quotient will be positive. Where the signs of dividend and divisor are unlike, the quotient will be negative. Thus,

$$(-4a) \div (-2a) = 2; \qquad 15x^2 \div (-3x) = -5x$$

The Law of Exponents in division is that the exponent of the quotient of two powers of the same quantity is equal to the exponent of the dividend minus the exponent of the divisor. Thus,

$$x^5 \div x^3 = x^{(5-3)} = x^2$$

$$x^2y^3 \div xy^2 = x^{(2-1)}y^{(3-2)} = xy$$

If the exponent of the divisor is larger than the exponent of the dividend, the exponent of the quotient will be negative. Thus,

$$x^3 \div x^5 = x^{(3-5)} = x^{-2}$$

$$a^4b^3 \div a^2b^6 = a^{(4-2)}b^{(3-6)} = a^2b^{-3}$$

A quantity having a negative exponent may also be expressed as one over that quantity to the same exponent made positive.

$$x^{-2} = \frac{1}{x^2} \qquad ab^{-4} = a \times \frac{1}{b^4} = \frac{a}{b^4}$$

When dividing one algebraic expression by another, arrange the terms of both the dividend and divisor according to the ascending or descending powers of any letter common to both. Divide the first term of the dividend by the first term of the divisor to obtain the first term of the quotient. Multiply each term of the divisor by this first term of the quotient and subtract the product from the dividend. In making this subtraction, like terms are set under like terms and unlike terms are set by themselves. The remainder is now considered to be a new dividend and is arranged according to the same ascending or descending powers of a given letter as the original dividend and the divisor. The first term of it is divided by the first term of the divisor to get the second term of the quotient. The subsequent operations are continued in the same order as before until either the remainder is zero or the quotient is found to include a fractional remainder. Thus,

Example 1: Divide $a^4 + a^2b^2 + b^4$ by $a^2 + ab + b^2$

Note: In setting down the dividend arranged in descending powers of a, it will facilitate the division if a space is left after the a^4 term and a space after the a^2b^2 term. Then, when the divisor is multiplied by the first term of the quotient to obtain the expression to be subtracted from the dividend, if an a^3 term appears in this expression, it can be set down in proper order, i.e., beneath the space after the a^4 term in the dividend and similarly if an a term appears it can be set down underneath the second space.

$$
\begin{array}{lllll}
\textit{Divisor} & & \textit{Dividend} & & \textit{Quotient} \\
a^2 + ab + b^2) & a^4 & + a^2b^2 & + b^4 & (a^2 - ab + b^2 \\
& -(a^4 + a^3b + a^2b^2) & & & \\
\hline
& - a^3b & & + b^4 & \\
& -(-a^3b - a^2b^2 - ab^3) & & & \\
\hline
& & a^2b^2 + ab^3 + b^4 & & \\
& & -(a^2b^2 + ab^3 + b^4) & & \\
\hline
\end{array}
$$

Ans., $a^2 - ab + b^2$

Example 2: Divide $a + b$ by $a - b$.

$$
\begin{array}{lll}
\textit{Divisor} & \textit{Dividend} & \textit{Quotient} \\
a - b) & a + b & (1 \\
& -(a - b) & \\
\hline
\textit{Remainder} & 2b &
\end{array}
$$

Ans., $1 + \dfrac{2b}{a - b}$

3. DIVISORS AND MULTIPLES

The Greatest Common Divisor or Highest Common Factor of two quantities is the continued product of all the prime factors which are common to both. This product is thus the expression of the highest degree which will divide each of the two quantities without remainder.

To Find the G. C. D. of Two Quantities, resolve each quantity into its prime factors and take the product of all of the latter which are common to both quantities.

Example: Find the G. C. D. of $a^2xy - b^2xy$ and $a^3x^3 - b^3x^3$.

Factoring both expressions

$$a^2xy - b^2xy = xy(a + b)(a - b)$$

$$a^3x^2 - b^3x^2 = x^2(a - b)(a^2 + ab + b^2)$$

Therefore G. C. D. $= x(a - b)$ which is the product of all the factors common to both.

When the two quantities cannot be readily resolved into their respective prime factors, arrange them with reference to the descending powers of some letter common to both and proceed as follows:

1. Divide both quantities by any factor which is common to all of their terms. (Such a factor, being common to both quantities is a factor of the Greatest Common Divisor.)
2. Multiply the larger quantity or divide the smaller quantity by some factor which will make the first term of the larger quantity exactly divisible by the first term of the smaller quantity. (If this division can be performed without first multiplying the larger quantity by some factor, omit this step.)
3. Divide the larger quantity by the smaller and continue the division until the highest exponent of the leading letter in the remainder is at least 1 less than highest exponent of the corresponding letter in the divisor.
4. Take the divisor just used, and set it down as a new dividend. Take the remainder and use it as a new divisor. (If necessary, multiply the dividend or divide the divisor by some factor which will make the first term of the dividend evenly divisible by the first term of the divisor.) Continue with the division until a remainder is found that is either 0 or that does not contain the leading letter of the divisor.

If the remainder is 0, the G. C. D. is equal to the last divisor multiplied by any factor employed in Step 1.

If the remainder is other than 0, then the G. C. D. is equal to the factor, if any, used in Step 1.

If there is no such factor and the remainder is other than 0, then the two quantities are prime to each other.

Example: Find the G. C. D. of $2a^3b + 10a^2b + 30ab + 36b$ and $2a^3b - 2a^2b + 6ab + 36b$

1. Here, $2b$ is a factor common to both quantities. Dividing both by this factor:

$$\frac{2a^3b + 10a^2b + 30ab + 36b}{2b} = a^3 + 5a^2 + 15a + 18$$

$$\frac{2a^3b - 2a^2b + 6ab + 36b}{2b} = a^3 - a^2 + 3a + 18$$

2. It is not necessary to multiply the larger quantity by some factor to make its first term exactly divisible by the first term of the smaller.
3. Dividing the larger quantity by the smaller:

$$\begin{array}{r|l} a^3 - a^2 + 3a + 18) & a^3 + 5a^2 + 15a + 18\,(1 \\ & \underline{-(a^3 - a^2 + 3a + 18)} \\ & 6a^2 + 12a \end{array}$$

4. The new dividend is now $a^3 - a^2 + 3a + 18$ and the new divisor is $6a^2 + 12a$.

Divide the new divisor by 6:

$$\frac{6a^2 + 12a}{6} = a^2 + 2a$$

Then:

$$\begin{array}{r|l} a^2 + 2a) & a^3 - a^2 + 3a + 18\,(a - 3 \\ & \underline{-(a^3 + 2a^2)} \\ & -3a^2 + 3a + 18 \\ & \underline{-(-3a^2 - 6a)} \\ & 9a + 18 \end{array}$$

The new dividend is now $a^2 + 2a$ and the new divisor is $9a + 18$.

Divide the new divisor by 9: $\dfrac{9a + 18}{9} = a + 2$

Then:

$$a + 2\overline{)\ a^2 + 2a\ }(a$$
$$\underline{-(a^2 + 2a)}$$
$$0$$

The G. C. D. is, then, the product of the last divisor and the factor used in Step 1.

$$\text{G. C. D.} = (a + 2)(2b) = 2ab + 4b$$

The Least Common Multiple of two quantities is the simplest quantity, i.e., the one of lowest degree and smallest numerical coefficient, which can be divided by each without a remainder. The L. C. M. of two quantities is thus equal to the product of those quantities, divided by their G. C. D. The L. C. M. of three or more quantities is found by taking the L. C. M. of two of them, then the L. C. M. of this result and the third, and so on.

To Find the L. C. M. of Two Quantities, either multiply their prime factors together, taking each prime factor the greatest number of times it enters into either of the quantities; or divide the product of the two quantities by their G. C. D.

Example: Find the L. C. M. of $a^3 - a^2b - ab^2 + b^3$ and $a^3 - b^3$.

Factoring both quantities:

$$a^3 - a^2b - ab^2 + b^3 = (a + b)(a - b)(a - b)$$

and

$$a^3 - b^3 = (a - b)(a^2 + ab + b^2)$$

Note that the factor $a - b$ appears twice in the first quantity. Hence:

$$\text{L. C. M.} = (a - b)(a - b)(a + b)(a^2 + ab + b^2)$$

4. FRACTIONS

To Reduce a Fraction to Its Lowest Terms, divide the numerator and denominator by their Greatest Common Divisor. Thus,

$$\frac{a^2 - b^2}{a^3 + b^3} = \frac{(a + b)(a - b)}{(a + b)(a^2 - ab + b^2)} = \frac{a - b}{a^2 - ab + b^2}$$

To Reduce a Mixed Quantity to an Improper Fraction, multiply the integral part by the denominator of the fraction and add the product to the numerator. Thus,

$$a - \frac{a-b}{c+d} = \frac{a(c+d) - (a-b)}{c+d} = \frac{ac + ad - a + b}{c+d}$$

To Convert Two or More Fractions to Equivalent Fractions Having the Least Common Denominator, i.e., the smallest denominator common to all of them, first find the Least Common Multiple of the denominators. Then multiply the numerator and denominator of each fraction by the quotient of this Least Common Multiple divided by the denominator of the fraction. Thus, for the fractions $\frac{2a}{3d}$, $\frac{3b}{4d^2}$ and $\frac{5c}{6d}$ the Least Common Multiple of their denominator is $12d^2$.

Hence: $\frac{12d^2}{3d} = 4d$ and $\frac{2a \times 4d}{3d \times 4d} = \frac{8ad}{12d^2}$

$\frac{12d^2}{4d^2} = 3$ and $\frac{3b \times 3}{4d^2 \times 3} = \frac{9b}{12d^2}$

$\frac{12d^2}{6d} = 2d$ and $\frac{5c \times 2d}{6d \times 2d} = \frac{10cd}{12d^2}$

The equivalent fractions are, therefore,

$$\frac{8ad}{12d^2}, \frac{9b}{12d^2} \quad \text{and} \quad \frac{10cd}{12d^2}$$

To Add Fractions, first reduce the fractions to equivalent fractions having a common denominator, preferably the lowest common denominator; then add the numerators and set their sum above the common denominator. Thus,

$$\frac{a}{b} + \frac{c}{d} + \frac{e}{f} = \frac{adf}{bdf} + \frac{cbf}{bdf} + \frac{ebd}{bdf} = \frac{adf + cbf + ebd}{bdf}$$

To Subtract Fractions, proceed as in addition, except that the numerator of the subtrahend is subtracted from that of the minuend. Thus,

$$\frac{3m + 4a}{3b} - \frac{m - a}{3c} = \frac{c(3m + 4a) - b(m - a)}{3bc}$$

$$= \frac{3mc + 4ac - mb + ab}{3bc}$$

To Multiply Fractions, multiply the numerators for the numerator of the product, and, similarly, the denominators for the denominator of the product. Thus,

$$\frac{m}{a} \times \frac{2na}{3c} \times -\left(\frac{3pb}{d}\right) = \frac{m \times 2na \times (-3pb)}{a \times 3c \times d}$$

$$= \frac{-6abmnp}{3acd} = -\frac{2bmnp}{cd}$$

(Note that minus sign of any fraction can be applied to *either* its numerator or denominator, but not to both.)

Division of Fractions. Dividing a/b by c is the same as multiplying a/b by $1/c$; similarly, dividing a/b by c/d is the same as multiplying a/b by d/c. Hence, to divide fractions, invert that fraction which is the divisor and proceed as in the multiplication of fractions. Thus,

$$\left(a - \frac{b}{c}\right) \div \frac{m}{n} = \left(a - \frac{b}{c}\right) \times \frac{n}{m} = \frac{ac - b}{c} \times \frac{n}{m} = \frac{acn - bn}{cm}$$

A Complex Fraction is one which has a fraction for the numerator, for the denominator, or for both. To reduce such a fraction to its simplest form, divide the numerator by the denominator, i.e., multiply the numerator by the reciprocal of the denominator. Thus,

$$\frac{\dfrac{a}{b + c}}{\dfrac{x}{y + z}} = \frac{a}{b + c} \times \frac{y + z}{x} = \frac{ay + az}{bx + cx}$$

Signs. As in the multiplication and division of integral quantities, like signs before two fractions to be multiplied or divided, give a product or quotient having a plus sign and unlike signs a product or quotient having a minus sign. In any fraction, the signs may be changed throughout *either* its numerator or denominator, if the sign before the fraction is also changed.

5. POWERS

A Power of a quantity or expression is the product obtained by using the quantity or expression as a multiplying factor a given number of times. Thus, a^4 is the fourth power of a or $a \times a \times a \times a$.

Involution is the operation of raising the quantity or expression to any required power.

Law of Exponents in Involution. The expression a^m is the mth power of a and means the product of a by itself m times. Similarly, $(a^m)^n$ is the nth power of a^m, or the continued product of a^m for n times, or a used as a factor $m \times n$ times. Again, $(ab)^m = ab$ to the mth power $= a$ to the mth power $\times\ b$ to the mth power $= a^m b^m$.

Law of Signs in Involution. Since involution is essentially a process of multiplication, it is evident that all of the powers of a positive quantity and all the even powers of a negative quantity are positive, while all the odd powers of a negative quantity are negative.

Summarizing as to the rules for powers:

Fundamental Rules:

$$(+a)^m = +a^m$$

$$(-a)^m = +a^m \text{ when } m \text{ is an even integer}$$

$$(-a)^m = -a^m \text{ when } m \text{ is an odd integer}$$

$$a^m \times a^n = a^{m+n}$$

$$a^m \div a^n = a^{m-n}$$

$$(a^m)^n = a^{mn} = (a^n)^m$$

$$\sqrt[n]{a^m} = a^{\frac{m}{n}}$$

Applications of these Rules:

$$a^{-m} = \frac{1}{a^m} = \left(\frac{1}{a}\right)^m; \qquad \frac{a}{a^m} = a^{1-m}; \qquad \frac{a^m}{a} = a^{m-1};$$

$$a^0 = 1, \text{ since } \frac{a^m}{a^m} = 1 = a^{m-m} = a^0$$

$$[(a^m)^n]^p = a^{mnp}; \qquad (a^{\frac{m}{n}})^n = a^m$$

$$\frac{a^{\frac{m}{n}}}{a} = a^{\frac{m}{n}-1} = a^{\frac{m-n}{n}}; \qquad \frac{a}{a^{\frac{m}{n}}} = a^{1-\frac{m}{n}} = a^{\frac{n-m}{n}};$$

$$a^m \times b^m = (ab)^m; \qquad \frac{a^m}{b^m} = \left(\frac{a}{b}\right)^m; \qquad \left(\frac{a}{b}\right)^{-m} = \left(\frac{b}{a}\right)^m$$

Powers of Polynomials. The power of a *binomial* is obtained in the same way as that of a *monomial*, that is by using it as a factor for the required number of times; the *Binomial Theorem* (which follows) makes this operation less laborious. The power of a *trinomial* or other polynomial is most readily obtained by reducing it to a binomial by substitution, and then applying the Binomial Theorem, as explained in the next section.

6. BINOMIAL THEOREM

If $a + b$ is raised to the fourth power, the resulting expression will be:

$$(a + b)^4 = a^4 + 4a^3b + \frac{4 \times 3}{1 \times 2}a^2b^2 + \frac{4 \times 3 \times 2}{1 \times 2 \times 3}ab^3 + \frac{4 \times 3 \times 2 \times 1}{1 \times 2 \times 3 \times 4}b^4$$

$$= a^4 + 4a^3b + 6a^2b^2 + 4ab^3 + b^4$$

Examination of the right-hand member of the first equation shows that the coefficients and exponents of the various terms follow the laws given below, and hence, for the nth power of $a + b$, we may write:

$$(a + b)^n = a^n + na^{n-1}b + \frac{n(n-1)}{1 \times 2}a^{n-2}b^2 + \frac{n(n-1)(n-2)}{1 \times 2 \times 3}a^{n-3}b^3 + \cdots + b^n$$

which is a general formula for finding the powers of binomials, called the *Binomial Theorem*. If b had been negative, it is evident that the sign of every term containing an odd power of b in the expansion of $(a - b)$ would be negative also.

The coefficient of the first term is 1; of the second term, $n/1 = n$; of the third term, $\frac{n(n-1)}{1 \times 2}$; of the fourth term, $\frac{n(n-1)(n-2)}{1 \times 2 \times 3}$, etc. The exponent of a is n in the first term, and this exponent decreases by 1 in each succeeding term until $a^0 = 1$ is reached in the last term. Conversely, the exponent of b is zero in the first term, and this exponent increases by 1 in each succeeding term, becoming n in the last term. The number of terms is thus $n + 1$. The series of coefficients from the middle to the final term is the same as that from the first to the middle terms, taken in reverse order.

To Find Any Term in the expansion of $(a + b)^n$, let r be the number of the required term and the latter will then be:

$$\frac{n(n-1)(n-2)\cdots(n-r+2)}{1 \times 2 \times 3 \cdots (r-1)} a^{n-r+1} b^{r-1}$$

If b is negative, as in $(a - b)^n$, the term will be the same as above, and will be positive or negative as r is odd or even, respectively.

The application of the Binomial Theorem can be extended to cover *any power of any polynomial.* This follows since a and b are used generally in the expansions as above to indicate either quantities or expressions. Thus, $(1 + 3x + 5y)^3 = [1 + (3x + 5y)]^3$. In the general formula a is replaced by 1 and b by $3x + 5y$, thus:

$$(1 + 3x + 5y)^3 = [1 + (3x + 5y)]^3 = 1 + 3 \times 1^2 \times (3x + 5y) + \frac{3 \times 2}{1 \times 2} \times 1 \times (3x + 5y)^2 + \frac{3 \times 2 \times 1}{1 \times 2 \times 3} (3x + 5y)^3$$

If the various terms of this expression are multiplied out, the resulting expression will be:

$$1 + 9x + 15y + 27x^2 + 90xy + 75y^2 + 27x^3 + 135x^2y + 225xy^2 + 125y^3$$

The use of the following table of coefficients will save the labor of computing them for the range which it covers.

Binomial Coefficients

	Terms												
Exponent	1	2	3	4	5	6	7	8	9	10	11	12	13
1	1	1											
2	1	2	1										
3	1	3	3	1									
4	1	4	6	4	1								
5	1	5	10	10	5	1							
6	1	6	15	20	15	6	1						
7	1	7	21	35	35	21	7	1					
8	1	8	28	56	70	56	28	8	1				
9	1	9	36	84	126	126	84	36	9	1			
10	1	10	45	120	210	252	210	120	45	10	1		
11	1	11	55	165	330	462	462	330	165	55	11	1	
12	1	12	66	220	495	792	924	792	495	220	66	12	1

The form of the expansion of $(a \pm b)^n$ by the Binomial Theorem, as given above, holds strictly for all cases when n is a positive integer, the terms then forming a finite series; but, when n is negative or fractional, the number of terms becomes infinite, furthermore, the theorem holds only when b is numerically less than a, thus making the series convergent. (For an explanation of convergent series see Section 12, Series, in this chapter.)

7. ROOTS

The Root of a number is one of the equal factors whose continued product is the number. Thus, $\sqrt{a} \times \sqrt{a} = a$, $\sqrt[3]{a} \times \sqrt[3]{a} \times \sqrt[3]{a} = a$, etc. Since, in algebraic multiplication, the exponents of the same quantity are added, $a = a^1 = a^{1/2} \times a^{1/2} = \sqrt{a} \times \sqrt{a}$, and hence, $\sqrt{a} = a^{1/2}$. Similarly, $\sqrt[3]{a} = a^{1/3}$, $\sqrt[4]{a} = a^{1/4}$, etc. Again, as in algebraic division, the exponents of the same quantity are subtracted, $1 \div \sqrt{a} = a^0 \div a^{1/2} = a^{0-1/2} = a^{-1/2}$, $1 \div \sqrt[3]{a} = a^0 \div a^{1/3} = a^{0-1/3} = a^{-1/3}$, etc.

Evolution is the operation of extracting any given root of a quantity, i.e., of resolving it into equal factors, the number of which is equal to the index of the root. Thus, the square root of a number is one of two equal factors; the cube root, one of three, etc. Since the nth root of a is a raised to the $1/n$th power, roots are essentially powers, and the laws stated previously as to the exponents and signs of powers apply also to roots, the difference being only the substitution of $1/m$ and $1/n$ for the exponents m and n, respectively, of the powers.

Since

$$\left.\begin{aligned} a^2 &= (+a) \times (+a) \\ a^2 &= (-a) \times (-a) \end{aligned}\right\} \quad \text{and} \quad \sqrt{+a^2} = +a \text{ or } -a = \pm a;$$

also,

$$a^3 = (+a) \times (+a) \times (+a) \qquad \text{and} \qquad \sqrt[3]{+a^3} = +a,$$

an even root of a positive quantity is positive or negative, and an odd root of a positive quantity is positive.

Again, since

$$\sqrt{-a^2} = \sqrt{a^2 \times (-1)} = \sqrt{a^2} \times \sqrt{-1} = a\sqrt{-1};$$

and

$$-a^3 = (-a) \times (-a) \times (-a) \qquad \text{and} \qquad \sqrt[3]{-a^3} = -a,$$

an even root of a negative quantity is what is known as an *imaginary quantity* (see Section 9 of this chapter) and an odd root of a negative quantity is negative.

To Find the Root of a Monomial, extract the root of any numerical factor and divide the exponents of the literal (letter) factors by the exponent of the root, prefixing the proper sign, as explained above.

Thus: $\sqrt{a^2} = \pm a$, $\sqrt[3]{a^6b^3} = a^2b$, and $\sqrt[4]{16a^4b^8} = \pm 2ab^2$.

Summarizing as to the rules for roots:

$$\sqrt[m]{a} = a^{\frac{1}{m}}; \qquad \sqrt[m]{\frac{1}{a}} = \frac{1}{\sqrt[m]{a}} = \frac{1}{a^{\frac{1}{m}}} = a^{-\frac{1}{m}};$$

$$\sqrt[2m]{+a} = \pm a^{\frac{1}{2m}}; \qquad \sqrt[2m+1]{+a} = +a^{\frac{1}{2m+1}};$$

$$\sqrt[2m]{-a} = (-a)^{\frac{1}{2}\times\frac{1}{m}} = \sqrt[m]{\sqrt{-a}} = \sqrt[m]{i\sqrt{a}} \ (\textit{imaginary});$$

$$\sqrt[2m+1]{-a} = \sqrt[2m+1]{a \times (-1)} = -\sqrt[2m+1]{a} = -a^{\frac{1}{2m+1}};$$

$$\sqrt[m]{a^m} = a; \qquad \sqrt[m]{ab} = \sqrt[m]{a} \times \sqrt[m]{b} = a^{\frac{1}{m}}b^{\frac{1}{m}} = (ab)^{\frac{1}{m}};$$

$$\sqrt[m]{\frac{a}{b}} = \frac{\sqrt[m]{a}}{\sqrt[m]{b}} = a^{\frac{1}{m}}b^{-\frac{1}{m}}; \qquad \sqrt[m]{a^n} = (\sqrt[m]{a})^n = a^{\frac{n}{m}} = \sqrt[\frac{m}{n}]{a};$$

$$\sqrt[m]{\sqrt[n]{a}} = \sqrt[m]{\frac{1}{a^n}} = a^{\frac{1}{mn}} = \sqrt[mn]{a} = \sqrt[n]{\sqrt[m]{a}}.$$

It will be noted that, if m is any number, $2m$ will be an even root and $2m + 1$ will be odd. The symbol i is used to denote the imaginary quantity $\sqrt{-1}$, see Section 9 of this chapter.

8. SURDS

An Irrational Quantity is one that cannot be expressed as an integer or as the quotient of two integers.

A Surd (irrational quantity or radical) is the root of a quantity which is an imperfect power of that root. Thus, $\sqrt{3}$ is a quadratic surd or radical of the second degree, $\sqrt[3]{3ab}$ is a radical of the third degree, and so on.

Similar Surds are those which are of the same degree and which have the same quantity under the radical sign, as $2a\sqrt{b}$ and $3c\sqrt{b}$.

A Binomial Surd is a binomial, one term at least of which is a surd, as $a + \sqrt{b}$, or $\sqrt{a} + \sqrt{b}$.

The principles employed in the multiplication, division, and transformation (change of form without change in value), etc., of surds are indicated by the following:

Addition: $a\sqrt{c} + b\sqrt{c} = (a + b)\sqrt{c}$.

Subtraction: $a\sqrt{c} - b\sqrt{c} = (a - b)\sqrt{c}$.

Multiplication: $a\sqrt[m]{c} \times b\sqrt[m]{d} = a \times b \times \sqrt[m]{c} \times \sqrt[m]{d} = ab\sqrt[m]{cd}$.

Division: $\dfrac{a\sqrt[m]{c}}{b\sqrt[m]{d}} = \dfrac{a}{b} \times \dfrac{\sqrt[m]{c}}{\sqrt[m]{d}} = \dfrac{a}{b} \times \sqrt[m]{\dfrac{c}{d}} = \dfrac{a}{b}\sqrt[m]{\dfrac{c}{d}}$.

Powers: $(a\sqrt[n]{b})^m = a^m(\sqrt[n]{b})^m = a^m\sqrt[n]{b^m}$.

Roots: $\sqrt[m]{a^m(\sqrt[n]{b^m})} = \sqrt[m]{a^m(\sqrt[n]{b})^m} = a\sqrt[n]{b}$.

Transformation:

$$C \times ab\sqrt[m]{K \times ac} = \sqrt[m]{C^m \times a^m b^m} \times \sqrt[m]{K \times ac} = \sqrt[m]{C^m K a^{m+1} b^m c}$$

$$\sqrt[m]{\sqrt[n]{a}} = \sqrt[mn]{a} = \sqrt[n]{\sqrt[m]{a}};\quad \sqrt[m]{\sqrt[n]{a^n}} = \sqrt[mn]{a^n} = \sqrt[m]{a}$$

9. IMAGINARY QUANTITIES

An Imaginary Quantity is an even root of a negative quantity, as $\sqrt{-1}$, $\sqrt{-a^2}$. The square root of -1, which is denoted by i, is the simplest imaginary quantity. Its powers are:

$$i = (\sqrt{-1})^1 = +\sqrt{-1} \qquad i^2 = (\sqrt{-1})^2 = -1$$

$$i^3 = (\sqrt{-1})^3 = -\sqrt{-1} \qquad i^4 = (\sqrt{-1})^4 = +1$$

$$i^5 = (\sqrt{-1})^5 = +\sqrt{-1} \qquad i^6 = (\sqrt{-1})^6 = -1, \text{etc.}$$

which form the repeating series $+\sqrt{-1}$, -1, $-\sqrt{-1}$, $+1$, etc.

The product of two imaginary factors is negative, if they have the same sign, and positive, if the signs differ. This reversal of the law of signs for surds is due to the imaginary factor i. Thus:

$$\begin{aligned}(+\sqrt{-a})(+\sqrt{-b}) &= (+\sqrt{a}\sqrt{-1})(+\sqrt{b}\sqrt{-1}) \\ &= +\sqrt{ab} \times i^2 = -\sqrt{ab}\end{aligned}$$

$$(-\sqrt{-a})(-\sqrt{-b}) = (-\sqrt{a}\sqrt{-1})(-\sqrt{b}\sqrt{-1})$$
$$= +\sqrt{ab} \times i^2 = -\sqrt{ab}$$
$$(+\sqrt{-a})(-\sqrt{-b}) = (+\sqrt{a}\sqrt{-1})(-\sqrt{b}\sqrt{-1})$$
$$= -\sqrt{ab} \times i^2 = +\sqrt{ab}$$
$$(-\sqrt{-a})(+\sqrt{-b}) = (-\sqrt{a}\sqrt{-1})(+\sqrt{b}\sqrt{-1})$$
$$= -\sqrt{ab} \times i^2 = +\sqrt{ab}$$

An imaginary monomial of the second degree always appears in the general form $\sqrt{-a^2}$, which may be factored into the typical form $\pm a\sqrt{-1}$. A complex imaginary quantity can also be reduced to the typical form $a \pm b\sqrt{-1}$, in which a and b are either whole numbers, fractions, or surds. Reduction to these typical forms gives the readiest means of performing operations on imaginary quantities. The product $(a + bi)(a - bi) = a^2 + b^2$. If $a + bi = 0$, $a = 0$ and $b = 0$. When $a + bi = c + di$, $a = c$ and $b = d$.

Examples:

1. $(a + b\sqrt{-1}) + (a - b\sqrt{-1}) = 2a$
2. $a + \sqrt{-b^2} + c + \sqrt{-d^2} = a + b\sqrt{-1} + c + d\sqrt{-1}$
   $= a + c + (b + d)\sqrt{-1}$
3. $(a + b\sqrt{-1}) \times (a - b\sqrt{-1}) = a^2 + b^2$
4. $(a + b\sqrt{-1}) \times (c + d\sqrt{-1}) = ac - bd + (ad + bc)\sqrt{-1}$
5. $\dfrac{1}{a + b\sqrt{-1}} + \dfrac{1}{a - b\sqrt{-1}} = \dfrac{a - b\sqrt{-1}}{a^2 + b^2} + \dfrac{a + b\sqrt{-1}}{a^2 + b^2}$
   $= \dfrac{2a}{a^2 + b^2}$

10. FACTORING

Factoring is the resolution of a complex expression into its factors. The factors of a monomial are ascertained by inspection. There are relatively few processes for finding the factors of a polynomial, although their application is varied. A table of useful factors is given on page **2**–17.

Table of Useful Factors

$$a^2 - b^2 = (a + b)(a - b)$$
$$a^2 + b^2 = (a + b\sqrt{-1})(a - b\sqrt{-1})$$
$$a^2 + b^2 - c^2 - d^2 + 2ab + 2cd = (a + b + c - d)(a + b - c + d)$$

$$a^3 - b^3 = (a - b)(a^2 + ab + b^2)$$
$$a^3 + b^3 = (a + b)(a^2 - ab + b^2)$$
$$a^3 + b^3 + c^3 - 3abc = (a + b + c)(a^2 + b^2 + c^2 - ab - bc - ac)$$

$$a^4 - b^4 = (a - b)(a^3 + a^2b + ab^2 + b^3)$$
$$a^4 - b^4 = (a + b)(a^3 - a^2b + ab^2 - b^3)$$
$$a^4 - b^4 = (a^2 - b^2)(a^2 + b^2) = (a + b)(a - b)(a + b\sqrt{-1})(a - b\sqrt{-1})$$
$$a^4 + b^4 = (a^2 + b^2\sqrt{-1})(a^2 - b^2\sqrt{-1})$$
$$a^4 + a^2b^2 + b^4 = (a^2 + ab + b^2)(a^2 - ab + b^2)$$

$$a^5 - b^5 = (a - b)(a^4 + a^3b + a^2b^2 + ab^3 + b^4)$$
$$a^5 + b^5 = (a + b)(a^4 - a^3b + a^2b^2 - ab^3 + b^4)$$

$$a^n - b^n = (a - b)(a^{n-1} + a^{n-2}b + a^{n-3}b^2 + \cdots + b^{n-1})$$
When n is an odd or even number
$$a^n - b^n = (a + b)(a^{n-1} - a^{n-2}b + a^{n-3}b^2 - \cdots - b^{n-1})$$
When n is an even number
$$a^n + b^n = (a + b)(a^{n-1} - a^{n-2}b + a^{n-3}b^2 - \cdots + b^{n-1})$$
When n is an odd number

$$x^2 + (a + b)x + ab = (x + a)(x + b)$$
$$x^2 + (a - b)x - ab = (x + a)(x - b)$$

11. EQUATIONS

An Equation is a statement of equality between two quantities. Thus, $x = a + b$ means that the quantity x is equal to the sum of the quantities a and b. The expressions connected by the sign of equality are called the *Members* of the equation. Since the members are equal, their equality will still hold if equal quantities are added to, or subtracted from both. Again, the equation will hold, if both members be multiplied or divided by the same quantity, if both be raised to the same power, or if the same root of each be found. Any term of either member can be transferred to the other, if its sign be changed, and, similarly, the signs of all terms on both sides may be changed.

These transformations are utilized to place all of the unknown terms on one side of an equation and the known terms on the other, or to convert an equation into more suitable form.

An Identical Equation is one in which the two members are identical in value, although their form may differ; in such cases, the *Sign of Identity* $\equiv$ sometimes replaces the sign of equality as $(a + b)^2 \equiv a^2 + 2ab + b^2$. Equations are classified according to the number of unknown quantities and according to their degree.

The Degree of an equation containing more than one unknown quantity, is the greatest sum of the exponents of the unknown quantities in any term. Thus, in the equation $x^2 + xy^2 + xy^3 + y^3 = 28$, the term in which the sum of the exponents is the greatest is xy^3. In this term, the sum of the exponents is $1 + 3 = 4$. Hence, this is an equation of the fourth degree. Equations of the first, second, third, and fourth degrees are called *Simple*, *Quadratic*, *Cubic* and *Biquadratic Equations*, respectively.

Simultaneous Equations are those in which the unknown quantities have the same values throughout the series of equations. Thus, if $x = 2$ and $y = 3$, then $x + y = 5$ and $3x - y = 3$ are simultaneous equations. To find the values of the unknown quantities, as many equations are required as there are unknown quantities.

To Solve a Simple Equation. If the equation contains one unknown, the first step is to clear the equation of fractions, if any, by multiplying each term by the Least Common Multiple of the denominators. Then, transpose all terms containing the unknown quantity to the left side of the equation and all other terms to the right side, combine like terms, and divide the equation by the coefficient of the unknown quantity. Thus, $2x + \frac{1}{3}x = 62 - \frac{1}{4}x$; multiplying by 12, which is the L. C. M. of the denominators, gives $24x + 4x = 744 - 3x$; transposing, the equation becomes $24x + 7x = 744$; whence, $31x = 744$ and $x = \frac{744}{31} = 24$.

Simultaneous simple equations containing two or more unknown quantities are solved by elimination, which is the process of combining the equations either by addition or subtraction, by substitution, or by comparison.

Take, for example, the two following simultaneous simple equations:

$$2x + 3y = 17 \tag{1}$$

$$3x - 2y = 6 \tag{2}$$

(*A*) *To eliminate by addition or subtraction*, Equation (1) is multiplied by some factor and Equation (2) is multiplied by some other factor such that either when the two equations are added together, or when one equation is subtracted from the other, one of the terms will be eliminated. Thus, if Equation (1) is multiplied by 2 and Equation (2) by 3 and then they are added together:

$$2(2x + 3y = 17) = 4x + 6y = 34 \tag{1a}$$

$$3(3x - 2y = \;\; 6) = \underline{9x - 6y = 18} \tag{2a}$$

$$13x \qquad = 52 \quad \text{and} \quad x = 4$$

the y term is eliminated and the resulting equation can be solved for x directly. If, on the other hand, Equation (1) were multiplied by 3 and Equation (2) by 2 and Equation (2) were subtracted from Equation (1):

$$3(2x + 3y = 17) = 6x + 9y = 51 \tag{1b}$$

$$2(3x - 2y = \;\; 6) = 6x - 4y = 12 \tag{2b}$$

$$6x + 9y = 51 \tag{1b}$$

$$\underline{-(6x - 4y = 12)} \tag{2b}$$

$$13y = 39 \quad \text{and} \quad y = 3$$

the x term is eliminated and the resulting equation can be solved for y directly.

(B) *To eliminate by substitution*, find the value of x in terms of y, or of y in terms of x, from one equation, and substitute this value in the other equation. Thus, in Equation (2):

$$x = \frac{2y + 6}{3} \tag{2c}$$

Substituting this value in Equation (1):

$$\frac{2(2y + 6)}{3} + 3y = 17 \tag{3}$$

$$4y + 12 + 9y = 51$$

$$13y = 39 \qquad y = 3$$

(C) *To eliminate by comparison*, find the value of x in terms of y, or of y in terms of x from both equations, and equate these values. Thus, from Equation (1)

$$x = \frac{17 - 3y}{2} \tag{1d}$$

and from the second:

$$x = \frac{2y + 6}{3} \qquad \text{(2d)}$$

Equating the right-hand side of each and solving for y

$$\frac{17 - 3y}{2} = \frac{2y + 6}{3} \qquad (4)$$

$$51 - 9y = 4y + 12$$

$$13y = 39 \qquad y = 3$$

Where there are three unknown quantities, the process is similar. Thus, let there be three equations involving x, y, and z. Eliminate z by combining Equations (1) and (2) and then (1) and (3). These eliminations give Equations (4) and (5) involving x and y only, from which, by elimination again, the values of these quantities can be found and substituted in either of the three original equations to find the value of z.

Quadratic Equations. A quadratic equation is an equation of the second degree, i.e., the highest power of the unknown quantity is the square. If, in an equation containing but one unknown quantity, both the first and second powers of the unknown quantity are present, the equation is a *complete quadratic*, as $x^2 - x = 12$. If the first power of the unknown quantity is lacking, the equation is an *incomplete quadratic*, as $6x^2 - 36 = 0$.

The *roots* of a quadratic equation are the values — rational, surd, or imaginary — of the unknown quantity which will satisfy the equation, i.e., will make the values of the two members equal. A quadratic equation always has two roots and two only.

General Form of a Quadratic Equation. Every quadratic equation containing but one unknown quantity can be reduced to the general form:

$$ax^2 + bx + c = 0$$

in which a, b and c are the known coefficients and x is the unknown quantity. (In the incomplete quadratic, the value of b is 0.)

The two roots, x' and x'', of this general equation are:

$$x' = \frac{-b + \sqrt{b^2 - 4ac}}{2a} \qquad \text{and} \qquad x'' = \frac{-b - \sqrt{b^2 - 4ac}}{2a}$$

or, as is sometimes written:

$$x = \frac{-b \pm \sqrt{b^2 - 4ac}}{2a}$$

Note: If the quadratic equation is in the form $ax^2 + bx = c$

then $$x = \frac{-b \pm \sqrt{b^2 + 4ac}}{2a}$$

It can be shown that if the quantity, $b^2 - 4ac$, under the radical sign is positive, the roots are real and unequal; if it is equal to 0, they are real, equal, and the value of each is $-\frac{b}{2a}$; if it is negative, they are imaginary and unequal; if it is a perfect square, they are rational and unequal; and, if it is not a perfect square, the roots are surds. Thus in the equation $4x^2 - 12x + 7 = 0$, $a = 4$, $b = -12$, $c = 7$, and $b^2 - 4ac = 144 - 4 \times 4 \times 7 = 32$, which is positive but not a perfect square. Hence the roots are real, unequal and both are surds.

Solving Quadratic Equations Containing One Unknown. If the equation is an incomplete quadratic, the first step is to simplify and collect like terms, divide by the coefficient of x and extract the square root of both members: Thus,

$$20(x^2 - 2) = 5(x^2 + 4) \quad (1)$$

$$20x^2 - 40 = 5x^2 + 20 \quad (2)$$

$$15x^2 = 60 \quad (3)$$

$$x^2 = 4 \quad (4)$$

$$x' = +2 \quad \text{and} \quad x'' = -2 \quad (5)$$

The principal method of solving a complete quadratic equation is that of *completing the square.* The principle governing this operation is shown by the expansion of $(x + b)^2$ into $x^2 + 2bx + b^2$. This is a perfect square and it will be noted that the third term is equal to the square of one-half the coefficient of the second term, or

$$b^2 = \left(\frac{2b}{2}\right)^2.$$

For example, take the fundamental equation $x^2 + px + q = 0$. Transposing the third term to the right-hand side of the equation it becomes $x^2 + px = -q$. The left-hand side of this equation is an imperfect square. To complete it the square of one-half the coefficient of the second term is added to both sides in order not to disturb their equality. This gives: $x + px + \frac{p^2}{4} = \frac{p^2}{4} - q$.

Extracting the square root of both sides:

$$x + \frac{p}{2} = \pm\sqrt{\frac{p^2}{4} - q} = \pm\tfrac{1}{2}\sqrt{p^2 - 4q}$$

and

$$x = -\frac{p}{2} \pm \tfrac{1}{2}\sqrt{p^2 - 4q}$$

When an equation is reduced to the form $x^2 + px + q = 0$, its roots can be found by substituting the values for the coefficients p and q in the equation for x just given, thus in the equation, $x^2 - 8x + 15 = 0$, $p = -8$ and $q = 15$, hence

$$x = -\frac{-8}{2} \pm \tfrac{1}{2}\sqrt{64 - 60} = 4 \pm 1,$$

hence $x' = 5$ and $x'' = 3$.

Factoring may sometimes be employed to determine the roots by inspection. For example, take $x^2 + 3x - 28 = 0$. Factoring this equation gives $(x + 7)(x - 4) = 0$. The required roots or values of x are, then, $x = -7$ or 4.

Although a quadratic has two roots and its solution therefore gives two answers, there is usually no difficulty in selecting the one which meets the specific conditions of a problem, where only one answer is applicable.

Solving of Quadratic Equations Containing Two Unknowns. Simultaneous quadratic equations involving two unknown quantities can be solved in a number of ways depending on their form. The following cases show general methods only.

1. *Substitution* may be employed when one equation is of the first degree and the other of the second. Thus, take $x + y = 7$ and $x^2 + 3xy + y^2 = 31$. From the first equation, $y = 7 - x$. Substituting this value in the second equation and simplifying, there results $x^2 - 7x = 18$, which is a complete quadratic that can be solved by methods given previously.

2. *Division* is often resorted to. As one of the simplest examples, take $x^2 - y^2 = 16$ and $x - y = 2$. Dividing the first equation by the second, member by member, there results $x + y = 8$ which, combined with the second equation, will give the values of x and y. Thus $x = 5$ and $y = 3$.

3. *Making y a Multiple of x.* This can be done when the two equations are of the second degree and both are homogeneous with respect to the unknown quantities. Thus, take $x^2 + xy - y^2 = 11$ and $4x^2 - 3xy - 2y^2 = 10$. For y substitute px in both equations, the value of p being yet unknown. Find the value of x in terms of p in each equation, and equate the two values, giving $12p^2 + 43p = 34$, a complete quadratic from which the two values of p, and hence of y in terms of x, can be found. Substitute the latter values in the first equation to obtain the two values of x and the corresponding values of y.

4. Given: $$x^2 + xy + y^2 = 84 \quad (1)$$

and $$x - \sqrt{xy} + y = 6 \quad (2)$$

Dividing (1) by (2): $$x + \sqrt{xy} + y = 14 \quad (3)$$

Adding (2) and (3) and dividing by 2: $$x + y = 10 \quad (4)$$

Substituting (4) in (2): $\sqrt{xy} = 4$ or $xy = 16$ and $y = 16/x$. Substituting the value of y in (4): $x + 16/x = 10$ or $x^2 - 10x = -16$, from which the values of x can be found and substituted in (4) to find those of y.

In solving equations containing roots of the unknown quantity or powers of that quantity which are higher than the square, it should be remembered that an equation is quadratic in type if it contains but two powers of the unknown quantity, one of which is the square of the other. Thus, $ax + b\sqrt{x} = c$ and $ax^4 + bx^2 = c$ are both of the quadratic form.

12. SERIES

A Series is a succession of terms whose formation is governed by the *Law of the Series*, each term being derived from one or more preceding terms by the provisions of this law. The Binomial Theorem expressed as a series is given in the table on page 2–25. Other series are arithmetical, geometrical, harmonical, recurring, exponential, logarithmic, trigonometric, etc.

Finite and Infinite Series. A series may be *finite* or *infinite* with regard to the number of its terms. The numerical values of many constants, such as π and e the base of the Naperian or natural system of logarithms, are expressed as *infinite* or *non-terminating* series, since either it is not possible to express them otherwise, or no *exact* form for the quantity is known and hence approximations by series must be used.

Sum of a Series. The *sum* of a series is the sum of its terms. The sum of a given series varies with the number of terms taken and may have a maximum value. Thus, the sum of *any* number of terms of the series $1 + \frac{1}{2} + \frac{1}{4} + \frac{1}{8} + \frac{1}{16} + \frac{1}{32} \cdots$ can never exceed 2. By adding more, more and more terms, it may be made to differ from 2 by a smaller and smaller amount. Hence, if a very great number of terms be taken, their sum will differ from 2 by a negligible quantity, and this result is expressed thus: the sum to infinity = 2. Similarly, the sum to infinity of $1 + \frac{1}{1} + \frac{1}{2!} + \frac{1}{3!} + \frac{1}{4!} \cdots + \frac{1}{n!}$ is $2.718281828 \cdots$, an approximate value which is correct to the last decimal place shown.

Convergent and Divergent Series. A series may be convergent or divergent with regard to the sum of its terms. A finite series is always convergent. Again, a convergent series is one the sum of n terms of which remains finite as n is made infinite. This characteristic is of fundamental importance, since *series which are not convergent*, that is divergent series, *cannot be used in practice as they give meaningless results*. It will be readily seen, for example, that by taking a sufficient number of terms of the series, $1 + 2 + 3 + 4 \ldots n$, the sum can be made greater than any assignable number. Such a series is called divergent because the sum of n terms of it, as n is made indefinitely large, diverges from or exceeds any number whatever.

The sum of n terms of $1 + \frac{1}{2} + \frac{1}{3} + \frac{1}{4} + \cdots + \frac{1}{n} \cdots$ as n is increased indefinitely, becomes greater than any assignable number, and hence this series is divergent; but $1 + \frac{1}{2^2} + \frac{1}{3^2} + \frac{1}{4^2} + \cdots + \frac{1}{n^2} + \cdots$ is convergent.

Important Series and Limits of Useful Application —1

Expansion of Series	Values for which Series is Convergent
Binomial Series $(a+b)^n = a^n + \frac{n}{1}a^{n-1}b + \frac{n(n-1)}{1\times 2}a^{n-2}b^2 + \frac{n(n-1)(n-2)}{1\times 2\times 3}a^{n-3}b^3 + \cdots \frac{n(n-1)\cdots(n-r+2)}{1\times 2\times 3\cdots(r-1)}a^{n-r+1}b^{r-1} + \cdots$ where r is the number of the term, as fourth, fifth, sixth, etc.	All finite values of a and b, if n is a positive whole number. All absolute (numerical) values of b less than a, if n is either negative or a fraction.
Special Binomial Series	
Square Root $\sqrt{1+x} = (1+x)^{1/2} = 1 + \frac{1}{2}x - \frac{1\times 1}{2\times 4}x^2 + \frac{1\times 1\times 3}{2\times 4\times 6}x^3 - \frac{1\times 1\times 3\times 5}{2\times 4\times 6\times 8}x^4 + \cdots$	All values of x between -1 and $+1$.
Cube Root $\sqrt[3]{1+x} = (1+x)^{1/3} = 1 + \frac{1}{3}x - \frac{1\times 2}{3\times 6}x^2 + \frac{1\times 2\times 5}{3\times 6\times 9}x^3 - \frac{1\times 2\times 5\times 8}{3\times 6\times 9\times 12}x^4 + \cdots$	All values of x between -1 and $+1$.
Exponential Series	
$e^x = 1 + \frac{x}{1} + \frac{x^2}{1\times 2} + \frac{x^3}{1\times 2\times 3} + \cdots \frac{x^n}{n!}$	All finite values of x.
$a^x = 1 + \frac{xA}{1} + \frac{x^2A^2}{1\times 2} + \frac{x^3A^3}{1\times 2\times 3} + \cdots \frac{x^nA^n}{n!}$ in which $A = \log_e a$	All finite values of x and a.
Logarithmic Series (for computation)	
$\log(1+x) = x - \frac{1}{2}x^2 + \frac{1}{3}x^3 - \frac{1}{4}x^4 + \cdots$	All values of x *between* -1 and $+1$.
$\log x = (x-1) - \frac{1}{2}(x-1)^2 + \frac{1}{3}(x-1)^3 - \cdots$	All positive values of x up to and including 2.
$\log x = \frac{x-1}{x} + \frac{1}{2}\left(\frac{x-1}{x}\right)^2 + \frac{1}{3}\left(\frac{x-1}{x}\right)^3 + \cdots$	All positive values of x less than $\frac{1}{2}$.
$\log x = 2\left[\left(\frac{x-1}{x+1}\right) + \frac{1}{3}\left(\frac{x-1}{x+1}\right)^3 + \frac{1}{5}\left(\frac{x-1}{x+1}\right)^5 + \cdots\right]$	Rapidly convergent and useful for positive values of x.

Important Series and Limits of Useful Application — 2

Expansion of Series	Values for which Series is Convergent
Trigonometric Series (for computations)	
$\sin x = x - \frac{x^3}{3!} + \frac{x^5}{5!} - \frac{x^7}{7!} + \cdots$	All finite values of x.
$\cos x = 1 - \frac{x^2}{2!} + \frac{x^4}{4!} - \frac{x^6}{6!} + \cdots$	All finite values of x.
$\tan x = x + \frac{1}{3}x^3 + \frac{2}{15}x^5 + \frac{17}{315}x^7 + \frac{62}{2835}x^9 + \frac{1382}{155925}x^{11} + \cdots$	All values of x between $-\frac{\pi}{2}$ and $\frac{\pi}{2}$ radians.
In the series for tan x, the coefficients follow no simple law, but depend upon Bernoulli's numbers. The series, as given above, is carried far enough for any practical purpose.	
Inverse Trigonometric Series	
$\sin^{-1} x = x + \frac{1}{2}\frac{x^3}{3} + \frac{1 \times 3}{2 \times 4}\frac{x^5}{5} + \frac{1 \times 3 \times 5}{2 \times 4 \times 6}\frac{x^7}{7} + \cdots$	All values of x^2 less than 1.
$\cos^{-1} x = \frac{\pi}{2} - \sin^{-1} x$	
$\tan^{-1} x = x - \frac{1}{3}x^3 + \frac{1}{5}x^5 - \cdots$	All values of x from -1 to $+1$ inclusive.
Sums of Powers of Natural Numbers	
$1 + 2 + 3 + 4 + \cdots n = \frac{1}{2}n^2 + \frac{1}{2}n$	
$1^2 + 2^2 + 3^2 + 4^2 + \cdots n^2 = \frac{1}{3}n^3 + \frac{1}{2}n^2 + \frac{1}{6}n$	*Note:* In each case, n is the last number in the series for which the sum is to be found.
$1^3 + 2^3 + 3^3 + 4^3 + \cdots n^3 = \frac{1}{4}n^4 + \frac{1}{2}n^3 + \frac{1}{4}n^2$	
$1^4 + 2^4 + 3^4 + 4^4 + \cdots n^4 = \frac{1}{5}n^5 + \frac{1}{2}n^4 + \frac{1}{3}n^3 - \frac{1}{30}n$	
$1^5 + 2^5 + 3^5 + 4^5 + \cdots n^5 = \frac{1}{6}n^6 + \frac{1}{2}n^5 + \frac{5}{12}n^4 - \frac{1}{12}n^2$	

Series Which May Be Convergent or Divergent. The series $1 + \frac{1}{2^x} + \frac{1}{3^x} + \frac{1}{4^x} + \cdots + \frac{1}{n^x} + \cdots$ may be either convergent or divergent depending upon the value of the variable x. If x is greater than 1, the series is convergent. If x is equal to or less than 1 the series is divergent. A large number of series are convergent for only certain values of the variable and when used for any other values of the variable will give results which have no meaning. For this reason, the limits within which the variable must lie in order to make the series convergent should always be stated.

Notation. In the preceding table the expansions of some of the more important series are given together with the values of x for which the series are convergent. All logarithms in these series are taken to the base e, and all angles are given in circular measure (radians). To change logarithms to or from base e, see page **5–14**; to convert radians to degrees, see page **27–142**.

13. PROGRESSIONS

An Arithmetical Progression is a series in which each term differs from the preceding one by a fixed amount called the *common difference, d*. If this common difference is added, as in the progression 1, 3, 5, 7 $\cdots$ in which d is 2, the progression is said to be *increasing*; if subtracted, as in the progression $\cdots$ 13, 10, 7, in which d is -3, the progression is *decreasing*.

In any section of an arithmetical progression, let the first and last terms considered be denoted by a and l, respectively; let d be the common difference, n the number of terms considered, and S the sum of n terms. Then, a and l are called the extremes and all terms between them are *arithmetical means*.

The general formulas for arithmetical progression are given in the table on page **2–28**. In these formulas d is positive in an increasing and negative in a decreasing progression. When any three of the five quantities listed in the above notation are given, the other two can be found by the formulas in the accompanying table.

Example: In an arithmetical progression, the first term is 4 and the last 39. If there are 8 terms, find their sum. Thus,

$$a = 4 \qquad l = 39 \qquad n = 8$$

according to the table of formulas on page **2–28**, when a, l and n are

Formulas for Arithmetical Progression

a = the first term considered
l = the last term considered
n = the number of terms
d = the common difference
S = the sum of n terms

To Find	Given	Use Equation
a	d l n	$a = l - (n-1)d$
	d n S	$a = \frac{S}{n} - \frac{n-1}{2} \times d$
	d l S	$a = \frac{d}{2} \pm \frac{1}{2}\sqrt{(2l+d)^2 - 8dS}$
	l n S	$a = \frac{2S}{n} - l$
d	a l n	$d = \frac{l-a}{n-1}$
	a n S	$d = \frac{2S - 2an}{n(n-1)}$
	a l S	$d = \frac{l^2 - a^2}{2S - l - a}$
	l n S	$d = \frac{2nl - 2S}{n(n-1)}$
l	a d n	$l = a + (n-1)d$
	a d S	$l = -\frac{d}{2} \pm \frac{1}{2}\sqrt{8dS + (2a-d)^2}$
	a n S	$l = \frac{2S}{n} - a$
	d n S	$l = \frac{S}{n} + \frac{n-1}{2} \times d$
n	a d l	$n = 1 + \frac{l-a}{d}$
	a d S	$n = \frac{d-2a}{2d} \pm \frac{1}{2d}\sqrt{8dS + (2a-d)^2}$
	a l S	$n = \frac{2S}{a+l}$
	d l S	$n = \frac{2l+d}{2d} \pm \frac{1}{2d}\sqrt{(2l+d)^2 - 8dS}$
S	a d n	$S = \frac{n}{2}[2a + (n-1)d]$
	a d l	$S = \frac{a+l}{2} + \frac{l^2 - a^2}{2d} = \frac{a+l}{2d}(l + d - a)$
	a l n	$S = \frac{n}{2}(a+l)$
	d l n	$S = \frac{n}{2}[2l - (n-1)d]$

Formulas for Geometrical Progression

a = the first term
l = the last (or nth) term
n = the number of terms
r = the ratio of the progression
S = the sum of n terms

To Find	Given	Use Equation
a	$l \quad n \quad r$	$a = \frac{l}{r^{n-1}}$
	$n \quad r \quad S$	$a = \frac{(r-1)S}{r^n - 1}$
	$l \quad r \quad S$	$a = lr - (r-1)S$
	$l \quad n \quad S$	$a(S-a)^{n-1} = l(S-l)^{n-1}$
l	$a \quad n \quad r$	$l = ar^{n-1}$
	$a \quad r \quad S$	$l = \frac{1}{r}[a + (r-1)S]$
	$a \quad n \quad S$	$l(S-l)^{n-1} = a(S-a)^{n-1}$
	$n \quad r \quad S$	$l = \frac{S(r-1)r^{n-1}}{r^n - 1}$
n	$a \quad l \quad r$	$n = \frac{\log l - \log a}{\log r} + 1$
	$a \quad r \quad S$	$n = \frac{\log[a + (r-1)S] - \log a}{\log r}$
	$a \quad l \quad S$	$n = \frac{\log l - \log a}{\log(S-a) - \log(S-l)} + 1$
	$l \quad r \quad S$	$n = \frac{\log l - \log[lr - (r-1)S]}{\log r} + 1$
r	$a \quad l \quad n$	$r = \sqrt[n-1]{\frac{l}{a}}$
	$a \quad n \quad S$	$r^n = \frac{Sr}{a} + \frac{a-S}{a}$
	$a \quad l \quad S$	$r = \frac{S-a}{S-l}$
	$l \quad n \quad S$	$r^n = \frac{Sr^{n-1}}{S-l} - \frac{l}{S-l}$
S	$a \quad n \quad r$	$S = \frac{a(r^n - 1)}{r-1}$
	$a \quad l \quad r$	$S = \frac{lr - a}{r-1}$
	$a \quad l \quad n$	$S = \frac{\sqrt[n-1]{l^n} - \sqrt[n-1]{a^n}}{\sqrt[n-1]{l} - \sqrt[n-1]{a}}$
	$l \quad n \quad r$	$S = \frac{l(r^n - 1)}{(r-1)r^{n-1}}$

known and S is to be found:

$$S = \frac{n}{2}(a + l) = \frac{8}{2}(4 + 39) = 172$$

A Geometrical Progression is a series in which each term is derived by multiplying the preceding term by a constant multiplier called the *ratio.* A geometrical progression is, therefore, a continued proportion in which each term is a *geometrical mean* or *mean proportional* between the term which precedes and that which follows it. When the ratio is greater than 1, that is, $r > 1$, the progression is increasing; when the ratio is less than 1, that is, $r < 1$, it is decreasing. Thus 4, 12, 36, 108 is an increasing geometrical progression with a ratio of 3, and $\cdots$ 128, 64, 32, 16 is a decreasing geometrical progression with a ratio of 1/2.

The general formulas for geometrical progression are given in the table on page **2**–29.

Example: In a geometrical progression the first term is 3, the last term is 6.2208 and the sum of the terms is 22.3248. Find the ratio. Thus

$$a = 3 \qquad l = 6.2208 \qquad S = 22.3248$$

According to the table of formulas on page **2**–29 when a, l and S are known and r is to be found

$$r = \frac{S - a}{S - l} = \frac{22.3248 - 3}{22.3248 - 6.2208}$$

$$r = \frac{19.3248}{16.1040} = 1.2$$

3

Geometry—Definitions and Principles Commonly Applied

1. DEFINITIONS

Lines and Angles

A point indicates only position and has neither length, breadth, nor thickness.

A line has only one dimension, *length*. It may be the boundary of a surface or that which separates one part of a surface from another.

A straight line is one that has the same direction throughout its entire length.

A curved line is one that has a continually changing direction.

A broken line is one that is composed of a succession of straight lines having different directions.

Two lines are said to be **parallel** with each other when they are the same distance apart at all points.

An angle is formed when two straight lines meet at a point. This point is called the vertex of the angle and the two lines are called the sides of the angle.

Adjacent angles are formed when one straight line meets another at some point between its ends.

Opposite angles are formed, as well as adjacent angles, when two straight lines intersect.

When one straight line meets another straight line to form adjacent angles that are equal, the two lines are said to be **perpendicular** to each other and the adjacent angles are called **right angles.**

An acute angle is less than a right angle.

An obtuse angle is greater than a right angle.

Two angles are said to be **complementary** when their sum is equal to a right angle.

Two angles are said to be **supplementary** when their sum is equal to two right angles.

Surfaces and Figures

A surface has two dimensions, length and breadth.

A plane surface is one in which any two points can be connected by a straight line that will lie wholly on it.

A curved surface is any surface no part of which is a plane surface.

A plane figure is any part of a plane surface bounded by lines.

If a plane figure is bounded by straight lines it is called a **polygon.**

The **side** of a polygon is any straight line forming part of its boundary.

The **perimeter** of a polygon is the sum of the straight lines forming its boundary.

The **name** of any polygon indicates its number of sides, thus:

Name	*Number of Sides*
Triangle	3
Quadrilateral	4
Pentagon	5
Hexagon	6
Heptagon	7
Octagon	8

A polygon is said to be **equilateral** if all its sides are equal.

A polygon is said to be **equiangular** if all its angles are equal.

A scalene triangle has no two of its sides equal.

An isosceles triangle has two of its sides equal.

An equilateral triangle has all of its sides equal.

A right-angled triangle has one right angle.

An oblique-angled triangle is any triangle that is not a right-angled triangle.

An acute-angled triangle has each of its angles less than a right angle.

An obtuse-angled triangle has one angle greater than a right angle.

Any side of a triangle may be considered to be its **base.** Usually the base is considered to be the bottom side or the side on which the triangle is supposed to rest.

The **altitude** of a triangle is a line drawn perpendicular to its base and from the base to the vertex of the angle opposite the base.

Two triangles are said to be **equal** when the sides of one are respectively equal to the sides of the other.

Two triangles are said to be **similar** when the angles of one are respectively equal to the angles of the other.

A circle is a plane figure bounded by a curved line every point on which is equidistant from a point called the center.

The **boundary** of a circle is called its **circumference.**

A diameter of a circle is a straight line drawn through the center and terminating at the circumference.

A radius of a circle is any straight line drawn from its center to its circumference.

The **arc** of a circle is any part of its circumference.

A **chord** of a circle is any straight line connecting two points on its circumference. It may also be defined as a straight line connecting the end points of an arc.

A secant is any straight line which passes through a circle intercepting its circumference at two points.

A segment of a circle is that part of its area included between an arc and its chord.

A sector of a circle is that part of its area included between an arc and two radii drawn from the ends of the arc.

A central angle is an angle having its vertex at the center of a circle.

An inscribed angle is formed by two chords which meet at some point on the circumference of a circle.

A tangent to a circle is a straight line which touches the circle at only one point. This point is called *the point of tangency.*

Two circles are said to be **tangent** to each other if their circumferences coincide at only one point.

Two or more circles are said to be **concentric** with each other when they have the same center.

A quadrant is an arc equal to one-quarter of the circumference of a circle. Four quadrants are formed when two diameters are drawn at right angles to each other.

Geometrical Propositions

1. Sum of three angles in a triangle.
2. Equal triangles—one side and two angles equal.
3. Equal triangles—two sides and included angle equal.
4. Equal triangles—corresponding sides equal.
5. Similar triangles—corresponding sides proportional.
6. Similar triangles—corresponding angles equal.

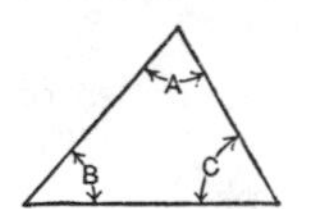

The sum of the three angles in a triangle always equals 180 degrees. Hence, if two angles are known, the third angle can always be found.

$$A + B + C = 180° \qquad A = 180° - (B + C)$$
$$B = 180° - (A + C) \qquad C = 180° - (A + B)$$

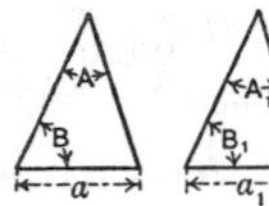

If one side and two angles in one triangle are equal to one side and similarly located angles in another triangle, then the remaining two sides and angle are also equal.

If $a = a_1$, $A = A_1$ and $B = B_1$, then the two other sides and the remaining angle are also equal.

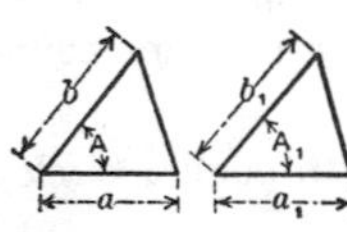

If two sides and the angle between them in one triangle are equal to two sides and a similarly located angle in another triangle, then the remaining side and angles are also equal.

If $a = a_1$, $b = b_1$ and $A = A_1$, then the remaining side and angles are also equal.

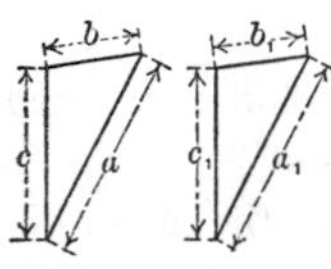

If the three sides in one triangle are equal to the three sides of another triangle, then the angles in the two triangles are also equal.

If $a = a_1$, $b = b_1$ and $c = c_1$, then the angles between the respective sides are also equal.

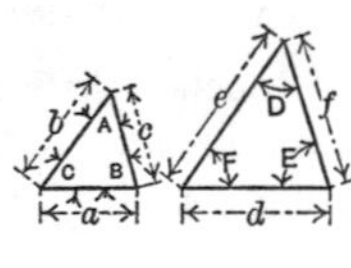

If the three sides of one triangle are proportional to corresponding sides in another triangle, then the triangles are called *similar*, and the angles in the one are equal to the angles in the other.

If $a : b : c = d : e : f$, then $A = D$, $B = E$ and $C = F$.

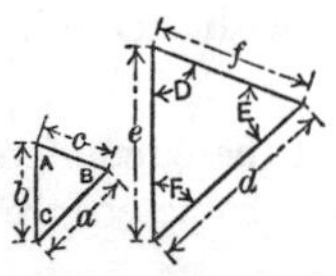

If the angles in one triangle are equal to the angles of another triangle, then the triangles are similar and their corresponding sides are proportional.

If $A = D$, $B = E$ and $C = F$, then $a : b : c = d : e : f$.

Geometrical Propositions

1. Equilateral triangle—three sides and three angles equal.
2. Line bisecting angle and opposite side of triangle.
3. Isosceles triangle—two sides and two angles equal.
4. Two angles and sides opposite them equal.
5. Bisector in an isosceles triangle.
6. Greater angle opposite greater side.

	If the three sides in a triangle are equal — that is, if the triangle is *equilateral* — then the three angles are also equal. Each of the three equal angles in an equilateral triangle is 60 degrees. If the three angles in a triangle are equal, then the three sides are also equal.
	A line which in an equilateral triangle bisects or divides any of the angles into two equal parts, bisects also the side opposite the angle and is at right angles to it. If line *AB* divides angle *CAD* into two equal parts, it also divides line *CD* into two equal parts and is at right angles to it.
	If two sides in a triangle are equal — that is, if the triangle is an *isosceles* triangle — then the angles opposite these sides are also equal. If side *a* equals side *b*, then angle *A* equals angle *B*.
	If two angles in a triangle are equal, then the sides opposite these angles are also equal. If angles *A* and *B* are equal, then side *a* equals side *b*.
	In an isosceles triangle, if a straight line is drawn from the point where the two equal sides meet, so that it bisects the third side or base of the triangle, then it also bisects the angle between the equal sides and is perpendicular to the base.
	In every triangle, that angle is greater which is opposite a longer side. — In every triangle, that side is greater which is opposite a greater angle. If *a* is longer than *b*, then angle *A* is greater than *B*. If angle *A* is greater than *B*, then side *a* is longer than *b*.

Geometrical Propositions

1. Sum of two sides of triangle is greater than third.
2. Square of hypotenuse equals sum of squares of sides.
3. Exterior angle equals sum of two interior angles.
4. Opposite angles are equal.
5. Corresponding angles are equal.
6. Sum of interior angles of four-sided figure.

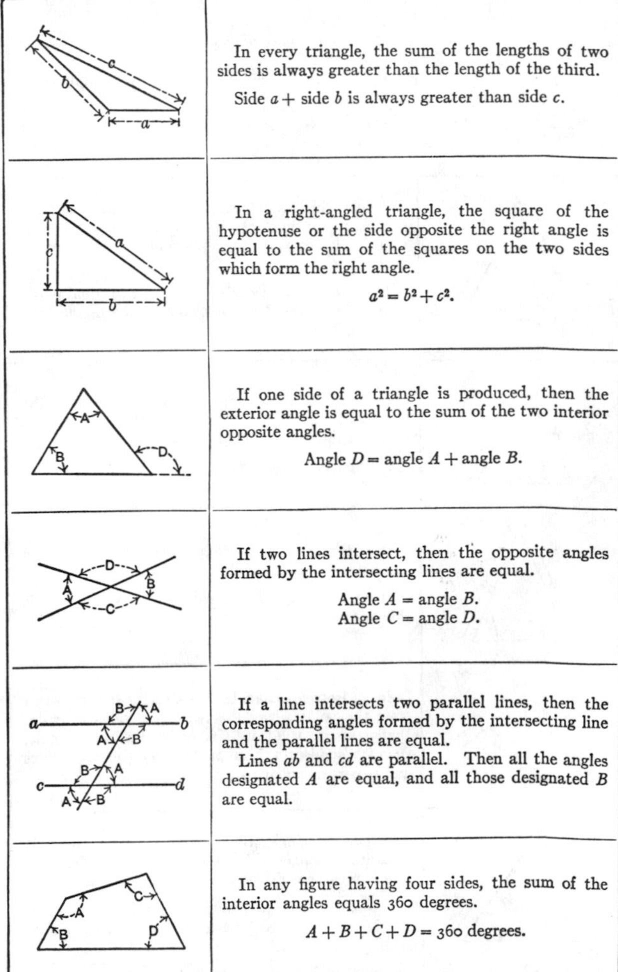

In every triangle, the sum of the lengths of two sides is always greater than the length of the third.

Side $a +$ side b is always greater than side c.

In a right-angled triangle, the square of the hypotenuse or the side opposite the right angle is equal to the sum of the squares on the two sides which form the right angle.

$$a^2 = b^2 + c^2.$$

If one side of a triangle is produced, then the exterior angle is equal to the sum of the two interior opposite angles.

Angle $D =$ angle $A +$ angle B.

If two lines intersect, then the opposite angles formed by the intersecting lines are equal.

Angle $A =$ angle B.
Angle $C =$ angle D.

If a line intersects two parallel lines, then the corresponding angles formed by the intersecting line and the parallel lines are equal.

Lines *ab* and *cd* are parallel. Then all the angles designated A are equal, and all those designated B are equal.

In any figure having four sides, the sum of the interior angles equals 360 degrees.

$$A + B + C + D = 360 \text{ degrees.}$$

Geometrical Propositions

1. Sides, angles and diagonals in parallelogram.
2. Areas of two parallelograms of equal base and height.
3. Areas of two triangles of equal base and height.
4. Diameter at right angles to chord.
5. Tangent and radius at point of tangency.
6. Two tangent circles and line through centers.

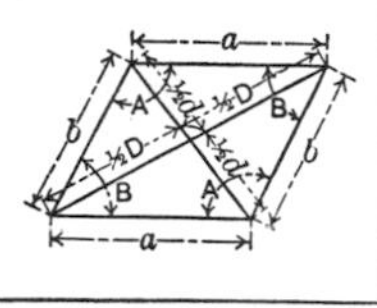	The sides which are opposite each other in a parallelogram are equal; the angles which are opposite each other are equal; the diagonal divides it into two equal parts. If two diagonals are drawn, they bisect each other.
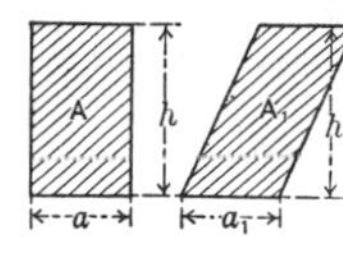	The areas of two parallelograms which have equal base and equal height, are equal. If $a = a_1$ and $h = h_1$, then area A = area A_1.
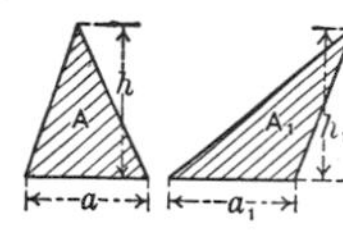	The areas of triangles having equal base and equal height are equal. If $a = a_1$ and $h = h_1$, then area A = area A_1.
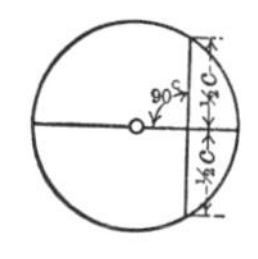	If a diameter of a circle is at right angles to a chord, then it bisects or divides the chord into two equal parts.
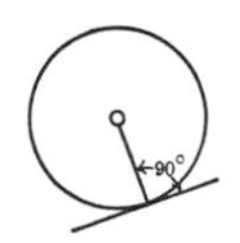	If a line is tangent to a circle, then it is also at right angles to a line drawn from the center of the circle to the point of tangency — that is, to a radial line through the point of tangency.
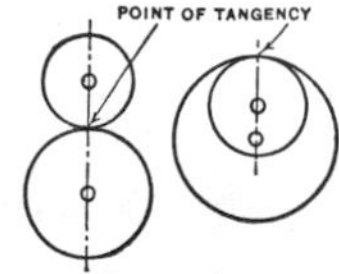	If two circles are tangent to each other, then the straight line which passes through the centers of the two circles must also pass through the point of tangency.

Geometrical Propositions

1. Two tangents to circle drawn from given point.
2. Angle between tangent and chord and angle at center.
3. Angle between tangent and chord and angle at periphery.
4. Angles subtended by common chord.
5. Angle at circumference and angle at center.
6. Size of angle when chord is greater or less than diameter.

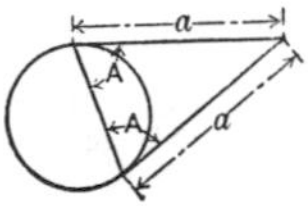 a, A, A, a	If from a point without a circle tangents are drawn to a circle, the two tangents are equal and make equal angles with the chord joining the points of tangency.
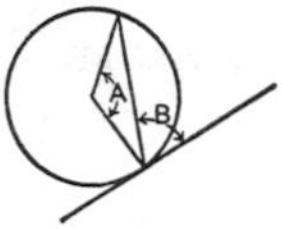 A, B	The angle between a tangent and a chord drawn from the point of tangency equals one-half the angle at the center subtended by the chord. Angle $B = \frac{1}{2}$ angle A.
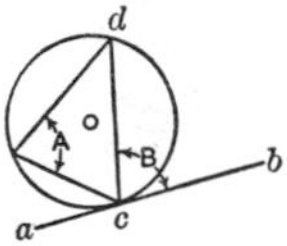 d, o, A, B, b, a, c	The angle between a tangent and a chord drawn from the point of tangency equals the angle at the periphery subtended by the chord. Angle B, between tangent ab and chord cd, equals angle A subtended at the periphery by chord cd.
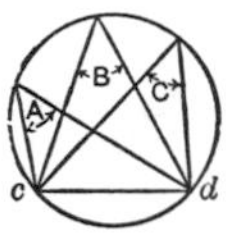 B, C, A, c, d	All angles having their vertex at the periphery of a circle and subtended by the same chord are equal. Angles A, B and C, all subtended by chord cd, are equal.
 A, B	If an angle at the circumference of a circle, between two chords, is subtended by the same arc as the angle at the center, between two radii, then the angle at the circumference is equal to one-half of the angle at the center. Angle $A = \frac{1}{2}$ angle B.
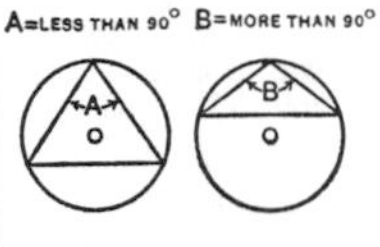 A=LESS THAN 90° B=MORE THAN 90°	An angle subtended by a chord in a circular segment larger than one-half the circle is an acute angle — an angle less than 90 degrees. An angle subtended by a chord in a circular segment less than one-half the circle is an obtuse angle — an angle greater than 90 degrees.

Geometrical Propositions

1. Two intersecting chords.
2. Tangent and secant drawn from given point.
3. Angles subtended by diameter.
4. Relation of length of arcs to corresponding angles.
5. Relation of length of arcs to length of radii.
6. Relation of circumference and area of circles to radii.

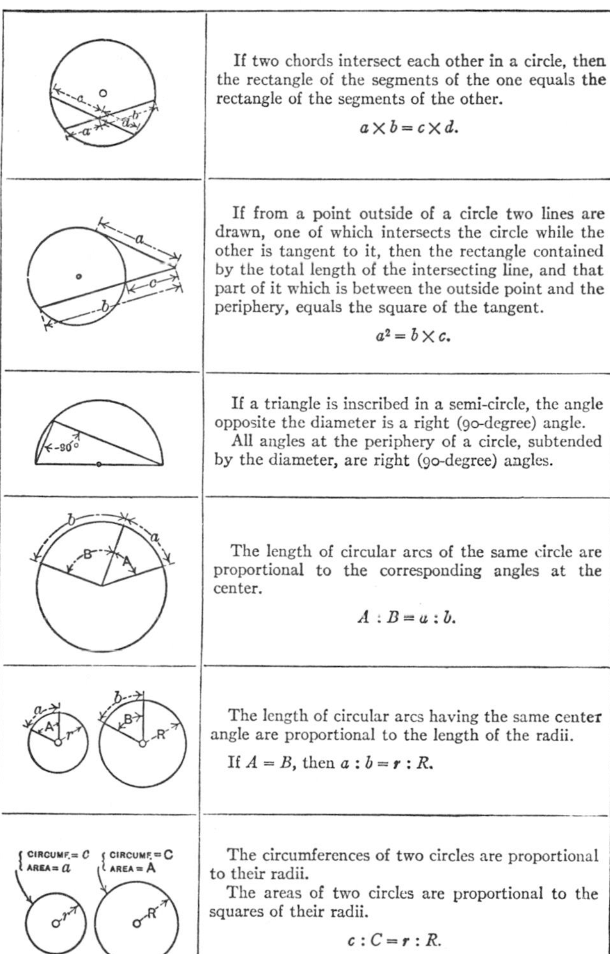

If two chords intersect each other in a circle, then the rectangle of the segments of the one equals the rectangle of the segments of the other.

$$a \times b = c \times d.$$

If from a point outside of a circle two lines are drawn, one of which intersects the circle while the other is tangent to it, then the rectangle contained by the total length of the intersecting line, and that part of it which is between the outside point and the periphery, equals the square of the tangent.

$$a^2 = b \times c.$$

If a triangle is inscribed in a semi-circle, the angle opposite the diameter is a right (90-degree) angle.

All angles at the periphery of a circle, subtended by the diameter, are right (90-degree) angles.

The length of circular arcs of the same circle are proportional to the corresponding angles at the center.

$$A : B = a : b.$$

The length of circular arcs having the same center angle are proportional to the length of the radii.

If $A = B$, then $a : b = r : R$.

The circumferences of two circles are proportional to their radii.

The areas of two circles are proportional to the squares of their radii.

$$c : C = r : R.$$

$$a : A = r^2 : R^2.$$

Geometrical Propositions

1. Two arcs equal when radii and chords are equal.
2. Line between centers at right angles to line between intersections.
3. Line drawn from circumference and perpendicular to diameter.
4. Chords at same distance from center of circle are equal.
5. Straight line between two sides and parallel to third side of triangle.

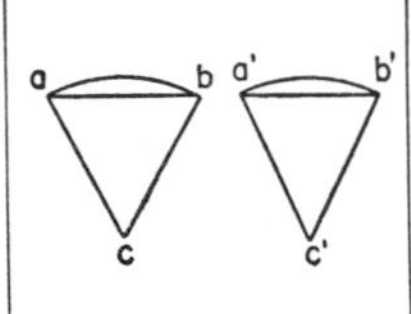

Two arcs are equal when the radius and chord of one are equal to the radius and chord of the other.

If $ac = a'c'$ and $ab = a'b'$ then
arc ab = arc $a'b'$.

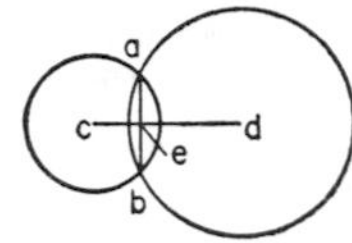

If two circles intersect each other, a line drawn between their centers bisects and is at right angles to a line drawn between the points of intersection.

$$ae = eb$$
$$\angle aed = \angle bed = \angle bec = \angle aec = 90^\circ$$

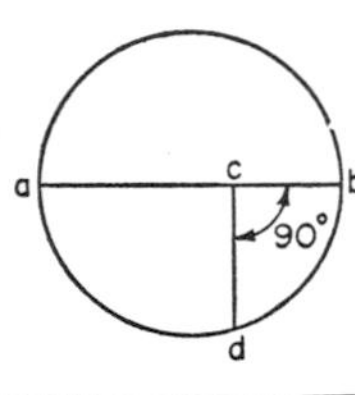

If from any point on the circumference of a circle, a line is drawn perpendicular to a diameter, this line will be a mean proportional between the two parts into which it divides the diameter.

$$ac : dc :: dc : cb$$

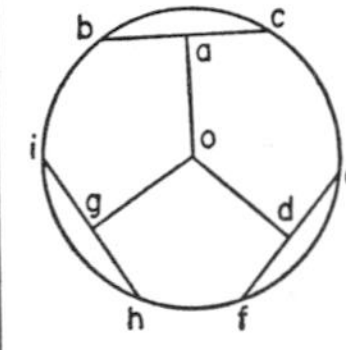

In the same circle or equal circles, chords that are the same distance from the center are equal.

If $oa = od = og$ then
$$bc = ef = hi$$

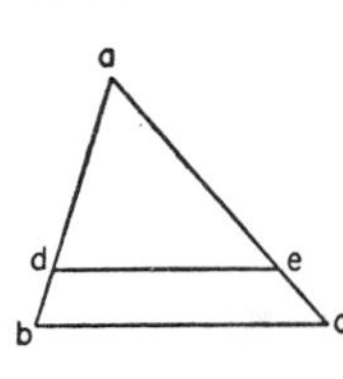

If a straight line is drawn between two sides of a triangle and parallel to the third side, it divides those sides proportionately.

If $de \parallel bc$ then
$$ad : db :: ae : ec$$

Geometrical Propositions

1. Angles equal when corresponding sides parallel or perpendicular.
2. Sum of angles at a point and on same side of straight line.
3. Sum of angles about a point.
4. Sum of interior angles of a polygon.
5. Angles inscribed in a semicircle.

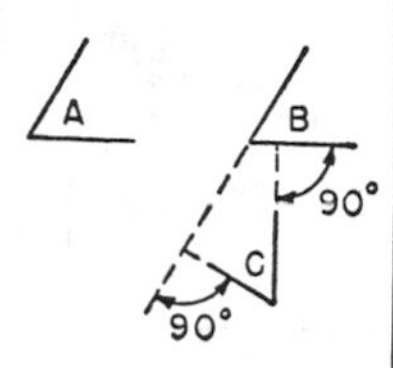	If two sides of one angle are respectively parallel to two sides of another angle, the angles are equal. If two sides of one angle are respectively perpendicular to two sides of another angle, the angles are equal. $A = B = C$
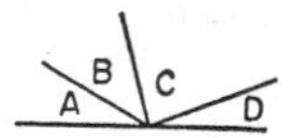	The sum of the angles formed by any number of straight lines on the same side of a straight line, meeting at a point is equal to two right angles or 180 degrees. $A + B + C + D = 180°$
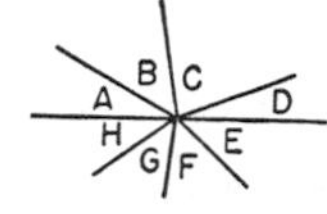	Through any given point, any number of straight lines may be drawn and the sum of all the angles formed about the point is equal to four right angles or 360 degrees $A + B + C + D + E + F + G + H = 360°$
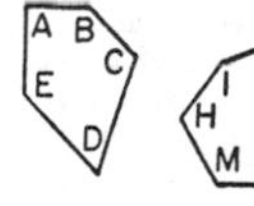	The sum of all the interior angles of any polygon is equal to the number of sides minus two, times 180 degrees. $A + B + C + D + E = (5 - 2) \times 180 = 540°$ $H + I + J + K + L + M = (6 - 2) \times 180 = 720°$
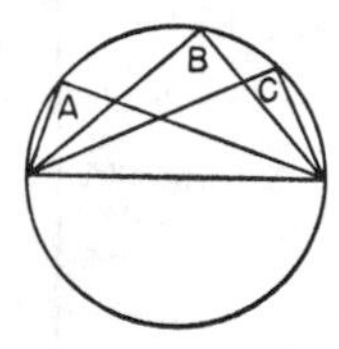	Any angle inscribed in a semicircle and intercepting a semicircumference is a right angle $A = B = C = 90°$

Geometrical Constructions

1. To divide a line into two equal parts.
2. To draw a perpendicular to a line at a point on the line.
3. To draw a perpendicular from a point at the end of a line.
4. To draw a perpendicular to a line from a given point.
5. To divide a straight line into a number of equal parts.

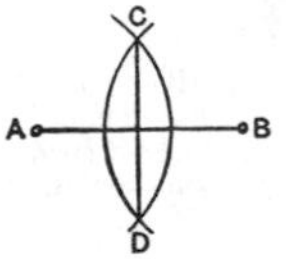

To divide a line *AB* into two equal parts: With the ends *A* and *B* as centers and a radius greater than one-half the line, draw circular arcs. Through the intersections *C* and *D*, draw line *CD*. This line divides *AB* into two equal parts and is also perpendicular to *AB*.

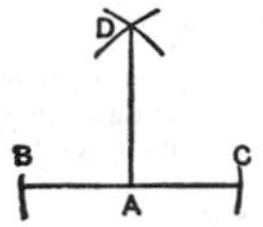

To draw a perpendicular to a straight line from a point *A* on that line: With *A* as a center and with any radius, draw circular arcs intersecting the given line at *B* and *C*. Then, with *B* and *C* as centers and a radius longer than *AB*, draw circular arcs intersecting at *D*. Line *DA* is perpendicular to *BC* at *A*.

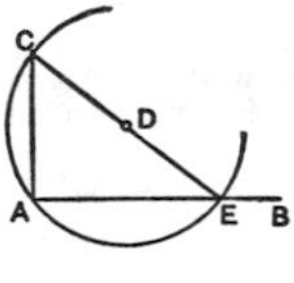

To draw a perpendicular line from a point *A* at the end of a line *AB*: With any point *D*, outside of the line *AB*, as a center, and with *AD* as a radius, draw a circular arc intersecting *AB* at *E*. Draw a line through *E* and *D* intersecting the arc at *C*; then join *AC*. This line is the required perpendicular.

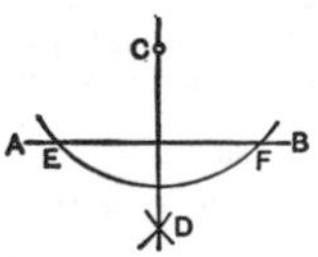

To draw a perpendicular to a line *AB* from a point *C* at a distance from it: With *C* as a center, draw a circular arc intersecting the given line at *E* and *F*. With *E* and *F* as centers, draw circular arcs with a radius longer than one-half the distance between *E* and *F*. These arcs intersect at *D*. Line *CD* is the required perpendicular.

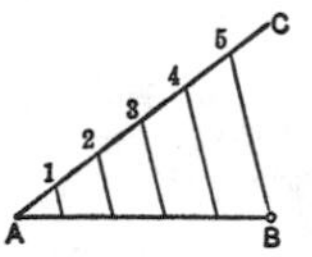

To divide a straight line *AB* into a number of equal parts: Let it be required to divide *AB* into five equal parts. Draw line *AC* at an angle with *AB*. Set off on *AC* five equal parts of any convenient length. Draw *B* 5 and then draw lines parallel with *B* 5 through the other division points on *AC*. The points where these lines intersect *AB* are the required division points.

Geometrical Constructions

1. To draw a straight line parallel to a given line.
2. To bisect a given angle.
3. To draw an angle equal to a given angle.
4. To lay out a 60-degree and a 30-degree angle.
5. To lay out a 45-degree angle.
6. To draw an equilateral triangle.

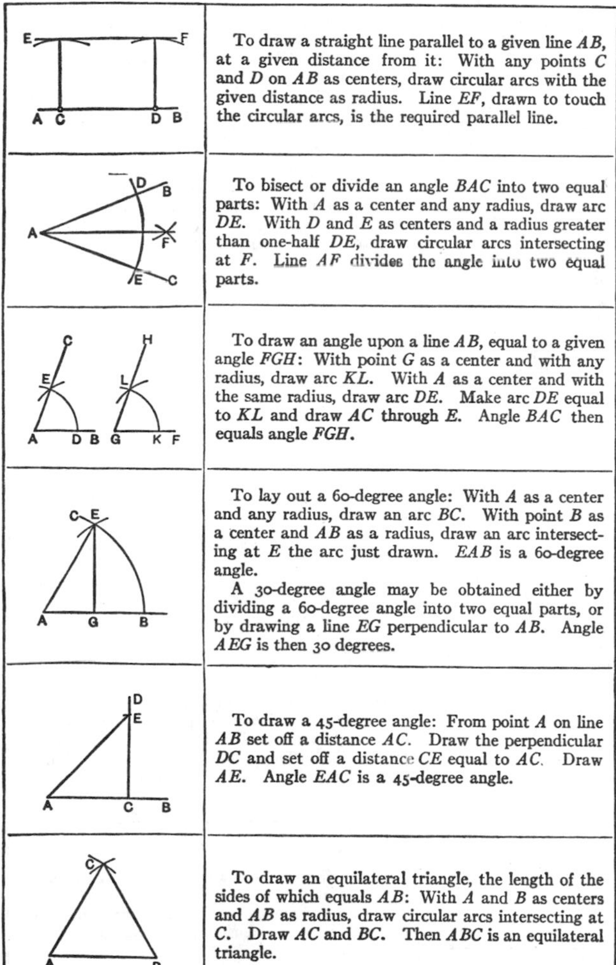

To draw a straight line parallel to a given line *AB*, at a given distance from it: With any points *C* and *D* on *AB* as centers, draw circular arcs with the given distance as radius. Line *EF*, drawn to touch the circular arcs, is the required parallel line.

To bisect or divide an angle *BAC* into two equal parts: With *A* as a center and any radius, draw arc *DE*. With *D* and *E* as centers and a radius greater than one-half *DE*, draw circular arcs intersecting at *F*. Line *AF* divides the angle into two equal parts.

To draw an angle upon a line *AB*, equal to a given angle *FGH*: With point *G* as a center and with any radius, draw arc *KL*. With *A* as a center and with the same radius, draw arc *DE*. Make arc *DE* equal to *KL* and draw *AC* through *E*. Angle *BAC* then equals angle *FGH*.

To lay out a 60-degree angle: With *A* as a center and any radius, draw an arc *BC*. With point *B* as a center and *AB* as a radius, draw an arc intersecting at *E* the arc just drawn. *EAB* is a 60-degree angle.

A 30-degree angle may be obtained either by dividing a 60-degree angle into two equal parts, or by drawing a line *EG* perpendicular to *AB*. Angle *AEG* is then 30 degrees.

To draw a 45-degree angle: From point *A* on line *AB* set off a distance *AC*. Draw the perpendicular *DC* and set off a distance *CE* equal to *AC*. Draw *AE*. Angle *EAC* is a 45-degree angle.

To draw an equilateral triangle, the length of the sides of which equals *AB*: With *A* and *B* as centers and *AB* as radius, draw circular arcs intersecting at *C*. Draw *AC* and *BC*. Then *ABC* is an equilateral triangle.

Geometrical Constructions

1. To draw arc of given radius through two given points.
2. To find center of circle or of arc of a circle.
3. To draw tangent to a circle at a given point.
4. To divide a circular arc into two equal parts.
5. To describe a circle about a triangle.
6. To inscribe a circle in a triangle.

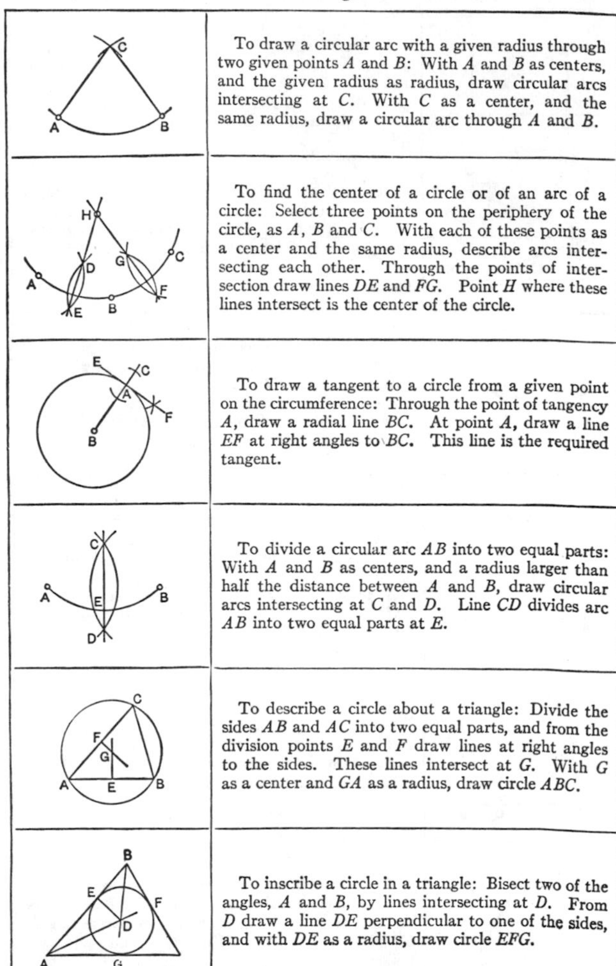

To draw a circular arc with a given radius through two given points *A* and *B*: With *A* and *B* as centers, and the given radius as radius, draw circular arcs intersecting at *C*. With *C* as a center, and the same radius, draw a circular arc through *A* and *B*.

To find the center of a circle or of an arc of a circle: Select three points on the periphery of the circle, as *A*, *B* and *C*. With each of these points as a center and the same radius, describe arcs intersecting each other. Through the points of intersection draw lines *DE* and *FG*. Point *H* where these lines intersect is the center of the circle.

To draw a tangent to a circle from a given point on the circumference: Through the point of tangency *A*, draw a radial line *BC*. At point *A*, draw a line *EF* at right angles to *BC*. This line is the required tangent.

To divide a circular arc *AB* into two equal parts: With *A* and *B* as centers, and a radius larger than half the distance between *A* and *B*, draw circular arcs intersecting at *C* and *D*. Line *CD* divides arc *AB* into two equal parts at *E*.

To describe a circle about a triangle: Divide the sides *AB* and *AC* into two equal parts, and from the division points *E* and *F* draw lines at right angles to the sides. These lines intersect at *G*. With *G* as a center and *GA* as a radius, draw circle *ABC*.

To inscribe a circle in a triangle: Bisect two of the angles, *A* and *B*, by lines intersecting at *D*. From *D* draw a line *DE* perpendicular to one of the sides, and with *DE* as a radius, draw circle *EFG*.

Geometrical Constructions

1. To describe a circle about and inscribe a circle in a square.
2. To inscribe a hexagon in a circle.
3. To describe a hexagon about a circle.
4. To describe an ellipse having given axes.
5. To construct an approximate ellipse.

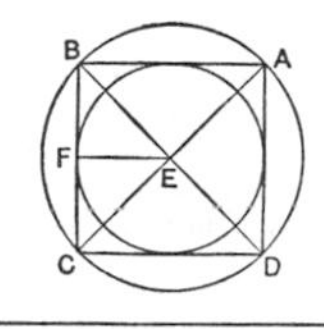	To describe a circle about a square and to inscribe a circle in a square: The center of both the circumscribed and inscribed circle is located at the point *E*, where the two diagonals of the square intersect. The radius of the circumscribed circle is *AE*, and of the inscribed circle, *EF*.
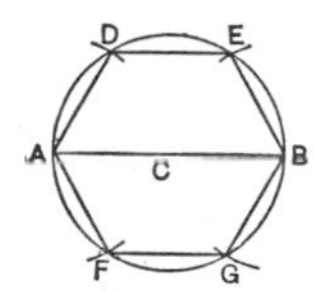	To inscribe a hexagon in a circle: Draw a diameter *AB*. With *A* and *B* as centers and with the radius of the circle as radius, describe circular arcs intersecting the given circle at *D, E, F* and *G*. Draw lines *AD, DE*, etc., forming the required hexagon.
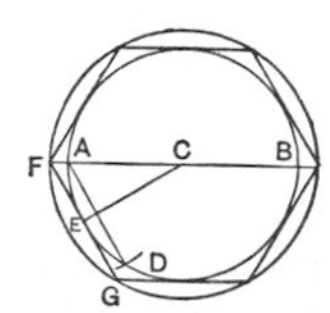	To describe a hexagon about a circle: Draw a diameter *AB*, and with *A* as a center and the radius of the circle as radius, cut the circumference of the given circle at *D*. Join *AD* and bisect it with radius *CE*. Through *E*, draw *FG* parallel to *AD* and intersecting line *AB* at *F*. With *C* as a center and *CF* as radius, draw a circle. Within this circle inscribe the hexagon as in the preceding problem.
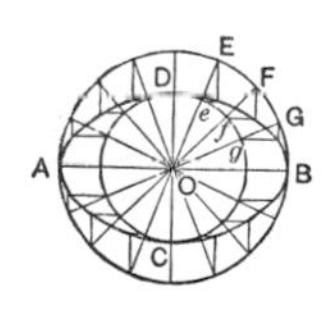	To describe an ellipse with the given axes *AB* and *CD*: Describe circles with *O* as a center and *AB* and *CD* as diameters. From a number of points, *E, F, G*, etc., on the outer circle draw radii intersecting the inner circle at *e, f, g*. From *E, F* and *G* draw lines perpendicular to *AB*, and from *e, f* and *g* draw lines parallel to *AB*. The intersections of these perpendicular and parallel lines are points on the curve of the ellipse.
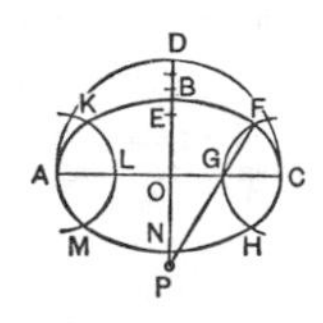	To construct an approximate ellipse by circular arcs: Let *AC* be the major axis and *BN* the minor. Draw half circle *ADC* with *O* as a center. Divide *BD* into three equal parts and set off *BE* equal to one of these parts. With *A* and *C* as centers and *OE* as radius, describe circular arcs *KLM* and *FGH*; with *G* and *L* as centers, and the same radius, describe arcs *FCH* and *KAM*. Through *F* and *G* draw line *FP*, and with *P* as a center draw the arc *FBK*. Arc *HNM* is drawn in the same manner.

Geometrical Constructions

1. To construct a parabola.
2. To construct a hyperbola.
3. To construct an involute.
4. To construct a helix.

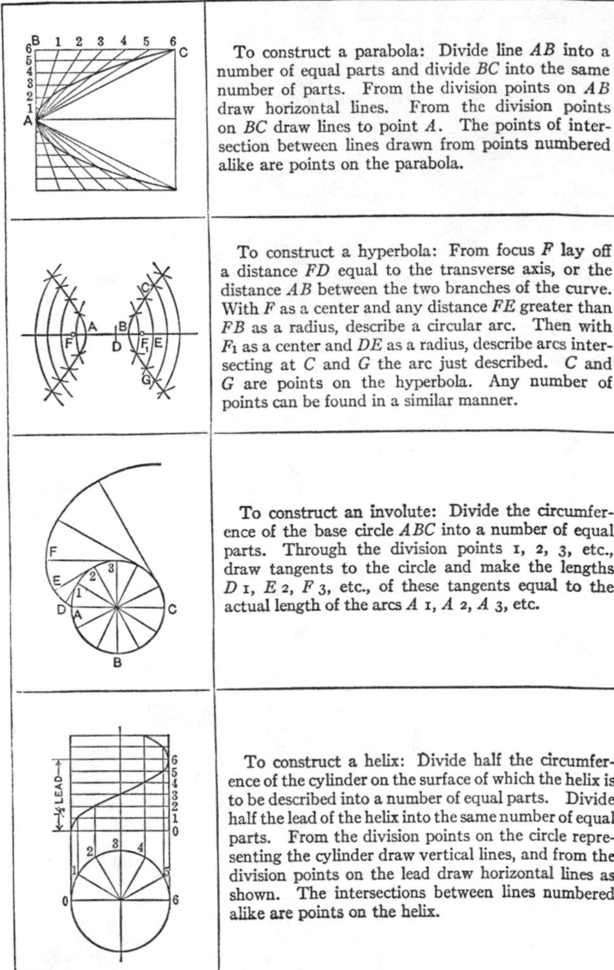

To construct a parabola: Divide line *AB* into a number of equal parts and divide *BC* into the same number of parts. From the division points on *AB* draw horizontal lines. From the division points on *BC* draw lines to point *A*. The points of intersection between lines drawn from points numbered alike are points on the parabola.

To construct a hyperbola: From focus *F* lay off a distance *FD* equal to the transverse axis, or the distance *AB* between the two branches of the curve. With *F* as a center and any distance *FE* greater than *FB* as a radius, describe a circular arc. Then with F_1 as a center and *DE* as a radius, describe arcs intersecting at *C* and *G* the arc just described. *C* and *G* are points on the hyperbola. Any number of points can be found in a similar manner.

To construct an involute: Divide the circumference of the base circle *ABC* into a number of equal parts. Through the division points 1, 2, 3, etc., draw tangents to the circle and make the lengths *D* 1, *E* 2, *F* 3, etc., of these tangents equal to the actual length of the arcs *A* 1, *A* 2, *A* 3, etc.

To construct a helix: Divide half the circumference of the cylinder on the surface of which the helix is to be described into a number of equal parts. Divide half the lead of the helix into the same number of equal parts. From the division points on the circle representing the cylinder draw vertical lines, and from the division points on the lead draw horizontal lines as shown. The intersections between lines numbered alike are points on the helix.

Geometrical Constructions

1. To draw tangent to given circle from given point.
2. To draw circular tangent to two non-parallel lines.
3. To draw outside tangent to two circles.
4. To draw inside tangent to two circles.
5. To draw circle of given size tangent to two circles.

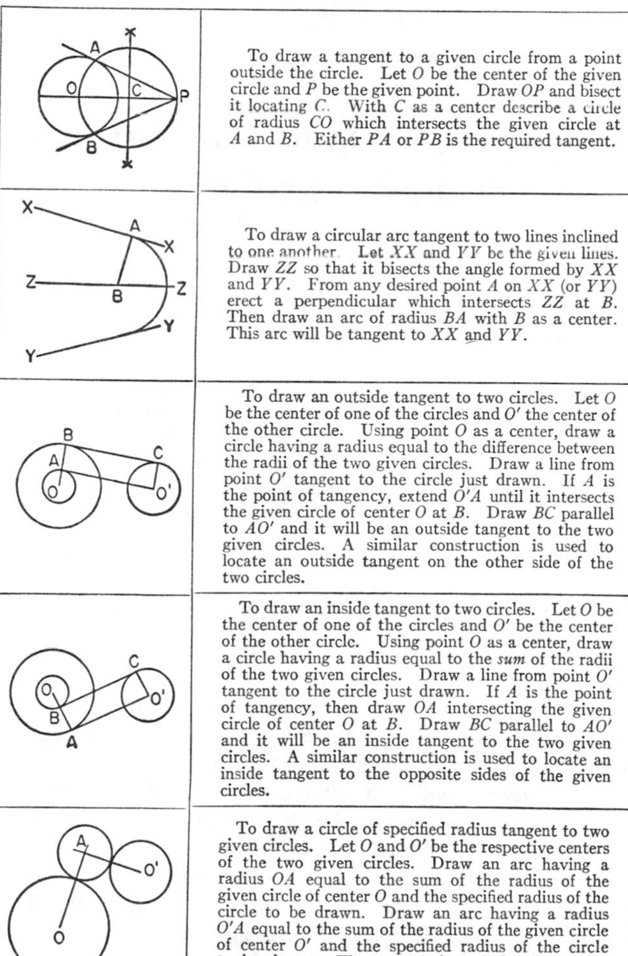

To draw a tangent to a given circle from a point outside the circle. Let O be the center of the given circle and P be the given point. Draw OP and bisect it locating C. With C as a center describe a circle of radius CO which intersects the given circle at A and B. Either PA or PB is the required tangent.

To draw a circular arc tangent to two lines inclined to one another. Let XX and YY be the given lines. Draw ZZ so that it bisects the angle formed by XX and YY. From any desired point A on XX (or YY) erect a perpendicular which intersects ZZ at B. Then draw an arc of radius BA with B as a center. This arc will be tangent to XX and YY.

To draw an outside tangent to two circles. Let O be the center of one of the circles and O' the center of the other circle. Using point O as a center, draw a circle having a radius equal to the difference between the radii of the two given circles. Draw a line from point O' tangent to the circle just drawn. If A is the point of tangency, extend $O'A$ until it intersects the given circle of center O at B. Draw BC parallel to AO' and it will be an outside tangent to the two given circles. A similar construction is used to locate an outside tangent on the other side of the two circles.

To draw an inside tangent to two circles. Let O be the center of one of the circles and O' be the center of the other circle. Using point O as a center, draw a circle having a radius equal to the *sum* of the radii of the two given circles. Draw a line from point O' tangent to the circle just drawn. If A is the point of tangency, then draw OA intersecting the given circle of center O at B. Draw BC parallel to AO' and it will be an inside tangent to the two given circles. A similar construction is used to locate an inside tangent to the opposite sides of the given circles.

To draw a circle of specified radius tangent to two given circles. Let O and O' be the respective centers of the two given circles. Draw an arc having a radius OA equal to the sum of the radius of the given circle of center O and the specified radius of the circle to be drawn. Draw an arc having a radius $O'A$ equal to the sum of the radius of the given circle of center O' and the specified radius of the circle to be drawn. The intersection of these two arcs locates the center of the circle to be drawn.

4

Trigonometry—Fundamentals Utilized in Solving Practical Problems

1. FUNCTIONS AND FORMULAS

Trigonometric Functions. In the table on page 4–2, the various trigonometric functions are defined and the formulas for each are given. It may be noted that the sine of an angle is equal to the cosine of its complement; the tangent of an angle is equal to the cotangent of its complement; and the secant of an angle is equal to the cosecant of its complement. Since, in a right-angled triangle, either of the acute angles is the complement of the other, the function of either of the acute angles is equal to the co-function of the other.

Sign of Trigonometric Functions. In the charts on pages 4–3 to 4–14 are shown the various functions drawn for angles in each of four quadrants of a circle. Since the radius of the circle is taken to be 1, the function in each case can be represented by a single line. The sign of each function is *plus* for angles located in two of the quadrants and *minus* for angles located in two of the quadrants. For example, from 0 to 90 degrees and from 90 degrees to 180 degrees the sine is positive. From 180 degrees to 270 degrees and from 270 degrees to 360 degrees the sine is negative. Also given in these charts are the functions of 90 degrees minus the angle, 90 degrees plus the angle, 180 degrees minus the angle, 180 degrees plus the angle, 270 degrees minus the angle, 270 degrees plus the angle, and 360 degrees minus the angle in terms of the angle.

Trigonometric Functions — Basic Formulas and Rules

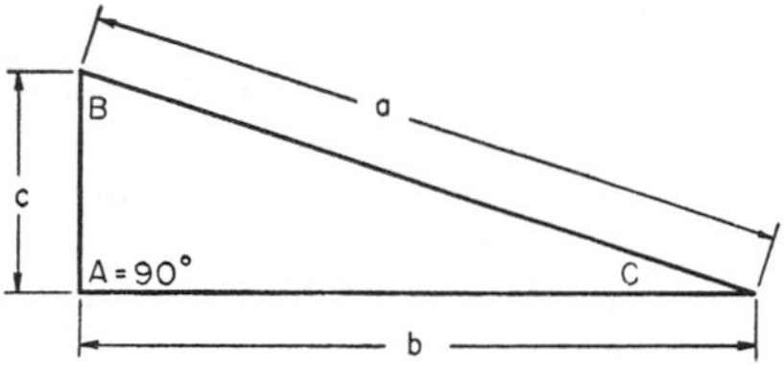

Formula	Rule
$\sin A = 1$ $\sin B = b \div a$ $\sin C = c \div a$	The *sine* of an angle equals the opposite side divided by the hypotenuse.
$\cos A = 0$ $\cos B = c \div a$ $\cos C = b \div a$	The *cosine* of an angle equals the adjacent side divided by the hypotenuse.
$\tan A = \infty$ $\tan B = b \div c$ $\tan C = c \div b$	The *tangent* of an angle equals the opposite side divided by the adjacent side.
$\cot A = 0$ $\cot B = c \div b$ $\cot C = b \div c$	The *cotangent* of an angle equals the adjacent side divided by the opposite side.
$\sec A = \infty$ $\sec B = a \div c$ $\sec C = a \div b$	The *secant* of an angle equals the hypotenuse divided by the adjacent side.
$\text{cosec}\ A = 1$ $\text{cosec}\ B = a \div b$ $\text{cosec}\ C = a \div c$	The *cosecant* of an angle equals the hypotenuse divided by the opposite side.
$\text{vers}\ A = 1$ $\text{vers}\ B = 1 - \cos B$ $\text{vers}\ C = 1 - \cos C$	The *versine* of an angle equals one minus the cosine of the angle.
$\text{covers}\ A = 0$ $\text{covers}\ B = 1 - \sin B$ $\text{covers}\ C = 1 - \sin C$	The *coversine* of an angle equals one minus the sine of the angle.

2. BASIC FACTS ABOUT THE SINE

Explanation of Diagrams. Each of the diagrams shown on page 4–4 consists of a circle, having a radius equal to 1, which is divided into four quadrants. In the first diagram a line *ob* has been drawn at angle A with the 0-degree line. In the succeeding diagrams *ob* is drawn at some multiple of 90 degrees plus or minus A with the 0-degree line, measuring in a counter-clockwise direction. In each case, the sine of the angle indicated is equal to $ab \div ob$. Since *ob* is a radius of the circle and is therefore equal to 1, the sine of the indicated angle is, in each case, equal to *ab*.

When Sine is Positive and When Negative. It will be noted from the smaller key diagram at the top-center of this figure, that the two upper quadrants have plus signs, indicating that the sines of all angles in these two quadrants are positive, and the two lower quadrants have minus signs, indicating that the sines of all angles in these two quadrants are negative.

Multiples of 90 degrees. It will also be seen that:

$$\text{Sine } 0° = 0 \qquad \text{Sine } 180° = 0$$

$$\text{Sine } 90° = 1 \qquad \text{Sine } 270° = -1$$

$$\text{Sine } 360° = 0$$

Multiples of 90 Degrees Plus or Minus a Given Angle. It is sometimes convenient to convert the function for an angle greater than 90 degrees to the equivalent function of an angle of less than 90 degrees. Since every angle greater than 90 degrees is equal to some multiple of 90 degrees, plus or minus some angle less than 90 degrees, the formulas derived from the eight diagrams shown, facilitate the conversion. Thus, it will be seen that:

$$\text{Sine } (90° - A) = \text{Cosine } A$$

$$\text{Sine } (90° + A) = \text{Cosine } A$$

$$\text{Sine } (180° - A) = \text{Sine } A$$

$$\text{Sine } (180° + A) = -\text{ Sine } A$$

$$\text{Sine } (270° - A) = -\text{ Cosine } A$$

$$\text{Sine } (270° + A) = -\text{ Cosine } A$$

$$\text{Sine } (360° - A) = -\text{ Sine } A$$

$$\text{Sine } (360° + A) = \text{Sine } A$$

$$\text{Sine } (-A) = -\text{ Sine } A$$

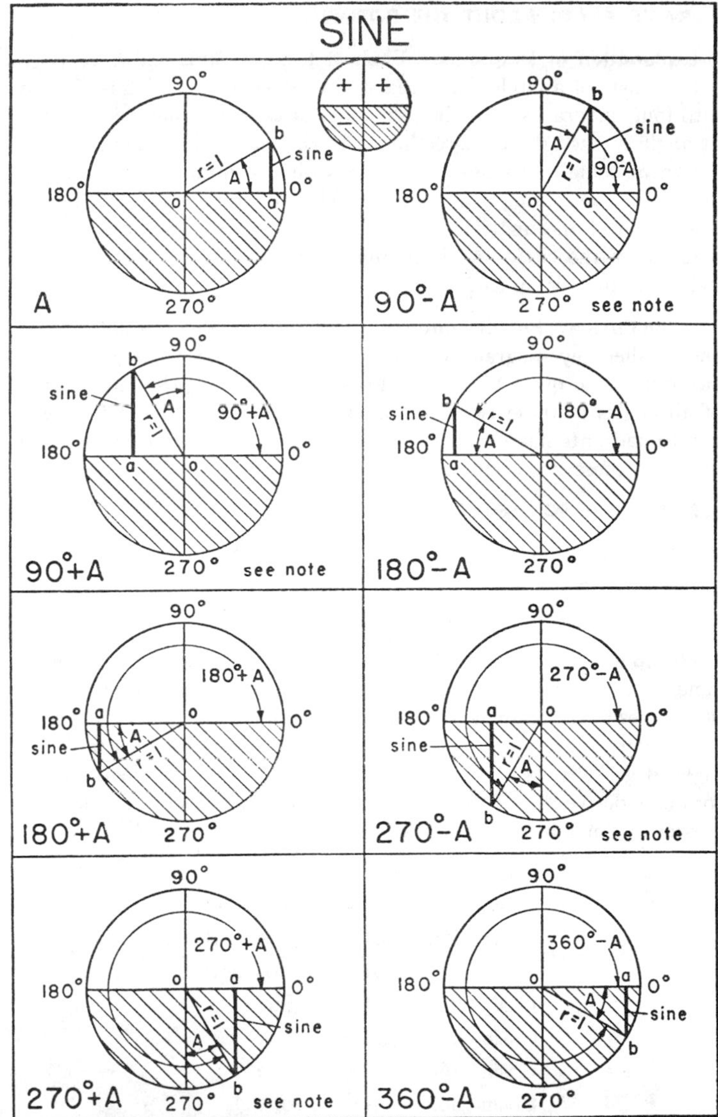

NOTE: For 90° – *A*, 90° + *A*, 270° – *A*, and 270° + *A*, angle *oba* equals angle *A*.

3. BASIC FACTS ABOUT THE COSINE

Explanation of Diagrams. Each of the diagrams shown on page 4–6 consists of a circle, having a radius equal to 1, which is divided into four quadrants. In the first diagram a line *ob* has been drawn at angle A with the 0-degree line. In the succeeding diagrams *ob* is drawn at some multiple of 90 degrees plus or minus A with the 0-degree line, measuring in a counter-clockwise direction. In each case, the cosine of the angle indicated is equal to $oa \div ob$. Since *ob* is a radius of the circle and is therefore equal to 1, the cosine of the indicated angle is, in each case, equal to *oa*.

When Cosine is Positive and When Negative. It will be noted from the smaller key diagram at the top-center of this figure, that the two right-hand quadrants have plus signs, indicating that the cosines of all angles in these two quadrants are positive, and the two left-hand quadrants have minus signs, indicating that the cosines of all angles in these two quadrants are negative.

Multiples of 90 degrees. It will also be seen that:

$$\text{Cosine } 0° = 1 \qquad \text{Cosine } 180° = -1$$
$$\text{Cosine } 90° = 0 \qquad \text{Cosine } 270° = 0$$
$$\text{Cosine } 360° = 1$$

Multiples of 90 Degrees Plus or Minus a Given Angle. It is sometimes convenient to convert the function of some angle greater than 90 degrees to the equivalent function of an angle less than 90 degrees. Since every angle greater than 90 degrees is equal to some multiple of 90 degrees, plus or minus some angle less than 90 degrees, the formulas derived from the eight diagrams shown, facilitate the conversion. Thus, it will be seen that:

$$\text{Cosine } 90° - A = \text{Sine } A$$
$$\text{Cosine } 90° + A = -\text{ Sine } A$$
$$\text{Cosine } 180° - A = -\text{ Cosine } A$$
$$\text{Cosine } 180° + A = -\text{ Cosine } A$$
$$\text{Cosine } 270° - A = -\text{ Sine } A$$
$$\text{Cosine } 270° + A = \text{Sine } A$$
$$\text{Cosine } 360° - A = \text{Cosine } A$$
$$\text{Cosine } 360° + A = \text{Cosine } A$$
$$\text{Cosine } -A = \text{Cosine } A$$

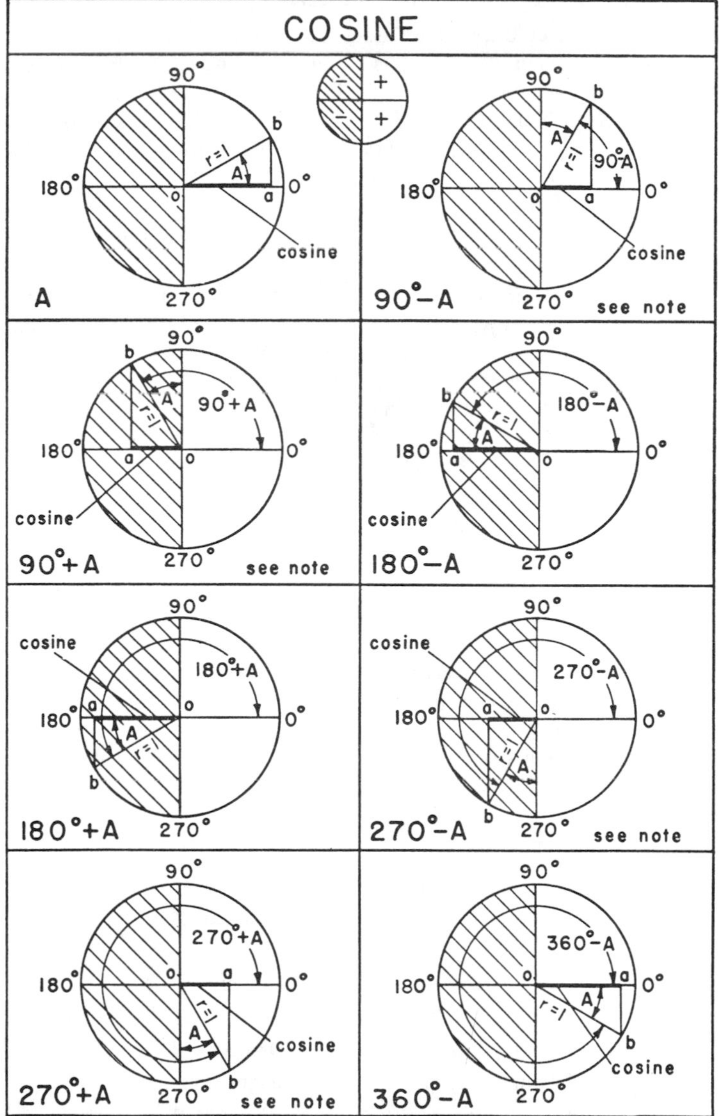

NOTE: For 90° − *A*, 90° + *A*, 270° − *A*, and 270° + *A*, angle *oba* equals angle *A*

4. BASIC FACTS ABOUT THE TANGENT

Explanation of Diagrams. Each of the diagrams shown on page 4–8 consists of a circle, having a radius equal to 1, which is divided into four quadrants. In the first diagram a line *od* has been drawn at angle A with the 0-degree line. In the succeeding diagrams *od* is drawn at some multiple of 90 degrees plus or minus A with the 0-degree line, measuring in a counter-clockwise direction. In each case, the tangent of the angle indicated is equal to $cd \div oc$. Since *oc* is a radius of the circle and is therefore equal to 1, the tangent of the indicated angle is, in each case, equal to *cd*.

When Tangent is Positive and When Negative. It will be noted from the smaller key diagram at the top-center of this figure, that the upper right and lower left-hand quadrants have plus signs, indicating that the tangents of all angles in these two quadrants are positive, and the upper left- and lower right-hand quadrants have minus signs, indicating that the tangents of all angles in these two quadrants are negative.

Multiples of 90 degrees. It will also be seen that:

$$\text{Tangent } 0° = 0 \qquad \text{Tangent } 180° = 0$$
$$\text{Tangent } 90° = \infty \qquad \text{Tangent } 270° = \infty$$
$$\text{Tangent } 360° = 0$$

Multiples of 90 Degrees Plus or Minus a Given Angle. It is sometimes convenient to convert the function for an angle greater than 90 degrees to the equivalent function of an angle of less than 90 degrees. Since every angle greater than 90 degrees is equal to some multiple of 90 degrees, plus or minus some angle less than 90 degrees, the formulas derived from the eight diagrams shown, facilitate the conversion. Thus, it will be seen that:

$$\text{Tangent } (90° - A) = \text{Cotangent } A$$
$$\text{Tangent } (90° + A) = -\text{ Cotangent } A$$
$$\text{Tangent } (180° - A) = -\text{ Tangent } A$$
$$\text{Tangent } (180° + A) = \text{Tangent } A$$
$$\text{Tangent } (270° - A) = \text{Cotangent } A$$
$$\text{Tangent } (270° + A) = -\text{ Cotangent } A$$
$$\text{Tangent } (360° - A) = -\text{ Tangent } A$$
$$\text{Tangent } (360° + A) = \text{Tangent } A$$
$$\text{Tangent } (-A) = -\text{ Tangent } A$$

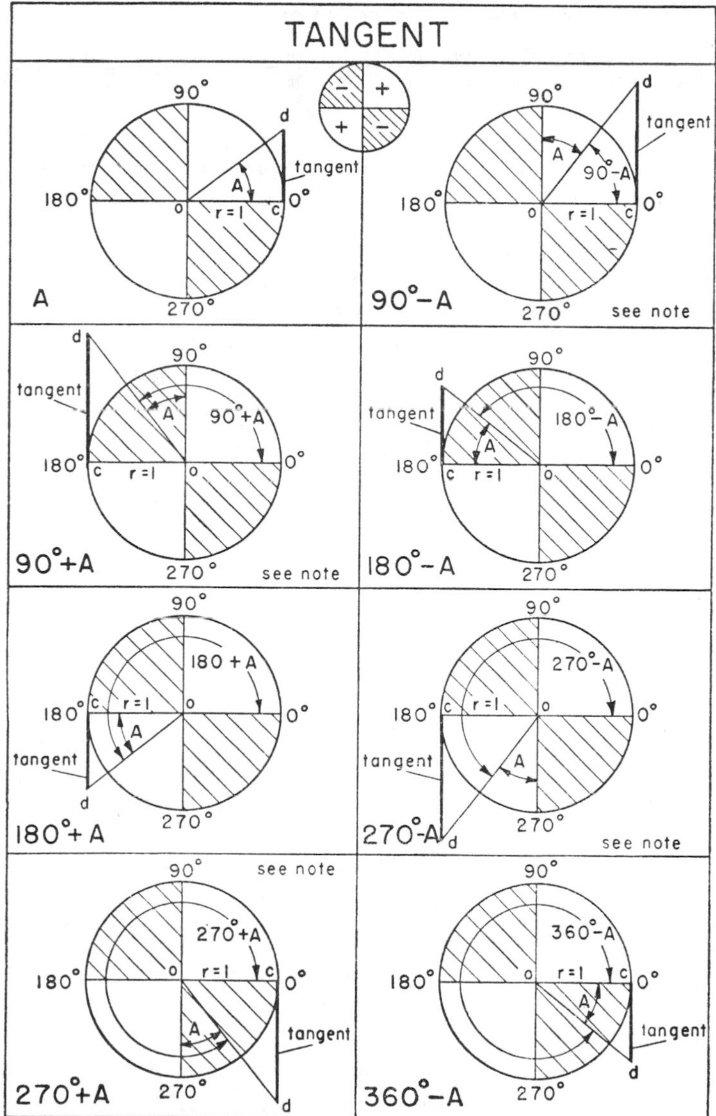

NOTE: For 90° – A, 90° + A, 270° – A and 270° + A, angle odc equals angle A.

5. BASIC FACTS ABOUT THE COTANGENT

Explanation of Diagrams. Each of the diagrams shown on page 4–10 consists of a circle, having a radius equal to 1, which is divided into four quadrants. In the first diagram a line *of* has been drawn at angle A with the 0-degree line. In the succeeding diagrams *of* is drawn at some multiple of 90 degrees plus or minus A with the 0-degree line, measuring in a counter-clockwise direction. In each case, the cotangent of the angle indicated is equal to $ef \div oe$. Since *oe* is a radius of the circle and is therefore equal to 1, the cotangent of the indicated angle is, in each case, equal to *ef*.

When Cotangent is Positive and When Negative. It will be noted from the smaller key diagram at the top-center of this figure, that the upper right and lower left-hand quadrants have plus signs, indicating that the cotangents of all angles in these two quadrants are positive, and the upper left and lower right-hand quadrants have minus signs, indicating that the cotangents of all angles in these two quadrants are negative.

Multiples of 90 Degrees. It will also be seen that:

$$\text{Cotangent } 0° = \infty \qquad \text{Cotangent } 180° = \infty$$
$$\text{Cotangent } 90° = 0 \qquad \text{Cotangent } 270° = 0$$
$$\text{Cotangent } 360° = \infty$$

Multiples of 90 Degrees Plus or Minus a Given Angle. It is sometimes convenient to convert the function for an angle greater than 90 degrees to the equivalent function of an angle of less than 90 degrees. Since every angle greater than 90 degrees is equal to some multiple of 90 degrees, plus or minus some angle less than 90 degrees, the formulas derived from the eight diagrams shown, facilitate the conversion. Thus, it will be seen that:

$$\text{Cotangent } (90° - A) = \text{Tangent } A$$
$$\text{Cotangent } (90° + A) = -\text{ Tangent } A$$
$$\text{Cotangent } (180° - A) = -\text{ Cotangent } A$$
$$\text{Cotangent } (180° + A) = \text{Cotangent } A$$
$$\text{Cotangent } (270° - A) = \text{Tangent } A$$
$$\text{Cotangent } (270° + A) = -\text{ Tangent } A$$
$$\text{Cotangent } (360° - A) = -\text{ Cotangent } A$$
$$\text{Cotangent } (360° + A) = \text{Cotangent } A$$
$$\text{Cotangent } (-A) = -\text{ Cotangent } A$$

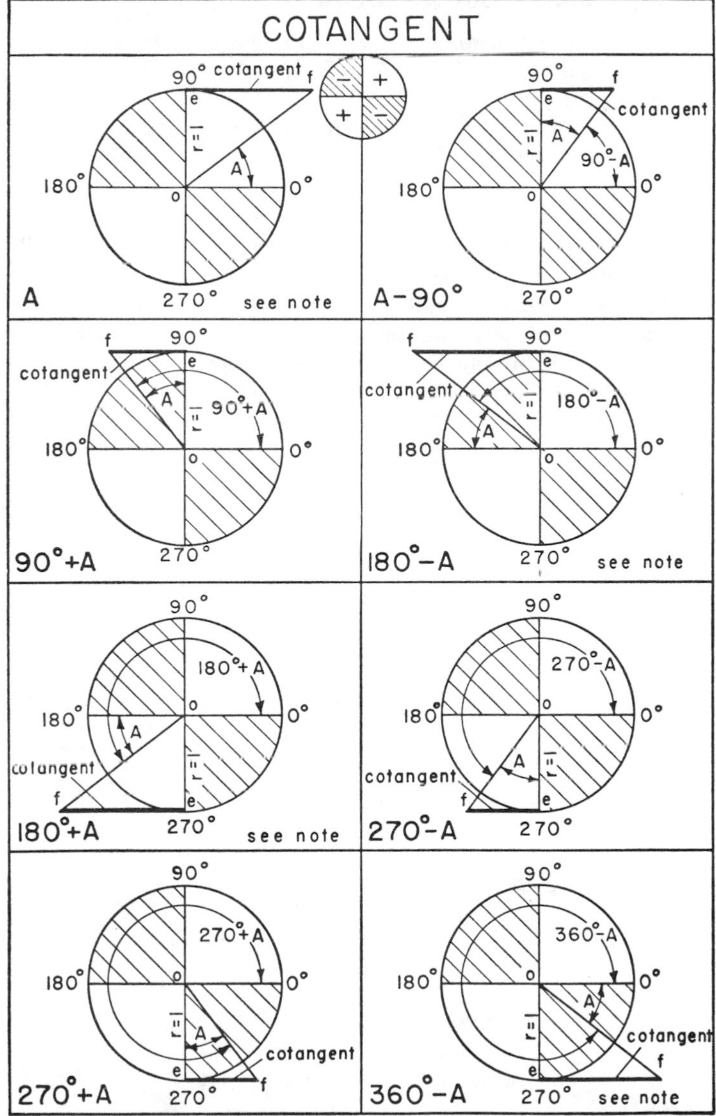

NOTE: For A, $180° - A$, $180° + A$, and $360° - A$, angle *ofe* equals angle A.

6. BASIC FACTS ABOUT THE SECANT

Explanation of Diagrams. Each of the diagrams shown on page 4–12 consists of a circle, having a radius equal to 1 which is divided into four quadrants. In the first diagram a line *od* has been drawn at angle A with the 0-degree line. In the succeeding diagrams *od* is drawn at some multiple of 90 degrees plus or minus A with the 0-degree line, measuring in a counter-clockwise direction. In each case, the secant of the angle indicated is equal to $od \div oc$. Since *oc* is a radius of the circle and is therefore equal to 1, the secant of the indicated angle is, in each case, equal to *od*.

When Secant is Positive and When Negative. It will be noted from the smaller key diagram at the top-center of this figure, that the two right-hand quadrants have plus signs, indicating that the secants of all angles in these two quadrants are positive and the two left-hand quadrants have minus signs, indicating that the secants of all angles in these two quadrants are negative.

Multiples of 90 Degrees. It will also be seen that:

$$\text{Secant } 0^\circ = 1 \qquad \text{Secant } 180^\circ = -1$$

$$\text{Secant } 90^\circ = \infty \qquad \text{Secant } 270^\circ = \infty$$

$$\text{Secant } 360^\circ = 1$$

Multiples of 90 Degrees Plus or Minus a Given Angle. It is sometimes convenient to convert the function for an angle greater than 90 degrees to the equivalent function of an angle of less than 90 degrees. Since every angle greater than 90 degrees is equal to some multiple of 90 degrees, plus or minus some angle less than 90 degrees, the formulas derived from the eight diagrams shown, facilitate the conversion. Thus, it will be seen that:

$$\text{Secant } (90^\circ - A) = \text{Cosecant } A$$

$$\text{Secant } (90^\circ + A) = -\text{ Cosecant } A$$

$$\text{Secant } (180^\circ - A) = -\text{ Secant } A$$

$$\text{Secant } (180^\circ + A) = -\text{ Secant } A$$

$$\text{Secant } (270^\circ - A) = -\text{ Cosecant } A$$

$$\text{Secant } (270^\circ + A) = \text{Cosecant } A$$

$$\text{Secant } (360^\circ - A) = \text{Secant } A$$

$$\text{Secant } (360^\circ + A) = \text{Secant } A$$

$$\text{Secant } (-A) = \text{Secant } A$$

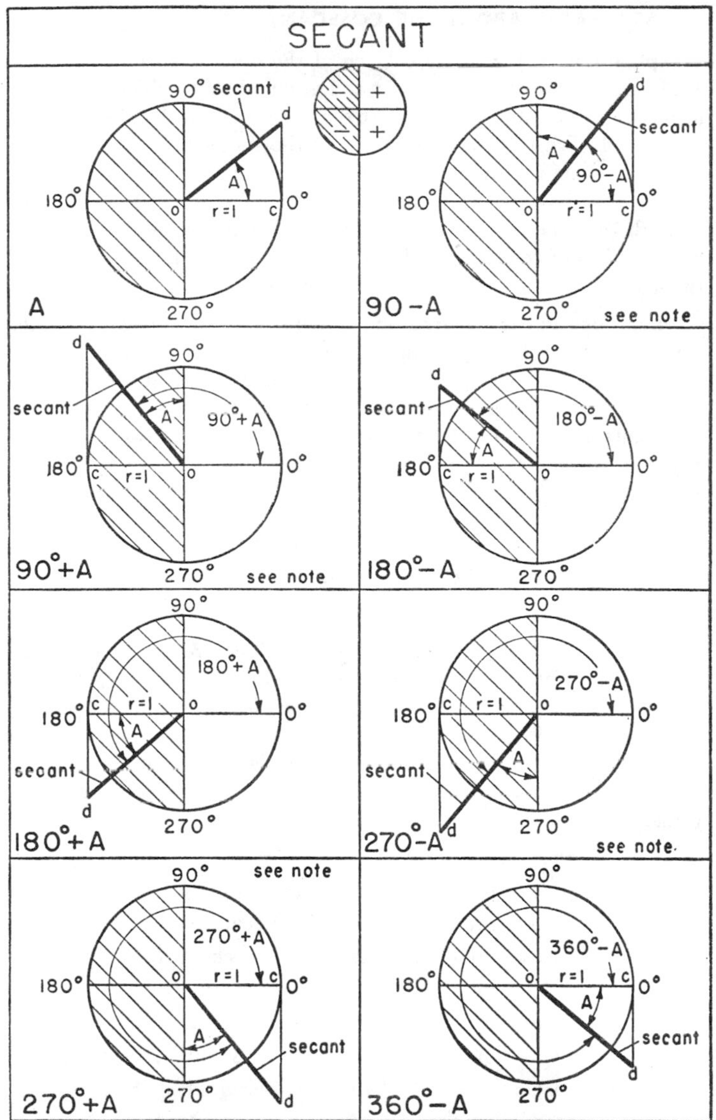

NOTE: For 90° − A, 90° + A, 270° − A, and 270° + A, angle odc equals angle A.

7. BASIC FACTS ABOUT THE COSECANT

Explanation of Diagrams. Each of the diagrams shown on page 4–14 consists of a circle, having a radius equal to 1, which is divided into four quadrants. In the first diagram a line *of* has been drawn at angle A with the 0-degree line. In the succeeding diagrams *of* is drawn at some multiple of 90 degrees plus or minus A with the 0-degree line, measuring in a counter-clockwise direction. In each case, the cosecant of the angle indicated is equal to $of \div oe$. Since *oe* is a radius of the circle and is therefore equal to 1, the cosecant of the indicated angle is, in each case, equal to *of*.

When Cosecant is Positive and When Negative. It will be noted from the smaller key diagram at the top-center of this figure, that the two upper quadrants have plus signs, indicating that the cosecants of all angles in these two quadrants are positive, and the two lower quadrants have minus signs, indicating that the cosecants of all angles in these two quadrants are negative.

Multiples of 90 Degrees. It will be seen that:

$$\text{Cosecant } 0^\circ = \infty \qquad \text{Cosecant } 180^\circ = \infty$$
$$\text{Cosecant } 90^\circ = 1 \qquad \text{Cosecant } 270^\circ = -1$$
$$\text{Cosecant } 360^\circ = \infty$$

Multiples of 90 Degrees Plus or Minus a Given Angle. It is sometimes convenient to convert the function for an angle greater than 90 degrees to the equivalent function of an angle of less than 90 degrees. Since every angle greater than 90 degrees is equal to some multiple of 90 degrees, plus or minus some angle less than 90 degrees, the formulas derived from the eight diagrams shown, facilitate the conversion. Thus, it will be seen that:

$$\text{Cosecant } (90^\circ - A) = \text{Secant } A$$
$$\text{Cosecant } (90^\circ + A) = \text{Secant } A$$
$$\text{Cosecant } (180^\circ - A) = \text{Cosecant } A$$
$$\text{Cosecant } (180^\circ + A) = -\text{ Cosecant } A$$
$$\text{Cosecant } (270^\circ - A) = -\text{ Secant } A$$
$$\text{Cosecant } (270^\circ + A) = -\text{ Secant } A$$
$$\text{Cosecant } (360^\circ - A) = -\text{ Cosecant } A$$
$$\text{Cosecant } (360^\circ + A) = \text{Cosecant } A$$
$$\text{Cosecant } (-A) = -\text{ Cosecant } A$$

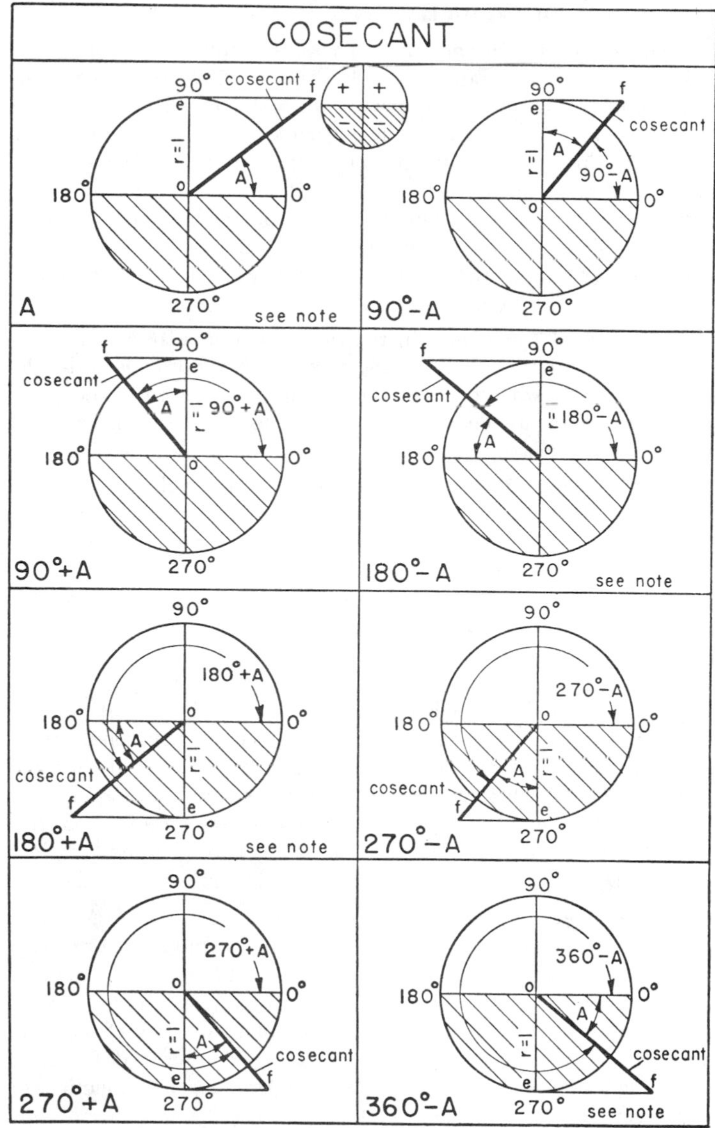

NOTE: For A, $180° - A$, $180° + A$, and $360° - A$, angle *ofe* equals angle A.

8. SOLUTION OF TRIANGLES

Application of Trigonometry. The most important application of trigonometry from a practical standpoint is in the solution of triangles, that is, in the finding of unknown sides or angles when certain sides or angles are known. The triangle is made up of six elements: three sides and three angles. If, in a right-angled triangle, one side and any one of remaining elements, other than the right-angle are known, the other elements can be found by the use of trigonometric functions. If in an oblique-angled triangle one side and any two of the remaining elements are known, the other elements can be found by the use of trigonometric functions.

Right-angled Triangles. In the chart on page 4–16 are given examples of the four types of problems in which right-angled triangles are solved: (1) when one side and the hypotenuse are known; (2) when two sides are known; (3) when one side and one angle are known; (4) when the hypotenuse and one angle are known.

Oblique-angled Triangles. The formulas and rules upon which the solution of oblique-angled triangles is based are given in the chart on page 4–18. It should be noted that in solving for a given angle where two angles are possible—one greater and one less than 90 degrees—the Law of Sines will not indicate whether the angle is greater or less than 90 degrees so that the Law of Cosines should be used, if this fact is not known.

In the charts on pages 4–19 to 4–22 are given four types of problems in which oblique-angled triangles are solved: (1) when three sides are known; (2) when two sides and the angle between them are known; (3) when two sides and the angle opposite one of them are known; and (4) when one side and two angles are known.

It should be noted that in type 3 problems there are four possible cases depending upon the relative sizes of the two sides and the product of one of them and the sine of the angle opposite the other.

General Trigonometric Formulas. In the tables on pages 4–23 and 4–24 are given certain general formulas which permit the substitution of one trigonometric function or combination of functions for another trigonometric function or combination of functions. These include functions of the sum and difference of two angles in terms of the functions of each angle; the functions of half-angles in terms of functions of the angle and the functions of double angles in terms of functions of the angle.

How to Solve Right-Angled Triangles

(*For Basic Formulas see Page* **4**–2)

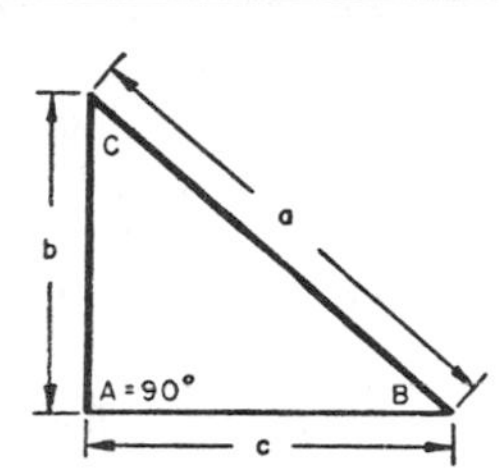

Example 1—One side and hypotenuse are known.

Given: $a = 5.91 \qquad b = 4.51$

To Find: Side c and Angles B and C.

Solution: $\sin B = \frac{b}{a} = \frac{4.51}{5.91} = 0.76311$

$$B = 49°\,44.4'$$

$$C = 90° - B = 90° - 49°\,44.4' = 40°\,15.6'$$

$$c = a \sin C = 5.91 \times 0.64626 = 3.82$$

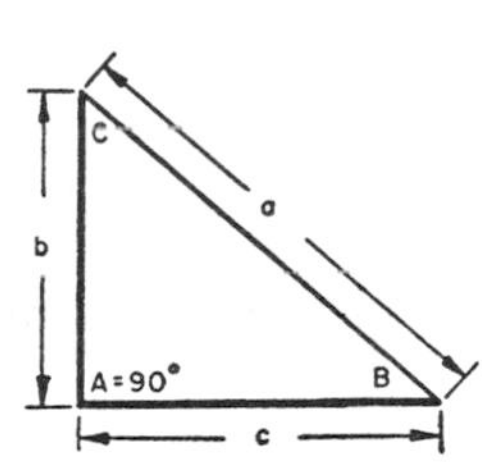

Example 2—Two sides are known.

Given: $b = 14.73 \qquad c = 8.04$

To Find: Hypotenuse a and angles B and C

Solution: $\tan B = \frac{14.73}{8.04} = 1.8321$

$$B = 61°\,22.4'$$

$$C = 90° - B = 90° - 61°\,22.4' = 28°\,37.6'$$

$$a = b \sec C = 14.73 \times 1.1392 = 16.78$$

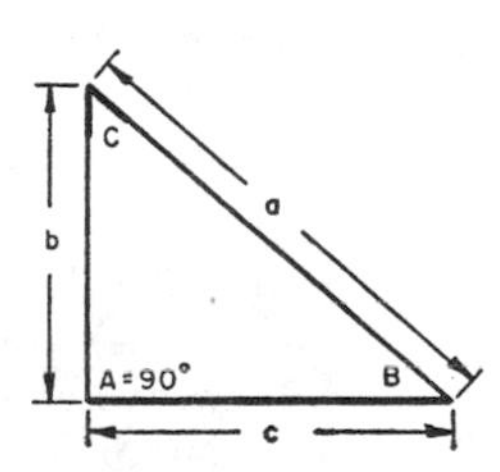

Example 3—One side and one angle other than 90° angle are known.

Given: $c = 7.98 \qquad C = 58°\,33'$

To Find: Side b, hypotenuse a and angle B.

Solution:

$$a = c \operatorname{cosec} C = 7.98 \times 1.1722 = 9.35$$

$$b = c \operatorname{cotan} C = 7.98 \times 0.61160 = 4.88$$

$$B = 90° - C = 90° - 58°\,33' = 31°\,27'$$

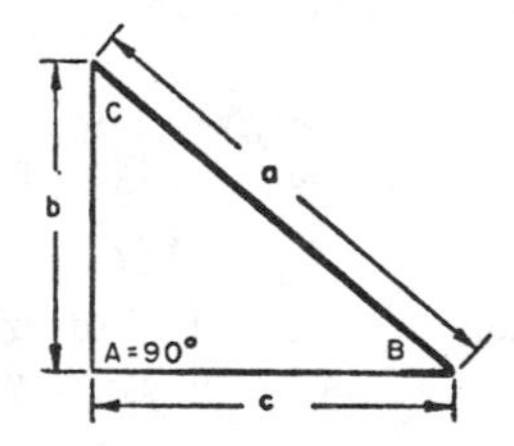

Example 4—Hypotenuse and one angle other than 90° angle are known.

Given: $a = 7.51 \qquad B = 54°\,51'$

To Find: Sides b and c and angle C.

Solution:

$$c = a \cos B = 7.51 \times 0.57572 = 4.32$$

$$b = a \sin B = 7.51 \times 0.81765 = 6.14$$

$$C = 90° - B = 90° - 54°\,51' = 35°\,9'$$

9. OTHER USEFUL FACTS

Inverse Trigonometric Functions. The equation $\sin A = x$ may be written inversely as $\text{arc} \sin x = A$ or $\sin^{-1} x = A$. Either of these two expressions is read: "The angle of which the sine is x equals A." This is called an inverse function. Other inverse functions are: *arc cosine, arc tangent; arc cotangent; arc secant;* and *arc cosecant.* These expressions provide a convenient short-hand way of designating an angle, the size of which is not known, but some function of which is known.

Computation of Sine and Cosine Values. The value of the sine of any angle A can be computed by the following series, using as many terms as are needed for the accuracy required:

$$\sin A = A - \frac{A^3}{3!} + \frac{A^5}{5!} - \frac{A^7}{7!} \cdots \frac{A^n}{n!}$$

where the value of A in radians is used in the right-hand side of the equation.

Similarly for the computation of the value of the cosine of any angle:

$$\cos A = 1 - \frac{A^2}{2!} + \frac{A^4}{4!} - \frac{A^6}{6!} \cdots \frac{A^n}{n!}$$

Since the values of tangent, cotangent, secant and cosecant of any angle may all be computed, if the values of its sine and cosine are known, it is possible to compute any value in the Trigonometric table, with the aid of the above series.

Functions of Small Angles. For angles up to 1 degree, the values of the sine, tangent and length of arc (measured in radians) are approximately equal to six decimal places. For angles between 1 and 2 degrees, the respective values are approximately equal to five decimal places. For angles between 2 and 3 degrees, the respective values are approximately equal to four decimal places.

Thus:

$\sin 1° = 0.017452$	$\sin 2° = 0.03490$	$\sin 3° = 0.0523$
$\tan 1° = 0.017455$	$\tan 2° = 0.03492$	$\tan 3° = 0.0524$
$\text{arc } 1° = 0.017453$	$\text{arc } 2° = 0.03491$	$\text{arc } 3° = 0.0524$

These approximate equalities are sometimes helpful in substituting one value for another in certain types of problems involving very small angles.

Oblique-angled Triangles — Basic Formulas

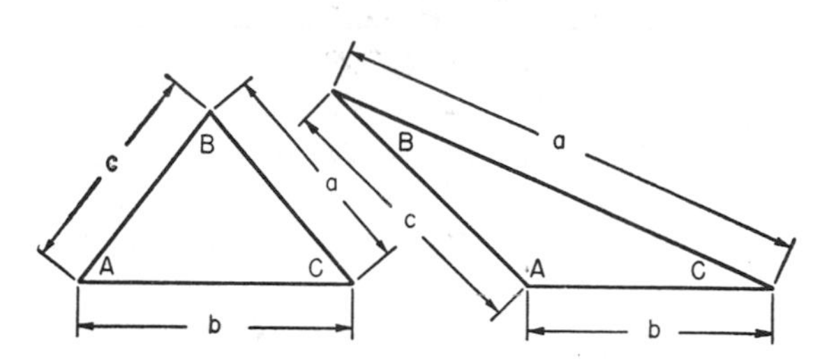

Formula	Rule
$\frac{a}{b} = \frac{\sin A}{\sin B}$ $\frac{b}{c} = \frac{\sin B}{\sin C}$ $\frac{a}{c} = \frac{\sin A}{\sin C}$	**Law of Sines:** (See also Note 1) In any triangle the ratio of any two sides is equal to the ratio of the sines of the respective angles opposite these sides.
$a^2 = b^2 + c^2 - 2bc \cos A$ $b^2 = a^2 + c^2 - 2ac \cos B$ $c^2 = a^2 + b^2 - 2ab \cos C$	**Law of Cosines:** (See also Note 2) In any triangle the square of any side is equal to the sum of the squares of the other two sides minus twice their product times the cosine of the included angle.

Note 1: If the *Law of Sines* is used to find an unknown angle in an oblique-angled triangle, the answer, which will always be a positive sine value, will not indicate whether the angle is greater or less than 90°. This is because for every positive sine value there is one equivalent angle between 0° and 90° and another equivalent angle between 90° and 180°.

Note 2: If the *Law of Cosines* is used to find an unknown angle, the resulting cosine value may be either positive or negative. A positive value indicates that the angle lies between 0° and 90°. A negative value indicates that it lies between 90° and 180°.

(For problems having two solutions, see also Example 3*c* on Page 4–22)

How to Solve Oblique-angled Triangles — 1

(For Examples See Opposite Page)

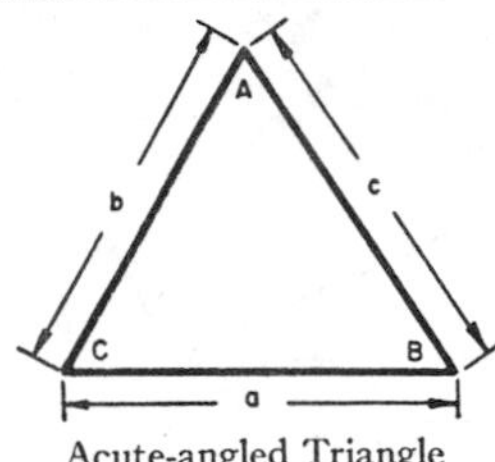

Acute-angled Triangle

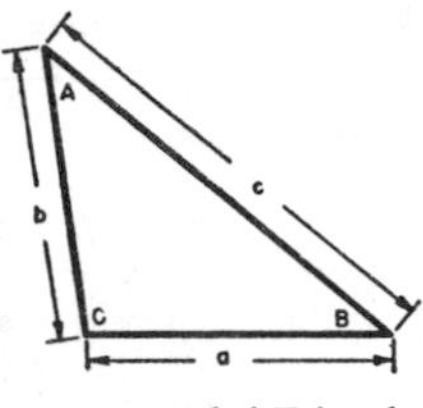

Obtuse-angled Triangle

Type 1—Three sides are known

Given: Sides a, b and c.

To Find: Angles A, B and C.

Solution:

$$a^2 = b^2 + c^2 - 2bc \cos A; \quad \cos A = \frac{b^2 + c^2 - a^2}{2bc}$$

$$b^2 = a^2 + c^2 - 2ac \cos B; \quad \cos B = \frac{a^2 + c^2 - b^2}{2ac}$$

or $$\frac{a}{\sin A} = \frac{b}{\sin B}; \quad \sin B = \frac{b}{a} \sin A$$

$$C = 180° - A - B$$

Note: In solving for the second angle it is preferable to use the Law of Sines, which involves less numerical calculation, if it is known whether the angle is acute or obtuse. If this is not known, the Law of Cosines must be used to determine this. (See both notes on Page 4–18.)

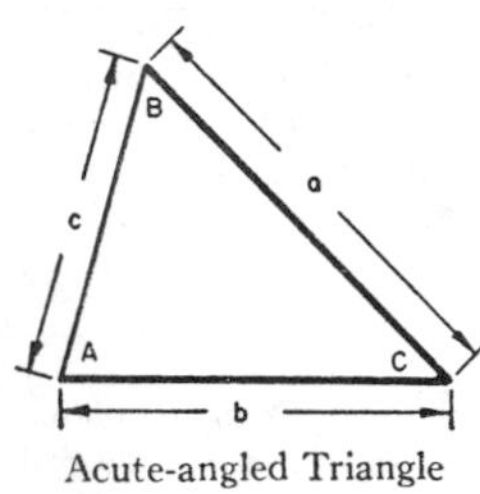

Acute-angled Triangle

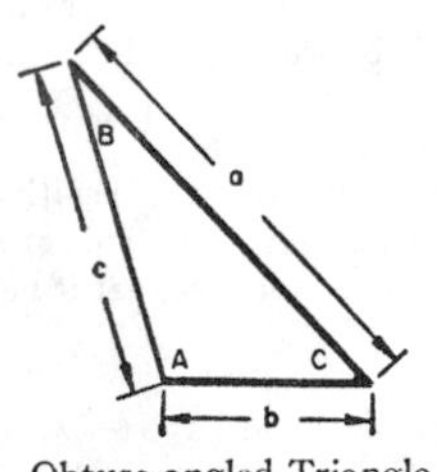

Obtuse-angled Triangle

Type 2—Two sides and angle included between them known

Given: Sides a and b and Angle C.

To Find: Side c and Angles A and B.

Solution:

$$c^2 = a^2 + b^2 - 2ab \cos C; \quad c = \sqrt{a^2 + b^2 - 2ab \cos C}$$

$$a^2 = b^2 + c^2 - 2bc \cos A; \quad \cos A = \frac{b^2 + c^2 - a^2}{2bc}$$

or $$\frac{a}{\sin A} = \frac{c}{\sin C}; \quad \sin A = \frac{a}{c} \sin C$$

$$B = 180° - A - C$$

Note: In solving for the first unknown angle, it is preferable to use the Law of Sines, which involves less numerical calculation, if it is known whether the angle is acute or obtuse. If this is not known, the Law of Cosines must be used to determine this. (See both notes on Page 4–18.)

How to Solve Oblique-angled Triangles — 2

(*For Basic Formulas See Page* 4–18)

Three sides are known

Example 1a

Given: $a = 5.36$; $b = 5.51$; $c = 5.18$

To Find: Angles A, B and C.

Solution:

$$\cos A = \frac{(5.51)^2 + (5.18)^2 - (5.36)^2}{2 \times 5.51 \times 5.18} = 0.49862 \qquad A = 60°\,5.5'$$

$$\sin B = \frac{5.51}{5.36} \times 0.86675 = 0.89102 \qquad B = 63°\,0.1'$$

$$C = 180° - 60°\,5.5' - 63°\,0.1' = 56°\,54.4'$$

Example 1b

Given: $a = 3.88$; $b = 6.87$; $c = 5.20$

To Find: Angles A, B and C.

Solution:

$$\cos A = \frac{(6.87)^2 + (5.20)^2 - (3.88)^2}{2 \times 6.87 \times 5.20} = 0.82833 \qquad A = 34°\,4.4'$$

$$\cos B = \frac{(3.88)^2 + (5.20)^2 - (6.87)^2}{2 \times 3.88 \times 5.20} = -0.12645 \qquad B = 97°\,15.9'$$

$$C = 180° - 34°\,4.4' - 97°\,15.9' = 48°\,39.7'$$

Two sides and angle included between them known

Example 2a

Given: $a = 6.42$; $b = 5.71$; $C = 46°\,58'$

To Find: Side c and angles A and B.

Solution:

$$c = \sqrt{(6.42)^2 + (5.71)^2 - 2 \times 6.42 \times 5.71 \times 0.68242} = 4.88$$

$$\sin A = \frac{6.42}{4.88} \times 0.73096 = 0.96194 \qquad A = 74°\,8.5'$$

$$B = 180° - 74°\,8.5' - 46°\,58' = 58°\,53.5'$$

Example 2b

Given: $a = 6.83$; $b = 3.18$; $C = 41°\,2'$

To Find: Side c and angles A and B.

Solution:

$$c = \sqrt{(6.83)^2 + (3.18)^2 - 2 \times 6.83 \times 3.18 \times 0.75433} = 4.90$$

$$\cos A = \frac{(3.18)^2 + (4.90)^2 - (6.83)^2}{2 \times 3.18 \times 4.90} = -0.40195 \qquad A = 113°\,42'$$

$$B = 180° - 41°\,2' - 113°\,42' = 25°\,16'$$

How to Solve Oblique-angled Triangles — 3

(For Examples See Opposite Page)

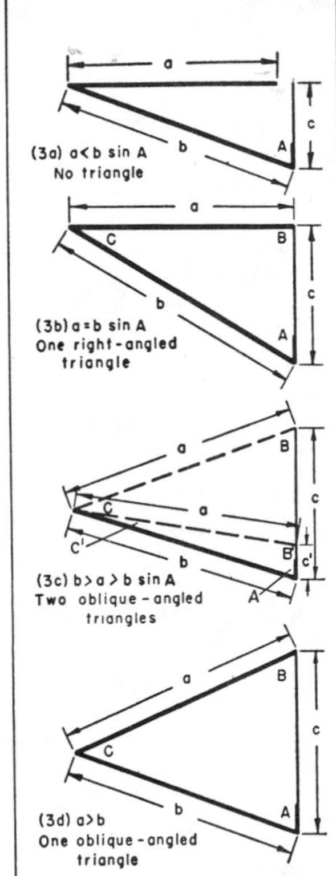

Type 3—Two sides and angle opposite one of them known

Given: Sides a and b and angle A.

To Find: Side c and angles B and C.

Solution:

$$\frac{a}{\sin A} = \frac{b}{\sin B}; \quad \sin B = \frac{b}{a} \sin A$$

$$C = 180° - A - B$$

$$\frac{a}{\sin A} = \frac{c}{\sin C}; \quad c = \frac{\sin C}{\sin A} \times a$$

Note: There are four cases:

(3a) When the given side opposite the given angle is less than the product of the other given side and the sine of the given angle, no triangle is possible.

(3b) When the given side opposite the given angle equals the product of the other given side and the sine of the given angle, one right-angled triangle is possible.

(3c) When the given side opposite the given angle is less than the other side but more than the product of the other given side and the sine of the given angle, two oblique-angled triangles are possible.

(3d) When the given side opposite the given angle is greater than the other given side, only one oblique-angled triangle is possible.

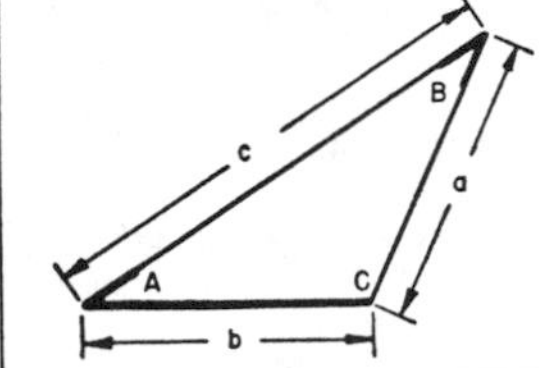

Type 4—One side and two angles known

Given: Side b and angles A and B.

To Find: Sides a and c and angle C.

Solution: $C = 180° - A - B$.

$$a = \frac{\sin A}{\sin B} \times b \qquad c = \frac{\sin C}{\sin B} \times b$$

How to Solve Oblique-angled Triangles — 4

(*For Basic Formulas See Page* 4–18)

Two sides and angle opposite one of them known

Example 3a

Given: $a = 2.59 \quad b = 5.50 \quad A = 68° 7'$

To Find: Side c and angles B and C.

Solution: $5.50 \times \sin 68° 7' = 5.50 \times 0.92974 = 5.11$.

Since in this case $a < b \sin A$, no triangle is possible.

Example 3b

Given: $a = 5.11 \quad b = 5.50 \quad A = 68° 7'$

To Find: Side c and angles B and C.

Solution: $5.50 \times \sin 68° 7' = 5.50 \times 0.92974 = 5.11$.

Since in this case $a = b \sin A$, this is a right-angled triangle.

Example 3c

Given: $a = 5.28 \quad b = 5.50 \quad A = 68° 7'$

To Find: Side c and angles B and C.

Solution: $5.50 \times \sin 68° 7' = 5.50 \times 0.92974 = 5.11$.

Since in this case $b > a > b \sin A$, there are two possible triangles.

$$\sin B = \frac{5.50}{5.28} \times 0.92974 = 0.96848$$

Since two triangles are possible: $B = 75° 34.6'$ or $B' = 104° 25.4'$.

If $B = 75° 34.6'$ then $C = 180° - 68° 7' - 75° 34.6' = 36° 18.4'$

and $$c = \frac{\sin 36° 18.4'}{\sin 68° 7'} \times 5.28 = \frac{0.59211}{0.92974} \times 5.28 = 3.36$$

If $B' = 104° 25.4$ then $C' = 180° - 68° 7' - 104° 25.4' = 7° 27.6'$

and $$c' = \frac{\sin 7° 27.6'}{\sin 68° 7'} \times 5.28 = \frac{0.12985}{0.92974} \times 5.28 = 0.74$$

Example 3d

Given: $a = 6.09 \quad b = 5.50 \quad A = 68° 7'$

To Find: Side c and angles B and C.

Solution: Since in this case $a > b$, there is only one triangle.

$$\sin B = \frac{5.50}{6.09} \times 0.92974 = 0.83966 \qquad B = 57° 6.2'$$

$$C = 180° - 68° 7' - 57° 6.2' = 54° 46.8' \qquad c = \frac{0.81695}{0.92974} \times 6.09 = 5.35$$

One side and two angles known

Example 4

Given: $b = 3.78 \quad A = 37° 20' \quad B = 34° 29'$

To Find: Sides a and c and Angle C.

Solution: $C = 180° - 37° 20' - 34° 29' = 108° 11'$

$$a = \frac{0.60645}{0.56617} \times 3.78 = 4.05 \qquad c = \frac{0.95006}{0.56617} \times 3.78 = 6.34$$

Important Trigonometric Formulas — 1

1. $\sin A = \dfrac{1}{\operatorname{cosec} A}$

2. $\cos A = \dfrac{1}{\sec A}$

3. $\tan A = \dfrac{\sin A}{\cos A} = \dfrac{1}{\cot A}$

4. $\cot A = \dfrac{\cos A}{\sin A} = \dfrac{1}{\tan A}$

5. $\sec A = \dfrac{1}{\cos A}$

6. $\operatorname{cosec} A = \dfrac{1}{\sin A}$

---

7. $\sin^2 A + \cos^2 A = 1$

7a. $\sin A = \sqrt{1 - \cos^2 A} = \dfrac{\tan A}{\sqrt{1 + \tan^2 A}} = \dfrac{1}{\sqrt{1 + \cot^2 A}}$

7b. $\cos A = \sqrt{1 - \sin^2 A} = \dfrac{\cot A}{\sqrt{1 + \cot^2 A}} = \dfrac{1}{\sqrt{1 + \tan^2 A}}$

---

8. $\sin (A + B) = \sin A \cos B + \cos A \sin B$

9. $\sin (A - B) = \sin A \cos B - \cos A \sin B$

10. $\cos (A + B) = \cos A \cos B - \sin A \sin B$

11. $\cos (A - B) = \cos A \cos B + \sin A \sin B$

---

12. $\sin A + \sin B = 2 \sin \tfrac{1}{2}(A + B) \cos \tfrac{1}{2}(A - B)$

13. $\sin A - \sin B = 2 \sin \tfrac{1}{2}(A - B) \cos \tfrac{1}{2}(A + B)$

14. $\cos A + \cos B = 2 \cos \tfrac{1}{2}(A + B) \cos \tfrac{1}{2} (A - B)$

15. $\cos A - \cos B = -2 \sin \tfrac{1}{2}(A + B) \sin \tfrac{1}{2}(A - B)$

---

16. $\sin A \sin B = \tfrac{1}{2} \cos (A - B) - \tfrac{1}{2} \cos (A + B)$

17. $\cos A \cos B = \tfrac{1}{2} \cos (A - B) + \tfrac{1}{2} \cos (A + B)$

18. $\sin A \cos B = \tfrac{1}{2} \sin (A - B) + \tfrac{1}{2} \sin (A + B)$

---

19. $\sin^2 A = \tfrac{1}{2}(1 - \cos 2A)$

20. $\cos^2 A = \tfrac{1}{2}(1 + \cos 2A)$

21. $\sin^2 A - \sin^2 B = \cos^2 B - \cos^2 A = \sin (A + B) \sin (A - B)$

22. $\cos^2 A - \sin^2 B = \cos^2 B - \sin^2 A = \cos (A + B) \cos (A - B)$

23. $\cos^2 A - \cos^2 B = \sin^2 B - \sin^2 A = -\sin (A + B) \sin (A - B)$

Important Trigonometric Formulas — 2

24. $\tan(A + B) = \dfrac{\tan A + \tan B}{1 - \tan A \tan B}$

25. $\tan(A - B) = \dfrac{\tan A - \tan B}{1 + \tan A \tan B}$

26. $\cot(A + B) = \dfrac{\cot A \cot B - 1}{\cot B + \cot A}$

27. $\cot(A - B) = \dfrac{\cot A \cot B + 1}{\cot B - \cot A}$

28. $\tan A + \tan B = \dfrac{\sin(A + B)}{\cos A \cos B}$

29. $\tan A - \tan B = \dfrac{\sin(A - B)}{\cos A \cos B}$

30. $\cot A + \cot B = \dfrac{\sin(B + A)}{\sin A \sin B}$

31. $\cot A - \cot B = \dfrac{\sin(B - A)}{\sin A \sin B}$

32. $\tan A \tan B = \dfrac{\tan A + \tan B}{\cot A + \cot B}$

33. $\cot A \cot B = \dfrac{\cot A + \cot B}{\tan A + \tan B}$

34. $\sin \frac{1}{2}A = \sqrt{\frac{1}{2}(1 - \cos A)} = \frac{1}{2}\sqrt{1 + \sin A} - \frac{1}{2}\sqrt{1 - \sin A}$

35. $\cos \frac{1}{2}A = \sqrt{\frac{1}{2}(1 + \cos A)} = \frac{1}{2}\sqrt{1 + \sin A} + \frac{1}{2}\sqrt{1 - \sin A}$

36. $\tan \dfrac{1}{2} A = \sqrt{\dfrac{1 - \cos A}{1 + \cos A}} = \dfrac{1 - \cos A}{\sin A} = \dfrac{\sin A}{1 + \cos A}$

37. $\cot \dfrac{1}{2} A = \sqrt{\dfrac{1 + \cos A}{1 - \cos A}} = \dfrac{1 + \cos A}{\sin A} = \dfrac{\sin A}{1 - \cos A}$

38. $\sin 2A = 2 \sin A \cos A$

39. $\cos 2A = \cos^2 A - \sin^2 A = 1 - 2\sin^2 A = 2\cos^2 A - 1$

40. $\tan 2A = \dfrac{2 \tan A}{1 - \tan^2 A} = \dfrac{2}{\cot A - \tan A}$

41. $\cot 2A = \dfrac{\cot^2 A - 1}{2 \cot A} = \dfrac{\cot A - \tan A}{2}$

42. $\sin 3A = 3 \sin A - 4 \sin^3 A$

43. $\cos 3A = 4 \cos^3 A - 3 \cos A$

44. $\sin 4A$
$= 8 \sin A \cos^3 A - 4 \sin A \cos A$

45. $\cos 4A = 8 \cos^4 A - 8 \cos^2 A + 1$

5

Basic Facts about Common and Hyperbolic Logarithms

1. NATURE OF LOGARITHMS

Logarithm calculations are based upon essentially simple principles which have to do with the relationship between an arithmetical series of numbers and a geometrical series of numbers, as shown by the following example:

(A)	0	1	2	3	4	5	6	7	8	9	10	11	12
(B)	1	2	4	8	16	32	64	128	256	512	1024	2048	4096

In row (A) is an arithmetic series of numbers having a *common difference* of 1 (each number is 1 greater than the number preceding it) and progressing from 0 to 12. In row (B) is a geometric series of numbers having a *ratio* of 2 (each number is 2 times the number preceding it) and progressing from 1 to 4096. It will be seen that any given number in row (A) is the power to which the number 2 must be raised to equal the number directly below it in row (B). Thus, below 5 in row (A) is 32 in row B since $2^5 = 32$. Similarly, below 0 in row (A) is 1 in row (B) since $2^\circ = 1$, and so on.

The numbers in row (A) are called *logarithms* and those in row (B) are called *antilogarithms*. Because these logarithms represent the powers to which the number 2 must be raised to equal the antilogarithms, they are said to have the *base* 2. Thus, a logarithm of a given number is the power to which the base of the logarithm must be raised to equal the given number.

2. USE OF LOGARITHMS

The advantage of logarithms is that they greatly facilitate the operations of multiplication, division, evolution or raising a number to a given power, and involution or finding a given root of a number. These operations are carried out by the use of logarithms in the same way regardless of the base of the system of logarithms used. Referring again to the series of logarithms to the base 2 as given in row (A) and the corresponding antilogarithms in row (B), these operations are carried out as follows:

(a) *To multiply* two numbers, say 16 and 256, which are given in row (B), *add* their corresponding logarithms in row (A), or $4 + 8 = 12$. Now, find the antilogarithm in row (B) which corresponds to the logarithm 12 in row (A). This is 4096, hence $16 \times 256 = 4096$.

(b) *To divide* one number by another, say 2048 by 32, *subtract* the logarithm of the divisor from the logarithm of the dividend or $11 - 5 = 6$. Now, find the antilogarithm in row (B) which corresponds to the logarithm 6 in row (A). This is 64, hence $2048 \div 32 = 64$.

Comparison of Arithmetical and Logarithmic Operations

Arithmetical Operation	Example	Logarithmic Operation	Example
Multiplication	$a \times b$	Addition	$\log a + \log b = \log (a \times b)$
Division	$a \div b$	Subtraction	$\log a - \log b = \log (a \div b)$
Square	a^2	Multiplication	$2 \times \log a = \log a^2$
Cube	a^3	Multiplication	$3 \times \log a = \log a^3$
*n*th Power	a^n	Multiplication	$n \times \log a = \log a^n$
Square Root	$\sqrt{a}$	Division	$\log a \div 2 = \log \sqrt{a}$
Cube Root	$\sqrt[3]{a}$	Division	$\log a \div 3 = \log \sqrt[3]{a}$
*n*th Root	$\sqrt[n]{a}$	Division	$\log a \div n = \log \sqrt[n]{a}$

(c) *To find a power* of a given number, say the cube of 16, *multiply* the logarithm of the number by the index of the power, or $4 \times 3 = 12$. Now, find the antilogarithm in row (B) which corresponds to the logarithm 12 in row (A). This is 4096, hence $16^3 = 4096$.

(d) *To find the root* of a given number, say the 5th root of 1024, *divide* the logarithm of the number by the index of the root, or $10 \div 5 = 2$. Now, find the antilogarithm in row (B) which corresponds to the logarithm 2 in row (A). This is 4, hence $\sqrt[5]{1024} = 4$.

3. COMMON OR BRIGGS LOGARITHMS

The logarithms commonly used in computations are called Common or Briggs logarithms and have 10 as a base. Using this series of logarithms the (A) and (B) rows shown in Section 1, Nature of Logarithms, would look like this:

(A)	0	1	2	3	4	5	6
(B)	1	10	100	1,000	10,000	100,000	1,000,000

and so on. This series of logarithms is used in exactly the same way as the series of logarithms to the base 2 previously described.

It will be noted that the logarithm or number in row (A) is the power to which 10 must be raised to equal the corresponding antilogarithm or number in row (B). A peculiarity of this series of logarithms that makes it convenient to use is that each logarithm as given in row (A) is equal to 1 less than the number of digits to the left of the decimal point of the corresponding antilogarithm in row (B). Thus, 100 has three digits and its log is 2; 1000 has four digits and its log is 3, and so on.

It can be also concluded that since the log of 10 is 1 and that of 100 is 2, the log of any number between 10 and 100 lies between 1 and 2 or, in other words, is equal to 1 plus a decimal value. This decimal part of the logarithm is called the *mantissa* and the whole number part, such as is given in row (A), is called the *characteristic*.

Another very useful feature of this series of logarithms to the base 10 is that the decimal part or mantissa of the logarithm of all numbers having the same digits in the same order, disregarding zeros which are not preceded *and* followed by any digit from 1 to 9 inclusive, is the same. Thus, the mantissa of the log of 2, 20, 200 or 0.2, 0.02, 0.002, etc., is .30103 to five places.

Referring again to the rows (A) and (B) for the series of logarithms to the base 10, it can be seen that if the logarithm of 1 is 0, then the logarithm of 2 is 0 + .30103 or 0.30103. Similarly, if the logarithm of 10 is 1, then the logarithm of 20 is 1 + .30103 or 1.30103. Since the characteristic or whole number part of a logarithm is easily determined by counting the number of digits to the left of the decimal point or the number of zeros directly to the right of the decimal point (as will be explained later) only a table of mantissas need be referred to in order to set down the logarithm of any number. This table of mantissas is what is commonly referred to as a table of logarithms.

Characteristics of Logarithms of Numbers Less than 1. So far, only the logarithms of numbers of 1 or greater have been dealt with. If an (A) and (B) row for decimal fractions were set up for logarithms to the base 10, it would look like this:

(A)	0	−1	−2	−3	−4	−5	−6
(B)	1	0.1	0.01	0.001	0.0001	0.00001	0.000001

Here again the numbers in row (A) are the logarithms (or the characteristics of the logarithms, more correctly stated) of the corresponding numbers in row (B) which are the antilogarithms. Also, these logarithms in row (A) are the powers to which the base 10 must be raised to equal the corresponding antilogarithm in row (B). Thus, the log of 0.1 is −1 since $10^{-1} = 1 \div 10 = 0.1$; and the log of 0.001 is −3 since $10^{-3} = 1 \div 10^3 = 0.001$, and so on. It can be readily seen that the numerical value of the log (or characteristic of the log) in row (A) is equal to 1 more than the number of zeros immediately following the decimal point. Thus, 0.1 has no zeros immediately following the decimal point (even if expressed 0.1000) and hence its logarithm is −1; 0.0001 has three zeros immediately following the decimal point and hence its logarithm is −4.

Since the log of 0.01 is −2 and the log of 0.1 is −1, the log of any number between 0.01 and 0.1 lies between −2 and −1. Thus, remembering that the mantissa of 2 is .30103, it can be seen that the log of 0.02 is .30103 on the positive side of −2, that is, it is −2 + .30103. This is written $\bar{2}.30103$ to indicate that the 2 is minus but the remainder of the logarithm is positive. It may also be written 0.30103 − 2 with the negative characteristic following the positive mantissa. In the series of logarithms to the base 10, all numbers between 0 and 1 have logarithms with negative characteristics and positive mantissas. A somewhat more convenient form of writing the logarithm of a number between 0 and 1 is obtained by *adding and subtracting* a whole number which makes the negative characteristic equal to −10. In the particular case just mentioned, that of log 0.02, the conversion is made as follows:

$$\begin{aligned} \log 0.02 &= \quad 0.30103 - 2 \\ \text{adding and subtracting } 8 &= +8 \qquad\quad - 8 \\ \text{also the } \log 0.02 &= \quad 8.30103 - 10 \end{aligned}$$

If the number were 0.002, for example, the conversion would be as follows:

$$\begin{aligned} \log 0.002 &= \quad 0.30103 - 3 \\ & \quad +7 \qquad\quad - 7 \\ \log 0.002 &= \quad 7.30103 - 10 \end{aligned}$$

4. USE OF TABLES OF LOGARITHMS

Finding a Logarithm. Tables of common logarithms are published in many books of mathematical functions; and many inexpensive calculators give logarithms and antilogarithms. With such calculators no tables are necessary. However, to provide an understanding of how tables are used, the following examples are given using portions of a table of logarithms. Suppose the number for which the logarithm is to be found is 4239. Looking down the first column headed N to 423 and then following this line over to the column headed 9 will be found the number 62726 which is the mantissa of the log of 4239. Since the given number is greater than 1 and has four digits to the left of the decimal point, its logarithmic characteristic is 3 and its complete logarithm is 3.62726.

Suppose the number for which the logarithm is to be found is 0.0008431. Referring again to the table and following down column N to 843 and over to the column headed 1 will be found the mantissa .92588. Since the given number is less than 1 and has three zeros to the immediate right of the decimal point, its logarithmic characteristic is −4 and its complete logarithm is $\bar{4}.92588$, or $0.92588 - 4$, or $6.92588 - 10$.

Finding an Antilogarithm. When a logarithm is given and the antilogarithm or number corresponding to it is to be found, the process is the reverse of that just described. Suppose the given logarithm is 2.58625. Referring to the tables, the mantissa .58625 will be found in the row opposite 385 and under the column headed

7. Since the characteristic of the logarithm is 2, the corresponding number or antilogarithm is greater than 1 and has three digits to the left of the decimal point, hence the number is 385.7.

How to interpolate to Find a Logarithm. If a number is given to more digits than are provided in the tables of logarithms, the logarithm of the number must be found by interpolation. Thus, the number 38233 has five digits, but only the first four can be found in the tables under the column headed *N* and the fourth digit indicated by the column headed *3*. This means that the log of 38230 can be found directly in the tables but the log of 38233 must be found by interpolation. The process of interpolation to find the desired logarithm is carried out in four steps.

Step 1. Find the nearest smaller number and the nearest larger number to the given number, which are given in the log tables. In this particular case, the nearest smaller number is 38230 and the nearest larger number is 38240.

Step 2. Find what proportion the difference between the given number and the next smaller number is to the difference between the next larger number and the next smaller number. In this particular case, this would be:

$$\frac{38233 - 38230}{38240 - 38230} = \frac{3}{10} = 0.3$$

If we think of going from 38230 to 38240 then 38233 is three-tenths of the way from 38230. For the purpose of practical interpolation, it may, therefore, be assumed without appreciable error, that the mantissa of the logarithm of 38233 is three-tenths of the way from the mantissa of the logarithm of 38230 to the mantissa of the logarithm of 38240, that is, it is equal to three-tenths of the difference between these two mantissas added to the smaller mantissa of the two.

Step 3. Find the difference between the logarithmic mantissa of the next larger number and that of the next smaller number.

$$\begin{aligned} \text{logarithmic mantissa } 38240 &= .58252 \\ \text{logarithmic mantissa } 38230 &= \underline{.58240} \\ & .00012 \end{aligned}$$

Step 4. Multiply the difference obtained in *Step* 3 by the decimal fraction obtained in *Step* 2 and add the result to the logarithmic mantissa of the next smaller number. This will give the mantissa of the logarithm of the given number. The characteristic of this logarithm is found by the rules previously stated and based upon counting the number of digits in the given number to the left of the decimal point, if the number is 1 or greater or counting the number of zeros immediately to the right of the decimal point, if the number is less than 1. In this particular case:

$$.00012 \times 0.3 = .000036 \text{ or } .00004 \text{ to five decimal places}$$

N.	L.	0	1	2	3	4	5	6	7	8	9
378		749	761	772	784	795	807	818	830	841	852
379		864	875	887	898	910	921	933	944	955	967
380		978	990	*001	*013	*024	*035	*047	*058	*070	*081
381	58	092	104	115	127	138	149	161	172	184	195
382		206	218	229	240	252	263	274	286	297	309
383		320	331	343	354	365	377	388	399	410	422
384		433	444	456	467	478	490	501	512	524	535
385		546	557	569	580	591	602	614	625	636	647
386		659	670	681	692	704	715	726	737	749	760
387		771	782	794	805	816	827	838	850	861	872
388		883	894	906	917	928	939	950	961	973	984
389		995	*006	*017	*028	*040	*051	*062	*073	*084	*095
390	59	106	118	129	140	151	162	173	184	195	207
391		218	229	240	251	262	273	284	295	306	318
392		329	340	351	362	373	384	395	406	417	428
415		805	815	826	836	847	857	868	878	888	899
416		909	920	930	941	951	962	972	982	993	*003
417	62	014	024	034	045	055	066	076	086	097	107
418		118	128	138	149	159	170	180	190	201	211
419		221	232	242	252	263	273	284	294	304	315
420		325	335	346	356	366	377	387	397	408	418
421		428	439	449	459	469	480	490	500	511	521
422		531	542	552	562	572	583	593	603	613	624
423		634	644	655	665	675	685	696	706	716	726
424		737	747	757	767	778	788	798	808	818	829
425		839	849	859	870	880	890	900	910	921	931
426		941	951	961	972	982	992	*002	*012	*022	*033
427	63	043	053	063	073	083	094	104	114	124	134
428		144	155	165	175	185	195	205	215	225	236
429		246	256	266	276	286	296	306	317	327	337
836		221	226	231	236	241	247	252	257	262	267
837		273	278	283	288	293	298	304	309	314	319
838		324	330	335	340	345	350	355	361	366	371
839		376	381	387	392	397	402	407	412	418	423
840		428	433	438	443	449	454	459	464	469	474
841		480	485	490	495	500	505	511	516	521	526
842		531	536	542	547	552	557	562	567	572	578
843		583	588	593	598	603	609	614	619	624	629
844		634	639	645	650	655	660	665	670	675	681
845		686	691	696	701	706	711	716	722	727	732
846		737	742	747	752	758	763	768	773	778	783
847		788	793	799	804	809	814	819	824	829	834
848		840	845	850	855	860	865	870	875	881	886
849		891	896	901	906	911	916	921	927	932	937
850		942	947	952	957	962	967	973	978	983	988

Multiplication by Logarithms. When two or more numbers are to be multiplied together, the procedure is to find the logarithm of each, add these logarithms together, and find the antilogarithm corresponding to their sum. This antilogarithm will be the required product.

Example 1: Find the product of 1496 × 312.7 × 1.431 × 18.7.

$$\begin{aligned} \log 1496 &= 3.17493 \\ \log 312.7 &= 2.49513 \\ \log 1.431 &= 0.15564 \\ \log 18.7 &= \underline{1.27184} \\ & 7.09754 \end{aligned}$$

$$\text{antilog } 7.09754 = 12{,}518{,}000$$

When any of the numbers to be multiplied together is less than 1, it is usually convenient to write such number in the form having the negative part of its characteristic equal to −10, as previously described under "Characteristics of Logarithms of Numbers less than 1."

N.	L.	0	1	2	3	4	5	6	7	8	9
271		297	313	329	345	361	377	393	409	425	441
272		457	473	489	505	521	537	553	569	584	600
273		616	632	648	664	680	696	712	727	743	759
274		775	791	807	823	838	854	870	886	902	917
275		933	949	965	981	996	*012	*028	*044	*059	*075
276	44	091	107	122	138	154	170	185	201	217	232
277		248	264	279	295	311	326	342	358	373	389
278		404	420	436	451	467	483	498	514	529	545
279		560	576	592	607	623	638	654	669	685	700
280		716	731	747	762	778	793	809	824	840	855
281		871	886	902	917	932	948	963	979	994	*010
282	45	025	040	056	071	086	102	117	133	148	163
283		179	194	209	225	240	255	271	286	301	317
284		332	347	362	378	393	408	423	439	454	469
285		484	500	515	530	545	561	576	591	606	621
286		637	652	667	682	697	712	728	743	758	773
287		788	803	818	834	849	864	879	894	909	924
288		939	954	969	984	*000	*015	*030	*045	*060	*075
289	46	090	105	120	135	150	165	180	195	210	225
290		240	255	270	285	300	315	330	345	359	374
291		389	404	419	434	449	464	479	494	509	523
292		538	553	568	583	598	613	627	642	657	672
293		687	702	716	731	746	761	776	790	805	820
294		835	850	864	879	894	909	923	938	953	967
295		982	997	*012	*026	*041	*056	*070	*085	*100	*114
296	47	129	144	159	173	188	202	217	232	246	261
297		276	290	305	319	334	349	363	378	392	407
298		422	436	451	465	480	494	509	524	538	553
299		567	582	596	611	625	640	654	669	683	698
300		712	727	741	756	770	784	799	813	828	842

Example 2: Find the product of 0.004947 × 28.44 × 0.912 × 6.385

$$\begin{aligned}
\log\ 0.004947 &= 7.69434 - 10 \\
\log 28.44 &= 1.45393 \\
\log\ 0.912 &= 9.95999 - 10 \\
\log\ 6.385 &= \underline{0.80516\qquad\ \ } \\
&\quad 19.91342 - 20
\end{aligned}$$

$$\text{antilog } 9.91342 - 10 = 0.81926$$

Division by Logarithms. When dividing one number by another, subtract the logarithm of the divisor from the logarithm of the dividend. The remainder will be the logarithm of the quotient.

Example 1: Find the quotient of 8319 ÷ 327.6

$$\begin{aligned}
\log 8319 &= 3.92007 \\
-\log\ 327.6 &= -\underline{2.51534} \\
&\quad\ 1.40473
\end{aligned}$$

$$\text{antilog } 1.40473 = 25.394$$

If the dividend or divisor consists of two or more numbers which are to be multiplied together, the logarithm of the product of these numbers is obtained by adding their respective logarithms together and then the logarithm of the divisor is subtracted from the logarithm of the dividend to get the logarithm of the quotient.

Example 2: Find the quotient of $\dfrac{84 \times 0.00796 \times 0.319}{436.2 \times 0.00051}$

$$\begin{aligned}
\log\ 84 &= 1.92428 \\
\log\ 0.00796 &= 7.90091 - 10 \\
\log\ 0.319 &= \underline{9.50379 - 10} \\
\text{log of dividend} &= 19.32898 - 20
\end{aligned}$$

$$\begin{aligned}
\log 436.2 &= 2.63969 \\
\log\ 0.00051 &= \underline{6.70757 - 10} \\
\text{log of divisor} &= 9.34726 - 10
\end{aligned}$$

$$\begin{aligned}
\text{log of dividend} &= 19.32898 - 20 \\
-\text{ log of divisor} &= -\underline{(9.34726 - 10)} \\
\text{log of quotient} &= 9.98172 - 10
\end{aligned}$$

$$\text{antilog } 9.98172 - 10 = 0.95878$$

This problem can be solved by the single operation of addition if cologarithms of the numbers in the divisor are used, as explained later under "Making Use of Cologarithms."

Both the positive and negative parts of the characteristic of the logarithm of the dividend should be, correspondingly, larger than the positive and negative parts of the characteristic of the logarithm of the divisor before the latter is subtracted from the former. If necessary, a suitable number is added to and subtracted from the characteristic of the logarithm of the dividend to bring this about. This facilitates the subtraction of the logarithm of the divisor.

Example 3: Find the quotient of 731 ÷ 4392

$$\begin{array}{rl} \log 731 = & 2.86392 \\ & \underline{+2 \qquad -2} \\ \log 731 = & 4.86392 - 2 \\ -\log 4392 = & \underline{-3.64266 \qquad\;} \\ & 1.22126 - 2 \end{array}$$

$$\text{antilog } \bar{1}.22126 = 0.16644$$

Example 4: Find the quotient of 411 ÷ 0.00527

$$\begin{array}{rl} \log 411 = & 2.61384 \\ & \underline{11 \qquad -11} \\ \log 411 = & 13.61384 - 11 \\ -\log 0.00527 = & -\underline{(7.72181 - 10)} \\ & 5.89203 - 1 \end{array}$$

$$\text{antilog } 4.89203 = 77{,}988$$

Obtaining the Powers of Numbers. To find a power of a given number, multiply the logarithm of the number by the index of the power and find the antilogarithm of this product.

Example 1: Find the value of 8.39^4

$$\begin{array}{rl} \log 8.39 = & 0.92376 \\ 4 \times \log 8.39 = & 3.69504 \\ \text{antilog } 3.69504 = & 4{,}955 \end{array}$$

Logarithms are particularly useful where decimal or fractional powers are involved.

Example 2: Find the value of $259^{2.61}$

$$\log 259 = 2.41330$$
$$2.61 \times \log 259 = 6.29871$$
$$\text{antilog } 6.29871 = 1{,}989{,}400$$

The multiplication of 2.61×2.41330 is carried out in the customary arithmetical way, only five decimal places being retained in the product.

When the number to be raised to a decimal power is less than 1, the logarithm of the number should be written so that the product of the negative part of its characteristic and the index of the power is a whole number.

Example 3: Find the value of $0.0318^{.46}$

The log of 0.0318 is to be written so that the product of the negative part of its characteristic and the decimal index .46 is a whole number. An easy way to do this is to make the negative part of this characteristic equal to −100.

$$\begin{array}{lrr}
\log 0.0318 = & 8.50243 & -\ \ 10 \\
 & +\ 90 & -\ \ 90 \\ \hline
\log 0.0318 & 98.50243 & -\ 100 \\
 & \times .46 & \\ \hline
 & 5\ 9101458 & \\
 & 39\ 400972 & \\ \hline
\log 0.0318^{.46} = & 45.3111178 & -\ \ 46
\end{array}$$

or when written to only five decimal places

$$\log 0.0318^{.46} = 45.31112 - 46 = 9.31112 - 10$$

$$\text{antilog } 9.31112 - 10 = 0.2047$$

Extracting Roots by Logarithms. Any given root of a number can be found by dividing the logarithm of the number by the index of the root and finding the antilogarithm of the resulting quotient.

Example 1: Find the value of $\sqrt[3]{3941}$

$$\log 3941 = 3.59561$$
$$\log \sqrt[3]{3941} = (\log 3941) \div 3 = 1.19854$$
$$\text{antilog } 1.19854 = 15.796$$

If the number of which the root is to be found is less than 1, its logarithm should be written such that the negative part of its characteristic is evenly divisible by the index of the root.

Example 2: Find the value of $\sqrt[4]{0.00313}$

$$\log 0.00313 = 7.49554 - 10$$

To make the negative part of the characteristic divisible by 4, 2 is added to and subtracted from the characteristic, giving:

$$\log 0.00313 = 9.49554 - 12$$
$$(\log 0.00313) \div 4 = 2.37388 - 3 = 9.37388 - 10$$
$$\text{antilog } 9.37388 - 10 = 0.23653$$

Example 3: Find the value of $\sqrt[0.7]{0.078}$

$$\begin{aligned} \log 0.078 &= 8.89209 - 10 \\ &\ +4 \qquad\quad -4 \\ \log 0.078 &= 12.89209 - 14 \end{aligned}$$

$$(\log 0.078) \div 0.7 = 18.41727 - 20 = 8.41727 - 10$$
$$\text{antilog } 8.41727 - 10 = 0.026138$$

Making Use of Cologarithms. The cologarithm of a number is the logarithm of the reciprocal of the number. In other words it is equal to the logarithm of 1 minus the logarithm of the number. Cologarithms enable division to be carried out by the addition of the logarithm of the dividend and the cologarithm of the divisor because the addition of a cologarithm is the same as the subtraction of a logarithm. The use of cologarithms is particularly convenient in computations involving two or more factors in the divisor. Only a single operation, that of adding the various logarithms and cologarithms is required, whereas, if only logarithms were employed, at least two additions, that of the logarithms of the factors in the dividend and that of the logarithms of the factors in the divisor and a subtraction of the latter sum from the former, would be required. This saving in time can be effected if a table of cologarithms is available so that they can be set down directly.

Note: A table of cologs is given in Chappell's "Five Figure Mathematical Tables" published by D. Van Nostrand Co., Inc. The rules for finding the characteristics of cologs and for interpolating in the table of mantissas of cologs are different from those for logs.

Example: Use logarithms and cologarithms to make the following computation:

$$x = \frac{25 \times 0.079 \times 8640}{136{,}000 \times 0.0001 \times 935}$$

Solution:

$$\begin{aligned}
\log 25 &= 1.39794 \\
\log 0.079 &= 8.89763 - 10 \\
\log 8640 &= 3.93651 \\
\text{colog } 136{,}000 &= 4.86646 - 10 \\
\text{colog } 0.0001 &= 4.00000 \\
\text{colog } 935 &= 7.02919 - 10 \\
\hline
\log x &= 30.12773 - 30 \\
\log x &= 0.12773 \\
x &= 1.3419
\end{aligned}$$

5. NATURAL LOGARITHMS

Other Systems of Logarithms. Since any positive number except 1 may be used as a base for a system of logarithms, there is no limit to the possible number of systems that might be devised. However, there are only two tables of logarithms in common use. One is the table of *Common* or *Briggs* logarithms previously described, in which the base number is 10. The other is the table of *Natural* (formerly also called *Hyperbolic*) logarithms in which the base, denoted as ***e***, is 2.7182818 . . .

Natural Logarithms. Natural logarithms are used in the calculations of pure mathematics and in certain theoretical formulas. For example, the mean effective pressure p in a steam engine cylinder is found by the formula

$$p = P(1 + \log_e R) \div R$$

in which P is the initial absolute pressure, R is the ratio of expansion and $\log_e R$ is the natural log of R.

The table of Natural Logarithms is given on pages **27**–12 to **27**–15. It will be noted that complete logarithms, that is characteristic and mantissa together, are shown.

Changing from One System of Logarithms to Another. If the logarithm of a given number, say to the base 10, is known, this logarithm can be converted to the logarithm of the number to some other base, say ***e***, by multiplying by a factor called a *modulus*. Let x equal the logarithm to the base 10 of a given number N, what is the modulus for converting x to the logarithm to the base e of N?

$$x = \log_{10} N \tag{1}$$

hence
$$N = 10^x \tag{2}$$

and
$$\log_e N = \log_e 10^x = x \log_e 10 \tag{3}$$

Substituting the equivalent of x from Equation (1):

$$\log_e N = \log_{10} N \times \log_e 10 \tag{4}$$

Thus, $\log_e 10$, which is equal to 2.3025851 is the required modulus.

If
$$\log_e N = \log_{10} N \times 2.3025851 \tag{5}$$

then
$$\log_{10} N = \log_e N \times \frac{1}{2.3025851} \tag{6a}$$

or
$$\log_{10} N = \log_e N \times 0.4342945 \tag{6b}$$

In other words, to convert the logarithm to the base 10 of any number to the logarithm to the base e of that number, multiply the logarithm to the base 10 of the number by 2.3025851.

To convert the logarithm to the base e of any number to the logarithm to the base 10 of that number, multiply the logarithm to the base e of the number by 0.4342945.

6

Right-angled Triangles—Finding Unknown Sides and Angles

The problems in this chapter illustrate how principles of geometry and trigonometry are applied in solving mechanical problems involving right-angled triangles. Since the right-angled triangle is one of the most common elements to be found in mechanical problems, the methods of solving such triangles are of basic importance. The formulas for finding unknown sides and angles which are applied in these problems are those given in Chapter 4. That chapter should be reviewed if these formulas are not clearly recalled as they will be used again and again throughout the manual. The problems presented in this present chapter are typical of those in which the right-angled triangle is an important element.

It will be noted that each of the problems presented has been given a general heading indicating the kind of a problem it is from a trigonometric standpoint. This permits the application of the method of solution outlined to other problems which, although different in detail, are basically similar.

Unknown Side of Right-angled Triangle

PROBLEM 1: To find X in Fig. 1. X is the distance from the outside of a shaft to the bottom of a keyway to which a cutter must be fed down to cut a keyway of width W and height H.

Analysis of Problem: This is essentially a problem in solving for an unknown side of a right-angled triangle when one side and the hypotenuse are known. As shown in the smaller diagram, the right-angled triangle to be solved has the known radius R as the hypotenuse, one-half of the width W has the known side, and the radius R minus the altitude A of the segment removed in

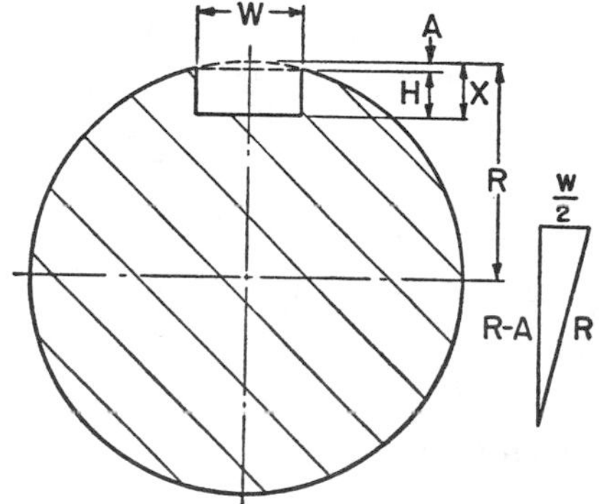

FIG. 1. To Find Depth X to Bottom of Keyway when Width W and Radius R are Known

cutting the keyway, as the unknown side. The second geometrical proposition on page **3–6** is the basis for setting up the first equation. This equation is solved for A. With H known, X can then be found.

Formula:

$$X = H + R - \sqrt{R^2 - \frac{W^2}{4}}$$

Derivation of Formula:

$$(R - A)^2 = R^2 - \left(\frac{W}{2}\right)^2 \tag{1a}$$

$$R - A = \sqrt{R^2 - \frac{W^2}{4}} \tag{1b}$$

$$A = R - \sqrt{R^2 - \frac{W^2}{4}} \tag{1c}$$

$$X = H + A \tag{2a}$$

$$X = H + R - \sqrt{R^2 - \frac{W^2}{4}} \tag{2b}$$

Example: Find the depth X to which a milling cutter must be sunk to cut a keyway having a width W of 7/16 inch and a depth H of 1/4 inch if the radius R of the shaft is 1 15/16 inches.

Solution:

$$X = 0.25 + 1.9375 - \sqrt{(1.9375)^2 - \frac{(0.4375)^2}{4}} \qquad (2b)$$

$$X = 0.25 + 1.9375 - \sqrt{3.7539 - 0.04785}$$

$$X = 0.25 + 1.9375 - \sqrt{3.7061}$$

$$X = 0.25 + 1.9375 - 1.9251$$

$$X = 0.2624 \text{ inch}$$

Unknown Side of Right-angled Triangle

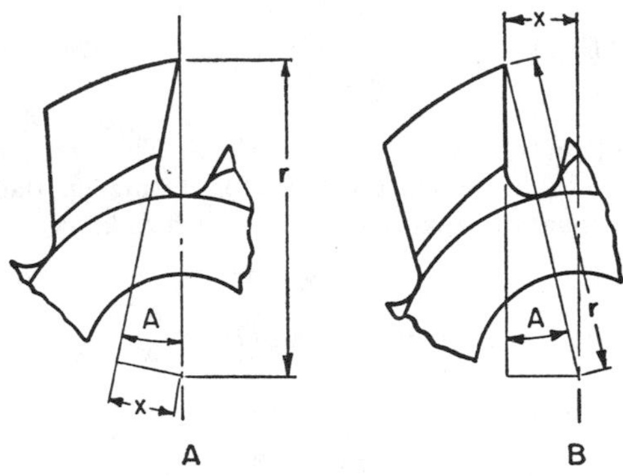

FIG. 2. To Find Horizontal Distance for Positioning Milling Cutter Tooth for Grinding Rake Angle A

PROBLEM 2: To find x in Fig. 2. In sharpening the teeth of thread milling cutters, if the teeth have rake, it is necessary to position each tooth for the grinding operation so that the outside tip of the tooth is at horizontal distance x from the vertical center line of the milling cutter as shown in Fig. 2B. What must this distance x be if the outside radius to the tooth tip is r and the rake angle is to be A?

Analysis of Problem: In Fig. 2A, it will be seen that, assuming the tooth has been properly sharpened to rake angle A, if a line is drawn extending the front edge of the tooth, it will be at a perpendicular distance x from the center of the cutter. Let the cutter now be rotated until the tip of the tooth is at a horizontal distance x from the vertical center line of the cutter as shown in Fig. 2B. It will be noted that an extension of the front edge of the cutter is still at perpendicular distance x from the center of the cutter indicating that the cutter face is parallel to the vertical center line or is itself vertical, which is the desired position for sharpening with a vertical wheel. Thus, x is the proper offset distance for grinding the tooth to rake angle A, if the radius to the tooth tip is r.

Formula: $x = r \sin A$

Derivation of Formula: Since r is the hypotenuse and x is one side of a right-angled triangle it can be readily seen that

$$x = r \sin A$$

It should be noted that as the face of each tooth is ground back by successive sharpenings, the radius to the tooth tip becomes less and less, since the radius to the outer edge of the cutter becomes less and less as it is swung from the front face of the tooth toward the back. This progressive decrease in outside radius of the tooth gives it the necessary clearance for the cut.

Thus as r becomes smaller, x should also theoretically be smaller, if rake angle A is to be maintained constant. In many cases, however, the change in rake angle A, if x is maintained constant, is so small as to be ignored over an appreciable number of successive sharpenings.

Example: What distance x off center must a $4\frac{1}{2}$-inch diameter cutter tooth be set, if it is to have a 3-degree rake angle? If this setting is maintained, what will the rake angle be if the cutter diameter has been reduced by successive sharpenings to 4 inches?

Solution: $A = 3°$ $r = 4.5 \div 2 = 2.25$ inches

$x = 2.25 \sin 3° = 2.25 \times 0.05234$

$x = 0.118$ inch

If $A' =$ rake angle for 4-inch cutter and a setting $x = 0.118$ is used, then

$$\sin A' = \frac{0.118}{2.00} = 0.0590$$

$$A' = 3°23'$$

Unknown Hypotenuse of Right-angled Triangle

PROBLEM 3: To compute *st* in Fig. 3. To obtain teeth with rake on milling cutters and reamers, the milling cutter or reamer blank must be offset from the radial position as shown. When the small-angle side of the double-angle fluting cutter is in line with *ot*, it is in position to cut a radial tooth. By moving it ahead of this position, as shown, negative rake is obtained. By moving it behind the radial position, positive rake is obtained. The problem is to compute the amount of horizontal offset *st* required to produce a negative rake angle *B* on the milling cutter teeth if the milling cutter blank radius *op* is known, as well as the small angle *A* of the cutter.

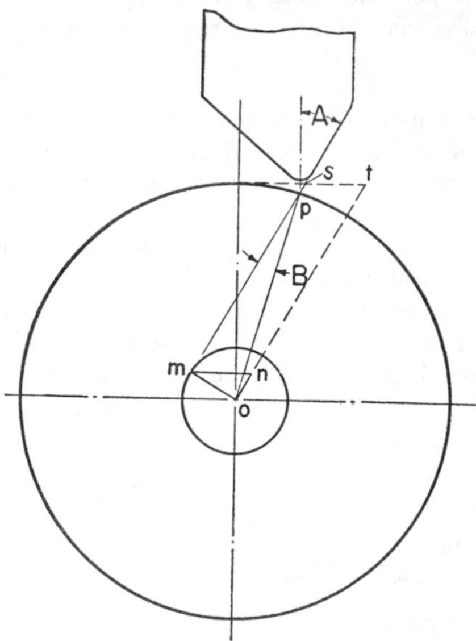

FIG. 3. To Find a Horizontal Offset *st* to Produce Negative Rake Angle *B* when Tool Angle *A* and Radius *op* are Known

Analysis of Problem: Line *ms* is an extension of the small angle side of the fluting cutter and is, therefore, parallel to *ot*. Draw a circle tangent to *ms* and having *O* as a center. Draw *mn* parallel with *st*. Then two right-angle triangles *omp* and *nom* have been formed. In the first, angle B and hypotenuse op are known, hence side mo which is common to the second triangle can be found. In the second triangle, angle omn is equal to angle A since their corresponding sides are perpendicular. Hence, with mo and angle omn known, side mn can be found and hence st, which is equal to mn.

Formula: $st = op \sin B \sec A$

Derivation of Formula: In right-angled triangle omp

$$om = op \sin B \quad (1)$$

$$\text{angle } nmo = \text{angle } A \quad (2)$$

In right-angled triangle nom

$$nm = om \sec nmo = om \sec A \quad (3)$$

but

$$nm = st \quad (4)$$

Substituting the equivalent of om from Equation (1) and the equivalent of nm from Equation (4) in Equation (3)

$$st = op \sin B \sec A \quad (5)$$

Note: A similar derivation for a positive rake angle shows that this formula also holds true where st is, in that case, the horizontal offset *behind* the radial position.

Example 1: Find the amount of offset required to produce a *negative* rake angle of 7 degrees when the radius of the blank is 3¼ inches and the small angle of the cutter is 30 degrees.

Solution: $A = 30°$; $B = 7°$; $op = 3.25$ inches;

$st = 3.25 \times \sin 7° \sec 30°$ (5)

$st = 3.25 \times 0.12187 \times 1.1547$

$st = 0.457$ inch *ahead* or, as shown in Fig. 3, to the *left* of the radial position

Example 2: Find the amount of offset required to produce a *positive* rake angle of 7 degrees when the radius of the blank is 3¼ inches and the small angle of the cutter is 30 degrees.

Solution: $A = 30°$; $B = 7°$; $op = 3.25$ inches;

$st = 3.25 \times \sin 7° \sec 30°$ (5)

$st = 0.457$ inch *behind* or, if shown in Fig. 3, to the *right* of the radial position

Two Right-angled Triangles with Common Hypotenuse

PROBLEM 4: To find angle X in Fig. 4. Given two lines ab and ac of known length at right angles to each other. An arc of known radius dg is drawn through point c with its center on a perpendicular to line ac at c. It is required to find the angle X between a line bf, drawn tangent to the given arc, and line ac.

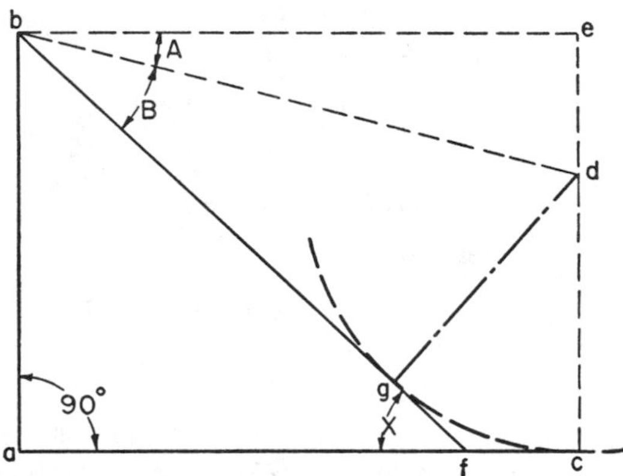

FIG. 4. To Find Angle X when Radius dg, and Dimensions ab and ac are Known

Analysis of Problem: This is a problem in solving a series of right-angled triangles. The first step is to complete the rectangle by drawing be equal and parallel to ac, and ec. Also draw line bd which forms the right triangle bde, in which two sides be and ed can be directly found from the data given. Then solve for the hypotenuse bd and angle A. The next step is to solve right-angled triangle bdg, in which hypotenuse bd has just been found and side dg is known, for angle B. (Note that bg is at right angles to gd, for a line drawn tangent to an arc is always at right angles to a radius of the arc drawn from the point of tangency.) Angle X is then equal to angles $A + B$.

Procedure: $be = ac$ (1)

$$ed = ec - dc \tag{2}$$

but $ba = ec$ and $dg = dc$

hence, $$ed = ab - dg \tag{3}$$

$$bd = \sqrt{\overline{be}^2 + \overline{ed}^2} \quad (4)$$

$$\tan A = \frac{ed}{be} \quad (5)$$

$$\sin B = \frac{dg}{bd} \quad (6)$$

$X = A + B$ (Geometrical Proposition No. 5 on page 3–6) (7)

Example 1: If $ab = 3.75$; $ac = 5$; and $dg = 2.5$, find angle X.

Solution:

$$be = 5 \quad (1)$$

$$ed = 3.75 - 2.5 = 1.25 \quad (3)$$

$$bd = \sqrt{(5)^2 + (1.25)^2} = 5.154 \quad (4)$$

$$\tan A = \frac{1.25}{5} = 0.25 \quad (5)$$

$$A = 14°2.2'$$

$$\sin B = \frac{2.5}{5.154} = 0.48507 \quad (6)$$

$$B = 29°1'$$

$$X = 14°2.2' + 29°1' = 43°3.2' \quad (7)$$

Example 2: If $ab = 5.85$; $ac = 4.5$; and $dg = 4.0$, find angle X.

Solution:

$$be = 4.5 \quad (1)$$

$$ed = 5.85 - 4.0 = 1.85 \quad (3)$$

$$bd = \sqrt{(4.5)^2 + (1.85)^2} = 4.865 \quad (4)$$

$$\tan A = \frac{1.85}{4.5} = 0.41111 \quad (5)$$

$$A = 22°20.9'$$

$$\sin B = \frac{4.0}{4.865} = 0.82220 \quad (6)$$

$$B = 55°18.4'$$

$$X = 22°20.9' + 55°18.4' = 77°39.3' \quad (7)$$

Two Right-angled Triangles with Common Hypotenuse

PROBLEM 5: To find angle X which is required for making the forming tool for the poppet valve head shown in Fig 5.

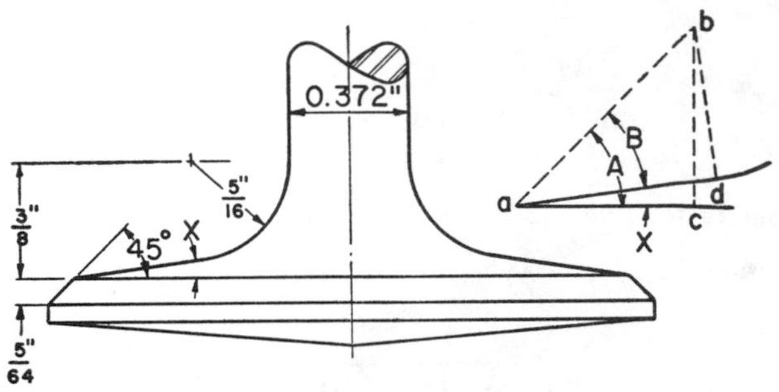

FIG. 5. To Find Angle X for Making Forming Tool for Poppet Valve Head when Angle A, Radius bd, and Height bc are Known

Analysis of Problem: Not all of the dimensions given are required to find angle X. Those which are needed are shown in the insert diagram as bc, bd and angle A. As can be seen, the solution of this problem involves two right-angled triangles with a common hypotenuse. The procedure is similar to that outlined in Problem 4.

Formulas: $X = A - B$; where $\sin B = \dfrac{bd \sin A}{bc}$

Derivation of Formula:

$$ab = bc \operatorname{cosec} A \qquad (1)$$

$$\sin B = \frac{bd}{ab} = \frac{bd}{bc \operatorname{cosec} A} = \frac{bd \sin A}{bc} \qquad (2)$$

$$X = A - B \qquad (3)$$

Example: Find angle X, if $bc = 3/8$ inch, $bd = 5/16$ inch and $A = 45$ degrees.

Solution: $bc = 0.375$; $bd = 0.3125$; $A = 45°$

$$\sin B = \frac{0.3125 \times \sin 45°}{0.375}$$

$$\sin B = \frac{0.3125 \times 0.70711}{0.375} = 0.58925$$

$$B = 36°6'$$

$$X = 45° - 36°6' = 8°54'$$

Two Right-angled Triangles with Common Hypotenuse

PROBLEM 6: To find angles X and Y in Fig. 6, of a special screw thread. The axial distances cf and cg from the center of a root radius to the center lines of adjoining crest radii are known, as is also the height fa and ge between the center of the root arc and the center of the crest arc. The root radius r and the crest radius R are known. It is required to find angles X and Y between each side of the thread flank and the vertical.

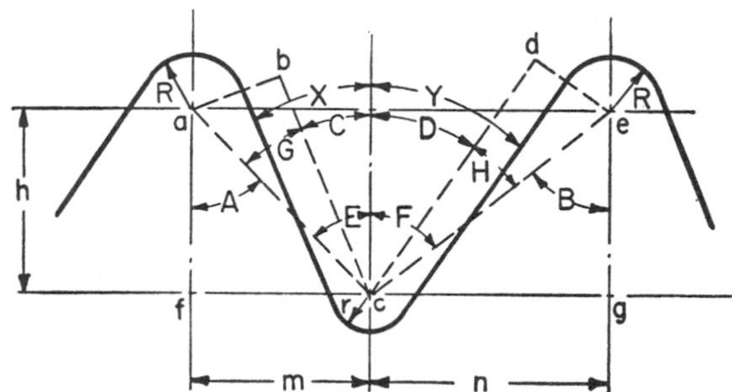

FIG. 6. To Find Angles X and Y of a Special Screw Thread when Radii r and R, Height h and Dimensions m and n are Known

Analysis of Problem: This is essentially a problem in solving two pairs of right-angled triangles, each of which has a common hypotenuse.

Draw bc parallel to the thread flank. Draw ac. Draw ab at right angles to bc, thus forming right-angled triangle abc. The hypotenuse ac of this triangle is also the hypotenuse of right-angled triangle afc. In triangle afc, two sides are known so that angle A and the hypotenuse can be determined. In triangle abc, side ab is equal to the sum of R and r. Hence with one side and the hypotenuse known, angle G can be determined. Angle E equals angle A (Geometrical Proposition No. 5 on page 3–6). Hence angle C can be found. But angle X equals angle C because both angles have one side common to each other and their other sides are parallel. Angle Y can be found in a similar manner. The detailed procedure for finding both X and Y is given below.

Procedure: To find angle X

$$\tan A = \frac{cf}{af} \tag{1}$$

$$ac = \frac{cf}{\sin A} \tag{2}$$

$$ab = r + R \tag{3}$$

$$\sin G = \frac{ab}{ac} \tag{4}$$

but
$$E = A \tag{5}$$

and
$$C = E - G = A - G \tag{6}$$

therefore
$$X = C = A - G \tag{7}$$

To find angle Y

$$\tan B = \frac{cg}{eg} = \frac{cg}{af} \tag{8}$$

$$ce = \frac{cg}{\sin B} \tag{9}$$

$$de = r + R \tag{10}$$

$$\sin H = \frac{de}{ce} \tag{11}$$

but
$$F = B \tag{12}$$

and
$$D = F - H = B - H \tag{13}$$

therefore
$$Y = D = B - H \tag{14}$$

Example: Let $cf = 0.25$; $cg = 0.375$; $af = 0.3125$; $r = 0.04$; and $R = 0.05$. Find angles X and Y.

Solution:
$$\tan A = \frac{0.25}{0.3125} = 0.80000 \tag{1}$$

$$A = 38°39.6'$$

$$ac = \frac{0.25}{0.62470} = 0.400 \quad (2)$$

$$ab = 0.04 + 0.05 = 0.09 \quad (3)$$

$$\sin G = \frac{0.09}{0.40} = 0.22500 \quad (4)$$

$$G = 13°0.2'$$

$$X = 38°39.6' - 13°0.2' = 25°39.4' \quad (7)$$

$$\tan B = \frac{0.375}{0.3125} = 1.2000 \quad (8)$$

$$B = 50°11.7'$$

$$ce = \frac{0.375}{0.76823} = 0.488 \quad (9)$$

$$de = 0.04 + 0.05 = 0.09 \quad (10)$$

$$\sin H = \frac{0.09}{0.488} = 0.18443 \quad (11)$$

$$H = 10°37.7'$$

$$Y = 50°11.7' - 10°37.7' = 39°34' \quad (14)$$

Three Right-angled Triangles with Common Side

PROBLEM 7: To find x in Fig. 7. This is a problem involving three right-angled triangles *ade*, *bde*, and *cde* having a common side *de*. Angles A and B are equal and known. Lengths *ab*, *bc*, and *cd* are known. It is required to find x, the length of side *de*.

Analysis of Problem: The objective is to set up an equation in which x is the only unknown value. The tangents of angles B, C, and D each have x as the only unknown factor. Furthermore, the tangent of angle $(D - C)$ is equal to the tangent of angle B, since $D - C$ equals A and A equals B. The tangent of $(D - C)$ can be expressed in terms of the tangent of D and the tangent of C. Thus, the desired equation with x as the only unknown can be established.

Formula:

$$x = \frac{o(m + n + o)(n + o)}{m - o}$$

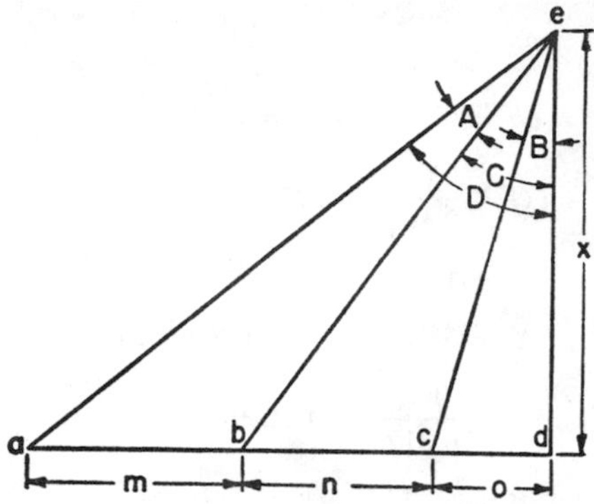

FIG. 7. To Find x, when Sides m, n and o and Angles A and B are Known

Derivation of Formula:

$$B = A = D - C \tag{1}$$

$$\tan B = \tan (D - C) \tag{2}$$

Substituting for tan $(D - C)$ in accordance with basic trigonometric formula 25, as given on page **4–24.**

$$\tan B = \frac{\tan D - \tan C}{1 + \tan D \tan C} \tag{3}$$

Substituting in Equation (3) for the tangents in terms of ratios of sides of the three right-angled triangles:

$$\frac{o}{x} = \frac{\dfrac{m + n + o}{x} - \dfrac{n + o}{x}}{1 + \dfrac{m + n + o}{x} \times \dfrac{n + o}{x}} \tag{4}$$

Simplifying numerator and denominator of the right-hand side of Equation (4):

$$\frac{o}{x} = \frac{\dfrac{m}{x}}{\dfrac{x^2 + (m + n + o)(n + o)}{x^2}} \tag{5}$$

Multiplying numerator and denominator of the right-hand side of Equation (5) by x:

$$\frac{o}{x} = \frac{m}{\dfrac{x^2 + (m+n+o)(n+o)}{x}} = \frac{mx}{x^2 + (m+n+o)(n+0)} \qquad (6)$$

Multiplying numerator of left-hand side of Equation (6) by denominator of right-hand side and setting this equal to product of denominator of left-hand side of (6) and numerator of right-hand side:

$$ox^2 + o(m+n+o)(n+o) = mx^2 \qquad (7)$$

$$mx^2 - ox^2 = o(m+n+o)(n+o) \qquad (8)$$

$$x^2(m-o) = o(m+n+o)(n+o) \qquad (9)$$

$$x^2 = \frac{o(m+n+o)(n+o)}{m-o} \qquad (10)$$

$$x = \sqrt{\frac{o(m+n+o)(n+o)}{m-o}} \qquad (11)$$

Example 1: If $m = 2.5$; $n = 2.1875$; and $o = 1.3125$, find x.

Solution:

$$x = \sqrt{\frac{1.3125(2.5 + 2.1875 + 1.3125)(2.1875 + 1.3125)}{2.5 - 1.3125}} \qquad (11)$$

$$x = \sqrt{\frac{1.3125 \times 6 \times 3.5}{1.1875}}$$

$$x = \sqrt{\frac{27.5625}{1.1875}} = \sqrt{23.2105} = 4.8177$$

Example 2: If $m = 5.86$; $n = 1.24$; and $o = 0.53$, find x.

Solution:

$$x = \sqrt{\frac{0.53(5.86 + 1.24 + 0.53)(1.24 + 0.53)}{5.86 - 0.53}} \qquad (11)$$

$$x = \sqrt{\frac{0.53 \times 7.63 \times 1.77}{5.33}}$$

$$x = \sqrt{1.343} = 1.16$$

Solving Right-angled Triangle When One Side, Part of Other Side and Part of Angle are Known

PROBLEM 8: To find *bd* and angle *E* in Fig. 8. This is a problem involving a right-angled triangle *abc*. Side *bc* is known. Part *ad* of side *ab* is known and part *A* of angle *B* is known. It is required to find part *bd* of side *ab*.

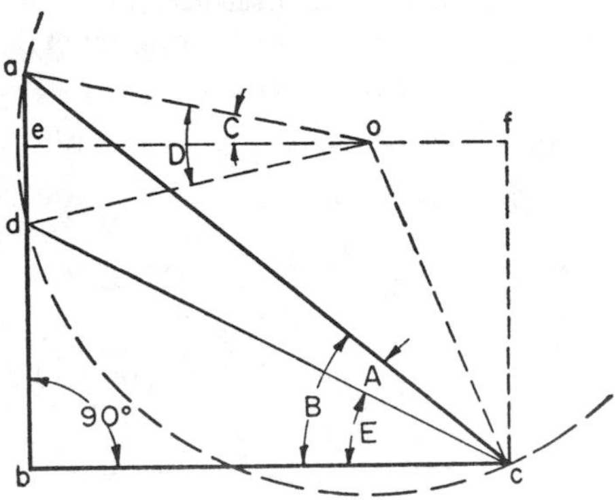

FIG. 8. To Find Part *bd* of Side *ab* and Angle *E* when *ad*, *bc* and Angle *A* are Known

Analysis of Problem: This is a problem in which it will be helpful to make use of another geometric figure, the circle, in solving for the unknown value. Three points determine the location and size of a circle so that a circle can be drawn through points *a*, *c* and *d* with its center at *o*. Then *oa*, *od*, and *oc* are its radii. Construct *eo* so that it is perpendicular to *ad*. Continue *eo* until it intersects a line drawn from *c* and parallel to *ab* at point *f*.

The arc *ad* subtends the angle *A* between chords *ac* and *cd*. It also subtends angle *D* between radii *oa* and *od*. According to the Geometric Proposition No. 5 given on page **3–8**, angle *A* equals one-half angle *D*. Since *oe* is drawn perpendicular to chord *ad* it bisects angle *D* and hence angle *C* equals angle *A*. With angle *C* and one-half of *ad* known, the radius of the circle can be determined, as well as distance *oe* from the center of the circle to chord *ad*. Since *ef* equals *bc*, which is known, *of* can be determined. Then *oc*, being a radius of the circle, is known so that *cf* can be found. Then *be* equals *cf* and knowing de, bd can be found. Knowing two sides of right-angled triangle, bcd, angle E can then be found.

Procedure:

$$C = \frac{D}{2} = A \quad (1)$$

$$ao = \frac{ae}{\sin C} = \frac{ad}{2 \sin C} = 0.5\, ad \operatorname{cosec} C \quad (2)$$

$$eo = ao \cos C \quad (3)$$

$$of = ef - eo = bc - eo \quad (4)$$

$$fc = \sqrt{\overline{oc}^2 - \overline{of}^2} = \sqrt{\overline{ao}^2 - \overline{of}^2} \quad (5)$$

$$bd = be - de = fc - de = fc - \frac{ad}{2} \quad (6)$$

$$\tan E = \frac{bd}{bc} \quad (7)$$

If ac were also required, then

$$ac = bc \sec B = bc \sec (A + E) \quad (8)$$

Example: Let $ad = 1.25$; $bc = 4.00$; and $A = 12°$. Find bd and angle E.

Solution:

$$C = 12° \quad (1)$$

$$ao = \frac{1.25}{2 \times 0.20791} = 3.006 \quad (2)$$

$$eo = 3.006 \times 0.97815 = 2.940 \quad (3)$$

$$of = 4.00 - 2.94 = 1.06 \quad (4)$$

$$fc = \sqrt{\overline{3.006}^2 - \overline{1.06}^2} = \sqrt{7.912} = 2.813 \quad (5)$$

$$bd = 2.813 - 0.625 = 2.188 \quad (6)$$

$$\tan E = \frac{2.188}{4.00} = 0.54700 \quad (7)$$

$$E = 28°40.7'$$

$$ac = bc \sec (12° + 28°40.7') \quad (8)$$

$$ac = 4.00 \times 1.3186 = 5.274$$

Solving Two Right-angled Triangles to Find Fourth Side of a Trapezoid

PROBLEM 9: To find the fourth side of a trapezoid when three sides and the altitude are given. In Fig. 9, if a, b and c are known, find side d. Similarly find a if b, c and d are known; b, if c, d and a are known; and c, if d, a and b are known.

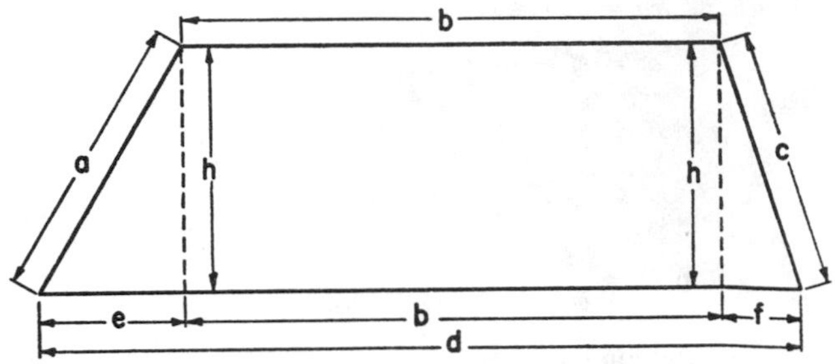

FIG. 9. To Find the Fourth Side of a Trapezoid when any Three Sides and Altitude h are Known

Analysis of Problem: If altitudes h are drawn from two corners of the trapezoid as shown, two right-angled triangles will be formed. Based on the theorem that the sum of the squares of the sides equals the square of the hypotenuse, relationships between a, e and h, and c, f and h can be established which lead to equations for a, b, c and d in terms of the known sides and altitude.

Formulas:

$$d = b + \sqrt{a^2 - h^2} + \sqrt{c^2 - h^2}$$

$$a = \sqrt{(d - b - \sqrt{c^2 - h^2})^2 + h^2}$$

$$b = d - \sqrt{a^2 - h^2} - \sqrt{c^2 - h^2}$$

$$c = \sqrt{(d - b - \sqrt{a^2 - h^2})^2 + h^2}$$

Derivation of Formulas:

$$e = \sqrt{a^2 - h^2} \tag{1}$$

$$f = \sqrt{c^2 - h^2} \tag{2}$$

$$d = b + e + f = b + \sqrt{a^2 - h^2} + \sqrt{c^2 - h^2} \tag{3}$$

$$b = d - e - f = d - \sqrt{a^2 - h^2} - \sqrt{c^2 - h^2} \quad (4)$$

$$a = \sqrt{e^2 + h^2} \quad (5)$$

but $$e = d - b - f = d - b - \sqrt{c^2 - h^2} \quad (6)$$

hence $$a = \sqrt{(d - b - \sqrt{c^2 - h^2})^2 + h^2} \quad (7)$$

$$c = \sqrt{f^2 + h^2} \quad (8)$$

but $$f = d - b - e = d - b - \sqrt{a^2 - h^2} \quad (9)$$

hence $$c = \sqrt{(d - b - \sqrt{a^2 - h^2})^2 + h^2} \quad (10)$$

Example: A. If $a = 6.6$; $b = 12.3$; $c = 5.9$; and $h = 5.6$, find d.

B. If $b = 4.1$; $c = 2.0$; $d = 5.8$; and $h = 1.8$, find a.

C. If $c = 10$; $d = 29$; $a = 11$ and $h = 9$, find b.

D. If $d = 23.2$; $a = 8.8$; $b = 16.4$; and $h = 7.2$, find c.

Solution:

A. $$d = 12.3 + \sqrt{6.6^2 - 5.6^2} + \sqrt{5.9^2 - 5.6^2} \quad (3)$$

$$d = 12.3 + \sqrt{43.56 - 31.36} + \sqrt{34.81 - 31.36}$$

$$d = 12.3 + 3.49 + 1.86 = 17.65 \text{ or } 17.6 \text{ rounded off}$$

B. $$a = \sqrt{(5.8 - 4.1 - \sqrt{2.0^2 - 1.8^2})^2 + 1.8^2} \quad (7)$$

$$a = \sqrt{(5.8 - 4.1 - 0.87)^2 + 1.8^2}$$

$$a = \sqrt{3.93} = 1.98 \text{ or } 2.0 \text{ rounded off}$$

C. $$b = 29 - \sqrt{11^2 - 9^2} - \sqrt{10^2 - 9^2} \quad (4)$$

$$b = 29 - 6.32 - 4.36 = 18.32 \text{ or } 18.3 \text{ rounded off}$$

D. $$c = \sqrt{(23.2 - 16.4 - \sqrt{8.8^2 - 7.2^2})^2 + 7.2^2} \quad (10)$$

$$c = \sqrt{(23.2 - 16.4 - 5.06)^2 + 7.2^2}$$

$$c = \sqrt{54.87} = 7.41 \text{ or } 7.4 \text{ rounded off}$$

7

Oblique-angled Triangles—Finding Unknown Sides and Angles

The oblique-angled triangle does not appear as frequently in mechanical problems as the right-angled triangle and hence the formulas for finding its unknown sides or angles are not as easily remembered. The two basic formulas for oblique-angled triangles are known as the Law of Sines and the Law of Cosines. These are given in Chapter 4.

An important and frequently puzzling kind of oblique-angled problem is to find one side when the other two sides and angle opposite one of them are known. This problem in its four possible cases is discussed in Chapter 4 (see Type 3 in the chart, page 4–21, entitled "How to Solve Oblique-angled Triangles").

The problems presented in this chapter illustrate the application of formulas representing the Law of Sines and the Law of Cosines, for finding of unknown sides or angles of oblique-angled triangles. Two of the problems involve the location of altitudes.

As in Chapter 6, these problems have been given general headings that are intended to extend their usefulness as models for the solution of other basically similar problems.

Determining Sides of Oblique-angled Triangle

PROBLEM 1: To find n and p in Fig. 1. If the difference d between the lengths of two sides of an oblique-angled triangle, the length of the third side m, and the angle opposite the third side M are known, what are the lengths n and p of the two unknown sides?

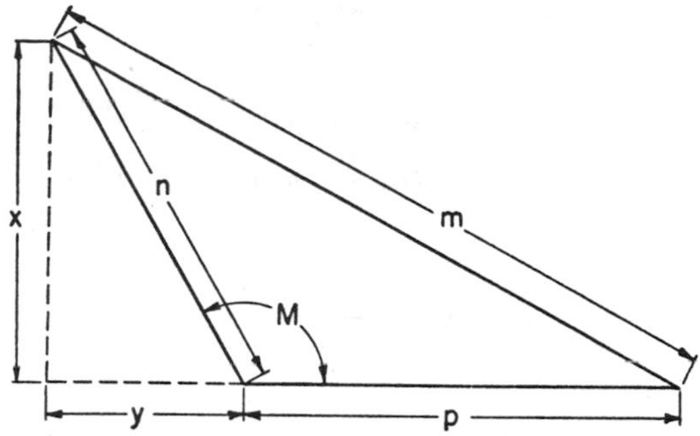

FIG. 1. To Find Sides n and p when Difference $(p - n)$, Side m and Angle M are Known

Analysis of Problem: This problem can be attacked in two ways. One is to use the Law of Cosines (page 4–18) to establish a relationship between known side m, known angle M and unknown sides n and p. Then substitute the equivalent of n in terms of p and d. The result will be a quadratic equation in terms of p that can be solved for p by the basic formula given on page 2–20 under General Form of a Quadratic Equation.

The other way is to draw lines x and y as shown, thus forming two right-angled triangles. Equations are now established for x, y, and n as two sides and hypotenuse of one of the right-angled triangles, and for x, $y + p$, and m as two sides and hypotenuse of the other right-angled triangle. By substituting in the second equation equivalents for $x^2 + y^2$, y, and n, a quadratic equation in terms of p is arrived at which can be solved as before.

Formulas:

$$p^2 + dp = \frac{m^2 - d^2}{2(1 - \cos M)} \quad \text{(to be solved as a quadratic for } p\text{)}$$

$$n = p + d$$

Derivation of Formula:

By First Method

$$\cos M = \frac{n^2 + p^2 - m^2}{2np} \tag{1}$$

but

$$n - p = d \quad \text{or} \quad n = p + d \tag{2}$$

hence
$$\cos M = \frac{(p + d)^2 + p^2 - m^2}{2p(p + d)} \tag{3a}$$

$$(2p^2 + 2pd) \cos M = p^2 + 2pd + d^2 + p^2 - m^2 \tag{3b}$$

$$2p^2 - 2p^2 \cos M + 2pd - 2\,pd \cos M = m^2 - d^2 \tag{3c}$$

$$2p^2(1 - \cos M) + 2pd(1 - \cos M) = m^2 - d^2 \tag{3d}$$

$$p^2 + pd = \frac{m^2 - d^2}{2(1 - \cos M)} \tag{3e}$$

By Second Method

$$x^2 + y^2 = n^2 \tag{1}$$

$$y = n \cos (180° - M) = -n \cos M \tag{2}$$

$$x + (y + p)^2 = m^2 \tag{3a}$$

$$x^2 + y^2 + 2yp + p^2 = m^2 \tag{3b}$$

Substituting the equivalents of $x^2 + y^2$ and y from Equations (1) and (2) in Equation (3b):

$$n^2 - 2np \cos M + p^2 = m^2 \tag{4}$$

but
$$n = p + d \tag{5}$$

hence
$$(p + d)^2 - 2p(p + d) \cos M + p^2 = m^2 \tag{6a}$$

or
$$p^2 + 2pd + d^2 - 2p^2 \cos M - 2pd \cos M + p^2 = m^2 \tag{6b}$$

$$2p^2 - 2p^2 \cos M + 2pd - 2pd \cos M = m^2 - d^2 \tag{6c}$$

$$2p^2(1 - \cos M) + 2pd(1 - \cos M) = m^2 - d^2 \tag{6d}$$

$$p^2 + pd = \frac{m^2 - d^2}{2(1 - \cos M)} \tag{6e}$$

Example: If $m = 14.2$, $d = 0.9$ and $M = 115°$, find n and p.

Solution:
$$p^2 + 0.9p = \frac{14.2^2 - 0.9^2}{2(1 + 0.42262)} \tag{6e}$$

(Note that since M lies between 90° and 180° its cosine is negative and, hence, $-\cos M$, as used in the above equation, is positive.)

$$p^2 + 0.9p = \frac{201.64 - 0.81}{2.84524} = 70.58$$

This is a quadratic equation in the form of $ap^2 + bp = c$ in which $a = 1$; $b = 0.9$ and $c = 70.58$.

$$p = \frac{-b \pm \sqrt{b^2 + 4ac}}{2a}$$

therefore
$$p = \frac{-0.9 \pm \sqrt{0.9^2 + 4 \times 1 \times 70.58}}{2 \times 1}$$

$$p = \frac{-0.9 \pm \sqrt{283.13}}{2} = \frac{-0.9 \pm 16.83}{2}$$

$$p = -8.86 \quad \text{or} \quad +7.96$$

Taking the plus value as being the one applicable to this problem:

$$n = 7.96 + 0.9 = 8.86 \tag{5}$$

Unknown Side of Oblique-angled Triangle

PROBLEM 2: To determine the depth of cut x in Fig. 2 required to machine a vee-shaped channel having known width w and sides inclined at known angles A and B respectively with the vertical, and having an arc of known radius r at the bottom of the vee.

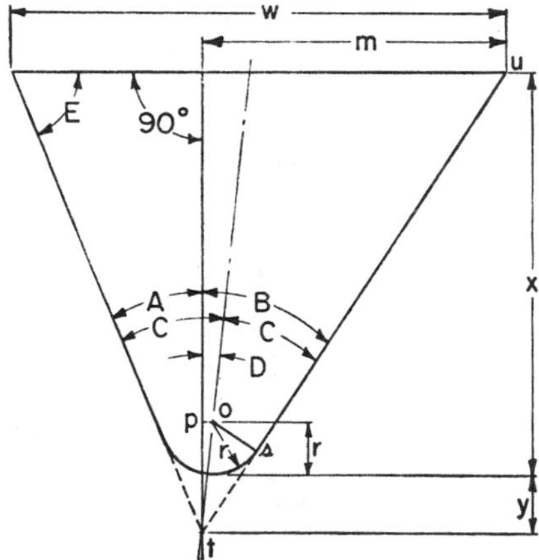

FIG. 2. To Find Depth of Cut x for Machining Vee-Shaped Channel when Angles A and B, Width W and Radius r are Known

Analysis of Problem: With the angles A and B and the width w known, side tu of the oblique-angled triangle can be found. Then vertical distance $(x + y)$ from the top of the vee to the point at which the sides would intersect if extended can be readily computed.

Knowing angles A and B, angle C, which is one-half of $A + B$, and angle D can be readily found. With angle C known and the radius r of the bottom arc, the distance to which is the hypotenuse of right-angled triangle ost can be computed. With to and angle D known, the distance tp which is one side of right-angled triangle tpo can be found. With tp and r known, y and x can now be determined.

Formula:

$$x = \frac{w \cos A \cos B}{\sin (A + B)} - r\left[\csc\left(\frac{A + B}{2}\right)\cos\left(\frac{B - A}{2}\right) - 1\right]$$

Derivation of Formula:

$$C = \frac{A + B}{2} \tag{1}$$

$$E = 90° - A \tag{2}$$

According to the Law of Sines (page 4–18) for obtuse-angled triangles:

$$\frac{tu}{\sin E} = \frac{w}{\sin (A + B)} \tag{3a}$$

$$tu = \frac{w \sin E}{\sin (A + B)} \tag{3b}$$

but

$$\sin E = \sin (90° - A) = \cos A \tag{4}$$

therefore

$$tu = \frac{w \cos A}{\sin (A + B)} \tag{5}$$

$$x + y = tu \cos B \tag{6}$$

Substituting the equivalent of tu from Equation (5)

$$x + y = \frac{w \cos A \cos B}{\sin (A + B)} \tag{7}$$

$$to = os \csc C = os \csc\left(\frac{A + B}{2}\right) \tag{8}$$

$$tp = to \cos D \tag{9}$$

but $$D = C - A = \frac{A + B}{2} - A = \frac{B - A}{2} \tag{10}$$

hence $$tp = to \cos\left(\frac{B - A}{2}\right) \tag{11}$$

Substituting the equivalent of *to* from Equation (8)

$$tp = os \csc\left(\frac{A + B}{2}\right) \cos\left(\frac{B - A}{2}\right) \tag{12a}$$

hence $$tp = r \csc\left(\frac{A + B}{2}\right) \cos\left(\frac{B - A}{2}\right) \tag{12b}$$

but $$y = tp - r \tag{13}$$

therefore $$y = r\left[\csc\left(\frac{A + B}{2}\right) \cos\left(\frac{B - A}{2}\right) - 1\right] \tag{14}$$

Substituting this value of y in Equation (7) and solving for x

$$x = \frac{w \cos A \cos B}{\sin (A + B)} - r\left[\csc\left(\frac{A + B}{2}\right) \cos\left(\frac{B - A}{2}\right) - 1\right] \tag{15}$$

Example: Find the depth of cut x for a vee-shaped channel having angle $A = 23°15'$, angle $B = 33°30'$, a width $w = 10.400$ and radius $r = 1.100$.

Solution: $$x = \frac{10.4 \cos 23°15' \cos 33°30'}{\sin (23°15' + 33°30')}$$

$$- 1.1\left[\csc\left(\frac{23°15' + 33°30'}{2}\right) \cos\left(\frac{33°30' - 23°15'}{2}\right) - 1\right] \tag{15}$$

$$x = \frac{10.4 \cos 23°15' \cos 33°30'}{\sin 56°45'} - 1.1(\csc 28°22.5' \cos 5°7.5' - 1)$$

$$x = \frac{10.4 \times 0.91879 \times 0.83388}{0.83629} - 1.1(2.1042 \times 0.99600 - 1)$$

$$x = 9.528 - 1.205 = 8.323$$

Location of Altitude in Oblique-angled Triangle

PROBLEM 3: To find distances x and y in Fig. 3 of the altitude h from the respective corners of an oblique-angled triangle, when sides a, b and c of the oblique-angled triangle are known.

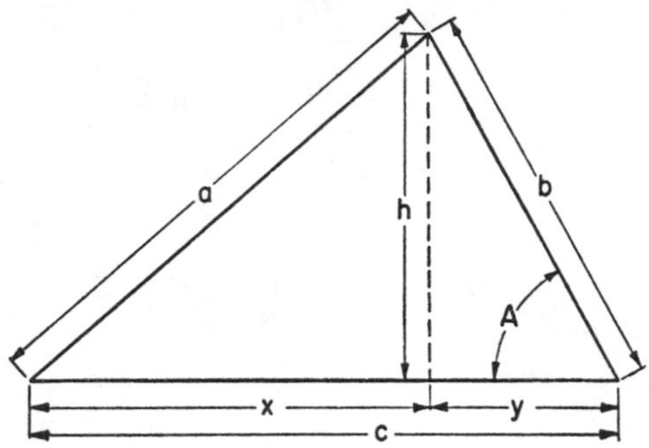

FIG. 3. To Find Distances x and y of Altitude h from Corners of Triangle when the Three Sides are Known

Analysis of Problem: This problem can be solved by using the Law of Cosines (page **4–18**) to find angle A and then finding y as one side of a right-angled triangle in which the hypotenuse and one angle are known. With y known, x is, of course, readily found.

A second method is to make use of the following geometrical proposition: *In any oblique-angled triangle where the three sides are known, the ratio of the length of the base to the sum of the other two sides equals the ratio of the difference between the lengths of the two sides to the difference between the lengths of the two parts of the base formed by the intersection of an altitude drawn from the vertex opposite the base.*

Thus: $$c : (a + b) :: (a - b) : (x - y)$$

or $$\frac{c}{a + b} = \frac{a - b}{x - y}$$

so that $$x - y = (a - b)(a + b) \div c$$

This will enable the value of $x - y$ to be determined. Since $x + y$ equals c which is known, the theorem that *the larger of two numbers is equal to one-half the sum of their sum and difference* can be used to find x.

Formula:

Method 1

$$y = \frac{b^2 + c^2 - a^2}{2c}$$

$$x = c - y$$

Method 2

$$x = \frac{c^2 + a^2 - b^2}{2c}$$

$$y = c - x$$

Derivation of Formulas:

Method 1

$$\cos A = \frac{b^2 + c^2 - a^2}{2bc} \tag{1}$$

$$y = b \cos A \tag{2}$$

hence

$$y = \frac{b(b^2 + c^2 - a^2)}{2bc} = \frac{b^2 + c^2 - a^2}{2c} \tag{3}$$

and

$$x = c - y \tag{4}$$

Method 2

$$\frac{c}{a+b} = \frac{a-b}{x-y} \tag{5a}$$

$$x - y = \frac{(a-b)(a+b)}{c} = \frac{a^2 - b^2}{c} \tag{5b}$$

$$x + y = c \tag{6}$$

$$x = \frac{(x+y) + (x-y)}{2} \tag{7}$$

hence

$$x = \frac{c + \dfrac{a^2 - b^2}{c}}{2} = \frac{c^2 + a^2 - b^2}{2c} \tag{8}$$

and

$$y = c - x \tag{9}$$

Example: If $a = 12.30$; $b = 9.20$; and $c = 13.80$, find x and y.

Solution:

By Method 1

$$y = \frac{9.20^2 + 13.80^2 - 12.30^2}{2 \times 13.80} \tag{3}$$

$$y = \frac{123.79}{27.60} = 4.49$$

$$x = 13.80 - 4.49 = 9.31 \tag{4}$$

By Method 2

$$x = \frac{13.80^2 + 12.30^2 - 9.20^2}{2 \times 13.80} \tag{8}$$

$$x = \frac{257.09}{27.60} = 9.31$$

$$y = 13.80 - 9.31 = 4.49 \tag{9}$$

Location of Altitude in Oblique-angled Triangle

PROBLEM 4: To find x and y in Fig. 4. Two sides a and b of an oblique-angled triangle are known and the amount of their overlap d when laid out on the third side c, as shown. It is required to find the length y of the altitude and its distance x from one corner of the triangle.

Analysis of Problem: The first method of solving this problem requires only an elementary knowledge of trigonometry and the use of trigonometrical functions. In the figure for Method 1, the angle A is found by using the formula for the cosine of an angle in an oblique-angled triangle when all three sides are known. (Side c can readily be found since a, b, and d are known.) With angle A known, x and y are readily found, since they constitute two sides of a right-angled triangle in which the hypotenuse b, as well as angle A, is known.

The second method, which avoids the use of trigonometry, is based on an arithmetical theorem that if the sum of two numbers and their difference be given, the greater of the two numbers is equal to one-half the sum of their sum and difference. In the figure for Method 2, $x + e$ and $x - e$ can be found from a, b, and d, so, according to this theorem, x can be found.

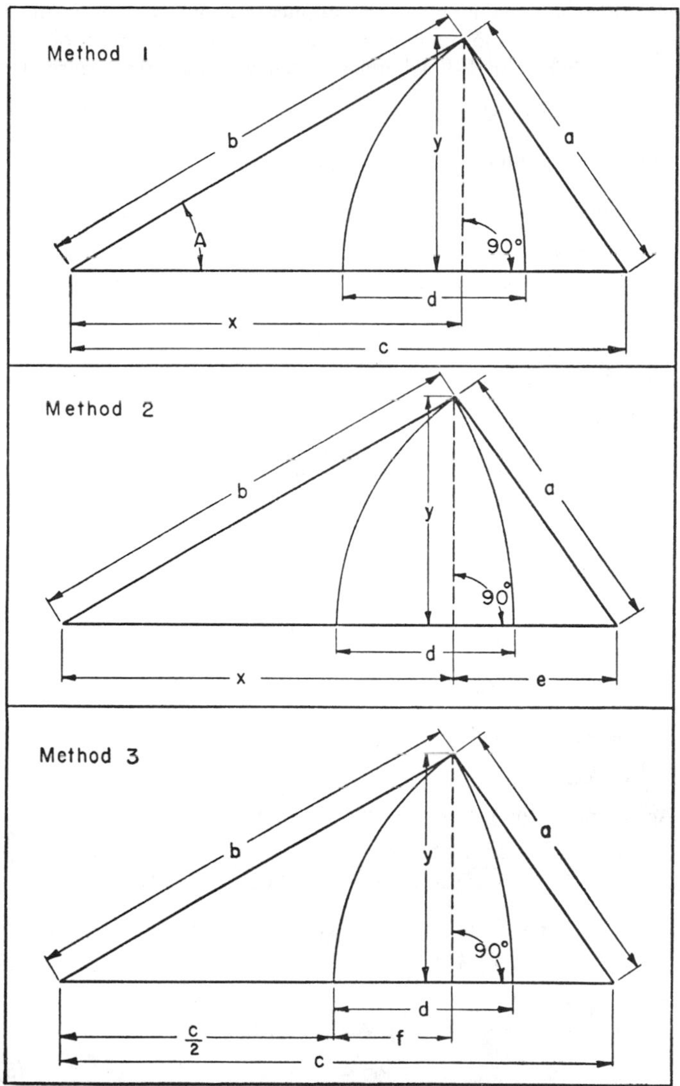

FIG. 4. To Find Altitude y and Its Distance x from One Corner of Triangle

The third method involves the application of a simple formula that is seldom given in handbooks, but is useful in solving problems involving oblique-angled triangles. Referring to the figure for Method 3, this formula is: $f = \dfrac{b^2 - a^2}{2c}$. With f known, x and y can be readily found.

Formulas:

By Method 1

$$x = b \cos A$$

$$y = b \sin A$$

where $$A = \text{arc cos} \frac{b^2 + (b + a - d)^2 - a^2}{2b(b + a - d)}$$

By Method 2

$$x = \frac{(x + e) + (x - e)}{2}$$

$$y = \sqrt{b^2 - x^2}$$

By Method 3

$$x = \frac{c}{2} + \frac{b^2 - a^2}{2c}, \text{ where } c = a + b - d$$

$$y = \sqrt{b^2 - x^2}$$

Derivation of Formulas:

By Method 1

According to the Law of Cosines (page **4–18**)

$$\cos A = \frac{b^2 + c^2 - a^2}{2bc} = \frac{b^2 + (b + a - d)^2 - a^2}{2b(b + a - d)} \tag{1}$$

$$x = b \cos A \tag{2}$$

$$y = b \sin A \tag{3}$$

By Method 2

$$x + e = a + b - d \tag{4}$$

$$y^2 = b^2 - x^2 \tag{5}$$

$$y^2 = a^2 - e^2 \tag{6}$$

therefore $$b^2 - x^2 = a^2 - e^2 \tag{7a}$$

and $$b^2 - a^2 = x^2 - e^2 \tag{7b}$$

or $$(b - a)(b + a) = (x - e)(x + e) \tag{7c}$$

so that $$x - e = \frac{(b - a)(b + a)}{x + e} \tag{7d}$$

According to the arithmetical theorem for sum and difference of two numbers:

$$x = \frac{(x + e) + (x - e)}{2} \tag{8}$$

$$y = \sqrt{b^2 - x^2} \tag{9}$$

By Method 3

$$b^2 - y^2 = \left(\frac{c}{2} + f\right)^2 \tag{10a}$$

$$b^2 - y^2 = \frac{c^2}{4} + cf + f^2 \tag{10b}$$

or $$4b^2 - 4y^2 = c^2 + 4cf + 4f^2 \tag{10c}$$

$$a^2 - y^2 = \left(\frac{c}{2} - f\right)^2 \tag{11a}$$

$$a^2 - y^2 = \frac{c^2}{4} - cf + f^2 \tag{11b}$$

$$4a^2 - 4y^2 = c^2 - 4cf + 4f^2 \tag{11c}$$

Subtracting Equation (11c) from Equation (10c)

$$4b^2 - 4a^2 = 8cf \tag{12a}$$

$$f = \frac{4b^2 - 4a^2}{8c} = \frac{b^2 - a^2}{2c} \tag{12b}$$

$$x = \frac{c}{2} + f = \frac{c}{2} + \frac{b^2 - a^2}{2c}, \text{ where } c = a + b - d \tag{13}$$

$$y = \sqrt{b^2 - x^2} \tag{14}$$

Example: Find x and y if $a = 6.81$; $b = 10.74$; and $d = 4.32$.

Solution:

By Method 1

$$\cos A = \frac{10.74^2 + (10.74 + 6.81 - 4.32)^2 - 6.81^2}{2 \times 10.74\,(10.74 + 6.81 - 4.32)} \qquad (1)$$

$$\cos A = \frac{115.3476 + 175.0329 - 46.3761}{284.1804}$$

$$\cos A = 0.85863$$

$$A = 30°50.2'$$

$$\sin A = 0.51259$$

$$x = 10.74 \times 0.85863 = 9.22 \qquad (2)$$

$$y = 10.74 \times 0.51259 = 5.51 \qquad (3)$$

By Method 2

$$x + e = 10.74 + 6.81 - 4.32 = 13.23 \qquad (4)$$

$$x - e = \frac{(10.74 - 6.81)(10.74 + 6.81)}{13.23} = \frac{68.9715}{13.23} \qquad (7d)$$

$$x - e = 5.21$$

$$x = \frac{13.23 + 5.21}{2} = 9.22 \qquad (8)$$

$$y = \sqrt{10.74^2 - 9.22^2} = \sqrt{30.3392} \qquad (6)$$

$$y = 5.51$$

By Method 3

$$c = 6.81 + 10.74 - 4.32 = 13.23$$

$$x = \frac{13.23}{2} + \frac{10.74^2 - 6.81^2}{2 \times 13.23} \qquad (13)$$

$$x = 6.615 + \frac{115.3476 - 46.3761}{26.46}$$

$$x = 6.615 + 2.607 = 9.22 \text{ (to two decimal places)}$$

$$y = \sqrt{10.74^2 - 9.22^2} = \sqrt{30.3392} = 5.51 \qquad (14)$$

Two Oblique-angled Triangles with Common Side

PROBLEM 5: To find angles A and B in Fig. 5. Two oblique triangles with a common side are given. The length of the common side ac, and the lengths of the two sides ab and ad are known. The angle C is known. It is also known that sides bc and dc are equal, although their lengths are not known. It is required to find angles A and B.

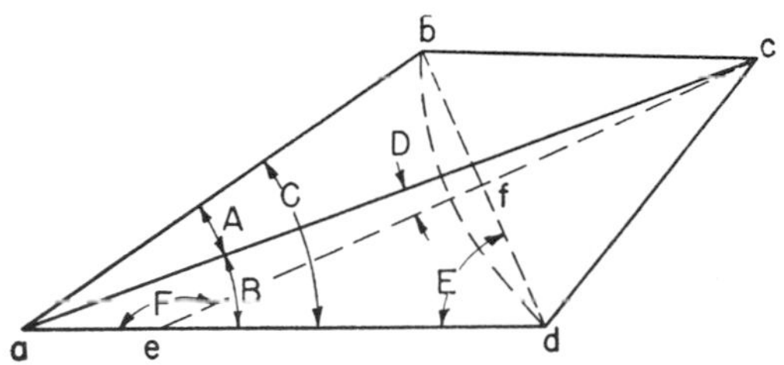

FIG. 5. To Find Unknown Angles A and B when Sides ab, ac and ad, and Angle C are Known

Analysis of Problem: In each of the two oblique triangles abc and adc, two sides are the only known factors, hence it is not possible to solve directly for the desired angles A and B. (In any oblique triangle at least two angles or two sides and an angle must be known before an unknown angle can be determined.)

The procedure, then, is to first construct an oblique triangle that can be solved with the elements given (sides ab and ad and angle C) and then use the new elements so determined in solving for angles A and B.

Procedure: Draw bd, making oblique triangle abd. Apply the formula given under Type 2 on page **4–19** for determining the third side of an oblique triangle when two sides and an included angle are known:

$$bd = \sqrt{\overline{ab}^2 + \overline{ad}^2 - 2ab \times ad \times \cos C} \qquad (1)$$

Draw ec at right angles to bd. Since bcd is an isosceles triangle (sides bc and dc being equal), a line drawn from the vertex perpendicular to the base will bisect the base, hence

$$fd = \frac{bd}{2} \qquad (2)$$

Next find angle E in triangle abd by applying The Law of Sines (page 4–18).

$$\frac{bd}{\sin C} = \frac{ab}{\sin E}$$

transposing,

$$\sin E = \frac{ab}{bd} \sin C \tag{3}$$

Since triangle efd is a right-angled triangle:

$$ed = \frac{fd}{\cos E} \tag{4}$$

It can be readily seen that:

$$ae = ad - ed \tag{5}$$

Angle F is equal to the sum of the two remote angles of triangle efd, or

$$F = 90° + E \tag{6}$$

Now angle D in triangle ace can be found by applying The Law of Sines.

$$\sin D = \frac{ae}{ac} \sin F \tag{7}$$

With angles D and F known, angles B and A can easily be determined:

$$B = 180° - D - F \tag{8}$$

$$A = C - B \tag{9}$$

Example: Let $ab = 3.31$; $ac = 5.25$; $ad = 3.56$; and $C = 35°$. Find angles A and B.

Solution: $bd = \sqrt{3.31^2 + 3.56^2 - 2 \times 3.31 \times 3.56 \times 0.81915}$ (1)

$$bd = \sqrt{4.325} = 2.08$$

$$fd = \frac{2.08}{2} = 1.04 \tag{2}$$

$$\sin E = \frac{3.31}{2.08} \times 0.57358 = 0.91199 \tag{3}$$

$$E = 65°46.9'$$

$$ed = \frac{1.04}{0.41022} = 2.535 \quad (4)$$

$$ae = 3.56 - 2.54 = 1.02 \quad (5)$$

$$F = 90° + 65°46.9' = 155°46.9' \quad (6)$$

$$\sin D = \frac{1.02}{5.25} \times 0.41022 = 0.07970 \quad (7)$$

$$D = 4°34.3'$$

$$B = 180° - 4°34.3' - 155°46.9' = 19°38.8' \quad (8)$$

$$A = 35° - 19°38.8' = 15°21.2' \quad (9)$$

8

Tapers—Figuring Angles, Diameters and Lengths

Tapering surfaces are used for certain machine parts and for the shanks of various tools such as twist drills, arbors, reamers, etc., to insure a tight fit, accurate centering or alignment of parts, quick assembly and easy operation. Such tapers may be of the "self-holding" or "quick-releasing" type. The shank of a drill or reamer is of the "self-holding" type and when seated offers considerable resistance to any force tending to rotate the tool in its socket. Examples of this type of taper are the Morse standard taper, used on the shanks of twist drills and a variety of other tools, which is approximately 5/8 inch per foot in most cases and the Brown and Sharpe taper, used for many arbors, collets and machine tool spindles, which is approximately 1/2 inch per foot in most cases. Another type of taper is the self-releasing type which also insures concentric seating of one part with relation to another but requires a positive locking device or drive to prevent slipping. An example of this type of taper is a milling machine spindle having a taper of 3 1/2 inches per foot or an included angle of over 16 degrees.

Aside from taper plug and ring gages which are used for repetitive checking, as in quantity inspection, various methods can be used to check or measure outside and inside tapers. To determine the amount of a given taper either in terms of the included angle (angle between the sides of the taper) or in terms of the amount of taper per inch, it is necessary to know the diameter of the tapered part at two different points and the distance between these points.

If the taper is on the outside of a part and the diameter of one end is known, as in Problem 4, it is only necessary to measure the distance over two spheres or rolls of known diameter when placed in contact with the tapered part at diametrically opposite points as shown. If no diameter of the tapered part is known, two sets of spheres or rolls of different diameters can be used as in Problem 5, or two sets of pins of the same diameter placed at two different points along the taper as shown in Problem 6. The distance between the two pairs of pins in this latter problem is established by using two sets of precision gage blocks to support the upper set of pins.

If the taper is an inside one, a sphere or spheres are used for checking. In Problem 8, the angle of taper is known and a sphere is used to check the end diameter of the tapered hole. In Problem 9, two spheres of different diameters are used to check the taper per inch and two diameters at specified points.

Other problems in this chapter show how the intersection of two tapers is located; how the angle of elevation for machining a tapered part when supported by a vee-block is computed; and how to find the width of intersection of a taper and an arc of known radius.

Here, again, although specific problems taken from the shop are presented, their headings indicate a general application to other problems in which the same basic elements are to be found.

Locating Intersection of Two Tapers

PROBLEM 1: To find ab in Fig. 1. This is a problem in determining the point of intersection of two tapers when one diameter of each is measured and the distance between the points of measurement is known. The distance mm' is the measured diameter of one taper, oo' is the measured diameter of the other, and co is the known distance between these two diameters. The taper per inch X of the long taper, which is equal to 2 tan A, is known and the taper Y of the short taper, which is equal to 2 tan B, is known. It is required to find the distance ab locating the intersection of the two tapers with respect to the point where diameter mm' is measured.

Analysis of Problem: It can be seen that cm can be expressed as one-half the difference between mm' and oo'. It can also be expressed as the sum of ab tan A and do tan B. The distance do can be expressed in terms of co and ab; tan A can be expressed in terms of X; and tan B in terms of Y. Hence, an equation can be set up with ab as the only unknown.

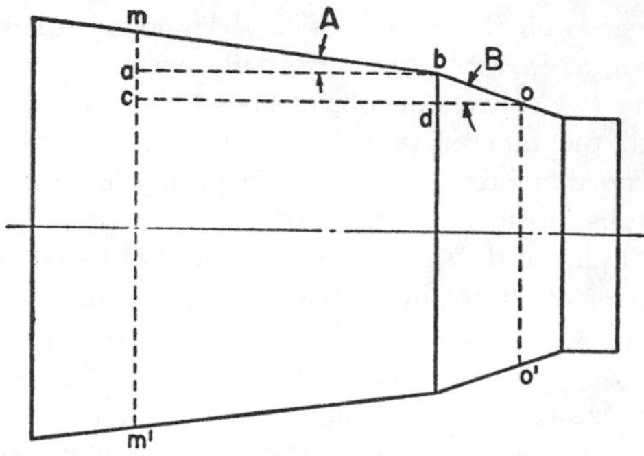

FIG. 1. To Find Distance ab Locating Intersection of Two Tapers when Diameters mm' and oo', Distance co between Them and Angles A and B are Known

Formula:

$$ab = \frac{mm' - oo' - coY}{X - Y}$$

Derivation of Formula:

$$am = ab \tan A \tag{1}$$

$$bd = do \tan B \tag{2}$$

but $bd = ac$

and $do = co - cd = co - ab$

Substituting in (2) $ac = (co - ab) \tan B \quad (3)$

but $cm = am + ac \quad (4)$

From (1) and (3)

$$cm = ab \tan A + (co - ab) \tan B \tag{5}$$

but $cm = \dfrac{mm' - oo'}{2} \quad (6)$

therefore $\dfrac{mm' - oo'}{2} = ab \tan A + (co - ab) \tan B \quad (7)$

$$mm' - oo' = 2ab \tan A + 2(co - ab) \tan B \tag{8}$$

but $X = 2 \tan A$ and $Y = 2 \tan B$

therefore $mm' - oo' = abX + (co - ab)Y \quad (9)$

$$mm' - oo' = abX + coY - abY \tag{10}$$

transposing

$$abX - abY = mm' - oo' \times coY \tag{11}$$

$$ab = \frac{mm' - oo' - coY}{X - Y} \tag{12}$$

Example: Let $mm' = 6.841$; $oo' = 3.798$; $co = 7.402$; $X = 0.371$; and $Y = 0.750$. Find ab.

Solution:

$$ab = \frac{6.841 - 3.798 - 7.402 \times 0.75}{0.371 - 0.750} = \frac{-2.508}{-0.379} \tag{12}$$

$$ab = 6.617$$

Locating Intersection of Two Tapers

PROBLEM 2: To find X in Fig. 2. The dimensions D and L and the angles α and β of the templet are known. On each side of the templet, the two tapering edges at known angles α and β with the horizontal are connected by a tangent arc of radius R. The distance X to be found extends from the left-hand edge of the templet to the point where the two tapering edges would intersect if continued.

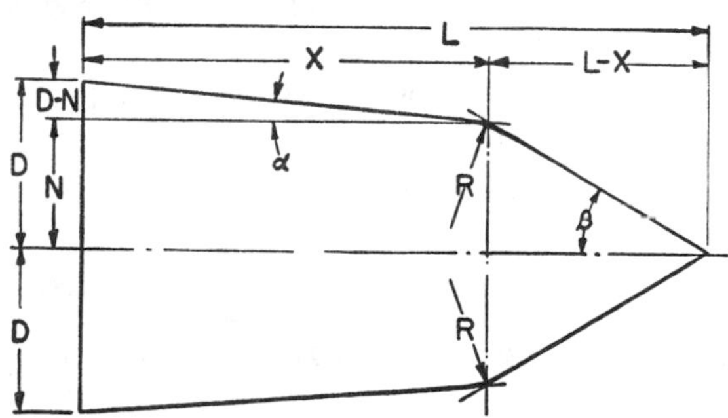

FIG. 2. To Find Distance X Locating Intersection of Two Tapers when Angles α and β, Radius D, and Length L are Known

Analysis of Problem: An expression for tan α in terms of D, N and X and one for tan β in terms of N, L and X can be readily established. Each of these expressions can be rearranged so that N

is on one side of the equation and the balance of the terms on the other. These two equivalents of N are now set equal to each other and the resulting equation is solved for X.

Formula: $$X = \frac{L \tan \beta - D}{\tan \beta - \tan \alpha}$$

Derivation of Formula:

$$\tan \alpha = \frac{D - N}{X}; \quad \text{hence } N = D - X \tan \alpha \tag{1}$$

$$\tan \beta = \frac{N}{L - X}; \quad \text{hence } N = (L - X) \tan \beta \tag{2}$$

Therefore

$$D - X \tan \alpha = (L - X) \tan \beta \tag{3a}$$

$$D - X \tan \alpha = L \tan \beta - X \tan \beta \tag{3b}$$

$$X \tan \beta - X \tan \alpha = L \tan \beta - D \tag{3c}$$

$$X = \frac{L \tan \beta - D}{\tan \beta - \tan \alpha} \tag{3d}$$

Example 1: Let $D = 3.16$; $L = 12.31$; $\alpha = 5°$; and $\beta = 30°$. Find X.

Solution: $$X = \frac{12.31 \tan 30° - 3.16}{\tan 30° - \tan 5°} \tag{3d}$$

$$X = \frac{12.31 \times 0.57735 - 3.16}{0.57735 - 0.08749} = \frac{3.947}{0.4899}$$

$$X = 8.06$$

Example 2: Let $D = 3.04$; $L = 21.19$; $\alpha = 7°$; and $\beta = 10°$. Find X.

Solution: $$X = \frac{21.19 \tan 10° - 3.04}{\tan 10° - \tan 7°}$$

$$X = \frac{21.19 \times 0.17633 - 3.04}{0.17633 - 0.12278} = \frac{0.696}{0.05355}$$

$$X = 13.00$$

Length of Intersecting Taper and Diameter at Intersection

PROBLEM 3: To find radius x and length y in Fig. 3. When two tapers are formed on a piece of work, it is sometimes necessary to determine the length of each taper and the diameter at the point where the two tapers join. The following dimensions are known: the radius a at the small end; the radius b at the large end; the angle A of taper at the large end; the angle B of taper at the small end; and the total length f of the piece. With radius x and length y found, the diameter at the intersection of the two tapers and the length of the small tapered section can be determined.

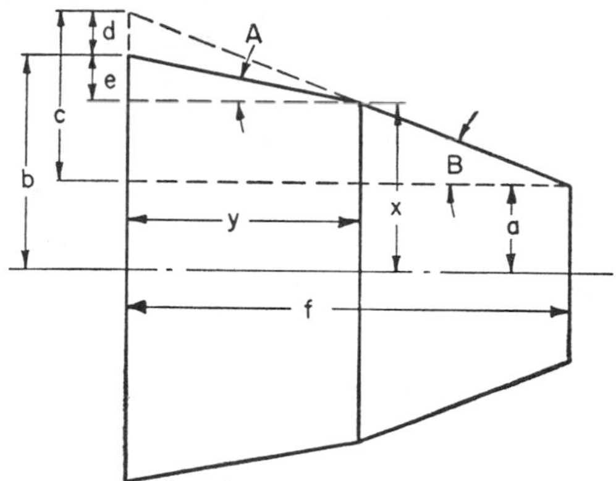

FIG. 3. To Find Radius x and Length y of Long Taper when Radii a and b, Angles A and B and Length f are Known

Analysis of Problem: With the addition of the construction lines shown in Fig. 3, two right-angled triangles are formed having a common side of length y. Two equations can thus be established for y, one in terms of angle A and side e, and the other in terms of angle B and side $(d + e)$. Setting the equivalents of y equal to each other it is then possible to solve for e. Knowing e, both x and y can be found.

Formulas:

$$y = \frac{(f \tan B + a - b) \cot B \cot A}{\cot A - \cot B}$$

$$x = b - \frac{(f \tan B + a - b) \cot B}{\cot A - \cot B}$$

Derivation of Formulas:

$$y = e \cot A \tag{1}$$

$$y = (e + d) \cot B \tag{2}$$

Setting these two equivalents of y equal to each other

$$e \cot A = (e + d) \cot B \tag{3a}$$

$$e \cot A = e \cot B + d \cot B \tag{3b}$$

$$e \cot A - e \cot B = d \cot B \tag{3c}$$

$$e = \frac{d \cot B}{\cot A - \cot B} \tag{3d}$$

but

$$d = c + a - b \tag{4}$$

and

$$c = f \tan B \tag{5}$$

therefore

$$d = f \tan B + a - b \tag{6}$$

and

$$e = \frac{(f \tan B + a - b) \cot B}{\cot A - \cot B} \tag{7}$$

Substituting this equivalent of e in Equation (1):

hence

$$y = \frac{(f \tan B + a - b) \cot B \cot A}{\cot A - \cot B} \tag{8}$$

$$x = b - e \tag{9}$$

therefore

$$x = b - \frac{(f \tan B + a - b) \cot B}{\cot A - \cot B} \tag{10}$$

Example: Let $a = 1.65$; $b = 3.97$; $f = 8.22$; $A = 10°30'$ and $B = 21°$. Find x and y.

Solution:

$$y = \frac{(8.22 \tan 21° + 1.65 - 3.97) \cot 21° \cot 10°30'}{\cot 10°30' - \cot 21°} \tag{8}$$

$$y = \frac{(8.22 \times 0.38386 + 1.65 - 3.97) \times 2.6051 \times 5.3955}{5.3955 - 2.6051}$$

$$y = \frac{11.8070}{2.7904} = 4.23$$

$$x = \frac{3.97 - (8.22 \times 0.38386 + 1.65 - 3.97) \times \cot 21°}{\cot 10°30' - \cot 21°} \tag{10}$$

$$x = 3.97 - \frac{2.1883}{2.7904}$$

$$x = 3.97 - 0.78 = 3.19$$

Checking Angle of Tapered Plug by Measurement over Pins

PROBLEM 4: To find x in Fig. 4. The width ab of the base of the tapered plug is known and the radius r of each of the two pins or cylindrical plugs. It is required to compute what the measurement x over the two pins should be for a given angle A.

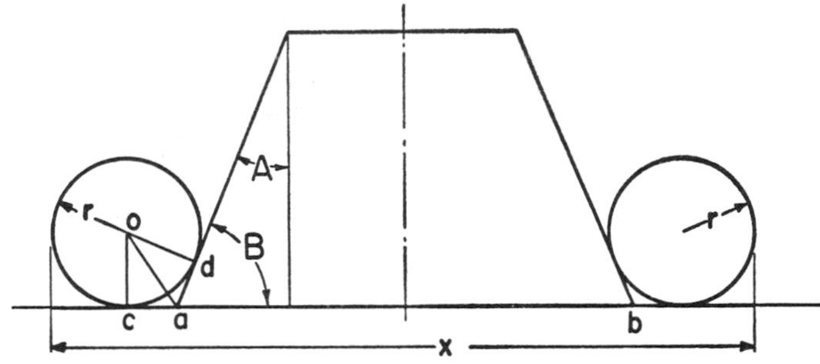

FIG. 4. To Compute Distance x Over Two Pins when Radius r, Diameter ab and Angle A are Known

Analysis of Problem: In order to find x when ab and the radius r of each pin are known, it is only necessary to compute ac. This can be done since oca is a right-angled triangle and angle coa can be shown to be equal to one-half angle B which, in turn, is equal to $90° - A$.

Formula: $$x = ab + 2r\left(\tan\frac{90° - A}{2} + 1\right)$$

Derivation of Formula:

$$\text{Angle } B = 90° - A \tag{1}$$

Angle cod = angle B since their respective sides are at right angles to each other.

$$\text{Angle } coa = \frac{\text{angle } cod}{2} \tag{2}$$

since a line drawn from the center of a circle to the point of intersection of two tangents to the circle bisects the angle formed by radii drawn from the two points of tangency.

Therefore $$\text{Angle } coa = \frac{B}{2} = \frac{90° - A}{2} \tag{3}$$

But $$ca = oc \tan coa \tag{4}$$

hence $$ca = r \tan \frac{90° - A}{2} \tag{5}$$

$$x = ab + 2ca + 2r \tag{6}$$

$$x = ab + 2r \tan \frac{90° - A}{2} + 2r \tag{7a}$$

$$x = ab + 2r\left(\tan \frac{90° - A}{2} + 1\right) \tag{7b}$$

Example: Let $ab = 3.210$; $r = 0.625$, and $A = 15°$. Find x.

Solution:

$$x = 3.210 + 2 \times 0.625\left(\tan \frac{90° - 15°}{2} + 1\right) \tag{7b}$$

$$x = 3.210 + 1.25\ (\tan 37°30' + 1)$$

$$x = 3.210 + 1.25 \times 1.76733 = 5.419$$

Checking Angle of Tapered Plug by Measurement Over Pins

PROBLEM 5: To find angle B in Fig. 5. The angle of a tapered plug may be checked by taking micrometer readings over pins or cylindrical plugs of two sizes. This method is particularly useful for checking the angle of a short taper such as that shown. In this problem, the diameters m and n of the two sizes of pins are known and, of course, the measurements r and s over these pins.

Analysis of Problem: This is a problem involving the solution of a right-angled triangle in which the two sides can be directly determined from the dimensions given. The triangle is abc and with its two sides known, angle A can be determined and then angle B.

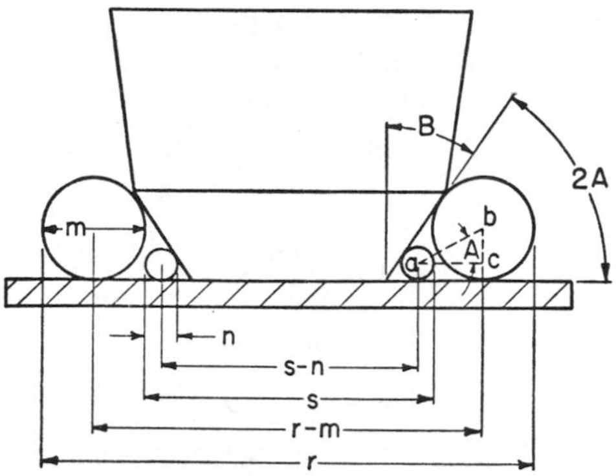

FIG. 5. To Compute Angle B when Measurements r and s are Taken and Diameters m and n are Known

Formula: $$B = 90° - 2A$$

where $$A = \text{arc cot}\, \frac{r - s}{m - n} - 1$$

Derivation of Formula:

$$bc = \frac{m - n}{2} \tag{1}$$

$$ac = \frac{(r - m) - (s - n)}{2} = \frac{(r - s) - (m - n)}{2} \tag{2}$$

$$\cot A = \frac{ac}{bc} = \frac{\dfrac{(r - s) - (m - n)}{2}}{\dfrac{m - n}{2}} \tag{3a}$$

$$\cot A = \frac{\dfrac{r - s}{2}}{\dfrac{m - n}{2}} - 1 = \frac{r - s}{m - n} - 1 \tag{3b}$$

$$B = 90° - 2A \tag{4}$$

Taper per inch = change in diameter per inch of length

tan B = change in radius per inch of length

$$\text{Taper per inch} = 2 \tan B \tag{5}$$

Example: Let $m = 0.875$; $n = 0.250$; $r = 4.083$; and $s = 2.345$. Find angle B and taper per inch.

Solution:

$$\cot A = \frac{4.083 - 2.345}{0.875 - 0.250} - 1 \tag{3b}$$

$$\cot A = \frac{1.738}{0.625} - 1 = 1.7808$$

$$A = 29°19'$$

$$B = 90° - 2 \times 29°19' \tag{4}$$

$$B = 90° - 58°38' = 31°22'$$

$$\text{Taper per inch} = 2 \times 0.60960 = 1.2192 \tag{5}$$

Checking Diameters of Tapered Plug Gage by Measurement Over Pins

PROBLEM 6: To find (1) the diameters $D(Not\ Go)$ and $D(Go)$ and (2) the diameters E, F and G in Fig. 6. The problem is to check a tapered plug gage by taking measurements over two pairs of pins or cylindrical plugs of equal diameter, the top pair being placed on two stacks of precision gage blocks of equal height and the lower pair resting on the surface plate supporting the tapered plug. The height L of the gage blocks, the radius R of the pins, the measurement M_1 over the top pair of pins and the measurement M_2 over the bottom pair of pins are known. Ordinarily, the diameters $D(Not\ Go)$ and $D(Go)$ are required, but diameters at other points such as E, which is in line with the top of the gage blocks; F, which is at some specified distance from either end of the plug; and G, which is the diameter of the small end of the plug, might also be required.

Analysis of Problem: The solution of this problem hinges on determining the horizontal distance X between the center of each pin and the plug being measured. To find X, one-half the angle of taper α is computed and from this angle β and hence X (which is one side of a right-angled triangle of which R is the other side) are found. With X known, the diameter E at distance L from the small end of the plug as well as diameter G at the small end of the plug can be determined. By adding to or subtracting from E or F the proper amount, as deter-

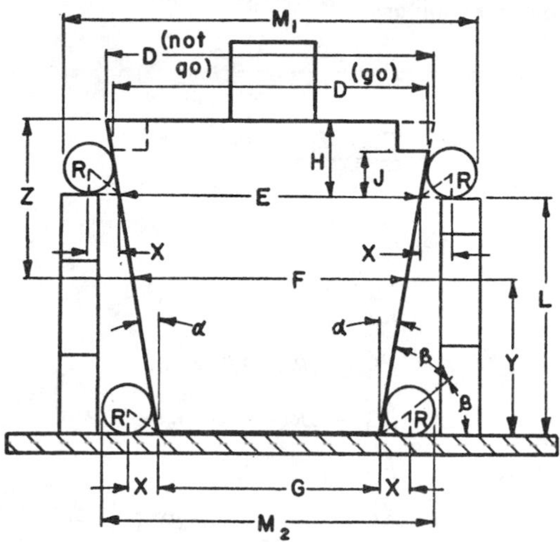

FIG. 6. To Check Diameters D (NOT GO) and D (GO) of Tapered Plug Gage by Measurements Over Pins

mined by the taper per inch and the vertical distance of the desired diameter from E or G, the other diameters are readily found. Formulas for each of these diameters are given below.

Formulas:

$$D(\textit{Not go}) = M_1 - 2R(1 + \cot B) + \frac{H(M_1 - M_2)}{L}$$

$$D(\textit{Go}) = M_1 - 2R(1 + \cot B) + \frac{J(M_1 - M_2)}{L}$$

$$E = M_1 - 2R(1 + \cot B)$$

$$F = M_1 - 2R(1 + \cot B) - \frac{(Z - H)(M_1 - M_2)}{L}$$

$$F = M_2 - 2R(1 + \cot B) + \frac{Y(M_1 - M_2)}{L}$$

$$G = M_2 - 2R(1 + \cot B)$$

Derivation of Formulas:

$$\text{Taper per inch} = T = \frac{M_1 - M_2}{L} \tag{1}$$

$$\tan \alpha = \frac{T}{2} = \frac{M_1 - M_2}{2L} \tag{2}$$

$$\beta = \frac{90° - \alpha}{2} \tag{3}$$

$$X = R \cot \beta \tag{4}$$

$$E = M_1 - 2R - 2R \cot \beta \tag{5a}$$

$$E = M_1 - 2R(1 + \cot \beta) \tag{5b}$$

$$D(Not\ Go) = E + HT \tag{6}$$

$$= M_1 - 2R(1 + \cot \beta) + \frac{H(M_1 - M_2)}{L} \tag{7}$$

$$D(Go) = E + JT \tag{8}$$

$$D(Go) = M_1 - 2R(1 + \cot \beta) + \frac{J(M_1 - M_2)}{L} \tag{9}$$

Similarly, the diameter G is found using measurement M_2.

$$G = M_2 - 2R - 2R \cot B \tag{10a}$$

$$G = M_2 - 2R(1 + \cot B) \tag{10b}$$

and any other diameter F at a distance Y from the small end of the plug:

$$F = G + YT \tag{11}$$

$$F = M_2 - 2R(1 + \cot B) + \frac{Y(M_1 - M_2)}{L} \tag{12}$$

or if diameter F is measured at a distance Z from the large end of the plug:

$$F = E - (Z - H)T \tag{13}$$

$$F = M_1 - 2R(1 + \cot B) - \frac{(Z - H)(M_1 - M_2)}{L} \tag{14}$$

Example: Let $L = 4.43$; $R = 0.5$; $H = 1.40$; $J = 0.58$; $M_1 = 7.71$; $M_2 = 6.20$; $Y = 2.95$ and $Z = 2.88$. Find $D(Not\ Go)$; $D(Go)$; E, F, and G.

Solution:

$$\tan \alpha = \frac{1}{2}\left(\frac{M_1 - M_2}{L}\right) = \frac{1}{2}\left(\frac{7.71 - 620}{4.43}\right)$$

$$\tan \alpha = \frac{0.34086}{2} = 0.17043$$

$$\tan \alpha = 9°40.3'$$

$$B = \frac{90° - 9°40.3'}{2} = 40°9.8' \qquad \cot B = 1.1848$$

Note that $T = \dfrac{M_1 - M_2}{L} = 0.34086$ as found above. This value will be used in the succeeding equations.

$$D(Not\ Go) = 7.71 - (2 \times 0.5)(1 + 1.1848) + 1.40 \times 0.34086$$

$$= 7.71 - 2.1848 + 0.4772 = 6.0024 \tag{7}$$

$$D(Go) = 7.71 - (2 \times 0.5)(1 + 1.1848) + 0.58 \times 0.34086 \tag{9}$$

$$D(Go) = 7.71 - 2.1848 + 0.1977 = 5.7229$$

$$E = 7.71 - (2 \times 0.5)(1 + 1.1848) = 5.5252 \tag{5b}$$

$$F = 7.71 - (2 \times 0.5)(1 + 1.1848) - \frac{(2.88 - 1.40)(7.71 - 6.20)}{4.43} \tag{14}$$

$$F = 7.71 - 2.1848 - 0.5045 = 5.0207$$

$$G = 6.20 - (2 \times 0.5)(1 + 1.1848) = 4.0152 \tag{10b}$$

Radius of Plug for Checking Tapered Hole

PROBLEM 7: To find radius *bd* in Fig. 7. A round plug when placed in a gage as shown is to be exactly flush with the top. Angles A and B are known and the width *ac* of the opening in the gage.

Analysis of Problem: Draw lines *ba* and *bc* from the center of the plug making the two right-angled triangles *abd* and *cbd*. Angles F

and C are equal and angles D and E are equal since a line drawn from the center of a circle to the intersection of two tangents to the circle will bisect the angle formed by the two tangents. Angle $(F + C)$ is equal to 90 degrees minus angle A; hence angle C can be found. Angle $(D + E)$ equals 90 degrees minus angle B; hence angle D can be found. With angles C and D known, an expression for radius bd can be established in terms of these two angles and ac.

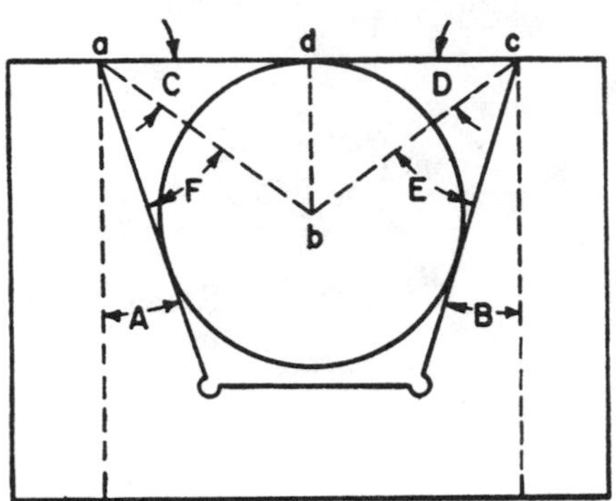

Fig. 7. To Find Radius bd of Round Plug when ac and Angles A and B are Known

Formula:

$$bd = \frac{ac}{\cot \frac{90° - A}{2} + \cot \frac{90° - B}{2}}$$

Derivation of Formula:

$$F = C \qquad D = E \tag{1) and (2}$$

$$F + C = 90° - A \tag{3}$$

$$2C = 90° - A \tag{4a}$$

$$C = \frac{90° - A}{2} \tag{4b}$$

$$D + E = 90° - B \tag{5}$$

$$2D = 90° - B \tag{6a}$$

$$D = \frac{90° - B}{2} \tag{6b}$$

$$\cot C = \cot \frac{90° - A}{2} = \frac{ad}{bd} \tag{7a}$$

$$ad = bd \cot \frac{90° - A}{2} \tag{7b}$$

$$\cot D = \cot \frac{90° - B}{2} = \frac{dc}{bd} \tag{8a}$$

$$dc = bd \cot \frac{90° - B}{2} \tag{8b}$$

$$ac = ad + dc \tag{9}$$

$$ac = bd \cot \frac{90° - A}{2} + bd \cot \frac{90° - B}{2} \tag{10a}$$

$$ac = bd \left(\cot \frac{90° - A}{2} + \cot \frac{90° - B}{2} \right) \tag{10b}$$

$$bd = \frac{ac}{\cot \frac{90° - A}{2} + \cot \frac{90° - B}{2}} \tag{10c}$$

Example: Let $A = 13°30'$. $B = 18°45'$ and $ac = 0.750$. Find bd.

Solution:

$$bd = \frac{0.750}{\cot \frac{90° - 13°30'}{2} + \cot \frac{90° - 18°45'}{2}} \tag{10c}$$

$$bd = \frac{0.750}{\cot \frac{76°30'}{2} + \cot \frac{71°15'}{2}} = \frac{0.750}{\cot 38°15' + \cot 35°37\tfrac{1}{2}'}$$

$$bd = \frac{0.750}{1.2685 + 1.3955}$$

$$bd = 0.2815$$

Checking Diameters of Tapered Hole by Measurement Over a Sphere

PROBLEM 8: To find diameters x and y in Fig. 8.

The angle of taper M, the length of taper d in the taper gage are known and a steel ball of known radius r is placed in the tapered hole and distance c determined by measurement.

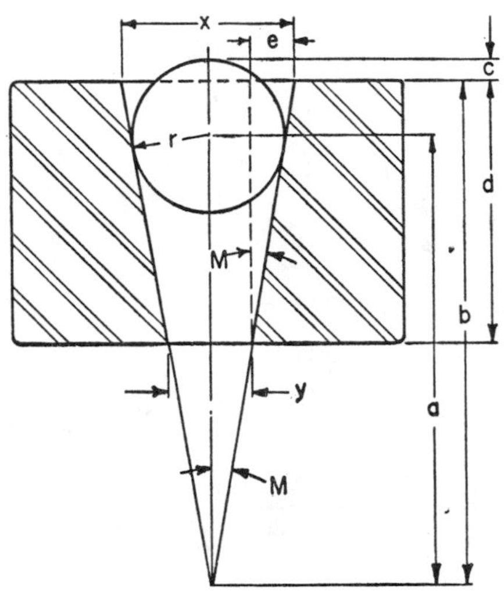

FIG. 8. To Find Diameters x and y of Tapered Hole when Radius r, Heights c and d, and Angle M are Known

Analysis of Problem: Since angle M and radius r are known, the theoretical distance a from the center of the ball to the apex of the taper can be computed. With a, d and c known, b can be found and hence x and y can be computed.

Note: To obtain an accurate answer, M and r must be known to very close limits.

Formulas: $x = 2r \sec M + 2 \tan M(r - c)$

$$y = x - 2d \tan M$$

Derivation of Formula:

$$a = r \operatorname{cosec} M \qquad (1)$$

$$b = a + r - c = r \text{ cosec } M + r - c \quad (2)$$

$$x = 2b \tan M \quad (3)$$

Hence $$x = 2(r \text{ cosec } M + r - c) \tan M \quad (4a)$$

Simplifying $$x = 2r \sec M + 2(r - c) \tan M \quad (4b)$$

To find y $$y = x - 2e \quad (5)$$

but $$e = d \tan M$$

Hence $$y = x - 2d \tan M \quad (6)$$

Example: Let $M = 9$ degrees; $d = 1.250$; $c = 0.250$ and $r = 0.500$. To find x and y.

Solution: $$x = 2 \times 0.500 \sec 9° + 2(0.500 - 0.250) \tan 9° \quad (4b)$$

$$x = 1.000 \times 1.0125 + 2 \times 0.250 \times 0.15838$$

$$x = 1.0125 + 0.0792 = 1.0917$$

$$y = 1.0917 - 2 \times 1.25 \times 0.15838 \quad (6)$$

$$y = 1.0917 - 0.3960 = 0.6957$$

Checking Diameters and Taper of Ring Gage

PROBLEM 9: To find diameters X and Y in Fig. 9 and the taper per inch. These dimensions of a ring gage, may be checked by using two balls of different known radii G and H that are placed within the gage as shown. The distance N from the top of the large ball to the top of the gage and the distance M from the top of the small ball to the top of the gage are found by measurement. Height O is known.

Analysis of Problem: Knowing the radii G and H of the two balls and measured distances M and N, it is possible to compute the distance ab between the centers of the two balls. Distance ac represents the difference between the radii of the two balls. Knowing ab and ac, angle α and hence the taper per inch can be found. The diameters D_1 and D_2 of the gage which pass through the centers of the large and small balls respectively, may now be found. Knowing these two diameters, the distances O, M, H and the taper per inch; the diameters X and Y at the large and small ends of the gage can be computed. If the top of the larger ball comes below the top of the gage, the sign of N is changed in the formulas given.

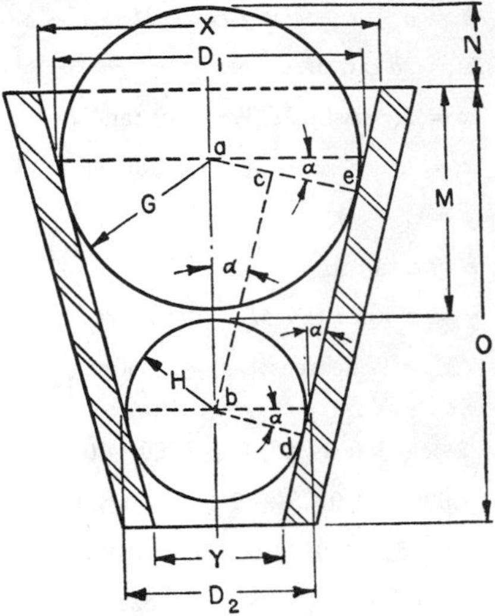

FIG. 9. To Find Diameters X and Y and Taper Per Inch of Ring Gage when Radii G and H and Heights M, N and O are Known

Formulas: Taper per inch $= 2 \tan \alpha$, where $\sin \alpha = \dfrac{G-H}{H+M+N-G}$

$$X = 2G \sec \alpha + 2(G - N) \tan \alpha$$

$$Y = 2H \sec \alpha - 2(O - M - H) \tan \alpha$$

Derivation of Formulas:

$$ab = H + M + N - G \qquad (1)$$

$$ac = ae - ce = ae - bd = G - H \qquad (2)$$

$$\sin \alpha = \frac{ac}{ab} = \frac{G - H}{H + M + N - G} \qquad (3)$$

$$2 \tan \alpha = \text{taper per inch} \qquad (4)$$

$$\frac{D_1}{2} = G \sec \alpha \qquad (5)$$

$$X = D_1 + (G - N)2 \tan \alpha \qquad (6)$$

$$X = 2G \sec \alpha + 2(G - N) \tan \alpha \tag{7}$$

$$\frac{D_2}{2} = H \sec \alpha \tag{8}$$

$$Y = D_2 - (O - M - H)2 \tan \alpha \tag{9}$$

$$Y = 2H \sec \alpha - 2(O - M - H) \tan \alpha \tag{10}$$

Example: Let $G = 2.91$; $H = 1.79$; $M = 4.43$; $N = 1.60$; $O = 8.43$. Find X and Y.

Solution: $$\sin \alpha = \frac{2.91 - 1.79}{1.79 + 4.43 + 1.60 - 2.91} = \frac{1.12}{4.91} \tag{3}$$

$$\sin \alpha = 0.22811 \qquad \alpha = 13°11.1'$$

$$\text{taper per inch} = 2 \tan \alpha = 2 \times 0.23427 = 0.46854 \tag{4}$$

$$X = 2 \times 2.91 \times 1.0271 + (2.91 - 1.60)0.46854 \tag{7}$$

$$X = 5.9777 + 0.6138 = 6.5915$$

$$Y = 2 \times 1.79 \times 1.0271 - (8.43 - 4.43 - 1.79)0.46854 \tag{10}$$

$$Y = 3.6770 - 1.0355 = 2.6415$$

Checking Tapered Slotted Recess

PROBLEM 10: To find M in Fig. 10. In checking a slotted recess having sides tapered at angle β with the center-line, a curved surface of radius R and chordal width $2A$, measurement M is taken between pins of diameter D located as shown. It is required to compute what M should be for any given angle, chordal width, and radius of the recess.

Analysis of Problem: The solution of this problem involves a series of steps, each of which consists in solving a right-angled triangle for an unknown side or angle. Considering these steps in the reverse order to which they are taken will show how some problems can be analyzed by working back from what is desired to what is known.

Thus, M can be found when *he* is known. To find *he* with *oe*, which is equal to R minus one-half D, known in right-angled triangle *ohe*, angle *hoe* must be found. By construction, angle *boh* equals known angle β, so that if angle *eob* is known, angle *hoe* can be found.

In triangle *oeb*, hypotenuse *oe* and section *ef* of side *eb* are known. If section *fb* of side *eb* can be found, angle *eob* can be determined. But *fb* equals *cd* and *cd* is one side of triangle *ocd* in which angle *cod* equals known angle β by construction. It is necessary, then, to find hypotenuse *oc* in order to solve this triangle for *cd*. But *oc* equals *ok* minus *ck*. Since *ok* is one side of triangle *olk* in which hypotenuse *ol* equals R and side *lk* equals A, *ok* can be found. Similarly, in triangle *clk*, angle β and *lk* are known so that *ck* can also be found, hence the series of required steps from the known values A, R and B to the desired measurement M is now complete.

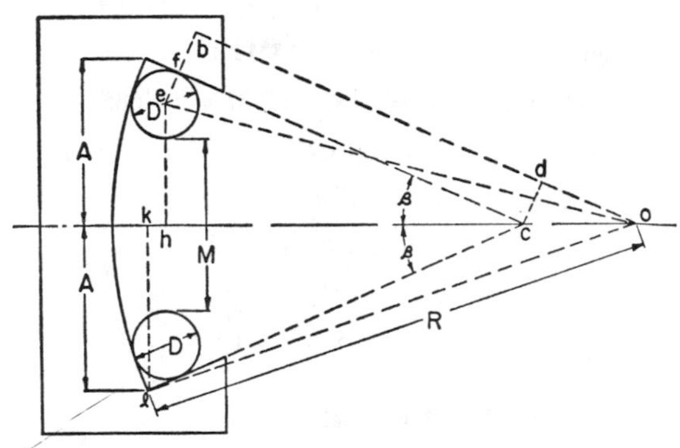

FIG. 10. To Compute Distance M between Pins when Diameter D, Distance A, Radius R, and Angle β are Known

Procedure: $$ok = \sqrt{\overline{ol}^2 - \overline{lk}^2} = \sqrt{R^2 - A^2} \quad (1)$$

$$ck = lk \cot \beta = A \cot \beta \quad (2)$$

$$oc = ok - ck = \sqrt{R^2 - A^2} - A \cot \beta \quad (3)$$

$$cd = oc \sin \beta = (\sqrt{R^2 - A^2} - A \cot \beta) \sin \beta \quad (4)$$

$$bf = cd = \sqrt{R^2 - A^2} \sin \beta - A \cos \beta \quad (5)$$

$$\sin eob = \frac{eb}{oe} = \frac{\frac{D}{2} + bf}{R - \frac{D}{2}} \quad (6a)$$

$$\sin eob = \frac{\frac{D}{2} + \sqrt{R^2 - A^2}\sin\beta - A\cos\beta}{R - \frac{D}{2}} \tag{6b}$$

$$\text{angle } hoe = \text{angle } \beta - \text{angle } eob \tag{7}$$

$$he = oe \sin hoe \tag{8a}$$

$$he = \left(R - \frac{D}{2}\right)\sin hoe \tag{8b}$$

$$M = 2he - D = 2\left(R - \frac{D}{2}\right)\sin hoe - D \tag{9}$$

Example: Let $A = 4.701$; $D = 1.900$; $R = 14.294$; and $\beta = 24°$.

Solution:

$$\sin eob = \frac{\frac{1.900}{2} + \sqrt{14.294^2 - 4.701^2}\sin 24° - 4.701\cos 24°}{14.294 - \frac{1.900}{2}} \tag{6b}$$

$$\sin eob = \frac{0.95 + 13.499 \times 0.40674 - 4.701 \times 0.91354}{14.294 - \frac{1.900}{2}}$$

$$\sin eob = \frac{2.146}{13.344} = 0.16082 \qquad eob = 9°15.3'$$

$$hoe = 24° - 9°15.3' = 14°44.7' \tag{7}$$

$$M = 2\left(14.294 - \frac{1.900}{2}\right)\sin 14°44.7' - 1.900 \tag{9}$$

$$M = 2 \times 13.344 \times 0.25452 - 1.900 = 4.893$$

Angle for Locating Tapered Part in Vee-Block

PROBLEM 11: To find angle X in Fig. 11A. When a cone-shaped or tapered part is supported in a vee-block for the purpose of grinding a horizontal surface such as wx or a vertical surface such as yz, it is necessary to tilt the vee-block to some angle X which will bring the axis of the cone-shaped or tapered part into a horizontal position.

The required angle X is a resultant of the half-angle N of the vee and the half-angle M of the cone-shaped piece, as shown in Fig. 11*B*.

Analysis of Problem: Since the value of angle X is not affected by whether the cone-shaped or tapered part is a full cone or a frustum of a cone, or by the longitudinal position of the cone-shaped part in the vee (providing it is fully supported), several possible diagrams might be drawn as a basis for the solution.

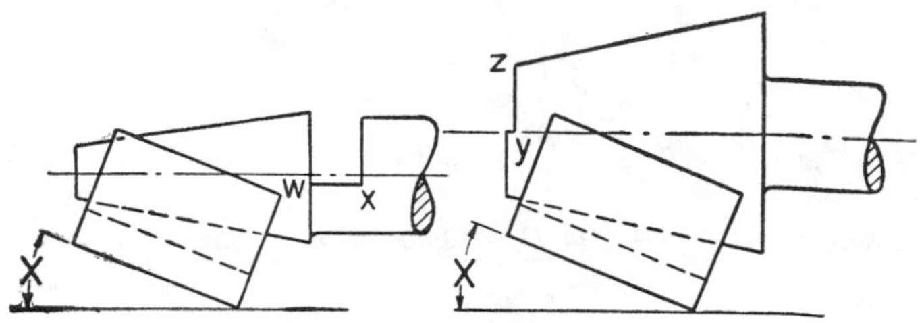

Fig. 11A. Cone-Shaped Part Supported in Vee-Block for Grinding Horizontal Surface *wx* or Vertical Surface *yz*

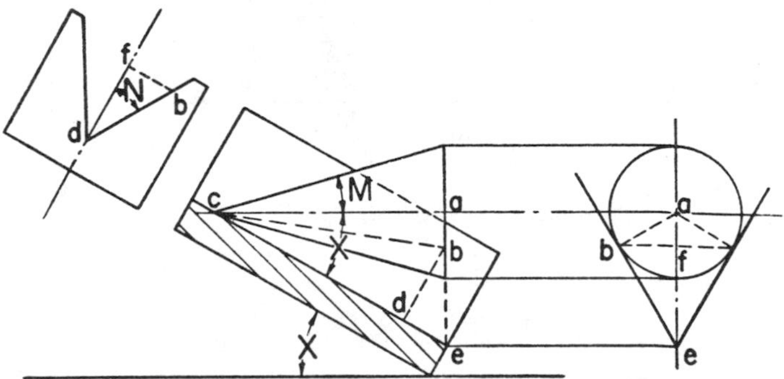

Fig. 11B. To Find Angle X to which Vee-Block Must be Raised to Make Cone Axis Horizontal when Angles M and N are Known

The diagram used in the solution given herein is shown in Fig. 11*B*. In this diagram a full cone resting in a vee-block is shown. It can be shown that if a full cone is supported in a vee, its apex will be somewhere along either the line of intersection of the two sides of the vee, or an extension of this line. In this diagram, the apex c of the cone is shown resting on the bottom of the vee. The angle

X is the angle through which that end of the vee-block toward the apex of the cone must be raised to place the axis of the cone in a horizontal position.

The first step in the solution of this problem is to establish that bce is a right-angled triangle. The second step is to establish an expression for ae in terms of proportions from similar triangles abe and bef. The third step is to establish an expression for ce in terms of proportions from similar triangles bce and bde. The final step is to express ae and ce as a function of angle X and their equivalents as functions of angles M and N.

Formula: $\sin X = \sin M \operatorname{cosec} N$

Derivation of Formula:

In right-angled triangle abe

$$\overline{ae}^2 = \overline{ab}^2 + \overline{be}^2 \tag{1a}$$

or

$$\overline{be}^2 = \overline{ae}^2 - \overline{ab}^2 \tag{1b}$$

In right-angled triangle abc

$$\overline{bc}^2 = \overline{ab}^2 + \overline{ac}^2 \tag{2}$$

In right-angled triangle ace

$$\overline{ce}^2 = \overline{ae}^2 + \overline{ac}^2 \tag{3}$$

Adding the corresponding sides of Equations (1b) and (2)

$$\overline{be}^2 + \overline{bc}^2 = \overline{ae}^2 + \overline{ac}^2 = \overline{ce}^2 \tag{4}$$

Hence bce is a right-angled triangle with the right-angle at b.

Since bf is an altitude of triangle abe, triangles abe and bfe are similar (three angles of one are equal to three angles of the other)

so that

$$\frac{ae}{ab} = \frac{be}{bf} \tag{5a}$$

or

$$ae = \frac{ab \times be}{bf} \tag{5b}$$

Since bd is an altitude of triangle bce, triangles bce and bde are similar

so that

$$\frac{ce}{bc} = \frac{be}{bd} \tag{6a}$$

or
$$ce = \frac{be \times bc}{bd} \tag{6b}$$

Dividing the corresponding sides of Equation (5b) by those of Equation (6b).

$$\frac{ae}{ce} = \frac{ab \times be}{bf} \times \frac{bd}{be \times bc} = \frac{ab \times bd}{bf \times bc} \tag{7}$$

But
$$\frac{ae}{ce} = \sin X \tag{8}$$

and
$$\frac{ab}{bc} = \sin M \tag{9}$$

and
$$\frac{bd}{bf} = \operatorname{cosec} N \tag{10}$$

Substituting these equivalents in Equation (7):

$$\sin X = \sin M \operatorname{cosec} N \tag{11}$$

When a 60-degree vee-block is used to support the cone or tapered part $N = 30$ degrees and cosec $N = 2$

or
$$\sin X = 2 \sin M \tag{12}$$

It is useful to remember, when the angle of taper is not known but the taper per inch is given, that:

$$2 \tan M = \text{taper per inch} \tag{13}$$

Example: In Fig. 11*B* if the angle of the vee is 60 degrees and the cone has a taper of 0.5161 inch per foot, through what angle must the block end be raised to place the cone axis in a horizontal position?

Solution:
$$N = \frac{60°}{2} = 30°$$

$$2 \tan M = \frac{0.5161}{12} = 0.04301 \tag{13}$$

$$\tan M = 0.02150$$

$$M = 1°14'$$

$$\sin X = 2 \times 0.02152 = 0.04304 \tag{12}$$

$$X = 2°28'$$

Width of Intersection of Taper and Circle

PROBLEM 12: To find *cd* and *oe* in Fig. 12. A part with sides which together have a known taper is placed on a circular disk of known radius so that the middle point of its lower edge is on the center of the disk. The length of this lower edge is known. It is required to find the chordal distance *cd* between the points of intersection. It is also required to find the vertical height *oe*.

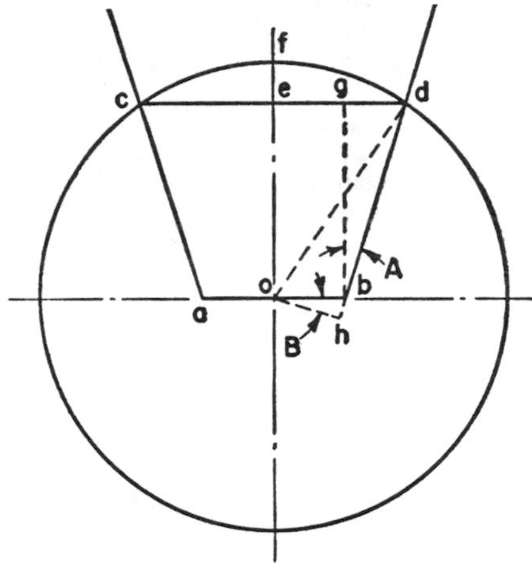

FIG. 12. To Find Chordal Distance *cd* and Vertical Depth *oe* when Radius *od*, Width *ab* and Taper are Known

Analysis of Problem: Draw the vertical line *bg*, the radius *od*, and a line *oh* perpendicular to an extension of *bd*. Thus, three right-angled triangles are formed: *obh*; *odh*; and *bdg*. Since the taper of side *bd* is known, the value of angle *A* can be determined from its tangent which is equal to this taper. Angles *A* and *B* are equal since their respective sides are at right angles to each other. In the right-angled triangle *obh*, the hypotenuse *ob* is known and angle *B* can be found directly so that *oh* and *bh* can be determined. With *od* and *oh* known in right-angled triangle *odh*, *dh* can be found, so that *bd* can now be determined. In right-angled triangle *bdg*, with *bd* and angle *A* known, *dg* and hence *ed* can be found. Knowing *ed*, *cd* can be determined. Knowing *bd* and angle *A*, *bg* and hence *oe* can be found.

Procedure:

$$\tan A = \frac{\text{taper of } ac + \text{taper of } bd}{2} \quad (1)$$

$$A = B \text{ (Geom. Prop. 1, page 3–11)} \quad (2)$$

$$bh = ob \sin B = \frac{ab}{2} \sin A \quad (3)$$

$$oh = ob \cos B = \frac{ab}{2} \cos A \quad (4)$$

$$dh = \sqrt{\overline{od}^2 - \overline{oh}^2} = \sqrt{\overline{of}^2 - \overline{oh}^2} \quad (5)$$

$$bd = dh - bh \quad (6)$$

$$dg = bd \sin A \quad (7)$$

$$ed = dg + ge = dg + ob \quad (8)$$

$$cd = 2ed \quad (9)$$

$$oe = bg = bd \cos A \quad (10)$$

Example: Let of = 3.592 inches and ab = 2.245 inches. The combined taper of ac and bd is 1 inch per foot. Find oe and cd.

Solution:

$$\tan A = \frac{1}{12} \div 2 = \frac{1}{24} = 0.04167 \quad (1)$$

$$A = 2°23.2' \text{ and } B = 2°23.2' \quad (2)$$

$$bh = \frac{2.245}{2} \times 0.04164 = 0.0467 \quad (3)$$

$$oh = \frac{2.245}{2} \times 0.99913 = 1.122 \quad (4)$$

$$dh = \sqrt{(3.592)^2 - (1.122)^2} = 3.412 \quad (5)$$

$$bd = 3.412 - 0.047 = 3.365 \quad (6)$$

$$dg = 3.365 \times 0.04164 = 0.140 \quad (7)$$

$$ed = 0.140 + \frac{2.245}{2} = 1.2625 \quad (8)$$

$$cd = 2 \times 1.2625 = 2.525 \quad (9)$$

$$oe = 3.365 \times 0.99913 = 3.362 \quad (10)$$

9

Compound-Angle Problems—How They are Analyzed

A compound angle may be defined as the resultant of two or more component angles located in two or more planes other than that in which the compound angle is located. The compound angle is said to be the resultant of the component angles since: (1) if any one of the component angles is changed, the compound angle is changed; and (2) if all of the component angles are known, the compound angle can be computed. Compound angles are frequently encountered in tool design and tool-making, in aircraft design and in miscellaneous classes of structural work.

The method of solving compound-angle problems is based on the use of trigonometry in finding unknown angles in adjacent triangles, that is, in triangles that have a common side. Although a compound angle may be the resultant of more than two angles, usually, compound-angle problems involve only two components. In the discussion which follows, the compound angle and each of the two component angles are located in separate triangles, at least two of which are adjacent to each other.

The following classification provides a basis for identifying a compound-angle problem so that the proper method of solution may be applied. In Chapter 10, specific compound-angle problems are analyzed and the various methods of solution are illustrated.

Careful study of both chapters should enable the reader to classify any one of the four types of compound angle problems which he may encounter. The method of solution called for can then be followed directly.

Types of Compound-Angle Problems. Four types of cases will be considered in which the compound angle is a resultant of two component angles.

Case A: As shown in Fig. 1, four right-angled triangles are adjacent to each other, forming a triangular pyramid. If an oblique angle is known in each of two of these triangles, any one of the oblique angles in the other two triangles is a compound angle of which the given two oblique angles are components.

Case B: As shown in Fig. 2, three right-angled triangles and one oblique-angled triangle are adjacent to each other, forming a triangular pyramid. If an oblique angle is known in each of two of the right-angled triangles, any one of the oblique angles in the third right-angled triangle or in the oblique-angled triangle is a compound angle of which the two given oblique angles are components.

Case C: As shown in Fig. 3, four right-angled triangles are adjacent to each other and to a rectangle, forming a quadrangular pyramid. If an oblique angle is known in each of two of the triangles, any one of the oblique angles in the other two triangles is a compound angle of which the two given angles are components.

Case D: As shown in Fig. 4, two triangular pyramids, each consisting of four right-angled triangles, are formed by passing a plane through opposite edges of a quadrilateral pyramid. Since one triangle is common to both triangular pyramids, there are actually only seven triangles and two of these, each of which is equal to one-half of the original base rectangle, are equal to each other.

If one oblique angle is known in each of two of these triangles which are not equal, then any other oblique angle in the five other triangles is a compound angle of which the two given angles are components.

General Method of Solving Case A: (Fig. 1)

Step 1. Express that side of the compound-angle triangle which is common to one of the component-angle triangles in terms of the component angle and that side of the component-angle triangle which is common to the other component-angle triangle.

Step 2. Similarly, express that side of the compound-angle triangle which is common to the other component-angle triangle in terms of this component angle and that side common to both component-angle triangles.

Step 3. Express the compound angle in terms of the two sides of its triangle for which equivalents were found in *Steps* 1 and 2.

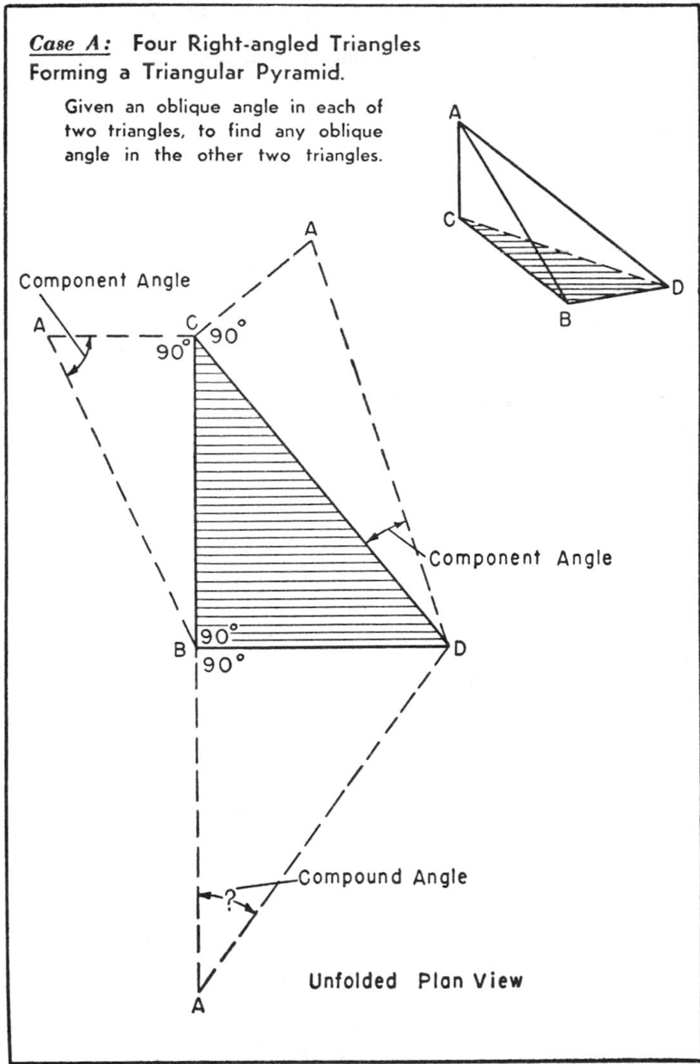

FIG. 1. First Type of Compound-angle Problem with Four Right-angled Triangles Adjacent to Each Other

Example: *Given:* Component angles BAC and ADC in Fig. 1.

To Find: Compound angle BAD.

Solution:

Step 1:
$$AB = AC \sec BAC \tag{1}$$

Step 2:
$$AD = AC \csc ADC \tag{2}$$

Step 3:
$$\cos BAD = \frac{AB}{AD} = \frac{AC \sec BAC}{AC \csc ADC} \tag{3a}$$

$$\cos BAD = \frac{\sec BAC}{\csc ADC}; \quad \text{or } \sec BAC \sin ADC; \quad \text{or } \frac{\sin ADC}{\cos BAC} \tag{3b}$$

General Method of Solving Case B: (Fig. 2) If the compound angle is in the third right-angled triangle, the procedure is the same as in Case A.

If the compound angle is in the oblique-angle triangle, the procedure is as follows:

Step 1. If not given, find the other two oblique angles at the same vertex of the pyramid as the compound angle. Let these be designated as angles M and N.

Step 2. Pass a plane through the apex of either of the other two angles in the compound-angle triangle and also through the apex of the pyramid at which all three right angles are located. (See DB in Fig. 2.) It can be shown that such a plane can be made perpendicular to that side of the compound-angle triangle opposite to the angle through which it is drawn. (In Fig. 2, DB is perpendicular to AC at E.) Then the compound-angle triangle will be divided into two right-angled triangles. Let that one of these two right-angled triangles in which the compound angle is located be designated as triangle X.

Step 3. Express that side of triangle X which is common to the triangle containing angle M in terms of angle M and the side common to triangles containing angles M and N, respectively.

Step 4. Express that side of triangle X which is common to the triangle containing angle N in terms of angle N and the side common to the triangles containing angles M and N, respectively.

Step 5. Express the compound angle in terms of the two sides of triangle X for which equivalents were found in *Steps* 3 and 4.

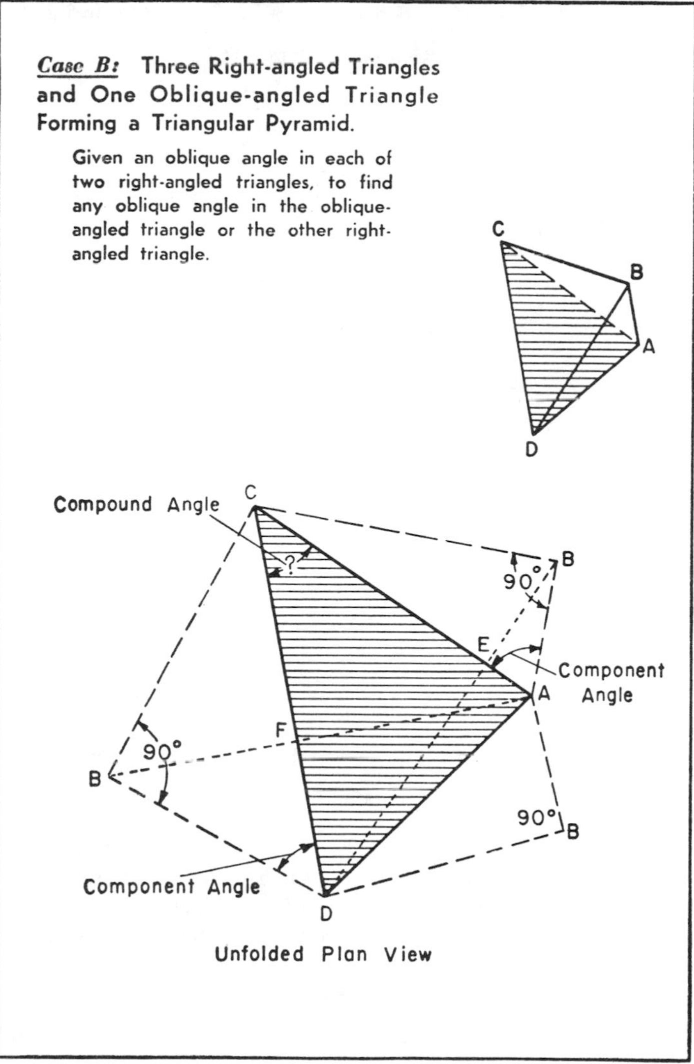

FIG. 2. Second Type of Compound-angle Problem with Three Right-angled and One Oblique-angled Triangles Adjacent to Each Other

Example 1: *Given:* Component angles BDC and BAC in Fig. 2.

To Find: Compound-angle ACD.

Solution:

Step 1.
$$BCD = 90^\circ - BDC \tag{1}$$
$$BCA = 90^\circ - BAC \tag{2}$$

Step 2. A plane represented by line DB is passed through the apex of angle ADC and perpendicular to side AC at E.

Step 3.
$$CE = CB \cos BCE \tag{3}$$

Step 4.
$$CD = CB \sec BCD \tag{4}$$

Step 5.
$$\cos ACD = \cos ECD = \frac{CE}{CD} = \frac{CB \cos BCE}{CB \sec BCD}$$
$$= \frac{CB \cos BCA}{CB \sec BCD} = \frac{\cos BCA}{\sec BCD} \tag{5}$$

Since
$$BCA = 90^\circ - BAC \tag{6}$$

and
$$\cos BCA = \sin BAC \tag{7}$$

Similarly
$$BCD = 90^\circ - BDC \tag{8}$$

and
$$\sec BCD = \csc BDC \tag{9}$$

Hence
$$\cos ACD = \frac{\sin BAC}{\csc BDC} = \sin BAC \sin BDC \tag{10}$$

Example 2: *Given:* Component angles BDC and BAC in Fig. 2.

To Find: Compound angle ADC.

Solution:

Step 1. Since angle ADB at the same vertex of the pyramid as the compound angle is not known and since it is in neither of the component-angle triangles, it will be found by the method employed in Case A, thus

$$AB = BC \cot BAC \tag{1}$$
$$BD = BC \cot BDC \tag{2}$$
$$\tan ADB = \frac{AB}{BD} = \frac{BC \cot BAC}{BC \cot BDC} \tag{3a}$$

$$\tan ADB = \frac{\cot BAC}{\cot BDC} = \cot BAC \tan BDC \tag{3b}$$

Step 2. A plane represented by line AB is passed through the vertex of angle CAD and perpendicular to side CD at F.

Step 3. $$DF = BD \cos BDF \tag{4}$$

Step 4. $$DA = BD \sec ADB \tag{5}$$

Step 5. $$\cos ADC = \cos ADF = \frac{DF}{DA} = \frac{BD \cos BDF}{BD \sec ADB} = \frac{\cos BDF}{\sec ADB} = \frac{\cos BDC}{\sec ADB} \tag{6a}$$

$$\cos ADC = \frac{\cos BDC}{\sec ADB} = \cos BDC \cos ADB, \tag{6b}$$

where $$\tan ABD = \cot BAC \tan BDC \tag{3b}$$

General Methods of Solving Case C: (Fig. 3) (*a*) *If the two component-angle triangles are adjacent to each other.*

Step 1. Express that side of the compound-angle triangle which is common to one of the component-angle triangles in terms of the component angle and that side which is common to both component-angle triangles.

Step 2. Express the base (that side adjacent to the rectangle) of the other component-angle triangle in terms of the component angle and that side which is common to both component-angle triangles.

Step 3. Express the compound angle in terms of the two sides of its triangle for which equivalents were found in *Steps* 1 and 2.

(*b*) *If the two component-angle triangles are opposite each other.*

Step 1. Express that side of the compound-angle triangle which is common to one of the component angle triangles in terms of the component angle and the base (that side adjacent to the rectangle) of the component-angle triangle.

Step 2. Express that side of the compound-angle triangle which is common to the other component-angle triangle in terms of the component angle and the base (that side adjacent to the rectangle) of this component-angle triangle.

Step 3. Express the compound angle in terms of the two sides of its triangle for which equivalents were found in *Steps* 1 and 2.

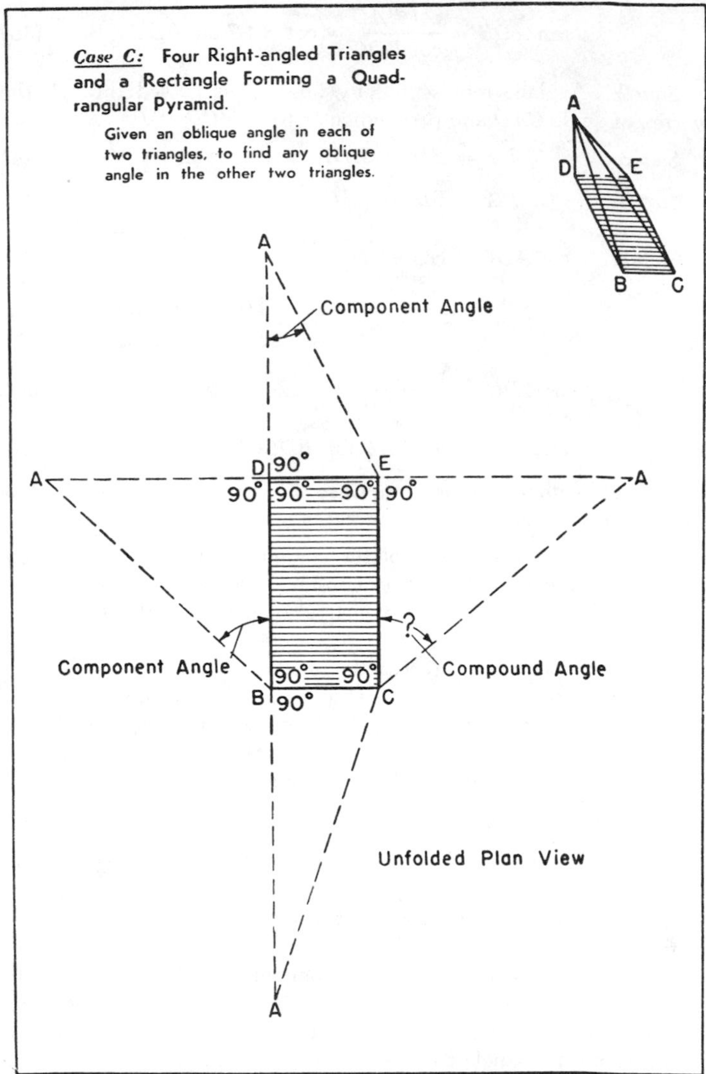

FIG. 3. Third Type of Compound-angle Problem with Four Right-angled Triangles Adjacent to Each Other and to Rectangle

Example 1: *Given:* Component angles DAE and ABD in Fig. 3.
To Find: Compound angle ECA.

Solution: It will be noted that the two component-angle triangles are adjacent to each other.

Step 1. $$AE = AD \sec DAE \tag{1}$$

Step 2. $$BD = AD \cot ABD \tag{2}$$

Step 3. $$\tan ACE = \frac{AE}{CE} \tag{3}$$

but $$BD = CE \tag{4}$$

hence $$\tan ACE = \frac{AE}{BD} = \frac{AD \sec DAE}{AD \cot ABD} \tag{5a}$$

$$\tan ACE = \frac{\sec DAE}{\cot ABD} = \sec DAE \tan ABD \tag{5b}$$

Example 2: *Given:* Component angles ACB and DAE in Fig. 3.
To Find: Compound angle ACE.

Solution:

Step 1. It will be noted that the two component-angle triangles are opposite each other.

$$AC = BC \sec ACB \tag{1}$$

Step 2. $$AE = DE \csc DAE \tag{2}$$

Step 3. $$\sin ACE = \frac{AE}{AC} = \frac{DE \csc DAE}{BC \sec ACB} \tag{3}$$

but $$BC = DE \tag{4}$$

hence $$\sin ACE = \frac{\csc DAE}{\sec ACB} = \csc DAE \cos ACB \tag{5}$$

General Method of Solving Case D: (Fig. 4) (*a*) *If the compound angle is in one of the base triangles and the two component-angle triangles are adjacent to each other.*

Step 1. Express the base of one component-angle triangle in terms of that component angle and the side common to both component-angle triangles.

Step 2. Express the base of the other component-angle triangle in terms of that component angle and the side common to both component-angle triangles.

Step 3. Express the compound angle in terms of the two sides for which equivalents were found in *Steps* 1 and 2.

(*b*) *If the compound angle is in one of the base triangles and the two component-angle triangles are opposite each other.*

Step 1. Find an oblique angle in one of the triangles adjoining both component-angle triangles by the procedure for Case C.

Step 2. Proceed as in (*a*).

(*c*) *If the compound angle is in the triangle common to both pyramids and the component-angle triangles have a common side with the compound-angle triangle.*

Step 1. Express the base of the compound-angle triangle in terms of the compound angle and the common side.

Step 2. Express the base of one of the component-angle triangles in terms of the component angle and the common side.

Step 3. Express the base of the other component-angle triangle in terms of the component angle and the common side.

Step 4. Since these three bases for which equivalents were found in *Steps* 1, 2, and 3 constitute a right-angled triangle, the sum of the squares of the two equivalents representing the two sides are set down as equal to the square of the equivalent representing the hypotenuse.

(*d*) *If the compound angle is in the triangle common to both pyramids and the component-angle triangles are adjacent but do not have a side in common with the compound-angle triangle.*

Step 1. Express that side of the compound-angle triangle which is adjacent to one of the component-angle triangles in terms of the component angle and the side common to both component-angle triangles.

Step 2. Express that side of the compound-angle triangle which is adjacent to the other component-angle triangle in terms of the component angle and the side common to both component-angle triangles.

Step 3. Express the compound angle in terms of the two sides for which equivalents were found in *Steps* 1 and 2.

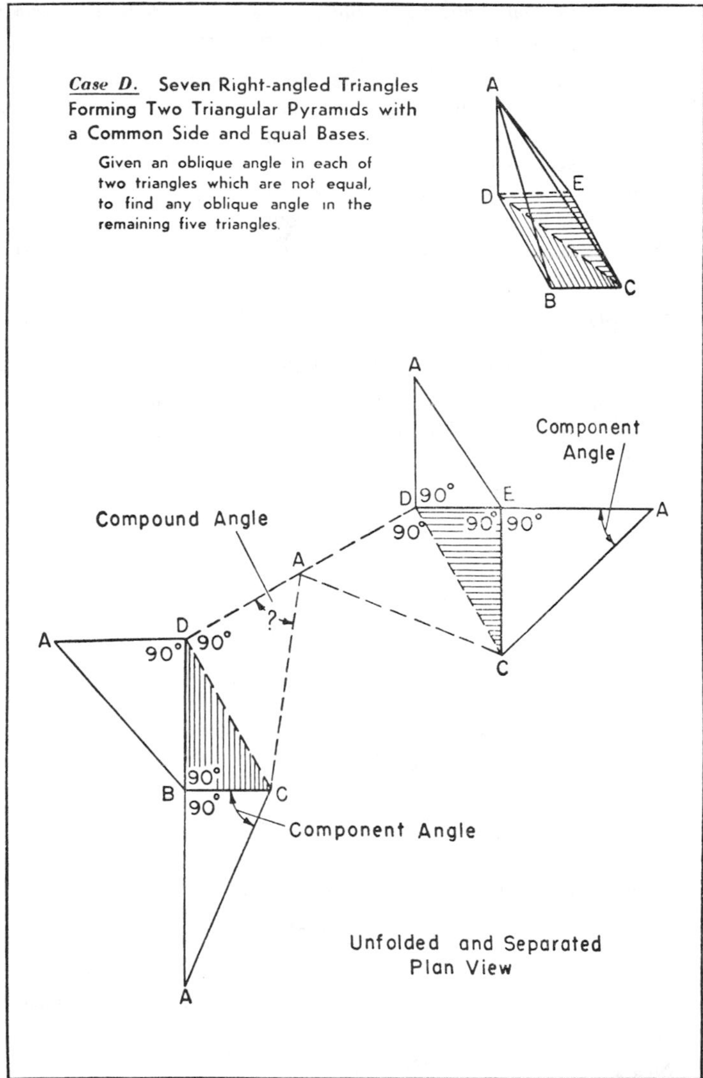

FIG. 4. Fourth Type of Compound-angle Problem with Seven Right-angled Triangles Adjacent to Each Other

(*e*) *If the compound angle is in the triangle common to both pyramids and the component-angle triangles are opposite each other.*

Step 1. Express that side of the compound-angle triangle which is adjacent to one of the component-angle triangles in terms of the component angle and the base of the component-angle triangle.

Step 2. Express that side of the compound-angle triangle which is adjacent to the other component-angle triangle in terms of the component angle and the base of the component angle triangle.

Step 3. Express the compound angle in terms of the two sides for which equivalents were found in *Steps* 1 and 2.

Example 1: *Given:* Component angles DAE and ACE in Fig. 4.

To find: Compound angle BDC.

Solution: It will be noted that the compound angle is in one of the base triangles and the two component-angle triangles are adjacent to each other.

Step 1. $$DE = AE \sin DAE \tag{1}$$

Step 2. $$CE = AE \cot ACE \tag{2}$$

Step 3. $$\tan BDC = \frac{BC}{BD} \tag{3}$$

but $$BC = DE \tag{4}$$

and $$BD = CE \tag{5}$$

hence $$\tan BDC = \frac{DE}{CE} = \frac{AE \sin DAE}{AE \cot ACE} \tag{6a}$$

$$\tan BDC = \frac{\sin DAE}{\cot ACE} = \sin DAE \tan ACE \tag{6b}$$

Example 2: *Given:* Component angles BAD and ECA in Fig. 4.

To find: Compound angle CDE.

Solution: It will be noted that the compound angle is in one of the base triangles and that the component-angle triangles are opposite each other.

Step 1. $$AB = BD \csc BAD \tag{1}$$

$$AC = CE \sec ECA \tag{2}$$

$$\cos BAC = \frac{AB}{AC} = \frac{BD \csc BAD}{CE \sec ECA} \tag{3}$$

but $$BD = CE \tag{4}$$

hence $$\cos BAC = \frac{\csc BAD}{\sec ECA} = \csc BAD \cos ECA \tag{5}$$

Step 2. $$BD = AB \sin BAD \tag{6}$$

$$BC = AB \tan BAC \tag{7}$$

$$\tan BCD = \frac{BD}{BC} = \frac{AB \sin BAD}{AB \tan BAC} \tag{8a}$$

$$\tan BCD = \frac{\sin BAD}{\tan BAC} = \sin BAD \cot BAC \tag{8b}$$

but $$CDE = BCD \tag{9}$$

therefore $$\tan CDE = \sin BAD \cot BAC \tag{10}$$

where $$\cos BAC = \csc BAD \cos ECA \tag{11}$$

Example 3: *Given:* Component angles BAC and ECA in Fig. 4.

To find: Compound angle DCA.

Solution: It will be noted that the compound angle is in the triangle common to both pyramids and the component-angle triangles have a common side with the compound-angle triangle.

Step 1. $$DC = AC \cos DCA \tag{1}$$

Step 2. $$BC = AC \sin BAC \tag{2}$$

Step 3. $$EC = AC \cos ECA \tag{3}$$

Step 4. $$\overline{DC}^2 = \overline{BD}^2 + \overline{BC}^2 \tag{4}$$

but $$BD = EC \tag{5}$$

hence $$\overline{DC}^2 = \overline{EC}^2 + \overline{BC}^2 \tag{6}$$

and $$\overline{AC}^2 \cos^2 DCA = \overline{AC}^2 \cos^2 ECA + \overline{AC}^2 \sin^2 BAC \tag{7a}$$

therefore $$\cos^2 DCA = \cos^2 ECA + \sin^2 BAC \tag{7b}$$

$$\cos DCA = \sqrt{\cos^2 ECA + \sin^2 BAC} \tag{8}$$

Example 4: *Given:* Component angles BAC and ABD in Fig. 4.

To find: Compound angle DCA.

Solution: It will be noted that the compound angle is in the triangle common to both pyramids and the component-angle triangles are adjacent but do not have a side in common with the compound-angle triangle.

Step 1. $$AC = AB \sec BAC \tag{1}$$

Step 2. $$AD = AB \sin ABD \tag{2}$$

Step 3. $$\sin DCA = \frac{AD}{AC} \tag{3}$$

therefore $$\sin DCA = \frac{AB \sin ABD}{AB \sec BAC} \tag{4a}$$

$$\sin DCA = \frac{\sin ABD}{\sec BAC} = \sin ABD \cos BAC \tag{4b}$$

Example 5: *Given:* Component angles BAD and ECA in Fig. 4.

To find: Compound angle DCA.

Solution: It will be noted that the compound angle is in the triangle common to both pyramids and that the component-angle triangles are opposite each other.

Step 1. $$AD = DB \cot BAD \tag{1}$$

Step 2. $$CA = EC \sec ECA \tag{2}$$

Step 3. $$\sin DCA = \frac{AD}{CA} \tag{3}$$

therefore $$\sin DCA = \frac{DB \cot BAD}{EC \sec ECA} \tag{4}$$

but $$DB = EC \tag{5}$$

hence $$\sin DCA = \frac{\cot BAD}{\sec ECA} = \cot BAD \cos ECA \tag{6}$$

10

Compound-angle Problems—Some Typical Examples

The following problems illustrate how the methods of solving adjacent right-angled and oblique-angled triangles for compound angles outlined in Chapter 9 can be applied to practical work. Having classified a problem the reader may wish to turn back to Chapter 9 and try his hand at deriving a formula for its solution. He can then check this against the derivation given.

Determining Angle for Grinding Precision Thread-cutting Tool

PROBLEM 1. To find the compound angle C of the precision thread-cutting tool shown in Fig. 1 if angle A and angle B are known. Angle A is measured in the plane of the top of the tool and angle C in a plane at right angles to the front edge of the tool. Angle B is the front clearance angle and is also the angle between the planes in which angle A and angle C are located, as shown.

Analysis of Problem: If that part of the tool bounded by lines *nm*, *mo*, *or*, *rm*, *on* and *rn* is considered, it will be found to be a triangular pyramid of which each side is a right-angled triangle. Angle $\frac{A}{2}$ is an oblique angle in one of the four adjacent triangles and angle B is an oblique angle in another one of the four adjacent triangles. Hence angle $\frac{C}{2}$ is a compound angle and angles $\frac{A}{2}$ and B are its components. This is solved as a Case A type of problem. (See method of solution outlined in Chapter 9.)

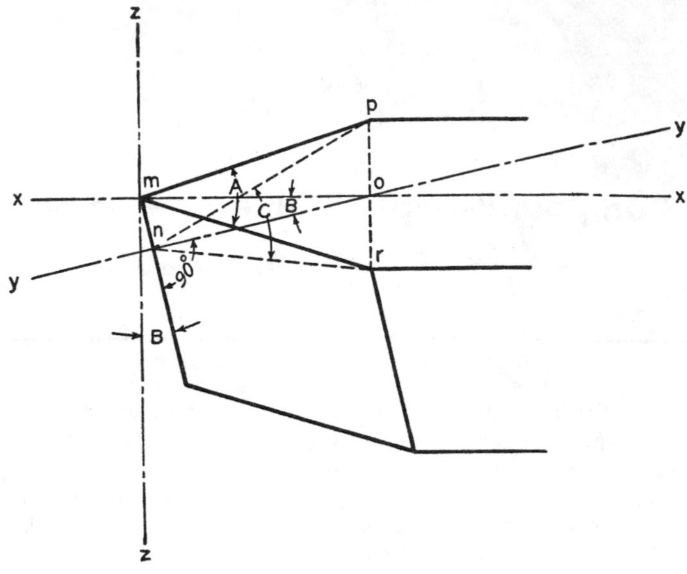

FIG. 1. To Find Compound Angle C in Precision Thread Cutting Tool when Angles A and B are Known

Formula: $$\tan\frac{C}{2} = \frac{\tan\frac{A}{2}}{\cos B} = \tan\frac{A}{2}\sec B$$

Derivation of Formula:

Step 1. $$no = mo \times \cos B \tag{1}$$

Step 2. $$ro = mo \times \tan\frac{A}{2} \tag{2}$$

Step 3. $$\tan\frac{C}{2} = \frac{ro}{no} = \frac{mo\tan\frac{A}{2}}{mo\cos B} \tag{3}$$

$$\tan\frac{C}{2} = \frac{\tan\frac{A}{2}}{\cos B} = \tan\frac{A}{2}\sec B \tag{4}$$

Example: Let A = 60 degrees and B = 15 degrees. Find compound angle C.

Solution:
$$\tan \frac{C}{2} = \tan \frac{60°}{2} \sec 15° \qquad (4)$$

$$\tan \frac{C}{2} = 0.57735 \times 1.0353 = 0.59773$$

$$\frac{C}{2} = 30°52.1'$$

$$C = 61°44.2'$$

The practical application of this formula is shown in Fig. 2 which illustrates how compound angle C is utilized. In making a holding block for locating the thread cutting tool at the correct angle for grinding, two faces of the block are machined at an angle with the center line equal to $C \div 2$ as shown. The slot into which the tool is inserted is cut into the block at an angle with a vertical plane parallel to the side of the block equal to the clearance angle B.

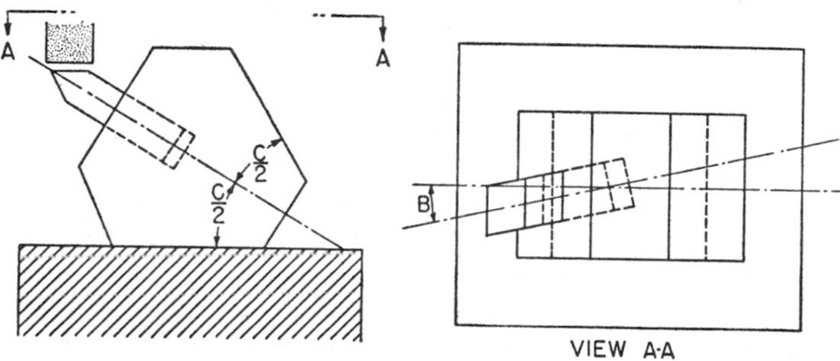

FIG. 2. Showing How Compound-angle C is Utilized in Making Holding Block for Grinding Operation

Fig. 3 shows a similar problem, except that a forming tool is involved. Here, the angle A in the top plane $x - x$ of the cutting tool is as shown, and for a given clearance angle B, the compound angle C in the plane $y - y$ at right angles to the inclined front edge is required. The same formula applies as in Problem 1 and the derivation is similar.

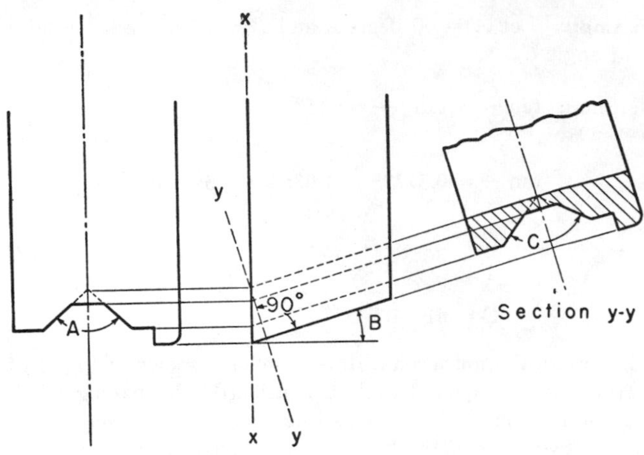

FIG. 3. Compound Angle C of Forming Tool is Found by Same Formula as for Thread Cutting Tool

Determining Position of Rectangular Block for Drilling Inclined Hole

PROBLEM 2. Find the angle R and C in Fig. 4 if angles A and B are known. The figure shown is a quadrangular pyramid having a rectangle as a base and four right-angled triangles as sides. When a plane is passed through *ony*, the figure becomes two triangular pyramids, each having four right-angled triangles as faces, one face being common to both pyramids and the two bases being equal.

Analysis of Problem: Angle R is in the base triangle of one of the pyramids and the two component angles are in triangles adjacent to each other, hence B is found according to the procedure for (*a*) in Case D. (See Chapter 9.)

Angle C is in the triangle common to both pyramids and, as before, the component angles are in triangles which have a common side, with the compound angle triangle, hence C is found according to the procedure for (c) in Case D. (See Chapter 9.)

Formulas: $\tan R = \tan A \cot B$

(where angle R is measured from the plane in which angle A is located)

$$\cot C = \sqrt{\cot^2 A + \cot^2 B}$$

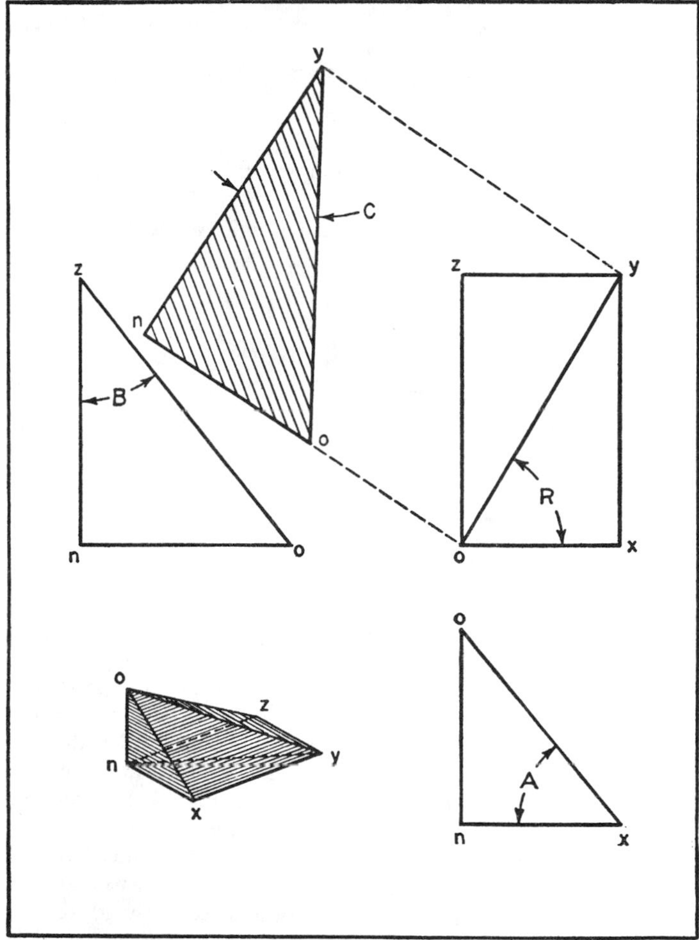

FIG. 4. Two Compound Angles R and C are to be Found in this Quadrangular Pyramid in which Angles A and B are Known

Derivation of Formula: To find angle R

Step 1. $$nz = on \times \cot B \tag{1}$$

Step 2. $$nx = on \times \cot A \tag{2}$$

Step 3. $$\tan R = \frac{xy}{nx} = \frac{nz}{nx} \tag{3}$$

$$\tan R = \frac{on \times \cot B}{on \times \cot A} \tag{4a}$$

$$\tan R = \frac{\cot B}{\cot A} = \tan A \cot B \tag{4b}$$

If angle R were measured from the plane in which angle B is located, then

$$\tan R = \frac{\cot A}{\cot B} = \cot A \tan B$$

To find angle C

Step 1. $$ny = on \times \cot C \tag{5}$$

Step 2. $$nx = on \times \cot A \tag{2}$$

Step 3. $$nz = on \times \cot B \tag{1}$$

Step 4. $$\overline{ny}^2 = \overline{nx}^2 + \overline{xy}^2 = \overline{nx}^2 + \overline{nz}^2 \tag{6}$$

hence $$\overline{on}^2 \cot^2 C = \overline{on}^2 \cot^2 A + \overline{on}^2 \cot^2 B \tag{7a}$$

and $$\cot^2 C = \cot^2 A + \cot^2 B \tag{7b}$$

$$\cot C = \sqrt{\cot^2 A + \cot^2 B} \tag{7c}$$

A practical compound-angle problem of the Case D type is the drilling of a hole in a rectangular plate for the insertion of a rod or pipe at an angle with the horizontal when the angles which the inserted rod or pipe make in the front and side elevations are given.

Four variations of this problem are shown in Figs. 5, 6, 7, and 8. As viewed from above, the rod or pipe will be inclined in one of the four quadrants formed by the horizontal and vertical center lines of the plate in the plan view. In each case it is required to find the angle of rotation R, which for all four variations is measured from a plane parallel to that in which angle B is located, and the angle of inclination C which is measured from the vertical.

Referring back to Fig. 4 showing the quadrangular pyramid, the edge oy might be considered to represent an inclined pipe or rod. In this case, however, angle C would be in the angle of elevation of the pipe or rod above the horizontal. In this case, too, the angle of rotation is measured from the plane in which angle A is located.

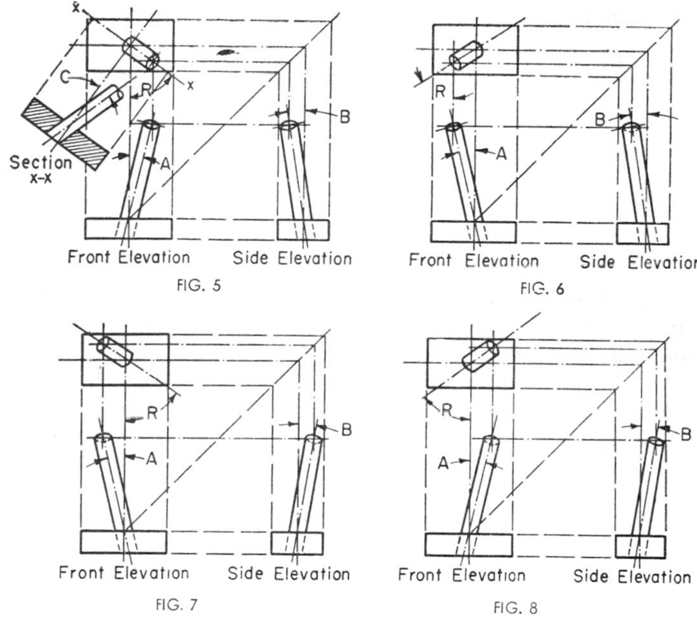

FIGS. 5–8. Four Variations of a Compound-angle Problem Involving the Drilling of a Hole in a Rectangular Block for an Inclined Rod or Pipe

Taking these differences into consideration, the formulas derived for R and C can be applied to the problem represented by Figs. 5–8.

Since angle R is measured from a plane parallel to that in which angle B is located and angles A and B are measured from the vertical:

$$\tan R = \tan (90° - B) \cot (90° - A) \qquad (8a)$$

and

$$\tan R = \cot B \tan A \qquad (8b)$$

Since angles A, B and C are measured from the vertical:

$$\cot^2 (90° - C) = \cot^2 (90° - A) + \cot^2 (90° - B) \qquad (9a)$$

and

$$\tan^2 C = \tan^2 A + \tan^2 B \qquad (9b)$$

$$\tan C = \sqrt{\tan^2 A + \tan^2 B} \qquad (9c)$$

Example: If in Fig. 5 the front angle of inclination A is 13 degrees and the side angle of inclination B is 9 degrees, find angles R and C.

$$\tan R = \cot B \tan A = \cot 9° \tan 13° \tag{8b}$$

$$\tan R = 6.3137 \times 0.23087 = 1.4576$$

$$\tan C = \sqrt{\tan^2 A + \tan^2 B} = \sqrt{\tan^2 13° + \tan^2 9°} \tag{9c}$$

$$\tan C = \sqrt{(0.23087)^2 + (0.15838)^2}$$

$$\tan C = \sqrt{0.07838} = 0.27996$$

$$C = 15°38.4'$$

Practical Application. Assume that a hinged angle-plate is to be employed in drilling the rectangular block. The block is located on the angle-plate with its front edge farthest from but parallel to the hinge if the front and side elevations are as in Figs. 5 and 6, or with the front edge nearest to and parallel with the hinge, if the front and side elevations are as shown in Figs. 7 and 8. Fig. 9 shows the initial position of the block on the plate.

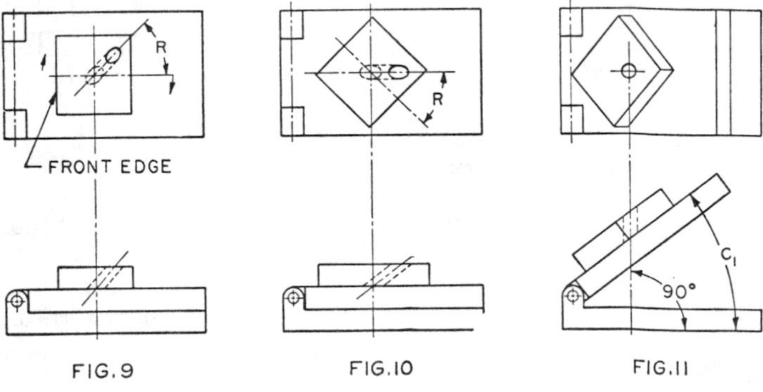

FIGS. 9–11. Three Steps in Positioning Block for Vertical Drilling of Hole in Which a Rod or Pipe is to be Inserted at a Required Angle

The block is next rotated clockwise on the angle-plate through the required angle of rotation R as shown in Fig. 10 if the front and side elevation angles are inclined as in Fig. 5 or 7; or counter-clockwise if these angles are inclined as in Fig. 6 or 8. The upper half of the angle-plate is then raised through angle C, as shown in Fig. 11, and the block will be in correct position for vertical drilling of the hole.

Finding Angles for Locating Block for Horizontal Machining of Inclined Face

PROBLEM 3. Find the angles R and C_1 if angles A and B in Fig. 12 are known.

The block shown has an inclined face which, if continued until it intersects the horizontal plane, makes a triangular pyramid, each face of which is a right-angled triangle. The angle that this inclined plane makes with the horizontal plane is measured in a plane which is at right angles to the intersection *vw*. The angle R is the angle between this vertical plane and the plane in which angle A is located.

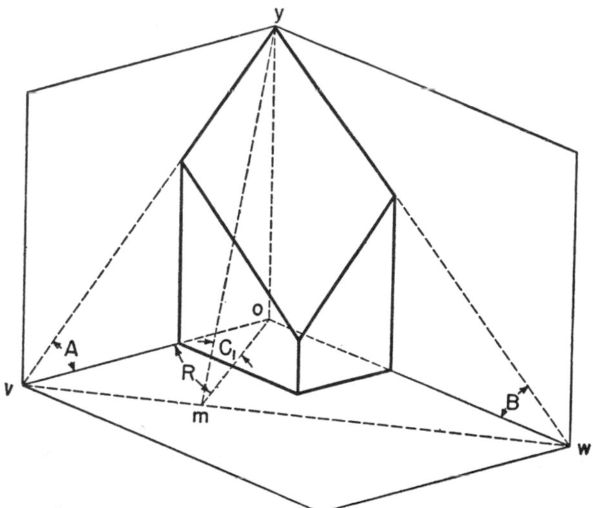

FIG. 12. Two Compound Angles R and C_1 are Required to Machine Inclined Face of this Block. Angles A and B are Known

Analysis of Problem: Referring to Fig. 12, it will be seen that angle *owv* equals angle R since their respective sides are at right angles to each other. Hence in triangular pyramid *yovw*, one oblique angle is known in each of two triangles and the oblique angle to be found is located in a third triangle. The angle to be found is thus a compound angle and the two given angles are its components. The procedure to be followed is that for Case A. (See Chapter 9.)

It will be noted from Fig. 12 that *yvom* is a triangular pyramid, each face of which is a right-angled triangle. In this pyramid, two oblique angles (A and R) in two different triangles are now known and the oblique angle C_1 to be found is located in a third triangle of the pyramid. Thus, C_1 is a compound angle of which angles A and R are components and the procedure to be followed is, again, that for Case A.

Formulas:

$$\text{Tan } R = \cot A \tan B$$

$$\text{Tan } C_1 = \tan A \sec R$$

Derivation of Formula: To find angle R

Step 1. $$ow = oy \cot B \qquad (1)$$

Step 2. $$ov = oy \cot A \qquad (2)$$

Step 3. $$\tan owv = \frac{ov}{ow} = \frac{oy \cot A}{oy \cot B} = \cot A \tan B \qquad (3)$$

but $$\text{angle } owv = \text{angle } R \qquad (4)$$

hence $$\tan R = \cot A \tan B \qquad (5)$$

To find angle C_1

Step 1. $$oy = ov \tan A \qquad (6)$$

Step 2. $$om = ov \cos R \qquad (7)$$

Step 3. $$\tan C_1 = \frac{oy}{om} = \frac{ov \tan A}{ov \cos R} = \tan A \sec R \qquad (8)$$

Example: If, in Fig. 12, A = 47 degrees 14 minutes and B = 38 degrees 10 minutes, find angles R and C_1.

Solution: $\tan R = \cot A \tan B = \cot 47°14' \tan 38°10'$ (5)

$$\tan R = 0.92493 \times 0.78598 = 0.72698$$

$$R = 36°1'$$

$$\tan C_1 = \tan A \sec R = \tan 47°14' \sec 36°1' \qquad (8)$$

$$\tan C_1 = 1.0812 \times 1.2361 = 1.3365 \qquad C_1 = 53°11.8'$$

Practical Application. Suppose that a rectangular block is to be machined to form an angular face, the component angles A and B of which are as given in Fig. 12. The angles R and C_1 are first computed, and the block is then placed on the angle-plate so that the side from which angle R is measured is parallel with the hinge axis as shown in Fig. 14. The block is then rotated on the face of the angle-plate through the angle $(90° - R)$ as shown in Fig. 15. This places the vertical plane in which the compound angle is measured at right angles to the hinge axis of the angle-plate as shown in Fig. 15. The upper face of the angle-plate is then tilted upward through angle C_1 as shown in Fig. 16, placing the plane in which the angular face is to be in a horizontal position.

Finding Angles for Locating Block for Vertical Machining of Inclined Face

PROBLEM 4. Find angle C_2 in Fig. 13 if angles A and B are given. The angle C_2 which the inclined face of the block makes with the vertical is measured in a plane which is vertical to line wy representing the intersection of the inclined face plane and the horizontal plane.

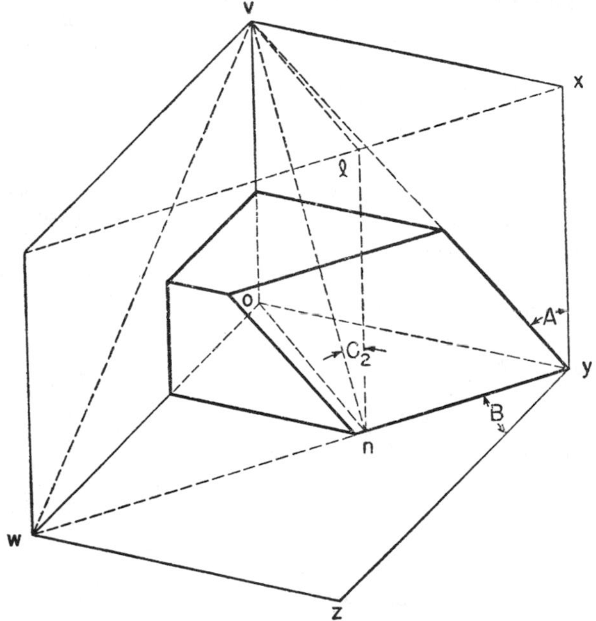

FIG. 13. In this Problem Both Component Angles A and B as well as Compound Angle C_2 are Measured from Vertical Planes

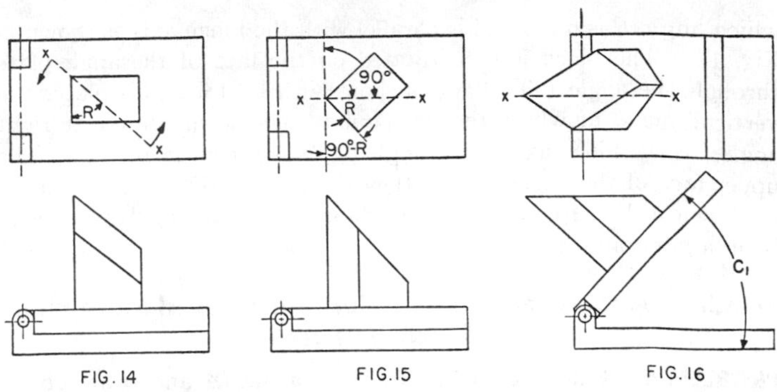

FIGS. 14–16. Three Steps in Positioning Block Shown in Fig. 12 for Horizontal Machining of Inclined Face

Analysis of Problem: It will be noted that angles A and C_2 are in triangles which form faces of a quadrangular pyramid *vlnyx*, all four sides of which are right-angled triangles and the base of which is a rectangle. If an equivalent of angle B can be found in one of the four triangles of this pyramid, the problem can be solved as a Case C type. Referring to Fig. 13 it will be noted that angle *noy* is equal to angle B since their corresponding sides are at right angles to each other. It can also be seen that angle *lvx* is equal to angle *noy* since each is the angle between planes *onlv* and *oyxv* as measured in a horizontal plane. Hence, angle B equals angle *lvx* and the problem is solved according to the procedure for (*a*) under Case C in which the component angle triangles are adjacent to each other. (See Chapter 9.)

Formula: $\tan C_2 = \tan A \cos B$

Derivation of Formula:

Step 1. $vl = vx \cos lvx$ (1)

Step 2. $xy = vx \cot A$ (2)

Step 3. $\tan C_2 = \dfrac{vl}{nl}$ (3)

but $nl = xy$ (4)

therefore $\tan C_2 = \dfrac{vl}{xy} = \dfrac{vx \cos lvx}{vx \cot A}$ (5a)

$$\tan C_2 = \frac{\cos lvx}{\cot A} \tag{5b}$$

but $$B = lvx \tag{6}$$

hence $$\tan C_2 = \frac{\cos B}{\cot A} = \tan A \cos B \tag{7}$$

Example. If, in Fig. 13, A equals 47 degrees 14 minutes and B equals 38 degrees 10 minutes, find the angle C_2.

$$\tan C_2 = \tan A \cos B = \tan 47°14' \cos 38°10' \tag{7}$$

$$\tan C_2 = 1.0812 \times 0.78622 = 0.85006$$

$$C_2 = 40°22'$$

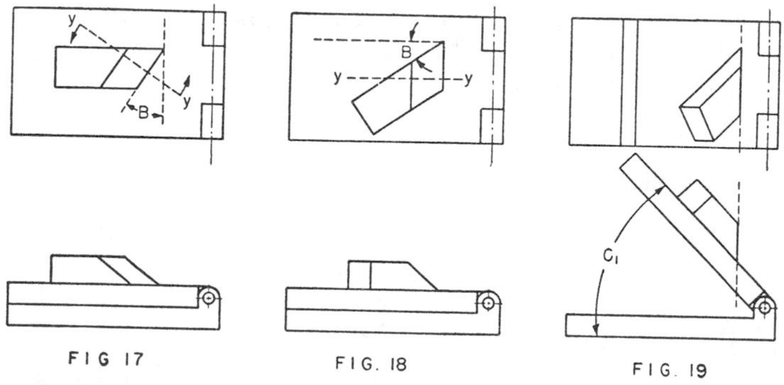

FIGS. 17–19. Three Steps in Positioning Block Shown in Fig. 13 for Vertical Machining of Inclined Face

Practical Application: If a rectangular block is to be machined to form an angular face and angle C_2 has been computed, the block is placed on an angle-plate with its sides parallel with the sides of the plate and at right angles to the hinge axis, as shown in Fig. 17. The block is then rotated on the face of the angle-plate through the angle of rotation (angle B) as shown in Fig. 18, so that the vertical plane y–y in which the compound angle C_2 is measured is at right angles to the hinge axis of the angle-plate. The upper face of the angle-plate is then tilted through angle C_2 as shown in Fig. 19, placing the plane in which the angular face is to lie in a vertical position.

Computing Angle Formed by Two Sides of Block in Plane of Inclined Face

PROBLEM 5. Given a rectangular block, one corner of which has been cut off to form a flat triangular surface *pvw*, as shown in Fig. 20. If angles A and B, formed by the edges of this surface and the sides of the block, with the horizontal are known, what is angle C?

Analysis of Problem: If that part of the block which has been cut away is restored, as indicated by the dotted lines in Fig. 20, a triangular pyramid *opvw* will be formed. It will be noted that three faces of this pyramid are right-angled triangles while the fourth, *pvw*, is an oblique-angled triangle. It will also be noted that angle *ovw* equals angle A and angle *opw* equals angle B. Hence, one oblique angle is known in each of the two right-angled triangles and an oblique angle is to be found in the oblique-angled triangle. The angle to be found is, therefore, a compound angle of which the two known angles are components. The problem is solved according to the procedure for Case B. (See Chapter 9.)

Formula: $\cos C = \sin A \sin B$

Derivation of Formula:

Step 1. $\text{angle } owv = 90° - ovw = 90° - A \quad (1)$

$\text{angle } owp = 90° - opw = 90° - B \quad (2)$

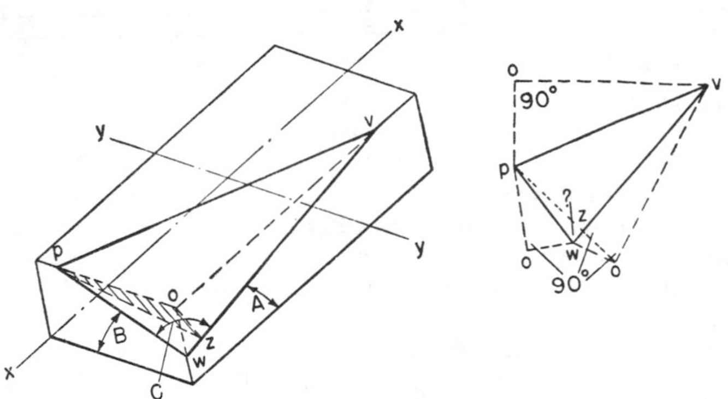

FIG. 20. (Left) Compound Angle C Formed on Inclined Face by Sides of Block is to be Found. (Right) Unfolded Plan View of Triangular Pyramid *pvwo* which Forms Corner Removed from Block

Step 2. Pass a plane *poz* through point *p* perpendicular to *wv* as shown in Fig. 20.

Step 3.

$$pw = ow \sec owp = ow \sec (90° - B) \tag{3a}$$

therefore

$$pw = ow \csc B \tag{3b}$$

Step 4.

$$wz = ow \cos owz = ow \cos owv = ow \cos (90° - A) \tag{4a}$$

therefore

$$wz = ow \sin A \tag{4b}$$

$$\cos C = \cos pwz = \frac{wz}{pw} \tag{5}$$

and $$\cos C = \frac{ow \sin A}{ow \csc B} = \frac{\sin A}{\csc B} = \sin A \sin B \tag{6}$$

Example: If, in Fig. 20, $A = 12°$ and $B = 19°$, find the compound angle C.

Solution:

$$\cos C = \sin A \sin B = \sin 12° \sin 19° \tag{6}$$

$$\cos C = 0.20791 \times 0.32557 = 0.06769$$

$$C = 86°7.1'$$

Finding Tool Angle for Cutting Tapered Hexagonal Hole

PROBLEM: To find angle $2A$ in Fig. 21. A tapered hole, with hexagonal sides has an angle of taper C in plane x–x as shown in the cross-sectional view of this plane. If the angle between any two adjacent sides of the hole as measured in a plane perpendicular to the axis of the hole is $2B$, what is the angle $2A$ required on the cutting face of a tool that is to machine the corners of the hole?

Analysis of Problem: The angle $2A$ is the angle between any two adjacent sides of the hole when measured in a plane perpendicular to these two sides as shown in the view in Fig. 21, taken perpendicular in line z–z. This is a compound angle problem in which angle B in a plane perpendicular to the axis of the hole and angle C in plane x–x are the known or component angles and angle A in plane z–z is the unknown or compound angle.

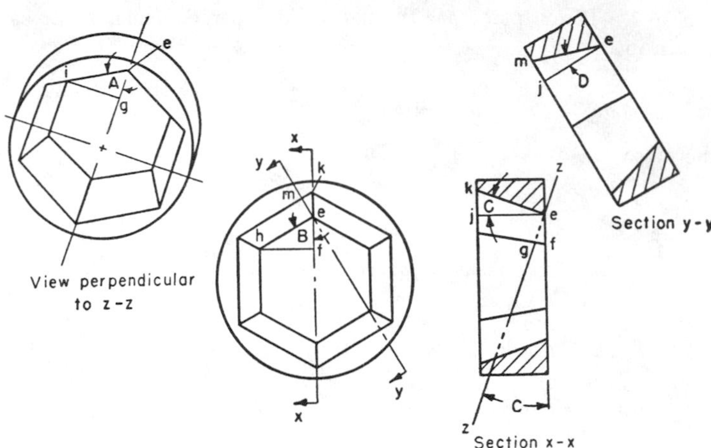

FIG. 21. To Find Tool Face Angle 2*A* for Cutting Corners of this Tapered Hexagonal Hole

In accordance with the general method for solving compound-angle problems explained, in Chapter 9, the first step is to establish the type of pyramid formed by the triangles in which the known and unknown angles appear. As can be seen from Fig. 21, one side of this pyramid is right-angled triangle *hef;* a second side is triangle *ieg;* and a third side is triangle *gef*. Not shown in Fig. 21 is the base *ihfg*, which is a rectangle and side *ihe* which is a right-angled triangle. This makes a quadrangular base pyramid as shown in Fig. 22. The component angle *B* and the compound angle *A* are located in the face triangles of the pyramid as shown. The other component angle *C* is equal to angle *gef*, as can be seen in section *x–x* of Fig. 21. As shown in Fig. 22, angle *gef*, is also in a face triangle of the pyramid. The procedure for finding angle *A* is as outlined under (*a*) in Case C, in Chapter 9.

In some instances the angle of taper is not given as *C* in plane *x–x* but as *D* in plane *y–y*, as shown in Fig. 21, which is at right angles to one of the sides of the hole. When this is the case, angle *C* has to be determined before angle *A* can be found. This is again a compound-angle problem. In Fig. 21, the three sides of triangle *mkj* are shown respectively in three separate views making it somewhat

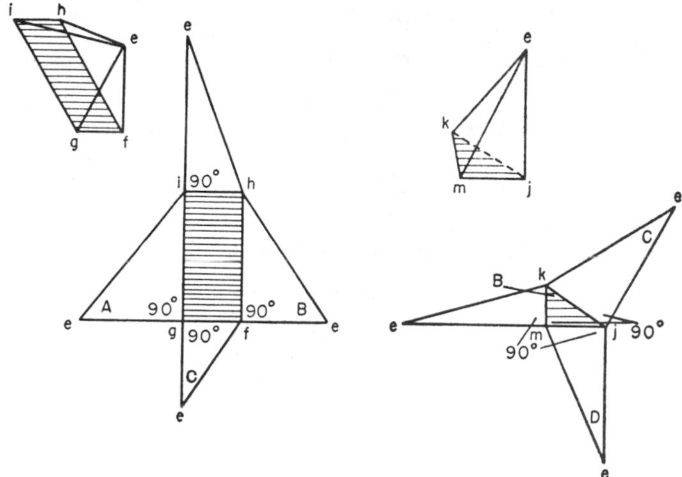

FIG. 22. (Left) Component Angles B and C and Compound Angle A are Located in Face Triangles of Quadrangular Pyramid. (Right) Component Angles B and D and Compound Angle C are Located in Faces of Triangular Pyramid

difficult to visualize. It lies, however, in the plane of angle B, hence angles mkj and B are equal, since their respective sides are parallel. Let angle mkj also be called angle B, then the known angles are angle B in right-angled triangle mkj; and angle D in right-angled triangle mje. The unknown angle is C in right-angled triangle ejk. The fourth face of the pyramid of which these three triangles are a part, is right-angled triangle mke. This pyramid is shown in Fig. 22. The component angles B and D and the compound-angle C are located in the triangles of the pyramid as shown. The procedure for finding angle B is as outlined under Case A in Chapter 9.

Formula: $$\tan A = \tan B \sec C$$

If D is given instead of C, then C is found by:

$$\tan C = \operatorname{cosec} B \tan D$$

Derivation of Formulas: Where B and C are known and A is to be found. (See Fig. 22.)

$$eg = ef \cos C \quad (1)$$

$$fh = ef \tan B \quad (2)$$

but $$gi = fh \tag{3}$$

hence $$gi = ef \tan B \tag{4}$$

$$\tan A = \frac{gi}{eg} = \frac{ef \tan B}{ef \cos C} \tag{5a}$$

$$\tan A = \frac{\tan B}{\cos C} = \tan B \sec C \tag{5b}$$

Where B and D are known and C is to be found. (See Fig. 22B.)

$$kj = mj \operatorname{cosec} B \tag{6}$$

$$je = mj \cot D \tag{7}$$

$$\tan C = \frac{kj}{je} = \frac{mj \operatorname{cosec} B}{mj \operatorname{cotan} D} \tag{8a}$$

$$\tan C = \operatorname{cosec} B \tan D \tag{8b}$$

Example: A tapered hole with hexagonal sides (Fig. 21) has an angle of taper in plane x–x equal to 10 degrees 2½ minutes. (1) Find the angle $2A$ required on the cutting face of the tool for machining the corners of the hole. (2) If this angle of taper of 10 degrees 2½ minutes is given as the angle in plane y–y, what would be the corresponding angle C of taper in plane x–x and what would the angle $2A$ then be?

Solution: (1) When $B = 60°$ and $C = 10°2.5'$

$$\tan A = \tan 60° \sec 10°2.5' \tag{5b}$$

$$\tan A = 1.7320 \times 1.0156 = 1.7590$$

$$A = 60°23' \qquad 2A = 120°46'$$

(2) When $B = 60°$ and $D = 10°2.5'$

$$\tan C = \operatorname{cosec} 60° \tan 10°2.5' \tag{8b}$$

$$\tan C = 1.1547 \times 0.17708 = 0.20447$$

$$C = 11°33.3'$$

$$\tan A = \tan 60° \sec 11°33.3' \tag{5b}$$

$$\tan A = 1.7320 \times 1.0207 = 1.7679$$

$$A = 60°30.3' \qquad 2A = 121°0.6', \text{ say } 121°1'$$

11

Arcs and Circles–Basic Concepts

Lengths and Locations of Tangential Lines

All circles have the very interesting characteristic that the ratio of their circumference to diameter is a constant for any size of circle. This ratio has been named with the Greek letter π or pi. The ratio π is an irrational number and is equal to 3.14159. . . to six significant figures. Pi is also the number of radians in half a circle, which is equal to 180°. This equality can be used to convert from radians to degrees

Basic formulas for any circle stem from the geometric constant relationship. Referring to Fig. 1 below, the circumference (C) = πD, or $2\pi R$, where D = diameter and R = radius. The length of part of a circumference, termed arc length or S, is found from $S = R\theta$, where θ = the number of radians corresponding to the small part of the circumference.

Another interesting feature of circles is their relationship to trigonometric functions, which sometimes are referred to as circular functions. Refer to the illustrations below, and to Chapter 4, pages 4–4 and 4–6. Note that ab = sin A and oa = cos A. The Pythagorean Theorem states that $oa^2 + ab^2 = r^2$. Consider a unit circle where $r = 1$, and substitute the trigonometric functions into the above equation. The result is $\sin^2 A + \cos^2 A = 1$. This well-known identity is shown as equation 7 on page 4–23, and is also a special case of the law of cosines (page 4–18), where the triangle considered is a right triangle.

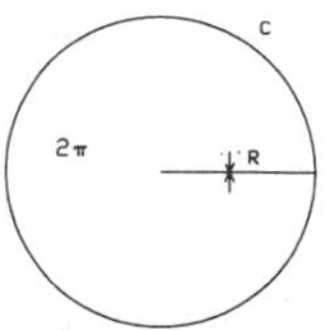

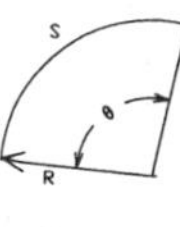

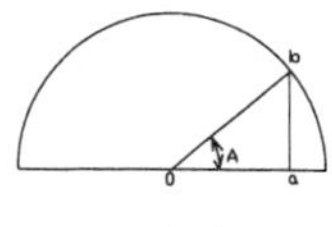

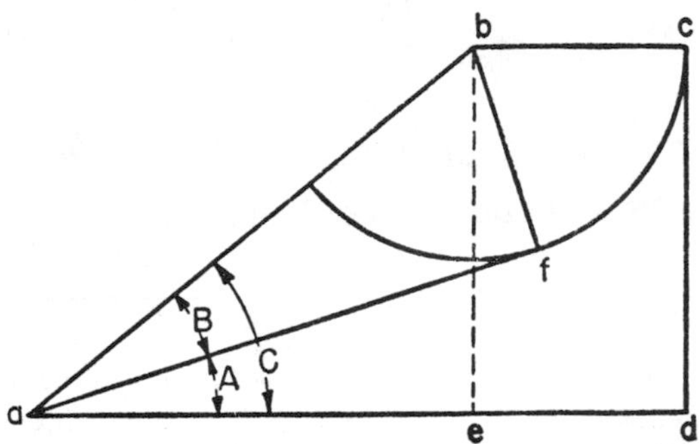

FIG. 2. To Find Angle *A* when *ad*, *dc* and *bc* are Known

The problem of finding the angular location or the length of a tangent to a given arc or arcs is frequently encountered in practical work. Since two points determine the position of a line, either one point and an arc or two arcs must be specified if the position of a tangent is to be established. The problems which follow illustrate a number of conditions in which the position or length of a tangent is to be determined when a point and an arc, or two arcs are given.

Angle of Inclination of Line Drawn from Given Point and Tangent to Given Arc

PROBLEM 1. To find angle *A* in Fig. 2 This is a problem involving two right-angled triangles with a common hypotenuse. Given figure *abcd* in which *ad*, *dc*, and *bc* are known. An arc is struck from point *c* with its center at *b*. A line is drawn from point *a* and tangent to this arc. It is required to find the angle *A* which this line makes with side *ad*.

Analysis of Problem: If angles *C* and *B* can be determined, angle *A* can be found. Draw *be* at right angles to *ad*. In right-angled triangle *abe* the two sides can be readily found from the given data and hence angle *C* can be found. Draw radius *bf* to the point of tangency of the line *af* with the arc. This forms the right-angled triangle *abf*. (A tangent to an arc is at right angles to that radius of the arc drawn from the point of tangency.) In this triangle, *ab* can be determined, since it is also the hypotenuse of triangle *abe*. One side is also known and hence angle *B* can be determined.

Procedure:

$$ae = ad - ed = ad - bc \quad (1)$$

$$be = cd \quad (2)$$

$$\tan C = \frac{be}{ae} = \frac{cd}{ad - bc} \quad (3)$$

$$ab = \sqrt{\overline{ae}^2 + \overline{be}^2} \quad (4)$$

$$\sin B = \frac{bf}{ab} = \frac{bc}{ab} \quad (5)$$

$$A = C - B \quad (6)$$

Example: Let $ad = 5.5$; $cd = 3$; and $bc = 1.625$. Find angle A.

Solution:

$$\tan C = \frac{3}{5.5 - 1.625} = \frac{3}{3.875} = 0.77419 \quad (3)$$

$$C = 37°44.8'$$

$$ab = \sqrt{(3.875)^2 + (3)^2} = \sqrt{24.016} \quad (4)$$
$$= 4.901$$

$$\sin B = \frac{1.625}{4.901} = 0.33156 \quad (5)$$

$$B = 19°21.8'$$

$$A = 37°44.8' - 19°21.8' = 18°23' \quad (6)$$

Length of Inside and Outside Tangents to Two Circles

PROBLEM 2: To find *de* in Figs. 2A and 2B. Given two circles of known radii *ad* and *be* and with the distance between their centers *ab* known. It is required to find the length *de* of a tangent to both circles.

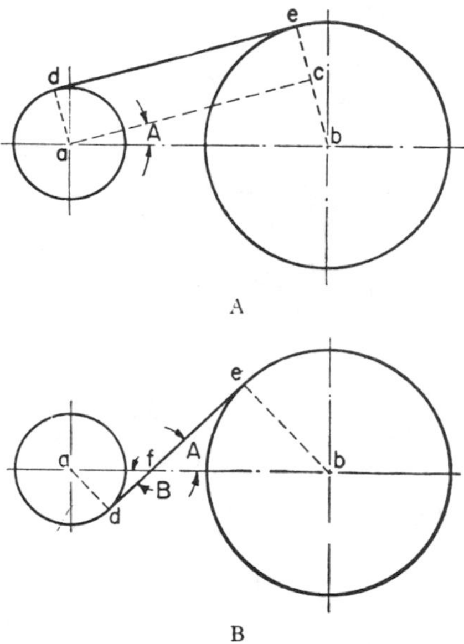

FIG. 3. To Find Tangent *de* when Radii *ad* and *be* and Distance *ab* are Known (A) Where Tangent is Outside. (B) Where Tangent is Inside

Analysis of Problem: There are two cases. The first case is illustrated by Fig. 3A where the tangent is "outside," that is, it does not cross the center line of the two circles. Draw *ac* parallel to *de*. According to the fifth geometrical proposition on page **3–7**, radius *be* is perpendicular to tangent *de*. Therefore *be* is also perpendicular to *ac* and *abc* is a right triangle. Since *ad* is also perpendicular to *de* it is parallel to *ce*. Therefore *ce* can be found directly and hence *bc*. Since *ab* is given, *ac* can then be determined and hence *de*.

The second case is illustrated by Fig. 3B where the tangent is "inside," that is, it crosses the center line of the two circles. This is a problem involving two right-angled triangles *adf* and *bef*. In triangle *bef* the sine of angle A can be expressed in terms of the known side *be* and the hypotenuse *bf*. In triangle *adf* the sine of angle B can be expressed in terms of the known side *ad* and the hypotenuse *af*. According to the fourth geometrical proposition on page 3–6, angles A and B are equal, hence the expression for sine B also equals sine A. It is now possible to express the sine of A in terms of the sum of *ad* and *be* and the sum of *af* and *bf*. Angle A can now be determined and hence *df* and *ef*.

Procedure: *Case* 1 (Fig. 3A)

$$bc = be - ce = be - ad \tag{1}$$

$$\sin A = \frac{bc}{ab} \tag{2}$$

$$de = ac = ab \cos A \tag{3}$$

Case 2 (Fig. 3B)

$$\sin A = \frac{be}{bf} \tag{4}$$

$$\sin B = \frac{ad}{af} \tag{5}$$

but

$$A = B$$

Therefore

$$\sin A = \frac{ad}{af} = \frac{be}{bf} = \frac{ad + be}{af + bf} = \frac{ad + be}{ab} \tag{6}$$

$$df = ad \cot B = ad \cot A \tag{7}$$

$$ef = be \cot A \tag{8}$$

$$de = df + ef \tag{9}$$

Example 1: (Fig. 3A) Let $ad = 1.35$; $be = 3.02$; and $ab = 6.33$. Find *de*.

Solution:

$$bc = 3.02 - 1.35 = 1.67 \tag{1}$$

$$\sin A = \frac{1.67}{6.33} = 0.26382 \tag{2}$$

$$A = 15°17.8'$$

$$de = 6.33 \times 0.96457 \tag{3}$$

$$de = 6.11$$

Example 2: (Fig. 3B) Let $ad = 1.35$; $be = 3.02$; and $ab = 6.33$. Find *de*.

Solution:

$$\sin A = \frac{1.35 + 3.02}{6.33} = 0.69036 \tag{6}$$

$$A = 43°39.5'$$

$$df = 1.35 \times 1.0480 = 1.415 \tag{7}$$

$$ef = 3.02 \times 1.0480 = 3.165 \tag{8}$$

$$de = 1.415 + 3.165 \tag{9}$$

$$de = 4.580$$

To Find Length and Angle of Inclination of Inside Tangent to Two Circles

PROBLEM 3: To find *ab* and angle *X* in Fig. 4. Two studs of known diameters are offset vertically and horizontally from each other. The horizontal distance *mc* and the vertical distance *cn* between their centers are known. A line *ab* is drawn tangent to these studs as shown. It is required to find the distance *ab* between the points of tangency and the angle *X* made by this line with the horizontal.

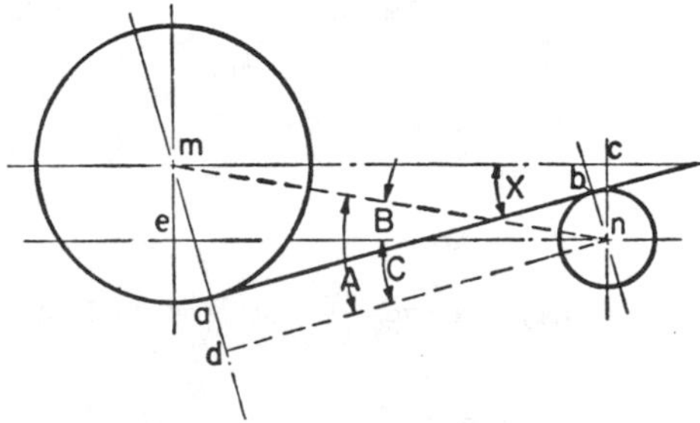

FIG. 4. To Find *ab* and Angle *X* when Center Distances *mc* and *cn* and Diameters *ma* and *nb* are Known

Analysis of Problem: Draw construction lines *mn*, connecting the center of the studs, and *dn* parallel to *ab*. Now, since *ab* is perpendicular to *ma* (fifth geometrical proposition on page 3–7) then *dn* is also perpendicular to *md*, and *mdn* is a right-angled triangle.

Since *bn* is perpendicular to *ab*, *ad* and *bn* are equal. Hence, *md* can be found directly. Distance *mn* can be found directly since *mc* and *cn* are known and *mcn* is a right-angled triangle. Knowing *mn* and *dm*, *dn* and consequently *ab* can be found. Angles X and C are equal since their corresponding sides are parallel. Angle C equals angle A minus angle B. Both angles A and B can be determined from the data given so that C and X can be found.

Procedure: $$\overline{mn}^2 = \overline{cm}^2 + \overline{cn}^2 \quad (1)$$

$$mn = \sqrt{\overline{cm}^2 + \overline{cn}^2} \quad (2)$$

$$md = ma + ad = ma + bn \quad (3)$$

$$\overline{dn}^2 = \overline{mn}^2 - \overline{md}^2 \quad (4)$$

therefore $$dn^2 = \overline{cm}^2 + \overline{cn}^2 - (ma + bn)^2 \quad (5)$$

$$dn = \sqrt{\overline{cm}^2 + \overline{cn}^2 - (ma + bn)^2} \quad (6)$$

$$ab = dn = \sqrt{\overline{cm}^2 + \overline{cn}^2 - (ma + bn)^2} \quad (7)$$

$$\tan A = \frac{md}{dn} = \frac{ma + bn}{ab} \quad (8)$$

$$\tan B = \frac{me}{en} = \frac{cn}{mc} \quad (9)$$

$$X = C = A - B \quad (10)$$

Example: Let $mc = 7.891$; $cn = 1.403$; $ma = 2.500$; and $nb = 0.875$. Find ab and angle X.

Solution:

$$ab = \sqrt{(7.891)^2 + (1.403)^2 - (2.500 + 0.875)^2} \quad (7)$$

$$ab = \sqrt{62.268 + 1.968 - 11.391}$$

$$ab = \sqrt{52.845} = 7.269$$

$$\tan A = \frac{2.500 + 0.875}{7.269} = 0.46430 \qquad A = 24°54.3' \quad (8)$$

$$\tan B = \frac{1.403}{7.891} = 0.17780 \qquad B = 10°4.9' \quad (9)$$

$$X = 24°54.3' - 10°4.9' = 14°49.4' \quad (10)$$

Angle of Inclination of Tangent to Two Arcs

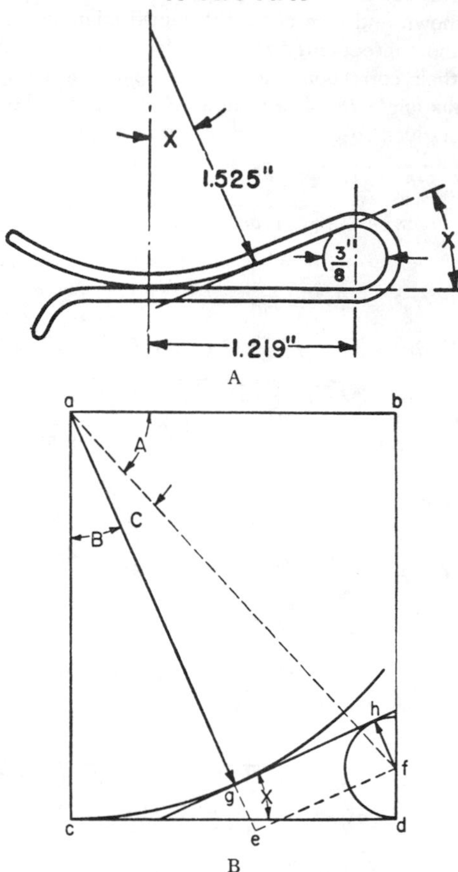

FIG. 5. (A) Side View of Spring Clamp. (B) Elements of Problem

PROBLEM 4: To find angle X in Fig. 5A. The side view of a spring clamp for which a perforating, embossing, and cutting-off die is to be produced is shown. The radii of the two arcs formed by the curved inner surface of the clamp and the horizontal distance between the centers of these arcs is given. It is required to find the angle which the inclined inner surface makes with the horizontal.

Analysis of Problem: It can be seen from Fig. 5B which shows the elements of this problem diagrammatically, that the solution of two right-angled triangles *abf* and *aef* with a common hypotenuse is involved. The given dimensions are *ag*, *cd*, and *fh*. Thus, two sides of triangle *abf* can be found directly from the given dimensions and hence angle *A* and the hypotenuse *af* can be found. But *af* is also the hypotenuse of triangle *aef*, of which side *ae* can be directly found from the dimensions originally given. Hence angle *C* can be found. Angle *B* is then easily determined and angle *X* equals angle *B* since the respective sides of these two angles are perpendicular to each other.

Procedure: It will be seen upon referring to Fig. 5B that:

$$ag = ac = bd;\ fh = df = eg;\ \text{and}\ ab = cd$$

hence

$$\tan A = \frac{bf}{ab} = \frac{bd - df}{ab} = \frac{ag - fh}{cd} \qquad (1)$$

$$af = \frac{bf}{\sin A} = \frac{ag - fh}{\sin A} \qquad (2)$$

$$\cos C = \frac{ag + eg}{af} = \frac{ag + fh}{af} \qquad (3)$$

$$B = 90^\circ - A - C \qquad (4)$$

$$X = B \qquad (5)$$

Example: In Fig. 5B, let $ag = 1.525$; $fh = 0.1875$; and $cd = 1.219$. Find angle X.

Solution: $\tan A = \dfrac{1.525 - 0.1875}{1.219} = \dfrac{1.3375}{1.219} = 1.0972$ (1)

$$A = 47^\circ 39.2'$$

$$af = \frac{1.525 - 0.1875}{0.73908} = \frac{1.3375}{0.73908} = 1.8097 \qquad (2)$$

$$\cos C = \frac{1.525 + 0.1875}{1.8097} = \frac{1.7125}{1.8097} = 0.94629 \qquad (3)$$

$$C = 18^\circ 51.8'$$

$$B = 90^\circ - 47^\circ 39.2' - 18^\circ 51.8' = 23^\circ 29' \qquad (4)$$

$$X = 23^\circ 29'$$

Angle of Inclination of Tangent to Two Arcs

PROBLEM 5: To find angle *A* in Fig. 6. Given the right-angled triangle *abc* with the two sides *ab* and *bc* known. An arc of known radius *ae* is drawn with its center on one end of the hypotenuse and an arc of known radius *cg* is drawn with its center on the other end of the hypotenuse. It is required to find the angle *A* between the line *fe* drawn tangent to the two arcs and the line *fh* drawn at right angles to side *bc*.

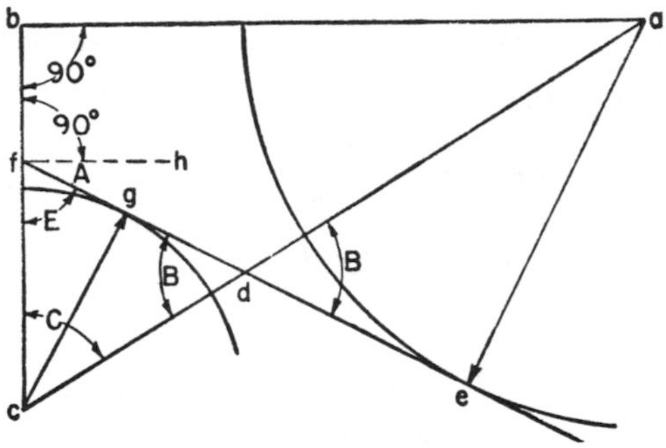

FIG. 6. To Find Angle *A* when Radii *ae* and *cg* and Sides *ab* and *bc* are Known

Analysis of Problem: If angles *B* and *C* can be determined, angle *E* can be found and hence angle *A*. Angle *C* can be found directly since sides *ab* and *bc* of right-angled triangle *abc* are known. Lines *ae* and *cg* are both perpendicular to *ef*. (A tangent to an arc is perpendicular to that radius of the arc which passes through the point of tangency.) Therefore *ade* is a right-angled triangle, as is *cdg*. Since these two triangles have two corresponding angles which are equal (their 90-degree angles and angles *B*), they are similar triangles.

The sum of the hypotenuses of these two similar triangles is also the hypotenuse of triangle *abc* and hence can be determined since sides *ab* and *bc* are given. Knowing this value and the length of one side in each of the two similar triangles, angles *B* can be determined, since the sine of angle *B* is equal to the sum of the sides opposite angles *B* divided by the sum of the hypotenuses. With angles *B* and *C* known, *E* is found and then *A*.

Procedure: $\tan C = \dfrac{ab}{bc}$ (1)

$$\sin B = \frac{ae}{ad} = \frac{gc}{dc} = \frac{ae + gc}{ad + dc}$$

since $ad + dc = ac$

and $ac = \dfrac{ab}{\sin C}$ (2)

then $\sin B = \dfrac{ae + gc}{ac} = \dfrac{(ae + gc)\sin C}{ab}$ (3)

$$E = 180° - B - C \quad (4)$$

$$A = 90° - E \quad (5)$$

Example: Given $ab = 3.75$; $bc = 3.1875$; $ae = 3.00$; and $gc = 1.625$. Find angle A.

Solution: $\tan C = \dfrac{3.75}{3.1875} = 1.1765$ (1)

$$C = 49°38.1'$$

$$\sin B = \frac{(3.00 + 1.625)0.76193}{3.75} \quad (3)$$

$$\sin B = 0.93971$$

$$B = 70°0.2'$$

$$E = 180° - 70°0.2' - 49°38.1' = 60°21.7' \quad (4)$$

$$A = 90° - 60°21.7' = 29°38.3' \quad (5)$$

Angle of Inclination of Tangent to Two Arcs

PROBLEM 6: To find angle A in Fig. 7 Given a right-angled triangle oec, the sides oe and ec being of known length. An arc of known radius og is drawn with its center at one end of the hypotenuse and another arc of known radius cb is drawn with its center at the other end of the hypotenuse. It is required to find angle A between a line ab drawn tangent to these two arcs and a line ac which forms an extension of one side of the triangle.

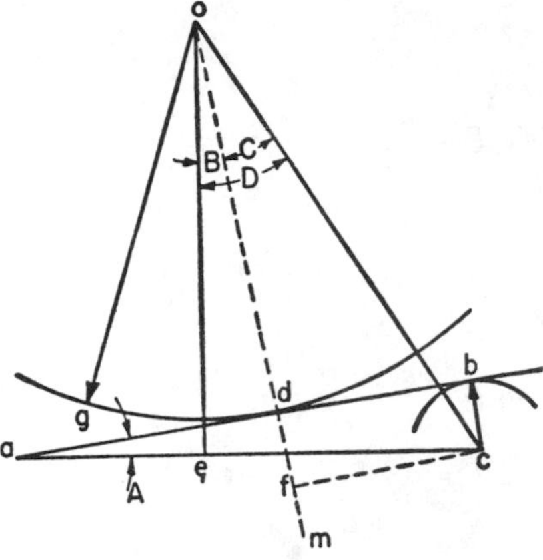

FIG. 7. To Find Angle A when Radii og and cb and Sides oe and ec are Known

Analysis of Problem: This is a problem in solving for the angles of two right-angled triangles with a common hypotenuse. One of these triangles oec is given. The other must be constructed. Draw om perpendicular to ab at the point of tangency d. Complete the triangle by drawing line cf at right angles to om.

Since ac is perpendicular to oe and ab is perpendicular to od, angle A is equal to angle B. (When the respective sides of two angles are perpendicular to each other, the angles are equal.)

Angle D is in right-angled triangle oec, of which two sides are known, hence angle D can be determined directly.

Angle C is in right-angled triangle ofc. The hypotenuse of this triangle can be determined as well as side of. Hence angle C can be found. Then, angle $B = D - C$.

Procedure: $\tan D = \dfrac{ec}{oe}$ (1)

$$oc = \frac{ec}{\sin D} \tag{2}$$

$$od = og \text{ and } df = bc$$

$$of = od + df = og + bc \tag{3}$$

$$\cos C = \frac{of}{oc} \tag{4}$$

$$B = D - C \tag{5}$$

Example: Let $og = 0.205$; $oe = 0.217$; $ec = 0.138$; and $bc = 0.040$. Find angle A.

Solution: $$\tan D = \frac{0.138}{0.217} = 0.63594 \tag{1}$$

$$D = 32°27.2'$$

$$oc = \frac{0.138}{0.53661} = 0.2572 \tag{2}$$

$$of = 0.205 + 0.040 = 0.245 \tag{3}$$

$$\cos C = \frac{0.245}{0.2572} = 0.95257 \tag{4}$$

$$C = 17°43'$$

$$B = 32°27.2' - 17°43' = 14°44.2 \tag{5}$$

Locating Second Tangent to a Circle at Given Angle with First

PROBLEM 7: To find the indexing angle X for positioning a round blank for a second milling cut to make a cam having straight sides which are at angle A with each other. The outside radius r_2 of the blank and the radius r_1 of the smaller circle to which the sides are tangent, are known.

Analysis of Problem: As shown in Figs. 8A, 8B, and 8C, there are three cases that may be considered: (1) when angle A is *less* than 180 degrees; (2) when angle A is *equal* to 180 degrees; and (3) when angle A is *greater* than 180 degrees. It will be found, upon analysis of each case that the same formula can be used to find angle X.

Formula: $$X = 2D + A - 180°$$

where $$D = \text{arc} \sin \frac{r_1}{r_2}$$

Derivation of Formula:

Case 1 (Fig. 8A) According to the third geometrical proposition on page 3–6.

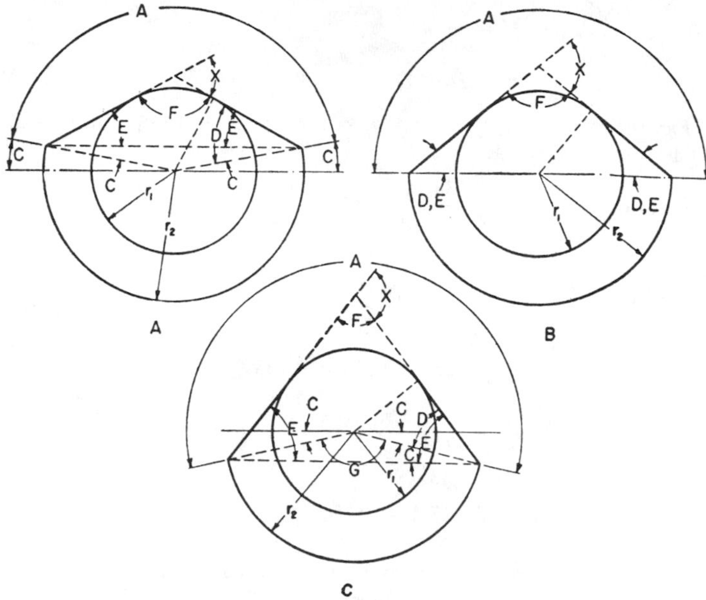

FIG. 8. To Find Indexing Angle X for Producing Angle A when Radii r_1 and r_2 are Known. (A) When A is Less than 180 Degrees. (B) When A Equals 180 Degrees. (C) When A is Greater than 180 Degrees

$$X = 2E \tag{1}$$

but

$$E = D - C \tag{2}$$

hence

$$X = 2D - 2C \tag{3}$$

$$C = \frac{180° - A}{2} \tag{4}$$

so that

$$X = 2D - 180° + A \tag{5}$$

where

$$\sin D = \frac{r_1}{r_2} \tag{6}$$

or

$$D = \text{arc sin } \frac{r_1}{r_2} \tag{7}$$

Case 2 (Fig. 8B)

$$X = 2E \tag{8}$$

In this case, angle C is equal to zero, so that

$$E = D \tag{9}$$

and

$$X = 2D \tag{10}$$

where

$$\sin D = \frac{r_1}{r_2} \tag{11}$$

or

$$D = \text{arc sin } \frac{r_1}{r_2} \tag{12}$$

Case 3 (Fig. 8C)

$$X = 2E \tag{13}$$

but

$$E = D + C \tag{14}$$

hence

$$X = 2D + 2C \tag{15}$$

$$C = \frac{A - 180°}{2} \tag{16}$$

so that

$$X = 2D + A - 180° \tag{17}$$

where

$$\sin D = \frac{r_1}{r_2} \tag{18}$$

or

$$D = \text{arc sin } \frac{r_1}{r_2} \tag{19}$$

It will be noted that Equation (17) is the same as Equation (5). If 180° is substituted for A in Equation (17) or Equation (5) the result will be Equation (10). Hence either Equations (17) or (5) may be used for all three cases.

Example: Find X if $r_1 = 1$ and $r_2 = 2$ when (1) $A = 140°$; (2) $A = 180°$; and (3) $A = 220°$.

Solutions: $\sin D = \frac{1}{2} = 0.5 \qquad D = 30°$

(1) $$X = 2 \times 30° + 140° - 180° = 20°$$

(2) $$X = 2 \times 30° + 180° - 180° = 60°$$

(3) $$X = 2 \times 30° + 220° - 180° = 100°$$

12

Arcs and Circles—Tangent to Circles and Lines

Another group of problems involving arcs and circles has to do with establishing their size and location when tangent to other arcs and lines. Three elements must be given if the location and size of the tangent arc is to be found. In the problems that follow the required arc is tangent to either a line, a point and an arc; two lines and an arc; one line and two arcs; or three circles or arcs.

Radius of Arc Tangent to a Given Arc and to a Given Line at a Given Point

PROBLEM 1: To find radius r in Fig. 1. In this problem two lines x–x and y–y at right angles to each other and an arc of radius R tangent to y–y at a distance A from x–x are given. It is required to find the radius r of a second arc which will be tangent to the first arc and also tangent to x–x at a distance B from y–y.

Analysis of Problem: If a line is drawn from the center a of the first arc through the center b of the second arc, it will intersect both arcs at d which is their point of tangency. If a center-line is now drawn through a and perpendicular to y–y and another center line is drawn through b and perpendicular to x–x, a right-angled triangle abc is formed. An equation in terms of r, R, A and B can then be set up based on the familiar geometrical theorem that in any right-angled triangle the square of the hypotenuse is equal to the sum of the squares of the other two sides.

Formula:

$$r = \frac{A^2 + B^2 - 2RB}{2A - 2R}$$

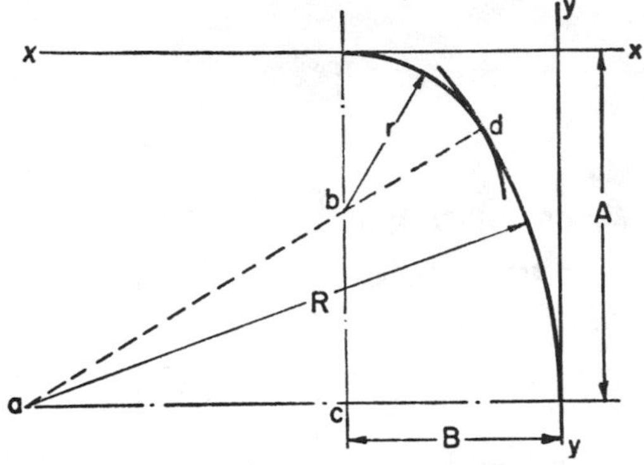

FIG. 1. To Find Radius r when Distances A and B and Radius R are Known

Derivation of Formula:

$$\overline{ab}^2 = \overline{bc}^2 + \overline{ac}^2 \qquad (1)$$

But

$$ab = R - r \qquad (2)$$

$$bc = A - r \qquad (3)$$

$$ac = R - B \qquad (4)$$

Therefore

$$(R - r)^2 = (A - r)^2 + (R - B)^2 \qquad (5a)$$

Squaring each term in Equation (5a) as indicated:

$$R^2 - 2Rr + r^2 = A^2 - 2Ar + r^2 + R^2 - 2RB + B^2 \qquad (5b)$$

Cancelling like terms and rearranging Equation (5b):

$$2Ar - 2Rr = A^2 + B^2 - 2RB \qquad (5c)$$

$$r(2A - 2R) = A^2 + B^2 - 2RB \qquad (5d)$$

$$r = \frac{A^2 + B^2 - 2RB}{2A - 2R} \qquad (5c)$$

Example: Find radius r when $A = 7.30$; $B = 4.48$; $R = 11.29$

Solution: $$r = \frac{(7.30)^2 + (4.48)^2 - 2 \times 11.29 \times 4.48}{2 \times 7.30 - 2 \times 11.29} \qquad (5c)$$

$$r = \frac{-27.798}{-7.98} = 3.483$$

Radius of Circle Tangent to Sides of Given Angle and to Given Circle Similarly Tangent

PROBLEM 2: To find radius *bc* in Fig. 2. Given two disks with centers at *b* and *d*. These disks are in contact with each other and two straight edges *ax* and *ay* are placed so that they are tangent to them both. The angle $2A$ formed by *ax* and *ay* and the radius *de* are known. It is required to find the radius *bc*.

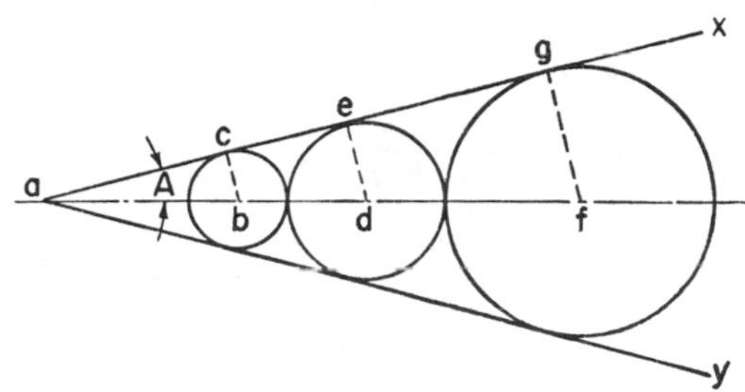

FIG. 2. To Find Radius *bc* when Angle $2A$ and Radius *de* are Known

Analysis of Problem: According to the fifth geometrical proposition on page 3–7, both *bc* and *de* are at right angles to *ax*. In triangle *ade* with angle A and *de* known, *ad* can be found. In triangle *abc*, the sine of A is expressed in terms of *bc* and *ab*. But *ab* can be expressed in terms of *ad*, *bc*, and *de*. Hence an equation can be established in which *bc* is the only unknown.

Formula:

$$bc = de \times \frac{1 - \sin A}{1 + \sin A}$$

Derivation of Formula:

$$\sin A = \frac{de}{ad} \tag{1}$$

$$ad = \frac{de}{\sin A} \tag{2}$$

$$\sin A = \frac{bc}{ab} = \frac{bc}{ad - bd} \tag{3}$$

but

$$bd = bc + de \tag{4}$$

hence
$$\sin A = \frac{bc}{ad - bc - de} \tag{5}$$

$$\sin A = \frac{bc}{\dfrac{de}{\sin A} - bc - de} \tag{6}$$

$$\sin A = \frac{bc \sin A}{de - bc \sin A - de \sin A} \tag{7}$$

Dividing both sides by $\sin A$:

$$1 = \frac{bc}{de - bc \sin A - de \sin A} \tag{8}$$

$$de - bc \sin A - de \sin A = bc \tag{9}$$

$$bc + bc \sin A = de - de \sin A \tag{10}$$

$$bc(1 + \sin A) = de(1 - \sin A) \tag{11}$$

$$bc = de \frac{1 - \sin A}{1 + \sin A} \tag{12}$$

Example: Let $de = 1$ and $A = 7°30'$. Find bc.

Solution:
$$bc = 1 \times \frac{1 - 0.13053}{1 + 0.13053} \tag{12}$$

$$bc = \frac{0.86947}{1.13053}$$

$$bc = 0.76908$$

Note: It can be shown, similarly, that if fg, the radius of a larger circle tangent to the given circle and to lines ax and ay, is to be determined, its formula will be:

$$fg = de \frac{1 + \sin A}{1 - \sin A} \tag{13}$$

Hence it will be seen upon comparison with Equation (12) that its value of fg will be the reciprocal of that for bc. From this it follows that the product of bc and fg is equal to the square of de. In other words the radius de of the given circle is a mean proportional between the radius bc of the smaller circle and the radius fg of the larger circle.

Radius of Circle Tangent to Given Circle and to Two Lines at Given Angle

PROBLEM 3: To find radius *gh* in Fig. 3. Two lines *oc* and *od* are drawn from the center of a circle of known radius *oj* at a given angle *A* with each other. Two lines *ac* and *af* are drawn respectively parallel to the lines *oc* and *od* and at a given distance *ab* from them. It is required to find the radius *gh* of the circle which will be externally tangent to the circle and to the lines *ae* and *af*. Two methods of solving this problem will be discussed.

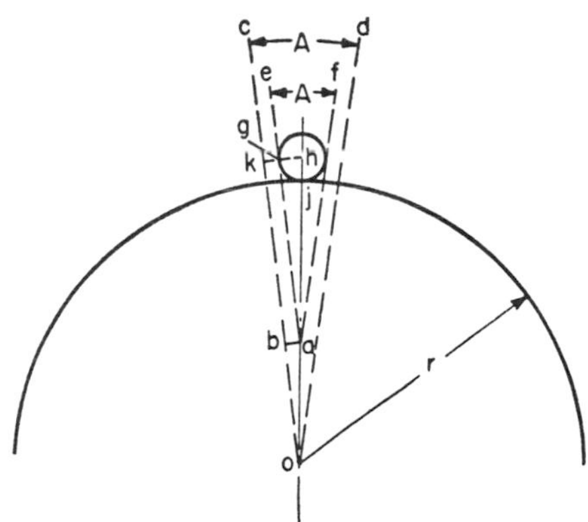

FIG. 3. To Find Radius *gh* when Radius *oj* Distance *ab* and Angle *A* are Known. A. First Method of Solution. B. Second Method of Solution

First Method

Analysis of Problem: In this diagram *gh* is perpendicular to *ea* (a radius drawn to point of tangency is perpendicular to tangent at that point) and *ab* is drawn perpendicular to *oc*. In the right-angled triangle *abo*, *oa* can be found since *ab* and angle $\frac{A}{2}$ are known. An expression for *aj* can then be established in terms of *oj* and *oa*. In right-angled triangle *agh* an expression for *gh* can be established in terms of angle $\frac{A}{2}$ and the equivalent of *ah*, expressed in terms of radius *oj* and distance *ab*.

Formula:

$$gh = \frac{oj \sin \frac{A}{2} - ab}{1 - \sin \frac{A}{2}}$$

Derivation of Formula: In the right-angled triangle *oab*

$$oa = ab \operatorname{cosec} \frac{A}{2} \tag{1}$$

$$aj = oj - oa \tag{2a}$$

$$aj = oj - ab \operatorname{cosec} \frac{A}{2} \tag{2b}$$

In the right-angled triangle *agh*

$$gh = ah \sin \frac{A}{2} \tag{3}$$

$$gh = (aj + jh) \sin \frac{A}{2} \tag{4}$$

$$gh = (aj + gh) \sin \frac{A}{2} \tag{5}$$

Substituting the equivalent of *aj* from Equation (2b)

$$gh = \left(oj - ab \operatorname{cosec} \frac{A}{2}\right) \sin \frac{A}{2} + gh \sin \frac{A}{2} \tag{6a}$$

$$gh = oj \sin \frac{A}{2} - ab \frac{\sin \frac{A}{2}}{\sin \frac{A}{2}} + gh \sin \frac{A}{2} \tag{6b}$$

$$gh - gh \sin \frac{A}{2} = oj \sin \frac{A}{2} - ab \tag{6c}$$

$$gh = \frac{oj \sin \frac{A}{2} - ab}{1 - \sin \frac{A}{2}} \tag{6d}$$

Example: In Fig. 3 let $oj = 2.00$; $ab = 0.21875$; and $A = 15°$. Find gh.

Solution:

$$gh = \frac{2 \sin \frac{15°}{2} - 0.21875}{1 - \sin \frac{15°}{2}} \qquad (6d)$$

$$gh = \frac{2 \times 0.13053 - 0.21875}{1 - 0.13053}$$

$$gh = \frac{0.04231}{0.86947} = 0.04866$$

Second Method

Analysis of Problem: Line hk is drawn perpendicular to oc forming right-angled triangle ohk. In this triangle hk can be expressed in terms of oh and angle $\frac{A}{2}$. By substituting the equivalents of hk and oh an expression for gh in terms of oj, ab and angle $\frac{A}{2}$ can be derived.

Formula:

$$gh = \frac{oj \sin \frac{A}{2} - ab}{1 - \sin \frac{A}{2}}$$

Derivation of Formula:

$$hk = oh \sin \frac{A}{2} \qquad (1)$$

Substituting equivalents of hk and oh:

$$kg + gh = (oj + jh) \sin \frac{A}{2} \qquad (2)$$

Substituting equivalents of kg and jh:

$$ab + gh = (oj + gh) \sin \frac{A}{2} \qquad (3a)$$

Transposing terms in (3a):

$$gh - gh \sin \frac{A}{2} = oj \sin \frac{A}{2} - ab \qquad (3b)$$

$$gh = \frac{oj \sin \frac{A}{2} - ab}{1 - \sin \frac{A}{2}} \qquad (3c)$$

Example: Let $oj = 5.41$; $ab = 0.85$; and $A = 30°$. Find gh.

Solution:

$$gh = \frac{5.41 \sin 15° - 0.85}{1 - \sin 15°} \qquad (3c)$$

$$gh = \frac{5.41 \times 0.25882 - 0.85}{1 - 0.25882}$$

$$gh = \frac{0.55}{0.74118} = 0.742$$

Radius of Arc Tangent to Given Circle and to Two Given Lines at Right Angles

PROBLEM 4: To find radius x in Fig. 4, which is a problem relating to the designing of dies for blanking a sheet metal piece. Given a circle of known radius r, the center of which is at a known distance

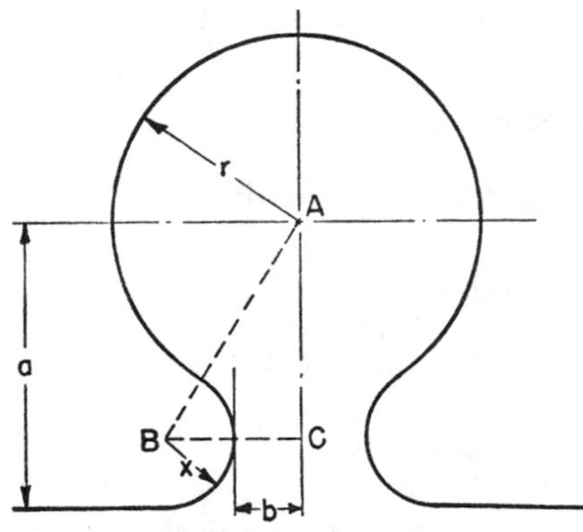

FIG. 4. To Find Radius x when Radius r and Distances a and b are Known

a from a given line, to find the radius x of a circle which will be tangent to the given circle and to the given line and will be so located that it is at a known distance b away from the vertical axis of the given circle or in other words tangent to a line drawn parallel to and at distance b from the vertical axis.

Analysis of Problem: Draw AB connecting the centers of the two circles and BC forming the right-angled triangle ABC. Then according to the familiar proposition "In any right-angled triangle the square of the hypotenuse equals the sum of the squares of the two sides," an equation can be established in terms of x and known values a, b, and r.

Formula: $x^2 + x(2b - 2a - 2r) + a^2 + b^2 - r^2 = 0$

To be solved as a quadratic for x.

Derivation of Formula:

$$\overline{AB}^2 = \overline{BC}^2 + \overline{AC}^2 \quad (1)$$

but

$$AB = r + x \quad (2)$$

$$BC = x + b \quad (3)$$

$$AC = a - x \quad (4)$$

Substituting these equivalents in Equation (1)

$$(r + x)^2 = (x + b)^2 + (a - x)^2 \quad (5a)$$

$$r^2 + 2rx + x^2 = x^2 + 2bx + b^2 + a^2 - 2ax + x^2 \quad (5b)$$

$$x^2 + 2bx - 2ax - 2rx + a^2 + b^2 - r^2 = 0 \quad (5c)$$

$$x^2 + x(2b - 2a - 2r) + a^2 + b^2 - r^2 = 0 \quad (5d)$$

This is a quadratic equation in the form given on page **2–20**. The easiest method of solution is to substitute the actual values of a, b, and r for each particular problem and then solve for x as shown in the following example.

Example: Let $a = 0.25$; $b = 0.09375$; and $r = 0.1875$. Find x.

Solution: $x^2 + x(0.1875 - 0.5 - 0.375) + (0.25)^2 + (0.09375)^2 - (0.1875)^2 = 0$ (5d)

$$x^2 - 0.6875x + 0.0625 + 0.00879 - 0.03516 = 0$$

$$x^2 - 0.6875x + 0.03613 = 0$$

This is a quadratic equation in the form given on page 2–20 in which $a = 1$; $b = -0.6875$; and $c = 0.03613$.

Solving according to the formula for the two roots given on page 2–21

$$x = \frac{-(-0.6875) \pm \sqrt{(-0.6875)^2 - 4 \times 1 \times 0.03613}}{2}$$

$$x = \frac{0.6875 \pm \sqrt{0.47266 - 0.14452}}{2}$$

$$x = \frac{0.6875 \pm \sqrt{0.32814}}{2}$$

$$x = \frac{0.6875 \pm 0.5728}{2} = 0.6301 \text{ or } 0.0573$$

It can be seen from Fig. 4 that 0.0573 is the correct answer in this case, since x must be less than a.

Radius of Circle Tangent to Given Arc, Its Radius and Line Drawn at Given Angle with Radius

PROBLEM 5: To find radius *ea* in Fig. 5. In a problem encountered in the design of a field pole tip for a piece of electrical equipment there is given an arc having a known radius *of*. At a known distance *od* from the center of the arc a line *bd* is drawn at right angles to *of*. Another line *og* is drawn from the center of the arc at a known angle A. It is required to find the radius *ea* of a circle that will be tangent to *og*, *bd*, and the given arc.

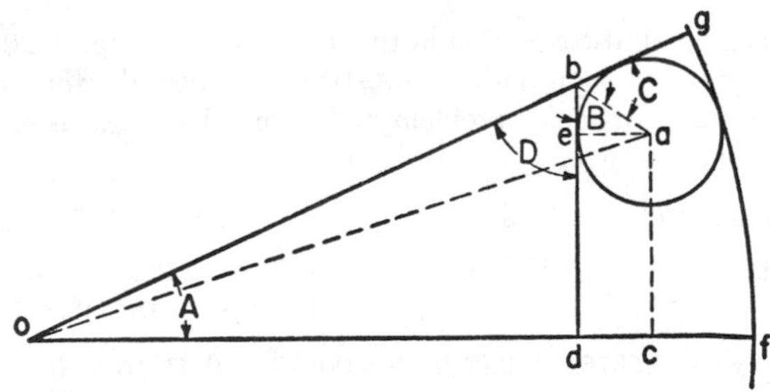

Fig. 5. To Find Radius *ea* when Radius *of*, Distance *od* and Angle A are Known

Analysis of Problem: Draw *oa* from the center of the arc to the center of the required circle. Draw *ac* parallel to *bd*, and *ae*, which is perpendicular to *bd* (Fifth geometrical proposition on page 3–7). Also draw *ba* from the intersection of the two tangents *og* and *bd* to the center of the required circle.

Then in right-angled triangle *aoc*, the familiar rule "the sum of the squares of the sides is equal to the square of the hypotenuse" can be used to set up an equation for *ac* in terms of the unknown radius of the circle and known values, since both *oa* and *oc* can be expressed in terms of the radius *ea* and given values.

In the right-angled triangle *abe*, an expression for *be* can be established in terms of angle B and side *ea*. Angle D equals 180 degrees minus angle $(B + C)$. But angles B and C are equal since a radius extended to the point of intersection of two tangents to a circle bisects the angle formed by these tangents. Angle D also equals 90 degrees minus angle A since they are both in right-angled triangle *obd*. By placing these two equivalents of angle D equal to each other, angle B can be found and hence *be* can be expressed in terms of radius *ea* as the only unknown.

In right-angled triangle *obd*, *bd* can be determined since *od* and angle A are known. It can be seen from the diagram that *bd* equals *be* plus *ac*. If, for each specific problem, the equivalents of *bd*, *be*, and *ac* are then substituted in this expression, a quadratic equation will be established which can be solved for *ea*. This is a much easier procedure than attempting to establish a general formula for *ea*, which is rather a cumbersome process.

Procedure: $$\overline{ac}^2 = \overline{oa}^2 - \overline{oc}^2 \quad (1)$$

$$\overline{ac}^2 = (of - ea)^2 - (od + ea)^2 \quad (2a)$$

$$ac = \sqrt{(of - ea)^2 - (od + ea)^2} \quad (2b)$$

$$be = ea \cot B \quad (3)$$

$$D = 180° - B - C \quad (4)$$

but $$B = C \quad (5)$$

hence $$D = 180° - 2B \quad (6)$$

also $$D = 90° - A \quad (7)$$

therefore $$90° - A = 180° - 2B \quad (8a)$$

$$2B = 90° + A \tag{8b}$$

$$B = \frac{90° + A}{2} \tag{8c}$$

Substituting this equivalent of B in Equation (3):

$$be = ea \cot \frac{90° + A}{2} \tag{9}$$

$$bd = od \tan A \tag{10}$$

$$bd = be + ed \tag{11}$$

$$bd = be + ac \tag{12}$$

Example: Let $od = 7$; $of = 10$; and $A = 33°38.6'$. Find radius ea.

Solution: $ac = \sqrt{(10 - ea)^2 - (7 + ea)^2}$ (2b)

$$ac = \sqrt{51 - 34ea}$$

$$be = ea \cot \frac{90° + 33°38.6'}{2} \tag{9}$$

$$be = ea \cot 61°49.3'$$

$$be = 0.53570ea$$

$$bd = 7 \times \tan 33°38.6' \tag{10}$$

$$bd = 7 \times 0.66549 = 4.6584$$

$$4.6584 = 0.53570ea + \sqrt{51 - 34ea} \tag{12}$$

$$4.6584 - 0.53570ea = \sqrt{51 - 34ea}$$

Squaring both sides of this equation

$$21.7007 - 4.9910ea + 0.28697ea^2 = 51 - 34ea$$

$$0.28697ea^2 + 29.0090ea - 29.2993 = 0$$

This is a quadratic equation in the form of $\boldsymbol{a}(ea^2) + \boldsymbol{b}(ea) + \boldsymbol{c} = 0$ (see formula on page **2–20**) where $\boldsymbol{a} = 0.28697$; $\boldsymbol{b} = 29.0090$; and $\boldsymbol{c} = -29.2993$.

$$ea = \frac{-29.0090 \pm \sqrt{(29.0090)^2 - 4 \times 0.28697 \times (-29.2993)}}{0.57394}$$

$$ea = \frac{-29.0090 \pm \sqrt{841.5221 + 33.6327}}{0.57394}$$

$$ea = \frac{-29.0090 \pm 29.5830}{0.57394} = \frac{0.5740}{0.57394} \quad \text{or} \quad -\frac{58.592}{0.57394}$$

Upon referring to Fig. 5 it will be seen at once that only the first value of *ea* applies to this problem.

$$ea = \frac{0.5740}{0.57394} = 1.000$$

Radius of Circle Tangent to Given Circle and to Two Given Lines

PROBLEM 6: To find radius *an* in Fig. 6. The diagram shows the conditions encountered in the design of a roller clutch. From a mathematical standpoint the problem is as follows. Given a circle of known radius *om*, to find the radius *an* of a circle which is tangent to the given circle and to two lines, one of which, *be*, is parallel to the horizontal axis *x–x* of the circle and at a given distance *bd* from it, and the other of which, *cb*, is at a given angle *A* with a line drawn parallel with the vertical axis *y–y*, and passes through point *b* on the first line at a given distance *be* from the *y–y* axis.

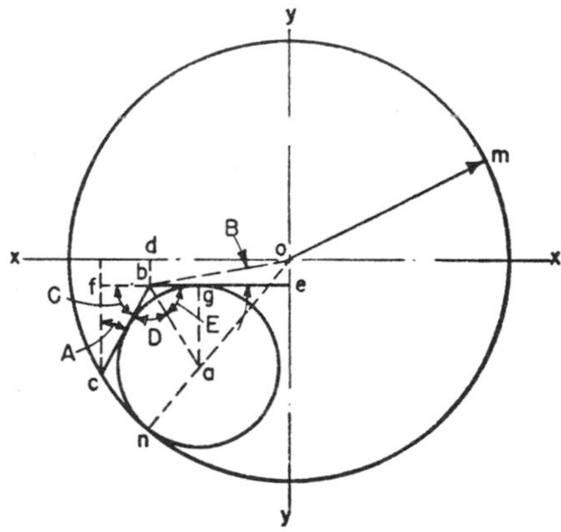

FIG. 6. To Find Radius *an* when Radius *om*, Distances *bd* and *be* and Angle *A* are Known

Analysis of Problem: In the right-angled triangle *obe*, *oe* and *be* are known, hence *ob* and angle B can be found. Since angle A is known in right-angled triangle *cfb*, angle C can be found and hence angle $(D + E)$. It can be shown that if two lines are drawn tangent to a circle, a line drawn from the center of the circle to the point of intersection of the two tangents will bisect the angle formed by them. Hence angle D equals angle E. It can be seen that *oa* equals *on*, which is known, minus *an* which is to be found. In right-angled triangle *abg*, *ab* can be expressed in terms of angle E and the unknown radius *an* of the smaller circle. Thus, with one side *ob* and one angle $(E + B)$ known in acute angle triangle *boa*, and with expressions for the other two sides *ab* and *ao* established in terms of the unknown radius *an*, a quadratic equation can be set up which can be solved for *an*.

Procedure: In right-angled triangle *bdo*

$$\overline{ob}^2 = \overline{bd}^2 + \overline{do}^2 \tag{1a}$$

$$ob = \sqrt{\overline{bd}^2 + \overline{do}^2} = \sqrt{\overline{bd}^2 + \overline{be}^2} \tag{1b}$$

In right-angled triangle *obe*

$$\tan B = \frac{oe}{be} = \frac{bd}{be} \tag{2}$$

In right-angled triangle *cfb*

$$C = 90° - A \tag{3}$$

$$D + E = 180° - C = 180° - 90° + A \tag{4a}$$

$$D + E = 90° + A \tag{4b}$$

$$D = E \tag{5}$$

since a line drawn from the center of a circle to the point of intersection of two tangents to the circle, bisects the angle formed by the tangents

hence

$$2E = 90° + A \tag{6a}$$

$$E = \frac{90° + A}{2} \tag{6b}$$

In right-angled triangle *abg*

$$ab = ag \operatorname{cosec} E \tag{7}$$

therefore $$ab = an \operatorname{cosec} E \tag{8}$$

$$ao = on - an \tag{9}$$

therefore $$ao = om - an \tag{10}$$

According to the Law of Cosines, page **4–18**, in acute-angled triangle *aob:*

$$ao = \sqrt{ab^2 + ob^2 - 2ab \times ob \times \cos (E + B)} \tag{11}$$

Rather than attempt to establish a general formula for *an* based on the above quadratic equation, it is easier to substitute the known values for each specific problem in this quadratic equation and then solve for *an* as illustrated in the following example.

Example: Let $be = 1.25$; $bd = 0.25$; $om = 2.00$; and $A = 30°$. Find radius *an*.

Solution: $$ob = \sqrt{(0.25)^2 + (1.25)^2} \tag{1b}$$

$$ob = \sqrt{0.0625 + 1.5625} = \sqrt{1.625}$$

$$ob = 1.275$$

$$\tan B = \frac{0.25}{1.25} = 0.2 \tag{2}$$

$$B = 11°18.6'$$

$$E = \frac{90° + 30°}{2} = 60° \tag{6b}$$

$$E + B = 71°18.6'$$

$$ab = an \operatorname{cosec} 60° = 1.1547\, an \tag{8}$$

$$ao = 2.00 - an \tag{10}$$

$$2.00 - an = \sqrt{(1.1547an)^2 + (1.275)^2 - 2 \times 1.1547an \times 1.275 \times 0.32045} \tag{11}$$

Squaring both sides of the above equation

$$4 - 4an \times \overline{an}^2 = 1.3333\overline{an}^2 + 1.625 - 0.94356an$$

$$0.3333\overline{an}^2 + 3.0564an - 2.375 = 0$$

This is a quadratic equation of the form illustrated by the formula on page **2–20**.

According to the formula for roots, on page 2–21

$$an = \frac{-3.0564 \pm \sqrt{(3.0564)^2 - 4 \times 0.3333 \times (-2.375)}}{0.6666}$$

$$an = \frac{-3.0564 \pm \sqrt{9.3416 + 3.1667}}{0.6666}$$

$$an = \frac{-3.0564 \pm \sqrt{12.508}}{0.6666} = \frac{-3.0564 \pm 3.5366}{0.6666}$$

$$an = \frac{0.4802}{0.6666} \quad \text{or} \quad -\frac{6.5930}{0.6666}$$

$$an = \frac{0.4802}{0.6666} = 0.7204$$

It is obvious that the negative value of *an* does not apply to this problem.

Radius and Center of Arc Tangent to Two Circles and Given Line

PROBLEM 7: To determine the radius R and center c of an arc that is to blend with two circles of known radii r_1 and r_2, respectively, as shown in Fig. 7. This arc is also to be tangent to a line *mn* which is at a known distance *de* from the center line of the two circles. The distance *ab* between the centers of the two circles is also known.

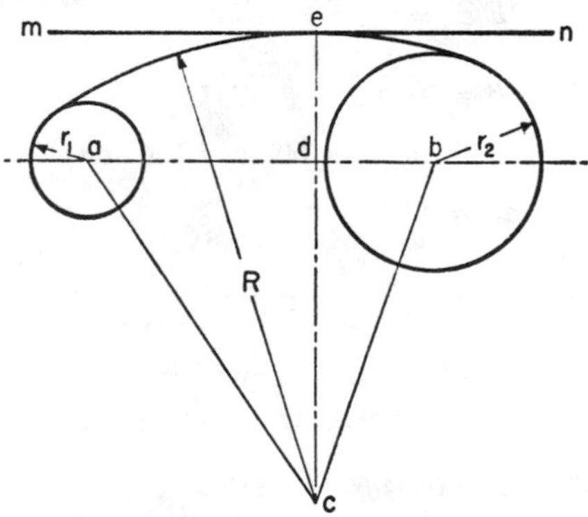

FIG. 7. To Find Radius R and Locate Center c when Radii r_1 and r_2 and Distances *de* and *ab* are Known

Analysis of Problem: Since *acd* is a right-angled triangle, the square of side *dc* can be expressed in terms of the square of the hypotenuse *ac* and the square of the side *ad*. Similarly the square of side *dc* can be expressed in terms of the square of the hypotenuse *bc* and the square of side *db*, since *bcd* is also a right-angled triangle.

If these two equivalents of *dc* are set equal to each other, an equation can be established in terms of R, r_1, r_2, *ab*, *ad* and *db*. Substituting the equivalent of *db* in terms of *ab* and *ad*, there results an equation with only two unknowns R and *ad*. This equation is now solved for *ad* in terms of R. A new equation for *ad* in terms of *ac*, which is $R - r_1$, and *dc*, which is $R - de$, is established and the equivalent of *ad* just found is substituted. There results a quadratic equation in terms of R and R^2 which can be readily solved.

Procedure:

$$\overline{dc}^2 = \overline{ac}^2 - \overline{ad}^2 \quad (1)$$

$$\overline{dc}^2 = \overline{bc}^2 - \overline{bd}^2 \quad (2)$$

therefore

$$\overline{ac}^2 - \overline{ad}^2 = \overline{bc}^2 - \overline{bd}^2 \quad (3a)$$

or

$$ac^2 - bc^2 = ad^2 - bd^2 \quad (3b)$$

factoring each side of Equation (3b)

$$(ac + bc)(ac - bc) = (ad + bd)(ad - bd) \quad (3c)$$

but

$$ac = R - r_1 \quad (4)$$

$$bc = R - r_2 \quad (5)$$

and

$$ad + bd = ab \quad (6)$$

Substituting these equivalents in Equation (3c)

$$(R - r_1 + R - r_2)(R - r_1 - R + r_2) = ab(ad - bd) \quad (7a)$$

$$(2R - r_1 - r_2)(r_2 - r_1) = ab(ad - bd) \quad (7b)$$

$$ad - bd = \frac{(2R - r_1 - r_2)(r_2 - r_1)}{ab} \quad (7c)$$

since

$$bd = ab - ad \quad (8)$$

then, substituting in Equation (7c)

$$ad - ab + ad = \frac{(2R - r_1 - r_2)(r_2 - r_1)}{ab} \quad (9a)$$

$$2ad = \frac{(2R - r_1 - r_2)(r_2 - r_1)}{ab} + ab \tag{9b}$$

$$ad = \frac{(2R - r_1 - r_2)(r_2 - r_1)}{2ab} + \frac{ab}{2} \tag{9c}$$

Having substituted the known values of r_1, r_2, and ab in Equation (9c), the resulting equivalent of ad is substituted in the following equation:

$$ad^2 = (R - r_1)^2 - (R - de)^2 \tag{10}$$

and this equation is then solved as a quadratic for R.

Example: In Fig. 7 let $r_1 = 12$, $r_2 = 23$, $ab = 73$ and $de = 27$. Find the radius R of an arc which will be tangent to both circles and also to line mn. Find also the location of the center of this arc with respect to point a.

Solution: $$ad = \frac{(2R - 12 - 23)(23 - 12)}{2 \times 73} + \frac{73}{2} \tag{9c}$$

$$ad = \frac{(2R - 35)11}{146} + \frac{73}{2} = \frac{22R - 385}{146} + 36.5$$

$$ad = 0.151R + 33.863$$

Substituting this equivalent of ad in Equation (10):

$$(0.151R + 33.863)^2 = (R - 12)^2 - (R - 27)^2 \tag{10}$$

$$0.0228R^2 + 10.23R + 1146.70 =$$
$$R^2 - 24R + 144 - R^2 + 54R - 729$$

$$0.0228R^2 - 19.77R + 1731.70 = 0$$

This is solved as a quadratic equation, so that if the coefficient of the first term is designated as a, of the second as b and of the third as c:

$$R = \frac{-b \pm \sqrt{b^2 - 4ac}}{2a}$$

or $$R = \frac{-(-19.77) \pm \sqrt{(-19.77)^2 - 4 \times 0.0228 \times 1731.70}}{2 \times 0.0228}$$

$$R = \frac{19.77 \pm \sqrt{390.85 - 157.93}}{0.0456}$$

$$R = \frac{19.77 \pm 15.26}{0.0456} = 768.2 \text{ or } 98.9$$

If $\quad R = 768.2$

then $ad = 0.151 \times 768.2 + 33.86 \qquad (9c)$

$ad = 116.00 + 33.86 = 149.86$

If $\quad R = 98.9$

then $ad = 0.151 \times 98.9 + 33.86 \qquad (9c)$

$ad = 14.93 + 33.86 = 48.79$

If the value of 768.2 for R is rejected as not suitable, then the center of the desired arc will be on a line drawn at right angles to ab at a distance 48.79 units from a and the radius R will be 98.9 units.

Location and Radius of Circle Tangent to Two Given Circles and Given Line

PROBLEM 8: To find radius r and center distances x and y in Fig. 8. Given two circles having known radii m and n, one of which is tangent to a reference line Z–Z and the other of which is located with its center E at a known distance b from this line. The distance a between centers of these two circles is also known. It is required to find the radius r of a circle which will be tangent to the two given circles and also to a line V–V which is parallel to and at a known distance d from the reference line Z–Z. It is also required to find the center distance x and y of this circle as shown.

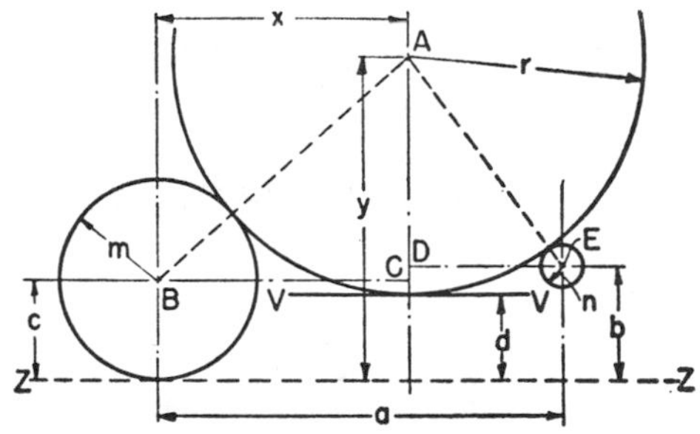

FIG. 8. To Find Radius r and Center Distances x and y when Radii m and n and Distances a, b and d are Known

Analysis of Problem: This is a problem involving two right-angled triangles ABC and ADE. Using the familiar theorem for right-angled triangles: "The square of the hypotenuse is equal to the sum of the squares of the two sides," two equations can be set up in terms of x, y, and r as the unknowns. Substituting for y the equivalent $r + d$, there remain two equations with two unknowns r and x. By manipulating these two equations, r can be eliminated leaving a quadratic equation in terms of x. Having found x, y and r can then be determined.

Procedure: In right-angled triangle ABC

$$\overline{AB}^2 = \overline{BC}^2 + \overline{AC}^2 \tag{1}$$

but
$$AB = r + m \tag{2}$$

and
$$BC = x \tag{3}$$

and
$$AC = y - m \tag{4}$$

since
$$y = r + d \tag{5}$$

$$AC = r + d - m \tag{6}$$

Then substituting equivalents in Equation (1)

$$(r + m)^2 = x^2 + (r + d - m)^2 \tag{7a}$$

$$r^2 + 2rm + m^2 = x^2 + r^2 + 2rd + d^2 - 2rm + m^2 - 2dm \tag{7b}$$

Simplifying

$$x^2 + 2rd - 4rm - 2dm + d^2 = 0 \tag{7c}$$

$$x^2 + r(2d - 4m) - 2dm + d^2 = 0 \tag{7d}$$

In right-angled triangle ADE

$$\overline{AE}^2 = \overline{AD}^2 + \overline{DE}^2 \tag{8}$$

but
$$AE = r + n \tag{9}$$

$$AD = y - b \tag{10}$$

$$DE = a - x \tag{11}$$

since
$$y = r + d \tag{5}$$

$$AD = r + d - b \tag{12}$$

Then substituting these equivalents in Equation (8)

$$(r + n)^2 = (r + d - b)^2 + (a - x)^2 \quad (13a)$$

$$r^2 + 2rn + n^2 = r^2 + 2rd + d^2 - 2rb + b^2 - 2db + a^2 - 2ax + x^2 \quad (13b)$$

Subtracting r^2 from both sides and rearranging terms

$$x^2 - 2ax + 2rd - 2rn - 2rb - 2db + a^2 + d^2 + b^2 - n^2 = 0 \quad (13c)$$

$$x^2 - x(2a) + r(2d - 2n - 2b) - 2db + a^2 + d^2 + b^2 - n^2 = 0 \quad (13d)$$

This problem is solved from this point on by substituting the specific values for a, b, d, and n in Equations (7d) and (13d) and then by manipulation getting rid of r, leaving a single quadratic equation in terms of x. These steps are illustrated in the following example.

Example: Let $m = 0.507$; $n = 0.109$; $a = 1.981$; $b = 0.523$; and $d = 0.417$. Find r, x, and y.

Solution:

$$x^2 + r(0.834 - 2.028) - 2 \times 0.417 \times 0.507 + (0.417)^2 = 0 \quad (7d)$$

$$x^2 - 1.194r - 0.423 + 0.174 = 0$$

$$x^2 - 1.194r - 0.249 = 0$$

$$x^2 - 3.962x + r(0.834 - 0.218 - 1.046) - 2 \times 0.417 \times 0.523 + (1.981)^2 + (0.417)^2 + (0.523)^2 - (0.109)^2 = 0 \quad (13d)$$

$$x^2 - 3.962x - 0.430r - 0.436 + 3.924 + 0.174 + 0.274 - 0.012 = 0$$

$$x^2 - 3.962x - 0.430r + 3.924 = 0 \quad (A)$$

$$x^2 - 1.194r - 0.249 = 0 \quad (B)$$

Multiply Equation (A) by 1.194 and Equation (B) by 0.430, then subtract (B) from (A)

$$\begin{array}{rrrr} 1.194x^2 & - 4.731x & - 0.513r & + 4.685 = 0 \\ - 0.430x^2 & & + 0.513r & + 0.107 = 0 \\ \hline 0.764x^2 & - 4.731x & & + 4.792 = 0 \end{array}$$

This is a quadratic equation in the form of the formula on page 2–20, in which $a = 0.764$; $b = -4.731$; and $c = 4.792$.

According to the formula for roots on page 2–21

$$x = \frac{+4.731 \pm \sqrt{(-4.731)^2 - 4 \times 0.764 \times 4.792}}{2 \times 0.764}$$

$$x = \frac{+4.731 \pm \sqrt{22.382 - 14.644}}{1.528}$$

$$x = \frac{4.731 \pm \sqrt{7.738}}{1.528} = \frac{4.731 \pm 2.782}{1.528}$$

$$x = \frac{7.513}{1.528} = 4.917 \quad \text{or} \quad x = \frac{1.949}{1.528} = 1.278$$

Since x must be less than a (see Fig. 8) the correct value for this problem is 1.278.

Substituting this value for x in Equation (B)

$$(1.278)^2 - 1.194r - 0.249 = 0$$

$$1.194r = 1.633 - 0.249$$

$$1.194r = 1.384$$

$$r = 1.159$$

Substituting this value for r in Equation (5)

$$y = 1.159 + 0.417$$

$$y = 1.576$$

Radius of Circles Externally Tangent and Internally Tangent to Three Given Circles

Problem 9: To find radius *od* in Fig. 9. Given three circles having radii of *ad*, *be*, and *cf*, respectively, and located at known center distances *ab*, *bc*, and *ca* from each other. It is required (1) to find the radius *od* of a circle (Fig. 9A) which will be tangent externally to these three circles; and (2) to find the radius *od* of a circle (Fig. 9B) which would be tangent internally to these three circles.

Analysis of Problem: Referring to Fig. 9A, equations can be set up according to the "Law of Cosines," as expressed by formula on page **4–18,** for the three obtuse-angled triangles *aob*, *boc*, and *aoc*. A fourth formula can be derived to show the relationships between angles A, B, and C, in terms of their cosines. If substitutions are now made in this fourth equation for the cosines of angles A, B, and C, in terms of their equivalents as given in the first three equations, a fifth equation can be set up. From this point on by manipulation and by substitution of terms, as shown in detail under "Derivation

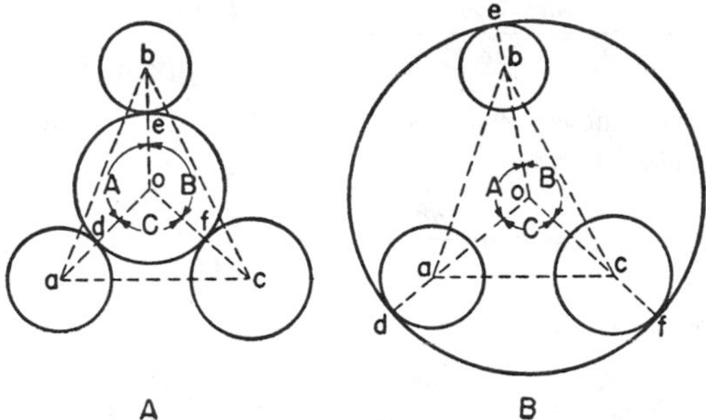

FIG. 9. To Find Radius *od* of a Circle that will be Tangent to Three Given Circles. A. When Tangent Externally. B. When Tangent Internally

of Formula," a quadratic equation in terms of the radius *od* of the circle to be found can be established. The two values obtained by the solution of this quadratic equation will be the respective radii of the two unknown circles in Fig. 9A and Fig. 9B.

Because of the number and complexity of the algebraic operations involved, many steps in the derivation of formula are shown, even where some rather obvious transformations are made.

Derivation of Formula: According to the "Law of Cosines," page 4–18 in obtuse-angled triangle *abo:*

$$\cos A = \frac{\overline{ao}^2 + \overline{bo}^2 - \overline{ab}^2}{2ao \times bo} \qquad (1)$$

but since
$$(ao - bo)^2 = \overline{ao}^2 - 2ao \times bo + \overline{bo}^2 \qquad (2)$$

then
$$(ao - bo)^2 + 2ao \times bo = \overline{ao}^2 + \overline{bo}^2 \qquad (3)$$

substituting this equivalent for $\overline{ao}^2 + \overline{bo}^2$ in Equation (1)

$$\cos A = \frac{2ao \times bo + (ao - bo)^2 - \overline{ab}^2}{2ao \times bo} \qquad (4a)$$

$$\cos A = 1 + \frac{(ao - bo)^2 - \overline{ab}^2}{2ao \times bo} \qquad (4b)$$

similarly
$$\cos B = \frac{\overline{bo}^2 + \overline{co}^2 - \overline{bc}^2}{2bo \times co} = 1 + \frac{(bo - co)^2 - \overline{bc}^2}{2bo \times co} \qquad (5)$$

$$\cos C = \frac{\overline{ao}^2 + \overline{co}^2 - \overline{ac}^2}{2ao \times co} = 1 + \frac{(ao - co)^2 - \overline{ac}^2}{2ao \times co} \tag{6}$$

Substituting the equivalents of *ao*, *bo*, and *co* in the numerators of Equations (4b), (5), and (6) referring to Fig. 9A

$$\cos A = 1 + \frac{(ad + do - be - eo)^2 - \overline{ab}^2}{2ao \times bo} \tag{7a}$$

but since $do = eo$, being radii of the same circle

$$\cos A = 1 + \frac{(ad - be)^2 - \overline{ab}^2}{2ao \times bo} \tag{7b}$$

similarly
$$\cos B = 1 + \frac{(be + eo - cf - fo)^2 - \overline{bc}^2}{2bo \times co} \tag{8a}$$

$$\cos B = 1 + \frac{(be - cf)^2 - \overline{bc}^2}{2bo \times co} \tag{8b}$$

and
$$\cos C = 1 + \frac{(ad + do - cf - co)^2 - \overline{ac}^2}{2ao \times co} \tag{9a}$$

$$\cos C = 1 + \frac{(ad - cf)^2 - \overline{ac}^2}{2ao \times co} \tag{9b}$$

To simplify subsequent manipulation of the terms in these three cosine equations, let:

$$X = (ad - be)^2 - \overline{ab}^2 \tag{10}$$

$$Y = (be - cf)^2 - \overline{bc}^2 \tag{11}$$

$$Z = (ad - cf)^2 - \overline{ac}^2 \tag{12}$$

Then substituting these equivalents in Equations (7b), (8b), and (9b):

$$\cos A = 1 + \frac{X}{2ao \times bo} \tag{13}$$

$$\cos B = 1 + \frac{Y}{2bo \times co} \tag{14}$$

$$\cos C = 1 + \frac{Z}{2ao \times co} \tag{15}$$

Turning now to the relationships between angles A, B, and C:

$$A + B + C = 360° \qquad (16a)$$

$$A = 360° - (B + C) \qquad (16b)$$

$$\cos A = \cos [360° - (B + C)] \qquad (17a)$$

therefore

$$\cos A = \cos (B + C) \qquad (17b)$$

According to Formula 10, page 4–23, Equation (17b) can be transformed into the following:

$$\cos A = \cos B \cos C - \sin B \sin C \qquad (18a)$$

$$\cos A - \cos B \cos C = -\sin B \sin C \qquad (18b)$$

Squaring both sides:

$$\cos^2 A - 2 \cos A \cos B \cos C + \cos^2 B \cos^2 C = \sin^2 B \sin^2 C \qquad (19)$$

According to Formula 7, page 4–23:

$$\sin^2 B = 1 - \cos^2 B \qquad (20)$$

$$\sin^2 C = 1 - \cos^2 C \qquad (21)$$

Substituting these equivalents in Equation (19):

$$\cos^2 A - 2 \cos A \cos B \cos C + \cos^2 B \cos^2 C = (1 - \cos^2 B)(1 - \cos^2 C) \qquad (22a)$$

Multiplying the two factors on the right-hand side of Equation (22a):

$$\cos^2 A - 2 \cos A \cos B \cos C + \cos^2 B \cos^2 C = 1 - \cos^2 B - \cos^2 C + \cos^2 B \cos^2 C \qquad (22b)$$

Subtracting $\cos^2 B \cos^2 C$ from both sides of Equation (22b) and moving all terms on the right side, except 1, to the left:

$$\cos^2 A + \cos^2 B + \cos^2 C - 2 \cos A \cos B \cos C = 1 \qquad (22c)$$

In this equation substitute the equivalents for $\cos A$, $\cos B$, and $\cos C$ as given in Equations (13), (14), and (15).

$$\left(1 + \frac{X}{2ao \times bo}\right)^2 + \left(1 + \frac{Y}{2bo \times co}\right)^2 + \left(1 + \frac{Z}{2ao \times co}\right)^2 - 2\left(1 + \frac{X}{2ao \times bo}\right)\left(1 + \frac{Y}{2bo \times co}\right)\left(1 + \frac{Z}{2ao \times co}\right) = 1 \qquad (23)$$

Equation (23) must be expanded and simplified. The *ao*, *bo*, and *co* in Fig. 9A must be replaced by the sum of their two smaller segments respectively. Substitute *do* for *eo* and *fo* because they are radii of the same circle in Fig. 9A. Combining terms then produces Equation (24)

Combining terms:

$$\overline{do}^2(X^2 + Y^2 + Z^2 - 2XZ - 2YZ - 2XY) + 2do[cf \times X^2 + ad \times Y^2 + be \times Z^2 - (be + cf)XZ - (ad + cf)XY - (ad + be)YZ] + \overline{cf}^2 \times X^2 + \overline{ad}^2 \times Y^2 + \overline{be}^2 \times Z^2 - 2be \times cf \times XZ - 2ad \times cf \times XY - 2ad \times be \times YZ - XYZ = 0 \quad (24)$$

Equation (24) is a quadratic equation in the form of that illustrated by the formula on page 2–20.

The solution of this problem is best carried forward form this point by substituting the equivalent numerical values for *ad*, *be*, *cf*, X, Y, and Z; these three latter values to be found by substituting numerical values for *ab*, *bc*, *ca*, *ad*, *be*, and *cf* in Equations (10), (11), and (12) in Equation (24), and then solving for *do* using the formula for roots on page 2–21.

Two values for *do* will be found by using this root formula. The smaller value is the radius of the circle which is externally tangent to the three given circles, Fig. 9A, while the larger value is the radius of the circle internally tangent to the three given circles, Fig. 9B.

Example: In Figs. 9A and 9B, let $ad = 6$; $be = 4$; $cf = 8$; $ab = 17$; $bc = 18$; and $ac = 19$. Find *do* in each case.

Solution:

$$X = (6 - 4)^2 - (17)^2 = 4 - 289 = -285 \quad (10)$$

$$Y = (4 - 8)^2 - (18)^2 = 16 - 324 = -308 \quad (11)$$

$$Z = (6 - 8)^2 - (19)^2 = 4 - 361 = -357 \quad (12)$$

Substituting numerical values in Equation (24)

$$\overline{do}^2[(-285)^2 + (-308)^2 + (-357)^2 - 2(-285)(-357) - 2(-308)(-357) - 2(-285)(-308)] + 2do[8(-285)^2 + 6(-308)^2 + 4(-357)^2 - (4 + 8)(-285)(-357) - (6 + 8)(-285)(-308) - (6 + 4)(-308)(-357)] + (8^2)(-285)^2 + (6)^2(-308)^2 + (4)^2(-357)^2 - (2)(4)(8)(-285)(-357) - (2)(6)(8)(-285)(-308) - (2)(6)(4)(-308)(-357) - (-285)(-308)(-357) = 0$$

$$-295{,}424\overline{do}^2 - 3{,}641{,}280do + 21{,}773{,}700 = 0$$

Dividing all terms by −295,424

$$\overline{do}^2 + 12.326do - 73.703 = 0$$

This is a quadratic equation in the form illustrated by the formula page **2**–20, in which $a = 1$; $b = 12.326$; and $c = -73.703$.

Therefore according to the formula for roots on page **2–21**:

$$do = \frac{-12.326 \pm \sqrt{(12.326)^2 - 4 \times 1 \times (-73.703)}}{2}$$

$$do = \frac{-12.326 \pm \sqrt{151.930 + 294.812}}{2}$$

$$do = \frac{-12.326 \pm 21.136}{2}$$

$$do = -16.731 \text{ or } 4.405$$

The value of 4.405 is the radius of the circle in Fig. 9A. The value 16.731 is the radius of the circle in Fig. 9B.

Radius of Circle Tangent to Three Other Circles

PROBLEM 10: To find radius x in Fig. 10. Given a circle with center at O and of known radius s and three inscribed circles which are tangent to each other. Two of these circles have their respective centers A and B located on a diameter of the circle in which they are inscribed. Their respective radii, t and r, are known. The radius x of the third circle is required.

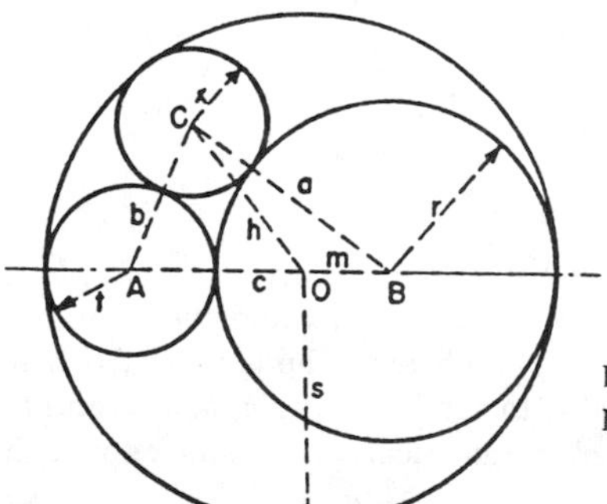

Fig. 10. To Find Radius x when Radii r, s and t are Known

Analysis of Problem: Draw AB, BC, AC, and CO connecting the respective centers of the four circles. Then in this diagram let: $a = BC$; $m = OB$; $b = AC$; $h = OC$; $c = AB$.

There are thus formed three oblique triangles: ABC, AOC, and COB. In triangle ABC an expression for the cosine of the angle at B can be set up in terms of a, b, and c. In triangle COB an expression for the cosine of the angle at B can be set up in terms of a, m, and h. These two equivalents of the cosine of angle B are then set equal to each other. The resulting equation is simplified but no general equation for x is set up.

This is a type of problem which is easier to solve by establishing an equation without the unknown x, and then substituting for each term its equivalent in terms of specific numerical values and x for a given set of data. This is clearly shown in the example.

Procedure: In obtuse-angled triangle ABC

$$\cos B = \frac{a^2 + c^2 - b^2}{2ac} \tag{1}$$

In obtuse-angled triangle COB

$$\cos B = \frac{a^2 + m^2 - h^2}{2am} \tag{2}$$

therefore

$$\frac{a^2 + c^2 - b^2}{2ac} = \frac{a^2 + m^2 - h^2}{2am} \tag{3}$$

multiplying both sides of Equation (3) by $2amc$

$$m(a^2 + c^2 - b^2) = c(a^2 + m^2 - h^2) \tag{4}$$

It can be seen that:

$$a = x + r \qquad h = s - x$$

$$b = x + t \qquad m = s - r$$

$$c = r + t$$

The process of attempting to substitute the equivalents of a, b, c, h, and m in Equation (4) in terms of r, s, t, and x and then working out a general formula in terms of x is cumbersome and of no particular value. The more direct method is to substitute for a, b, c, h, and m in Equation (4) in terms of known numerical values and x for each particular problem.

Example: Let $r = 2$, $s = 3$, and $t = 1$. Find x.

Solution:

$$a = x + 2 \qquad h = 3 - x$$

$$b = x + 1 \qquad m = 3 - 2$$

$$c = 2 + 1$$

$$1[(x + 2)^2 + 9 - (x + 1)^2] = 3[(x + 2)^2 + 1 - (3 - x)^2] \quad (4)$$

$$x^2 + 4x + 4 + 9 - x^2 - 2x - 1 = 3x^2 + 12x + 12 + 3 - 27 + 18x - 3x^2$$

Combining terms and simplifying:

$$-28x = -24$$

$$x = \frac{-24}{-28} = \frac{6}{7}$$

$$x = 0.857$$

13

Arcs and Circles—Fixed by Lines and Points

The third group of problems relating to arcs and circles deals with cases where they are fixed by lines and points. If either three tangential lines, two tangential lines and a point, one tangential line and two points, three points, or, if the radius of the circle is known a tangential line and a point, two lines or two points are known, the size and position of the arc or circle is determined. Each of the following problems illustrates an instance where one of above combination of elements is known and the size or location of the arc or circle is to be found. Several ways of analyzing and solving a given problem are often possible, as illustrated by the different methods presented in Problems 4 and 5. Such possibilities offer a challenge to the ingenuity of those working with mechanical problems.

Height of Arc Tangent to Two Straight Lines at Given Angle

PROBLEM 1: To find height *mp* in Fig. 1. If an arc of known radius *on* is drawn tangent to two straight lines that are at a known angle *A* with each other, it is required to find the height *mp* of the arc.

Analysis of Problem: As can be seen, the height *mp* is equal to the radius of the arc *op* minus *om*. Since *om* is one side of a right-angled triangle in which the hypotenuse *on* is known and one angle *B* can be found, *om* and hence *mp* can be found.

Formula: $mp = on\left(1 - \sin\frac{A}{2}\right)$

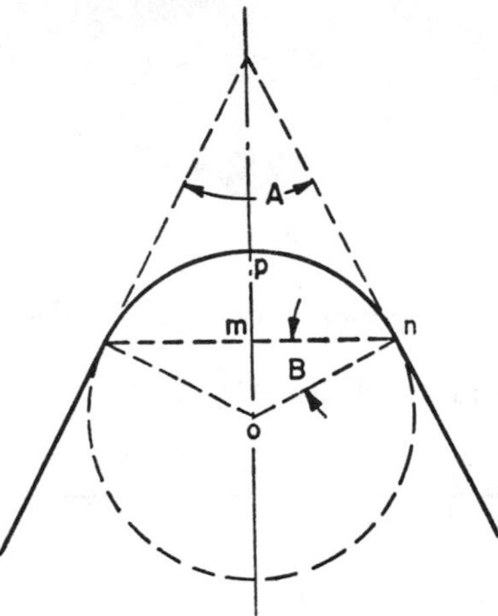

FIG. 1. To Find Height mp when Radius on and Angle A are Known

Derivation of Formula:

$$mp = op - om = on - om \tag{1}$$

$$om = on \sin B \tag{2}$$

but $$B = \frac{A}{2}$$

since their corresponding sides are mutually perpendicular;

hence, $$om = on \sin \frac{A}{2} \tag{4}$$

and $$mp = on - on \sin \frac{A}{2} = on\left(1 - \sin \frac{A}{2}\right) \tag{5}$$

Example: A Whitworth screw thread having 3½ threads per inch has a crest radius of 0.03924 inch and an included angle of 55 degrees. Find the depth of rounding for making a lap for the crests.

Solution:

$on = 0.03924 \qquad A = 55° \qquad mp =$ depth of rounding

$$mp = 0.03924\left(1 - \sin \frac{55°}{2}\right) \tag{5}$$

$$mp = 0.03924(1 - 0.46175) = 0.021121 \text{ inch}$$

Diameter of Tapered End to be Rounded to Given Radius

PROBLEM 2: To find diameter *bc* in Fig. 2. The end of a tapered section of a round rod has been machined to a spherical form. The radius *de* at the rounded end and the angle of taper *A* are known. It is required to find the diameter *bc* at the end of the rod before it was machined to a spherical form.

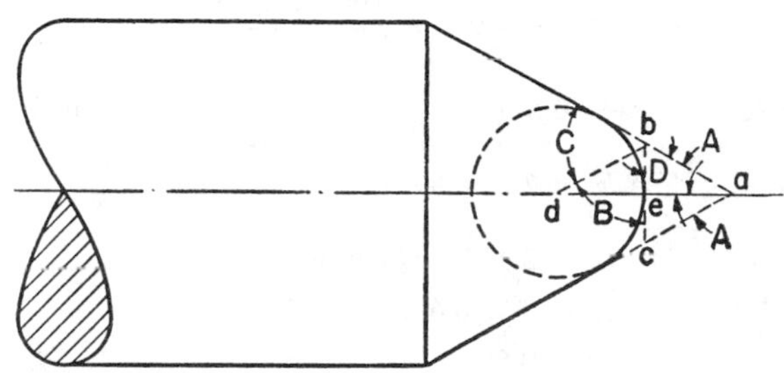

FIG. 2. To Find Diameter *bc* when Radius *de* and Angle of Taper *A* are Known

Analysis of Problem: Extend the tapered surface lines until they intersect at point *a*. Draw *db* from the center *d* of the spherical surface to the intersection of diameter *bc* with the upper tapered surface line. Since angle *A* is known, angle *D* can be found. When angle *D* is known, angle $(C + B)$ can be found. But angles *B* and *C* are equal since it can be shown that a radius of a circle extended to the point of intersection of two tangents to that circle will bisect the angle formed by the two tangents. In right-angled triangle *bde* with *de* and angle *B* known, *be* and hence *bc* can be found.

Formula:

$$bc = 2de\left(\cot\frac{90° + A}{2}\right)$$

Derivation of Formula:

$$D = 90° - A \tag{1}$$

$$B + C = 180° - D = 180° - (90° - A) \tag{2}$$

$$= 90° + A \tag{3}$$

but

$$B = C \tag{4}$$

hence

$$2B = 90° + A \tag{5a}$$

$$B = \frac{90° + A}{2} \tag{5b}$$

In right-angled triangle *bde*:

$$be = de \cot B \tag{6}$$

therefore

$$be = de \cot \frac{90° + A}{2} \tag{7}$$

and

$$bc = 2\, de \cot \frac{90° + A}{2} \tag{8}$$

Example: Let $de = 0.697$ and angle $A = 30°$. Find diameter *bc*.

Solution:

$$bc = 2 \times 0.697 \cot \frac{90° + 30°}{2} \tag{8}$$

$$bc = 2 \times 0.697 \cot 60°$$

$$bc = 2 \times 0.697 \times 0.57735$$

$$bc = 0.805$$

Radius of Circle Tangent to Two Lines and Passing Through Given Point

PROBLEM 3: To find radius r in Fig. 3. A circle is to be drawn tangent to two sides of a square and also passing through the point of intersection of the other two sides of the square. The length a of the sides of the square is known. It is required to find the radius r of the circle.

Analysis of Problem: This is a relatively simple problem involving two right-angled triangles. The derivation of the formula requires no explanation.

Formula:

$$r = \frac{a}{1 + \cos 45°}$$

Derivation of Formula:

$$a = r + r \cos 45° \tag{1a}$$

$$a = r(1 + \cos 45°) \tag{1b}$$

hence

$$r = \frac{a}{1 + \cos 45°} \tag{1c}$$

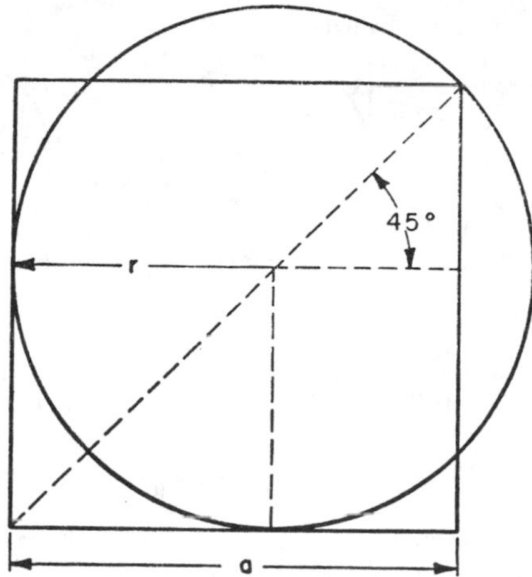

FIG. 3. To Find Radius r when Length a of Side of Square is Known

Example: Let $a = 4.372$. Find radius r.

Solution:

$$r = \frac{4.372}{1 + 0.70711} = 2.561 \tag{1c}$$

Radius of Circle Tangent to Two Lines and Passing Through a Given Point

PROBLEM 4: To find radius r in Fig. 4. A block of known dimensions m and n is placed against two surfaces at right angles to each other as shown. It is required to find radius r of a disk which will be tangent to these two surfaces and at the same time will make contact with the corner of the block.

Three methods of solving this problem will be discussed. The first method, which is the simplest, is based on the familiar geometrical proposition that the sum of the squares of two sides of a right triangle is equal to the square of the hypotenuse. The second method is based upon the general formula for a circle in analytical geometry. The third method is based on the geometrical proposition that if from a point outside a circle a secant and a tangent are drawn, the product of the entire length of the secant and that part of it which is outside the circle is equal to the square of the tangent.

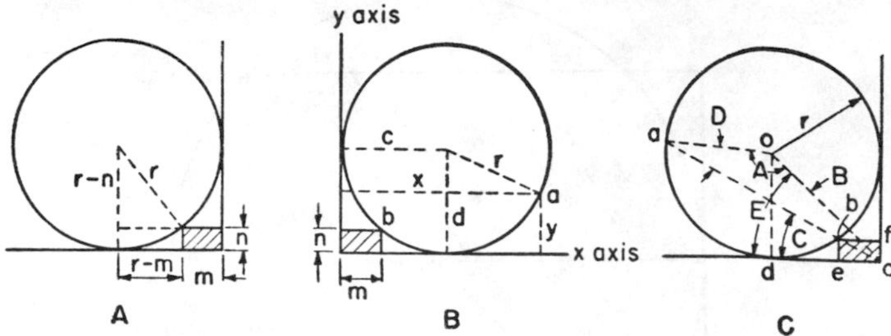

FIG. 4. To Find Radius r when Dimensions m and n are Known. (A) By First Method. (B) By Second Method. (C) By Third Method

First Method (Fig. 4A)

Analysis of Problem: Draw the radius r to the point of contact with the block. Draw a vertical line from the point of tangency with the horizontal surface to the center of the disk. Draw a horizontal line from the point of contact between block and disk until it intersects the vertical line which passes through the center of the disk. A right-angled triangle is thus formed having sides equal to $r - n$ and $r - m$ and a hypotenuse equal to r. An expression can thus be set up for r in terms of m and n which are known.

Formula: $$r = m + n \pm \sqrt{2mn}$$

Derivation of Formula:

$$r^2 = (r - m)^2 + (r - n)^2 \tag{1}$$

Expanding (1)

$$r^2 = r^2 - 2rm + m^2 + r^2 - 2rn + n^2 \tag{2a}$$

Rearranging (2a)

$$r^2 - 2rm - 2rn + m^2 + n^2 = 0 \tag{2b}$$

$$r^2 - 2(m + n)r + m^2 + n^2 = 0 \tag{2c}$$

Equation (4) is a quadratic in the form of that illustrated by the formula on page 2–20.

Thus, $$ar^2 + br + c = 0 \tag{3}$$

where $a = 1;\ b = -2(m + n);\ c = (m^2 + n^2)$

Substituting these values in the formula for roots on page **2–21**:

$$r = \frac{-b \pm \sqrt{b^2 - 4ac}}{2a} \tag{4}$$

$$r = \frac{+2(m + n) \pm \sqrt{4m^2 + 8mn + 4n^2 - 4m^2 - 4n^2}}{2} \tag{5}$$

$$r = \frac{2(m + n) \pm 2\sqrt{2mn}}{2} \tag{6}$$

$$r = m + n \pm \sqrt{2mn} \tag{7}$$

Example: In Fig. 4A, let $m = 0.5$ and $n = 0.25$. Find r.

Solution:

$$r = 0.5 + 0.25 \pm \sqrt{2 \times 0.5 \times 0.25} \tag{7}$$

$$r = 0.75 \pm 0.5$$

$$r = 1.25 \quad \text{or} \quad 0.25$$

It can be seen from the diagram that r must be greater than n; hence the value 0.25 does not apply to this problem and the correct value must be 1.25.

Second Method (Fig. 4B)

Analysis of Problem: Draw the radius r to some point a on the circumference of the disk. Let the coordinates of this point be x and y and the coordinates of the center of the circle be c and d. Then, according to analytic geometry, the general equation for this circle would be $(x - c)^2 + (y - d)^2 = r^2$. Since c and d are both radii of the disk, r can be substituted for them in this general equation which is now an expression for locating a point anywhere on the circumference of the disk in terms of x, y, and r. Take the specific point b where the disk is in contact with the block. The values of the coordinates x and y for point b are m and n, respectively. Since m and n are known, this equation can be used to find r.

Formula:

$$r = m + n \pm \sqrt{2mn}$$

Derivation of Formula:

$$(x - c)^2 + (y - d)^2 = r^2 \tag{1}$$

Substituting r for d and c

$$(x - r)^2 + (y - r)^2 = r^2 \tag{2}$$

Substituting m for x and n for y

$$(m - r)^2 + (n - r)^2 = r^2 \tag{3a}$$

Expanding Equation (3a)

$$m^2 - 2mr + r^2 + n^2 - 2nr + r^2 = r^2 \tag{3b}$$

Rearranging terms

$$r^2 - 2(m + n)r + m^2 + n^2 = 0 \tag{3c}$$

This equation is a quadratic in the form of that illustrated by the formula on page **2–20**.

Thus,

$$ar^2 + br + c = 0 \tag{6}$$

where $a = 1$; $b = -2(m + n)$; $c = m^2 + n^2$.

Solving in a similar manner as in Equations (6), (7), and (8) under the first method:

$$r = m + n \pm \sqrt{2mn}$$

The actual solution of a specific problem is exactly the same as that illustrated in the example under the first method.

Third Method (Fig. 4C)

Analysis of Problem: Draw the diagonal cb and extend it until it intersects the edge of the disk at a. Draw radius oa and line oc. Draw radius od to the point of tangency of the disk with the horizontal surface.

In right-angled triangle ocd, both od and dc are equal to r, hence this is an isosceles triangle and angle E equals 45 degrees. Angle C can be found, since in right-angled triangle bce, be and ec are known. Hence angle B can be found. Since od and dc are both equal to r, oc is equal to $r\sqrt{2}$. The relative lengths of oc and oa are now known so that with angle B known in obtuse-angled triangle aoc, angle D can be found.

According to the second geometrical proposition on page **3–9**, the product of ac and bc is equal to $\overline{dc}^2$. Since dc is equal to r, ac can be expressed in terms of r and bc. Using the Law of Sines for an obtuse-angled triangle, ac can be expressed in terms of ao and angles B and A. Since ao is a radius, r can be substituted for it. These two equivalents of ac can then be set equal to each other and an equation will be established with r as the only unknown.

Formula:

$$r = \frac{\sin A \sqrt{\overline{be}^2 + \overline{ec}^2}}{\sin B}$$

Derivation of Formula:

$$\tan C = \frac{be}{ce} \tag{1}$$

$$B = E - C \tag{2}$$

$$oc = \sqrt{\overline{od}^2 + \overline{dc}^2} \tag{3}$$

$$oc = \sqrt{r^2 + r^2} \tag{4a}$$

$$oc = r\sqrt{2} \tag{4b}$$

According to the Law of Sines, in obtuse-angled triangle oac:

$$\frac{oc}{\sin D} = \frac{oa}{\sin B} \tag{5a}$$

$$\sin D = \frac{oc \sin B}{oa} \tag{5b}$$

Substituting equivalents of oc and oa:

$$\sin D = \frac{r\sqrt{2} \sin B}{r} \tag{6}$$

$$\sin D = \sqrt{2} \sin B$$

$$A = 180° - B - D \tag{7}$$

According to the second geometrical proposition on page 3–9:

$$ac \times bc = \overline{dc}^2 \tag{8a}$$

$$ac = \frac{\overline{dc}^2}{bc} = \frac{r^2}{bc} \tag{8b}$$

but

$$bc = \sqrt{\overline{be}^2 + \overline{ec}^2} \tag{9}$$

hence
$$ac = \frac{r^2}{\sqrt{\overline{be}^2 + \overline{ec}^2}} \tag{10}$$

In obtuse angle triangle *oac*:

$$\frac{ac}{\sin A} = \frac{ao}{\sin B} \tag{11a}$$

$$ac = \frac{ao \sin A}{\sin B} = \frac{r \sin A}{\sin B} \tag{11b}$$

From Equations (10) and (11b)

$$\frac{r^2}{\sqrt{\overline{be}^2 + \overline{ec}^2}} = \frac{r \sin A}{\sin B} \tag{12a}$$

$$r^2 = \frac{r \sin A \times \sqrt{\overline{be}^2 + \overline{ec}^2}}{\sin B} \tag{12b}$$

$$r = \frac{\sin A \times \sqrt{\overline{be}^2 + \overline{ec}^2}}{\sin B} \tag{12c}$$

Example: In Fig. 4C let $ec = 0.5$ and $be = 0.25$. Find r.

Solution: $\tan C = \dfrac{0.25}{0.5} = 0.5$ (1)

$$C = 26°33.9'$$

$$B = 45° - 26°33.9' = 18°26.1' \tag{2}$$

$$\sin D = \sqrt{2} \sin B \tag{6}$$

$$\sin D = \sqrt{2} \sin 18°26.1'$$

$$\sin D = 1.4142 \times 0.31623 = 0.44721$$

$$D = 26°33.9'$$

$$A = 180° - 18°26.1' - 26°33.9' = 135° \tag{7}$$

$$r = \frac{0.70711 \times \sqrt{0.5^2 + 0.25^2}}{0.31623} \tag{12c}$$

$$r = \frac{0.70711 \times 0.559}{0.31623} = 1.25$$

Radius of Circle Tangent to Two Lines and Passing Through a Given Point

PROBLEM 5: To find radius *om*. In Fig. 5 is shown a circular disk which is in contact with a vertical and a horizontal surface. A gage block of known height *ab* is so located on the horizontal surface that its upper edge is in contact with the circumference of the disk as shown. The distance *bc* of the front edge of the gage block from the vertical surface is then measured. From these two known dimensions, *ab* and *bc*, it is required to determine the radius of the disk. Three methods of solution will be described.

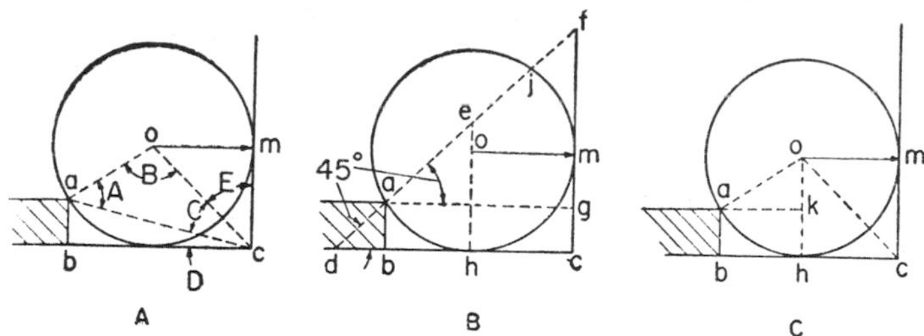

FIG. 5. To Find Radius *om* when Height *ab* and Distance *bc* are Known. (A) First Method. (B) Second Method. (C) Third Method

First Method (Fig. 5A)

Analysis of Problem: Draw *ao* from the point of contact of disk and block to the center of the disk. Draw *oc* from the center of the disk to the intersection of the horizontal and vertical surfaces. In right-angled triangle *abc*, *ab* and *bc* are known. Hence *ac* and angle *D* can be determined. In right-angled triangle *omc*, *om* and *mc* are equal, hence angle *E* equals 45 degrees. Therefore angles *C* plus *D* equal 45 degrees, and hence angle *C* can be determined. In right-angled triangle *omc*, the length of the hypotenuse *oc* is equal to $\sqrt{2}$ times either *om* or *mc* since these sides are equal. Since *om* is also equal to *oa*, the relative length of *oa* and *oc* can be established. Knowing this and the value of angle *C*, angle *A* can be determined by means of the Law of Sines. Angle *B* can now be found and again using this same formula with *ac* and angles *B* and *C* known, the radius *oa* can be determined.

Procedure:

$$\tan D = \frac{ab}{bc} \tag{1}$$

$$ac = ab \operatorname{cosec} D \tag{2}$$

$$oc = \sqrt{2om} = 1.414om$$

$$oc = 1.414oa \tag{3}$$

$$C = 45° - D \tag{4}$$

$$\frac{oc}{\sin A} = \frac{oa}{\sin C} \tag{5a}$$

$$\sin A = \frac{oc \sin C}{oa} \tag{5b}$$

Substituting for *oc*:

$$\sin A = \frac{1.414oa \times \sin C}{oa} = 1.414 \sin C \tag{6}$$

$$B = 180° - A - C \tag{7}$$

$$\frac{oa}{\sin C} = \frac{ac}{\sin B} \tag{8a}$$

$$oa = \frac{ac \sin C}{\sin B} \tag{8b}$$

Example: In Fig. 5A, let $ab = 0.25$ and $bc = 1.00$. Find radius *om*.

Solution:

$$\tan D = \frac{0.25}{1.00} = 0.25 \tag{1}$$

$$D = 14°2.2'$$

$$ac = 0.25 \times 4.1229 = 1.0307 \tag{2}$$

$$C = 45° - 14°2.2' = 30°57.8' \tag{4}$$

$$\sin A = 1.414 \times 0.51449 = 0.72749 \tag{6}$$

$$A = 46°40.6'$$

$$B = 180° - 46°40.6' - 30°57.8' = 102°21.6' \tag{7}$$

$$oa = om = \frac{1.0307 \times 0.51449}{0.97682} = 0.543 \tag{8b}$$

Second Method (Fig. 5B)

Analysis of Problem: Through point *a* draw *df* at an angle of 45 degrees with the horizontal surface. Draw the vertical line *eh* through the center of the disk *o*. It can be shown that if a circle is tangent to the two sides of an isosceles right-angled triangle the two parts of the hypotenuse lying outside of the circle will be equal. Hence *ad* equals *jf* and *dj* equals *af*. In right-angled triangle *afg*, *af* can be found since *ag* is equal to *bc* which is known. Similarly, in right-angled triangle *dba*, *da* can be found since *ab* is known. Then making use of Geometrical Proposition 2 on page 3–9 an expression can be set up for *dj* in terms of *dh* and *da*. With *dc* and *dh* known, *hc*, which is equal to the radius of the disk, can be found.

Formula:

$$om = ab + bc - \sqrt{2bc \times ab}$$

Derivation of Formula:

$$af = ag \sec 45° = 1.4142ag \tag{1}$$

$$dj = af = 1.4142ag = 1.4142bc \tag{2}$$

$$da = ab \operatorname{cosec} 45° = 1.4142ab \tag{3}$$

According to Geometrical Proposition 2, page 3–9:

$$dj \times da = dh^2 \tag{4}$$

Substituting equivalents of *dj* and *da*:

$$1.4142bc \times 1.4142ab = dh^2 \tag{5}$$

or $$dh = \sqrt{1.4142bc \times 1.4142ab}$$

but $$hc = dc - dh \tag{6}$$

and $$dc = db + bc = ab + bc \tag{7}$$

hence $$hc = ab + bc - \sqrt{1.4142bc \times 1.4142ab} \tag{8}$$

$$om = hc = ab + bc - \sqrt{2bc \times ab} \tag{9}$$

Example: In Fig. 5B let $ab = 0.25$ and $bc = 1.00$. Find *om*.

Solution: $$om = 0.25 + 1.00 - \sqrt{2 \times 1.00 \times 0.25} \tag{9}$$

$$om = 1.25 - \sqrt{0.50} = 1.25 - 0.707 = 0.543$$

Third Method (Fig. 5C)

Analysis of Problem: This method shows how the problem can be solved entirely by algebra, without the use of trigonometrical formulas or tables.

Draw *oc* from the center of the disk to the intersection of the horizontal and vertical surfaces, and *oa* from the center of the disk to the point of contact of disk and gage block. Draw the vertical line *oh* and the horizontal line *ak*.

Then *ako* is a right-angled triangle and the familiar "sum of the squares of the sides is equal to the square of the hypotenuse" theorem can be used to establish an equation for *ao*, *ak* and *ko*. It will be noted from Fig. 5C, that *ak* can be expressed in terms of known distance *bc* and the radius of the disk and also that *ko* can be expressed in terms of known distance *ab* and the radius of the disk. When these substitutions are made in the equation for *ao*, *ak* and *ko*, a quadratic equation in terms of the radius of the disk and known values *ab* and *bc* is established which can be solved by means of the quadratic formula for roots on page **2–21**.

Formula:

$$\overline{om}^2 - (2bc + 2ab)om + \overline{bc}^2 + \overline{ab}^2 = 0$$

To be solved as a quadratic for *om*.

Derivation of Formula: In right-angled triangle *ako*:

$$\overline{oa}^2 = \overline{ak}^2 + \overline{ok}^2 \qquad (1)$$

but

$$ak = bh = bc - hc \qquad (2)$$

and

$$ok = oh - kh = oh - ab \qquad (3)$$

Substituting these equivalents in Equation (1):

$$\overline{oa}^2 = (bc - hc)^2 + (oh - ab)^2 \qquad (4)$$

$$om = oa = oh = hc \qquad (5)$$

Substituting *om* in Equation (4)

$$\overline{om}^2 = (bc - om)^2 + (om - ab)^2 \qquad (6a)$$

$$\overline{om}^2 = \overline{bc}^2 - 2bc \times om + \overline{om}^2 + \overline{om}^2 - 2ab \times om + \overline{ab}^2 \qquad (6b)$$

$$\overline{om}^2 - (2bc + 2ab)om + \overline{bc}^2 + \overline{ab}^2 = 0 \qquad (6c)$$

This is a quadratic equation in terms of $\overline{om}$ which can be solved by the quadratic formula for roots on page **2–21**.

Example: In Fig. 5C let $ab = 0.25$ and $bc = 1.00$. Find om.

Solution: $om^2 - (2 \times 1 + 2 \times 0.25)om + 1^2 + 0.25^2 = 0$

This is a quadratic equation in the form (see page 2–20)

$$a(om)^2 + b(om) + c = 0$$

in which $a = 1$; $b = -2.5$; and $c = 1.0625$

Since $$om = \frac{-b \pm \sqrt{b^2 - 4ac}}{2a}$$

then $$om = \frac{2.5 \pm \sqrt{6.25 - 4 \times 1 \times 1.0625}}{2 \times 1}$$

$$om = \frac{2.5 \pm \sqrt{2.00}}{2} = \frac{3.914}{2} \quad \text{or} \quad \frac{1.086}{2}$$

$$om = 1.957 \quad \text{or} \quad 0.543$$

Since, as can be seen from Fig. 5C, om must be less than bc, the value 1.957 is not applicable to this problem. Therefore $om = 0.543$.

Radius of Circle which will Pass Through Two Given Points and be Tangent to Given Line

PROBLEM 6: To find radius *of* in Fig. 6. Given two lines *am* and *aw* at a known angle *A* to each other, the problem is to construct a circle which will be tangent to *am* and which will pass through points *c* and *e* on *aw*. The vertical distances *de* and *bc* are known. The problem is, then, to find the radius *of* of the desired circle.

Analysis of Problem: Draw the two radii *oc* and *of* and the chord *cf*. Draw *og* perpendicular to *cf*. Since in right-angled triangle *ade* angle *A* and *ed* are known, *ae* can be found. In right-

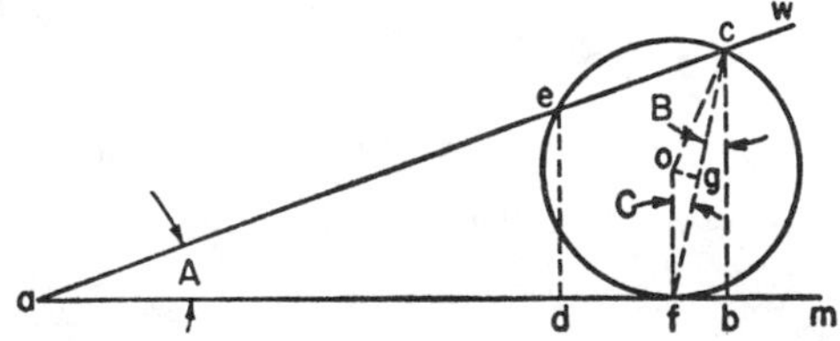

FIG. 6. To Find Radius *of* when Distances *de* and *bc* and Angle *A* are Known

angled triangle *abc*, angle A and bc are known so that ac and ab can be found. Then according to Geometrical Proposition 2 on page 3–9, af can be found. With af and ab known, fb is determined. Then, in right-angled triangle *bcf*, fb and bc are known so that angle B and cf can be determined. According to the fifth geometrical proposition on page 3–6, angles B and C are equal. If og is at right angles to chord cf it bisects this chord according to Geometrical Proposition 4 on page 3–7. Hence, with fg and angle C known in right-angled triangle *ofg*, of can be determined.

With the radius of the circle known, its center can be located by striking two arcs having this radius from points e and c as centers. The point of intersection of these two arcs is the center of the desired circle.

Procedure:

$$ae = ed \operatorname{cosec} A \tag{1}$$

$$ac = bc \operatorname{cosec} A \tag{2}$$

$$ab = bc \cot A \tag{3}$$

$$ae \times ac = \overline{af}^2 \text{ (Geometrical Prop. 2, page 3–9)} \tag{4}$$

$$af = \sqrt{ae \times ac} \tag{5}$$

$$bf = ab - af \tag{6}$$

$$\tan B = \frac{bf}{bc} \tag{7}$$

$$C = B \tag{8}$$

$$cf = bc \sec B \tag{9}$$

$$of = gf \sec C \tag{10}$$

$$2gf = cf \text{ (Geometrical Prop. 4, page 3–7)} \tag{11}$$

$$of = \frac{cf}{2} \sec C \tag{12}$$

Example: Let $A = 23°$; $de = 3.8$; $bc = 5.1$. Find of.

Solution:

$$ae = 3.8 \times 2.5593 = 9.73 \tag{1}$$

$$ac = 5.1 \times 2.5593 = 13.05 \tag{2}$$

$$ab = 5.1 \times 2.3558 = 12.01 \tag{3}$$

$$af = \sqrt{9.73 \times 13.05} = \sqrt{126.98} = 11.27 \quad (4)$$

$$bf = 12.01 - 11.27 = 0.74 \quad (5)$$

$$\tan B = \frac{0.74}{5.1} = 0.14510 \quad (7)$$

$$B = 8°15.4'$$

$$C = 8°15.4' \quad (8)$$

$$cf = 5.1 \times 1.0104 = 5.15 \quad (9)$$

$$of = \frac{5.15}{2} \times 1.0104 = 2.60 \quad (12)$$

Finding Center of Arc Tangent to a Given Line and Passing Through a Given Point

PROBLEM 7: In views A and B of Fig. 7 are shown a line *bd* drawn through point *d*, which is at a known vertical distance from point *o*, and at a known angle *D* with the horizontal; and point *a* which is located at known horizontal distance *oa* from point *o*. It is required to find the location of the center *p* of an arc of given radius *pe*, which will pass through point *a* and be tangent to line *bd*.

Analysis of Problem: The location of point *p* is established if *ip* and *oi* can be determined. There are two methods of procedure.

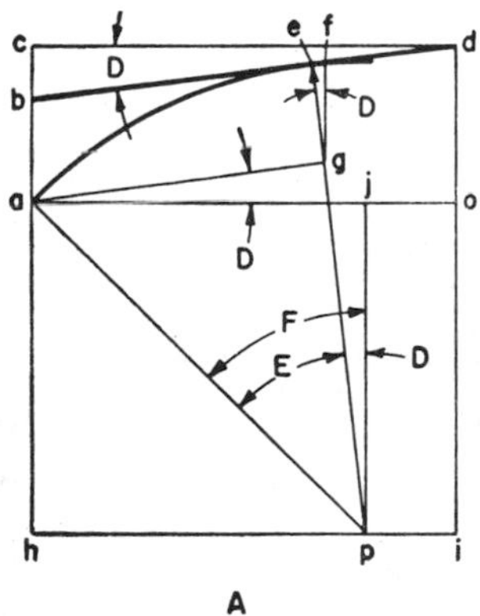

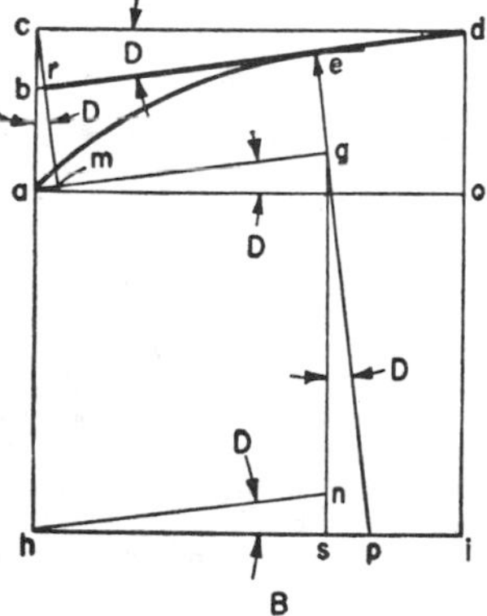

FIG. 7. To Locate Center *p* when Distances *od* and *cd*, Radius *pe*, and Angle *D* are Known. (A) By First Method. (B) By Second Method

By one method, referring to Fig. 7A, if angle F can be found, then jp and ja can be determined, since ap is a radius of the arc and is, therefore, known. With ja and jp known, oi is known and ip can be found directly.

By the second method, referring to Fig. 7B, if rm and ag can be found, then hn is known and sp and ns can be determined. Both oi and ip can then be found.

Both methods will now be worked out step by step.

Formulas:

Method 1.

$$ip = cd - pe \sin F$$

$$oi = pe \cos F$$

where $$F = D + \text{arc} \cos \left(\frac{pe - od \cos D + cd \sin D}{pe} \right)$$

Method 2.

$$ip = cd - \sqrt{eg(2pe - eg)} \cos D - (pe - eg) \sin D$$

$$oi = (pe - eg) \cos D - \sqrt{eg(2pe - eg)} \sin D$$

where $$eg = od \cos D - cd \sin D$$

Derivation of Formulas: *Method* 1. Let construction lines be drawn as shown in Fig. 7A.

$$bc = cd \tan D \tag{1}$$

$$ab = ac - bc \tag{2}$$

hence $$ab = ac - cd \tan D \tag{3}$$

$$ge = gf \cos D \tag{4}$$

but $$gf = ab \tag{5}$$

hence $$ge = ab \cos D = (ac - cd \tan D) \cos D \tag{6a}$$

or $$ge = ac \cos D - cd \sin D \tag{6b}$$

$$pg = pe - ge \tag{7}$$

hence $$pg = pe - ac \cos D + cd \sin D \tag{8}$$

but $$ac = od \tag{9}$$

therefore $$pg = pe - od \cos D + cd \sin D \tag{10}$$

$$\cos E = \frac{pg}{pa} = \frac{pg}{pe} = \frac{pe - od \cos D + cd \sin D}{pe} \tag{11}$$

$$hp = aj = pa \sin F = pe \sin F \tag{12}$$

$$ip = hi - hp = cd - pe \sin F \tag{13}$$

$$oi = jp = pa \cos F = pe \cos F \tag{14}$$

where

$$F = D + E = D + \text{arc cos} \left(\frac{pe - od \cos D + cd \sin D}{pe} \right) \tag{15}$$

Method 2. Let construction lines be drawn as in Fig. 7B.

$$cr = cd \sin D \tag{1}$$

$$cm = ca \cos D = od \cos D \tag{2}$$

$$rm = cm - cr \tag{3}$$

hence

$$rm = od \cos D - cd \sin D \tag{4}$$

$$eg = rm = od \cos D - cd \sin D \tag{5}$$

The formula for one-half the length of the chord of a circular segment in terms of the altitude and radius, if ag = one-half the length of the chord; eg = the altitude, and pe = the radius, is:

$$ag = \sqrt{eg(2pe - eg)} \tag{6}$$

but

$$hn = ag = \sqrt{eg(2pe - eg)} \tag{7}$$

$$ip = hi - hp = cd - hp \tag{8}$$

and

$$hp = hs + sp \tag{9}$$

hence

$$ip = hi - hs - sp = cd - hs - sp \tag{10}$$

but

$$hs = hn \cos D \tag{11}$$

therefore

$$hs = \sqrt{eg(2pe - eg)} \cos D \tag{12}$$

$$sp = (pe - eg) \sin D \tag{13}$$

so that

$$ip = cd - \sqrt{eg(2pe - eg)} \cos D - (pe - eg) \sin D \tag{14}$$

$$oi = ah = gs - ns \tag{15}$$

but $$gs = (pe - eg)\cos D \tag{16}$$

and $$ns = hn \sin D = \sqrt{eg(2pe - eg)}\,\sin D \tag{17}$$

hence $$oi = (pe - eg)\cos D - \sqrt{eg(2pe - eg)}\,\sin D \tag{18}$$

Example: If the vertical distance of point *d* from *o* is $2\frac{11}{16}$ inches; the angle *D* which line *bd* makes with the horizontal is 10 degrees; the horizontal distance of point *a* from *o* is 7 inches; and the radius *pe* of the arc is 8 inches, find the location of the center of the arc with reference to point *o*.

$$od = 2.6875; \quad cd = 7; \quad pe = 8; \quad D = 10°$$

Method 1. (Fig. 7A)

$$F = 10° + \text{arc cos}\left(\frac{8 - 2.6875 \cos 10° + 7 \sin 10°}{8}\right) \tag{15}$$

$$\text{arc cos}\left(\frac{8 - 2.6875 \times 0.98481 + 7 \times 0.17365}{8}\right) = \frac{6.56887}{8}$$

$$\text{arc cos}\,\frac{6.56887}{8} = \text{arc cos}\,0.82111 = 34°48.24'$$

hence $$F = 10° + 34°48.24' = 44°48.24'$$

$$ip = 7 - 8 \sin 44°48.24' \tag{13}$$

$$ip = 7 - 8 \times 0.70468 = 1.3626 \text{ inches}$$

$$oi = 8 \times \cos 44°48.24' \tag{14}$$

$$oi = 8 \times 0.70952 = 5.6762 \text{ inches}$$

Method 2. (Fig. 7B)

$$eg = 2.6875 \times 0.98481 - 7 \times 0.17365 = 1.43113 \tag{5}$$

$$ip = 7 - (\sqrt{20.84995} \times 0.98481) - (6.56887 \times 0.17365) \tag{14}$$

$$ip = 7 - 4.49681 - 1.14068 = 1.3625 \text{ inches}$$

(Note slight difference from value obtained by Method 1 due to use of five-place trigonometrical tables.)

$$oi = (6.56887 \times 0.98481) - \sqrt{20.84995} \times 0.17365 \tag{18}$$

$$oi = 6.46909 - 0.79292 = 5.6762 \text{ inches}$$

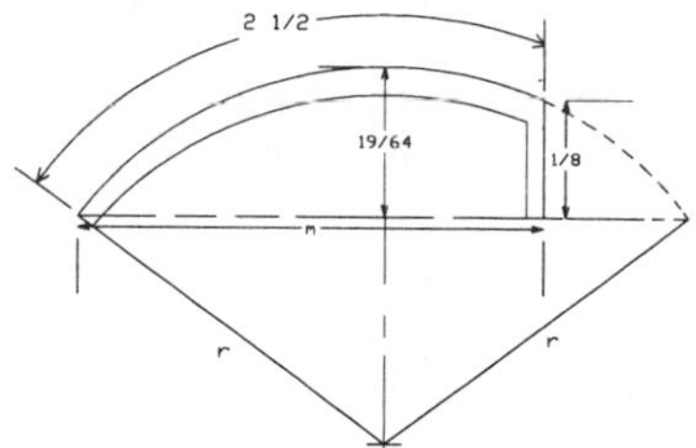

Calculating the Dimensions of a Spring

Problem 8: In the design of the flat spring shown in Fig. 8, it was required to find the dimensions r and m. From the figure we obtain the following relationships:

$$r - r\cos\alpha = \frac{19}{64} \quad (1) \qquad r - r\cos\theta = \frac{11}{64} \quad (2) \qquad r(\alpha + \theta) = 2.5 \quad (3)$$

Dividing Equation (3) by Equation (1) gives

$$\frac{\alpha + \theta}{1 - \cos\alpha} = \frac{160}{19} \text{ and } 1 - \cos\alpha = 2\sin^2\frac{\alpha}{2}, \text{ then}$$

$$\alpha + \theta = \frac{320}{19}\sin^2\frac{\alpha}{2} \qquad (4)$$

Dividing Equation (2) by Equation (1) gives

$$\frac{1 - \cos\theta}{1 - \cos\alpha} = \frac{11}{19} = \frac{2\sin^2\frac{\theta}{2}}{2\sin^2\frac{\alpha}{2}}, \quad \text{or,} \quad \sin\frac{\theta}{2} = \sqrt{\frac{11}{19}}\sin\frac{\alpha}{2}$$

let $\quad K = \sqrt{\frac{11}{19}}$, then $\theta = 2\sin^{-1}(k\sin\frac{\alpha}{2})$

From Equation (4) we get $\alpha + 2\sin^{-1}(k\sin\frac{\alpha}{2}) = \frac{320}{19}\sin^2\frac{\alpha}{2}$ (5)

let $\quad x = \sin\frac{\alpha}{2}$, then $\sin^{-1}x + \sin^{-1}kx = \frac{160}{19}x^2$ (6)

Equation (6) may be solved by the method of successive approximations:

When x is small, $\sin^{-1} x = x + \frac{x^3}{6} + \frac{3x^5}{40} \ldots$

When x is small we may use $\sin^{-1} = x$ and $\sin^{-1} kx = kx$. Then, from

Equation (6) we get $1\,k = \frac{160}{19}x.\ \ k = \sqrt{\frac{11}{19}} = 0.76$

Then $\quad 1 + \frac{76}{100} = \frac{160}{19}x.\ x = \frac{176 \times 19}{100 \times 160} = \frac{20}{100}.\ \ x = 0.209$

or $\quad \sin\frac{\alpha}{2} = x = 0.209,\ \frac{\alpha}{2} = 12°\ 3'\ 50''$ and $\alpha = 24°\ 7'\ 40''$

$$\sin\frac{\theta}{2} = \sqrt{\frac{11}{19}}\sin\frac{\alpha}{2} = 0.76 \times 0.209$$

$$\sin\frac{\theta}{2} = 0.15884 \quad \frac{\theta}{2} = 9°\ 8'\ 23''$$

$\theta = 18°\ 16'\ 46''$. $\alpha = 24°\ 7'\ 40''$. $\alpha + \theta = 42°\ 24'\ 26''$

From (3) $r = \frac{2.5}{\alpha + \theta} = \frac{2.5}{42°\ 24'\ 26''} = \frac{2.5}{0.7401146} = 3.378$

$$b = r\sin\alpha = 3.3777 \sin 24°\ 7'\ 40'' = 1.3807$$

$$a = r\sin\theta = 3.3777 \sin 18°\ 16'\ 46'' = 1.0594$$

$$m = a + b = 2.4401$$

The above is an approximate solution and could be more accurate. Equation 5 is exact and can be used to evaluate these results. When $\frac{\alpha}{2} = 0.209$ Equation (5) yields 0.737408 > 0.735680. By trial and error, a better approximation is $\sin\frac{\alpha}{2}$ = 0.2104 and Equation (5) yields 0.7455344 > 0.7455688.

This answer is definitely more accurate and yields the following results: $\alpha = 0.423968$ radians $= 24.29$ degrees

$\theta = 0.321564$ radians $= 18.42$ degrees

$r = 3.353$

$b = 1.379$

$a = 1.060$

$m = 2.439$

The corrections in a, b, and m are small, but the correction in r is 0.7 per cent, which is significant.

Radius of Cutter to Produce a Convex or Concave Surface Without Undercutting

Problem 9: Cams and other components having convex or concave curves are often produced by numerically controlled machinery, tracer-type duplicating equipment, or other, generating-type machinery.

In the case of cams which transmit motion utilizing a roller follower, undercutting of the cam profile at convex and concave points on the cam profile can occur if the cutter radius does not bear a certain relationship to the radius of the cam follower roller and to the radius of curvature at the point on the cam being considered. When undercutting occurs, the cam will not transmit to the roller follower the intended movements. In figures 9(a), 9(b), and 9(c) are shown the three possible conditions resulting from the cutting of a convex curve on a cam or other product.

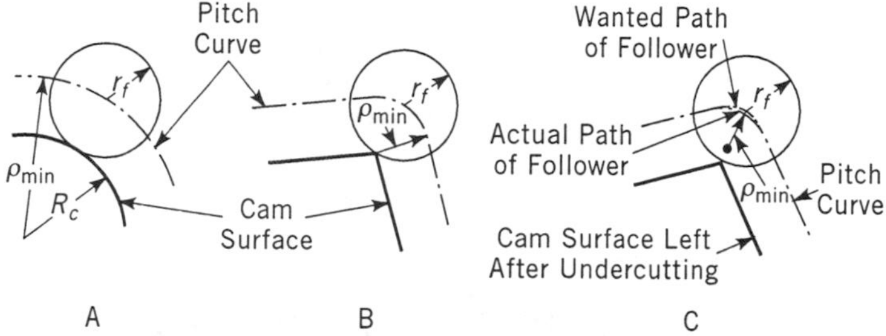

Radius of Curvature.— The minimum radius of curvature of a cam should be kept as large as possible (1) to prevent undercutting of the convex portion of the cam and (2) to prevent too high surface stresses. Figure 9 shows how undercutting occurs and how it can be prevented.

In Fig. 9a, the radius of curvature of the path of the follower is ρ_{min} and the cam will at that point have a radius of curvature $R_c = \rho_{min} - r_f$.

In Fig. 9b, $\rho_{min} = r_f$ and $R_c = 0$. Therefore, the actual cam will have a sharp corner which, in most cases, will result in too high surface stresses.

In Fig. 9c, is shown the case where $\rho_{min} < r_f$. This case is not possible because undercutting will occur and the actual motion of the roller follower will deviate from the desired one as shown.

Undercutting cannot occur at the *concave* portion of a cam profile (working surface), but caution should be exerted in not making the radius of curvature equal to the radius of the roller follower. This condition would occur if there is a cusp on the displacement diagram, which of course, should be avoided. To enable milling or grinding of *concave* portions of the cam profile, the radius of curvature of concave portions of the cam, $R_c = \rho_{min} + r_f$, must be larger than the radius of the cutter to be used.

14

Vees and Arcs—Checking with Cylinders and Spheres

One of the most accurate methods of checking the angle or width of a vee or conical opening, or the radius or diameter of an arc or circular surface, is by measurement over cylinders or spheres placed in contact with the surface of the vee or arc to be checked. The cylinder or sphere will make line or point contact with the sides of the vee or arc and the location of these lines or points of contact can be determined mathematically.

The problems which follow illustrate some of the ways in which cylinders and spheres are used for this type of checking and show in detail how the desired formula or method of procedure is worked out and applied.

Checking Width of V-shaped Slot by Measurement Over Round Pin

PROBLEM 1: To find D in Fig. 1. One method of checking the width of a V-shaped slot or a conical opening is to place a pin of sufficient size in the V (a ball is used for a conical opening) so that it protrudes above the surface. Then by measuring the vertical distance ef from the horizontal surface at the edge of the V-shaped slot or conical opening to the top of the pin or ball, the corresponding width of slot can be computed and checked to see if it is within allowable limits. To use this method of checking, however, the angle of the V or cone, the diameter of pin or ball and the measured height ef must be known very accurately. It is also required that the center line of the V or cone be at right angles to the horizontal surface.

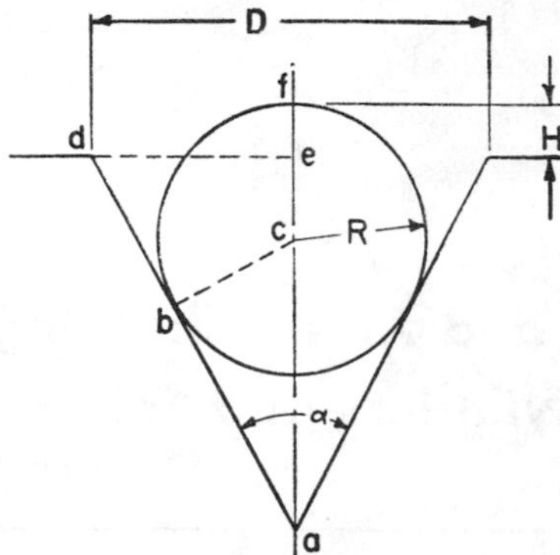

FIG. 1. To Find Diameter D when Distance ef and Radius R of Pin are Known

Analysis of Problem: Draw cb from the center of the pin to the point of tangency with the side of the block. According to the fifth geometrical proposition on page 3–7, angle cba is a right angle, and in the right-angled triangle abc, angle bac and side bc are known, hence ac can be found. If the center line through the pin or ball that passes through the intersection of the sides of the V (or the apex of the cone) is at right angles to the horizontal surface, then ade is a right-angled triangle and an expression for de, which is equal to $\frac{D}{2}$, in terms of ae and angle $\frac{\alpha}{2}$ can be found. But ae equals ac plus cf minus ef, all of which are now known, hence $\frac{D}{2}$ can be computed. The computed value of D is then compared with the specified value.

Formula: $D = 2 \tan \frac{\alpha}{2}\left(R \csc \frac{\alpha}{2} + R - H\right)$

Derivation of Formula:

In right-angle triangle abc

$$ac = bc \csc \frac{\alpha}{2} \tag{1}$$

In right-angle triangle ade

$$ae = de \cot \frac{\alpha}{2} \tag{2}$$

but $$ef = ac + cf - ae \tag{3}$$

therefore $$ef = bc \csc \frac{\alpha}{2} + cf - de \cot \frac{\alpha}{2} \tag{4}$$

since $$ef = H;\ bc = R;\ cf = R \text{ and } de = \frac{D}{2}$$

then $$H = R \csc \frac{\alpha}{2} + R - \frac{D}{2} \cot \frac{\alpha}{2} \tag{5a}$$

Transposing terms

$$\frac{D}{2} \cot \frac{\alpha}{2} = R \csc \frac{\alpha}{2} + R - H \tag{5b}$$

$$\frac{D}{2} = \frac{R \csc \frac{\alpha}{2} + R - H}{\cot \frac{\alpha}{2}} \tag{5c}$$

$$D = 2\left(R \csc \frac{\alpha}{2} + R - H\right) \tan \frac{\alpha}{2} \tag{5d}$$

Example: Specifications call for a conical hole having a diameter D at the widest point of 2.735 ± 0.002 inches. In checking the hole with a ball of 1.8750 diameter, the height H measures 0.3739 inch. The angle α of the hole being checked is known to be 56°30′. Is the diameter D within the specified limits?

Solution:

$H = 0.3739$

$\alpha = 56°30'$

$R = 0.9375$

$D = 2 \tan 28°15' \ (0.9375 \csc 28°15' + 0.9375 - 0.3739)$ (5d)

$D = 2 \times 0.53732 \ (0.9375 \times 2.1127 + 0.9375 - 0.3739)$

$D = 1.0746 \ (1.9807 + 0.9375 - 0.3739)$

$D = 2.734$ inches

Hence, the hole being checked is within the allowable limits of 2.735 ± 0.002 inches.

Checking Location of Round Workpiece in V-block

PROBLEM 2: To find height h in Fig. 2. Using a standard plug gage of radius R, it is required to check a work-holding V-block to determine at what height h the center of a workpiece of known radius r will be located.

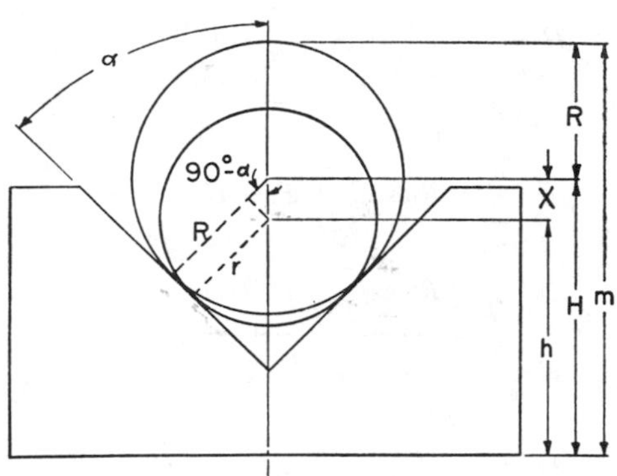

FIG. 2. To Find Height h when Radii R and r are Known

Analysis of Problem: With the standard plug gage seated in the V-block, the overall height m can be measured by means of a height gage and, knowing R, then H can be readily found. Next, the vertical distance X between the center of the workpiece when in position and the center of the plug gage when in position can be computed since X is the hypotenuse of a right-angled triangle in which one side and one angle are known.

Formula: $h = m - R - (R - r) \csc \alpha$

Derivation of Formula:

$$H = m - R \quad (1)$$

$$h = H - X = m - R - X \quad (2)$$

$$X = (R - r) \sec (90° - \alpha) \quad (3)$$

but $$\sec (90° - \alpha) = \csc \alpha \quad (4)$$

therefore $$X = (R - r) \csc \alpha \quad (5)$$

and substituting this equivalent of X in Equation (2):

$$h = m - R - (R - r) \csc \alpha \qquad (6)$$

Example: Given a V-block designed to hold a workpiece of 1.064 inch diameter, to determine height h if, with a plug gage of 1.3125 inch diameter the overall height m measures 2.0391 inches. The half-angle α of the V-block equals 45 degrees.

Solution: $m = 2.0391$; $r = 0.532$; $R = 0.6562$; $\alpha = 45°$

$$h = 2.0391 - 0.6562 - (0.6562 - 0.532)\ 1.4142 \qquad (6)$$

$$h = 2.0391 - 0.6562 - 0.1756$$

$$h = 1.2073 \text{ inches}$$

Radius of Round Plug for Checking Taper Gage

PROBLEM 3: To find radius r in Fig. 3. A round plug when placed in a tapered gage is tangent to three sides as shown. The width a of the bottom of the gage and the angle A, formed by the tapered side and a vertical line, are known. It is required to find the radius r of the plug.

Analysis of Problem: Draw a line from the corner of the gage, formed by the tapered side and the bottom of the gage, to the center of the disk. Draw another line from the center of the disk to the

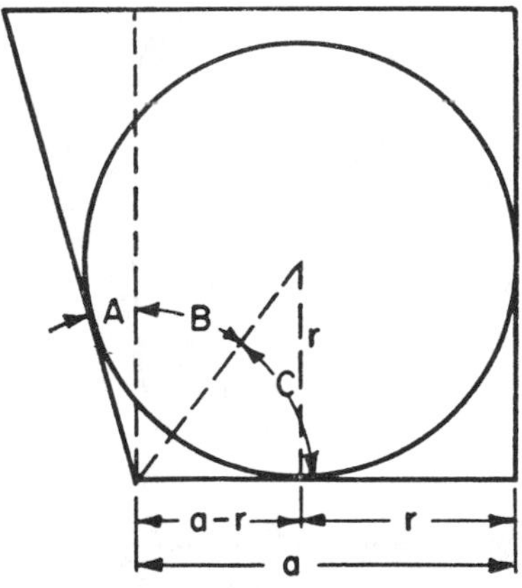

FIG. 3. To Find Radius r when Width a and Angle A are Known

point of tangency with the bottom of the gage. Angle $(A + B)$ equals angle C since it can be shown that a line drawn from the center of a circle to the point of intersection of two tangents to the circle bisects the angle formed by the two tangents. But angle $(B + C)$ is a right angle and angle A is known. Hence angle C can be found. An expression for the tangent of angle C in terms of r and a can now be set up and since a is known, r can be found.

Formula:

$$r = \frac{a \tan\left(45° + \frac{A}{2}\right)}{1 + \tan\left(45° + \frac{A}{2}\right)}$$

Derivation of Formula:

$$A + B + C = 90° + A \quad (1)$$

$$C = A + B \quad (2)$$

substituting for $A + B$ in (1)

$$C + C = 90° + A \quad (3a)$$

$$C = \frac{90° + A}{2} \quad (3b)$$

$$\tan C = \tan \frac{90° + A}{2} = \frac{r}{a - r} \quad (4a)$$

$$(a - r) \tan\left(45° + \frac{A}{2}\right) = r \quad (4b)$$

$$a \tan\left(45° + \frac{A}{2}\right) - r \tan\left(45° + \frac{A}{2}\right) = r \quad (4c)$$

$$a \tan\left(45° + \frac{A}{2}\right) = r + r \tan\left(45° + \frac{A}{2}\right) \quad (4d)$$

$$a \tan\left(45° + \frac{A}{2}\right) = r\left[1 + \tan\left(45° + \frac{A}{2}\right)\right] \quad (4e)$$

$$r = \frac{a \tan\left(45° + \frac{A}{2}\right)}{1 + \tan\left(45° + \frac{A}{2}\right)} \quad (4f)$$

Example: Let $a = 2$ and $A = 15°$. Find r.

Solution:

$$r = \frac{2 \tan\left(45° + \frac{15°}{2}\right)}{1 + \tan\left(45° + \frac{15°}{2}\right)} \tag{4f}$$

$$r = \frac{2 \tan 52\frac{1}{2}°}{1 + \tan 52\frac{1}{2}°} = \frac{2 \times 1.3032}{1 + 1.3032}$$

$$r = 1.1316$$

Measuring Sharp V-Thread Over Three Wires

PROBLEM 4: To find M in Fig. 4 for a sharp V-thread in which the thread angle β, the major diameter D, and the full height of the thread H are known. The measurement M is made over three pins of equal and known radius R, placed in the thread grooves as shown.

Analysis of Problem: If the lead angle of the thread is so small that it can be ignored, or, in other words, the thread groove is considered to be annular, and if the sides of the thread are considered to be straight, as shown in Fig. 4, the following analysis is valid.

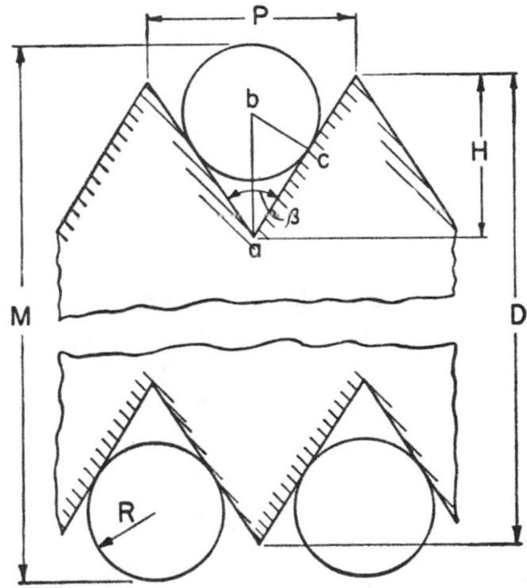

FIG. 4. To Check Measurement M when Major Diameter D and Height H are Known

It will be seen from Fig. 4 that the measurement M is equal to the major diameter D minus twice the full thread height H plus twice the distance ab from the bottom of the V to the center of the pin plus twice the radius of the pin. If radius bc is drawn to the point of tangency of the pin with the side of the thread, it forms a right angle (fifth geometrical proposition on page 3–7) and in the right-angled triangle abc, one side and one angle are known, hence ab can be found and M can be computed.

The resulting formula will be approximate, since the lead angle has not been taken into account, but it is sufficiently accurate, however, to check single-thread screws having a sharp V-thread unless exceptional accuracy is required.

(*Note:* For a discussion of various formulas for finding the measurement M over wires for screw threads having an appreciable lead angle, see Chapter 16, Approximate Formulas.)

Formula:

$$M = D - 2H + 2R\left(\csc\frac{\beta}{2} + 1\right)$$

Derivation of Formula:

$$M = D - 2H + 2ab + 2R \qquad (1)$$

$$\text{angle } bac = \frac{\beta}{2} \qquad (2)$$

$$ab = bc \csc\frac{\beta}{2} = R \csc\frac{\beta}{2} \qquad (3)$$

$$M = D - 2H + 2R \csc\frac{\beta}{2} + 2R \qquad (4)$$

Example: A sharp V-thread has 7 threads per inch, a major diameter of $1\frac{1}{4}$ inches, a full depth of thread of 0.1237 inch and a thread angle of 60 degrees. If three wires having a diameter of 0.090 inch are used, what will be the measurement M over these wires?

Solution:

$$M = 1.25 - 2 \times 0.1237 + 2 \times 0.0450 \csc\frac{60°}{2} + 2 \times 0.0450 \quad (4)$$

$$M = 1.25 - 0.2474 + 0.09 \times 2.0000 + 0.09$$

$$M = 1.2726 \text{ inches}$$

Checking V-shaped Groove by Measurement over Pins

PROBLEM 5: To find x in Fig. 5. Given a forming tool of the shape shown, it is required to compute the over-all measurement x over pins of radius r, if dimensions a and r and angles α and β are known.

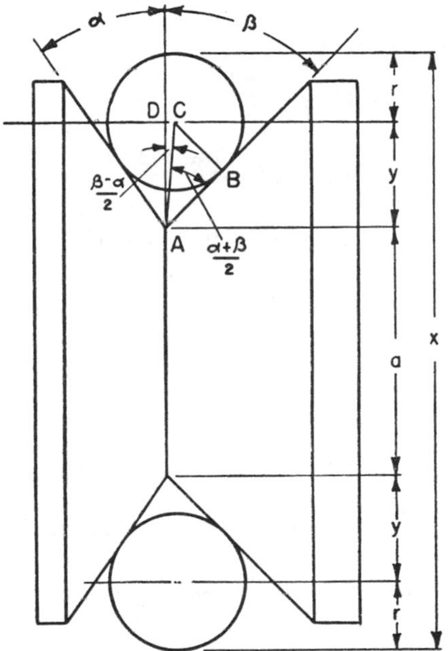

FIG. 5. To Check Measurement x when Radius r, Distance a and Angles α and β are Known

Analysis of Problem: Since a and r are known, the problem is to determine y. If AC is a line drawn from the bottom of the V to the center of the upper pin and CB is a line drawn from the center of this pin to its point of tangency with the side of the V, then a right-angled triangle is formed in which one side, CB, is known and one angle, CAB, can be determined. A line drawn from the center of a circle to the point of intersection of two tangents to the circle bisects the angle made by them. Hence, angle $CAB = \frac{\alpha + \beta}{2}$. Therefore AC can be found. Angle DAC can now be found and with AC known in right-angled triangle ADC, AD, which is equal to y can be found.

Formula: $x = a + 2r\left(\csc\frac{\alpha + \beta}{2}\cos\frac{\beta - \alpha}{2} + 1\right)$

Derivation of Formula:

$$\angle CAB = \frac{\alpha + \beta}{2} \tag{1}$$

$$AC = CB\csc\frac{\alpha + \beta}{2} \tag{2}$$

$$\angle DAC = \angle DAB - \angle CAB \tag{3}$$

$$\angle DAC = \beta - \frac{\alpha + \beta}{2} \tag{4a}$$

$$= \frac{2\beta - \alpha - \beta}{2} = \frac{\beta - \alpha}{2} \tag{4b}$$

$$AD = AC\cos\frac{\beta - \alpha}{2} \tag{5a}$$

Substituting the equivalent of AC given in Equation (2):

$$AD = CB\csc\frac{\alpha + \beta}{2}\cos\frac{\beta - \alpha}{2} \tag{5b}$$

but $\quad y = AD \quad$ and $\quad r = CB$

therefore

$$y = r\csc\frac{\alpha + \beta}{2}\cos\frac{\beta - \alpha}{2} \tag{6}$$

Since

$$x = a + 2y + 2r \tag{7}$$

$$x = a + 2r\csc\frac{\alpha + \beta}{2}\cos\frac{\beta - \alpha}{2} + 2r \tag{8a}$$

$$x = a + 2r\left(\csc\frac{\alpha + \beta}{2}\cos\frac{\beta - \alpha}{2} + 1\right) \tag{8b}$$

Example: Let $r = 0.500$; $a = 1.824$; $\alpha = 35°$; and $\beta = 45°$. Find x.

Solution:

$$x = 1.824 + 1\left(\csc\frac{35° + 45°}{2}\cos\frac{45° - 35°}{2} + 1\right) \tag{8b}$$

$$x = 1.824 + \csc 40° \cos 5° + 1$$

$$x = 1.824 + 1.5557 \times 0.99619 + 1$$

$$x = 1.824 + 1.550 + 1 = 4.374$$

Radius of Roll for Checking Radius of a Given Arc

PROBLEM 6: To find radius r in Fig. 6. In making a gage having an arc of given radius a and height of arc c, it was required to determine the size of a roll or disk, for checking the curvature of this arc, that would be tangent to extensions of the bottom and end of the gage as well as the arc.

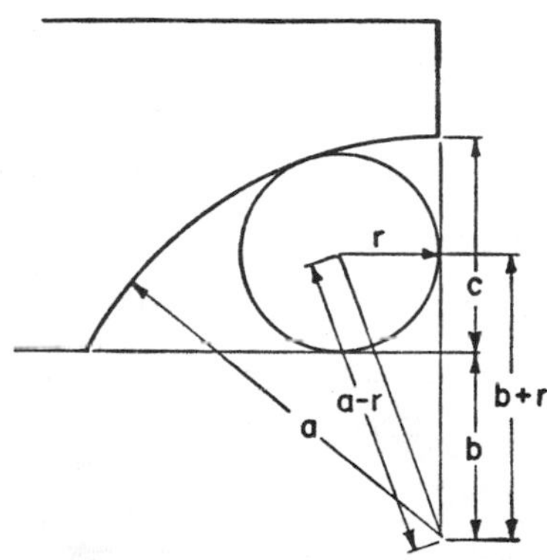

FIG. 6. To Find Radius r when Radius a and Height c are Known

Analysis of Problem: This is a problem involving the use of the theorem for right-angled triangles that the square of the hypotenuse equals the sum of the squares of the other two sides. Using this theorem, an equation in terms of a, b, and r can be set up. Since a is given and b equals $a - c$, the equation can be solved for r.

Formula: $r = -(2a - c) \pm \sqrt{2a(2a - c)}$

Derivation of Formula:

$$(a - r)^2 = r^2 + (b + r)^2 \quad (1a)$$

Squaring all terms as indicated:

$$a^2 - 2ar + r^2 = r^2 + b^2 + 2br + r^2 \quad (1b)$$

Simplifying by placing all terms including r on one side of the equation:

$$r^2 + 2ar + 2br = a^2 - b^2 \quad (1c)$$

Factoring both sides of the equation:

$$r^2 + 2r(a + b) = (a + b)(a - b) \quad (1d)$$

Adding $(a + b)^2$ to both sides, thus completing the square on the left-hand side:

$$r^2 + 2r(a + b) + (a + b)^2 = (a + b)(a - b) + (a + b)^2 \quad (2a)$$

$$[r + (a + b)]^2 = (a + b)(a - b + a + b) \quad (2b)$$

Extracting square root of both sides

$$r + a + b = \pm\sqrt{2a(a + b)} \quad (3a)$$

$$r = -(a + b) \pm \sqrt{2a(a + b)} \quad (3b)$$

Substituting $a - c$ for b

$$r = -(2a - c) \pm \sqrt{2a(2a - c)} \quad (4)$$

Example: Let $a = 7.20$ and $c = 3.88$. Find r.

Solution:

$$r = -(2 \times 7.20 - 3.88)$$

$$\pm\sqrt{2 \times 7.20\ (2 \times 7.20 - 3.88)} \quad (4)$$

$$r = -10.52 \pm \sqrt{151.4880}$$

$$r = -10.52 \pm 12.31$$

$$r = 1.79 \text{ or } -22.83$$

Since the minus answer has no significance in this particular problem, the positive value of r is taken as the one required. It can also be seen from Fig. 6 that r must be numerically less than c.

Radius of Arc by Measurement Over Rolls

PROBLEM 7: To find radius R in Figs. 7A and 7B. Large radius gages of the type shown can be checked by measurement L over two rolls with the gage resting on the rolls as shown. The diameter of the rolls D is known as is also the height H of the top of the arc above the surface plate in Fig. 7A.

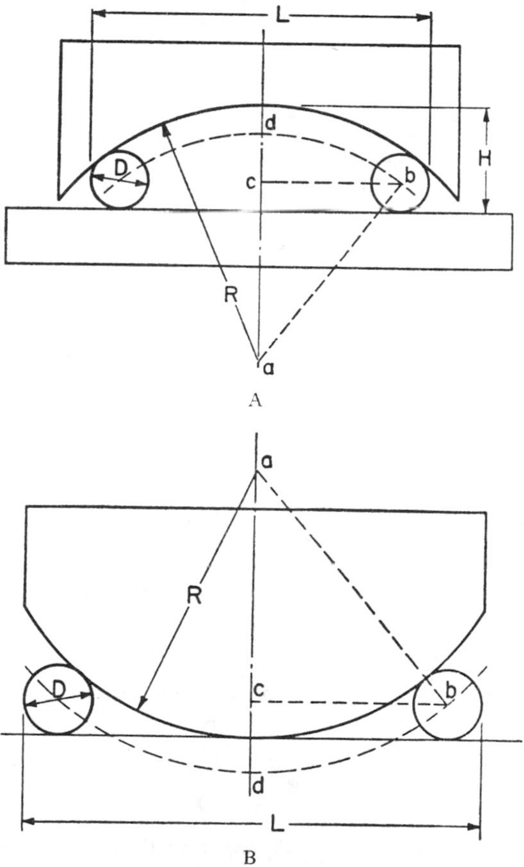

FIG. 7. To Find Radius R when Measurement L, Height H and Diameter D are Known. (A) For Concave Surface. (B) For Convex Surface

Analysis of Problem: (Fig. 7A) Since L and D are known, cb can be found. Also, knowing H and D, cd can be found. With cb and cd known, ab can be found by means of two similar triangles in which cb is a common side as indicated in Fig. 7C by drawing a semicircle with a as a center and ad as a radius. Connecting point b with point e, two similar right-angled triangles cdb and cbe are formed. Returning to Fig. 7A, if ab and D are known, R can be found.

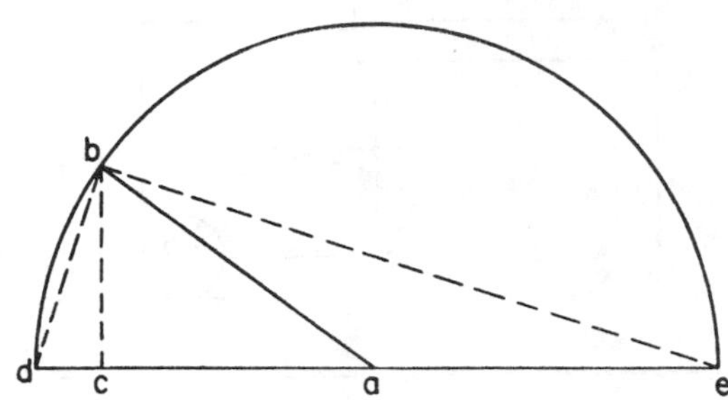

FIG. 7C. Using Point d in Fig. 7A as a Center, a Semi-Circle is Drawn through d and b and da is Extended to e

(Fig. 7B) The solution is similar. The distances cb and cd are readily found, since L and D are known. From these two distances, ab is computed on the basis of similar triangles, as before. R is then readily found.

Formulas:

$$R = \frac{(L - D)^2}{8(H - D)} + \frac{H}{2} \quad \text{(For Fig. 7A)}$$

$$R = \frac{(L - D)^2}{8D} \quad \text{(For Fig. 7B)}$$

Derivation of Formulas:

Case 1: For concave surface as shown in Fig. 7A.

$$cb = \frac{L - D}{2} \tag{1}$$

$$dc = H - D \tag{2}$$

In Fig. 7C, angle dbc = angle bec since their corresponding sides are at right angles to each other, hence triangle dbc is similar to triangle cbe.

$$\frac{ce}{cb} = \frac{cb}{dc} \tag{3a}$$

and

$$ce = \frac{\overline{cb}^2}{dc} \tag{3b}$$

but

$$ad = \frac{ce + dc}{2} \tag{4}$$

hence

$$ad = \frac{\frac{\overline{cb}^2}{dc} + dc}{2} = \frac{\overline{cb}^2 + \overline{dc}^2}{2dc} \tag{5}$$

and

$$R = ad + \frac{D}{2} = \frac{\overline{cb}^2 + \overline{dc}^2}{2dc} + \frac{D}{2} \tag{6}$$

Substituting the equivalents of cb and dc from Equations (1) and (2):

$$R = \frac{\frac{(L - D)^2}{4} + (H - D)^2}{2(H - D)} + \frac{D}{2} \tag{7a}$$

$$R = \frac{(L - D)^2}{8(H - D)} + \frac{H - D}{2} + \frac{D}{2} \tag{7b}$$

$$R = \frac{(L - D)^2}{8(H - D)} + \frac{H}{2} \tag{7c}$$

Case 2: For convex surface as shown in Fig. 7B.

$$cb = \frac{L - D}{2} \tag{8}$$

$$dc = D \tag{9}$$

As has been shown in Fig. 7A:

$$ad = \frac{\overline{cb}^2 + \overline{dc}^2}{2dc} \tag{5}$$

Referring again to Fig. 7B:

$$R = ad - \frac{D}{2} \tag{10}$$

Substituting the equivalent of *ad* from Equation (5):

$$R = \frac{cb^2 + dc^2}{2dc} - \frac{D}{2} \tag{11}$$

Substituting the equivalents of *cb* and *dc* from Equations (8) and (9):

$$R = \frac{\frac{(L - D)^2}{4} + D^2}{2D} - \frac{D}{2} \tag{12a}$$

$$R = \frac{(L - D)^2}{8D} + \frac{D}{2} - \frac{D}{2} \tag{12b}$$

$$R = \frac{(L - D)^2}{8D} \tag{12c}$$

Example 1: In Fig. 7A let $L = 17.80$; $D = 3.20$ and $H = 5.72$. Find R.

Solution: $$R = \frac{(17.80 - 3.20)^2}{8(5.72 - 3.20)} + \frac{5.72}{2} = \frac{(14.60)^2}{8 \times 2.52} + 2.86 \tag{7c}$$

$$R = \frac{213.16}{20.16} + 2.86 = 13.43$$

Example 2: In Fig. 7B let $L = 22.28$ and $D = 3.40$. Find R.

Solution: $$R = \frac{(22.28 - 3.40)^2}{8 \times 3.40} = \frac{356.45}{27.20} \tag{12c}$$

$$R = 13.1$$

Diameter of Arc by Measurement Over Rolls

PROBLEM 8: To determine the diameter D of the split liner shown in Fig. 8 by measurement over pins of known diameter d.

Analysis of Problem: As can be seen from Fig. 8, a right-angled triangle is formed by sides B and $R - r$ and hypotenuse $R + r$. Using the familiar "sum of the squares of the sides equals the square of the hypotenuse" equation, R is found in terms of B and r. Since B is unknown, its equivalent in terms of A and r is substituted and R, which is one-half D, can now be determined.

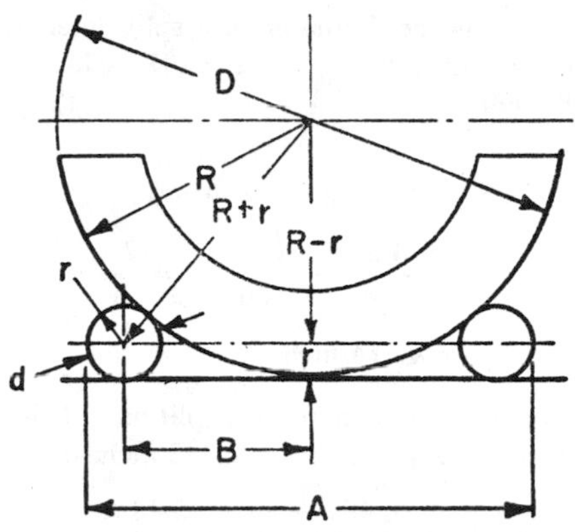

FIG. 8. To Find Diameter D when Measurement A and Diameter d are Known

Formula: $D = \dfrac{(A - d)^2}{4d}$, where A is the measurement over pins.

Derivation of Formula:

$$(R - r)^2 + B^2 = (R + r)^2 \tag{1a}$$

$$R^2 - 2Rr + r^2 + B^2 = R^2 + 2Rr + r^2 \tag{1b}$$

hence

$$B^2 = 4Rr \tag{2a}$$

and

$$R = \frac{B^2}{4r} = \frac{B^2}{2d} \tag{2b}$$

but

$$B = \frac{A}{2} - r = \frac{A - 2r}{2} = \frac{A - d}{2} \tag{3}$$

hence

$$R = \frac{(A - d)^2}{8d} \tag{4}$$

or

$$D = \frac{(A - d)^2}{4d} \tag{5}$$

It should be noted that if the size of the liner permits 1-inch diameter pins to be used, this equation can be simplified to:

$$D = \frac{(A - 1)^2}{4} \tag{6}$$

Example 1: What is the diameter of a split liner if the measurement over pins as shown in Fig. 8 is 3.297 inches? The pins are 0.460 inch in diameter.

Solution:

$$A = 3.297 \qquad d = 0.460$$

$$D = \frac{(3.297 - 0.460)^2}{4 \times 0.460} = \frac{2.837^2}{1.84}$$

$$D = 4.374 \text{ inches}$$

Example 2: What is the diameter of a split liner if the measurement over 1-inch pins as shown in Fig. 8 is 6.482 inches?

Solution:

$$D = \frac{(6.482 - 1)^2}{4} = \frac{5.482^2}{4}$$

$$D = 7.513 \text{ inches}$$

15

Circular Segments—Finding Unknown Elements

The circular segment, a familiar geometrical figure bounded by an arc and its subtending chord, is an important factor in many mechanical problems. Usually two of the elements—length of arc, length of chord, height of arc, radius, and subtending angle—of the circular segment are given and one of the remaining elements is to be found.

One of the most common problems is to find the radius of an arc, when the length of chord and height of the arc are known. Where chordal and height measurements can be taken of a circular surface, the center of which is either not accessible or not readily located, the radius or diameter can be readily computed. The second problem in this chapter shows how a vernier caliper equipped with a sliding block can be used for this purpose.

Also included in this chapter is a method of dividing the circumference of a circle into a given number of equal parts by means of chordal measurements. (Although not based on the use of circular segments, another method of spacing a circle for use with the jig boring machine is also presented for comparison.) The method of finding the radius of a curve for a given "degree of curvature" is given, because of its general interest and to further illustrate the importance of the circular segment as a problem factor.

Several methods of computing the area of circular segments are given in Chapter 16, Approximate Formulas. The formulas used for finding the various elements of circular segments are shown together in the mensuration tables in Chapter 27.

Length of Chord When Radius and Height of Arc are Known

PROBLEM 1: To find the length of chord *cd* in Fig. 1. The diameter *ab* of the flywheel is given and the distance *oe* that the center of its shaft is above the floor level. It is required to find the length of the opening in the floor that would permit the flywheel to be mounted in this position and turn freely.

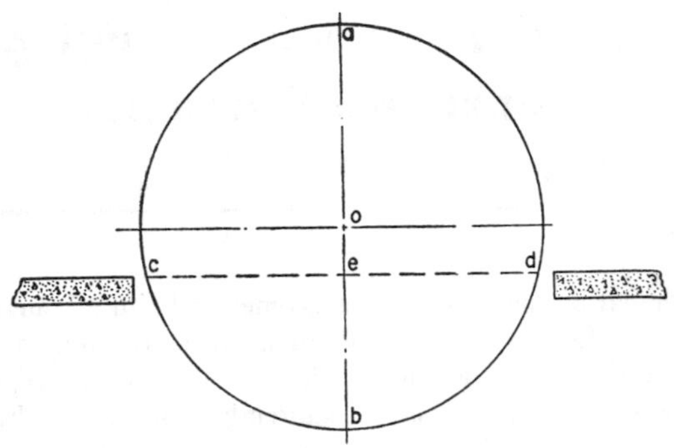

FIG. 1. To Find Length of Floor Opening When Diameter *ab* and Height *oe* are Given.

Analysis of Problem: The solution of this problem is based upon the geometric theorem that when two chords of a circle intersect, or a chord and a diameter, the product of the segments of one equals the product of the segments of the other. The length of *ab* is known, as is also *oe*, which is given, hence the length of the two segments *ae* and *be* can be directly determined. Since chord *cd* is horizontal, it is bisected by the vertical diameter *ab*. Hence *ce* equals *ed* and an equation can be set up with only one unknown.

Formula: $cd = \sqrt{\overline{ab}^2 - 4\overline{oe}^2}$

Derivation of Formula:

$$ae = ao + oe = \frac{ab}{2} + oe \qquad (1)$$

$$be = bo - oe = \frac{ab}{2} - oe \qquad (2)$$

$$ae \times be = \frac{\overline{ab}^2}{4} - \overline{oe}^2 = ce \times de = \overline{ce}^2 \qquad (3)$$

$$ce = \sqrt{\frac{\overline{ab}^2}{4} - \overline{oe}^2} \qquad (4)$$

$$cd = 2ce = 2\sqrt{\frac{\overline{ab}^2}{4} - \overline{oe}^2} = \sqrt{\overline{ab}^2 - 4\overline{oe}^2} \qquad (5)$$

Example: Let the diameter *ab* equal 16 feet and the distance *oe* of the center of the flywheel shaft above the floor level equal 3 feet. Find length of opening in floor.

Solution:

$$cd = \sqrt{16^2 - 4 \times 3^2} = \sqrt{220} \qquad (5)$$

$$cd = 14.832 \text{ feet}$$

An opening of at least 15 feet in length in the floor would be required to allow for clearance.

Finding Diameter of Arc When Length of Chord and Height of Segment are Known

PROBLEM 2: To find the diameter of a round work-piece where the center is not accessible but part of the circumference can be reached so that a chordal measurement can be taken with a vernier caliper. This caliper is equipped with a sliding block A between the two measuring jaws. The distance from the end of this block to the end of either jaw is 1.000 inch as shown in Fig. 2A. What will the diameter D of the work be for a given chordal measurement C as shown?

Analysis of Problem: As shown in Fig. 2B, this is a problem in which the length C and height H of a circular segment are known and the radius R is to be found.

Formula:

$$D = \left(\frac{C}{2}\right)^2 + 1$$

Derivation of Formula:

$$R^2 = \left(\frac{C}{2}\right)^2 + (R - H)^2 \qquad (1a)$$

$$R^2 = \left(\frac{C}{2}\right)^2 + R^2 - 2RH + H^2 \qquad (1b)$$

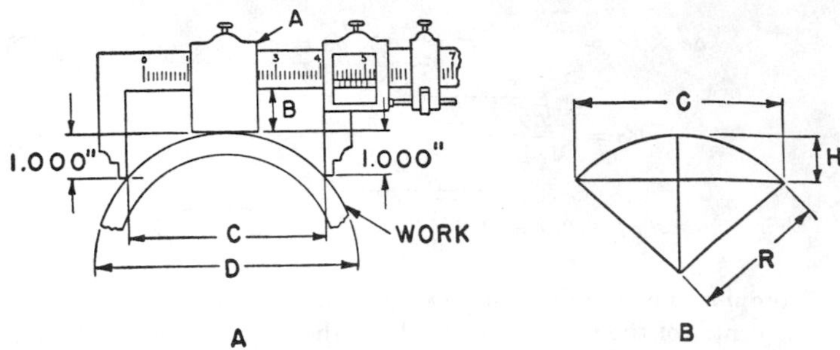

FIG. 2A. Vernier Caliper Equipped with Sliding Block for Taking Chordal Measurements of Large Diameter Work. B. Radius R is to be Found When Length C and Height H are Known.

If $H = 1.000$ inch and $D = 2R$

$$R^2 = \left(\frac{C}{2}\right)^2 + R^2 - D + 1 \tag{1c}$$

$$D = \left(\frac{C}{2}\right)^2 + 1 \tag{1d}$$

Example: If the reading taken with the vernier caliper as shown in Fig. 2A is 4.472 inches, what is the diameter of the work-piece?

Solution:

$$D = \left(\frac{4.472}{2}\right)^2 + 1.000 \tag{1d}$$

$$D = 5.000 + 1.000 = 6.000 \text{ inches}$$

Computing Diameter of Circle from Chordal Length and Central Angle

PROBLEM 3: To find *cd* in Fig. 3. Given two points *a* and *b* on the circumference of a circle which are equidistant from a diameter *cd*. If the length of chord *ac* or *bc* is known together with the central angle X between radii *oa* and *ob*, how can the length of diameter *cd* be found?

Analysis of Problem: This is the problem which would be presented if chordal measurements were taken over two spline or gear teeth nearest to being diametrically opposite each other where the total number of teeth were odd. The central angle X would then equal 360 degrees divided by the number of teeth.

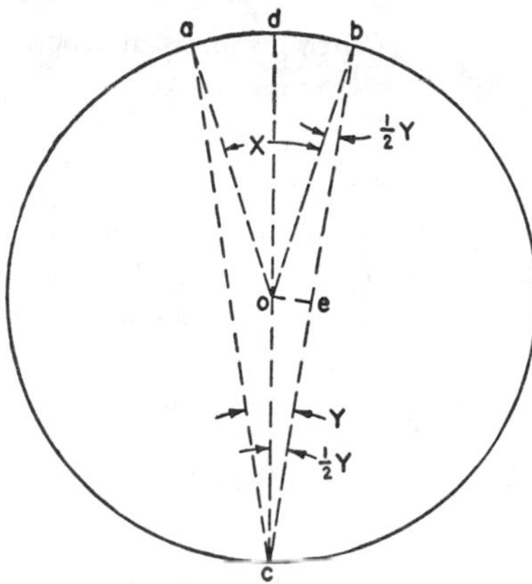

FIG. 3. To Find Diameter *cd* When Length of Chord *ac* or *bc* and Angle *X* are Known

As can be seen from Fig. 3, if angle $\frac{1}{2}Y$ can be determined, then radius *ob* can be found, since *be* is equal to one-half *bc*. According to the fifth geometrical proposition on page **3–8**, an angle at the circumference of a circle between two chords is equal to one-half the angle between two radii if both angles are subtended by the same arc. Thus *Y* can be found and the diameter determined.

Formula:

$$cd = bc \sec \frac{X}{4}$$

Derivation of Formula:

In right-angled triangle *obe:*

$$ob = eb \sec \frac{Y}{2} \qquad (1a)$$

or

$$2ob = 2eb \sec \frac{Y}{2} \qquad (1b)$$

but

$$2eb = bc \qquad (2)$$

and

$$2ob = cd \qquad (3)$$

hence

$$cd = bc \sec \frac{Y}{2} \qquad (4)$$

$$Y = \frac{X}{2} \qquad \text{(Fifth geometrical prop., p. 3–8)} \tag{5}$$

therefore

$$cd = bc \sec \frac{Y}{4} \tag{6}$$

Example: If a spline has 27 teeth and the chordal measurement over two teeth which are nearest to being diametrically opposite each other is 4.825 inches, what is the outside diameter of the spline?

Solution:

$$X = \frac{360°}{27} = 13°20'$$

$$cd = 4.825 \sec \frac{13°20'}{4} \tag{6}$$

$$cd = 4.825 \times 1.0017 = 4.833 \text{ inches}$$

Note: Ordinarily, it is not the outside diameter, but the pitch diameter of a spline or gear which is checked. This is done by placing wires or rolls of suitable size in the two tooth spaces which are diametrically opposite each other, if the spline or gear has an even number of teeth; or nearest diametrically opposite each other if the spline or gear has an odd number of teeth. Measurement is then made over the two wires and the pitch diameter is computed, or the measurement itself is compared with the computed measurement for the correct pitch diameter. This latter measurement for gears may be taken from gear checking tables such as appear in MACHINERY'S Handbook.

Length of Crank for Given Distance and Position of Travel

PROBLEM 4: To find the length *ac* of crank *C* in Fig. 4. In the link mechanism shown, crank *C* oscillates between the extreme positions shown, carrying the end of link *M* above and below the center line of travel *x–x* of slide *S*. The height *cf* of the pivot point of the crank above this center line is known, as well as the horizontal distance of travel *da*. It is required to find the length *ca* of the crank which will cause the end of link *M* to rise above center line *x–x* the same maximum distance that it falls below. In other words, *bf* is to equal *ef*.

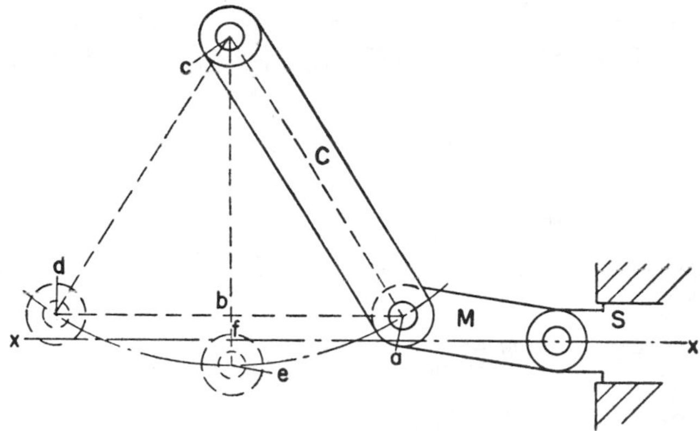

FIG. 4. To Find Length *ac* of Crank Such That Rise *bf* Above and Fall *ef* Below Center Line *x–x* are Equal

Analysis of Problem: This problem can be readily solved with the aid of the familiar rule that in any right-angled triangle the sum of the squares of the sides equals the square of the hypotenuse. In right-angled triangle *abc*, side *ab* is known; side *bc* can be expressed in terms of *cf*, which is known, and *bf*; and hypotenuse *ac* can be expressed in terms of *cf* and *ef*. Since *bf* and *ef* are equal, an equation can be set up with one unknown. With *ef* or *bf* determined, *ac* can be found.

Formula:

$$ac = cf + \frac{\overline{ad}^2}{16cf}$$

Derivation of Formula:

$$\overline{ab}^2 + \overline{bc}^2 = \overline{ac}^2 \qquad (1)$$

but $$bc = cf - bf \qquad (2)$$

and $$ac = cf + ef \qquad (3)$$

since $$bf = ef \qquad (4)$$

then $$ac = cf + bf \qquad (5)$$

Substituting for *bc* and *ac* in Equation (1):

$$\overline{ab}^2 + (cf - bf)^2 = (cf + bf)^2 \qquad (6a)$$

Expanding this equation:

$$\overline{ab}^2 + \overline{cf}^2 - 2bf \times cf + \overline{bf}^2 = \overline{cf}^2 + 2bf \times cf + \overline{bf}^2 \tag{6b}$$

Simplifying

$$\overline{ab}^2 = 4bf \times cf \tag{6c}$$

$$bf = \frac{\overline{ab}^2}{4cf} \tag{6d}$$

since $ab = \frac{ad}{2}$ then $\overline{ab}^2 = \frac{\overline{ad}^2}{4}$ (7)

and

$$bf = \frac{\overline{ad}^2}{16cf} \tag{8}$$

Substituting for bf in Equation (5):

$$ac = cf + \frac{\overline{ad}^2}{16cf} \tag{9}$$

Example: Let $ad = 8.64$ and $cf = 7.61$. Find ac.

Solution:

$$ac = 7.61 + \frac{(8.64)^2}{16 \times 7.61} \tag{9}$$

$$ac = 7.61 + 0.61 = 8.22$$

Dividing a Circle into Equal Number of Parts

PROBLEM 5: To lay out a given number of points that are equally spaced around a circle of given radius where a rotary table with precise angular adjustments is not available.

Methods: Two methods will be described. One method is based on the laying out of chordal dimensions and is therefore adapted to locating points around a circle by the use of dividers. The other method utilizes horizontal and vertical or co-ordinate dimensions and is primarily adapted to locating points around a circle on the jig-boring machine.

Application: The basis of the first method is the table shown in Fig. 5A from which the lengths of the longest chord A, the next longest chord B and the shortest chord C, may be found for dividing a circle of any diameter into any given number of divisions. Starting at any point on the circle, two other points can be located using

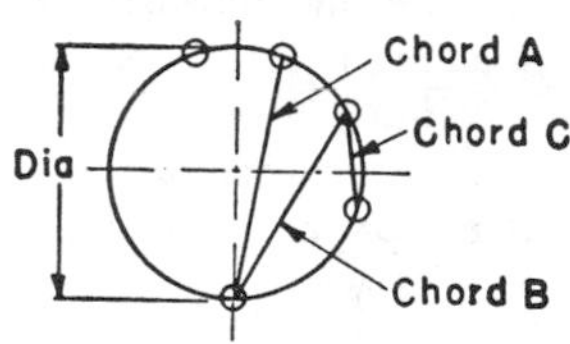

Number of Divisions	Chords, Inches A	B	C	Number of Divisions	Chords, Inches A	B	C
1	0.00000	0.00000	0.00000	26	1.00000	0.99270	0.12053
2	1.00000	1.00000	1.00000	27	0.99831	0.98480	0.11609
3	0.86603	0.86603	0.86603	28	1.00000	0.99371	0.11196
4	1.00000	0.70710	0.70710	29	0.99853	0.98682	0.10811
5	0.95106	0.58778	0.58778	30	1.00000	0.99452	0.10452
6	1.00000	0.86603	0.50000	31	0.99871	0.98847	0.10116
7	0.97493	0.78183	0.43388	32	1.00000	0.99518	0.09801
8	1.00000	0.92388	0.38268	33	0.99886	0.98982	0.09505
9	0.89481	0.86603	0.34202	34	1.00000	0.99573	0.09226
10	1.00000	0.95105	0.30901	35	0.99899	0.99095	0.08963
11	0.98982	0.90963	0.28173	36	1.00000	0.99619	0.08715
12	1.00000	0.96592	0.25881	37	0.99910	0.99192	0.08480
13	0.99271	0.93501	0.23931	38	1.00000	0.99658	0.08257
14	1.00000	0.97492	0.22251	39	0.99918	0.99269	0.08046
15	0.99452	0.95105	0.20791	40	1.00000	0.99691	0.07845
16	1.00000	0.98078	0.19509	41	0.09926	0.99338	0.07654
17	0.99573	0.96182	0.18374	42	1.00000	0.99720	0.07473
18	1.00000	0.98480	0.17364	43	0.99933	0.99355	0.07299
19	0.99658	0.96940	0.16459	44	1.00000	0.99745	0.07133
20	1.00000	0.98768	0.15643	45	0.99939	0.99452	0.06975
21	0.99720	0.97492	0.14904	46	1.00000	0.99766	0.06824
22	1.00000	0.99010	0.14231	47	0.99944	0.99497	0.06679
23	0.99766	0.97908	0.13616	48	1.00000	0.99785	0.06540
24	1.00000	0.99144	0.13052	49	0.99948	0.99539	0.06406
25	0.99803	0.98228	0.12533	50	1.00000	0.99802	0.06279

To Find Chordal Lengths *A, B* and *C*

The lengths of chords A, B and C given in this table are for a diameter of 1. To find the required chords A, B and C for dividing a circle of given diameter into a given number of divisions, multiply the values shown in the table by the length of the diameter.

For a greater number of divisions N than 50, the following formulas can be used to find the constants A, B and C for a circle having a diameter of 1: $A = \cos 90° \div N$ (odd number of divisions); $A = 1$ (even number of divisions); $B = \cos 270° \div N$ (odd number of divisons); $B = \cos 180° \div N$ (even number of divisions); $C = \sin 180° \div N$ (odd and even number of divisions).

FIG. 5A. Chordal Lengths for Dividing Circumferences of Circles.

dimension A and two more using dimension B. Dimension C can be used to locate the remaining points or, if a large number of points are to be located, a few points can be located by chord C and then chords A and B can be used from one of these points. The reason for again using chords A and B instead of repeatedly using chord C for a large number of spacings is to reduce the cumulative error.

The basis of the second method is the table shown in Fig. 5B from which horizontal and vertical distances of the various points from the horizontal and vertical center lines of the circle or from other established points can be computed for any diameter of circle. These "coordinate" distances are used to locate the jig-boring head in the equally spaced positions around the circle.

Example 1: Compute the necessary chordal dimensions for dividing a circle of 3.26-inch diameter into 17 equal divisions using the first method as given in table shown in Fig. 5A.

Solution: According to this table, for 17 divisions:

$$A = 0.99573 \times \text{diameter}$$

$$B = 0.96182 \times \text{diameter}$$

$$C = 0.18374 \times \text{diameter}$$

or

$$A = 0.99573 \times 3.26 = 3.246 \text{ inches}$$

$$B = 0.96182 \times 3.26 = 3.136 \text{ inches}$$

$$C = 0.18374 \times 3.26 = 0.599 \text{ inch}$$

Example 2: Find the chordal lengths A, B, and C for dividing a circle of 5.39-inch diameter into 51 equal divisions as shown in Fig. 5A.

Solution: According to this table, for 51 divisions the lengths of A, B, and C can be found by the following formulas:

$$A = \cos \frac{90°}{N} \times \text{diameter}$$

$$B = \cos \frac{270°}{N} \times \text{diameter}$$

$$C = \sin \frac{180°}{N} \times \text{diameter}$$

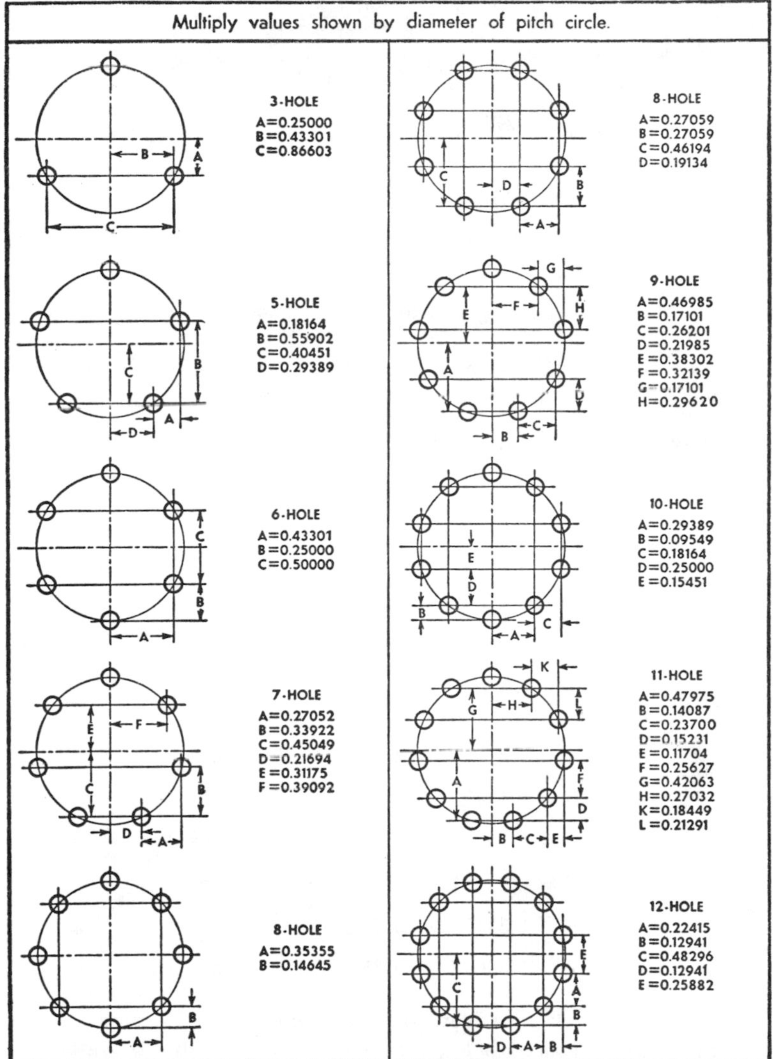

FIG. 5B. Coordinates for Locating Equally Spaced Holes in Jig Boring.

or
$$A = \cos\frac{90°}{51} \times 5.39 = \cos 1°45.9' \times 5.39$$

$$A = 0.99952 \times 5.39 = 5.387 \text{ inches}$$

$$B = \cos\frac{270°}{N} \times 5.39 = \cos 5°17.6' \times 5.39$$

$$B = 0.99573 \times 5.39 = 5.367 \text{ inches}$$

$$C = \sin\frac{180°}{51} \times 5.39 = \sin 3°31.76' \times 5.39$$

$$C = 0.06156 \times 5.39 = 0.3318 \text{ inch}$$

Example 3: Find the coordinates of locating seven equally spaced points around a circle of 11.35-inch diameter as shown by Fig. 5B.

Solution: According to this table:

$$A = 0.27052 \times \text{diameter}$$

$$B = 0.33922 \times \text{diameter}$$

$$C = 0.45049 \times \text{diameter}$$

$$D = 0.21694 \times \text{diameter}$$

$$E = 0.31175 \times \text{diameter}$$

$$F = 0.39092 \times \text{diameter}$$

or

$$A = 0.27052 \times 11.35 = 3.0704 \text{ inches}$$

$$B = 0.33922 \times 11.35 = 3.8501 \text{ inches}$$

$$C = 0.45049 \times 11.35 = 5.1131 \text{ inches}$$

$$D = 0.21694 \times 11.35 = 2.4623 \text{ inches}$$

$$E = 0.31175 \times 11.35 = 3.5384 \text{ inches}$$

$$F = 0.39092 \times 11.35 = 4.4369 \text{ inches}$$

While either of these methods can be used to divide a circle or locate equally spaced holes, an auxiliary rotary table provides a more direct method. With a rotary table, the holes are spaced by precise angular movements, after adjustment to the required radius.

Radius of Curve for Given Degree of Curvature

PROBLEM 6: To find the radius r in Fig. 6. In railroad practice the term "degree of curvature" is the angle of an arc subtended by a chord of 100 feet in length. It is required to find the radius r of a given curve when the degree of curvature A is known.

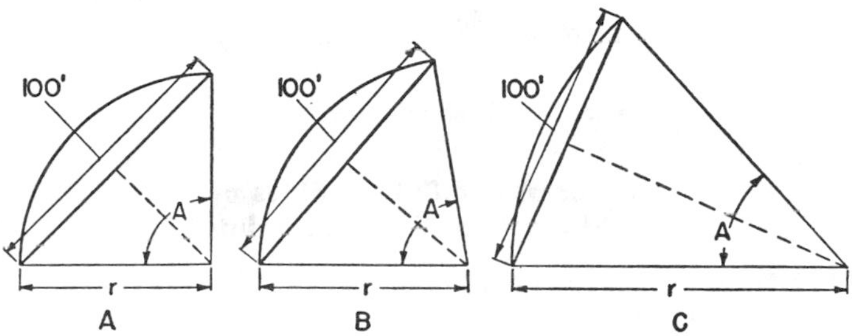

FIG. 6. To Find Radius of Curvature r When Degree of Curvature A is Known

Analysis of Problem: This problem can be readily solved by drawing a line perpendicular to the chord and passing through the vertex of the known angle. This line bisects the chord (see fourth geometrical proposition on page **3**–7) and the central angle. Thus two right triangles are formed in which one side and one acute angle are known. The other side r can then be readily found.

Formula:

$$r = 50 \operatorname{cosec} \frac{A}{2}$$

Derivation of Formula:

$$\operatorname{cosec} \frac{A}{2} = \frac{r}{50} \qquad r = 50 \operatorname{cosec} \frac{A}{2} \qquad \text{(1a) and (1b)}$$

Example 1: In Fig. 6A let $A = 100$ degrees. Find r.

Solution: $$r = 50 \operatorname{cosec} \frac{100°}{2} = 50 \operatorname{cosec} 50° \qquad \text{(1b)}$$

$$r = 50 \times 1.3054 = 65.27 \text{ feet}$$

Example 2: In Fig. 6B let $A = 65$ degrees. Find r.

Solution:

$$r = 50 \operatorname{cosec} \frac{65°}{2} = 50 \operatorname{cosec} 32°30' \qquad (1b)$$

$$r = 50 \times 1.8611 = 93.06 \text{ feet}$$

Example 3: In Fig. 6C let $A = 43°$

Solution:

$$r = 50 \operatorname{cosec} \frac{43°}{2} = 50 \operatorname{cosec} 21°30' \qquad (1b)$$

$$r = 50 \times 2.7285 = 136.42 \text{ feet}$$

Center of Arc Cutting Sides of Right Angle at Specified Points

PROBLEM 7: To find *ce* and *af* in Fig. 7. An arc of radius *oc* cuts the sides of right angle *mbn* at points *a* and *c*. Dimensions *ab* and *bc* are known. It is required to find the location of the center *o* of the arc from the sides of the right angle in terms of distance *ce* and distance *af*, if radius *oc* is known.

Analysis of Problem: It will be noted that *ce* forms one side of right-angled triangle *oce* in which the hypotenuse *oc* is known. If

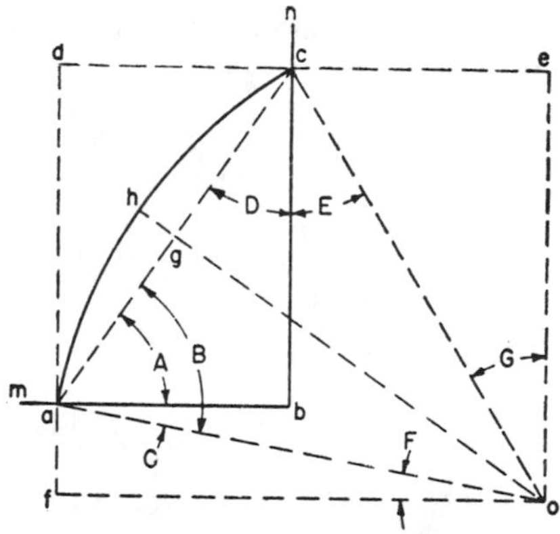

FIG. 7. To Find Location of Center *o* of Arc having Known Radius *oc* When Dimensions *ab* and *bc* are Known

angle G can be determined then ce can be found. Similarly af is one side of right-angled triangle oaf in which the hypotenuse oa is known. If F can be found then af can be determined.

The first step is to solve the right-angled triangle abc for angles A and D and hypotenuse ac. Since radius oh is drawn at right angles to chord ac it bisects it at g forming right-angled triangle aog. Knowing ao and ag, angle B can be found and hence angle C. Angle B equals angle $(D + E)$ since they are base angles of an isosceles triangle. Hence with angle D known, angle E can be found. Then angle C equals angle F and angle E equals angle G, hence af and ce can now be determined.

Procedure:

$$\tan A = \frac{bc}{ab} \quad (1)$$

$$D = 90° - A \quad (2)$$

$$ac = \frac{bc}{\sin A} \quad (3)$$

According to the fourth geometrical proposition on page **3–7**, since oh is drawn perpendicular to ac then:

$$ag = \frac{ac}{2} \quad (4)$$

$$\cos B = \frac{ag}{oa} = \frac{ag}{oc} \quad (5)$$

$$C = B - A \quad (6)$$

Since aoc is an isosceles triangle:

$$B = D + E$$

$$E = B - D \quad (7)$$

According to the fifth geometrical proposition on page **3–6**, it will be seen that:

$$F = C$$

and

$$G = E \quad (9)$$

Since aof is a right-angled triangle:

$$af = oa \sin F = oc \sin F \quad (10)$$

Since *coe* is a right-angled triangle:

$$ce = oc \sin G \tag{11}$$

Example: Let $ab = 1.375$; $bc = 1.850$; and $oc = 3.128$. Find af and ce.

Solution:
$$\tan A = \frac{1.850}{1.375} = 1.3455 \tag{1}$$

$$A = 53°22.8'$$

$$D = 90° - 53°22.8' = 36°37.2' \tag{2}$$

$$ac = \frac{1.850}{0.80261} = 2.305 \tag{3}$$

$$ag = \frac{2.305}{2} = 1.1525 \tag{4}$$

$$\cos B = \frac{1.1525}{3.128} = 0.36845 \tag{5}$$

$$B = 68°22.8'$$

$$C = 68°22.8' - 53°22.8' = 15° \tag{6}$$

$$E = 68°22.8' - 36°37.2' = 31°45.6' \tag{7}$$

$$F = 15° \tag{8}$$

$$G = 31°45.6' \tag{9}$$

$$af = 3.128 \times 0.25882 = 0.810 \tag{10}$$

$$ce = 3.128 \times 0.52636 = 1.646 \tag{11}$$

Locating Intersecting Points of Arc and Straight Line

PROBLEM 8: To find x and y in Fig. 8. Given a rectangular opening of length b which is to be enlarged by cutting a segment of a circle having radius r and height h into one side of the opening as shown. The distance a of the center of the arc of the segment from one end of the block is also given. It is required to find the distances x and y from the ends of the opening to the points of intersection of the arc with the side of the opening.

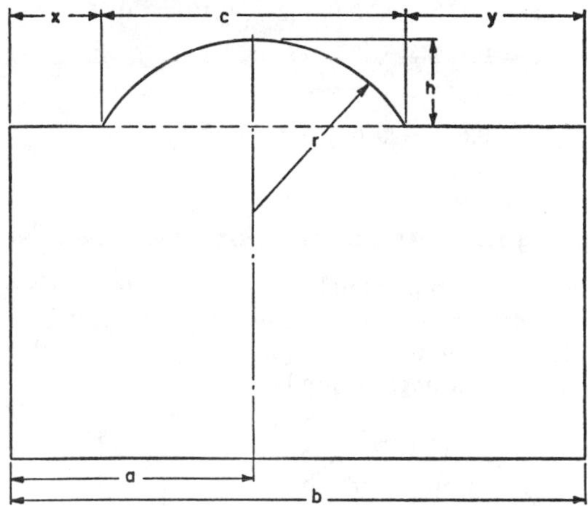

FIG. 8. To Find Distances x and y When Lengths a and b, Radius r and Height h are Known

Analysis of Problem: It is only required to find the length c of the chord of the segment, then x and y can be readily determined.

Formulas:

$$x = a - \sqrt{h(2r - h)}$$

$$y = b - a - \sqrt{h(2r - h)}$$

Derivation of Formulas: The formula for length of the chord c of a segment in terms of its height h and radius r is:

$$c = 2\sqrt{h(2r - h)} \quad (1)$$

$$x = a - \frac{c}{2} \quad (2)$$

therefore

$$x = a - \sqrt{h(2r - h)} \quad (3)$$

$$y = b - a - \frac{c}{2} \quad (4)$$

and

$$y = b - a - \sqrt{h(2r - h)} \quad (5)$$

Example: Let $r = 0.101$; $h = 0.051$; $a = 0.152$; and $b = 0.625$. Find x and y.

Solution:

$$x = 0.152 - \sqrt{0.051\,(2 \times 0.101 - 0.051)} \qquad (3)$$

$$x = 0.152 - \sqrt{0.0077}$$

$$x = 0.152 - 0.088 = 0.064$$

$$y = 0.625 - 0.152 - 0.088 \qquad (5)$$

$$y = 0.385$$

Dividing Head Angle for Milling V-shaped Teeth

PROBLEM 9: To determine angle B in Fig. 9B to which a dividing head would be set for milling V-shaped teeth in a clutch of the type shown in Fig. 9A with a double-angle cutter, when number of teeth n, radius r, and cutter angle D are known.

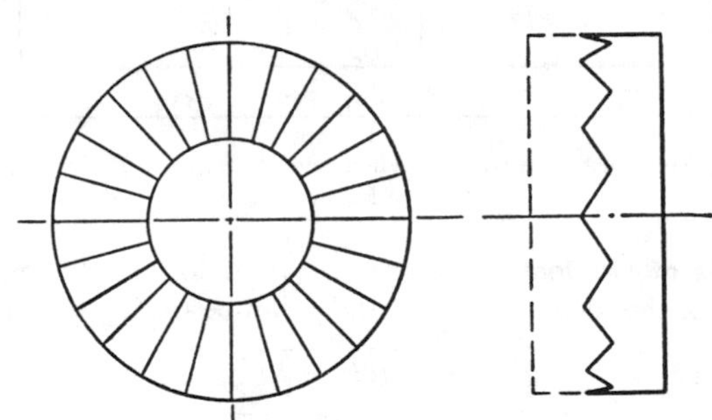

FIG. 9A. Clutch with V-shaped Teeth to be Milled on Dividing Head

Analysis of Problem: In Fig. 9B is shown a single tooth space having an angular width C in a clutch of radius r. The angle of the V-shaped cutter is shown as D. In order that the cutter shall move across the clutch blank at angle A with the face of the clutch blank so that it will just graze this face as it leaves the blank on the opposite side, the dividing head on which the blank is mounted must be inclined at angle B from the vertical.

If the number of teeth n is known, C can be determined. With r also known, a, which is one side and b, the other, of a right-angled triangle having known angle $\frac{C}{2}$ can be found. With the cutter angle D and also a known, d is computed. With r, b and d known, angle A and hence angle B, which is the complement of A, can be found, since $r + b$ is the hypotenuse and d one side of a right-angled triangle.

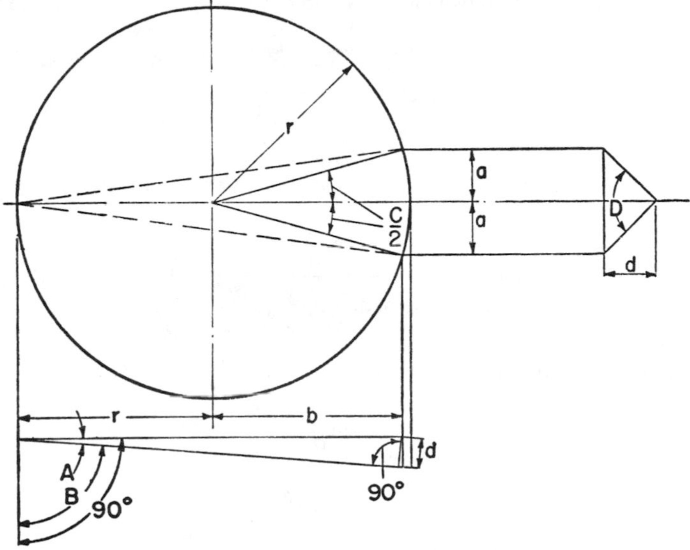

FIG. 9B. To Determine Angle B When Number of Teeth n Radius r and Cutter Angle D are Known. The Depth d of the Cutter is Shown Somewhat Enlarged at the Right. It Actually Equals Depth of Cut d Shown at the Bottom

Formula: $$\cos B = \tan \frac{90°}{n} \times \cot \frac{D}{2}$$

where B = index-head angle; n = number of teeth in clutch; and D = cutter angle.

Derivation of Formula: $$a = r \sin \frac{C}{2} \qquad (1)$$

$$d = a \cot \frac{D}{2} \qquad (2)$$

$$b = a \cot \frac{C}{2} \qquad (3)$$

$$\sin A = \frac{d}{r + b} \qquad (4)$$

Substituting the equivalents of d and b from Equations (2) and (3)

$$\sin A = \frac{a \cot \frac{D}{2}}{r + a \cot \frac{C}{2}} \tag{5}$$

Substituting the equivalent of a from Equation (1) in Equation (5)

$$\sin A = \frac{r \sin \frac{C}{2} \cot \frac{D}{2}}{r + r \sin \frac{C}{2} \cot \frac{C}{2}} \tag{6a}$$

Simplifying by cancellation of r and combining $\sin \frac{C}{2} \cot \frac{C}{2}$

$$\sin A = \frac{\sin \frac{C}{2} \cot \frac{D}{2}}{1 + \cos \frac{C}{2}} \tag{6b}$$

Substituting the equivalents of $\sin \frac{C}{2}$ (Equation 38, page 4–24) and of $1 + \cos \frac{C}{2}$ in terms of one-half this angle (Equation 39, page 4–24).

$$\sin A = \frac{2 \sin \frac{C}{4} \cos \frac{C}{4} \cot \frac{D}{2}}{2 \cos^2 \frac{C}{4}} \tag{6c}$$

Simplifying by dividing numerator and denominator by $2 \cos \frac{C}{4}$ and substituting for the remaining $\dfrac{\sin \frac{C}{4}}{\cos \frac{C}{4}}$

$$\sin A = \tan \frac{C}{4} \cot \frac{D}{2} \tag{6d}$$

But

$$C = \frac{360°}{n}; \quad \text{hence} \quad \frac{C}{4} = \frac{90°}{n} \tag{7}$$

and $$\sin A = \cos (90° - A) = \cos B \tag{8}$$

Substituting the equivalents of $\frac{C}{4}$ and $\sin A$

$$\cos B = \tan \frac{90°}{n} \cot \frac{D}{2} \tag{9}$$

Example: Find the angle B at which the dividing head is set to cut ten teeth with a cutter having an angle of 60°.

Solution: $n = 10$ and $D = 60°$

For this type of computation it is helpful to use logarithms, thus

$$\log \cos B = \log \tan \frac{90°}{10} + \log \cot \frac{60°}{2} \tag{9}$$

$$\log \cos B = \log \tan 9° + \log \cot 30°$$

$$\begin{aligned} \log \tan 9° &= 9.19971 - 10 \\ \log \cot 30° &= \underline{10.23856 - 10} \\ \log \cos B &= 19.43827 - 20 \end{aligned}$$

$$B = 74°4'41''$$

Calculating the Radius of a Spherical Segment

A sheet metal stamping in the form of a spherical segment with a flange is shown in Fig. A. Designing the forming die for this part requires a value for radius *R*. This dimension can be calculated from the data in Fig. B, with some algebraic manipulation.

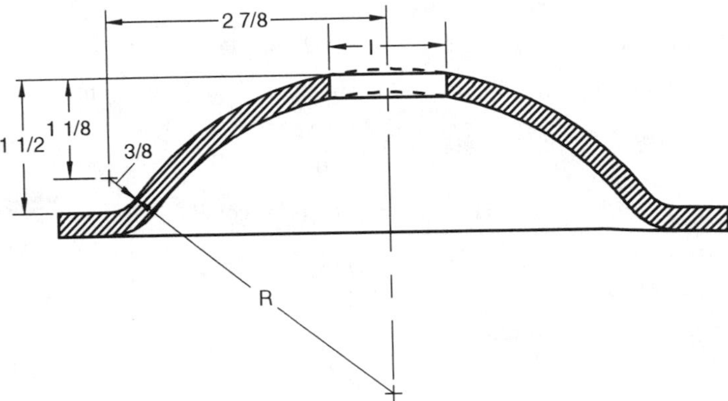

Fig. A. Flanged spherical segment for which radius *R* must be calculated.

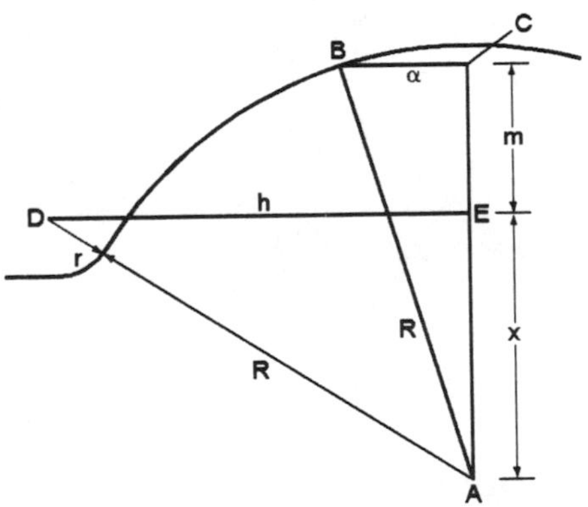

Fig. B. Diagram used in calculating radius *R*

Solution:

To avoid the use of fractions and large numbers, all dimensions are first multiplied by 8, so that $h = 23$, $m = 9$, $r = 3$, and $a = 4$.

In triangle ABC,

$$R^2 = a^2 + (x + m)^2 = 4^2 + (x + 9)^2 = x^2 + 18x + 97 \qquad (1)$$

In triangle ADE,

$$(R + r)^2 = x^2 + h^2$$

$$(R + r)^3 = x^2 + 23^2$$

$$(R^2 + 6R - 520 - x^2 = 0 \qquad (2)$$

Eliminating x^2 between Equations (1) and (2) and solving for x,

$$x = \frac{423 - 6R}{18} \qquad (3)$$

Substituting Equation (3) in Equation (2)

$$288R^2 + 7020R - 347{,}409 = 0$$

Dividing through by 9,

$$32R^2 + 780R - 38{,}601 = 0$$

Using the quadratic formula,

$$R = \frac{-780 \pm \sqrt{(780)^2 - 4(32)(-38{,}601)}}{2 x 32}$$

Dividing numerator and denominator of the right hand member by 4 to reduce the size of the numbers handled,

$$R = \frac{-195 \pm \sqrt{(195)^2 + 8 x 38{,}601}}{16}$$

$$= \frac{-195 \pm 346{,}833}{16} = \frac{-195 \pm 588{,}925}{16} = \frac{393{,}925}{16} = 24.6203$$

Since 24.6203 is 8 times the true value of R,

$$R = \frac{24.6203}{8} = 3.0775$$

16

Approximate Formulas—Why and How They Are Used

In certain problems, approximate formulas may sometimes be employed to advantage in place of theoretically exact formulas. The advantage arises from the fact that the approximate formulas provide the degree of accuracy called for in the answer while at the same time they are either simpler to apply, require less numerical calculation or involve less possibility of error.

In deciding whether or not to use an approximate formula, the accuracy required in the answer is, of course, an all-important factor. If the accuracy of each formula under consideration were known and if the accuracy of each remained more or less constant regardless of the conditions of the problem, the decision of which formula to use would be a relatively simple matter. Frequently the accuracies of the formulas under consideration are not known, however, and when determined, they will be found to vary, depending upon the conditions of the problem.

Only where a given problem is to be solved again and again with different sets of data does it usually prove worthwhile to make any elaborate determination of the accuracies of applicable formulas and the variations of these accuracies with the data to be used. Such a determination, if carefully made and the results set up in tabular form, can be a valuable time- and error-saving aid to repetitive calculations.

Exact Formulas for Area of Circular Segment. Two problems will be considered to show the selection and use of approximate formulas. The first problem is that of calculating the area of the segment of a circle, as indicated by the shaded portion in Fig. 1 when only the height h and length of chord c are known.

The exact formula for finding this area A is:

$$A = \tfrac{1}{2}[r\ell - c(r - h)] \qquad (1a)$$

where r = radius of the arc bounding the segment and ℓ = length of this arc.

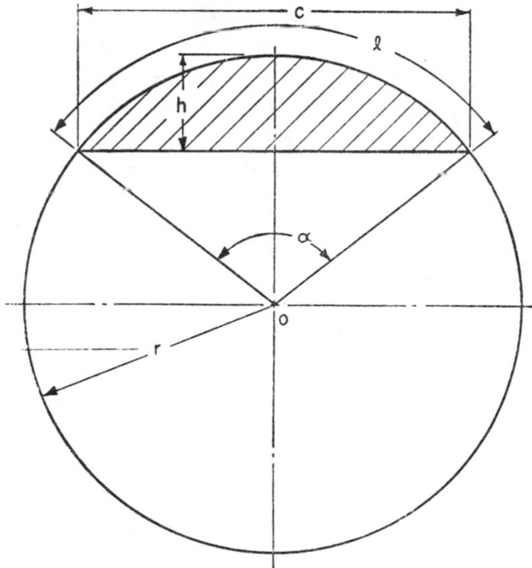

FIG. 1. To Find Area of Circular Segment When Height h and Length of Chord c are Known

In itself, this is a relatively simple formula but before it can be used, both r and ℓ have to be computed by the formulas: $r = \dfrac{c^2 + 4h^2}{8h}$; $\ell = r\hat{\alpha}$ where $\hat{\alpha}$ is in radians and $\sin \dfrac{\alpha}{2} = \dfrac{c}{2r}$.

The accuracy of the result depends upon the accuracy with which ℓ is calculated so that considerable numerical work is involved before A is found. A more convenient formula, also exact and equivalent to Formula (1a) is:

$$A = \frac{r^2}{2}(\hat{\alpha} - \sin\alpha) \qquad (1b)$$

in which r and α are computed as indicated with Formula (1a). In Formula (1b), $\hat{\alpha}$ is expressed in radians.

Approximate Formulas for Area of Circular Segment: Three different approximate formulas that may be used for calculating the area of a circular segment are given below. The accuracy of each is related to the ratio of the length of chord c to the height of segment h in the specific problem to which they are applied.

An approximate formula which gives an error of about 0.1 per cent or less for circular segments ranging almost up to a semi-circle is:

$$A = \frac{4h^2}{3}\sqrt{\frac{c^2}{4h^2} + 0.392} \quad \text{(approx.)} \qquad (2)$$

An approximate formula which gives an error of about 0.1 per cent when the c to h ratio is 3 or greater is:

$$A = \frac{2ch}{3} + \frac{h^3}{2c} \quad \text{(approx.)} \qquad (3)$$

An approximate formula which is more accurate than Formula (3) for segments having a c to h ratio of from 2.0 to 2.2 is:

$$A = \frac{h^3}{2c} + 0.6604ch \quad \text{(approx.)} \qquad (4)$$

For segments very close to a semi-circle, its accuracy is about that of Formula (2).

The following table shows a comparison of area values computed by Formulas (1b), (2), (3) and (4) for segments ranging from a semi-circle down to one having a c to h ratio of 40. The absolute and the percentage errors are given in each case. This table may be used as a guide in selecting the formula needed to obtain the accuracy required in a given problem.

Example 1: Compute the area of a circular segment having a chord c that is 40.25 inches long and a height h that is 20 inches, to an accuracy of three significant figures.

The first step is to find the ratio of c to h.

$$\frac{c}{h} = \frac{40.25}{20} = 2.012+$$

Table 1. Comparison of Formulas for Areas of Circular Segments

Values of c and h	Ratio of c to h	Area by Formula 1b	Area by Formula 2	Area by Formula 3	Area by Formula 4
$c = 10;\ h = 5$	2	39.2699	39.3277	39.5833	39.2700
Error		*0.0000*	*+0.0578*	*+0.3134*	*+0.0001*
			+0.15%	*+0.80%*	*0.00%*
$c = 10;\ h = 4.75$	2.1	36.8053	36.8449	37.0253	36.7276
Error		*0.0000*	*+0.0396*	*+0.2200*	*−0.0777*
			+0.11%	*+0.60%*	*−0.21%*
$c = 10;\ h = 4.5$	2.22	34.4103	34.4350	34.5562	34.2742
Error		*0.0000*	*+0.0247*	*+0.1459*	*−0.1361*
			+0.07%	*+0.42%*	*−0.40%*
$c = 10;\ h = 4$	2.25	29.8200	29.8244	29.8667	29.6160
Error		*0.0000*	*+0.0044*	*+0.0467*	*−0.2040*
			+0.02%	*+0.16%*	*−0.68%*
$c = 10;\ h = 3$	3.33	21.3736	21.3647	21.3500	21.1620
Error		*0.0000*	*−0.0089*	*−0.0236*	*−0.2116*
			−0.04%	*−0.11%*	*−0.99%*
$c = 10;\ h = 2$	5	13.7507	13.7451	13.7333	13.6080
Error		*0.0000*	*−0.0056*	*−0.0174*	*−0.1427*
			−0.04%	*−0.13%*	*−1.04%*
$c = 10;\ h = 1$	10	6.7197	6.7187	6.7167	6.6540
Error		*0.0000*	*−0.0010*	*−0.0030*	*−0.0657*
			−0.02%	*−0.04%*	*−0.98%*
$c = 10;\ h = 0.5$	20	3.3400	3.3399	3.3396	3.3082
Error		*0.0000*	*−0.0001*	*−0.0004*	*−0.0318*
			0.00%	*−0.01%*	*−0.95%*
$c = 10;\ h = 0.25$	40	1.6676	1.6675	1.6675	1.6518
Error		*0.0000*	*−0.0001*	*−0.0001*	*−0.0158*
			−0.01%	*−0.01%*	*−0.95%*

Referring to the table showing the accuracy of various formulas, it will be noted that for segments having $\frac{c}{h}$ ratios close to 2 that Formulas

(2) and (4) give the closest approximations. Since Formula (4) involves less computation, it will be used.

$$A = \frac{20^3}{2 \times 40.25} + 0.6604 \times 40.25 \times 20 \qquad (4)$$

$$A = 631.001 \text{ square inches}$$

The area as computed by the exact Formula (1b) is:

$$A = 631.160 \text{ square inches}$$

Hence it will be noted that the area as computed by Formula (4) is accurate to three significant figures, as specified in the problem.

Example 2: Find the area to three-place accuracy of a circular segment having a chordal length c of 54 inches and a height h of 6 inches.

The ratio of chordal length to height is $54 \div 6 = 9$. Referring to the table showing the accuracy of various formulas, it will be noted that for a chordal length to height ratio of 10, both Formulas (2) and (3) give results that are accurate to three significant figures. Formula (3), however, involves less numerical computation and will, therefore, be used.

$$A = \frac{2 \times 54 \times 6}{3} + \frac{6^3}{2 \times 54} \qquad (3)$$

$$A = \frac{648}{3} + \frac{216}{108} = 218 \text{ square inches}$$

The area to one decimal place as given by the exact formula is 218.1 square inches so that the answer given by the approximate formula is correct to three significant figures.

Checking Screw Thread Pitch Diameters. The second problem to be considered is that of checking the pitch diameter E of a screw thread by means of measurement over wires. The wires are held in the thread grooves as shown in Figs. 2A and 2B, and a measurement M over the wires is taken with some form of micrometer. This measured value of M is checked against a computed value of M for the thread in question. Any error in the pitch diameter will be indicated, if the computed value of M is exact, by a difference between the computed and the measured values of M. Since an error of one ten-thousandth of an inch in the pitch diameter may be of importance

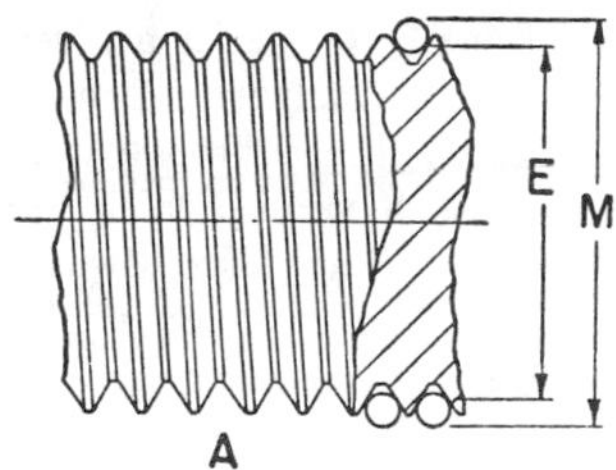

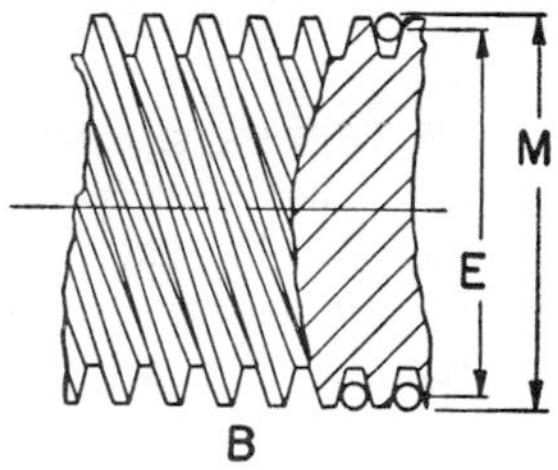

FIG. 2. To Compute What Measurement *m* Over Three Wires for a Given Pitch Diameter *e* Should be: A. For American Standard Thread, B. For Acme Standard Thread

in one case, such as that of a threaded plug gage, while in another case only errors of a thousandth of an inch or more are important, what is meant by a *computed value* of M is a value accurate to the degree required by the thread measurement in question. This may be to four significant figures, that is, to one ten-thousandth inch for a thread having an M value of less than one inch, or to one-thousandth inch for a thread having an M value of one inch or more but less than ten inches, for example; or it may be to five significant figures, that is, to one ten-thousandth of an inch for an M value of one inch or more but less than ten inches, and so on.

Choice of Wire Size a Factor. The size of wire used in measuring any given thread has an important effect on the accuracy of the computed measurement M. Thus, it would be preferable to have the wires always make contact with the sides of the thread groove at the pitch line or mid-slope. Then the measurement of the pitch diameter would be least affected by any error in the thread angle.

Actually it is difficult to determine the exact size of wire which would make contact right on the pitch line of a given thread. A theoretically correct solution for finding the *exact* size for pitch line contact involves the use of cumbersome indeterminate equations with solutions by successive trials. As a practical matter, then, wire sizes are chosen so that contact is made *near* the pitch line.

The position of any measuring wire of given diameter in a thread groove is determined not only by the pitch of the thread and the thread angle, but also by the lead angle. From a theoretical standpoint, therefore, it would be necessary to have a different size of wire for each combination of lead angle, thread angle and pitch as presented by a given thread. In commercial practice, however, the lead angle is ignored in determining the thread size. Thus, for a given pitch and thread angle, a size of wire is used which would make contact with the sides at the pitch line, if the thread groove were annular—that is, if it had no lead angle. Such wires (which are commercially available) are commonly called "best size" wires, although if "best size" means pitch line contact, it is evident that for a given pitch, the *true* "best size" will vary for every change in lead angle.

The use of commercial "best size" wires is justified by the fact that for most commercial threads, particularly those with a 60-degree thread angle, the effect of the lead angle on the position of the wire in the thread groove may be ignored.

In Acme and worm threads of multiple form (having two or more starts) the lead angles are usually quite large and must be taken into account if an accurate value of M is to be obtained. For threads of this type which must be accurate but not to the greatest degree of precision, a comparatively simple formula is available, as will be pointed out in the following discussion of formulas for computing M. This simple formula is made possible by using a *special size* of wire which is found by a formula in which the lead angle of the thread in question is taken into account.

Thus there are two formulas in practical use for determining wire size. The first is for the commercial "best size" of wire which makes contact at the pitch line only if the thread groove has a zero lead angle:

$$W = \frac{T}{\cos A} = T \sec A \tag{5}$$

The second is for the special size of wire which, even in cases of appreciably large lead angles, makes contact near the pitch line:

$$W = \frac{T \cos B}{\cos A_n} = T \cos B \sec A_n \tag{6}$$

In these two formulas, W = wire diameter; T = width of thread in axial plane at the pitch diameter, or one-half the pitch; B = lead angle of thread at pitch diameter; A = one-half included thread angle in the axial plane; and A_n = one-half included thread angle in the normal plane or plane perpendicular to sides of the thread.

Comparison of Thread Measurement Formulas: Several formulas of varying degrees of accuracy and complexity can be used to compute measurement M for any given thread. The relative accuracy of some of them depends upon the thread angle and lead angle of the thread in question. The absolute error resulting from certain of these formulas also varies with the diameter of the wires placed in the thread grooves.

These formulas for the computation of M may be classified into three groups. Those in the first group do not take the lead angle into account and, therefore, where the lead angle is an appreciable factor, that is, if it is anything but a very small angle, they give the least accurate results.

In the second group are approximate formulas which do take the lead angle of the thread into account but which are not suitable for extreme accuracy where the lead angle is of appreciable size.

In the third group are the so-called exact formulas which are the most complex as well. These are applicable to threads of low or high lead angle and are used for theoretical purposes and for the most precise classes of thread checking, particularly where the higher lead angles are involved.

A general formula of the first type which does not take the lead angle of the thread into account is:

$$M = E - T \cot A + W(1 + \operatorname{cosec} A) \qquad (7)$$

where M = measurement over wires; E = pitch diameter of thread; T = width of thread groove in axial plane at diameter E (or one-half the pitch P); W = wire diameter, which is usually the commercial "best size"; and A = one-half the included thread angle in an axial plane. This formula is accurate for checking practically all of the standard 60-degree *single* screw threads because of the low lead angles involved.

Two formulas of the second group which, although approximate, do take into account the effect of the lead angle are:

$$M = E - T \cot A + W(1 + \operatorname{cosec} A + 0.5 \tan^2 B \cos A \cot A) \qquad (8)$$

where B = lead angle at the pitch diameter and the remaining notation is the same as for Formula (7).

$$M = E + W(1 + \sin A_n), \text{ where } W = \frac{T \cos B}{\cos A_n} \qquad (9 \text{ and } 9a)$$

where A_n = one-half the included thread angle in the normal plane, that is, in a plane perpendicular to the sides of the thread.

Formula (8) is used by the Bureau of Standards in preference to

Formula (7) when the value of that part of the formula, $0.5W \tan^2 B \cos A \cot A$, exceeds 0.00015, as in the case of medium and large lead angles, particularly in combination with small thread angles, such as in the Acme thread.

Formula (9), which is known as the Buckingham Simplified Formula, gives about the same accuracy as Formula (8) for low lead angles, but is more accurate for the higher lead angles. It has the disadvantage that a special size of wires must be used in the measurement. This size is determined by Formula (9a) which takes the lead angle into account.

The third type of formula which takes the lead angle into account and gives a high degree of accuracy even for large lead angles is:

$$M = \frac{2R_b}{\cos G} + W \tag{10}$$

where

$$R_b = \frac{E}{2}\cos F; \quad \tan F = \frac{\tan A}{\tan B};$$

$$\text{inv } G = \frac{T_a}{E} + \text{inv } F + \frac{W}{2R_b \cos H_b} - \frac{\pi}{S};$$

$$T_a = \frac{T}{\tan B}; \quad \tan H_b = \cos F \tan H;$$

and

$$H = 90^\circ - B$$

In this formula, M is the measurement over wires; W is the diameter of the wires, which may be the commercial "best size," since in this formula wire contact near the pitch line is not required; and R_b and G are found by the succeeding formulas. In these succeeding formulas, E is the pitch diameter, A is one-half the thread angle, B is the lead angle, S is the number of "starts" or threads on a multiple worm or screw (2 for double thread, 3 for triple thread, etc.), and T is the width of the thread in the axial plane at the pitch diameter $= 0.5 \times$ pitch of thread.

This formula, known as the Buckingham Involute Helicoid Formula, is somewhat cumbersome to apply because of the number of subsidiary formulas which must be employed to compute the values of R_b and G. It also requires access to a table of involute functions to determine the involute of angle F and the angle corresponding to involute G. It does, however, provide a direct and very accurate solution.

Table 2 gives a comparison of the computed values of M as determined by Formulas (7), (8), (9) and (10) for American Standard 60-degree and Acme 29-degree threads of various lead angles. In each case, the value of M as computed by Formula (10) is taken to be correct to five decimal places. The errors in the values of M computed by the other formulas were found by comparison with the Formula (10) values.

It will be noted that for both American Standard and Acme threads the error in M as computed by Formula (7) increases rapidly as the lead angle increases in size. For the American Standard thread, the error in M as computed by Formula (8) is the same as in that computed by Formula (9). For the Acme Standard thread, the error in M as computed by Formula (8) increases rather rapidly for the higher lead angles, whereas that in the values computed by Formula (9) increases much more slowly throughout the lead angle range shown.

For Acme threads having lead angles of 9.8497 degrees and larger, values of M in Table 2 were computed by Formulas (7) and (8) using two sizes of wires. That is, for each thread, M was computed for the commercial "best size" wire, which would make contact with the flanks of the thread at the pitch line if the thread had a zero degree lead angle and for a special size of wire which is computed by Formula (9a) that takes the lead angle into account and will be referred to hereafter as the "Buckingham size."

If the lead angle of a thread is increased while the pitch is maintained the same, the width of the thread groove at the midpoint or pitch line as measured in a normal plane, perpendicular to the sides of the thread, becomes smaller and smaller. Hence, it can be seen that the commercial "best size" of wire will make contact with the flanks of the thread at points that are farther and farther away from the pitch line as the lead angle becomes larger. If contact at or near the pitch line is a prerequisite of accurate results in the use of the approximate Formulas (7), (8) or (9), then it might be expected that greater errors would be introduced when the so-called "best size" wires were used, than when wires of the size computed by Formula (9a) which make contact nearer the pitch line were used.

Table 2 shows that in the case of Formula (7), it makes comparatively little difference whether the "best wire size" or the "Buckingham wire size" is used. This is probably due to the fact that Formula (7) does not take the lead angle into account. In the

Table 2. Comparison of Computations of Measurement of Screw Threads Over Three Wires to Show Effect of Lead Angle and Wire Diameter on Accuracy of Four Formulas

Thread Angle	Lead Angle	No. of Starts	Pitch Diam.	Wire Diam.	By Formula 7		By Formula 8		By Formula 9		By Formula 10
					M	Error	M	Error	M	Error	
60°	2.1827°	1	1.8557	0.12817	2.04778	0.00005	2.04792	0.00009	2.04792	0.00009	2.04783
60°	4.3591°	2	1.8557	0.12782	2.04674	0.00048	2.04730	0.00008	2.04730	0.00008	2.04722
60°	6.5230°	3	1.8557	0.12725	2.04502	0.00121	2.04627	0.00004	2.04627	0.00004	2.04623
29°	3.3123°	1	1.3750	0.12888	1.53528	0.00080	1.53609	0.00001	1.53610	0.00002	1.53608
29°	6.6025°	2	1.3750	0.12820	1.53188	0.00313	1.53510	0.00009	1.53510	0.00009	1.53501
29°	9.8497°	3	1.3750	0.12709*	1.52634	0.00696	1.53351	0.00021	1.53346	0.00016	1.53330
29°	9.8497°	3	1.3750	0.12911†	1.53643	0.00698	1.54371	0.00030	—	—	1.54341
29°	19.1494°	6	1.3750	0.12156*	1.49872	0.02619	1.52616	0.00125	1.52541	0.00050	1.52491
29°	19.1494°	6	1.3750	0.12911†	1.53643	0.02625	1.56557	0.00289	—	—	1.56268
29°	26.9895°	12	1.8750	0.11430*	1.96247	0.05185	2.01796	0.00364	2.01497	0.00065	2.01432
29°	26.9895°	12	1.8750	0.12911†	2.03643	0.05251	2.09911	0.01017	—	—	2.08894
29°	34.1789°	16	1.8750	0.10575*	1.91977	0.08229	2.01104	0.00898	2.00287	0.00081	2.00206
29°	34.1789°	16	1.8750	0.12911†	2.03643	0.08135	2.14787	0.03009	—	—	2.11778

* Buckingham (Formula 3a) Wire Size
† Commercial "Best Size" Wire based on Zero Helix Angle.

case of Formula (8), however, much more accurate results are obtained in the computation of M for the higher lead angles if the Buckingham wire size is used instead of the "best wire size." This is undoubtedly because the "Buckingham wire size" is computed on the basis of contact near the pitch line after taking the lead angle into account.

Several conclusions may be drawn from the data, shown in Table 2, that have a bearing on the selection of an approximate formula for a given thread.

1. Formula (7) is satisfactory for computing M in the case of American Standard, Acme and worm threads of low lead angles.

2. Formulas (8) and (9) are more accurate than Formula (1) in all cases except for very low lead angles in the American Standard thread where Formula (7) gives sufficient accuracy.

3. For threads having the higher lead angles the best approximate formula is Formula (9).

4. For threads having moderate or large lead angles, more accurate results will be obtained by using the "Buckingham wire sizes" than if the "best size" wires are employed in the measurement of M.

5. Where Formulas (7), (8) or (9) are not sufficiently accurate for computing M, Formula (10) may be employed.

Three considerations govern the choice of an approximate formula for computing M: (a) the probable error that would result; (b) the difficulty of application; (c) whether a commercial wire size must be used or a special size may be obtained.

The probable error can be roughly estimated by using Table 2 as a guide. Take the error shown for the formula under consideration and the given lead angle, interpolating if the given lead angle is not close to any shown in the table. Multiply this error by the ratio of the wire size to be used for measuring the given thread and that given in the table, interpolating if necessary. (*Note:* In the case of Formula (8), both numerator and denominator of this fractional ratio should be either the "best wire size" or the "Buckingham wire size" if the lead angle is 10 degrees or more. Below 10 degrees for Formula (8) and for all estimates in the case of Formula (7) it is not important whether both numerator and denominator are standard or special wire sizes.) The resulting error will be roughly what may be expected if the formula in question is applied to the problem at hand. The actual procedure is outlined in the examples which follow.

As far as difficulty is concerned, Formula (10) undoubtedly involves the most work. Formula (8) is the next in difficulty and Formula (9) is the next, since, although the formula itself is the simplest of the four, a special wire size must be computed.

In the examples which follow, six-place trigonometric and involute values have been used to insure the desired accuracy in the computations.*

Example 1: Compute the measurement M over three wires of a No. 12 American Standard, 60-degree single thread having a pitch diameter of 0.1889 inch and 24 threads per inch. The measuring wires are to be of the "best size" for this pitch of thread or 0.02406 inch diameter. The computed measurement M is to be accurate within ±0.0001 inch.

Procedure: If Table 2 is to be used as a guide in selecting the formula to be used, the first step is to find the lead angle B of this thread.

$$\tan B = \frac{\text{lead}}{\pi \times \text{pitch diameter}} = \frac{\frac{1}{24}}{3.141593 \times 0.1889}$$

$$\tan B = \frac{1}{14.242726} = 0.070211$$

$$B = 4.0162°$$

Referring now to Table 2, it will be noted that for an American Standard 60-degree screw thread having a lead angle of about 4.4 degrees, the use of Formula (7) results in a value of M having an error of 0.00048 inch when a wire size of 0.12782 inch is used. In this example, due to the smaller pitch, the wire size is to be 0.02406 inch; hence the error that would result if Formula (7) were used is estimated as

$$\frac{0.024}{0.128} \times 0.00048 = 0.00009 \text{ inch}$$

which is within the accuracy requirements specified. Hence, Formula (7) will be employed:

* Two tables of trigonometric and involute functions which give data to at least six places are: "Mathematical Tables"—Vol. I of "Manual of Gear Design" by Earle Buckingham, published by The Industrial Press; and "Involutometry and Trigonometry" by Werner F. Vogel, published by The Michigan Tool Co.

$$M = E - T \cot A + W(1 + \operatorname{cosec} A)$$

$$E = 0.1889 \qquad W = 0.02406$$

$$T = \frac{P}{2} = \frac{1}{24 \times 2} \qquad A = \frac{60°}{2} = 30°$$

$$M = 0.1889 - \frac{1}{48} \times 1.732051 + 0.02406(1 + 2)$$

$$M = 0.1889 - 0.03608 + 0.07218 = 0.22500$$

(As checked by Formula (10), $M = 0.22508$, indicating an error of 0.00008 inch, which is within the required accuracy limits specified.)

Example 2: Compute the measurement over three wires of a $\frac{5}{16}$-inch General Purpose Acme *double* thread having 14 threads per inch and a pitch diameter of 0.2768 inch. The computed measurement M is to be accurate within ±0.0001 inch. The commercial "best size" of wires for this pitch of thread, which are 0.03689 inch in diameter are to be used.

Procedure: The lead angle B of the thread is first computed:

$$\tan B = \frac{\text{lead}}{\pi \times \text{pitch diameter}} = \frac{\frac{2}{14}}{3.141593 \times 0.2768}$$

$$\tan B = 0.164280$$

$$B = 9.3292°$$

Reference to Table 2 shows that for an Acme screw thread having a lead angle of about 9.8 degrees the value of M as computed by Formula (7) using the best wire size of 0.12911 inch is in error by about 0.0070 inch. The anticipated error in M if Formula (7) were used for this problem would be slightly less than the ratio of the wire diameters times this error:

$$\frac{0.037}{0.129} \times 0.0070 = 0.0020 \text{ inch}$$

indicating that Formula (7) would not be accurate enough.

Referring again to Table 2, the error in M for Acme thread having a lead angle of about 9.8 degrees as computed by Formula (8) using the best wire size of 0.12911 inch is 0.00030 inch. The anticipated error in M if Formula (8) were used for this problem would be slightly less than:

$$\frac{0.037}{0.129} \times 0.00030 = 0.00009 \text{ inch}$$

which is within the accuracy requirements specified. Hence, Formula (8) will be employed.

$$M = E - T \cot A + W(1 + \operatorname{cosec} A + 0.5 \tan^2 B \cos A \cot A)$$

$$E = 0.2768 \text{ inch} \qquad A = 14\tfrac{1}{2}°$$

$$T = \frac{P}{2} = \frac{1}{14 \times 2} \qquad B = 9.3292°$$

$$W = 0.03689$$

$$M = 0.2768 - \tfrac{1}{28} \times 3.866713 + 0.03689(1 + 3.993929 + 0.5 \times 0.026988 \times 0.968148 \times 3.866713)$$

$$M = 0.2768 - 0.13810 + 0.18609 = 0.32479 \text{ inch}$$

(As checked by Formula (10), $M = 0.32470$ inch, indicating an error of 0.00009 inch, which is within the required accuracy limits specified.)

Example 3: It is required to find the measurement M over three wires for an Acme worm thread having a pitch diameter of 3.25 inches, 2 threads per inch and 9 starts. The size of wires to be used is optional. The computed value of M is to be accurate within ±0.001 inch.

Procedure: The first step is to find the lead angle B:

$$\tan B = \frac{\text{lead}}{\pi \times \text{pitch diameter}} = \frac{9 \times \frac{1}{2}}{3.141593 \times 3.25}$$

$$\tan B = \frac{4.5}{10.21018} = 0.440737$$

$$B = 23.7849°$$

If a "best wire size" is used, W will be 0.25822 inch. If a "Buckingham size" wire is to be used, it must be computed by Formula (9a)

$$W = \frac{T \times \cos B}{\cos A_n}$$

where

$$T = \frac{P}{2} = \frac{0.5}{2} = 0.25 \text{ inch};\ B = 23.7849°;\ \text{and } \tan A_n = \tan A \cos B$$

$$\tan A_n = 0.258618 \times 0.915066$$

$$\tan A_n = 0.236652$$

$$A_n = 13.3142°$$

$$W = \frac{0.25 \times \cos 23.7849°}{\cos 13.3142°} = \frac{0.25 \times 0.915066}{0.973122} = 0.23508$$

Referring to Table 2, it will be noted that for a lead angle of about 19 degrees and a wire size of 0.12156 inch, Formula (7) gives a value for M with an error of 0.02619. With wires about twice this size, the error would be at least doubled so that Formula (7) is not accurate enough. If Formula (8) were used, the error would be at least twice that shown for a lead angle of about 19 degrees and a "Buckingham wire size" of 0.12156 inch, which would be 0.0025 inch, so that Formula (8) is not accurate enough.

Suppose it is assumed that Formula (9) will be used, what is the estimated error? The error shown under Formula (9) M values for 19.1494° is 0.00050 inch and for 26.9895° is 0.00065 inch. If we interpolate roughly between 19° and 27° the corresponding error for 24° (23.7849° rounded off) would be 0.00059 inch. Similarly, if we roughly interpolate between the two Buckingham wire sizes shown, namely 0.12156 inch and 0.11430 inch, to obtain an approximate size for 24° lead angle and pitch corresponding to that of the 19° and 27° threads shown in the table, we get 0.119 inch so that the estimated error would be

$$0.00059 \times \frac{0.235}{0.119} = 0.001 \text{ inch}$$

to the nearest thousandth, indicating that Formula (9) will give a sufficiently accurate result in this problem. Hence, Formula (9) will be used:

$$M = E + W(1 + \sin A_n)$$

where

$$E = 3.25 \text{ inches; } A_n = 13.3142°\text{; and } W = 0.23508$$

$$M = 3.25 + 0.23508(1 + 0.230291)$$

$$M = 3.25 + 0.289 = 3.539 \text{ inches}$$

(As checked by Formula (10), M = 3.538 inches, indicating an error of 0.001 inch, which is within the accuracy limits specified.)

Example 4: It is required to find the measurement M over three wires for an Acme triple worm thread having a pitch diameter of 2 inches and 1½ threads per inch. The size of wires to be used is optional. The computed value of M is to be accurate within ±0.0001 inch.

Procedure: The lead angle B is first found:

$$\tan B = \frac{\text{lead}}{\pi \times \text{pitch diameter}} = \frac{3 \times \frac{2}{3}}{3.141593 \times 2.00}$$

$$\tan B = \frac{1}{3.141593} = 0.318310$$

$$B = 17.6568°$$

If a "best size" wire is used, it will be 0.34430 inch in diameter. The "Buckingham size" would be somewhat smaller. Referring to Table 2 it will be found that Formula (9) gives the most accurate results of the first three formulas for lead angles between 9.8497° and 19.1494°, the error for measurement with wires of 0.12709-inch diameter and 0.12156-inch diameter, respectively, being 0.00016 inch for the smaller lead angle and 0.00050 inch for the larger lead angle. Since the wires to be used in this problem are appreciably larger than those just referred to in the table, the resulting error will certainly be greater than 0.0001 inch. Hence it is necessary to use Formula (10). Since Formula (10) is accurate whether standard or special wire size is used, provided contact is made by the wires on the sides of the thread, the commercial "best size" of wires will be used.

$$E = 2 \text{ inches} \qquad A = 14.5°$$

$$T = \tfrac{1}{3} \text{ inch} \qquad B = 17.6568°$$

$$S = 3 \qquad W = 0.34430 \text{ inch}$$

The first step is to find angle F:

$$\tan F = \frac{\tan A}{\tan B} = \frac{\tan 14.5°}{\tan 17.6568°} = \frac{0.258618}{0.318310}$$

$$\tan F = 0.812472 \qquad F = 39.0929°$$

The second step is to compute R_b:

$$R_b = \frac{E}{2} \cos F = \tfrac{2}{2} \times 0.776124 = 0.776124$$

The third step is to compute T_a:

$$T_a = \frac{T}{\tan B} = \frac{0.333333}{0.31830} = 1.047196$$

The fourth step is to find angle H_b:

$$H = 90° - B = 90° - 17.6568° = 72.3432°$$

$$\tan H_b = \tan H \cos F = 3.141593 \times 0.776124$$

$$\tan H_b = 2.438265$$

$$H_b = 67.7001$$

The fifth step is to find angle G:

$$\text{Inv } G = \frac{T_a}{E} + \text{inv } F + \frac{W}{2R_b \cos H_b} - \frac{\pi}{S}$$

Note that *inv* stands for involute function, the value of which for a given angle is found by referring to a table of involute functions which is similar in arrangement to, and may be included with, a table of trigonometric functions. See tables of involute functions referred to in footnote on page 16–13.

$$\text{Inv } G = \frac{1.047196}{2} + 0.130172 + \frac{0.34430}{2 \times 0.776124 \times 0.379454} - \frac{3.141593}{3}$$

$$\text{Inv } G = 0.523598 + 0.130172 + 0.584544 - 1.047198$$

$$\text{Inv } G = 0.191116$$

$$G = 43.5869°$$

The final step is to compute M:

$$M = \frac{2R_b}{\cos G} + W$$

$$M = \frac{2 \times 0.776124}{0.724329} + 0.34430 = 2.487315$$

$M = 2.4873$ to the nearest ten-thousandth

Finding the Area of an Irregular Plane Figure. The area of an irregular plane figure can be computed by a variety of methods, some of which make use of analytical geometry or the calculus. Two methods which are relatively simple, in that they involve only measurements of a graphical construction and arithmetical computation based on these measurements, are given below. These methods are approximate, but it is possible to obtain a considerable degree of accuracy if they are used with proper care.

Trapezoidal Rule. In Fig. 3 is shown a polygon which is composed of a series of right trapezoids, i.e., a series of four-sided figures having two sides parallel and two sides not parallel. The area of this figure

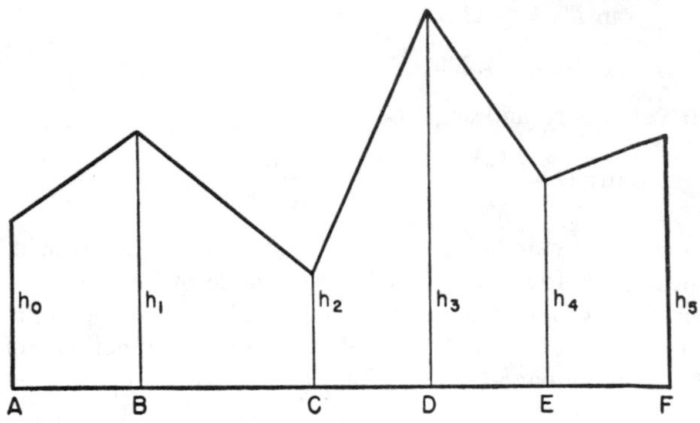

FIG. 3. To Find Area of Polygon. Vertical Lines are Drawn to Form Series of Right Trapezoids Having Two Sides Parallel and Two Sides Not Parallel.

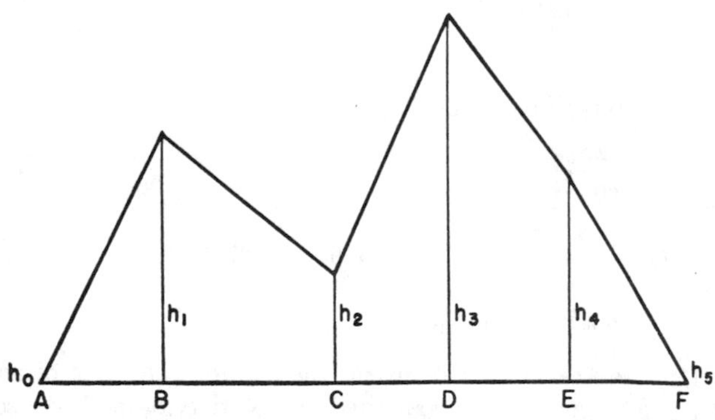

FIG. 4. Polygon in which Two End Areas are Triangles is Modification of that Shown in Fig. 3

can be computed by the following equation:

$$\text{Area} = \tfrac{1}{2}(AB \times h_0 + AC \times h_1 + BD \times h_2 + CE \times h_3 + DF \times h_4 + EF \times h_5) \quad (11)$$

When the shape of the polygon is of the type shown in Fig. 4, the two end areas being triangles, h_0 and h_5 are equal to zero and the formula becomes:

$$\text{Area} = \tfrac{1}{2}(AC \times h_1 + BD \times h_2 + CE \times h_3 + DF \times h_4) \quad (12)$$

When the horizontal distances between the vertical lines h_0, h_1, h_2, etc., are equal, that is $AB = BC = CD = DE = EF = d$, then the formula becomes:

$$\text{Area} = d(\tfrac{1}{2}h_0 + h_1 + h_2 + h_3 + h_4 + \tfrac{1}{2}h_5) \quad (13)$$

In any polygon, a straight line can be drawn between two opposite corners and perpendiculars erected on this line to the remaining corners of the polygon as shown in Fig. 5. This forms two figures of the type shown in Fig. 4. The area of this polygon is then found by computing the area of each figure according to Formula (12) and then adding these two areas together.

Example: In Fig. 5 let $h_1 = 4.4$; $h_2 = 2$; $h_3 = 4.8$; $h_4 = 2.1$; $h_5 = 2.1$; $h_6 = 3.7$; $h_7 = 3.5$; $h_8 = 2.8$; $h_9 = 1.6$; $h_{10} = 2.0$; $AB = 0.8$; $BC = 1.4$; $CD = 1.1$; $DE = 1.0$; $EF = 0.9$; $FG = 0.7$; $GH = 3.3$; $HI = 0.7$; $IJ = 0.2$; $JK = 2.2$; and $KL = 1.0$. Find the area A of this polygon.

Let X_1 = area of upper part of figure
X_2 = area of lower part of figure

Then $X_1 = \tfrac{1}{2}(AF \times h_1 + EI \times h_2 + FL \times h_3)$ (12)

$$= \tfrac{1}{2}(0.8 + 1.4 + 1.1 + 1.0 + 0.9)4.4 + (0.9 + 0.7 + 3.3 + 0.7)2 + (0.7 + 3.3 + 0.7 + 0.2 + 2.2 + 1.0)4.8$$

$$= \tfrac{1}{2}(5.2 \times 4.4 + 5.6 \times 2 + 8.1 \times 4.8)$$

$$X_1 = \tfrac{1}{2}(22.88 + 11.20 + 38.88) = 36.48$$

$$X_2 = \tfrac{1}{2}(AC \times h_{10} + BD \times h_9 + CG \times h_8 + DH \times h_7 + GJ \times h_6 + HK \times h_5 + JL \times h_4) \quad (12)$$

$$= \tfrac{1}{2}(0.8 + 1.4)2.0 + (1.4 + 1.1)1.6 + (1.1 + 1.0 + 0.9 + 0.7)2.8 + (1.0 + 0.9 + 0.7 + 3.3)3.5 + (3.3 + 0.7 + 0.2)3.7 + (0.7 + 0.2 + 2.2)2.1 + (2.2 + 1.0)2.1$$

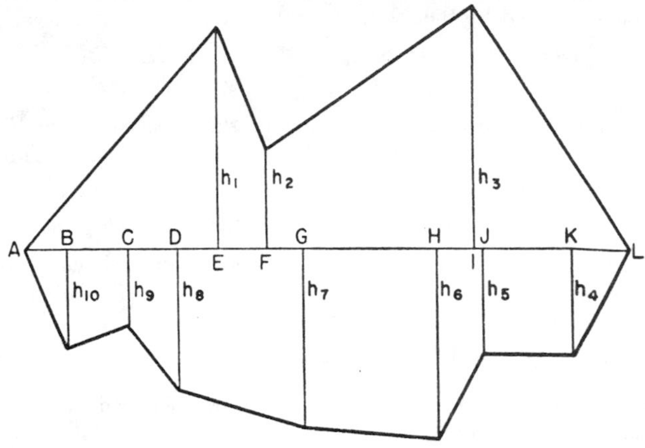

FIG. 5. Showing How Any Polygon can be Converted Into Two Figures of the Type Shown in Fig. 4 for Computation of Area

$$X_2 = \tfrac{1}{2}(2.2 \times 2.0 + 2.5 \times 1.6 + 3.7 \times 2.8 + 5.9 \times 3.5 + 4.2 \times 3.7 + 3.1 \times 2.1 + 3.2 \times 2.1)$$
$$= \tfrac{1}{2}(4.40 + 4.00 + 10.36 + 20.65 + 15.54 + 6.51 + 6.72)$$
$$X_2 = 34.09$$
$$\text{Area} = X_1 + X_2 = 36.48 + 34.09 = 70.57$$

In any irregularly shaped figure such as that shown in Fig. 6, a straight line can be drawn through the figure and perpendiculars erected upon it as shown. Each of the two parts into which the original figure has been divided by the straight line can now be treated as if it were composed of a series of trapezoids. If the boundary of the original figure is curved, then the figures formed by the erection of ordinates are actually not trapezoids since all of their sides are not straight lines and hence this assumption can only lead to an approximation of the area. An increasingly accurate result may be obtained, however, by placing the ordinates closer and closer together, thus increasing the number of "trapezoids."

If the periphery of the figure is a smooth curve, as it is in the upper part of Fig. 6, the ordinates may be erected at regular intervals and Formula (13) used.

If the periphery of the figure is composed of a number of distinct

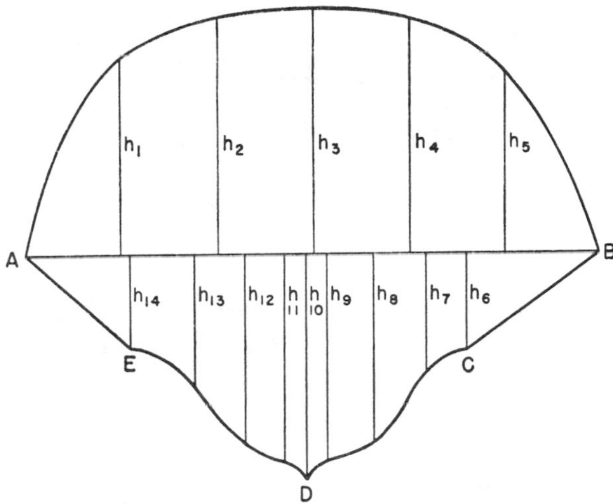

FIG. 6. To Find Area of Irregularly Shaped Figure. Vertical Lines are Drawn to Form Series of Approximate Trapezoids

curves with marked boundary points, as at C, D and E in Fig. 6, somewhat greater accuracy will be assured if ordinates are erected to intersect these points as well as points in between. As many additional ordinates as may be desired can be erected along the original dividing line and here again, the greater the number of ordinates, the greater the accuracy on the computed area. If ordinates are erected to intersect given points on the periphery of the figure, they will probably be unevenly spaced along the dividing line so that Formula (12) will be used. The solution is the same as that outlined in detail in the previous problem.

When the same ordinates extend through both upper and lower parts of the figure as in Fig. 7, the entire area is computed at once using the total lengths of the ordinates in the Trapezoidal formula.

Simpson's Rule. Another method of finding the area of irregular plane figures is by using Simpson's Rule. As in the Trapezoidal Method, a line is drawn connecting two points (usually about opposite each other) on the periphery of the figure such as line AB in Fig. 7. This line is then divided into any *even* number of *equal* parts (the greater the number, the greater the accuracy of the result) and

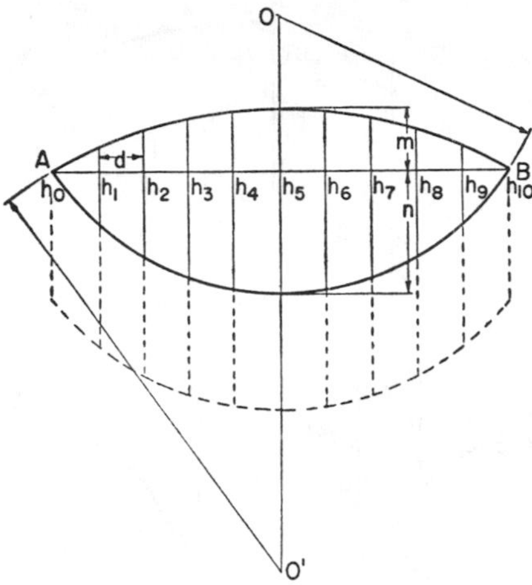

FIG. 7. Showing Application of Simpson's Rule for Finding Irregular Areas

through the points of division perpendiculars or *ordinates* are drawn. These ordinates are lettered h_0, h_1, h_2, h_3, etc., as in Fig. 7, except that in this case h_0 and h_{10} are equal to 0. The length of each ordinate is then measured very carefully. The ordinates having odd subscripts (h_1, h_3, h_5, etc.) are added together and multiplied by 4. Those which have even subscripts (h_2, h_4, h_6, etc., but not including the two end ordinates) are added together and multiplied by 2. These two products plus the end ordinates are multiplied by one-third the horizontal distance d between any two successive ordinates to obtain the area. Thus in Fig. 7,

$$\text{Area} = \frac{d}{3}\left[h_0 + h_{10} + 4(h_1 + h_3 + h_5 + h_7 + h_9) + 2(h_2 + h_4 + h_6 + h_8)\right] \quad (14)$$

The same formula is applied if the lower part of the figure is as shown by the dotted lines in Fig. 7. The only difference is that h_0 and h_{10} are no longer equal to 0.

Example: In Fig. 7 is shown an area made up of two circular segments. Let $h_0 = 0$; $h_1 = 0.820$; $h_2 = 1.350$; $h_3 = 1.718$; $h_4 = 1.910$; $h_5 = 1.98$; $h_6 = 1.910$; $h_7 = 1.718$; $h_8 = 1.350$; $h_9 = 0.820$; $h_{10} = 0$; and $d = 0.498$.

Find the area by Simpson's Rule; by the Trapezoidal Rule; and by computation of the area of the two circular segments.

By Simpson's Rule:

$$\text{Area} = \frac{d}{3}[h_0 + h_{10} + 4(h_1 + h_3 + h_5 + h_7 + h_9) + 2(h_2 + h_4 + h_6 + h_8)] \quad (14)$$

$$\text{Area} = \frac{0.498}{3}[0 + 0 + 4(0.820 + 1.718 + 1.980 + 1.718 + 0.820) + 2(1.350 + 1.910 + 1.910 + 1.350)]$$

$$\text{Area} = 0.166\,[(4 \times 7.056) + (2 \times 6.520)]$$

$$\text{Area} = 0.166 \times 41.264 = 6.850$$

By Trapezoidal Rule:

$$\text{Area} = d(\tfrac{1}{2}h_0 + h_1 + h_2 + h_3 + h_4 + h_5 + h_6 + h_7 + h_8 + h_9 + \tfrac{1}{2}h_{10}) \quad (13)$$

$$\text{Area} = 0.498(0 + 0.820 + 1.350 + 1.718 + 1.910 + 1.980 + 1.910 + 1.718 + 1.350 + 0.820 + 0)$$

$$\text{Area} = 0.498 \times 13.576 = 6.761$$

By Calculation of Areas of Circular Segments:

If the altitude m of the top segment is 0.66 and the length of its chord $10 \times d$ is equal to 4.98, the area may be found by Formula (3) given in the first part of this chapter:

$$\text{Area} = \frac{2 \times 10d \times m}{3} + \frac{m^3}{2 \times 10d} \quad (3)$$

$$\text{Area} = \frac{2 \times 4.98 \times 0.66}{3} + \frac{0.66^3}{2 \times 4.98}$$

$$\text{Area} = 2.1912 + 0.0289 = 2.2201$$

If the altitude n of the bottom segment is 1.32 and the length of its chord $10 \times d$ is equal to 4.98, the area is found by this same formula:

$$\text{Area} = \frac{2 \times 10d \times n}{3} + \frac{n^3}{2 \times 10d} \quad (3)$$

$$\text{Area} = \frac{2 \times 4.98 \times 1.32}{3} + \frac{1.32^3}{2 \times 4.98}$$

$$\text{Area} = 4.3824 + 0.2309 = 4.6133$$

$$\text{Total Area} = 2.220 + 4.613 = 6.833$$

Computing Periphery of an Ellipse. The periphery p of an ellipse can be computed approximately by means of the following formula:

$$p = \frac{\pi}{2}(D + d)k \tag{15}$$

where p = periphery

D = long diameter or major axis

d = short diameter or minor axis

k = a constant

n = ratio of difference to sum of long and short diameters $= \dfrac{D - d}{D + d}$

Values for the constant k corresponding to different values of n are given in Table 3.

In some instances, as for example, when n is known to more than two decimal places, it may be desirable to compute k. In such cases the value of k for values of n up to 0.70 can be computed correctly to five significant figures by the following formula:

$$k = \frac{64 - 3n^4}{64 - 16n^2} \tag{16}$$

The value of k for values of n from 0.70 to 1.00 can be computed correctly to five significant figures by the following formula:

$$k = 0.147202 + 5.355178n - 13.197449n^2 + 16.892262n^3 - 10.613854n^4 + 2.689901n^5 \tag{17}$$

In most instances, however, it will be sufficiently accurate to determine k by interpolation in the table when n is known to more than two decimal places. This method is used in the following example.

Example: Find the periphery of an ellipse having a long diameter D of 7 feet 5 inches and a short diameter d of 9 inches.

$$n = \frac{D - d}{D + d} = \frac{89 - 9}{89 + 9} = \frac{80}{98} = 0.8163$$

Table 3. Ratios *n* and Corresponding Values of *k*

n	k	n	k	n	k	n	k
0	1	0.25	1.0157	0.50	1.0635	0.75	1.1465
0.01	1.0000	0.26	1.0170	0.51	1.0662	0.76	1.1506
0.02	1.0001	0.27	1.0183	0.52	1.0688	0.77	1.1548
0.03	1.0002	0.28	1.0197	0.53	1.0716	0.78	1.1591
0.04	1.0004	0.29	1.0211	0.54	1.0743	0.79	1.1634
0.05	1.0006	0.30	1.0226	0.55	1.0772	0.80	1.1678
0.06	1.0009	0.31	1.0242	0.56	1.0801	0.81	1.1723
0.07	1.0012	0.32	1.0258	0.57	1.0830	0.82	1.1768
0.08	1.0016	0.33	1.0274	0.58	1.0860	0.83	1.1815
0.09	1.0020	0.34	1.0291	0.59	1.0891	0.84	1.1862
0.10	1.0025	0.35	1.0309	0.60	1.0922	0.85	1.1909
0.11	1.0030	0.36	1.0327	0.61	1.0954	0.86	1.1958
0.12	1.0036	0.37	1.0345	0.62	1.0987	0.87	1.2007
0.13	1.0042	0.38	1.0364	0.63	1.1020	0.88	1.2057
0.14	1.0049	0.39	1.0384	0.64	1.1053	0.89	1.2108
0.15	1.0056	0.40	1.0404	0.65	1.1088	0.90	1.2160
0.16	1.0064	0.41	1.0425	0.66	1.1123	0.91	1.2213
0.17	1.0072	0.42	1.0446	0.67	1.1158	0.92	1.2266
0.18	1.0081	0.43	1.0468	0.68	1.1194	0.93	1.2321
0.19	1.0090	0.44	1.0490	0.69	1.1231	0.94	1.2376
0.20	1.0100	0.45	1.0513	0.70	1.1268	0.95	1.2433
0.21	1.0111	0.46	1.0536	0.71	1.1306	0.96	1.2490
0.22	1.0121	0.47	1.0560	0.72	1.1345	0.97	1.2549
0.23	1.0133	0.48	1.0585	0.73	1.1384	0.98	1.2609
0.24	1.0145	0.49	1.0610	0.74	1.1424	0.99	1.2670
0.25	1.0157	0.50	1.0635	0.75	1.1465	1.00	1.2732

From the accompanying table it will be noted that when $n = 0.81$, $k = 1.1723$ and when $n = 0.82$, $k = 1.1768$. Thus, at this point in the table a difference in n of 0.01 is equivalent to a difference in k of 0.0045. Since $n = 0.8163$, a difference in n of $0.8163 - 0.8100 = 0.0063$ is equal to a difference in k of $\frac{0.0063}{0.0100} \times 0.0045$, or 0.0028. Hence, $k = 1.1723 + 0.0028 = 1.1751$.

$$\text{Thus, } p = \frac{\pi}{2}(89 + 9)1.1751 = 180.89 \text{ inches} = 15 \text{ feet } 0.89 \text{ inch} \qquad (15)$$

17

Problems in Mechanics—Where Forces Are in Equilibrium

What might be called "refresher problems" are presented in this and the two succeeding chapters. The purpose is to highlight and clarify, by examples worked out in detail, some of the methods which may be employed in solving problems involving forces acting on bodies which are assumed to be absolutely rigid (mechanics) and forces acting on non-rigid bodies (strength of materials). No attempt has been made to give a comprehensive review of problems in these fields nor to include any detailed treatment of basic theory underlying their solution. For information of this character, the reader is referred to standard textbooks on Mechanics and Strength of Materials. The intent is rather to present a selection of typical problems that will be helpful to readers who wish to brush up on methods of analysis and solution.

In each of these chapters, the first problems are relatively simple and serve to illustrate the principles involved and the methods of approach. They "pave the way" for the more difficult problems that follow. A clearer grasp of the methods of analysis and solution will, therefore, result if these problems are studied in the sequence presented.

In this chapter, the problems deal with forces acting upon bodies which are in equilibrium. A body is said to be in equilibrium when the resultant of the forces acting upon it is equal to zero. Such a body is either in a state of rest or is moving in a straight line at a constant velocity. It is assumed that the reader has some knowledge of the use of vector diagrams of forces as a basis for the mathematical solution of problems of this type.

Simple Lever

PROBLEM 1: What is the total pressure P (Fig. 1) that can be carried by the safety valve if the weight W_1 of the valve arm, the balance weight W_2 and the distances a, b and c from the pivot point of the arm are known.

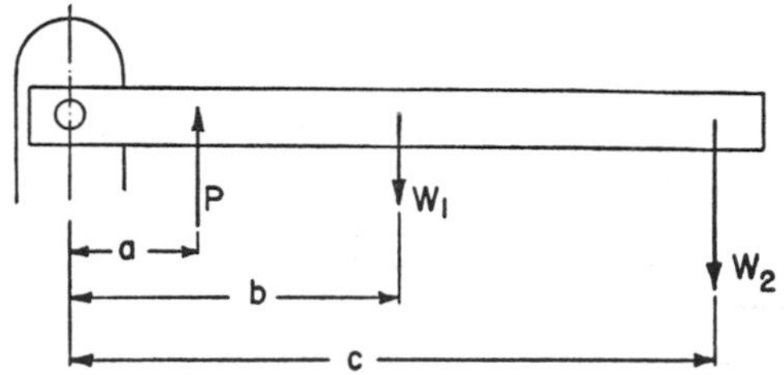

FIG. 1. To Find Upward Pressure on Safety Valve If Weights W_1 and W_2 and Distances a, b, and c are Known

Analysis of Problem: This is a relatively simple problem involving the equality of moments. If the safety valve arm is to be in equilibrium, the sum of the moments of W_1 and W_2 about the pivot point must equal the moment of P about this point. P can be found by dividing the sum of the moments by the moment-arm a of P.

Formula:

$$P = \frac{W_1 \times b + W_2 \times c}{a}$$

Derivation of Formula: The moment of a force about a given point is equal to the product of the magnitude of the force and the perpendicular distance between the line of action of the force and the point, so that:

$$P \times a = W_1 \times b + W_2 \times c \qquad (1a)$$

$$P = \frac{W_1 \times b + W_2 \times c}{a} \qquad (1b)$$

Example: In Fig. 1 let $W_1 = 8$ pounds, $W_2 = 25$ pounds, $a = 7\frac{1}{2}$ inches, $b = 15$ inches and $c = 34$ inches. Find P.

Solution:

$$P = \frac{8 \times 15 + 25 \times 34}{7.5} = \frac{970}{7.5} = 129.3 \text{ pounds} \qquad (1b)$$

Compound Levers

PROBLEM 2: In the system of compound levers shown in Fig. 2, which represents the working mechanism of a platform scales, it is required to find the ratio of the scale weight G to the platform load A. The distances n, o, p, q, r, s and t are known.

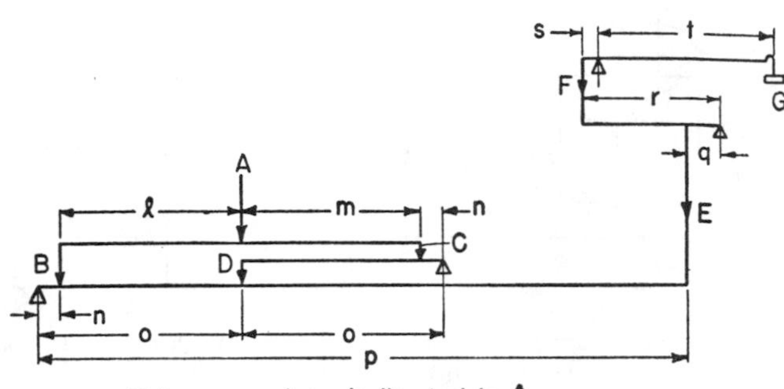

FIG. 2. To Find Ratio of Scale Weight G to Platform Load A When Distances n, o, p, q, r, s, and t are Known

Analysis of Problem: Beginning with either the weight G or the load A and utilizing the principle of moments that for a body at rest, the sum of the moments about a given point of forces tending to produce rotation in one direction is equal to the sum of the moments about the same point of forces tending to produce rotation in the opposite direction, the effect of either can be traced throughout the system. Note that when each lever is considered by itself, the direction of the forces acting on it should be taken as if it were in balance, as shown under Derivation of Formula.

Formula:

$$A = \frac{p}{n} \times \frac{r}{q} \times \frac{t}{s} \times G$$

Derivation of Formula: Taking moments about the fulcrum points:

$$F \times s = G \times t \qquad \text{hence} \qquad F = \frac{t}{s} G \tag{1}$$

Whereas in Equation (1) F is considered to be acting *downward* in opposition to G; in Equation (2) F is considered to be acting *upward* in opposition to E.

$$E \times q = F \times r \qquad \text{hence} \qquad E = \frac{r}{q} F \tag{2}$$

Substituting the equivalent of F in Equation (2) as indicated by Equation (1)

$$E = \frac{r}{q} \times \frac{t}{s} G \tag{3}$$

If E is considered to be acting in opposition to D and B

$$D \times o + B \times n = Ep \tag{4}$$

Assuming that C opposes D

$$C \times n = D \times o \tag{5}$$

Substituting in Equation (4) the equivalent of E as obtained from Equation (3) and the equivalent of $D \times o$ as obtained from Equation (5)

$$C \times n + B \times n = p \times \frac{r}{q} \times \frac{t}{s} G \tag{6a}$$

$$C + B = \frac{p}{n} \times \frac{r}{q} \times \frac{t}{s} G \tag{6b}$$

If there were a pivot point at B

$$A \times \ell = C \times (\ell + m) \qquad C = \frac{\ell}{\ell + m} A \tag{7}$$

If there were a pivot point at C

$$A \times m = B \times (\ell + m) \qquad B = \frac{m}{\ell + m} A \tag{8}$$

Substituting the equivalents for B and C obtained from Equations (7) and (8) in Equation (6b)

$$\frac{\ell}{\ell + m} A + \frac{m}{\ell + m} A = \frac{p}{n} \times \frac{r}{q} \times \frac{t}{s} G \tag{9a}$$

$$\frac{\ell + m}{\ell + m} A = A = \frac{p}{n} \times \frac{r}{q} \times \frac{t}{s} G \tag{9b}$$

From this equation it can be seen that ℓ or m has no effect on the relationship between A and G, or, in other words, the position of the load on the platform does not affect the weight needed to balance it. In working out this problem, it has been assumed that the weight of the levers themselves has been counterbalanced.

Example: In Fig. 2 let $n = 6$, $o = 48$, $p = 150$, $q = 8$, $r = 32$, $s = 4$ and $t = 40$ inches. Find G if A is 500 pounds.

Solution:
$$500 = \frac{150}{6} \times \frac{32}{8} \times \frac{40}{4} G = 25 \times 4 \times 10G \qquad (9b)$$

$$G = \frac{500}{1000} = 0.5 \text{ pound}$$

Forces Not Acting at Common Point

PROBLEM 3: Find the reactive forces P_x, F_x and F_y at the supporting points of the jib crane shown in Fig. 3 if the load L carried by the crane, the weight W of the crane itself, the distances a and b and the height h are known.

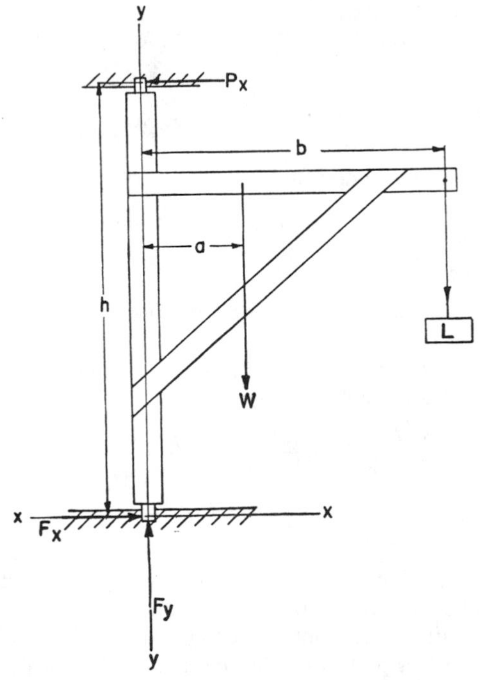

Fig. 3. To Find Reactive Forces P_x, F_x, and F_y Acting on Jib Crane

Analysis of Problem: This is a problem involving the equilibrium of several forces whose lines of action do not pass through a common point. The equilibrium of these forces is established, if the sum of their x-components is equal to zero, the sum of their y-components is equal to zero, and the sum of their moments about a given point is equal to zero. Draw the x-axis through that point in the lower supporting pin of the crane through which the horizontal reactive force F_x acts and the y-axis through that point in the same pin through which the vertical reactive force F_y acts. Then F_y must equal W plus L and F_x must equal P_x. By taking moments about the point of intersection of the x- and y-axes, the value of P_x, and hence F_x, can be found.

In this problem the forces are acting in lines parallel to either the x- or y-axes. Where a force is not parallel to one of these two axes, its x component is found by multiplying it by the cosine of the angle which its line of action makes with the x-axis and, similarly, its y-component is found by multiplying it by the cosine of the angle which its line of action makes with the y-axis.

Formulas:

$$F_y = W + L$$

$$F_x = P_x = \frac{W \times a + L \times b}{h}$$

Derivation of Formulas:

$$F_y = W + L \tag{1}$$

$$F_x = P_x \tag{2}$$

$$W \times a + L \times b = P_x \times h \tag{3a}$$

$$P_x = \frac{W \times a + L \times b}{h} \tag{3b}$$

Example: In Fig. 3 let $W = 1750$ pounds, $L = 1500$ pounds, $a = 3$ feet 6 inches, $b = 10$ feet and $h = 15$ feet 10 inches. Find P_x, F_x and F_y.

Solution: $F_y = 1750 + 1500 = 3250$ pounds

$$F_x = P_x = \frac{1750 \times 42 + 1500 \times 120}{190} = 1334 \text{ pounds} \tag{3b}$$

Forces Not Acting in Same Plane

PROBLEM 4: In the stiff-leg derrick shown in Fig. 4A, the load L, the length k of the boom, the angle C (between the boom and the horizontal), the height h of the mast above the pivot point of the boom, and the height m of this pivot point above the ground are known. The boom W, the boom supporting cable X and the mast are in the same vertical plane. The angle A of leg Y and the horizontal and the angle B of leg Z and the horizontal are known, as well as angles D and E which lie between the vertical planes respectively passing through each leg and the mast and the vertical plane of the mast and the boom. It is required to find the total stresses in boom W, supporting cable X and legs Y and Z.

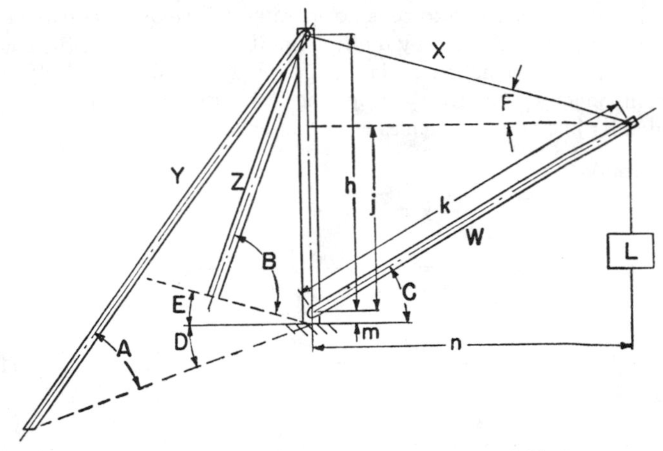

FIG. 4A. To Find Total Stresses in Boom W, Supporting Cable X, and Legs Y and Z, When Load L, Length k, Angle C, Height h and Height m are Known

Analysis of Problem: The angle F made by the boom supporting cable with the horizontal is first determined and a free body diagram, (a diagram showing all the forces acting on a given point or part) Fig. 4B, of the forces acting on the pivot point of the supporting cable, the boom and the load cable is drawn. Since this diagram shows forces acting in a vertical plane, x- and z-reference axes are used. Based on the fact that if this pivot point is in equilibrium, the sums of the x- and z-components respectively must equal zero, two simultaneous equations can be established and solved for X and W.

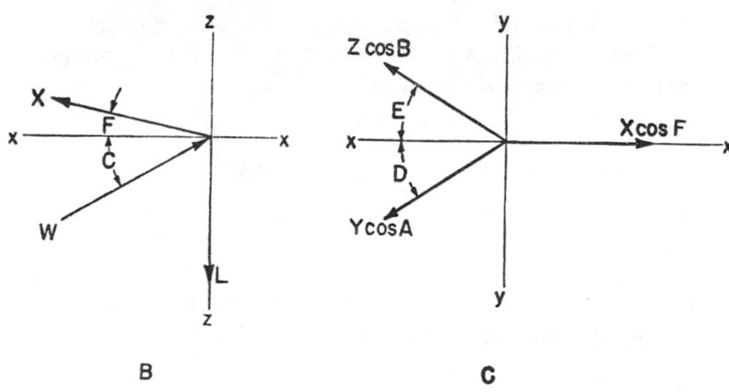

FIG. 4B. Free Body Diagram Showing Forces Acting on Pivot Point of Supporting Cable, Boom and Load. C. Free Body Diagram Showing Forces Acting on Pivot Point of Two Legs, Mast, and Supporting Cable

A free body diagram, Fig. 4C, is next drawn showing the forces acting in a horizontal plane on the pivot point of the two legs, the mast, and the supporting cable. It will be noted that the forces shown in this diagram are horizontal components of forces X, Y and Z. Since this diagram shows forces acting in a horizontal plane, x and y axes are used. Based on the fact that if this pivot point is in equilibrium, the sums of the x- and y-components of these forces must respectively equal zero, two simultaneous equations can be established and solved for Y and Z.

Procedure:

Referring to Fig. 4A. Establishing equations for unknown distances j and n and unknown angle F:

$$j = k \sin C \tag{1}$$

$$n = k \cos C \tag{2}$$

$$\tan F = \frac{h - j}{n} \tag{3}$$

Referring to Fig. 4B. Establishing equations for z-components of forces X and W and force L and for x-components of forces X and W.

$$X \sin F + W \sin C - L = 0 \tag{4}$$

$$X \cos F - W \cos C = 0 \tag{5}$$

Referring to Fig. 4C. Establishing equations for x-components of forces $Y \cos A$ and $Z \cos B$ and force $X \cos F$ and for y-components of forces $Y \cos A$ and $Z \cos B$:

$$Z \cos B \cos E + Y \cos A \cos D - X \cos F = 0 \qquad (6)$$

$$Z \cos B \sin E - Y \cos A \sin D = 0 \qquad (7)$$

Example: In Fig. 4A, the load L weighs 4200 pounds; the length k of the boom is 32 feet; the height h of the mast is 23 feet; the angle C is 29 degrees; the angle A is 41°; the angle B is 39°; the angle D is 44°; and the angle E is 46°. What are the total stresses W in the boom; X in the supporting cable; and Y and Z in the legs?

Solution:

$$j = 32 \sin 29° \qquad (1)$$

$$j = 32 \times 0.48481 = 15.51 \text{ feet}$$

$$n = 32 \cos 29° \qquad (2)$$

$$n = 32 \times 0.87462 = 27.99 \text{ feet}$$

$$\tan F = \frac{23 - 15.51}{27.99} = 0.26760 \qquad (3)$$

$$F = 14°58.9'$$

$$X \times 0.25851 + W \times 0.48481 - 4200 = 0 \qquad (4)$$

$$X \times 0.96607 - W \times 0.87462 = 0 \qquad (5)$$

Multiplying Equation (4) by 0.96607 and Equation (5) by 0.25851 and solving for W

$$\begin{array}{r} 0.24974X + 0.46836W = 4057.2 \\ -(0.24974X - 0.22610W = 0) \\ \hline 0.69446W = 4057.2 \end{array}$$

$$W = 5842.2 \text{ pounds}$$

Substituting this value of W in Equation (5) and solving for X

$$0.96607X = 5842.2 \times 0.87462$$

$$X = \frac{5109.7}{0.96607} = 5289.2 \text{ pounds}$$

$$Z \cos 39° \cos 46° + Y \cos 41° \cos 44° - 5289.2 \times \cos 14°58.9' = 0 \qquad (6)$$

$$Z \times 0.77715 \times 0.69466 + Y \times 0.75471 \times 0.71934 - 5289.2 \times 0.96607 = 0$$

$$0.53986Z + 0.54289Y = 5109.7$$

$$Z \cos 39° \sin 46° - Y \cos 41° \sin 44° = 0 \qquad (7)$$

$$Z \times 0.77715 \times 0.71934 - Y \times 0.75471 \times 0.69466 = 0$$

$$0.55904Z - 0.52427Y = 0$$

Multiplying the simplified Equation (6) by 0.55904 and the simplified Equation (7) by 0.53986 and solving for Y

$$\begin{array}{r} 0.30180Z + 0.30350Y = 2856.5 \\ -(0.30180Z - 0.28303Y = 0) \\ \hline 0.58653Y = 2856.5 \end{array}$$

$$Y = \frac{2856.5}{0.58653} = 4870.2 \text{ pounds}$$

Substituting this value of Y in the simplified Equation (7) and solving for Z

$$0.55904Z - 0.52427 \times 4870.2 = 0$$

$$Z = \frac{2553.3}{0.55904} = 4567.3 \text{ pounds}$$

Rounded off to three figures, the required values are:

$W = 5840$ pounds $\qquad Y = 4870$ pounds

$X = 5290$ pounds $\qquad Z = 4570$ pounds

Sliding on Inclined Plane

PROBLEM 5: Find the force P (Fig. 5) required (a) to just prevent motion of a block of weight W down a plane inclined to the horizontal at angle α and (b) to just start motion up the plane if the coefficient of static friction μ is known.

Analysis of Problem: In Fig. 5, F_1 represents the frictional force acting to oppose motion of the block down the plane. It will be noted that this force is inclined at angle ϕ with a line drawn vertical to the sliding surface of the block. The tangent of this angle, called the angle of friction, is equal to the coefficient of static friction. This coefficient usually represents the maximum static friction, which is present just before motion takes place. F_2 represents the frictional force acting to oppose motion of the block up the plane. It also is at an angle ϕ with the line drawn vertical to the sliding surface. Although both F_1 and F_2 are shown in Fig. 5, only one of these two forces is acting at any given time, F_1 when motion is impending down the plane and F_2 when motion is impending up the plane.

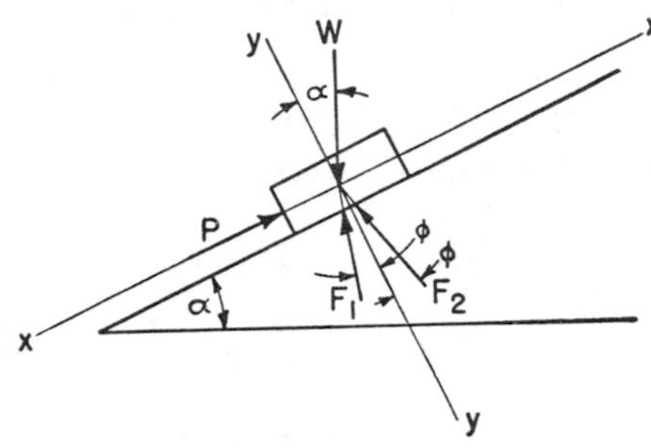

FIG. 5. To Find Force P Required to (a) Prevent Motion of Block Down Plane and (b) Start Motion of Block up Plane When Weight W, Angle α and Coefficient of Static Friction μ are Known

For forces in equilibrium whose lines of action intersect at a common point, the sum of the x-components and the sum of the y-components of these forces must respectively equal zero. See Analysis of Problem 3 for method of finding x- and y-components of forces not parallel to either axis.

Draw the x-axis through the line of action of force P and draw the y-axis at right angles to this through the intersection of the lines of action of forces W, P, F_1 and F_2.

(a) *Motion impending down the plane.* The equation of y-components permits F_1 to be determined. With F_1 known, the equation of x-components permits P to be found.

(b) *Motion impending up the plane.* The equation of y-components permits F_2 to be determined. With F_2 known, the equation of x-components permits P to be found.

Formulas:

(a) $$P = W(\sin \alpha - \mu \cos \alpha)$$

(b) $$P = W(\sin \alpha + \mu \cos \alpha)$$

Derivation of Formulas:

(a) *For motion impending down the plane*

Sum of y-components equals 0

$$F_1 \cos \phi - W \cos \alpha = 0 \tag{1a}$$

$$F_1 = \frac{W \cos \alpha}{\cos \phi} \tag{1b}$$

Sum of x-components equals 0

$$P + F_1 \sin \phi - W \sin \alpha = 0 \tag{2}$$

therefore
$$P + \frac{W \cos \alpha}{\cos \phi} \sin \phi - W \sin \alpha = 0 \tag{3a}$$

$$P = W \sin \alpha - W \cos \alpha \tan \phi \tag{3b}$$

But
$$\mu = \tan \phi \tag{4}$$

hence
$$P = W(\sin \alpha - \mu \cos \alpha) \tag{5}$$

(b) *For motion impending up the plane*

Sum of y-components equals 0

$$F_2 \cos \phi - W \cos \alpha = 0 \tag{6a}$$

$$F_2 = \frac{W \cos \alpha}{\cos \phi} \tag{6b}$$

Sum of x-components equals 0

$$P - W \sin \alpha - F_2 \sin \phi = 0 \tag{7}$$

Substituting the equivalent of F_2 from Equation (6b)

$$P - W \sin \alpha - \frac{W \cos \alpha}{\cos \phi} \sin \phi = 0 \tag{8a}$$

$$P = W \sin \alpha + W \cos \alpha \tan \phi \tag{8b}$$

$$P = W(\sin \alpha + \mu \cos \alpha) \tag{9}$$

Example: If $W = 43$ pounds; $\alpha = 25°$; and $\mu = 0.25$. Find P for motion impending (a) down the plane and (b) up the plane.

Solution:

(a) $P = 43(0.4226 - 0.25 \times 0.9063) = 8.43$ pounds (5)

(b) $P = 43(0.4226 + 0.25 \times 0.9063) = 27.92$ pounds (9)

Lifting Weight on Jackscrew

PROBLEM 6: How great a force Q must be applied to a lever passing through the head of a jackscrew, and at a distance R from the center line of the screw to lift a weight W, if the screw has a single square thread of pitch radius r and lead ℓ, and the coefficient of sliding friction for the screw is μ.

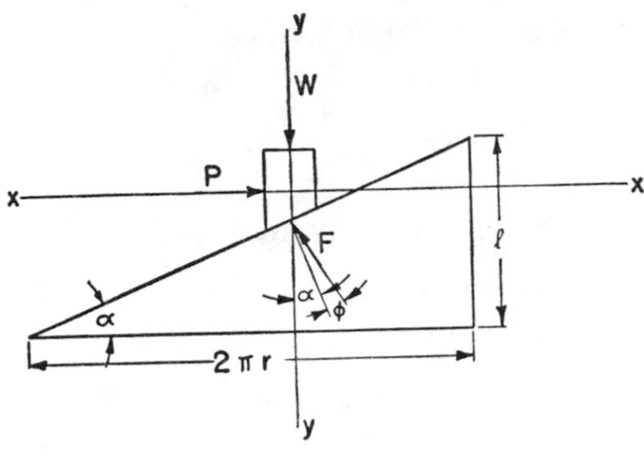

FIG. 6. To Find Force Q Required to Lift Weight W on Jack-screw When Pitch Radius r, Lead ℓ, Length R, and Coefficient of Friction μ are Known

Analysis of Problem: If the weight W is moving steadily upward without acceleration, the forces acting upon the jackscrew at any given moment may be considered to be in equilibrium, and a force diagram constructed accordingly. If the thread of the jackscrew could be "unrolled" it would, in effect, form an inclined plane. For a length of the screw thread equal to one turn, the base of the inclined plane would be equal to the pitch circumference of the screw or, $2\pi r$ in length, and the height would be equal to the lead ℓ of the screw. The diagram, Fig. 6, shows the weight W being pushed up this inclined plane by force P and being resisted in its upward motion by frictional force F. The angle of friction ϕ formed by the force F with a line at right angles to the inclined plane is readily found since its tangent is equal to the coefficient of sliding friction μ. Draw the y-axis through the line of action of weight W and the x-axis through the line of action of force P. Then, the sum of the x-components and the sum of the y-components of these three forces must equal zero. The y-component equation is set up first and solved for F.

The x-component equation is set up next and solved for P. P represents the amount of force required to be applied at the pitch radius r of the screw.

According to the principle of moments, the force Q times its distance R from the axis of the screw must equal the force P times its distance r from the screw axis. Hence, an equation can be established and solved for Q.

Formula: $$Q = \frac{rW}{R} \tan (\phi + \alpha)$$

where $\tan \phi = \mu$ and $\tan \alpha = \dfrac{\ell}{2\pi r}$

Derivation of Formula:

$$F \cos (\phi + \alpha) = W \tag{1a}$$

$$F = \frac{W}{\cos (\phi + \alpha)} \tag{1b}$$

$$P = F \sin (\phi + \alpha) \tag{2}$$

Substituting the equivalent of F from Equation (1b) in Equation (2)

$$P = W \frac{\sin (\phi + \alpha)}{\cos (\phi + \alpha)} = W \tan (\phi + \alpha) \tag{3}$$

$$Q \times R = P \times r \tag{4a}$$

$$Q = \frac{P \times r}{R} \tag{4b}$$

Substituting the equivalent of P from Equation (3) in Equation (4b)

$$Q = \frac{r}{R} \times W \tan (\phi + \alpha) \tag{5}$$

where $$\tan \phi = \mu \tag{6}$$

and $$\tan \alpha = \frac{\ell}{2\pi r} \tag{7}$$

Example: In this problem, let $W = 2000$ pounds; $R = 4$ feet; $r = 1\frac{3}{8}$ inches; $\ell = \frac{1}{2}$ inch; and $\mu = 0.15$.

Solution: $\tan \phi = 0.15$ (6)

$$\phi = 8°32'$$

$$\tan \alpha = \frac{0.5}{2 \times \pi \times 1.375} = 0.0579 \qquad (7)$$

$$\alpha = 3°19'$$

$$Q = \frac{1.375 \times 2000}{48} \times \tan 11°51' \qquad (5)$$

$$Q = 57.29 \times 0.2098 = 12.02 \text{ pounds}$$

Lifting Weight by Means of Wedge

PROBLEM 7: Determine the horizontal force P required to be imposed on a wedge of weight W_2 and having a supporting surface at angle α with the horizontal if it is to just start the upward movement of a block of weight W_1, sliding between two vertical surfaces as shown in Fig. 7A. The coefficient of friction μ is assumed to be the same for all surfaces in contact.

Analysis of Problem: In order to solve this problem, it is necessary to consider the forces acting on the block separately from those acting on the wedge as shown respectively in Figs. 7B and 7C.

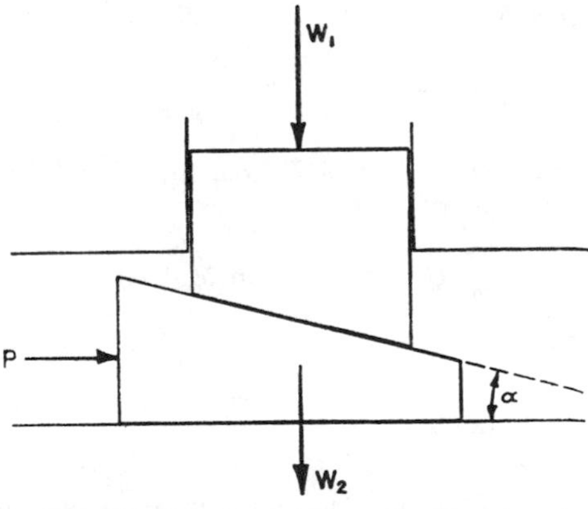

FIG. 7A. To Find Force P Imposed on Wedge of Weight W_2 to Start Movement of Block of Weight W_1 if Angle α and Coefficient of Friction μ are Known

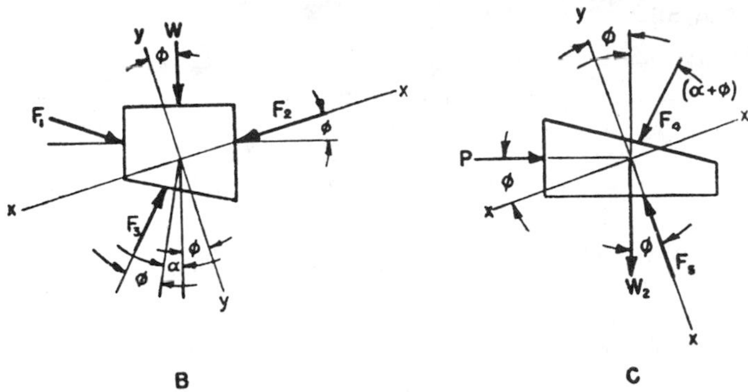

FIG. 7B. Free Body Diagram Showing Forces Acting on Block. C. Free Body Diagram Showing Forces Acting on Wedge

In Fig. 7B, three forces F_1, F_2 and F_3 are shown since there are three surfaces of the block in contact with other surfaces. However, the tendency of force P is to push the block in the direction which would create or increase the frictional resistance F_2 and to lessen or remove entirely any initial friction as represented by F_1. Since the block is considered to be free to slide easily in a vertical direction when no sideways pressure is exerted against it, F_1 will be considered as equal to 0. With F_1 eliminated, there are two unknowns F_2 and F_3, the latter representing the frictional force exerted between the wedge and the block contacting surfaces. If the forces acting on the block are in equilibrium, as they would be just before motion impends, the sum of their x-components and the sum of their y-components should equal zero. By drawing the x-axis through F_2 as shown, F_2 will have no y-component and in an equation of y components F_3 will be the only unknown and can be solved for.

In Fig. 7C, the weight W_2 of the wedge is known and F_4 is numerically equal to F_3 although opposite in direction. F_5 is the frictional force acting against the bottom surface of the wedge. Here, again, if these forces are in equilibrium, as they must be just before motion impends, the sum of their x-components and the sum of their y-components must be equal to 0. By taking the y-axis through F_5 as shown, the x-component of F_5 will be eliminated and in an equation of x-components P will be the only unknown and can be solved for.

Formula: $P = W_1 \tan (2\phi + \alpha) + W_2 \times \mu$

Derivation of Formula:

$$\tan \phi = \mu \tag{1}$$

$$F_3 \cos (2\phi + \alpha) = W_1 \cos \phi \tag{2a}$$

$$F_3 = \frac{W_1 \cos \phi}{\cos (2\phi + \alpha)} \tag{2b}$$

$$F_4 = F_3 \tag{3}$$

$$P \cos \phi = F_4 \sin (2\phi + \alpha) + W_2 \sin \phi \tag{4a}$$

$$P = \frac{F_4 \sin (2\phi + \alpha)}{\cos \phi} + \frac{W_2 \sin \phi}{\cos \phi} \tag{4b}$$

Substituting the equivalent of F_3 from Equation (2b) for F_4 in Equation (4b)

$$P = \frac{W_1 \cos \phi \sin (2\phi + \alpha)}{\cos \phi \cos (2\phi + \alpha)} + W_2 \tan \phi \tag{5a}$$

$$P = W_1 \tan (2\phi + \alpha) + W_2 \tan \phi \tag{5b}$$

$$P = W_1 \tan (2\phi + \alpha) + W_2 \times \mu \tag{5c}$$

Example: In Fig. 7A let $W_1 = 750$ pounds, $W_2 = 10$ pounds, $\mu = 0.30$, and $\alpha = 15$ degrees. Find P.

Solution: $\tan \phi = 0.30 \qquad \phi = 16°42'$ (1)

$$P = 750 \tan (2 \times 16°42' + 15°) + 10 \times 0.30 \tag{5c}$$

$$P = 750 \tan 48°24' + 3.0$$

$$P = 750 \times 1.1263 + 3.0 = 847.7 \text{ pounds}$$

Supporting Weight by Rope Wound About Steel Drum

PROBLEM 8: How many turns N of a rope about a steel drum having a rough surface will be required to support a freely suspended weight of W pounds if a pull of F pounds is exerted on the free end of the rope coming from the drum. Let μ be the coefficient of friction between rope and drum.

Analysis: According to a well-known formula for rope friction, the ratio of greater tension on one end of a rope or belt in contact with a pulley or drum having a rough surface which offers resistance to turning to the lesser tension on the other end is equal to e, the base of natural logarithms which is 2.7183, raised to a power having an exponent which is the product of the coefficient of friction μ between rope and drum and the angle of contact α in radians. Since the pull on one end of the rope and the weight on the other are known, together with the coefficient of friction, this formula can be used to solve for α which is the length of contact of the rope with the drum in radians. By dividing this value by 2π, which is the number of radians in 360 degrees, the number of turns needed to be wound about the drum will be obtained.

Formula:

$$N = \frac{\log W - \log F}{2\pi\mu \log e}$$

Derivation of Formula:

$$\frac{W}{F} = e^{\mu\alpha} \tag{1}$$

Taking the log (to the base 10) of both sides of this equation:

$$\log\left(\frac{W}{F}\right) = \log\,(e^{\mu\alpha}) \tag{2a}$$

or

$$\log W - \log F = \mu\alpha \log e \tag{2b}$$

Solving this equation for α:

$$\alpha = \frac{\log W - \log F}{\mu \log e} \tag{2c}$$

but

$$N = \frac{\alpha}{2\pi} \tag{3}$$

hence

$$N = \frac{\log W - \log F}{2\pi\mu \log e} \tag{4}$$

Example: How many turns of rope about a roughened steel drum will be required to support a weight of 2500 pounds if the pull exerted on the free end is 50 pounds? Assume the coefficient of friction between drum and rope to be 0.29.

Solution: $W = 2500$; $F = 50$; $\mu = 0.29$

$$N = \frac{\log 2500 - \log 50}{2 \times 3.1416 \times 0.29 \times \log e}$$

$$N = \frac{3.39794 - 1.69897}{2 \times 3.1416 \times 0.29 \times 0.43429}$$

$$N = \frac{1.69897}{0.7913} = 2.15 \text{ turns}$$

or say about $2\frac{1}{4}$ turns.

Friction Drive Transmission of Power

PROBLEM 9: Find the total normal pressure P in pounds required between a leather-faced crown wheel of diameter d inches and a cast-iron disk, if H horsepower at N revolutions per minute is to be transmitted. The coefficient of friction is μ.

Analysis of Problem: Knowing the horsepower to be transmitted, the revolutions per minute and the diameter of the crown-wheel, the tangential load T can be found. The total normal pressure P is then equal to the tangential load divided by the coefficient of friction.

Formula:

$$P = \frac{H \times 33{,}000 \times 12}{N \times \pi \times d \times \mu}$$

Derivation of Formula:

Converting horsepower to inch-pounds per minute and dividing by velocity of rotation of leather face of crown wheel in inches per minute.

$$T = \frac{H \times 33{,}000 \times 12}{N \times \pi \times d} \tag{1}$$

$$P = \frac{T}{\mu} \tag{2}$$

Therefore

$$P = \frac{H \times 33{,}000 \times 12}{N \times \pi \times d \times \mu} \tag{3}$$

Example: Find the normal pressure required between a cast-iron disk and leather-faced crown wheel to transmit 10 horsepower at 900 revolutions per minute, if the diameter of the crown wheel is 16 inches. Assume the coefficient of friction to be 0.31.

Solution:

$$P = \frac{10 \times 33{,}000 \times 12}{900 \times 16 \times 3.1416 \times 0.31} \tag{3}$$

$$P = \frac{3{,}960{,}000}{14{,}024} = 282 \text{ pounds}$$

Friction Clutch Transmission of Power

PROBLEM 10: How much longitudinal force F in pounds is required to hold the contact faces of the cone clutch shown in Fig. 8A tightly enough together so that P horsepower can be transmitted at N revolutions per minute? The small radius r_1 and the large radius r_2 of the cone contact surface are known, and the angle α of this surface with the axis of the shaft. Let μ be the coefficient of friction between the two surfaces in contact.

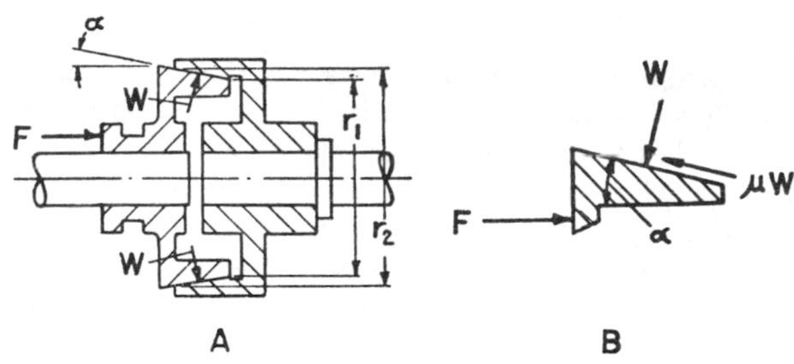

FIG. 8A. Friction Clutch for Which Longitudinal Force F is Required to Permit Transmission of P Horsepower at N Revolutions Per Minute. B. Free Body Diagram Showing Three Forces Acting on Section of Inner Clutch Member

Analysis of Problem: After the mean radius r of the driving clutch contact surface is found, and knowing its speed in revolutions per minute, the peripheral velocity in feet per minute at this radius can be computed. With the horsepower and velocity in feet per minute known, the tangential force T at this radius is found. From the tangential force and the coefficient of friction, the normal force is then found. From the normal force, the force of friction and the angle of the clutch surface with the shaft axis, the axial force is computed.

Formula:

$$F = \frac{63{,}023P\ (\sin\alpha + \mu\cos\alpha)}{\mu r N}$$

Derivation of Formula:

$$r = \frac{r_1 + r_2}{2} \text{ inches} \tag{1}$$

If r is in inches, the peripheral velocity V is

$$V = \frac{2\pi\, rN}{12} \text{ feet per minute} \tag{2}$$

$$T = \frac{P \times 33{,}000}{V} = \frac{12 \times 33{,}000 \times P}{2\pi\, rN} \text{ pounds} \tag{3a}$$

$$T = \frac{63{,}023P}{rN} \text{ pounds} \tag{3b}$$

$$W = \frac{T}{\mu} = \frac{63{,}023P}{\mu rN} \text{ pounds} \tag{4}$$

As indicated in Fig. 8B, which shows a section of the inner clutch member, there are three forces acting on this member. F is the longitudinal thrust tending to wedge this member into closer contact with the other clutch member, W is the reacting force exerted by the other clutch member normal to the contacting surfaces, and μW is the force of friction which tends to resist transverse motion of the two faces. If this clutch member is in equilibrium, the sum of the horizontal forces acting on it must equal zero:

$$F - W \sin \alpha - \mu W \cos \alpha = 0 \tag{5a}$$

so that

$$F = W \sin \alpha + \mu W \cos \alpha \tag{5b}$$

$$F = W (\sin \alpha + \mu \cos \alpha) \tag{5c}$$

Substituting the equivalent of W from Equation (4) in Equation (5c):

$$F = \frac{63{,}023P\ (\sin \alpha + \mu \cos \alpha)}{\mu rN} \text{ pounds} \tag{6}$$

Example: How much longitudinal force F would be required to hold the two faces of a cone clutch in contact without slipping if 5 horsepower is to be transmitted at 350 R.P.M.? The mean radius r of the clutch is 15 inches and the angle α of the clutch face with the axis is 10 degrees. Assume the coefficient of friction between the clutch faces μ to be 0.25.

Solution: $$F = \frac{63{,}023 \times 5 \times (\sin 10° + 0.25 \cos 10°)}{0.25 \times 15 \times 350} \tag{6}$$

$$F = \frac{315{,}115 \times 0.4198}{0.25 \times 5250} = 101 \text{ pounds, approx.}$$

Differential Chain Hoist

PROBLEM 11: Determine the load W that can be lifted by the differential chain hoist shown in Fig. 9 by the application of a given force F if the radii R and r of the two pulleys fastened together are known.

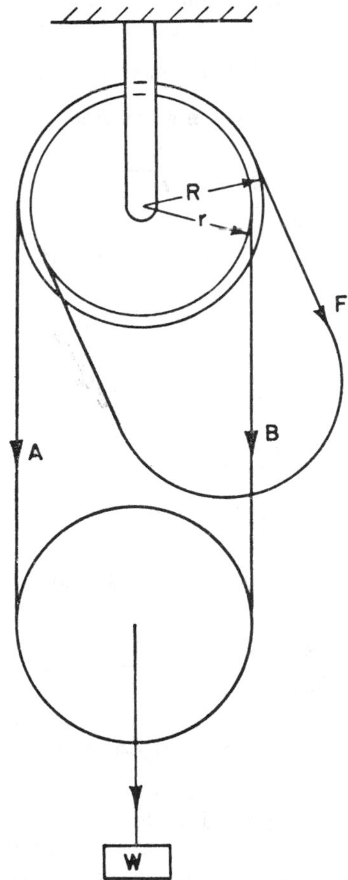

FIG. 9. To Find Load W That can be Lifted on Differential Chain Hoist by Force F if Radii R and r are Known

Analysis of Problem: This is a problem in the equilibrium of parallel forces. (Although F is not shown parallel to W for the sake of a clearer diagram, it can actually be considered as acting parallel to W since it may be applied at any point where the chain is tangent to the larger pulley.) Hence the sum of the moments of the given forces about a common axis must be equal to zero. Let the center of the two upper pulleys be the common axial point about which the moments are taken. The forces A and B imposed by weight W are each equal to $\frac{W}{2}$. Since A, B and moment arms R and r are known, an equation can be established in which F is the only unknown.

Formula:

$$W = \frac{2F \times R}{R - r}$$

Derivation of Formula:

$$A \times R = B \times r + F \times R \tag{1}$$

But

$$A = B = \frac{W}{2} \tag{2}$$

Hence

$$\frac{W}{2} \times R = \frac{W}{2} \times r + F \times R \tag{3a}$$

$$\frac{W}{2}(R - r) = F \times R \tag{3b}$$

$$W = \frac{2FR}{R - r} \tag{4}$$

Example: In Fig. 9 let $F = 35$ pounds and $R = 9\frac{1}{4}$ inches; $r = 8\frac{3}{4}$ inches. Find W.

Solution: $$W = \frac{2 \times 35 \times 9.25}{9.25 - 8.75} = \frac{647.5}{0.5} = 1295 \text{ pounds} \tag{4}$$

Balancing of Couples

PROBLEM 12: To find the forces X acting at the pitch line of the gear in Fig. 10 having a pitch diameter b which will balance the two couples produced by the forces F_1 acting at the pitch diameter of the gear having a pitch diameter a and that produced by forces F_2 acting at the pitch diameter of the gear having a pitch diameter c if all three gears are keyed to the same shaft.

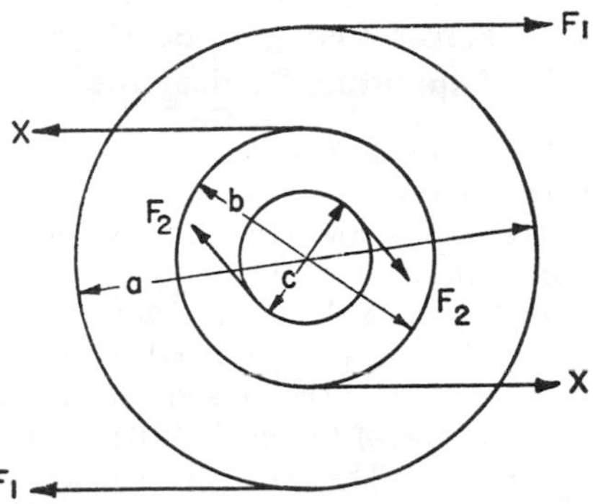

FIG. 10. To Find Forces X Acting at Known Pitch Diameter b which will Balance Two Couples Produced by Forces F_1 Acting at Known Pitch Diameter a and Forces F_2 Acting at Known Pitch Diameter c

Analysis of Problem: The couple produced by forces F_1 and that produced by forces F_2 act to produce clockwise rotation of the gears; hence their combined effect is a couple having a moment equal to the sum of the moments of these two couples. To balance this combined effect, forces X must be such as to produce an equivalent couple tending to cause counterclockwise rotation. Dividing this equivalent couple by the moment arm or distance between forces X which is b gives the amount of X.

Formula: $$X = \frac{F_1a + F_2c}{b}$$

Derivation of Formula:

The couple produced by forces F_1 is equal to $F_1 \times a$. (1)

The couple produced by forces F_2 is equal to $F_2 \times c$. (2)

The couple produced by forces X is equal to $X \times b$. (3)

Hence: $$Xb = F_1a + F_2c \quad (4a)$$

$$X = \frac{F_1a + F_2c}{b} \quad (4b)$$

Example: In Fig. 9 let $F_1 = 360$ pounds and $F_2 = 201$ pounds. Let $a = 8.6$ inches, $b = 4.8$ inches and $c = 2.4$ inches. Find X.

Solution: $$X = \frac{360 \times 8.6 + 201 \times 2.4}{4.8} = 745.5 \text{ pounds} \quad (4b)$$

Forces Acting on Bearings Supporting Vertical Shaft

PROBLEM 13: To find the forces S_x, S_y, T_x, T_y and T_z acting at side and end bearings supporting a vertical shaft on which are keyed two V-belt sheaves as shown in Fig. 11A. The weights E of the smaller sheave, F of the shaft and G of the larger sheave are known. The heights m and p of the two sheaves and n and r of the bearings S and T above the end bearing point of the shaft are also known. The pitch diameters k of the larger sheave and ℓ of the smaller sheave are also given, as well as three of the four belt tension forces A, B, C and D as shown in Fig. 11B. The shaft is in equilibrium and frictional forces are to be ignored.

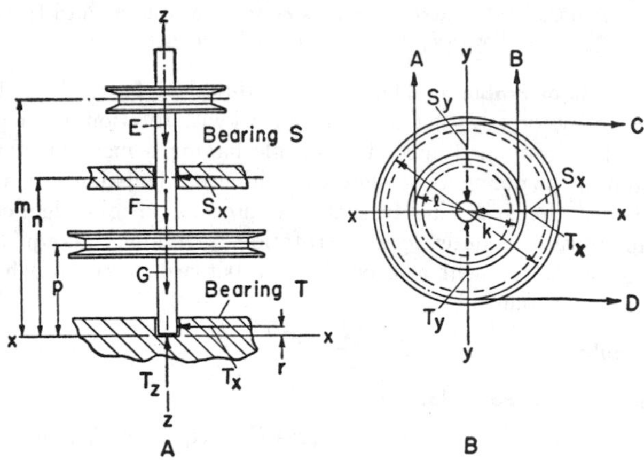

FIG. 11. To Find Forces S_x, S_y, T_x, T_y and T_z When Weights E, F and G, Heights m, p, n and r, Pitch Diameters k and l Shown in View A, and Belt Tension Forces A, B, C and D Shown in View B are Known

Analysis of Problem: Since this is a problem involving forces in three dimensions, three reference axes x, y and z will be established. The horizontal x-axis will be taken through the bottom bearing point of the shaft and is shown in Figs. 11A and B. The horizontal y-axis will be taken through the center-line of the shaft and is shown in Fig. 11B. Since the shaft is in equilibrium, that is, it is either standing still or rotating at a constant velocity, the sums of the forces acting on the shaft and parallel to the x-, y- and z-axes, respectively, are equal to zero and the sums of the moments of all forces acting on the shaft about the x-axis, the y-axis and the z-axis, respectively, are equal to zero. Hence, six simultaneous equations can be established expressing these six equalities and the unknown forces, D, S_x, S_y, T_x, T_y and T_z found algebraically.

Procedure: Let S_x and S_y be the forces exerted by bearing S on the shaft in directions parallel to the x- and y-axes, respectively, and T_x, T_y and T_z, the forces exerted by bearing T on the shaft in directions parallel to the x-, y- and z-axes, respectively.

Then, if

(1) Sum of forces acting parallel to the x-axis equals zero:

$$C + D - S_x - T_x = 0 \tag{1}$$

(2) Sum of forces acting parallel to the y-axis equals zero:

$$A + B + T_y - S_y = 0 \tag{2}$$

(3) Sum of forces acting parallel to the z-axis equals zero:

$$E + F + G - T_z = 0 \tag{3}$$

(4) Sum of moments of all forces about x-axis equals zero:

$$A \times m + B \times m - S_y \times n + T_y \times r = 0 \tag{4}$$

(5) Sum of moments of all forces about y-axis equals zero:

$$C \times p + D \times p - S_x \times n - T_x \times r = 0 \tag{5}$$

(6) Sum of moments of all forces about z-axis equals zero:

$$A \times \frac{\ell}{2} + C \times \frac{k}{2} - B \times \frac{\ell}{2} - D \times \frac{k}{2} = 0 \tag{6}$$

Example: In Fig. 11A and 11B, if belt tensions $A = 50$ pounds; $B = 90$ pounds and $C = 120$ pounds; weights $E = 15$ pounds; $F = 45$ pounds and $G = 35$ pounds; pitch diameters $k = 13.6$ inches and $\ell = 8.6$ inches; and heights $m = 72$ inches; $n = 48$ inches; $p = 28$ inches; and $r = 1$ inch, what are the values of belt tension D and bearing reactions S_x, S_y, T_x, T_y and T_z if the shaft is in equilibrium?

Solution: Setting down the six equations of equilibrium:

$$120 + D - S_x - T_x = 0 \tag{1}$$

$$50 + 90 + T_y - S_y = 0 \tag{2}$$

$$15 + 45 + 35 - T_z = 0 \tag{3}$$

$$50 \times 72 + 90 \times 72 - S_y \times 48 + T_y \times 1 = 0 \qquad (4)$$

$$120 \times 28 + D \times 28 - S_x \times 48 - T_x \times 1 = 0 \qquad (5)$$

$$50 \times \frac{8.6}{2} + 120 \times \frac{13.6}{2} - 90 \times \frac{8.6}{2} - D \times \frac{13.6}{2} = 0 \qquad (6)$$

Solving equation (6) for D:

$$215 + 816 - 387 - 6.8D = 0 \qquad (6)$$

$$6.8D = 644$$

$$D = 94.7 \text{ pounds}$$

Substituting this value of D in Equations (1) and (5) and solving for S_x and T_x:

$$3360 + 2651.6 - 48S_x - T_x = 0 \qquad (5)$$

$$-(120 + 94.7 - S_x - T_x = 0) \qquad (1)$$

$$3240 + 2556.9 - 47S_x = 0$$

$$47S_x = 5796.9$$

$$S_x = 123.3 \text{ pounds}$$

$$120 + 94.7 - 123.3 - T_x = 0 \qquad (1)$$

$$T_x = 91.4 \text{ pounds}$$

Using Equations (2) and (4), solve for S_y and T_y:

$$3600 + 6480 - 48S_y + T_y = 0 \qquad (4)$$

$$10080 - 48S_y + T_y = 0 \qquad (4)$$

$$-(140 - S_y + T_y = 0) \qquad (2)$$

$$9940 - 47S_y = 0$$

$$47S_y = 9940$$

$$S_y = 211.5 \text{ pounds}$$

$$140 - 211.5 + T_y = 0 \qquad (2)$$

$$T_y = 71.5 \text{ pounds}$$

Solving Equation (3) for T_z:

$$T_z = 15 + 45 + 35 = 95 \text{ pounds} \qquad (3)$$

18

Problems in Mechanics—Where Forces Are Not in Equilibrium

Where forces are not in equilibrium, the body on which they act is either being subjected to a change in linear velocity or a change in direction of motion or both. In such problems, the factor of acceleration must be dealt with. This factor frequently proves to be a troublesome one due to a lack of clear understanding of just what acceleration is. In a general way, it might be said that acceleration is a measure of how fast a body is being speeded up; how rapidly it is being slowed down; or how rapidly its direction of motion is being changed. More precisely, acceleration of a body is change in its velocity for a given unit of time, usually taken to be 1 second.

A factor of acceleration frequently encountered in mechanics is that due to gravity. This is the acceleration which would be imparted to a freely falling body by action of the force of gravity, assuming the fall of the body were not retarded in any way by air resistance. It is commonly designated by the symbol g and is taken to be 32.2 or 32.16 feet per second per second at sea level and a latitude of 41 degrees.

Another factor which may be a source of difficulty is that of mass. Wherever mass appears in mechanical problems involving forces not in equilibrium, it is taken to be the weight of the body in question divided by acceleration due to gravity. Thus, if W is the weight of the body in pounds; g is acceleration due to gravity in feet per second per second; and M is the mass of the body; $\frac{W}{g}$ may be substituted for M wherever M appears in the problem.

For a discussion of the difference between mass and weight, the reader is referred to Chapter 26.

As in the previous chapter, the reader is assumed to have some knowledge of the use of vector diagrams of forces as a basis for the mathematical solution of problems of this type.

Velocity and Time of Freely Falling Body

Problem 1: If an object of weight w falls through a vertical distance h to the ground, how long a time t is required for it to reach the ground, and with what velocity v does it strike?

Analysis of Problem: Neglecting air resistance, the length of time of fall and the velocity of impact are independent of the weight of the object. The time of fall is equal to the distance h divided by the average velocity. The only acceleration acting on the body is that of gravity, which is constant and known, therefore the maximum velocity at the end of the fall is equal to the acceleration due to gravity g times the time t. Since the acceleration is uniform, its velocity increases uniformly from zero during the time of fall and the average velocity may be taken as one-half the maximum velocity. Thus, two equations may be established which can be solved for v and t.

Formulas: $t = \sqrt{\frac{2h}{g}}$; $v = \frac{2h}{t}$

Derivation of Formulas:

A freely-falling body without air friction, complies with the following equations from physics and calculus:

$$\text{Acceleration} = A = g$$

$$\text{Velocity} = V = gT \quad (1)$$

$$\text{Displacement downward} = Y = \tfrac{1}{2}gT^2 \quad (2)$$

In these equations, Y is measured positive downward and the body starts at time $T = 0$, with zero initial velocity. The constant downward acceleration of gravity = g = 32.16 ft per second squared (s^2).

Substitute equation (1) into equation (2)

$$Y = \tfrac{1}{2}VT \qquad (3)$$

When the falling body falls a distance h, the corresponding time interval is t. Substitute these values into general equation (3).

$$h = \tfrac{1}{2}vt, \text{ which can be rearranged to } v = 2h/t \qquad (4)$$

Substitute t and h into cquation (2)

$$h = \tfrac{1}{2}gt^2, \text{ which can be rearranged to } t = \sqrt{2h/g} \qquad (5)$$

Substituting this equivalent for t in equation (5)

Example: How long is required for any object to fall a distance h of 300 feet neglecting air resistance? Assume $g = 32.16$ feet per second per second. What is the velocity of impact?

Solution: $$t = \sqrt{\frac{2 \times 300}{32.16}} = \sqrt{18.66} = 4.32 \text{ seconds} \qquad (6c)$$

$$v = \frac{2 \times 300}{4.32} = 139 \text{ feet per second, approx.} \qquad (5)$$

Force and Velocity of Impact of Freely Falling Body

PROBLEM 2: A projectile of weight w is fired *vertically* into the air. It strikes the ground $2t$ seconds after leaving the gun and buries itself in the ground to a depth d. Neglecting air resistance, what height h did the projectile reach? What was the velocity v when it struck the ground? What was the force f with which it struck the ground?

Analysis of Problem: The length of time required for the projectile to reach its greatest height is equal to one-half the time of flight or t. The time of fall, therefore, also equals t. Since during its fall, acceleration is due entirely to gravity, it is a constant, known value. The velocity v at the end of time t when the projectile strikes the ground is equal to the product of acceleration due to gravity g and the time t. (This is also equal to the muzzle velocity.) The distance traveled during the time of fall is the product of the average velocity and time. Since the velocity starts from zero and increases uniformly during the time of fall, the average velocity may be taken as one-half the maximum velocity which is v. The distance of fall is, of course, the maximum height of travel. The energy E with which the projectile strikes the ground is the product of one-half its mass m and the square of its final velocity v at the instant of impact. It also equals its weight w times its height of fall h. The average force with which it strikes the ground is equal to the energy of impact E divided by the distance of travel d, into the ground.

Formulas: $v = gt; \qquad h = \frac{v}{2}t; \qquad f = \frac{wh}{d}.$

Derivation of Formulas:

$$v = gt \tag{1}$$

$$h = \frac{v}{2}t \tag{2}$$

$$E = \tfrac{1}{2}mv^2 \tag{3}$$

but

$$m = \frac{w}{g}$$

therefore

$$E = \frac{wv^2}{2g} = \frac{wvt}{2} = wh \tag{4}$$

hence

$$f = \frac{E}{d} = \frac{wh}{d} \tag{5}$$

Example: A projectile weighing 250 pounds strikes the ground 25 seconds after being fired vertically, and buries itself 5 feet 6 inches into the ground. Find v, h, and f. Assume $g = 32.16$ feet per second per second.

Solution: $d = 5.5$ feet; $w = 250$ pounds; $t = \frac{25}{2} = 12.5$ seconds.

$$v = 32.16 \times 12.5 = 402 \text{ feet per second} \tag{1}$$

$$h = \frac{402}{2} \times 12.5 = 2{,}512 \text{ feet} \tag{2}$$

$$f = \frac{250 \times 2512}{5.5} = 114{,}200 \text{ pounds} \tag{5}$$

Caution: Most physical systems will start to deform in a somewhat elastic manner and plastic deformation will follow if the deformation is large. The maximum force experienced by the system will then be larger than the value given by the above equations if the total deformation is the same. Diagram A below shows the constant force analysis given by the above equations. Diagram B shows the situation for a combination of elastic and plastic deflections. The force equation for the condition of diagram B is:

$$F = \frac{wh}{\left(\frac{D_e}{2} + D_p\right)} \qquad \text{where } d = D_e + D_p$$

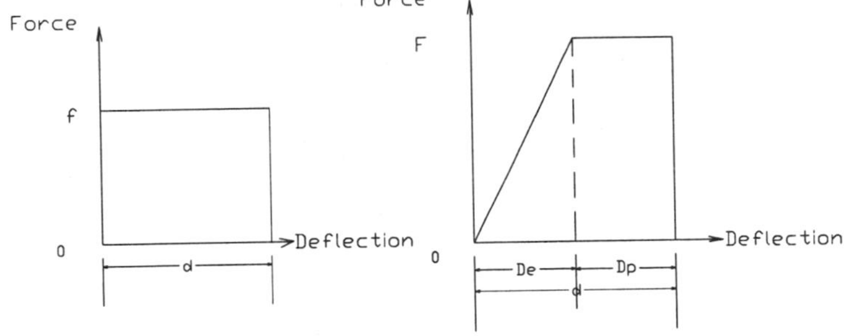

Force of Impact of Drop Hammer

PROBLEM 3: What is the average force F with which a drop hammer of weight W that falls through a distance h strikes the work, if it compresses it by an amount d.

Formula: $F = \frac{Wh}{d}$; where h and d are in the same units of measurement.

Derivation of Formula: The kinetic energy E delivered by the hammer as it strikes the work is equal to its potential energy at the beginning of its fall which, in turn, is equal to its weight times the height of the fall:

$$E = W \times h \tag{1}$$

The amount of average force with which it strikes the work, assuming no energy losses, is equal to the energy delivered to the work divided by the amount of compression or distance through which the force acts.

$$F = \frac{E}{d} \tag{2}$$

Substituting the equivalent of E from Equation (1):

$$F = \frac{Wh}{d} \tag{3}$$

Example 1: Find the force F imparted by the blow of a drop hammer if it weighs 500 pounds, drops a distance of 4½ feet, and compresses the material struck by 0.42 inch.

Solution: $W = 500$ pounds; $h = 4.5 \times 12 = 54$ inches; $d = 0.42$ inch

$$F = \frac{500 \times 54}{0.42} = 64{,}300 \text{ pounds.} \tag{3}$$

Example 2: Find force F if the drop hammer weighs 350 pounds and drops 3 feet to compress the material struck by 0.125 inch.

Solution: $W = 350$ pounds; $h = 3 \times 12 = 36$ inches; $d = 0.125$ inch.

$$F = \frac{350 \times 36}{0.125} = 100{,}800 \text{ pounds}$$

Caution: see the remarks at the end of Problem 2.

Unequal Weights Supported by Pulley

PROBLEM 4: Two weights W_1 and W_2 are attached to a rope passing over a pulley as shown in Fig. 1A. Weight W_1 is heavier than weight W_2 so that motion is in the direction indicated by the arrow on the pulley. Neglecting friction and the weights of pulley and rope, what is the acceleration a with which the weight W_1 moves downward? What is the tension F in the rope? What is the upward force P exerted by the support to which the pulley is attached?

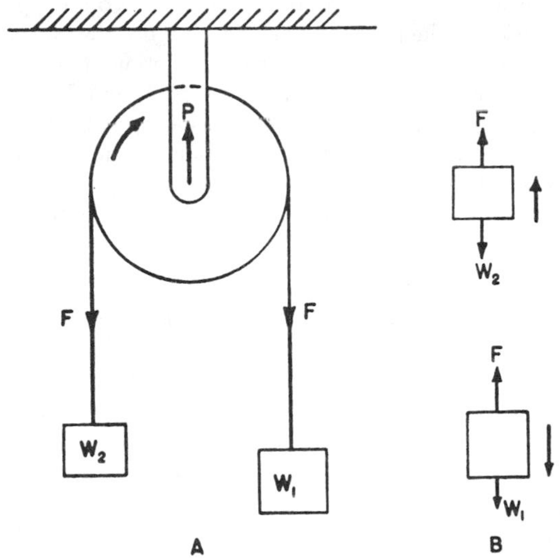

FIG. 1. (A) To Find Tension F, Upward Force P and Acceleration a When Weights W_1 and W_2 are Known. (B) Free Body Diagrams Showing Forces Acting on Weights W_1 and W_2

Analysis of Problem: The first step is to draw two diagrams showing the forces acting on weights W_1 and W_2 respectively, and the direction of motion as shown in Fig. 1B. If a body is considered to be rigid and to be moving in a straight line without rotation, that is, all points of the body have the same acceleration, then the acceleration of the body in feet per second per second is equal to the force (or resultant of several forces) in pounds acting on it divided by its mass (weight in pounds divided by acceleration due to gravity in feet per second per second). The resultant R_1 of the forces F and W_1 acting on the weight W_1 is their algebraic sum since they are both acting vertically. The mass of W_1 is equal to its weight W_1 divided by the acceleration g due to gravity. Hence, an equation can be established for a in terms of F, W_1, and g. But since both a and F are unknown, this equation cannot be solved by itself. Hence, a similar equation for a is established in terms of F, W_2, and g, since the acceleration of W_2 is also equal to a. These two equations are now solved simultaneously for a and F.

The downward forces acting on the pulley are, therefore, F and F acting as tension in the rope at either side. Since the pulley does not move in a vertical direction, i.e., it has no motion of translation but simply rotates on a fixed axis, the forces F, F and P may be considered in equilibrium. These three forces are parallel to each other but their lines of action do not pass through a common point, hence their algebraic sum and the algebraic sum of their moments must, respectively, equal zero. Setting up these two equations will enable the upward force P exerted by the pulley support to be computed and the condition of equilibrium to be checked.

Formulas: $F = \dfrac{2W_1W_2}{W_1 + W_2}$; $\quad a = g \times \dfrac{W_1 - W_2}{W_1 + W_2}$;

$$P = 2F = \frac{4W_1W_2}{W_1 + W_2};$$

where all forces are in pounds and all accelerations in feet per second per second.

Derivation of Formulas:

$$R_1 = W_1 - F \tag{1a}$$

hence

$$F = W_1 - R_1 \tag{1b}$$

but

$$R_1 = M_1a = \frac{W_1}{g}a \tag{2}$$

Substituting equivalent of R_1 in Equation (1b)

$$F = W_1 - \frac{W_1}{g}a \tag{3}$$

$$R_2 = F - W_2 \tag{4a}$$

hence $$F = R_2 + W_2 \tag{4b}$$

but $$R_2 = M_2 a = \frac{W_2}{g} a \tag{5}$$

Substituting the equivalent of R_2 from Equation (5) in Equation (4b)

$$F = \frac{W_2}{g} a + W_2 \tag{6}$$

Equating the right-hand sides of Equations (3) and (6)

$$W_1 - \frac{W_1}{g} a = \frac{W_2}{g} a + W_2 \tag{7a}$$

$$\frac{W_2}{g} a + \frac{W_1}{g} a = W_1 - W_2 \tag{7b}$$

$$a\left(\frac{W_2 + W_1}{g}\right) = W_1 - W_2 \tag{7c}$$

$$a = g \times \frac{W_1 - W_2}{W_1 + W_2} \tag{7d}$$

Substituting the equivalent of a from Equation (7d) in Equation (6)

$$F = \frac{W_2}{g} \times g\left(\frac{W_1 - W_2}{W_1 + W_2}\right) + W_2 \tag{8a}$$

$$F = W_2\left(\frac{W_1 - W_2}{W_1 + W_2}\right) + W_2 = W_2\left(\frac{W_1 - W_2}{W_1 + W_2} + 1\right) \tag{8b}$$

$$F = W_2\left(\frac{W_1 - W_2 + W_1 + W_2}{W_1 + W_2}\right) = W_2\left(\frac{2W_1}{W_1 + W_2}\right) \tag{8c}$$

$$F = \frac{2W_1W_2}{W_1 + W_2} \tag{8d}$$

$$P = F + F = 2F \tag{9}$$

If radius of pulley is r and vertical moment axis is taken through left-hand rope, then

$$P \times r = F \times 2r \tag{10}$$

or $P = 2F$ as before

Example: Find a, F and P if $W_1 = 75$ pounds and $W_2 = 46$ pounds. Assume $g = 32.2$ feet per second per second.

Solution:

$$F = \frac{2 \times 75 \times 46}{75 + 46} = 57.0 \text{ pounds} \tag{8d}$$

$$a = 32.2 \times \frac{75 - 46}{75 + 46} = 0.24 \times 32.2 \tag{7d}$$

$$a = 7.73 \text{ feet per second per second}$$

$$P = 2F = 57 \times 2 = 114 \text{ pounds} \tag{9}$$

Deceleration of Load Being Lowered on Hoisting Rope

PROBLEM 5: A load L is being lowered at a uniform rate by a hoisting rope which has a maximum safe working load rating S. How rapidly can the descent of this load be decelerated d without exceeding the safe load limit?

Analysis of Problem: A slowing down or deceleration of the descending load causes additional tension on the cable due to the inertia of the load. Neglecting the weight of the cable, the total tension on the cable is, then, the weight L of the load plus the inertia force F. This latter is equal to the mass of the load M times the deceleration d. Since the weight of the load is known, its mass is readily computed by dividing by the acceleration due to gravity g. Now F is found by subtracting L from the safe working load S. With F and M known, d can be readily found.

Formula: $d = \dfrac{(S - L)g}{L}$; where all forces are in pounds, and acceleration and deceleration are in feet per second per second.

Derivation of Formula:

If the tension in the cable is equal to S:

$$S = F + L \tag{1a}$$

and

$$F = S - L \tag{1b}$$

but

$$F = dM = d \times \frac{L}{g} \tag{2}$$

Equating the left-hand sides of Equations (2) and (1b):

$$\frac{dL}{g} = S - L \tag{3a}$$

and

$$d = \frac{(S - L)g}{L} \tag{3b}$$

Example: If the load is 4 tons, and a 1-inch plow steel hoisting rope having a maximum safe working load of 6.6 tons is used, what is the maximum permissible deceleration d if the safe working load is not to be exceeded? Assume that $g = 32.2$ feet per second per second.

Solution: $L = 8000$ pounds $S = 13{,}200$ pounds

$$d = \frac{13{,}200 - 8000}{8000} \times 32.2 = 20.9 \text{ feet per second per second} \tag{3b}$$

or, in other words, the velocity of descent cannot be decreased at a rate greater than 20.9 feet per second per second, without the maximum safe load being exceeded in the hoisting cable.

Acceleration of Elevator and Load Pressure on Floor

PROBLEM 6: A load of weight W is being carried upward in an elevator of weight W_e. If the upward force F exerted on the hoisting cable is known, what is the rate of upward acceleration a of elevator and load if friction and weight of the elevator cable are neglected? What is the downward pressure P_d of the load on the floor of the elevator?

Analysis of Problem: Since there are several forces acting on the elevator and its load, i.e., the upward force exerted on the elevator cable, the downward force of gravity acting on the elevator and the downward force of gravity acting on the load, the force tending to produce the upward acceleration a will be considered as the resultant R of these three forces.

The three forces all act in a vertical direction, so that their resultant R is equal to their algebraic sum. The total mass involved is the sum of the mass of the elevator M_e and that of the load M. (It should be remembered that the mass of an object is equal to its weight divided by the acceleration g due to gravity.)

It is assumed in this problem that all points of the elevator and load have the same acceleration and that both are rigid bodies moving in the same straight line without rotation. Under these conditions, the basic formula: *Force* (in pounds) is equal to *mass* (weight in pounds divided by acceleration due to gravity in feet per second per second) times *acceleration* in feet per second per second, can be applied. With force R and total mass $(M_e + M)$ known, acceleration a is readily found.

The same basic equation can be applied to solving the second part of the problem. Since the upward acceleration a of the load is now known, as well as its mass M, the force U acting upon it to produce this upward acceleration can be readily determined. This force U is the resultant of the upward force exerted on the load, which is the pressure P_u of the floor of the elevator on the load and the downward force of gravity W acting on the load. Since U and W are known, P_u can be found. According to Newton's third law, action and reaction forces are equal and oppositely directed, hence the downward pressure P_d of the load on the floor of the elevator is equal to the upward pressure P_u of the floor on the load. Hence P_d is now known.

Formulas:

$$a = g\left(\frac{E}{W + W_e} - 1\right)$$

where a and g are in feet per second per second; and F, W, and W_e are in pounds.

$$P_d = W\left(\frac{a}{g} + 1\right)$$

where a and g are in feet per second per second; and P_d and W are in pounds.

Derivation of Formulas:

To find the upward acceleration a:

$$R = F - (W + W_e) \tag{1}$$

$$R = (M + M_e)a \tag{2a}$$

$$a = \frac{R}{M + M_e} = \frac{R}{\dfrac{W}{g} + \dfrac{W_e}{g}} \tag{2b}$$

$$a = \frac{Rg}{W + W_e} \tag{2c}$$

Substituting the equivalent of R from Equation (1) in Equation (2b)

$$a = \frac{[F - (W + W_e)]g}{W + W_e} = \frac{Fg}{W + W_e} - \frac{(W + W_e)g}{W + W_e} \tag{3a}$$

$$a = \frac{Fg}{W + W_e} - g = g\left(\frac{F}{W + W_e} - 1\right) \tag{3b}$$

To find the downward pressure P_d

$$U = P_u - W \tag{4}$$

$$U = Ma = \frac{W}{g} \times a \tag{5}$$

Substituting the equivalent for U from Equation (5) in Equation (4) and solving for P_u

$$P_u = \frac{Wa}{g} + W = W\left(\frac{a}{g} + 1\right) \tag{6}$$

But $$P_d = P_u \qquad \text{(Newton's third law)} \tag{7}$$

Hence $$P_d = W\left(\frac{a}{g} + 1\right) \tag{8}$$

Example: A load of 800 pounds is being carried upward by an elevator weighing 1750 pounds. If the lifting force exerted on the elevator cable is 3000 pounds, find the rate of upward acceleration and the pressure exerted by the load on the floor of the elevator.

Solution: $W = 800$; $W_e = 1750$; $F = 3000$; and assume that $g = 32.2$ ft. per second per second.

$$a = 32.2\left(\frac{3000}{800 + 1750} - 1\right) = 32.2 \times (1.176 - 1) \tag{3b}$$

$$a = 0.176 \times 32.2 = 5.67 \text{ feet per second per second}$$

$$P_d = 800\left(\frac{5.67}{32.2} + 1\right) = 800(0.176 + 1) \tag{8}$$

$$P_d = 800 \times 1.176 = 940.8 \text{ pounds}$$

Acceleration of Loaded Truck

PROBLEM 7: What is the amount of pull F in pounds which is required to:

(A) Bring load W, in Fig. 2A, which is carried on a small four-wheel truck of weight w and having solid rubber tires from a standing position to a velocity of v miles per hour in t seconds if the truck is to move along a concrete runway? (The bearing friction of the truck wheels is to be ignored.)

(B) To keep the load W moving steadily at velocity v?

(C) To accelerate the truck from a standing position at a rate which will just cause the load to tip backwards.

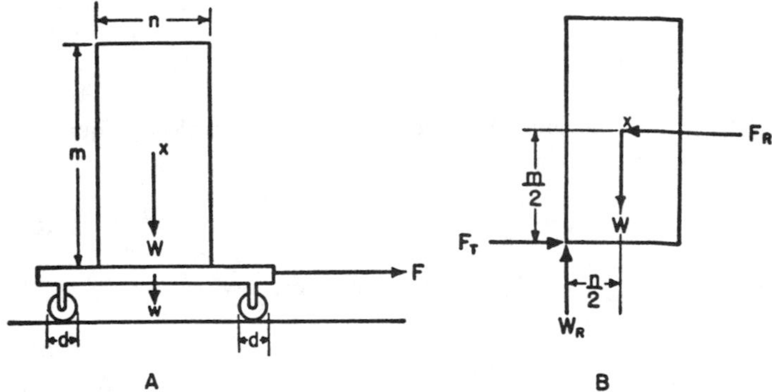

FIG. 2A. To Find Pull F Required for Different Conditions of Motion When Load W, Weight w, and Dimensions m, n and d are Known. B. Free Body Diagram Showing Forces Acting on Load When it is About to Tip Backwards

Analysis of Problem: (A) The force required to bring the load up to the required velocity must overcome the inertia effect due to the acceleration of the combined mass of load and truck and also the frictional resistance resulting from the movement of the solid rubber tires over the concrete floor.

Knowing the initial velocity of the load, which is zero, and the final velocity v, together with the elapsed time t during which the change in velocity is to occur, the acceleration a is computed. The force F_1 required to overcome the inertia effect is, then, equal to the sum of mass M of the load and mass m of the truck, multiplied by the acceleration.

Knowing the weight $\frac{1}{4}(W + w)$ supported by each wheel of the truck, the diameter d of each wheel in inches, and the coefficient of rolling friction expressed in inches for solid rubber tires on a concrete floor, the force F_2 required to overcome this frictional resistance is computed. The force F is, then, the sum of F_1 and F_2.

(B) If the load is not to be accelerated, but is to be moved at a constant velocity, there is no inertia effect to be overcome and the only force required is that needed to overcome the rolling friction, or F equals F_2.

(C) If the truck is to be accelerated so rapidly that the load is about to tip backwards (assuming that the frictional resistance between it and the truck is great enough to prevent it from slipping), a force F_T representing the horizontal force exerted by the truck on the load may be assumed to be applied at the rear edge of the block as shown in Fig. 2B, which is a free body diagram showing the forces acting on the block. Similarly, the reaction W_R of the floor of the truck to weight W of the load may also be considered as being applied to this rear edge. The inertia effect due to the acceleration of the load may be represented by force F_R acting through the center of gravity of the load and opposite to F_T.

Since F_R is equal to F_T and W_R is equal to W, it can be seen from Fig. 2B that the load is being acted upon by two couples. One of these whose moment is $F_T \times \frac{m}{2}$ is tending to tip the load backwards, while the other, whose moment is $W \times \frac{n}{2}$ is tending to resist this tipping of the load. When F_T is large enough so that the moments of the two couples are equal, tipping is about to occur.

After having found F_T, and hence F_R, the acceleration a of the load W required to produce an inertia effect equal to F_R is computed. Now, with acceleration a and the combined mass of load and truck $(M + m)$ known, the force F_1 required to overcome the inertia effect of this mass can be found. F is then equal to the sum of F_1 and F_2 as in (A).

Formulas:

$$\text{(A)}\quad F = (W + w)\left(\frac{1.47v}{gt} + \frac{0.08}{d}\right) \text{ pounds}$$

$$\text{(B)}\quad F = \frac{0.08\ (W + w)}{d} \text{ pounds}$$

$$(C)\ F = (W + w)\left(\frac{n}{m} + \frac{0.08}{d}\right) \text{ pounds}$$

Derivation of Formulas: For Condition (A) where v is in miles per hour; t is in seconds; and F_1, F_2, W and w are in pounds:

$(v - 0)$ is changed to feet per second by multiplying by 5280 and dividing by 3600.

$$a = \frac{(v - 0) \times 5280}{t \times 3600} \text{ ft. per sec. per sec.} \tag{1a}$$

$$a = 1.47\frac{v}{t} \text{ ft. per sec. per sec.} \tag{1b}$$

$$F_1 = (M + m)a = \frac{W + w}{g} \times a \tag{2}$$

$$F_1 = \frac{W + w}{g} \times 1.47\frac{v}{t} = \frac{1.47\ (W + w)v}{gt} \text{ pounds} \tag{3}$$

Resistance to rolling, or the force F_2 required to overcome this resistance, in pounds, is equal to the total weight of the rolling body, times the coefficient of rolling friction, in inches, divided by the radius of the wheels in inches. Assuming 0.04 inch as a reasonable coefficient of rolling friction for solid rubber tires on a cement surface:

$$F_2 = \frac{(W + w)0.04}{\frac{d}{2}} = \frac{0.08(W + w)}{d} \text{ pounds} \tag{4}$$

then

$$F = F_1 + F_2 \tag{5a}$$

$$F = \frac{1.47(W + w)v}{gt} + \frac{0.08(W + w)}{d} \text{ pounds} \tag{5b}$$

$$F = (W + w)\left(\frac{1.47v}{gt} + \frac{0.08}{d}\right) \text{ pounds} \tag{5c}$$

For Condition (B) where v is in miles per hour; t is in seconds; and F_2, W, and w are in pounds:

$$F = F_2 = \frac{0.08(W + w)}{d} \text{ pounds} \tag{6}$$

For Condition (C) where n and m are in feet; F_1, F_3, W, and w are in pounds:

$$F_T \times \frac{m}{2} = W_R \times \frac{n}{2} \text{ pounds-feet} \tag{7}$$

$$F_T = W_R \times \frac{n}{m} \text{ pounds}$$

but $$W = W_R \quad \text{and} \quad F_R = F_T \qquad \text{(8) and (9)}$$

hence $$F_R = W \times \frac{n}{m} \text{ pounds} \tag{10}$$

The next step is to find the acceleration required to produce this inertia reaction F_R:

$$a = \frac{F_R}{M} = \frac{F_R}{\frac{W}{g}} = \frac{F_R g}{W} \tag{11}$$

Substituting the equivalent of F_R from Equation (10)

$$a = \frac{Wn}{m} \times \frac{g}{W} = \frac{ng}{m} \tag{12}$$

Next, the force required to give the load and truck this acceleration is found:

$$F_1 = (M + m)\,a = \frac{(W + w)a}{g} \text{ pounds} \tag{13}$$

Substituting the equivalent of a from Equation (12)

$$F_1 = \frac{W + w}{g} \times \frac{ng}{m} = \frac{(W + w)n}{m} \tag{14}$$

but $$F = F_1 + F_2$$

hence $$F = \frac{(W + w)n}{m} + \frac{0.08(W + w)}{d} \tag{15a}$$

$$F = (W + w)\left(\frac{n}{m} + \frac{0.08}{d}\right) \tag{15b}$$

Example: In Fig. 2A let the load W = 1500 pounds; the weight w of the truck = 500 pounds; the height m of the load = 6 feet; the length n of the load = $2\frac{1}{2}$ feet; and the diameter d of the truck wheels = 6 inches.

(A) Find the force F required to bring the load from a standing position to a velocity of 10 miles per hour in a time t of 5 seconds.

(B) Find the force F required to maintain the velocity of the load at 10 miles per hour.

(C) Find the force F required to accelerate the load so that tipping is about to occur.

Assume acceleration g due to gravity to be 32.2 feet per second.

Solution:

$$\text{(A)}\quad F = (1500 + 500)\left(\frac{1.47 \times 10}{32.2 \times 5} + \frac{0.08}{6}\right) \qquad \text{(5c)}$$

$$F = 2000 \times 0.1046 = 209.2 \text{ pounds}$$

$$\text{(B)}\quad F = \frac{0.08(1500 + 500)}{6} = 26.7 \text{ pounds} \qquad \text{(6)}$$

$$\text{(C)}\quad F = (1500 + 500)\left(\frac{2.5}{6} + \frac{0.08}{6}\right) \qquad \text{(15b)}$$

$$F = 2000 \times 0.43 = 860 \text{ pounds}$$

Elevation of Track Curve for Given Speed

PROBLEM 8: To find the height BC (Fig. 3) that the outer rail of a railroad track of standard gage (width AC) would have to be raised for a given curve of radius r and train speed v so that there would be no flange pressure of the outer wheels against the rail.

Analysis of Problem: Knowing the distance AC between rail centers, the problem is to find angle y so that BC can be computed. There are two forces acting on the moving body—its weight W and the centrifugal force F. The resultant of these two forces is shown as R in Fig. 3. When the outer rail is elevated so that a line AC drawn parallel with the top of the rails is perpendicular to the line of action of resultant R, there will be no sideways pressure of the outer wheel flanges against the rail. With the rails elevated as shown, the angle x between resultant R and the vertical is equal to angle y since their corresponding sides are at right angles to each other. Since W is known and F can be computed in terms of known velocity v, weight W, acceleration due to gravity g, and radius of curve r, then angle x and hence angle y can be found.

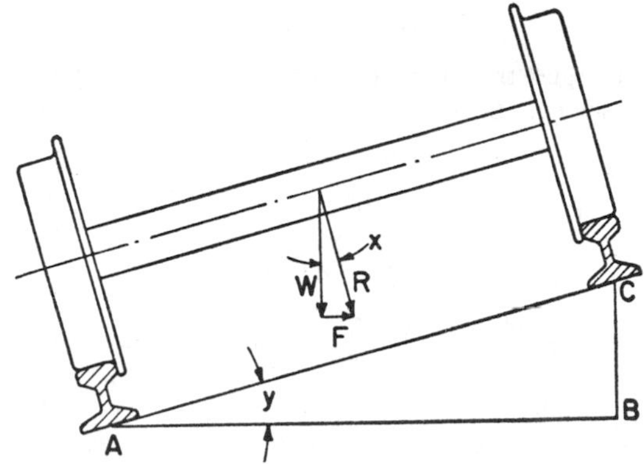

FIG. 3. To Find Height BC for Elevation of Outer Rail if Radius r of Track Curve, Train Speed v, and Track Width AC are Known

Formula:

$$BC = AC \times \frac{v^2}{gr};$$

where g is acceleration due to gravity and v, g, and r are all in units of feet and seconds or in units of inches and seconds.

Derivation of Formula:

$$\tan x = \frac{F}{W}, \text{ where } F \text{ and } W \text{ are in feet} \quad (1)$$

According to the formula for centrifugal force:

$$F = \frac{Wv^2}{gr} \quad (2)$$

where W is in pounds, v is in feet per second, g is in feet per second per second and r is in feet.

Substituting this equivalent for F in Equation (1)

$$\tan x = \frac{\frac{Wv^2}{gr}}{W} = \frac{v^2}{gr} \quad (3)$$

It will be noted that angle x is independent of the weight of the moving body. Hence, height BC is not affected by the weight W. This is indicated by the following equations:

$$\text{angle } y = \text{angle } x \quad (4)$$

$$BC = AC \sin y = AC \sin x \tag{5}$$

At small angles the tangent and sine are approximately equal so that tangent x may be substituted for sin x in Equation (5)

$$BC = AC \tan x \tag{6}$$

Substituting the equivalent of tan x from Equation (3)

$$BC = AC \times \frac{v^2}{gr} \tag{7}$$

Example: What should be the elevation of the outer rail above the inner rail if it is to prevent outward flange pressure of trains running at 50 miles per hour around an 8-degree curve? The standard gage center distance between rails is 4 feet 8½ inches.

Solution: According to Problem 6 in Chapter 15, the radius r of an 8-degree curve is found as follows:

$$r = 50 \text{ cosec } \frac{8^\circ}{2} = 50 \text{ cosec } 4^\circ$$

$$r = 50 \times 14.335 = 716.75 \text{ feet}$$

$$v = \frac{50 \times 5280}{60 \times 60} = 73.33 \text{ feet per second}$$

$$AB = 56.5 \text{ inches}$$

$$BC = \frac{56.5 \times (73.33)^2}{32.2 \times 716.75} = \frac{303{,}817}{23{,}079} = 13.2 \text{ inches}$$

Kinetic Energy of Flywheel

PROBLEM 9: A power press is being used for a blanking operation, the energy for which is supplied as a result of a decrease in the speed of rotation of a flywheel. The flywheel has a radius of gyration of k inches and weighs W pounds. E inch-tons of energy are required in the blanking operation. What must be the velocity of rotation N in revolutions per minute of the flywheel just before the blanking operation takes place? The ratio m of the decrease in velocity of rotation to the velocity just before blanking is given.

Analysis of Problem: During the "off" part of the power press working cycle, the velocity of the flywheel is increased so that it "stores up" kinetic energy. When the blanking operation begins, resistance to rotation of the flywheel develops so that it tends to slow down and give up some of this kinetic energy which is utilized in performing the blanking operation. The amount of kinetic energy which the flywheel loses is proportional to the difference in squares of its velocity of rotation in feet per second at the beginning and end of the blanking operation. This velocity of rotation is the velocity at the radius of gyration of the flywheel. (*Note:* Where the mass of a flywheel is largely concentrated in its rim, the mass of its hub and spokes is frequently ignored. The mean radius of the rim is then used as a close approximation to the radius of gyration.) Since in this problem, the decrease in velocity is given as a definite proportion of the initial velocity, the final velocity can be expressed in terms of the initial velocity. An equation can then be established for the loss of kinetic energy during the blanking operation in which the initial velocity is the only unknown.

Formula: $$N = \frac{11{,}860}{k}\sqrt{\frac{E}{W[1 - (1 - m)^2]}}$$

Derivation of Formula: Let V_1 = velocity of flywheel in feet per second immediately before the blanking operation takes place and V_2 = velocity of flywheel in feet per second at the end of the blanking operation. Then, according to the requirements of this problem:

$$\frac{V_1 - V_2}{V_1} = m \qquad (1)$$

$$V_2 = V_1 - mV_1 = (1 - m)V_1 \qquad (2)$$

$$V_2^2 = (1 - m)^2 V_1^2 \qquad (3)$$

The loss in kinetic energy is equal to one-half the mass of the flywheel $\frac{W}{g}$ times the difference of the squares of initial and final velocities; where W is in pounds and g is acceleration due to gravity in feet per second per second.

$$E \text{ (foot-pounds)} = \frac{W(V_1^2 - V_2^2)}{2g} \qquad (4)$$

Converting this equation to inch-tons

$$E \text{ (inch-tons)} = \frac{W(V_1^2 - V_2^2) \times 12}{2g \times 2000} = \frac{W(V_1^2 - V_2^2)}{333g} \qquad (5)$$

Substituting the equivalent of V_2 from Equation (3) in Equation (5)

$$E \text{ (inch-tons)} = \frac{W[V_1^2 - (1 - m)^2 V_1^2]}{333g} \tag{6a}$$

$$E \text{ (inch-tons)} = \frac{WV_1^2[1 - (1 - m)^2]}{333g} \tag{6b}$$

But

$$V_1 = \frac{N \times 2\pi k}{12 \times 60} = 0.00873Nk \text{ feet per second} \tag{7}$$

and

$$V_1^2 = 0.0000762N^2k^2 \tag{8}$$

Substituting this equivalent of V_1^2 in Equation (6b)

$$E \text{ (inch-tons)} = \frac{0.0000762WN^2k^2[1 - (1 - m)^2]}{333g} \tag{9a}$$

If g is taken to be 32.2 feet per second per second:

$$E \text{ (inch-tons)} = \frac{WN^2k^2[1 - (1 - m)^2]}{140{,}700{,}000} \tag{9b}$$

Rearranging this equation to solve for N:

$$N = \frac{11{,}860}{k}\sqrt{\frac{E}{W[1 - (1 - m)^2]}} \tag{10b}$$

Example: A blanking operation requiring 3 inch-tons of energy is to be performed on a power press having a flywheel with a radius of gyration of 15 inches and a weight of 1000 pounds. If the allowable decrease in velocity of the flywheel during the blanking operation is 10 per cent, what must be its initial velocity in revolutions per minute?

Solution: $m = 0.10$; $E = 3$ inch-tons; $k = 15$ inches; $W = 1000$ pounds

$$N = \frac{11{,}860}{15}\sqrt{\frac{3}{1000[1 - (1 - 0.10)^2]}}$$

$$N = 790.7\sqrt{0.0158}$$

$$N = 790.7 \times 0.126$$

$N = 99.6$ or 100 R.P.M. approximately

Forces Acting on Porter Type Governor

PROBLEM 10: To find the R.P.M. of the Porter type governor in Fig. 4A for a specified distance h. The weight W of the sleeve; the weight L of each ball; the length f of each upper arm; the length g of each lower arm; and the distance s of the lower pivot points from the governor axis are known. At a certain speed of rotation, each ball will rotate in a plane which is at a vertical distance h from the upper pivot point. What is the speed of rotation N for a given distance h?

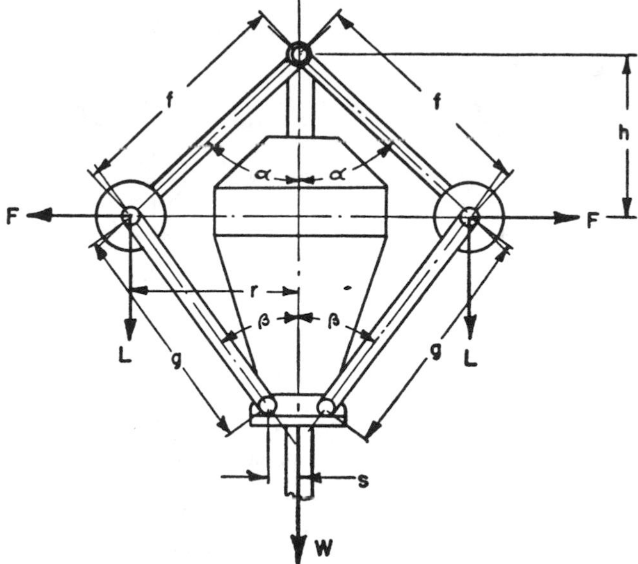

FIG. 4A. To Find Speed of Rotation N for Given Distance h When Weights W and L, and Lengths f, g and s are Known

Analysis of Problem: The first objective is to find the centrifugal force F acting on each ball which, in conjunction with the other forces acting on balls and sleeve, causes the balls to rotate in a plane at distance h from the upper pivot point. Since the balls are not moving up or down, the forces acting on them and on the sleeve are in equilibrium. By drawing a free-body diagram of the forces acting on the sleeve, as shown in Fig. 4B, and taking the sum of the vertical components of these forces as equal to 0, the tension T_2 in each of the lower arms can be found. Next, a free-body diagram of the forces acting on one of the balls, as shown at lower left in Fig. 4C, is drawn. By taking the sum of the vertical components of these forces, with T_2 now known, T_1 can be found. By taking the sum of the horizontal components of these forces with T_1 known, F can be found. F, being a centrifugal force, can be expressed in terms of speed of rotation N, radius r, weight L, and acceleration due to gravity g. This substitution is made and the resulting equation is solved for N.

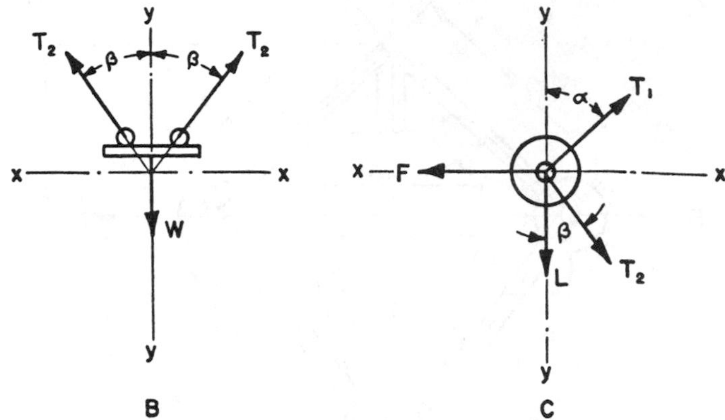

FIG. 4B. Free Body Diagram of Forces Acting on Sleeve. C. Free Body Diagram of Forces Acting on Ball Weight

Formula:

$$N = 187.7\sqrt{\frac{\frac{W}{2}\left(1 + \frac{\tan\beta}{\tan\alpha}\right) + L}{Lh}}$$

$$\text{where } \alpha = \text{arc sin } \frac{r}{f} \text{ and } \beta = \text{arc sin } \frac{r - s}{g}$$

Derivation of Formula: Let α be the angle between each upper arm and the vertical, and similarly, β the angle between each lower arm and the vertical, as shown in Fig. 4A. Then, taking the vertical components acting on the sleeve as shown in Fig. 4B,

$$2T_2 \cos\beta - W = 0 \tag{1a}$$

$$T_2 = \frac{W}{2\cos\beta} \tag{1b}$$

And taking the vertical components acting on one of the balls as shown in Fig. 4C

$$T_1 \cos\alpha - T_2 \cos\beta - L = 0 \tag{2a}$$

$$T_1 = \frac{T_2 \cos\beta + L}{\cos\alpha} \tag{2b}$$

Substituting the equivalent of T_2 from Equation (1b)

$$T_1 = \frac{\frac{W}{2} + L}{\cos\alpha} \tag{3}$$

Taking the horizontal components acting on one of the balls:

$$F - T_1 \sin\alpha - T_2 \sin\beta = 0 \tag{4a}$$

$$F = T_1 \sin\alpha + T_2 \sin\beta \tag{4b}$$

Substituting the equivalents of T_1 from Equation (3) and of T_2 from Equation (1b) in Equation (4b)

$$F = \frac{\frac{W}{2} + L}{\cos\alpha} \sin\alpha + \frac{W}{2\cos\beta} \sin\beta \tag{5a}$$

$$F = \left(\frac{W}{2} + L\right) \tan\alpha + \frac{W}{2} \tan\beta \tag{5b}$$

If v is the velocity of rotation in feet per second, M is the mass of one ball, L is the weight of one ball in pounds, and g is acceleration due to gravity in feet per second per second; the centrifugal force F is given by the equation:

$$F = M\frac{v^2}{r} = \frac{Lv^2}{gr} \quad \text{pounds} \tag{6}$$

but

$$v = \frac{2\pi rN}{60} = \frac{rN}{9.55} \quad \text{feet per second} \tag{7}$$

where r is in feet.

Therefore, if g is taken as 32.2 feet per second per second, and substituting the equivalent of v from Equation (7) in Equation (6):

$$F = \frac{LrN^2}{9.55^2 \times 32.2} = \frac{LrN^2}{2937} \tag{8}$$

Substituting the equivalent of F from Equation (8) in Equation (5b):

$$\frac{LrN^2}{2937} = \left(\frac{W}{2} + L\right) \tan \alpha + \frac{W}{2} \tan \beta \tag{9a}$$

$$\frac{LrN^2}{2937} = \left(\frac{W}{2} + L + \frac{W \tan \beta}{2 \tan \alpha}\right) \tan \alpha \tag{9b}$$

Substituting the equivalent $\frac{r}{h}$ for $\tan \alpha$ (see Fig. 4A) outside of the parentheses and moving all terms except N^2 to the right-hand side of the equation:

$$N^2 = \frac{2937}{L \times r} \times \frac{r}{h}\left[\frac{W}{2}\left(1 + \frac{\tan \beta}{\tan \alpha}\right) + L\right] \tag{10a}$$

$$N^2 = \frac{2937\left[\frac{W}{2}\left(1 + \frac{\tan \beta}{\tan \alpha}\right) + L\right]}{Lh} \tag{10b}$$

$$N = 54.19\sqrt{\frac{\frac{W}{2}\left(1 + \frac{\tan \beta}{\tan \alpha}\right) + L}{Lh}} \tag{10c}$$

If h is in inches, then

$$N = 187.7\sqrt{\frac{\frac{W}{2}\left(1 + \frac{\tan \beta}{\tan \alpha}\right) + L}{Lh}} \tag{11}$$

Angle α can be found by the equation

$$\sin \alpha = \frac{r}{f} \tag{12}$$

Angle β can be found by the equation

$$\sin \beta = \frac{r - s}{g} \tag{13}$$

Example: In Fig. 4A, W = 100 pounds; L = 18 pounds; f = 10 inches; g = 10 inches; and s = 1 inch. Find N if h = 7⅛ inches.

Solution: $r = \sqrt{f^2 - h^2} = \sqrt{100 - 50.77} = 7.017$ inches.

$$\sin \alpha = \frac{7.017}{10} = 0.7017 \qquad \alpha = 44^\circ\ 34' \qquad (12)$$

$$\sin \beta = \frac{7.017 - 1}{10} = 0.6017 \qquad \beta = 36^\circ\ 59\tfrac{1}{2}' \qquad (13)$$

$$N = 187.7\sqrt{\frac{50\left(1 + \dfrac{\tan 36^\circ\ 59.5'}{\tan 44^\circ\ 34'}\right) + 18}{18 \times 7.125}} \qquad (11)$$

$$N = 187.7\sqrt{\frac{50\left(1 + \dfrac{0.75332}{0.98499}\right) + 18}{18 \times 7.125}}$$

$$N = 187.7\sqrt{\frac{106.24}{128.25}} = 187.7\sqrt{0.8284}$$

$$N = 187.7 \times 0.910 = 171 \text{ R.P.M.}$$

19

Factors in Design—Efficiency, Load Torque, Strength, Power

One of the most common types of mechanical problems encountered, especially in the field of machine design, is that of finding the load or torque to which a given part or machine is subjected, or of computing its strength, power capacity, or efficiency. Problems of this type arise in a multitude of forms and frequently involve factors which cannot be accurately evaluated or measured. Many of them fall under the heading "Strength of Materials." They deal with shafts, gears, bearings, screw threads, springs, pipes, cylinders, etc.

Because of the nature of these problems, rigid mathematical analysis and derivation of formulas cannot always be relied upon to open the way to their solution. Where such an analysis is possible, it may require a more extensive theoretical knowledge than the practical man has at his command. Empirical formulas are employed, therefore, for solving many of these problems. Their application is usually quite simple and direct, but good judgment is required in selecting the right formula for the problem at hand and in interpreting the results obtained.

Space does not permit more than a few such formulas to be mentioned and even these cannot be given a detailed and extensive treatment. Nevertheless, a careful study of the problems presented should provide a clearer understanding of the methods of approach in this field.

For additional discussion of the nature and use of empirical formulas, the reader is referred to Chapter 24.

Size of Shaft for Given Power and Load Requirements

PROBLEM 1: Find the diameter d of a shaft required to transmit P horsepower at N revolutions per minute if the type of steel for the shaft and the kind of load are specified. In this particular case, bending load is not a factor.

Analysis of Problem: The formula below is commonly used to find the diameter of shafting for given power transmission requirements where the bending load is not a factor. Since steady torsional loads cause failure in shear, the basis for the design is the yield strength in shear which is usually taken as 60 per cent of the yield strength in tension. When endurance or shock loads are encountered it has been recommended that the design be based on the endurance limit in shear which is taken as one-quarter of the ultimate tensile strength. For ordinary service, a factor of safety of 3 is used. For shock loads, a higher safety factor is called for. The size of this safety factor depends upon the intensity and frequency of the shock and its selection is based upon the experience and judgment of the designer.

Formula:

$$d = \sqrt[3]{\frac{321{,}000P}{NS}}$$

where S = allowable shearing stress of steel used.

Example: Find the diameter d of a shaft required to transmit 125 H.P. at 500 R.P.M. if (a) a steady torsional load is to be carried; and (b) if a rather severe shock load is to be carried. The steel to be used is SAE 1040 hot rolled with an ultimate tensile strength of 94,000 pounds per square inch and a yield point in tension of 58,000 pounds per square inch.

Solution: Let Y_t = yield point in tension
Y_s = yield point in shear
T_s = ultimate tensile strength
L_s = endurance limit in shear

(a) For a steady torsional load

$Y_s = 0.60 Y_t = 0.60 \times 58{,}000 = 34{,}800$ pounds per square inch

Using a factor of safety of 3

$$S = \frac{Y_s}{3} = \frac{34{,}800}{3} = 11{,}600 \text{ pounds per square inch}$$

allowable stress

$$d = \sqrt[3]{\frac{321{,}000 \times 125}{500 \times 11{,}600}} = \sqrt[3]{6.918}$$

$d = 1.905$ inches. Probably a $1\frac{15}{16}$-inch or a 2-inch shaft would be used.

(b) For rather severe shock load

$$L_s = \frac{T_s}{4} = \frac{94{,}000}{4} = 23{,}500 \text{ pounds per square inch}$$

Using a factor of safety of 4

$$S = \frac{L_s}{4} = \frac{23{,}500}{4} = 5875 \text{ pounds per square inch}$$

$$d = \sqrt[3]{\frac{321{,}000 \times 125}{500 \times 5875}} = \sqrt[3]{13.660}$$

$d = 2.391$ inches. Probably a $2\frac{7}{16}$-inch or a $2\frac{1}{2}$-inch shaft would be used.

Shaft Torques for Given Horsepower and Speed

PROBLEM 2: To find the torques exerted by Shafts 1, 2 and 3 (Fig. 1) when the horsepower delivered to Shaft 1, the speed of rotation of Shaft 1, the sizes of Gears *A*, *B*, *C* and *D* and the percentage of power lost in transmission between shafts are given.

Analysis of Problem: Knowing the speed of rotation of Shaft 1, the speeds of rotation of Shafts 2 and 3 can be calculated as being inversely proportional to the number of respective gear teeth. The amounts of power delivered to Shafts 2 and 3 are readily found after allowance for power lost. Knowing the horsepower and speed for each shaft, the torque exerted by each can be directly computed.

Formula: $\text{Torque} = \dfrac{63{,}000 \text{ H.P.}}{\text{R.P.M.}}$ pound-inches

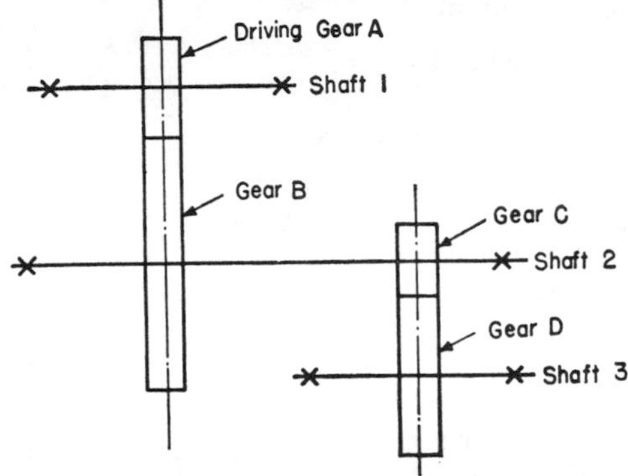

FIG. 1. To Find Torques Exerted by Shafts 1, 2 and 3

Example: Suppose that 40 H.P. at 1500 R.P.M. is delivered to Shaft 1. Gear A has 18 teeth; Gear B, 45 teeth; Gear C, 12 teeth; and Gear D, 24 teeth. The loss of power between Shafts 1 and 2 is assumed to be 5 per cent and between Shafts 3 and 4 also 5 per cent. What are the torques in pound-inches exerted by Shafts 1, 2 and 3?

Solution:

$$\text{Torque} = \frac{\text{H.P.} \times 33{,}000 \times 12}{2 \times \pi \times \text{R.P.M.}} = \frac{63{,}000 \times \text{H.P.}}{\text{R.P.M.}} \text{ pound-inches}$$

where H.P. = horsepower
R.P.M. = revolutions per minute
33,000 = no. of ft. pounds per minute in 1 H.P.
12 = no. of inches in 1 foot
2π = factor to convert work per revolution into torque

$$\text{Torque on Shaft 1} = \frac{63{,}000 \times 40}{1500} = 1680 \text{ pound-inches} \quad (1)$$

$$\frac{\text{R.P.M. Shaft 2}}{\text{R.P.M. Shaft 1}} = \frac{\text{Size Gear } A}{\text{Size Gear } B} \quad (2a)$$

$$\frac{\text{R.P.M. Shaft 2}}{1500} = \frac{18}{45} \quad (2b)$$

$$\text{R.P.M. Shaft 2} = \frac{18 \times 1500}{45} = 600 \text{ R.P.M.} \quad (2c)$$

$$\text{H.P. delivered to Shaft 2} = \text{95 per cent of H.P. delivered to Shaft 1} \tag{3a}$$

$$\text{H.P. Shaft 2} = 0.95 \times 40 = 38 \text{ H.P.} \tag{3b}$$

$$\text{Torque Shaft 2} = \frac{63{,}000 \times 38}{600} = 3990 \text{ pound-inches} \tag{4}$$

$$\frac{\text{R.P.M. Shaft 3}}{\text{R.P.M. Shaft 2}} = \frac{\text{Size Gear } C}{\text{Size Gear } D} \tag{5a}$$

$$\frac{\text{R.P.M. Shaft 3}}{600} = \frac{12}{24} \tag{5b}$$

$$\text{R.P.M. Shaft 3} = \frac{12}{24} \times 600 = 300 \text{ R.P.M.} \tag{5c}$$

$$\text{H.P. delivered to Shaft 3} = 0.95 \times 38 = 36.1 \text{ H.P.} \tag{6}$$

$$\text{Torque Shaft 3} = \frac{63{,}000 \times 36.1}{300} = 7581 \text{ pound-inches} \tag{7}$$

Bearing Loads for Two Spur Gears in Mesh

PROBLEM 3: Determine the amount and direction of the bearing loads of spur gears 1 and 2 (Fig. 2) if the horsepower delivered to gear 1, the speed and direction of rotation, and the pitch diameter of gear 1 and the pressure angles of gears 1 and 2 are known.

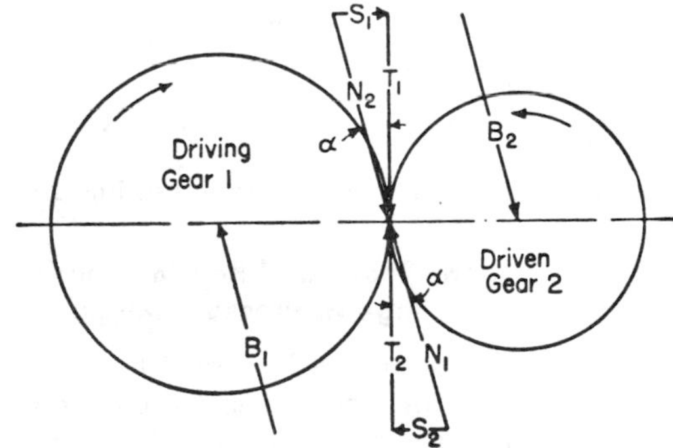

FIG. 2. To Determine Amount and Direction of Bearing Loads of Spur Gears 1 and 2

Analysis of Problem: With the amount of horsepower applied to gear 1 and its rotational speed known, the torque exerted by it can be computed as in Problem 2. Knowing the torque and the pitch diameter of gear 1, the tangential driving force, T_1, at this pitch diameter can be computed. Knowing the tangential resisting load T_2 (which will be equal and opposite to T_1) and the pressure angle, α, the total load, N_2, normal to the teeth of gear 2 can be computed. The bearing load, B_2, of gear 2 will then be equal and parallel to N_2 and in the same direction. The total load, N_1, normal to the teeth of gear 1 will be equal and opposite to N_2. The total load, B_1, on the bearing of gear 1 will be equal and parallel to and in the same direction as N_1.

Example: Determine the bearing loads, B_1 and B_2, if 5 H.P. is supplied to gear 1 which rotates in the direction shown in Fig. 2 at 1500 R.P.M. The pitch radius, R_1, of gear 1 is 8.6 inches. Both gears have a pressure angle, α, of $14\frac{1}{2}$ degrees.

Solution:

$$\text{Torque (gear 1)} = \frac{63{,}000 \times \text{H.P.}}{\text{R.P.M.}} = \frac{63{,}000 \times 5}{1500} \tag{1}$$

$$= 210 \text{ pound-inches}$$

$$T_1 = \frac{\text{Torque}}{\text{Pitch Radius}} = \frac{210}{8.6} = 24.4 \text{ pounds} \tag{2}$$

$$N_2 = T_1 \sec \alpha = 24.4 \times \sec 14\tfrac{1}{2}° \tag{3}$$

$$N_2 = 24.4 \times 1.0329 = 25.2 \text{ pounds}$$

$$N_1 = N_2 = 25.2 \text{ pounds but in opposite direction} \tag{4}$$

$$B_1 = N_1 = 25.2 \text{ pounds; } B_2 = N_2 = 25.2 \text{ pounds} \tag{5}$$

The directions of these bearing loads are shown in Fig. 1.

Change in Gear Tooth and Bearing Loads for Given Change in Pressure Angle

PROBLEM 4: Determine the change in normal tooth load N_1 and bearing load B_1 (Fig. 2) if the horsepower delivered to gear 1, the speed and direction of rotation of gear 1, the pitch diameters of gears 1 and 2 and the change in pressure angle α for the two gears are known.

Analysis of Problem: The solution of this problem proceeds along the same lines as that of Problem 3. The tangential load T_1 is computed and then the normal tooth load N_1 and bearing load B_1 for the two values of pressure angle α.

Example: Determine the change in normal tooth load and bearing load for a change in pressure angle α from $14\frac{1}{2}°$ to 20°. 25 H.P. is supplied to gear 1 (which rotates at 1200 R.P.M.). The pitch radius of gear 1 is 11.3 inches.

Solution: Torque gear $1 = \dfrac{63{,}000 \times 25}{1200} = 1312.5$ pound-inches (1)

$$T_1 = \frac{1312.5}{11.3} = 116.2 \text{ pounds} \quad (2)$$

$$N_1 = B_1 = T_1 \sec 14\tfrac{1}{2}° = 116.2 \times 1.0329 = 120.0 \text{ pounds} \quad (3)$$

$$N_1 = B_1 = T_1 \sec 20° = 116.2 \times 1.0642 = 123.7 \text{ pounds} \quad (4)$$

Thus a change in pressure angle from $14\frac{1}{2}$ degrees to 20 degrees increases the normal tooth load by 3.7 pounds in this case.

Size of Shaft When Gear Locations and Power Requirements Are Known

PROBLEM 5: To find the diameter d of a shaft (Fig. 3A) which is to carry two spur gears located at distance b apart and distances a and c, respectively, from the supporting shaft bearings. The horsepower P transmitted to gear B and its speed n in revolutions per minute are given. The weights of the two gears are known. Frictional losses are to be neglected.

Analysis of Problem: With the horsepower transmitted to gear B and its speed of rotation known, the torque which it exerts on the shaft can be computed.

With the torque known, the bearing load of gear B can be found. Since the shaft is rotating at a constant velocity, the applied torque of gear B and the resisting torque of gear C must be equal. (If this were not so, the shaft would either be accelerating or decelerating.) With the torque of gear C known, its bearing load is computed. With the bearing loads and weights of the two gears known, the loads on the supporting shaft bearings are found and the maximum bending moment M on the shaft due to these factors can be determined.

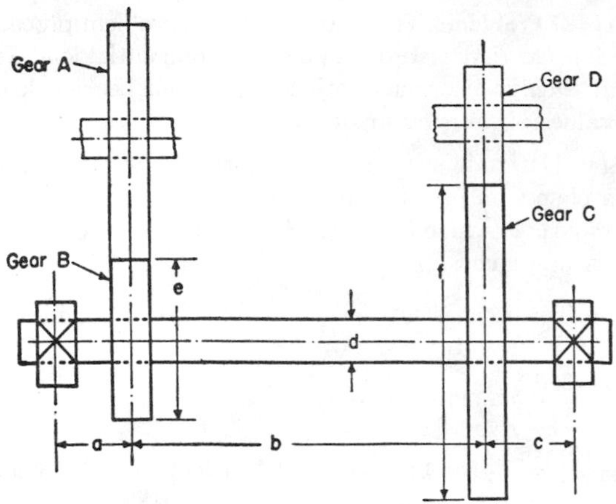

FIG. 3A. To Find Diameter d of Shaft Carrying Two Spur Gears Located as Shown, for Given Horsepower and Speed

The proper diameter of the shaft is then computed by means of the following formula for shafting subjected to both bending and torsional moments. This formula is given in the American Standard Association's Code for the Design of Transmission Shafting, where:

d = outside diameter of shaft in inches;

K_m = combined shock and fatigue factor to be applied in every case to the computed bending moment (for rotating shafts, K_m = 1.5 for gradually applied or steady loads; 1.5 to 2 for suddenly applied loads and minor shocks only; 2 to 3 for suddenly applied loads and heavy shocks);

K_t = combined shock and fatigue factor to be applied in every case to the computed torsional moment (for rotating shafts and gradually applied or steady loads $K_t = 1$; for suddenly applied loads and minor shocks only $K_t = 1$ to 1.5; for suddenly applied loads and heavy shocks $K_t = 1.5$ to 3);

M = maximum bending moment in inch pounds;

n = revolutions per minute;

P = maximum number of horsepower to be transmitted by the shaft;

p_t = maximum shearing stress in pounds per square inch (the maximum shearing stress p_t, under combined load = 8000 pounds per square inch for "commercial steel" shafting without allowance for keyways, and 6000 pounds per square inch with allowance for keyways. p_t = 30 per cent of the elastic limit in tension, but not more than 18 per cent of the ultimate tensile strength for shafting steel purchased under definite physical specifications);

Formula: If a solid circular shaft is subjected to combined torsion and bending:

$$d = \sqrt[3]{\frac{16}{\pi p_t}\sqrt{(K_m M)^2 + \left(\frac{396{,}000K_t P}{2\pi n}\right)^2}} \quad (1)$$

Example: Gear B in Fig. 3A has a pitch diameter of 10 inches and is rotating at 1200 R.P.M. Two hundred horsepower is being transmitted to it from gear A. Gear C has a pitch diameter of 20 inches. Both gears have 20-degree standard full-depth teeth. The distance a from the left-hand shaft bearing to gear B is 5 inches and similarly the distance from the right-hand shaft bearing to gear C is also 5 inches. The distance b between gears B and C is 23 inches. Gear B weighs 56 pounds and gear C, 174 pounds. What should be the diameter d of the shaft to carry a steady load? Use a maximum shearing stress which makes allowance for keyways. (p_t = 6000 p.s.i.)

Solution: Referring to Fig. 3B, let:

Speed of gears B and C = n = 1200 R.P.M.
Horsepower = P = 200 H.P.
Torque of gear B = t_B = ?
Torque of gear C = t_C = ?
Tangential load on teeth of gear B = T_B = ?
Tangential load on teeth of gear C = T_C = ?
Separating load on gear B = S_B
Separating load on gear C = S_C
Bearing load of gear B = B_B = ?
Bearing load of gear C = B_C = ?
Pitch diameter of gear B = D_B = 10 inches
Pitch diameter of gear C = D_C = 20 inches
Weight of gear B = W_B = 56 pounds
Weight of gear C = W_C = 174 pounds
Reacting upward force at left-hand shaft bearing = R_{LV} = ?
Reacting upward force at right-hand shaft bearing: R_{RV} = ?
Pressure angle = α = 20 degrees

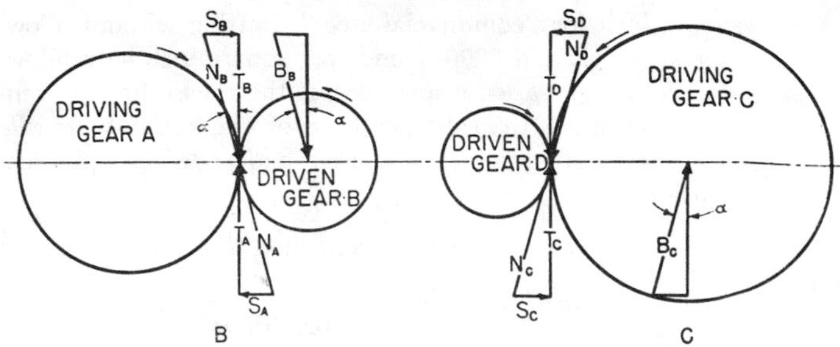

Fig. 3B. Diagram Showing Direction of Tooth and Bearing Loads for Gears A and B. C. Diagram Showing Tooth and Bearing Loads for Gears C and D

Then,

$$t_B = \frac{63{,}000 \times P}{n} = \frac{63{,}000 \times 200}{1200} = 10{,}500 \text{ pound-inches}$$

$$T_B = t_B \div \frac{D_B}{2} = 10{,}500 \div 5 = 2100 \text{ pounds}$$

It can be seen from Fig. 3B that $B_B \cos \alpha = T_B = 2100$ pounds

$$T_C = t_C \div \frac{D_C}{2}; \qquad \text{but } t_C = t_B, \text{ so that}$$

$$T_C = 10{,}500 \div 10 = 1050 \text{ pounds}$$

It can be seen from Fig. 3C that $B_C \cos \alpha = T_C = 1050$ pounds.

Next, the vertical force acting on the shaft at the left-hand bearing is found by taking moments about the right-hand bearing. (How this is done was shown in Problem 1, Chapter 17.)

$$R_{LV}(a + b + c) = W_B(b + c) + B_B \cos \alpha(b + c) + W_C(c) - B_C \cos \alpha(c)$$

$$R_{LV} = (W_B + B_B \cos \alpha)\left(\frac{b + c}{a + b + c}\right) + (W_C - B_C \cos \alpha) \times \left(\frac{c}{a + b + c}\right)$$

$$R_{LV} = (56 + 2100)\left(\frac{28}{33}\right) + (174 - 1050) \times \left(\frac{5}{33}\right)$$

$$R_{LV} = 1829 - 133 = 1696 \text{ pounds, upward}$$

Similarly, taking moments about the left-hand bearing,

$$R_{RV} = (W_B + B_B \cos \alpha)\left(\frac{a}{a + b + c}\right) + (W_C - B_C \cos \alpha) \times \left(\frac{a + b}{a + b + c}\right)$$

$$R_{RV} = (56 + 2100)\left(\frac{5}{5 + 23 + 5}\right) + (174 - 1050) \times \left(\frac{28}{33}\right)$$

$$R_{RV} = 327 - 743 = -416 \text{ pounds, downward}$$

The horizontal reacting force R_{LH} at the left-hand shaft bearing and that at the right-hand shaft bearing R_{RH} are now computed. These forces are due to the tooth load components S_B and S_C.

It can be seen from Figs. 3B and 3C that:

$$S_B = T_B \tan \alpha = 2100 \times 0.36397 = 764 \text{ pounds}$$

$$S_C = T_C \tan \alpha = 1050 \times 0.36397 = 382 \text{ pounds}$$

Taking moments about the right- and left-hand bearings, respectively,

$$R_{LH} = S_B \times \frac{b + c}{a + b + c} + S_C \times \frac{c}{a + b + c}$$

$$= 764 \times \frac{28}{33} + 382 \times \frac{5}{33} = 706 \text{ pounds, and, similarly,}$$

$$R_{RH} = S_C \times \frac{b + a}{a + b + c} + S_B \times \frac{a}{a + b + c} = 440 \text{ pounds}$$

The load on the left-hand bearing, R_L, and that on the right-hand bearing, R_R, is found by vector addition of the vertical and horizontal bearing load components as in Problem 7.

$$R_L = \sqrt{(R_{LV})^2 + (R_{LH})^2} = \sqrt{(1696)^2 + (706)^2} = 1837 \text{ pounds}$$

$$R_R = \sqrt{(R_{RV})^2 + (R_{RH})^2} = \sqrt{(-416)^2 + (440)^2} = 605 \text{ pounds}$$

The bending moment M_B on the shaft at gear B is equal to the reactive force at the left-hand shaft bearing times the distance a.

$$M_B = R_L \times a = 1864 \times 5 = 9320 \text{ pound-inches}$$

Similarly,

$$M_C = R_R \times c = 605 \times 5 = 3025 \text{ pound-inches}$$

The maximum bending moment on the shaft is therefore at gear *B*. Substituting the known values in Formula (1):

$$d = \sqrt[3]{\frac{16}{\pi \times 6000}\sqrt{(1.5 \times 9320)^2 + \left(\frac{396{,}000 \times 1 \times 200}{2\pi \times 1200}\right)^2}}$$

$$= 2\tfrac{3}{8} \text{ inches, approximately}$$

Gear Bearing Loads When Shaft Centers Are in Line

PROBLEM 6: To find the amount and direction of the bearing loads of spur gears 1, 2 and 3 which are located with the centers of their respective shafts in a straight line as shown in Fig. 4 if the horsepower delivered to gear 1, the speed, direction of rotation and the pitch diameter of gear 1, and the pressure angles of gears 1, 2 and 3 are known.

Analysis of Problem: The solution of this problem is similar to that of Problem 3. Gear 2, being an idling gear, merely serves to provide for the proper direction of rotation of driven gear 3 to which it transmits the full amount of horsepower received from gear 1

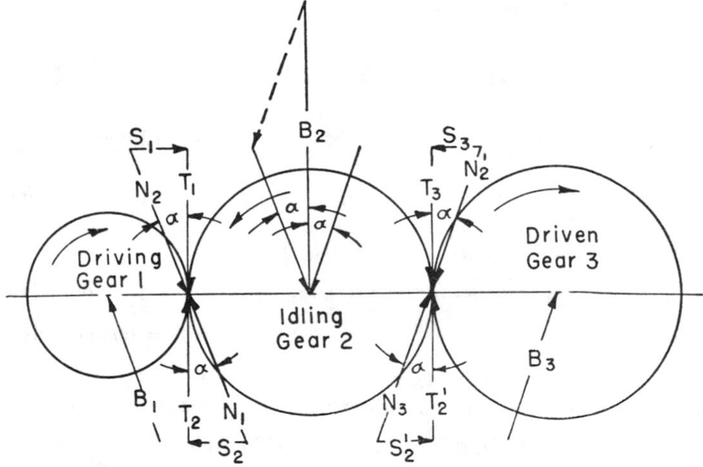

FIG. 4. To Find Amount and Direction of Bearing Loads of Spur Gears 1, 2 and 3 Having Shaft Centers in Line

(neglecting losses). The tangential driving force and resisting load is, therefore, the same between gears 2 and 3 as between gears 1 and 2. The loads, N_1 and N_3 normal to the teeth of gears 1 and 3 are in the directions shown and the loads, B_1 and B_3, on the bearings of gears 1 and 3 are equal and parallel to these respective normal tooth loads. The load, B_2, on the bearing of idling gear 2 is the vector sum of the two components respectively equal in magnitude and direction to loads N_2 and N_2' normal to the teeth of gear 2 as shown in Fig. 4.

Example: Determine the amount of bearing loads B_1, B_2 and B_3 if 15 H.P. is supplied to gear 1 which rotates in the direction shown in Fig. 4 at 900 R.P.M. The pitch radius R_1 of gear 1 is 8.5 inches. All three gears have a pressure angle α of 20 degrees.

Solution:

$$\text{Torque (gear 1)} = \frac{63{,}000 \times \text{H.P.}}{\text{R.P.M.}} \tag{1}$$

$$\text{Torque (gear 1)} = \frac{63{,}000 \times 15}{900} = 1050 \text{ pound-inches}$$

$$T_1 = \frac{\text{Torque}}{R_1} = \frac{1050}{8.5} = 123.5 \text{ pounds} \tag{2}$$

$$N_2 = T_1 \sec \alpha = 123.5 \times 1.0642 = 131.4 \text{ pounds} \tag{3}$$

$$N_1 = N_2 = 131.4 \text{ pounds but in the direction shown in Fig. 4} \tag{4}$$

$$B_1 = N_1 = 131.4 \text{ pounds} \tag{5}$$

$$T_3 = T_1 = 123.5 \text{ pounds} \tag{6}$$

$$T_2' = T_3 = 123.5 \text{ pounds but in the direction shown in Fig. 4} \tag{7}$$

$$N_3 = T_2' \sec \alpha = 123.5 \times 1.0642 = 131.4 \text{ pounds} \tag{8}$$

$$B_3 = N_3 = 131.4 \text{ pounds} \tag{9}$$

$$B_2 = \text{Vector sum of } N_2 \text{ and } N_2' \tag{10}$$

$$B_2 = N_2 \cos \alpha + N_2' \cos \alpha$$

$$B_2 = 131.4 \times 0.93969 + 131.4 \times 0.93969$$

$$B_2 = 123.5 + 123.5 = 247 \text{ pounds}$$

Gear Bearing Loads When Shaft Centers Are Not in Line

PROBLEM 7: Determine the amount and direction of bearing loads B_1, B_2 and B_3 for spur gears D, E and F respectively, Fig. 5A, if the horsepower delivered to gear D, the speed and direction of rotation of gear D, the pitch diameter of gear D, the pressure angle of gears D, E and F, the percentage of horsepower delivered to gear E that is transmitted to its shaft and the angle between the line connecting the centers of gears D and E and the line connecting the centers of gears E and F are known.

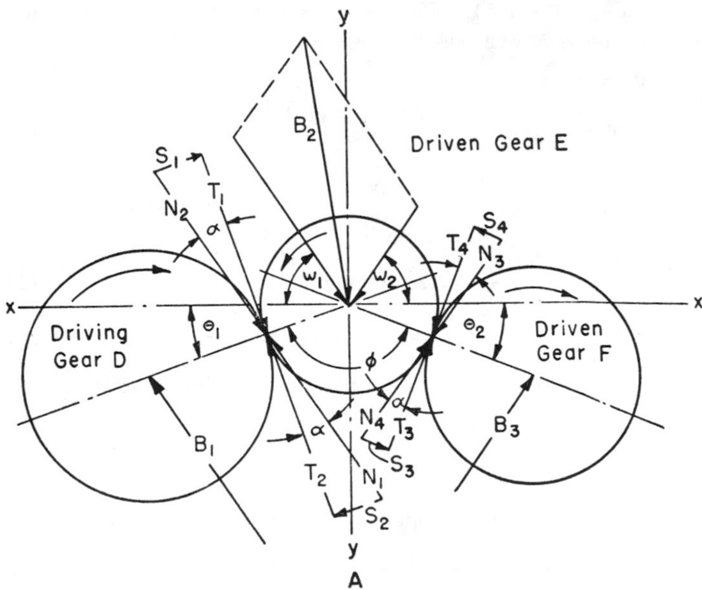

FIG. 5A. To Determine Amount and Direction of Bearing Loads for Spur Gears D, E and F with Shaft Centers not in Line

Analysis of Problem: The solution of this problem is similar to that of Problem 6, with the exception that determination of the bearing load of gear E is made somewhat more difficult, because the centers of gears D, E and F are not on the same straight line.

The tangential loads T_1 and T_2 and the normal tooth loads N_1 and N_2 at the point of mesh of gears D and E are computed as in Problem 6, also the bearing load B_1 of the driving gear D.

Since a given percentage of the power delivered to driven gear E is transmitted to its shaft and the balance is transmitted to gear F, neglecting losses, the tangential forces T_3 and T_4 are in the same proportion to the tangential forces T_1 and T_2 as the amount of power transmitted from gear E to gear F is to the amount of power transmitted from gear D to gear E. Having found T_3 and T_4, the normal tooth loads N_3 and the bearing load B_3 of gear F are computed. The bearing load B_2 is the vector sum of two components respectively equal in magnitude and direction to N_2 and N_3, as shown in Fig. 5B.

Example: Find the bearing loads B_1, B_2 and B_3 in Fig. 5A if 18 horsepower is supplied to gear D which rotates at 1200 R.P.M. and has a pitch diameter of 11.4 inches; the pressure angle of all three gears is 20 degrees; 40 per cent of the power delivered to gear E is transmitted to its shaft; and the angle ϕ between a line connecting the centers of gears D and E and a line connecting the centers of gears E and F is equal to 140 degrees.

Solution: 1. *Computing bearing load B_1 of gear D:*

$$\text{Torque } D \text{ to } E = \frac{63{,}000 \times \text{H.P.}}{\text{R.P.M.}} = \frac{63{,}000 \times 18}{1200}$$

$$= 945 \text{ pound-inches} \quad (1)$$

$$T_1 = \frac{\text{Torque}}{\text{Pitch radius}} = \frac{945}{5.7} = 165.8 \text{ pounds} \quad (2)$$

$$N_2 = T_1 \sec \alpha = 165.8 \times \sec 20^\circ$$
$$= 165.8 \times 1.0642 \quad (3)$$

$$N_2 = 176.4 \text{ pounds}$$

$$N_1 = N_2 = 176.4 \text{ pounds} \quad (4)$$

$$B_1 = N_1 = 176.4 \text{ pounds} \quad (5)$$

2. *Computing bearing load B_3 of gear F:*

Since 60 per cent of the power delivered to gear E is transmitted to gear F, tangential force T_3 is 60 per cent of T_1.

$$T_3 = 0.60 \times T_1 = 0.60 \times 165.8 = 99.4 \text{ pounds} \quad (6)$$

$$N_4 = T_3 \sec \alpha = 99.4 \sec 20^\circ = 99.4 \times 1.0642 \quad (7)$$

$N_4 = 105.8$ pounds

$B_3 = N_4 = 105.8$ pounds in the direction shown in Fig. 5A (8)

$N_3 = N_4 = 105.8$ pounds in the direction shown in Fig. 5A (9)

3. *Computing bearing load B_2 of gear E:*

Draw an arbitrary horizontal reference axis x–x and a vertical reference axis y–y through the center of gear E in Fig. 5A. Let it be assumed that the angle θ_1 of the line drawn through the centers of gears D and E with reference to the horizontal axis is 19 degrees. (Actually this horizontal reference line can be drawn at any convenient angle without affecting the procedure.) Hence T_1 which is at right angles to this line is at the same angle of 19 degrees with the vertical axis. Let it also be assumed that the angle θ_2 of the line drawn through the centers of gears E and F with reference to the horizontal axis is 21 degrees. Hence T_4 which is at right angles to this line is at the same angle of 21 degrees with the vertical axis.

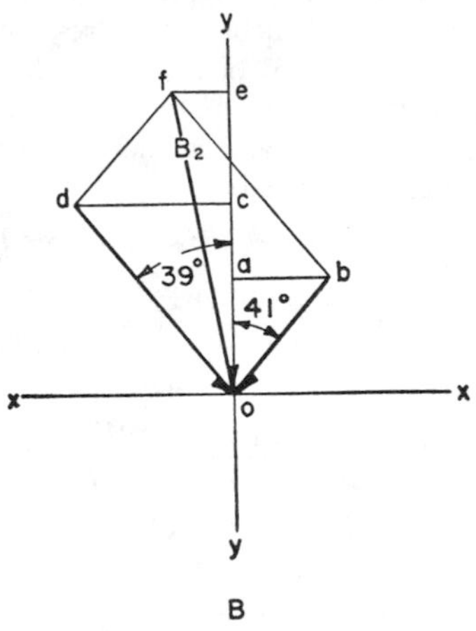

Fig. 5B. Diagram Showing Components of Bearing Load B_2

N_2 is therefore at an angle equal $\alpha + \theta_1$ or 39 degrees while N_3 is at an angle equal to $\alpha + \theta_2$ or 41 degrees with the vertical axis.

Referring now to Fig. 5B, one component of B_2 is *do* which is parallel to N_2. It is, therefore, at an angle of 39 degrees with the vertical as shown. Similarly the other component of B_2 which is *bo* being parallel to N_3 is at an angle of 41 degrees with the vertical as shown.

The vertical component *co* of *od* is equal to *od* cos 39° as shown in Fig. 5B. The vertical component *ao* of *bo* is equal to *bo* cos 41°. Since these vertical components both have a downward direction, they are added together to obtain the vertical component *eo* of B_2.

The horizontal component *dc* of *do* is equal to *do* sin 39° while the horizontal component *ba* of *bo* is equal to *bo* sin 41°. Since these two horizontal components are acting in opposite directions, the smaller is *subtracted* from the larger to obtain the horizontal component *fe* of B_2. Since B_2 is the hypotenuse of a right-angled triangle of which sides *fe* and *eo* are known, B_2 can be readily determined.

$$co = do \cos 39° = N_2 \cos 39° = 176.4 \times 0.77715 = 137.1 \qquad (10)$$

$$ao = bo \cos 41° = N_3 \cos 41° = 105.8 \times 0.75471 = 79.8 \qquad (11)$$

$$eo = ao + co = 137.1 + 79.8 = 216.9 \qquad (12)$$

$$dc = do \sin 39° = N_2 \sin 39° = 176.4 \times 0.62932 = 111.0 \qquad (13)$$

$$ba = bo \sin 41° = N_3 \sin 41° = 105.8 \times 0.65606 = 69.4 \qquad (14)$$

$$fe = de - ba = 111.0 - 69.4 = 41.6 \qquad (15)$$

$$B_2 = \sqrt{\overline{eo}^2 + \overline{fe}^2} = \sqrt{216.9^2 + 41.6^2} \qquad (16)$$

$$B_2 = \sqrt{48776} = 220.8 \text{ pounds in the direction shown in Fig. 5A.}$$

Efficiency of Square Threaded Screw and Nut

PROBLEM 8: What is the efficiency of a screw and nut with square thread if the helix angle α of the thread and the coefficient of friction μ are known?

Analysis of Problem: Efficiency is the ratio of energy output to energy input for a given period of time. It has been shown in Problem 6, Chapter 17, that the force P required at the pitch circumference of a screw to lift a weight W when the helix angle of the thread α and the angle of friction ϕ are known is:

$$P = W \tan (\phi + \alpha) \qquad (1)$$

Hence the energy input required when the screw is turned one complete revolution is equal to P times distance traveled which is the circumference of the pitch circle $2\pi r$.

The work done is equal to the product of the weight W and the distance it is raised in one revolution which is equal to l. Knowing energy input and work output for one revolution and neglecting any additional frictional losses such as may occur between a thrust collar at the top of the jackscrew and its bearing surfaces, the efficiency can be readily determined.

Formula: $E = \dfrac{\tan \alpha}{\tan (\alpha + \phi)}$, where $\phi = \arctan \mu$

Derivation of Formula: Referring to Fig. 6 in Chapter 17:

$$P = W \tan (\phi + \alpha) \tag{1}$$

Let E_i equal energy required to turn screw through one revolution and E_o equal work performed by screw in one revolution. Then

$$E_i = P \times 2\pi r = W \times 2\pi r \times \tan (\phi + \alpha) \tag{2}$$

$$E_o = W \times l \tag{3}$$

but

$$l = 2\pi r \times \tan \alpha \tag{4}$$

hence

$$E_o = W \times 2\pi r \times \tan \alpha \tag{5}$$

$$\text{efficiency} = \frac{E_o}{E_i} = \frac{W \times 2\pi r \times \tan \alpha}{W \times 2\pi r \times \tan (\phi + \alpha)} = \frac{\tan \alpha}{\tan (\phi + \alpha)} \tag{6}$$

Example: What is the efficiency of a jackscrew having a single square thread of ½ inch pitch and a pitch radius of 1⅜ inches when the coefficient of friction is 0.15? It is assumed that the jackscrew has a ball bearing thrust collar and hence the friction at the collar may be neglected in this calculation.

Solution:

$$\tan \phi = 0.15 \qquad \phi = 8°32'$$

$$\tan \alpha = \frac{0.5}{2\pi \times 1.375} = 0.0579; \; \alpha = 3°19'$$

$$\phi + \alpha = 8°32' + 3°19' = 11°51'$$

$$\text{efficiency} = \frac{\tan 3°19'}{\tan 11°51'} = \frac{0.0579}{0.2098} \times 100 = 27.6 \text{ per cent}$$

Load Capacity of Helical Spring

PROBLEM 9: What is the load capacity P in pounds of a helical spring if the diameter d in inches of the spring wire, the outside diameter D_o in inches of the spring, and the allowable stress S in pounds per square inch are given.

Analysis of Problem: The formula below is used to determine the loads on helical compression or tension springs required to produce given tensile or compressive stresses. When the stress value used in this formula is the allowable stress, the result is the maximum safe load.

Formula:

$P = \dfrac{\pi S d^3}{8DK}$, where D is the mean spring diameter in inches and K is a constant called the Wahl factor, the value of which depends on the ratio C of the mean spring diameter to the spring wire diameter. This factor tends to compensate for errors in spring equations which take into account torsional stress only, particularly where the ratio of spring diameter to wire diameter is low.

Example: What is the maximum safe load P of a carbon-steel helical compression spring having an outside diameter D of 5 inches and a wire diameter d of 0.5 inch, if it is intended for average service?

Solution: The allowable working stress S for a helical compression spring of 0.5 inch wire diameter in average service is given in MACHINERY'S HANDBOOK as 52,000 pounds per square inch.

The first step is to find the mean spring diameter D.

$$D = D_o - d = 5 - 0.5 = 4.5 \text{ inches}$$

The next step is to compute the ratio C of the mean spring diameter to wire diameter and find the corresponding value of K.

$$C = \frac{4.5}{0.5} = 9,$$

the corresponding value of Wahl factor K is given in MACHINERY'S HANDBOOK as 1.16.

With all the values now known, the allowable safe load P is computed:

$$P = \frac{3.1416 \times 52{,}000 \times 0.5^3}{8 \times 4.5 \times 1.16} = 489 \text{ pounds}$$

Total Stress in Flywheel Rim

PROBLEM 10: To find the total stress F in the rim of a rotating flywheel, if the density of the rim is k pounds per cubic inch; the width, a inches; the depth b inches; the mean radius R feet; and the speed of rotation is N revolutions per minute. The effect of the flywheel spokes is to be neglected.

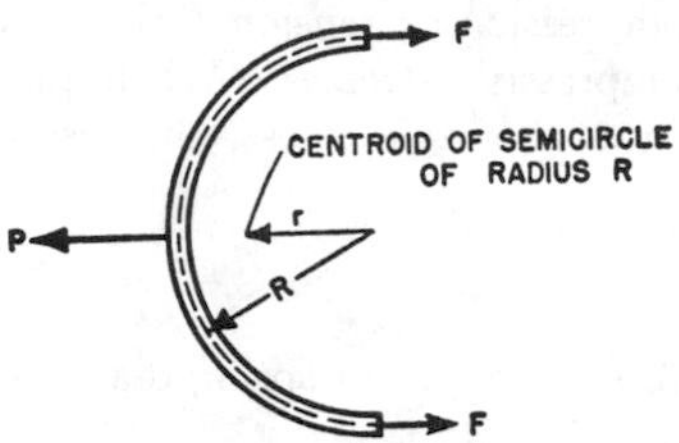

FIG. 6. Diagram Showing Forces Acting on One-Half of Rim of Flywheel and Location of Centroid of Semi-Circle of Radius R

Analysis of Problem: The stress in the rim of the flywheel is caused by the centrifugal force exerted on the rim as it rotates. For the purpose of analysis, the flywheel rim will be considered as composed of two equal and semi-circular parts. The centrifugal force P, which tends to separate these parts and thus causes the stress F in the rim, as shown in Fig. 6, will then be computed.

The formula for centrifugal force acting upon a body rotating about a fixed axis, in this case one-half the flywheel rim, is

$$P = \frac{W\omega^2 r}{g}$$

where P = centrifugal force in pounds; W = weight of body in pounds; ω = angular velocity of rotation in radians or 2π times the revolutions per second; r = radius from axis of rotation to center of gravity of the body in feet; and g = acceleration due to gravity in feet per second per second.

Since the dimensions of the flywheel rim and the density of the material composing the rim are known, its weight can be readily found. The angular velocity can be computed with the revolutions per minute known. There remains the radius to the center of gravity to be found. The thickness of the rim is relatively small compared with its radius. Therefore, the center of gravity of one-half the rim may be taken to be the center of gravity or, as it is more correctly called, the centroid of a semi-circle having a radius equal to the mean radius R of the flywheel rim. The distance from the center of a semi-circle to its centroid is equal to twice its radius divided by π. Thus, r can be expressed in terms of the mean radius R of the flywheel and all the elements necessary for the solution of the centrifugal force equation for P are known.

Formulas:

$$W = \frac{37.7 \times R \times a \times b \times k}{k} \text{ pounds}$$

$$F = \frac{0.0035WN^2R}{g} \text{ pounds}$$

where g is acceleration due to gravity in feet per second per second, and W is the weight of one-half the flywheel rim in pounds.

Derivation of Formula: The weight of one-half the rim is equal to its volume (mean circumference times area of cross section) in cubic inches multiplied by its density k.

$$W = \pi R \times 12 \times a \times b \times k = 37.7Rabk \tag{1}$$

The equation for the centrifugal force P exerted on one-half the rim is

$$P = \frac{W\omega^2 r}{g} \tag{2}$$

but

$$\omega = \frac{2\pi N}{60} \text{ radians per second} \tag{3}$$

If r is taken as the distance from the axis of rotation to the centroid of a semi-circle of radius R, then

$$r = \frac{2R}{\pi} \text{ feet} \tag{4}$$

Substituting these equivalents of ω and r in Equation (2)

$$P = \frac{W}{g} \times \left(\frac{2\pi N}{60}\right)^2 \times \frac{2R}{\pi} \tag{5a}$$

$$P = \frac{0.007WN^2R}{g} \tag{5b}$$

As shown in Fig. 6, the centrifugal force P is opposed by a tensile force F acting at each end of the half-rim and representing the stress in the rim, so that

$$2F = P \qquad \text{and} \qquad F = \frac{P}{2} \qquad \text{(6a and b)}$$

Hence,

$$F = \frac{0.007WN^2R}{2g} = \frac{0.0035WN^2R}{g} \qquad (7)$$

Example: Find the total stress in the rim of a cast-iron flywheel having a mean diameter of 5 feet, a width of 2 inches, a depth of 4 inches and a rotational speed of 200 revolutions per minute.

Solution: $R = \frac{5}{2} = 2.5$ feet; $a = 2$ inches; $N = 200$ R.P.M.; $b =$ 4 inches.

MACHINERY'S HANDBOOK gives the density of cast iron as 0.26 pound per cubic inch, hence $K = 0.26$.

$$W = 37.7 \times 2.5 \times 2 \times 4 \times 0.26 = 196 \text{ pounds} \qquad (1)$$

Acceleration due to gravity g is taken to be 32.2 feet per second per second.

$$F = \frac{0.0035 \times 196 \times 200^2 \times 2.5}{32.2} = 2130 \text{ pounds} \qquad (7)$$

Allowable Torque to be Applied to Bolt of Given Size

PROBLEM 11: It is required to find the maximum amount of torque which may be applied to a bolt without producing a combined tensile stress which is beyond the allowable limit.

Analysis of Problem: If the effective length (length equivalent to a lever arm) of a wrench is L and the force exerted in turning the nut is F, as shown in Fig. 7, the actual work W done in turning the wrench is equal to force times distance traveled, or $W = 2\pi L \times F$ per revolution. In one revolution the bolt moves axially a distance equal to its pitch p. If the average tensile pull on the bolt during this axial movement equal to p is M, the *useful* work done in moving the bolt is equal to force times distance, or $M \times p$ per revolution. Due to friction between the face of the nut and the surface on which it bears and also the thread friction, the total amount of work done in turning the wrench is equal to the useful work times the ratio $(p + D)/p$, where D is the nominal diameter of the bolt, hence

$$\frac{M \times p(p + D)}{p} = 2\pi L \times F$$

$$L \times F = \frac{M \times (p + D)}{2\pi} \text{ pound-inches}$$

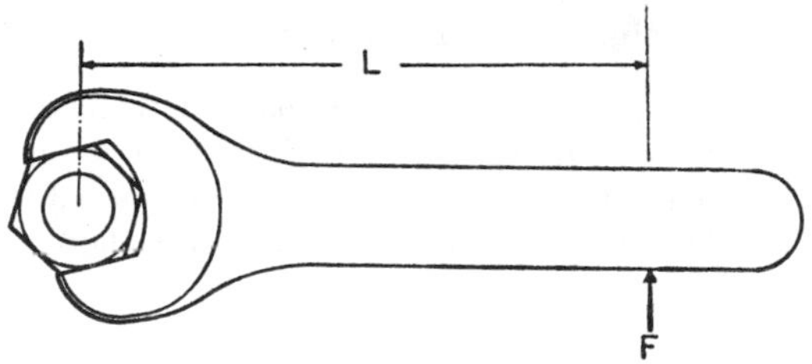

FIG. 7. To Find Maximum Torque Applied to Bolt by Wrench Having Effective Length L

Since the number of threads per inch N in the bolt is equal to $1/p$, the following formula is obtained by multiplying both numerator and denominator by N:

$$L \times F = \frac{M(1 + DN)}{2\pi N} \text{ pound-inches}$$

If S is the allowable stress and A is the area of the bolt cross-section taken at the root of the thread, then the maximum allowable pull on the bolt is found by the formula:

$$M = S \times A \text{ pounds}$$

and

$$LF = \frac{SA(1 + DN)}{2\pi N} \text{ pound-inches}$$

Example: Given a bolt made of annealed medium carbon steel that is ⅜ inch in diameter, has 16 threads per inch and a basic root area of 0.068 square inch. The allowable stress is 12,000 pounds per square inch. How much torque can be safely applied in tightening this bolt? What force would be necessary to produce this torque using a wrench with an effective length of 8 inches?

Solution: $$LF = \frac{12{,}000 \times 0.068(1 + 0.375 \times 16)}{2 \times 3.1416 \times 16}$$

$$LF = \frac{816 \times 7}{100.53} = 56.8 \text{ pound-inches}$$

$$F = \frac{56.8}{8} = 7.1 \text{ pounds applied to 8-inch wrench}$$

Determining Dimensions of Foundation Bolt

PROBLEM 12: To determine the dimensions of a foundation bolt.

A foundation bolt, Fig. 8, is to be designed with a lower end enlarged and slotted to receive a cotter-pin. It is required to compute the necessary dimensions so that the lower end of the bolt and also the cotter-pin will have the same strength as a section of the bolt taken through the root of the thread.

Analysis of Problem: Failure of the bolt may occur in several ways:

(a) It may rupture at the root of the thread as shown at B in Fig. 8.

(b) It may rupture at some section taken through the slotted section as at C.

(c) It may fail by crushing of the metal at the bottom of the keyway as at E.

(d) It may fail by the vertical shearing of that part of the bolt directly below the keyway as shown at F.

(e) The key may fail by shearing where it enters and leaves the bolt as shown at H and H_1.

(f) The key may fail by crushing at the contact surface between it and the keyway as shown at E.

The requirements are that the strengths at each of these locations C, E, F and H-H_1 should be equal to the strength at the root of the thread B. To determine the dimensions which will satisfy these requirements, it is necessary to compute the axial load required to produce failure at B and to set this equal to each of the maximum axial load equations for the conditions (b), (c), (d), (e) and (f) enumerated above. By solving these various equations the required dimensions are found.

(a) *Failure by rupture at root of thread.* Maximum axial load W that can be imposed before failure occurs at root of thread equals

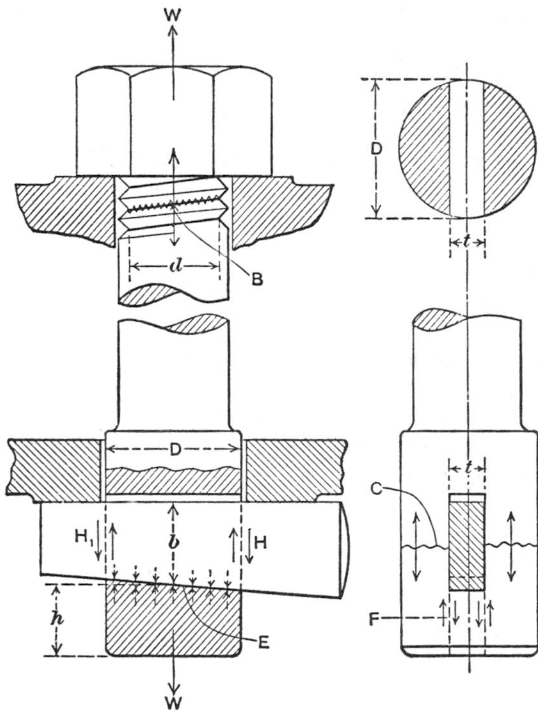

FIG. 8. To Determine Dimensions of Foundation Bolt Having Lower End Enlarged and Slotted to Receive a Cotter Pin

the ultimate tensile strength of the bolt S_t times the cross sectional area A_r at root of thread. Let d equal diameter at root of thread.

$$W = S_t \times A_r = S_t \times \frac{\pi d^2}{4} \tag{1}$$

(b) *Failure by rupture of slotted section of bolt.* Maximum axial load W that can be imposed is equal to the ultimate tensile strength S_t of the bolt times the area A_s of the cross-section of the bolt at this point. If D equals diameter of enlarged section of the bolt and t equals width of slot, then for this purpose it may be considered that the area removed by the slot is a rectangle of length D and width t.

$$W = S_t \times A_s = S_t\left(\frac{\pi D^2}{4} - D \times t\right) \tag{2}$$

(c) *Failure by crushing at bottom of keyway.* Maximum axial load W that can be imposed is equal to ultimate compressive strength S_c of the bolt times area of bottom of keyway. Here again, the bottom of the keyway may be considered to be a rectangle of length D and width t.

$$W = S_c \times D \times t \tag{3}$$

(d) *Failure by vertical shearing of that portion of bolt below keyway.* Maximum axial load that can be imposed is equal to ultimate shearing strength S_s of the bolt times twice the area of a vertical section directly below one side of the keyway. Let h equal the mean height of the section below the keyway.

$$W = S_s \times 2D \times h \tag{4}$$

(e) *Failure by shearing of key where it enters and leaves bolt.* Maximum axial load W that can be imposed is equal to the ultimate shearing strength of key S_s' times the sum of the vertical cross-sectional areas of the key where it enters and leaves the bolt, or times the equivalent, which is twice the mean vertical cross-sectional area of that portion of the key within the slot. Let b equal height of key at middle of keyway.

$$W = S_s' \times 2b \times t \tag{5}$$

(f) *Failure by crushing of the key.* Maximum axial load W that can be imposed is equal to the ultimate compressive stress of the key S_c' times the area of bottom or top of key.

$$W = S_c' \times D \times t \tag{6}$$

Procedure: The first step is to find the value of W which will cause rupture at the root of the thread.

$$W = S_t \times \frac{\pi d^2}{4} \tag{1}$$

Next, find the diameter of the bolt D at the enlarged section. Set the right-hand side of Equation (1) equal to the right-hand side of Equation (6)

$$S_t \times \frac{\pi d^2}{4} = S_c' \times Dt \tag{7a}$$

$$Dt = \frac{S_t}{S_c'} \times \frac{\pi d^2}{4} \tag{7b}$$

Then set the right-hand side of Equation (1) equal to the right-hand side of Equation (2)

$$S_t \times \frac{\pi d^2}{4} = S_t\left(\frac{\pi D^2}{4} - Dt\right) \tag{8a}$$

Divide both sides of this equation by S_t and place Dt alone on the left-hand side:

$$Dt = \frac{\pi D^2}{4} - \frac{\pi d^2}{4} \tag{8b}$$

Equating the right-hand sides of Equations (7b) and (8b)

$$\frac{S_t}{S_c'} \times \frac{\pi d^2}{4} = \frac{\pi D^2}{4} - \frac{\pi d^2}{4} \tag{9}$$

Dividing both sides of the equation by $\frac{\pi}{4}$ and placing D^2 alone on the left-hand side:

$$D^2 = \frac{S_t}{S_c'} \times d^2 + d^2 \tag{9a}$$

$$D = d\sqrt{\frac{S_t}{S_c'} + 1} \tag{9b}$$

With D, W and S_c' known, the thickness t of the key can be found from rearrangement of Equation (6)

$$t = \frac{W}{S_c'D} \tag{10}$$

With D, W and S_s known, the height h of the section below the slot can be found from a rearrangement of Equation (4):

$$h = \frac{W}{2S_s \times D} \tag{11}$$

With t, W and S_s' known, the height b of the key can be found from a rearrangement of Equation (5)

$$b = \frac{W}{2S_s' \times t} \tag{12}$$

Example: A 3-inch foundation bolt is to have an enlarged end to receive a cotter-pin. The bolt is made from cold worked steel with an ultimate tensile and compressive strength of 75,000 pounds per square inch and an ultimate shearing strength of 55,000 pounds per square inch. The cotter-pin is made of hardened steel with an ultimate compressive strength of 120,000 pounds per square inch and an ultimate shearing strength of 90,000 pounds per square inch.

Solution: $S_t = S_c = 75{,}000$; $S_c' = 120{,}000$; $S_s = 55{,}000$; $S_s' = 90{,}000$; $d = 2.675$ inches for a 3-inch bolt

$$W = 75{,}000 \times \frac{3.1416 \times 2.675^2}{4} \tag{1}$$

$$W = 75{,}000 \times 5.62 = 421{,}500 \text{ pounds}$$

$$D = 2.675\sqrt{\frac{75{,}000}{75{,}000} + 1} = 2.675\sqrt{2} \tag{9b}$$

$$D = 2.675 \times 1.414 = 3.782 \text{ inches}$$

$$t = \frac{421{,}500}{120{,}000 \times 3.782} = 0.929 \text{ inch} \tag{10}$$

$$h = \frac{421{,}500}{2 \times 55{,}000 \times 3.782} = 1.013 \text{ inches} \tag{11}$$

$$b = \frac{421{,}500}{2 \times 90{,}000 \times 1.857} = 1.261 \text{ inches} \tag{12}$$

Power Transmitting Capacity of Spur Gear

PROBLEM 13: What is the safe horsepower transmitting capacity of a steel spur gear having a pitch diameter D, a face width F, number of teeth N, and a pressure angle ϕ, if it is to run at a given R.P.M.? Assume that tooth wear is not a factor that has to be taken into consideration in this case.

Analysis of Problem: The above formula which gives the safe horsepower transmitting capacity, is a modification of the Lewis formula and is based upon the beam strength of the gear teeth.

In this formula, S_s is the allowable static unit stress in pounds per square inch for the gear material, the value of which is obtained by reference to a handbook; F is the face width of the gear in inches; V is the velocity of rotation of the gear in feet per minute at the pitch diameter; Y is a tooth form factor obtained by reference to a handbook; P is the diametral pitch of the gear; and G is a factor which is taken to be $55 \times (600 + V)$ for cut gears of ordinary accuracy, $27 \times (1200 + V)$ for cut gears of the better grades, and $423 \times (78 + V)$ for very accurate gears and pitch-line velocities of 4000 feet per minute or higher.

If tooth wear were an important consideration, the power transmitting capacity both with reference to wear and to tooth strength should be determined; then the smaller of the two values would be used.

Formula:

$$\text{H.P.} = \frac{S_s F V Y}{P G}$$

Example: What is the safe horsepower transmitting capacity of an SAE 1045 steel spur gear which has a pitch diameter of 16 inches, a face width of 3 inches, 64 teeth and a pressure angle of 14½ degrees, assuming that the gear is to run at a velocity of 120 revolutions per minute? The gear has been cut with ordinary accuracy. Assume safe static stress for this steel to be 40,000 pounds per square inch.

Solution: $S_s = 40{,}000$ pounds per square inch

$F = 3$ inches

$$V = \frac{\pi D \times \text{R.P.M.}}{12} = \frac{3.1416 \times 16 \times 120}{12} = 502.66 \text{ f.p.m.}$$

$Y = 0.360$ (From Machinery's Handbook)

$$P = \frac{N}{D} = \frac{64}{16} = 4$$

$$G = 55 \times (600 + V) = 55 \times (600 + 502.66) = 60{,}646$$

$$\text{H.P.} = \frac{40{,}000 \times 3 \times 502.66 \times 0.360}{4 \times 60{,}646} = 90$$

Bursting Pressure of Pipe

PROBLEM 14: To find the bursting pressure of a pipe when the wall thickness T and outside diameter D of a pipe as well as the ultimate tensile strength S of the material of which it is made are given.

Analysis of Problem: The bursting pressure can be found by an empirical formula, known as *Barlow's formula*, which for internal fluid pressures gives results within safe limits for pipes of all practical thickness ratios, i.e., ratios of wall thickness to outside diameter. The value of S to be used in this formula has been determined by actual tests to be 40,000 pounds per square inch for butt-welded and 50,000 pounds per square inch for lap-welded pipe.

Formula:

$$P = \frac{2T \times S}{D}$$

Example: What is the bursting pressure of lap-welded American Standard 4-inch pipe, Schedule 80?

Solution: The outside diameter of this pipe is given in MACHINERY'S HANDBOOK as 4.5 inches and the nominal wall thickness as 0.337 inch. Thus,

$$P = \frac{2 \times 0.337 \times 50{,}000}{4.5} = 7489 \text{ pounds per square inch, say}$$

7500 pounds per square inch

Thickness of Thin Cylindrical Shell for Given Pressure

PROBLEM 15: To find the required thickness t of a thin cylindrical shell subject to internal fluid pressure if the unit pressure of the fluid, the internal radius r of the shell, and the allowable working stress S of the material of which the shell is made are given.

Analysis of Problem: According to Pascal's law, the pressure exerted at any point by a fluid at rest is equally transmitted in all directions and is normal to the surface on which it acts.

In Fig. 9 is shown a cross-section perpendicular to the axis of the cylindrical shell. The arrows AB, HD and FE represent the pressure p which is radial, or normal to the unit of area at B, D and E, respectively. If the pressure is great enough it may be assumed that it will

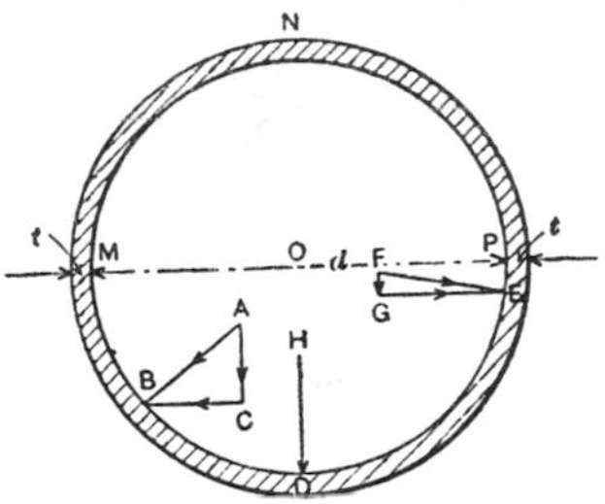

FIG. 9. Cross-Section of Cylindrical Shell Showing Forces Exerted on it by Internal Fluid Pressure

separate one-half of the shell from the other half. Suppose it separates the upper half *MNP* from the lower half *MDP*. Since the forces acting downward are equal to those acting upward, it will suffice to determine the downward pressure. At *D* the unit pressure acts entirely downward; at *B* it can be resolved into two components, one *AC* acting downward and the other *CB* acting horizontally. At *E* the vertical component *FG* is quite small, while at *P* it would equal *O*.

It can be shown by the calculus that the sum of these vertical downward acting components acting on the lower half of the shell is equal to $p \times 2r \times l$ where l is the length of the shell in the axial direction. This downward pressure is resisted by the unit strength of the material of the shell multiplied by the area of the ruptured surface or $2t \times l \times S_t$ in which $2t \times l$ is the area and S_t is the ultimate tensile strength of the material. Setting these two forces equal to each other

$$2tlS_t = 2prl \tag{1a}$$

$$t = \frac{pr}{S_t} \tag{1b}$$

If now, instead of the ultimate tensile strength S_t, the allowable working stress S of the material is substituted in Equation (1b) the required thickness t for a given fluid pressure p and internal radius r will be given.

This formula is for cases where t is equal to, or less than $0.1r$. Where computation by this formula shows that t should be greater than $0.1r$ the value of t should be recalculated using the formula for thick-walled cylinders given in Problem 16.

Formula:

$$t = \frac{pr}{S}$$

Example: Find the thickness of a cylindrical tank having an inside diameter of 6 feet if it is to be filled with a gas having a pressure of 80 pounds per square inch and is to be constructed of rolled-steel plate having an allowable working stress of 12,000 pounds per square inch for steady pressures.

Solution:

$$t = \frac{80 \times 3 \times 12}{12{,}000}$$

$$t = 0.240 \text{ inch, say } \tfrac{1}{4} \text{ inch}$$

Thickness of Thick-Walled Cylinder for Given Pressure

PROBLEM 16: To find the required thickness T of a high pressure cylinder wall when the inner radius R of the cylinder, the unit pressure P of the contained fluid, and the maximum allowable fiber stress S of the material are given.

Analysis of Problem: An empirical formula known as Lamé's formula is generally used in the design of thick-walled cylinders. It is valid only for internal pressure and cannot be applied in the accurate designing of a part such as a ram or plunger which is exposed to external pressure.

Formula:

$$T = R\left(\sqrt{\frac{S+P}{S-P}} - 1\right)$$

Example: Find the thickness of a cast iron cylinder to withstand a pressure of 1500 pounds per square inch. The inside diameter of the cylinder is to be 12 inches. The pressure of the contained fluid is not subject to frequent variation.

Solution: According to MACHINERY'S HANDBOOK, the allowable fiber stress for cast iron when subjected to steady or gradually applied pressures is from 3500 to 4000 pounds per square inch. For this problem assume the higher stress value, thus:

$$R = 6; \qquad S = 4000; \qquad P = 1500$$

$$T = 6\left(\sqrt{\frac{4000 + 1500}{4000 - 1500}} - 1\right)$$

$$T = 6 \times (1.483 - 1) = 2.90 \text{ inches}$$

Indicated Horsepower of Steam Engine

PROBLEM 17: What is the indicated horsepower of a steam engine having a length of stroke L and a piston area A if the mean effective pressure P in pounds per square inch and the number of strokes of the piston per minute are known.

Derivation of Formula: The average total force applied to the piston during any one stroke is equal to the mean effective pressure multiplied by the area of the piston. The work in foot-pounds done per stroke is equal to the average total force in pounds multiplied by the distance traveled or length of stroke in feet of the piston. The indicated power of the engine in foot-pounds per minute is equal to the work done per stroke multiplied by the number of strokes per minute. This product divided by the number of foot-pounds per minute in one horsepower is equal to the indicated horsepower.

Formula:

$$\text{H.P.} = \frac{PLAN}{33{,}000}$$

Example: What is the indicated horsepower of a steam engine in which the mean effective pressure is 110 pounds per square inch; the piston is 16 inches in diameter and has a stroke of 20 inches; and the engine flywheel has a speed of 80 R.P.M. The piston makes two strokes for each revolution of the flywheel.

Solution:

$$P = 110 \text{ lbs. per sq. in.} \qquad A = \frac{\pi \times 16^2}{4} = 201.06 \text{ sq. in.}$$

$$L = \frac{20}{12} = 1.67 \text{ feet} \qquad N = 2 \times 80 = 160 \text{ strokes per min.}$$

$$\text{H.P.} = \frac{110 \times 1.67 \times 201.06 \times 160}{33{,}000} = 179.1$$

Size of Steam Cylinder and Length of Stroke for Given Horsepower

PROBLEM 18: Find the cylinder diameter D and length of stroke L of a steam engine to deliver a given horsepower, H.P. The mean steam pressure P, the number of strokes per minute N, and the desired ratio k of length of stroke to cylinder diameter are given.

Analysis of Problem: Knowing the formula for indicated horsepower of a steam engine as given in Problem 17 and the ratio of length of stroke to cylinder diameter, an equation can be established with D as the only unknown, which can be solved for readily.

Formula:

$$D = 79.6 \sqrt[3]{\frac{\text{H.P.}}{kPN}} \text{ inches}$$

$$L = kD \text{ inches}$$

Derivation of Formula:

$$\text{H.P.} = \frac{PLAN}{33{,}000} \tag{1a}$$

$$LA = \frac{33{,}000 \text{ H.P.}}{PN} \tag{1b}$$

but

$$L = k \times \frac{D}{12} \text{ feet, where } D \text{ is in inches} \tag{2}$$

and

$$A = \frac{\pi D^2}{4} = 0.7854D^2 \tag{3}$$

Substituting the equivalents of L and A from Equations (2) and (3) in Equation (1b)

$$\frac{kD}{12} \times 0.7854D^2 = \frac{33{,}000 \text{ H.P.}}{PN} \tag{4a}$$

$$D^3 = \frac{504{,}202 \text{ H.P.}}{kPN} \tag{4b}$$

$$D = \sqrt[3]{\frac{504{,}202 \text{ H.P.}}{kPN}} = 79.6 \sqrt[3]{\frac{\text{H.P.}}{kPN}} \text{ inches} \tag{5}$$

$$L = kD \text{ inches} \tag{6}$$

Example: Find the length of stroke and diameter of the cylinder of a steam engine that is to deliver 175 horsepower if the mean steam pressure is 90 pounds per square inch and the number of strokes per minute is 120. The ratio of the length of stroke to the diameter of the cylinder is to be 1.5.

Solution: H.P. = 175; $P = 90$ pounds per sq. in.; $N = 120$ strokes per min.; $k = 1.5$.

$$D = 79.6\sqrt[3]{\frac{175}{1.5 \times 90 \times 120}} \tag{5}$$

$$D = 79.6\sqrt[3]{0.010802}$$

$$D = 79.6 \times 0.2211 = 17.60 \text{ inches}$$

$$L = 1.5 \times 17.60 = 26.40 \text{ inches} \tag{6}$$

20

Gear Ratios—Obtaining by Exact or Approximate Gear Combinations

In the design of machines and mechanical devices, certain speed relationships may be required between moving parts such as shafts. In some cases, these ratios must be exact; in others, approximate ratios are satisfactory. Such ratios may be obtained by using gears having different numbers of teeth, either in single or multiple pairs.

Where the required ratio is given as a common fraction having as numerator and denominator numbers for which gears with an equivalent number of teeth are available, say, $\frac{60}{47}$; $\frac{28}{56}$; or $\frac{72}{87}$; there is no ratio problem presented. A gear having 60 teeth and another having 47 teeth will provide the first ratio. Any combination of two available gears having a ratio of 1 : 2 will be satisfactory for the second ratio, and so on.

If the numbers in the numerator and denominator of the required ratio are too small, as $\frac{5}{8}$; $\frac{6}{11}$; $\frac{3}{13}$; to represent gears having satisfactory numbers of teeth, the numerator and denominator may be multiplied by some common factor to obtain gear teeth numbers of suitable size, as $\frac{5 \times 4}{8 \times 4} = \frac{20}{32}$; $\frac{6 \times 5}{11 \times 5} = \frac{30}{55}$. Or, if the numerators and denominators are too large, an attempt should be made to divide them both by a common factor, as $\frac{266 \div 7}{399 \div 7} = \frac{38}{57}$. If numerator and denominator are numbers for which factors are not readily discerned, the Table of Prime Numbers and Factors that is given in Chapter 27 will prove helpful. Thus, if the required ratio is $\frac{3645}{4221}$, referring to this table it will be found that 3 is a factor of 3645, or 3645 ÷

3 = 1215. Referring again to the table it will be found that 3 is a factor of 1215 or 1215 ÷ 3 = 405. It will be seen that 5 is a factor of 405, or 405 ÷ 5 = 81, and similarly 81 ÷ 9 = 9. Hence, the factors of 3645 are 3 × 3 × 5 × 9 × 9. Similarly, 4221 ÷ 3 = 1407; 1407 ÷ 3 = 469; 469 ÷ 7 = 67. Hence, the factors of 4221 are 3 × 3 × 7 × 67. The required ratio might then be written:

$$\frac{3645}{4221} = \frac{3 \times 3 \times 5 \times 9 \times 9}{3 \times 3 \times 7 \times 67}$$

Multiplying the first three factors of the numerator together; the last two factors of the numerator together; and the first three factors of the denominator together gives:

$$\frac{3645}{4221} = \frac{45 \times 81}{63 \times 67}$$

which ordinarily would be a suitable combination of two pairs of gears, from the standpoint of numbers of teeth.

So far, only ratios which can be obtained exactly by one or more pairs of gears have been considered. Much greater difficulty is faced when the fractional ratio cannot be broken down into factors in both numerator and denominator which combine to give suitable gear combinations. By "suitable" is meant gears which do not have too large or too small a number of teeth. In such cases, some gear combination which gives a ratio that approximates the required ratio must be used. Another difficulty is faced when the ratio is given as a decimal fraction and some equivalent common fraction which either equals or approximates it, must be found.

It is here that the method of continued fractions will be found useful in discovering a series of common fractions which are successively closer approximations to the given fraction, whether a common fraction or a decimal fraction. Where a table of gear ratios is not available, such a method permits the finding of several possible combinations of gears which offer different degrees of approximation to the required ratio.

This and other methods of establishing combinations which approximate a gear ratio will be taken up and illustrated in the discussion and problems which follow.

Method of Continued Fractions. Occasionally in dealing with a fraction which is cumbersome, that is, it has a large number of digits in the numerator and in the denominator; or a fraction, the numerator and denominator of which cannot be factored, it is desirable to substitute some other fraction approximately equal to it which is simpler or which can be factored satisfactorily. The method of continued fractions provides a means of computing a series of fractions, each of which is a closer approximation to the original fraction than the ones preceding it in the series.

Method of Computing Series. The following example gives the successive steps of this method in detail:

Example: The fraction $\frac{2153}{9277}$ is both cumbersome and non-factorable. It is required to find a simpler fraction with numerator and denominator that are factorable and which is approximately equal to $\frac{2153}{9277}$.

Step 1: Divide both numerator and denominator of this fraction by the numerator.

$$\frac{2153 \div 2153}{9277 \div 2153} = \frac{1}{4\frac{665}{2153}}$$

Step 2: Divide both numerator and demoninator of the fraction $\frac{665}{2153}$ by its numerator.

$$\frac{665 \div 665}{2153 \div 665} = \frac{1}{3\frac{158}{665}}$$

Step 3: Divide the numerator and denominator of the fraction $\frac{158}{665}$ by its numerator.

$$\frac{158 \div 158}{665 \div 158} = \frac{1}{4\frac{33}{158}}$$

Step 4: Divide the numerator and denominator of the fraction $\frac{33}{158}$ by its numerator.

$$\frac{33 \div 33}{158 \div 33} = \frac{1}{4\frac{26}{33}}$$

Step 5: Similarly,

$$\frac{26 \div 26}{33 \div 26} = \frac{1}{1\frac{7}{26}}$$

Step 6: Similarly,

$$\frac{7 \div 7}{26 \div 7} = \frac{1}{3\frac{5}{7}}$$

Step 7: Similarly,

$$\frac{5 \div 5}{7 \div 5} = \frac{1}{1\frac{2}{5}}$$

Step 8: Similarly,

$$\frac{2 \div 2}{5 \div 2} = \frac{1}{2\frac{1}{2}}$$

Step 9: Similarly,

$$\frac{1 \div 1}{2 \div 1} = \frac{1}{2}$$

Step 10: The successive fractions which have been computed in *Steps* 1 through 9 are now set down as follows without the fractional part of the denominators.

$$\cfrac{1}{4 + \cfrac{1}{3 + \cfrac{1}{4 + \cfrac{1}{4 + \cfrac{1}{1 + \cfrac{1}{3 + \cfrac{1}{1 + \cfrac{1}{2 + \cfrac{1}{2}}}}}}}}}$$

A more convenient arrangement is as follows:

$$\frac{1}{4+}\,\frac{1}{3+}\,\frac{1}{4+}\,\frac{1}{4+}\,\frac{1}{1+}\,\frac{1}{3+}\,\frac{1}{1+}\,\frac{1}{2+}\,\frac{1}{2}$$

These several elements make up what is called a *continued fraction.* By means of these elements it is possible to compute a series of fractions, each of which will be successively a closer approximation of the original fraction $\frac{2153}{9277}$. Because these computed fractions approach closer and closer to a definite value, they constitute what is called a *convergent series* and each of the fractions in this series is called a *convergent.*

The following steps show in detail how these convergents are computed.

Step 11: Set up a table as shown below with spaces for the denominator of each element in the continued fraction given in *Step* 10, plus two empty spaces at the beginning. Since there are nine denominators, there will be eleven spaces and these should be numbered 1 through 11, as shown. Provide a row of spaces for the convergents to be computed.

	1	2	3	4	5	6	7	8	9	10	11
Denominator			4	3	4	4	1	3	1	2	2
Convergent											

Step 12: In column 1 of the convergent row, insert $\frac{1}{0}$. In column 2 of this same row, insert $\frac{0}{1}$. These are two arbitrary convergents, the first equal to infinity and the second equal to zero which are always employed to facilitate the computation. The way in which they are used is indicated in the next step.

	1	2	3	4	5	6	7	8	9	10	11
Denominator			4	3	4	4	1	3	1	2	2
Convergent	$\frac{1}{0}$	$\frac{0}{1}$									

Step 13: The third convergent will now be computed. To find the numerator of the third convergent, multiply the denominator in column 3 by the numerator of the second convergent and add the numerator of the first convergent. Thus,

$$4 \times 0 + 1 = 1$$

To find the denominator of the third convergent, multiply the denominator in column 3 by the denominator of the second convergent and add the denominator of the first convergent, thus

$$4 \times 1 + 0 = 4$$

The third convergent is thus $\frac{1}{4}$.

The table with all computed convergents will now be shown and should be referred to in following the succeeding steps which give the computations for each convergent successively.

	1	2	3	4	5	6	7	8	9	10	11
Denominator			4	3	4	4	1	3	1	2	2
Convergent	$\frac{1}{0}$	$\frac{0}{1}$	$\frac{1}{4}$	$\frac{3}{13}$	$\frac{13}{56}$	$\frac{55}{237}$	$\frac{68}{293}$	$\frac{259}{1116}$	$\frac{327}{1409}$	$\frac{913}{3934}$	$\frac{2153}{9277}$

Step 14: The fourth convergent is computed in a similar manner. To find the numerator, multiply the denominator in column 4 by the numerator of the third convergent and add the numerator of the second convergent.

$$3 \times 1 + 0 = 3$$

To find the denominator of the fourth convergent, multiply the denominator in column 4 by the denominator of the third convergent and add the denominator of the second convergent.

$$3 \times 4 + 1 = 13$$

The fourth convergent is, therefore, $\frac{3}{13}$

Step 15: Similarly, the fifth convergent is computed:

$$\frac{4 \times 3 + 1}{4 \times 13 + 4} = \frac{13}{56}$$

Step 16: Similarly, the sixth convergent is computed:

$$\frac{4 \times 13 + 3}{4 \times 56 + 13} = \frac{55}{237}$$

Step 17: Similarly for the seventh convergent:

$$\frac{1 \times 55 + 13}{1 \times 237 + 56} = \frac{68}{293}$$

Step 18: Similarly for the eighth convergent:

$$\frac{3 \times 68 + 55}{3 \times 293 + 237} = \frac{259}{1116}$$

Convergent	Decimal Equivalent	Difference
$\frac{1}{0}$	∞	∞
$\frac{0}{1}$	0	0.2320793
$\frac{1}{4}$	0.25	0.0179207
$\frac{3}{13}$	0.2307692	0.0013101
$\frac{13}{56}$	0.2321428	0.0000635
$\frac{55}{237}$	0.2320675	0.0000118
$\frac{68}{293}$	0.2320819	0.0000026
$\frac{259}{1116}$	0.2320788	0.0000005
$\frac{327}{1409}$	0.2320795	0.0000002
$\frac{913}{3934}$	0.2320793	0.0000000
$\frac{2153}{9277}$	0.2320793	0

Step 19: Similarly for the ninth convergent:

$$\frac{1 \times 259 + 68}{1 \times 1116 + 293} = \frac{327}{1409}$$

Step 20: Similarly for the tenth convergent:

$$\frac{2 \times 327 + 259}{2 \times 1409 + 1116} = \frac{913}{3934}$$

Step 21: Similarly for the eleventh convergent:

$$\frac{2 \times 913 + 327}{2 \times 3934 + 1409} = \frac{2153}{9277}$$

which is the original fraction and is a check on the accuracy of the previous computations; in other words if the convergents have been computed correctly, the last one will be exactly equal to the original fraction.

The table on page **20–7** shows the decimal equivalents of these convergents and the respective differences between them and the decimal equivalent of the original fraction.

$$\frac{2153}{9277} = 0.2320793$$

It will be noted that the seventh, eighth, ninth, and tenth convergents are equal to the original fraction $\frac{2153}{9277}$ to five decimal places. If this accuracy is sufficient for the purpose at hand, then there is a choice of four possible fractions: $\frac{68}{293}$; $\frac{259}{1116}$; $\frac{327}{1409}$; and $\frac{913}{3934}$ to be used in place of the original.

The second of these can be factored as follows:

$$\frac{259}{1116} = \frac{37 \times 7}{31 \times 36} = \frac{37 \times 7}{62 \times 18} = \frac{37 \times 7}{93 \times 12} \text{ etc.}$$

The fourth of these can be factored as follows:

$$\frac{913}{3934} = \frac{83 \times 11}{281 \times 14}$$

The accuracy of each of these convergents is obvious upon comparison of its decimal equivalent with that of the original fraction. It is also possible to determine the range of error in each convergent without comparing decimal equivalents, as will be shown.

Intermediate convergents to those in the original series may sometimes be required. Two methods of finding these will be explained.

Accuracy of Any Convergent. Since the series of convergents computed by the method of continued fractions are alternately larger and smaller than the original fraction, the original fraction lies somewhere between each pair of successive convergents and the error of either is less than the difference between them.

The difference between any two successive convergents is equal to 1 divided by the product of the denominators of the two convergents. Thus, in the series just computed, the difference between the two convergents $\frac{55}{237}$ and $\frac{68}{293}$ is equal to $\frac{1}{237 \times 293} = \frac{1}{69{,}441}$. Hence, the error of the convergent $\frac{55}{237}$ is less than $\frac{1}{69{,}441}$. Since the intervals between successive convergents become smaller as the series progresses, it may be said that:

> *The error of any convergent is less than the difference between it and the next following convergent, such difference being equal to 1 divided by the products of the denominators of the two convergents.*

Computing Additional Intermediate Convergents. Where the method of computing continued fractions just described fails to give a ratio that is conveniently factorable, additional convergents intermediate between any two successive convergents in the original series may be found by one of two methods.

For the sake of illustration, let it be supposed that intermediate convergents between the ninth convergent, $\frac{327}{1409}$ and the tenth, $\frac{913}{3934}$ in the series just computed are to be found. In the following step-by-step explanations, the ninth convergent, representing that convergent of the chosen pair which comes first in the series, will be designated as convergent *A* and the tenth convergent or that which follows immediately after it in the series, as convergent *B*.

Method 1

Step 1: Multiply the numerator or denominator (whichever is larger) of convergent *B* by a factor which will be designated as *k*. Assuming that any intermediate convergent to be found by this method is to be factored, if possible, and that each of the factors cannot be larger than a certain permissible maximum, the value of *k* is dependent upon the product of these maximum factors. Thus, in this particular example, let it be assumed that the convergent to be used will have two factors in the numerator and two in the denominator. Let it be assumed further that each of these factors represents a gear, the maximum number of teeth of which can be 120. If each of the two factors in the numerator or denominator is 120, then their product is 14,400 which represents the maximum value of the numerator or denominator of any convergent that can be used in this example.

To find k, subtract from this maximum value the numerator or denominator, whichever is larger, of convergent A and divide the remainder by the corresponding numerator or denominator, as the case may be, of convergent B. k is equal to the resulting quotient if it is a whole number; if not, to the next smallest whole number.

Thus,
$$\frac{14400 - 1409}{3934} = 3.3$$

The next smallest whole number is 3 which is, then, the value of k.

Step 2: Multiply the numerator and denominator of convergent B by k and then add the numerator and denominator, respectively, of convergent A.

$$\frac{913 \times 3 + 327}{3934 \times 3 + 1409} = \frac{3066}{13211}$$

This is the first intermediate convergent.

Step 3: To find successive intermediate convergents, subtract the numerator of convergent B from the numerator of the first intermediate convergent just determined and the denominator of convergent B from the corresponding denominator. Thus,

$$\frac{3066 - 913}{13211 - 3934} = \frac{2153}{9277}$$

This subtraction may be carried on in a series of successive steps, each remainder of which will be an intermediate convergent, until the remainder is equal to the A convergent.

$$\frac{2153 - 913}{9277 - 3934} = \frac{1240}{5343}$$

$$\frac{1240 - 913}{5343 - 3934} = \frac{327}{1409}$$

which is the A convergent.

In the particular example worked out, the intermediate convergents computed between convergents A and B and a comparison of their decimal values is given below:

$$\text{Convergent } A = \frac{327}{1409} = 0.2320795$$

$$\text{1st Inter. Convergent} = \frac{1240}{5343} = 0.2320794$$

$$\text{2nd Inter. Convergent} = \frac{2153}{9277} = 0.2320793$$

$$\text{3rd Inter. Convergent} = \frac{3066}{13211} = 0.2320793$$

$$\text{Convergent } B = \frac{913}{3934} = 0.2320793$$

Still more intermediate convergents may be obtained by adding the numerators and denominators of any two of the above convergents, provided that the resulting numerator or denominator does not exceed the maximum allowable for a factorable convergent, in this case, 14,400. Thus,

$$\frac{327 + 1240}{1409 + 5343} = \frac{1567}{6752} = 0.2320794$$

Method 2

Step 1: Multiply the numerator or denominator (whichever is larger) of convergent A by a factor which will be designated as m. The factor m is found by subtracting from the maximum allowable value of the numerator or denominator in the desired factorable convergent, in this case 14,400, the numerator or denominator, whichever is larger, of convergent B and dividing the remainder by the corresponding numerator or denominator of convergent A. m is equal to the resulting quotient, if it is a whole number; if not, to the next smallest whole number, thus

$$\frac{14400 - 3934}{1409} = 7.4$$

m equals the next smallest whole number, or 7.

Step 2: Multiply the numerator and denominator of convergent *A* by *m* and then add the numerator and denominator, respectively, of convergent *B*

$$\frac{327 \times 7 + 913}{1409 \times 7 + 3934} = \frac{3202}{13797}$$

This is the first intermediate convergent.

Step 3: To find successive intermediate convergents, subtract the numerator of convergent *A* from the numerator of the intermediate convergent just determined and the denominator of convergent *A* from the corresponding denominator. This subtraction is continued until the remainder equals convergent *B*. Thus,

$$\frac{3202 - 327}{13797 - 1409} = \frac{2875}{12388} = 0.2320794$$

$$\frac{2875 - 327}{12388 - 1409} = \frac{2548}{10979} = 0.2320794$$

$$\frac{2548 - 327}{10979 - 1409} = \frac{2221}{9570} = 0.2320794$$

$$\frac{2221 - 327}{9570 - 1409} = \frac{1894}{8161} = 0.2320794$$

$$\frac{1894 - 327}{8161 - 1409} = \frac{1567}{6752} = 0.2320794$$

$$\frac{1567 - 327}{6752 - 1409} = \frac{1240}{5343} = 0.2320794$$

$$\frac{1240 - 327}{5343 - 1409} = \frac{913}{3934} = 0.2320793$$

which is convergent *B*.

Additional intermediate convergents can be obtained by adding the numerators and denominators of any two successive convergents as before.

Finding an approximate gear ratio by continuous fractions is illustrated in Problem 1. The methods of finding approximate ratios by proportional subtraction and proportional addition are explained and illustrated in Problems 2 and 3. Problem 4 shows how to find the new gear ratio for a given percentage change in driven gear speed.

Applying Method of Continued Fractions to Find Approximate Gear Ratio

PROBLEM 1: To find the change gears *C* and *E* in Fig. 1 required on a simple geared lathe, if the lead-screw has five threads per inch; gear *A* has one-half as many teeth as gear *B* and the thread to be cut has a pitch of 4.5 millimeters.

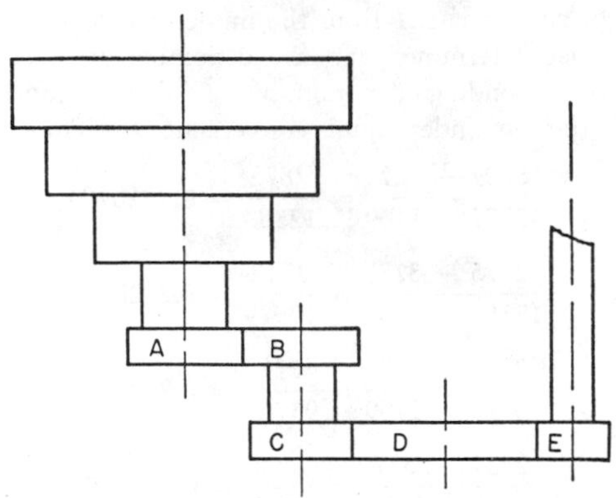

FIG. 1. To Find Change Gears *C* and *E* for Simple Geared Lathe for Cutting Thread of Given Pitch

Analysis of Problem: It is first necessary to express the pitch of the thread to be cut and the pitch of the lead-screw in the same units. This may be done by reducing 4.5 millimeters to inches (4.5 millimeters × 0.03937 inch per millimeter = 0.177165 inch). The pitch of the lead-screw = 1 ÷ 5 = 0.2 inch.

If gear *B* has twice as many teeth as gear *A* and it is assumed that the ratio of $\frac{C}{E} = 1$ (gear *D* only being an idler and not affecting the ratio) then, for every revolution of the spindle, the lead-screw will revolve one-half revolution and hence it will advance the tool carriage $0.2 \times \frac{1}{2} = 0.1$ inch. But it is desired to advance the tool carriage 0.177165 inch for each revolution of the lathe spindle in order to produce a thread having that pitch, hence gear *E* must be smaller than gear *C* in the ratio of $\frac{0.100000}{0.177165}$ or $\frac{E}{C} = \frac{0.100000}{0.177165}$.

In this problem, the method of continued fractions will be used to find a pair of gears C and E having this ratio.

Formula:

$$p = P \times \frac{A}{B} \times \frac{C}{E} \tag{1}$$

where p = pitch of thread to be cut

P = pitch of lead-screw

A, B, C and E = teeth numbers of gears shown in Fig. 1.

Solution: Substituting the known values for p, P and $\frac{A}{B}$ in Equation (1)

$$0.177165 = 0.2 \times \frac{1}{2} \times \frac{C}{E} \tag{1a}$$

$$\frac{C}{E} = \frac{0.177165}{0.100000} \tag{1b}$$

or

$$\frac{E}{C} = \frac{0.100000}{0.177165}$$

Multiply numerator and denominator of this ratio by 1,000,000 to remove the decimal points.

$$\frac{0.100000 \times 1{,}000{,}000}{0.177165 \times 1{,}000{,}000} = \frac{100{,}000}{177{,}165}$$

Now using the method of continued fractions outlined on pages 20–3 to 20–8, the convergents or simpler fractions approximating this ratio will be computed.

(1) $\frac{100{,}000}{177{,}165} = \cfrac{1}{1\,\frac{77{,}165}{100{,}000}}$; (2) $\frac{77{,}165}{100{,}000} = \cfrac{1}{1\,\frac{22{,}835}{77{,}165}}$

(3) $\frac{22{,}835}{77{,}165} = \cfrac{1}{3\,\frac{8{,}660}{22{,}835}}$; (4) $\frac{8{,}660}{22{,}835} = \cfrac{1}{2\,\frac{5{,}515}{8{,}660}}$

(5) $\frac{5{,}515}{8{,}660} = \cfrac{1}{1\,\frac{3{,}145}{5{,}515}}$; (6) $\frac{3{,}145}{5{,}515} = \cfrac{1}{1\,\frac{2{,}370}{3{,}145}}$

(7) $$\frac{2,370}{3,145} = \frac{1}{1\frac{775}{2,370}};$$ (8) $$\frac{775}{2,370} = \frac{1}{3\frac{45}{775}}$$

(9) $$\frac{45}{775} = \frac{1}{17\frac{10}{45}};$$ (10) $$\frac{10}{45} = \frac{1}{4\frac{5}{10}}; \frac{5}{10} = \frac{1}{2}$$

Expressing this ratio in terms of a continuous fraction:

$$\frac{100,000}{177,165} = \frac{1}{1} + \frac{1}{1} + \frac{1}{3} + \frac{1}{2} + \frac{1}{1} + \frac{1}{1} + \frac{1}{1} + \frac{1}{3} + \frac{1}{17} + \frac{1}{4} + \frac{1}{2}$$

Computing the fractional convergents and setting them down in tabular form.

	1	2	3	4	5	6	7	8	9	10	11	12	13
Denominators			1	1	3	2	1	1	1	3	17	4	2
Convergents	$\frac{1}{0}$	$\frac{0}{1}$	$\frac{1}{1}$	$\frac{1}{2}$	$\frac{4}{7}$	$\frac{9}{16}$	$\frac{13}{23}$	$\frac{22}{39}$	$\frac{35}{62}$	$\frac{127}{225}$	$\frac{2,194}{3,887}$	$\frac{8,903}{15,773}$	$\frac{20,000}{35,433}$

It will be seen that $$\frac{20,000 \times 5}{35,433 \times 5} = \frac{100,000}{177,165}$$

the original ratio, which is a check on the accuracy of the series of convergent fractions.

If gears of not more than, say, 120 teeth are available, then the fraction closest to the desired ratio which represents gears that could be used is $\frac{35}{62}$ or a gear C of 62 teeth and a gear E of 35 teeth.

Substituting these values for C and E in Equation (1) and solving for p:

$$p = 0.2 \times \frac{1}{2} \times \frac{62}{35}$$

$$p = 0.177143 \text{ inch}$$

$$p = \frac{0.177143}{0.03937} = 4.4994 \text{ millimeters}$$

which would be correct within six ten-thousandths of a millimeter.

Obtaining Approximate Ratio by Proportional Subtraction

PROBLEM 2: Determine the index gears and feed gears for cutting a helical gear on a hobbing machine having a fixed ratio between the feed and index gears, there being no differential mechanism. The number of teeth N to be cut on the helical gear; the normal diametral pitch P_n of the helical gear; the lead L; the feed G per revolution of the helical gear; the helix angle A; the machine constant K; the feed constant M; and the number of threads n on the hob are known.

Analysis of Problem: In gear hobbing machines not equipped with a differential, the accuracy of the helix angle or lead on the helical gear being cut depends upon the fixed relation between the index and feed gears. Even a slight error in the index gear ratio may cause an excessive error in the lead of the gear being cut. Where the ratio of the index gears is not exact, the ratio of the feed gears is modified slightly to compensate for this.

Procedure: The desired ratio R of the index gears is first found using the formula:

$$R = \frac{L \div F}{(L \div F) \pm 1} \times \frac{K_n}{N} = \frac{\text{Driving gears}}{\text{Driven gears}} \tag{1}$$

The minus sign in the denominator is used when gear and hob are of the same "hand," the plus sign, when of opposite "hand." The resulting fraction will probably have a decimal figure in the numerator and also in the denominator. One way to obtain a ratio closely approximating the desired ratio is to add a suitable decimal to the numerator and denominator to convert them to whole numbers. An attempt is then made to factor numerator and denominator into suitable gear sizes. If this cannot be done, another decimal number may be added to both numerator and denominator or one can be subtracted and the new whole numbers checked for factors. When a common fraction is found having factorable numerator and denominator that are suitable from the standpoint of gear sizes, it is converted to a decimal and this new ratio R' is compared with the required ratio R to determine the amount of error.

Since even a small error in the ratio of index gears will cause an appreciable change in the lead, the next step is to modify the feed by whatever slight amount is necessary to compensate for this error.

In order to compute the new ratio of feed gears which will provide the necessary correction, the following formula is used to first find the new feed F' per revolution of the helical gear for the new ratio R'.

$$F' = \frac{L(NR' - K_n)}{NR'} \qquad (2)$$

The ratio of feed gears Q' required to produce this feed is found by the formula:

$$Q' = \frac{F'}{M} \qquad (3)$$

To find an available combination of feed gears which will have a ratio closely approximating Q', several methods may be used. For this particular problem the method of obtaining an approximate ratio by proportional subtraction will be employed.

Step 1. Multiply numerator and denominator of the ratio (since ratio is a decimal, consider denominator to be 1) by some trial factor which will make both numerator and denominator quite large. For decimal ratios of 0.1 to 10, trial factors of 1000, 2000, 2500, 3000, 4000, 5000, 6000, 7500, 8000, 9000 and 10,000 are suggested.

Step 2. Round off the decimal in the numerator of the resulting fraction to the nearest whole number and compare the resulting ratio with that required.

Step 3. If the ratio which results in *Step* 2 is not close enough to the required ratio of feed gears, set down the original numerator and denominator of the fraction, side by side, and subtract successively one-hundredth of each until the numerator reaches a value which has a suitable decimal portion for rounding off, that is the decimal is either close to 0 or close to 1.

Step 4. Compare the resulting ratio Q'' with the required ratio Q' and check numerator and denominator for factors that would make suitable gear sizes.

If the resulting ratio is satisfactory, the new obtainable feed F'' is computed by the formula

$$F'' = Q'' \times M \qquad (4)$$

The lead L' produced in the helical gear when index gears having ratio R' and a feed F'' is used as found by the formula:

$$L' = \frac{F''NR'}{NR' - K_n} \qquad (5)$$

The final step is to check this lead L' with the required lead L.

Example: Find the index and feed gears for cutting a helical gear on a hobbing machine having a fixed ratio between feed and index gears, there being no differential mechanism. The helical gear is to have 48 teeth; a normal diametral pitch of 10; and a lead of 44.0894 inches. The feed per revolution of the helical gear is 0.035 inch; the helix angle is 20 degrees; the machine constant is 30; the feed gear constant is 0.075; and the number of threads on the hob is 1. The gear and hob are of the same hand.

Solution: $N = 48$; $P_n = 10$; $L = 44.0894$; $A = 20°$; $F = 0.035$; $K = 30$; $M = 0.075$; and $n = 1$.

The ratio R of the index gears is first found:

$$R = \frac{44.0894 \div 0.035}{(44.0894 \div 0.035) - 1} \times \frac{30 \times 1}{48} \qquad (1)$$

$$R = \frac{1259.697}{1258.697} \times \frac{5}{8} = 0.62549654$$

This is a preliminary ratio that will be changed slightly to suit available change gears. First try adding 0.303 to numerator and denominator of the fraction $\frac{1259.697}{1258.697}$ to obtain whole numbers which are factorable, thus

$$\frac{1259.697 + 0.303}{1258.697 + 0.303} = \frac{1260}{1259}$$

Referring to the table of prime numbers and factors in Chapter 27, it will be found that 1259 is a prime number and hence cannot be factored.

Next try adding 1.303 to numerator and denominator:

$$\frac{1259.697 + 1.303}{1258.697 + 1.303} = \frac{1261}{1260}$$

Using the table of prime numbers and factors:

$$\frac{1261}{13} = 97$$

$$\frac{1260}{5} = \frac{252}{3} = \frac{84}{3} = \frac{28}{4} = 7$$

Hence, $$\frac{1261}{1260} = \frac{13 \times 97}{5 \times 3 \times 3 \times 4 \times 7} = \frac{13 \times 97}{12 \times 105}$$

Let this new ratio be designated as R',

then $$R' = \frac{13 \times 97}{12 \times 105} \times \frac{5}{8} = \frac{65 \times 97}{96 \times 105} = 0.62549603$$

$$\begin{aligned} \text{Required ratio } R &= 0.62549654 \\ \text{Obtainable ratio } R' &= \underline{0.62549603} \\ \text{Error in ratio } R' &= 0.00000051 \end{aligned}$$

Since even this small error in the ratio of index gears will cause an appreciable change in the lead, the next step is to modify the feed by whatever slight amount is necessary to compensate for this index gear ratio error of 0.00000051.

Using the new ratio 0.62549603, the new feed F' and corresponding feed gear ratio Q' are computed:

$$F' = \frac{44.0894(48 \times 0.62549603 - 30 \times 1)}{48 \times 0.62549603} = 0.034964 \text{ inch} \quad (2)$$

$$Q' = \frac{F'}{M} = \frac{0.034964}{0.075} = 0.4661867 \quad (3)$$

Using the method of proportional subtraction to find a common fraction closely approximating 0.4661867:

Step 1. $$\frac{0.4661867 \times 10{,}000}{1 \times 10{,}000} = \frac{4661.867}{10{,}000}$$

Step 2. $\dfrac{4662}{10{,}000} = 0.4662$. Assuming that this is not close enough to 0.4661867, then

Step 3.

Numerators	*Denominators*
4661.867	10,000
− 46.619	− 100
4615.248	9900
− 46.619	− 100
4568.629	9800
− 46.619	− 100
4522.010	9700

Since the decimal 0.010 is quite close to 0, this last numerator will be rounded off giving the fraction $\frac{4522}{9700}$.

Step 4. $$\frac{4522}{9700} = 0.4661856$$

Assuming this to be sufficiently close to 0.4661867, the numerator and denominator will now be factored:

$$\frac{4522}{2} = \frac{2261}{7} = \frac{323}{17} = 19$$

or
$$\frac{4522}{9700} = \frac{2 \times 7 \times 17 \times 19}{97 \times 100} = \frac{38 \times 119}{97 \times 100}$$

Using this approximate ratio, the obtainable feed F'' now becomes:

$$F'' = \frac{38 \times 119}{97 \times 100} \times 0.075 = \frac{0.075 \times 38 \times 119}{97 \times 100} = 0.034964 \qquad (4)$$

The lead L' obtained when R' and F'' are used is now computed and compared with the required lead L.

$$L' = \frac{0.034964 \times 48 \times 0.62549603}{48 \times 0.62549603 - 30 \times 1} = 44.0898 \qquad (5)$$

The lead error is then 44.0898 − 44.0894 = 0.0004 inch.

Obtaining Approximate Ratio by Proportional Addition

PROBLEM 3: To find a combination of four gears that will have a ratio close to a given decimal ratio R.

Procedure: In the previous problem, proportional *subtraction* was used to find an approximate ratio. In this case, proportional *addition* will be employed.

Step 1. As in Problem 2, multiply numerator and denominator of the ratio to be approximated by some suitable factor.

Step 2. If the resulting decimal part of the numerator is sufficiently close to 0 or 1, round off the numerator and check the resulting ratio against the required ratio.

Step 3. If the ratio obtained in *Step* 2 is not close enough, set down the original numerator and denominator side by side and *add* one hundredth of each until a suitable decimal value for rounding off is obtained in the numerator.

Step 4. Convert the resulting common fraction back to a decimal and compare with the required ratio. If satisfactory, then check numerator and denominator of this common fraction for suitable factors.

Example: Find two pairs of gears having a ratio that will be within 0.00001 of the ratio 0.62549654.

Solution:

Step 1. $$R' = \frac{0.62549654 \times 5000}{1 \times 5000} = \frac{3127.4827}{5000}$$

Step 2. $\frac{3127}{5000} = 0.625400$ which is not close enough.

Step 3.

Numerators	*Denominators*
3127.4827	5000
+ 31.2748	+ 50
3158.7575	5050
+ 31.2748	+ 50
3190.0323	5100

Since .0323 is close to 0, this numerator will be rounded off giving the fraction $\frac{3190}{5100}$.

Step 4. $\frac{3190}{5100} = 0.625490$ which is within 0.0001 of the required ratio.

Factoring:
$$\frac{3190}{2} = \frac{1595}{5} = \frac{319}{11} = 29$$

$$\frac{5100}{2} = \frac{2550}{2} = \frac{1275}{5} = \frac{255}{5} = \frac{51}{3} = 17$$

so that

$$R' = \frac{2 \times 5 \times 11 \times 29}{2 \times 2 \times 5 \times 5 \times 3 \times 17} = \frac{55 \times 58}{60 \times 85} = \frac{110 \times 29}{75 \times 68}\text{; etc.}$$

Finding New Gear Ratio for Fixed Center Distance and Given Change in Driven Gear Speed

PROBLEM 4: To replace a pair of gears which are located on shafts having a fixed center distance C, as shown in Fig. 2, by a new pair of gears which will change the speed of the driven shaft by a given per cent k. Let t be the number of teeth in driving gear to be replaced and T be the number of teeth in driven gear to be replaced. Let P equal diametral pitch of these two gears. Let t_1 equal number of teeth in new driving gear and T_1 equal number of teeth in new driven gear.

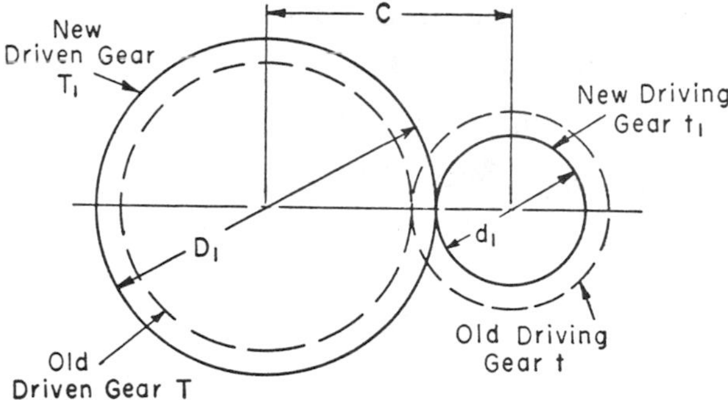

FIG. 2. To Replace Gears T and t by Gears T_1 and t_1 so that Speed of Driven Shaft Will be Changed by Given Per Cent

Analysis of Problem: The first step is to find the ratio R of the number of teeth in the *new* driven gear to the number of teeth in the *new* driving gear. If the speed of the driven shaft is to be *increased* by k per cent and the number of teeth were to remain the same, then the number of teeth on the driving gear would have to be increased by k per cent. Similarly, if the speed of the driven shaft were to be decreased by k per cent, the number of teeth in the driving gear would have to be decreased by k per cent, assuming the driven gear to remain the same. In either case, knowing the three factors T, t, and k, R can be found. With the new ratio, the center distance and the diametral pitch known, the number of teeth on the new driving gear can be found and also the number of teeth on the new driven gear can be determined.

Formula:

$$t_1 = \frac{2 \times C \times P}{R + 1}$$

$$T_1 = 2 \times C \times P - t_1$$

where

$$R = \frac{T}{t \pm kt}$$

Derivation of Formula:

$$R = \frac{T}{t \pm kt} = \frac{T_1}{t_1} \qquad (1)$$

If D_1 = pitch diameter of new driven gear
d_1 = pitch diameter of new driving gear

then $PD_1 = T_1$ (2)

and $Pd_1 = t_1$ (3)

or $$PD_1 + Pd_1 = T_1 + t_1 \qquad (4a)$$

or $$P(D_1 + d_1) = T_1 + t_1 \qquad (4b)$$

but $$D_1 + d_1 = 2C \qquad (5)$$

hence $$2CP = T_1 + t_1 \qquad (6)$$

$$R + 1 = \frac{T_1}{t_1} + 1 = \frac{T_1 + t_1}{t_1} \qquad (7)$$

hence $$t_1 = \frac{T_1 + t_1}{R + 1} \qquad (8)$$

Substituting the equivalent of $T_1 + t_1$ from Equation (6)

$$t_1 = \frac{2CP}{R + 1} \qquad (9)$$

$$T_1 = T_1 + t_1 - t_1 \qquad (10a)$$

Substituting the equivalent of $T_1 + t_1$ from Equation (6)

$$T_1 = 2CP - t_1 \qquad (10b)$$

Example: The driving and driven gears of a machine have 20 and 76 teeth, respectively. The center distance between shafts is 6 inches and the diametral pitch 8. Using the same center distance and gears of the same diametral pitch, how many teeth should there be in the driving and driven gears in order to obtain an increase in the speed of the driven gear of 20 per cent?

Solution: $t = 20$; $T = 76$; $C = 6$; $P = 8$; $k = 0.20$.

$$R = \frac{76}{20 + 20 \times 0.20} = 3.17 \text{ approx.} \qquad (1)$$

$$t_1 = \frac{2 \times 6 \times 8}{3.17 + 1} = 23 +, \text{ say } 23 \text{ teeth} \qquad (9)$$

$$T_1 = 2 \times 6 \times 8 - 23 = 73 \text{ teeth} \qquad (10b)$$

The original speed ratio was:

$\frac{20}{76} = 0.263158$ revolution of driven gear per revolution of driver.

The new speed ratio is:

$\frac{23}{73} = 0.315068$ revolution of driven gear per revolution of driver.

$$\frac{0.315068}{0.263158} = 1.197$$

or an increase of speed in the driven gear of 19.7 per cent, if the speed of the driver remains constant.

Simplified Change Gear Calculations

Most systematic methods for determining a set of four change-gears to produce a specified gear ratio require considerable calculation and pencil work. This work can usually be eliminated by using a calculator or computer, in conjunction with a simple trial and error procedure based on the usual formula expressing the relationship between a gear ratio and the numbers of teeth in the gears:

$$\text{Gear ratio } R = \frac{\text{Driving gear} \times \text{driving gear}}{\text{Driven gear} \times \text{driven gear}}$$

or

Gear ratio R × product of driven gears = product of driving gears

Method:

1. Set the desired ratio R in the memory of a calculator

2. By repeated addition of this ratio R (the equivalent of multiplication) find a multiplier such that the product M × R is a whole numnber I, or nearly so.

3. Convert the fraction $\frac{I}{M}$ to the desired change gears.

As a simple example of applying the method, consider the ratio 0.750. Here, 0.750 is set on the keyboard and by repeated addition the fraction $\frac{3}{4}$ is determined as follows:

R		M		I	
0.750	×	1	=	0.750	
	×	2	=	1.500	
	×	3	=	2.250	
	×	4	=	3.000	⇐ whole number

$$\text{Therefore } 0.750 = \frac{I}{M} = \frac{3}{4}$$

This fraction is next converted to provide the driving and riven gears, an operation familiar to those who make gear-ratio calculations.

$$0.750 = \frac{3}{4} = \frac{3 \times 800}{4 \times 800} = \frac{2400}{3200} = \frac{40 \times 60}{50 \times 64}$$

As a second example, gears for a ratio of 0.750176 will be determined:

0.750176 × 1400 = 1050.246400
× 1401 = 1050.996576
× 1402 = 1051.746752
× 1403 = 1052.496928
× etc. =
× 1413 = 1059.998688
× etc. =
× 1421 = 1066.000096

In the sequence above, the ratio 0.750176 was multiplied first by 1400, a number selected at random, and repeated addition was applied until 1066.000096 was obtained. It is evident that this number is very close to 1066, so that 0.750176 is very nearly represented by the fraction , $\frac{1066}{1421}$ and is within 0.0000001 of being equal to 0.750176.

If such a high degree of accuracy were not required, the fraction $\frac{1060}{1413}$ might have been considered. $\frac{1060}{1413} = 0.7501769$, which is within 0.0000009 of 0.750176. However, the factors of the denominator of this fraction, 1413, are 9 and 157, but because 157 is greater than the largest generally-available change-gear, this fraction could not be used.

The factors of the fraction $\frac{1066}{1421}$ are $\frac{26 \times 41}{29 \times 49}$. Each of these factors is within the standard range of change gears (20 to 120).

21

Finding Out What Happens in Planetary Gearing

Planetary gearing is a special type of mechanism which provides, in compact form, a means of obtaining a very small or very large change in angular velocity of driven shaft as compared with driving shaft. The distinctive characteristic of planetary gearing is that some of the gears turn on movable centers while others turn on fixed centers. The gears that turn on movable centers are called *planet gears* and those that turn on fixed centers, *sun gears*.

The planetary gearing mechanism may consist of external spur gears, of external and internal spur gears, of racks and spur gears, or of bevel gears. That member which receives motion from outside the mechanism is called the *driver*; that member from which motion is taken outside the mechanism is called the *follower*; that member which carries one or more bearing pins about which the planet gears rotate is called the *train arm*. In addition, there is usually one member which is maintained in a fixed position.

The objective in solving planetary gearing problems is usually to determine the number of turns of the driver required to produce one turn of the follower or, vice versa, the number of turns of the follower produced by one turn of the driver. Two methods of solution will be explained and illustrated by application to various problems.

The first method, called the *analytical method*, is used to *accurately* determine the ratio between rotational speeds of driver and follower shafts. The second method, called the *graphical method*, is used to *approximately* determine the ratio between rotational speeds of driver and follower shafts and is frequently employed to obtain a quick check on the results obtained by the first method.

Analytical Method of Solution. In this method, the revolution of the planet wheel axis or pin through 360 degrees in a clockwise direction is taken to be the reference basis for the analysis. The direction of rotation and number of turns made by the driver and by the follower during this single turn of the planet wheel axis are then determined and compared. The step-by-step procedure is as follows:

Step 1. The entire mechanism is considered to be locked together and the whole device is rotated one turn in a clockwise position around that axis about which the planet gear pin is carried.

Step 2. The planet gear pin has now made one revolution in a clockwise direction as required, but the fixed gear has also been rotated the same amount. Hence, the various elements of the mechanism are considered to be again free to turn; the train arm which carries the planet gear pin is now held stationary; and the fixed gear is rotated back in a counter-clockwise direction one turn to its original position.

Step 3. The number of revolutions and direction of rotation of the driver caused by returning the fixed gear to its original position in *Step* 2 are then noted. Let this be called X.

Step 4. The number of revolutions and direction of rotation of the follower caused by returning the fixed gear to its original position in *Step* 2 are also noted. Let this be called Y.

Step 5. In *Step* 1 the driver and follower were both rotated one turn in a clockwise direction. In *Step* 2 the driver makes X revolutions and the follower Y revolutions. The total number of revolutions made by the driver is $1 \pm X$ depending upon whether the driver rotates in a clockwise (plus) or counter-clockwise (minus) direction during *Step* 2. The total number of revolutions made by the follower is $1 \pm Y$ depending upon whether the follower rotates in a clockwise (plus) or counter-clockwise (minus) direction during *Step* 2. The number of turns of the driver for each turn of the follower is, therefore, expressed by the equation:

$$N = \frac{1 \pm X}{1 \pm Y}$$

Graphical Method of Solution: The graphical method is the same in principle for all planetary gearing problems but varies somewhat in the procedure to be followed depending upon whether the mechanism consists (a) of spur gears having the same circular pitch; (b) of spur gears having different circular pitches; or (c) of bevel gears.

The procedure to be described at this point is applicable to spur gears having the same circular pitch. The modifications of this method required for the other two types of problems will be presented when these problems are specifically discussed.

For the purpose of more clearly outlining this method, reference is made to a typical planetary gearing mechanism having gears of the same circular pitch. This mechanism is diagrammatically shown in Fig. 1.

Step 1. Draw a line *ab* tangent to the pitch circle of the driver, if it is a gear, and at the point *a* where the center line of the train arm intersects this pitch circle. (If the driver is the train arm, this line is drawn through the axis of the outermost bearing pin carried by the train arm and at right angles to the axis of the train arm.)

This line may be of any convenient length and represents the tangential linear velocity of the point from which it is drawn.

If the driver is a gear (as in Fig. 1) it will usually be found that point *a* on the pitch circle of the driver coincides with a point on the pitch circle of a planet gear which either itself engages the fixed gear in the train or is keyed to the same shaft as another planet gear which engages the fixed gear. (If the driver is a train arm, this point *a* will be found to coincide with the axis of a planet gear which engages the fixed gear.) In either case *ab* can also be taken to represent the linear velocity of point *a* on that planet gear which engages the fixed gear.

Step 2. Since there is no sliding of the planet gear on the fixed gear, that point *c* at which their pitch circles are in contact may be considered as having momentarily a linear velocity of zero. If a line is drawn from *b* to *c*, then, measuring at right angles to the center line of the train arm, or in other words parallel to *ab*, the distance to line *bc* of any point on this planet gear which also coincides with the train arm center line represents the linear velocity of that point.

Step 3. When the follower is the train arm, such a line is now drawn from point *p*, which is the center of the pin, carried by the train arm, on which the planet gear revolves. This line is drawn parallel to *ab* until it intersects *bc* at *e*. (If the follower is a gear, point *p* is taken to be the point of tangency between the pitch circles of the planet gear and the follower.) Line *pe* represents the velocity of point *p* on the planet gear. It also represents the velocity of coinciding point *p* on the follower.

Step 4. A line is now drawn from point *m*, which is the fixed center of the follower and therefore has zero linear velocity, through point *e* until it intersects line *ab* at *f*.

The distance to *mf*, if it is measured parallel to *ab*, of any point on the follower which coincides with the axis of the train arm, represents the velocity of the follower at a radius equal to the distance of that point from the axis *m*. Hence *af* represents the velocity of the follower at radius *am*.

Step 5. But *ab* represents the linear velocity of the driver at radius *am*. Hence, the value of *N* which is the number of turns of the driver required to produce one turn of the follower is found by the formula:

$$N = \frac{ab}{af}$$

Notation for Planetary Gearing Problems. In the problems that follow, solutions will be worked out by both the analytical and graphical methods. The notation used in these problems will be:

N = number of turns of driver to one of the follower or driven member;

N' = number of turns of follower to one of driver = $1 \div N$;

N_1 = number of turns of driver to one complete revolution of planet gear axis;

N_2 = number of turns of follower to one complete revolution of planet gear axis;

D = diameter of pitch circle of driver, if driver is a gear; (The driver, or the follower, may be the "train arm," and not one of the gears, according to the data of a problem.)

D_1 = diameter of pitch circle of follower, if follower is a gear;

D_2 = diameter of pitch circle of fixed gear;

D_3, D_4, etc. = diameters of pitch circles of planetary gears;

T = number of teeth in driver, if driver is a gear;

T_1 = number of teeth in follower, if follower is a gear;

T_2 = number of teeth in a fixed gear; and

T_3, T_4, etc. = number of teeth in planetary gears.

Single Planet Gear with Train Arm Acting as Follower

PROBLEM 1: As shown in Fig. 1, this planetary gearing mechanism has a fixed external gear, an internal gear as a driver, a single planet gear, and a train arm acting as follower.

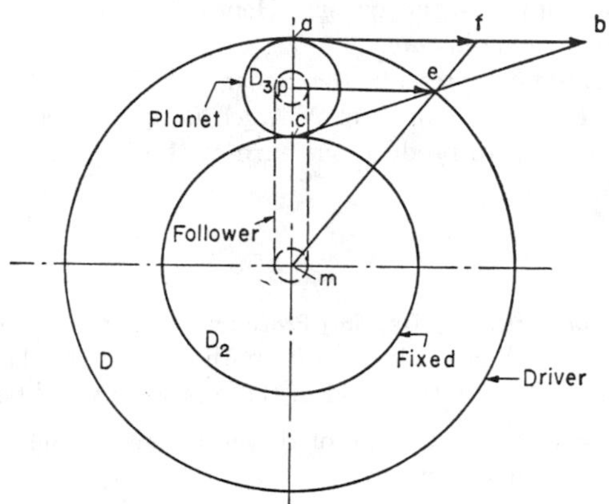

FIG. 1. Single Planet Gear with Train Arm Acting as Follower

Formulas: $N = 1 + \dfrac{D_2}{D}$ (By Analytical Method)

$N = \dfrac{ab}{af}$ (By Graphical Method)

Derivation of Formulas: *By the Analytical Method*

Step 1. Entire mechanism locked together and rotated once about m.

Step 2. Fixed sun gear D_2 rotated back one turn in counter-clockwise direction, with train arm held in fixed position.

Step 3. The number of revolutions X made by the driver in *Step* 2 is found as follows. When the fixed gear D_2 is rotated one turn in a counter-clockwise direction, planet gear D_3 is caused to rotate in a clockwise direction. The number of turns it makes is equal to the pitch diameter of the fixed gear divided by the pitch diameter of the planet gear (or the number of teeth of the fixed gear divided by the number of teeth of the planet gear), thus:

$$\text{No. turns of } D_3 = \frac{D_2}{D_3}$$

The number of turns of the driver for each turn of planet gear D_3 with which it meshes is equal to the pitch diameter D_3 divided by the pitch diameter D (or the number of teeth of the planet gear divided by the number of teeth of the driver).

$$\text{No. turns } D \text{ for one turn } D_3 = \frac{D_3}{D}$$

Hence the number of turns X of the driver equals the number of turns of D_3 times $\frac{D_3}{D}$, or

$$X = \frac{D_2}{D_3} \times \frac{D_3}{D} = \frac{D_2}{D}$$

Step 4. Since the follower is the train arm, it makes no revolutions during *Step* 2, hence $Y = 0$.

Step 5. It can be seen from Fig. 1 that during *Step* 2 the driver will rotate in a clockwise or positive direction, therefore

$$N = \frac{1 + X}{1}$$

or

$$N = 1 + \frac{D_2}{D}$$

By the Graphical Method (See Fig. 1). The graphical solution of this particular problem is given in the introductory text of this chapter under the heading "Graphical Method of Solution."

Example 1: In Fig. 1 if $D = 115$, $D_2 = 65$ and $D_3 = 25$, find N by the analytical method.

Solution: $N = 1 + \frac{65}{115} = 1\frac{13}{23}$ turns of the driver for each turn of the follower.

Example 2: In Fig. 1, if $D = 60$, $D_2 = 36$, and $D_3 = 12$, find N by the analytical method.

Solution: $N = 1 + \frac{36}{60} = 1\frac{3}{5}$ turns of the driver for each turn of the follower.

Two Planet Gears with Train Arm Acting as Follower and External Gear as Driver

PROBLEM 2: As shown in Fig. 2, this planetary gearing mechanism has two external spur gears as planet gears, a fixed internal spur gear, an external spur gear acting as driver and the train arm acting as follower. If the pitch diameters D, D_2, D_3 and D_4 are given, it is required to find the number of turns N of the driver required to produce one turn of the follower.

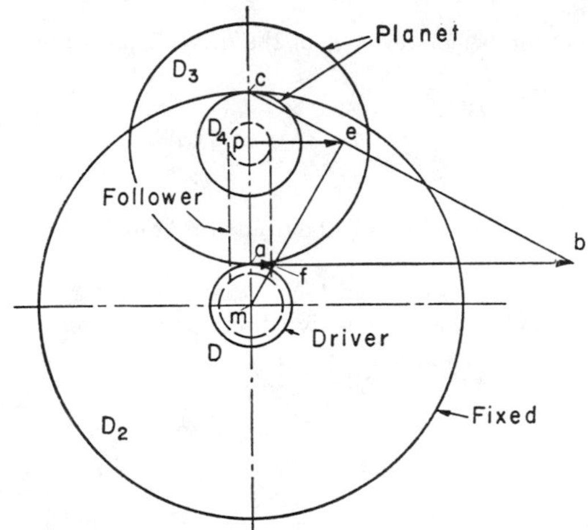

FIG. 2. Two Planet Gears with Train Arm Acting as Follower and External Gear as Driver

Formulas: $N = 1 + \frac{D_2}{D_4} \times \frac{D_3}{D}$ (By Analytical Method)

$N = \frac{ab}{af}$ (By Graphical Method)

Derivation of Formulas: *By the Analytical Method*

Step 1. Entire mechanism locked together and rotated clockwise once about m.

Step 2. Fixed sun gear D_2 rotated back one turn in counter-clockwise direction, with train arm held in fixed position.

Step 3. The number of revolutions X made by the driver in *Step* 2 is found as follows. When the fixed gear D_2 is rotated one turn in a counter-clockwise direction, planet gear D_4 is also caused to rotate in a counter-clockwise direction. The number of turns it makes is equal to the pitch diameter of the fixed gear divided by the pitch diameter of the planet gear (or the number of teeth of the fixed gear divided by the number of teeth of the planet gear) thus

$$\text{No. turns of } D_4 = \frac{D_2}{D_4}$$

Since planet gear D_3 is keyed to the same shaft as D_4 the number of turns it makes is the same as that made by D_4, or

$$\text{No. of turns } D_3 = \text{No. of turns } D_4 = \frac{D_2}{D_4}$$

The number of turns made by the driver D, for each turn of planet gear D_3 with which it meshes is equal to the pitch diameter D_3 divided by the pitch diameter D (or the number of teeth of the planet gear divided by the number of teeth of the driver), thus:

$$\text{No. of turns } D \text{ for one turn } D_3 = \frac{D_3}{D}$$

Hence the number of turns X of the driver equals the number of turns of D_3 times $\frac{D_3}{D}$, or

$$X = \frac{D_2}{D_4} \times \frac{D_3}{D}$$

Step 4. Since the follower is the train arm, it makes no revolutions during *Step* 2, hence $Y = 0$.

Step 5. It can be seen from Fig. 2 that during *Step* 2 the driver will rotate in a clockwise or positive direction, therefore

$$N = \frac{1 + X}{1}$$

or

$$N = 1 + \frac{D_2}{D_4} \times \frac{D_3}{D}$$

By the Graphical Method (See Fig. 2)

Step 1. Draw ab of any convenient length tangent to pitch circle of driver at intersection with center line of train arm to represent linear velocity of a on driver. It will be seen that ab also represents the velocity of a on planet gear D_4 if it had the same diameter as D_3.

Step 2. Point c at which D_4 makes contact with the fixed gear may be considered as a momentary pivot point and therefore has a linear velocity of zero. A line is drawn from b to c.

Step 3. Draw a line from p (which is center of pin about which planet gear rotates) parallel to ab until it intersects bc at e. Line pe then represents the velocity of point p on the planet gears D_3 and D_4. It also represents the velocity of point p on the follower which is the train arm.

Step 4. Point m on the train arm being the pivot point or axis may be considered as having zero velocity. A line is drawn from m to point e. It intersects ab at f. Hence af represents the velocity of the follower or train arm at radius ma.

Step 5. Since ab represents the velocity of the driver at radius ma

$$N = \frac{ab}{af}$$

Example 1: In Fig. 2 if $D = 8$; $D_2 = 36$; $D_3 = 20$ and $D_4 = 8$, find N by the analytical method.

Solution: $N = 1 + \frac{D_2}{D_4} \times \frac{D_3}{D}$

$$N = 1 + \frac{36}{8} \times \frac{20}{8} = 1 + \frac{720}{64}$$

$N = 1 + 11\frac{1}{4} = 12\frac{1}{4}$ turns of driver for each turn of follower.

Example 2: In Fig. 2 if $D = 12$; $D_2 = 60$; $D_3 = 32$ and $D_4 = 16$, find N by the analytical method.

Solution: $N = 1 + \frac{60}{16} \times \frac{32}{12}$

$N = 1 + 10 = 11$ turns of driver for each turn of follower.

Two Planet Gears with Train Arm Acting as Follower and Internal Gear as Driver

PROBLEM 3: As shown in Fig. 3 two external spur planet gears mesh with an external spur fixed and driving gear respectively. It is required to find the number of turns N of the driver required to produce one turn of the follower if the respective pitch diameters D, D_2, D_3 and D_4 are known.

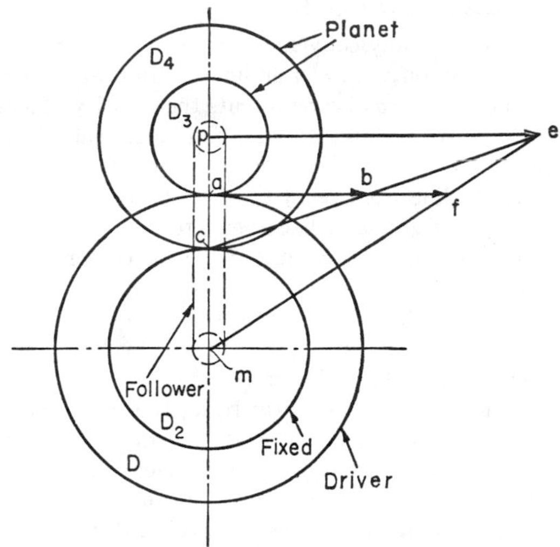

FIG. 3. Two Planet Gears with Train Arm Acting as Follower and Internal Gear as Driver

Formulas: $N = 1 - \frac{D_2}{D_4} \times \frac{D_3}{D}$ (By Analytical Method)

$N = \frac{ab}{ab}$ (By Graphical Method)

Derivation of Formulas: *By the Analytical Method*

Step 1. Entire mechanism is locked together and rotated clockwise about m.

Step 2. Fixed gear is rotated back one turn in counter-clockwise direction with train arm held in fixed position.

Step 3. $X = \frac{D_2}{D_4} \times \frac{D_3}{D}$ turns of driver during *Step* 2.

Step 4. $Y = 0$ turns of follower during *Step* 2.

Step 5. It can be seen from Fig. 3 that during *Step* 2 the driver will rotate in a counter-clockwise or negative direction, therefore

$$N = \frac{1 - X}{1} \quad \text{or} \quad N = 1 - \frac{D_2}{D_4} \times \frac{D_3}{D}$$

By the Graphical Method (See Fig. 3)

Step 1. Draw ab, of any convenient length tangent to pitch circle of driver at intersection with center line of train arm to represent velocity of a on driver. ab also represents the velocity of a on planet gear with which driver is tangent and a on planet gear which meshes with fixed gear.

Step 2. Draw a line from point c, which is point of tangency of pitch circles of planet gear and fixed gear, to point b.

Step 3. Draw a line from point p, which is center of pin about which both planet gears rotate parallel to ab until it intersects an extension of bc at e. Line pe then represents the velocity of point p on both planet gears. It also represents the velocity of point p on the follower which is the train arm.

Step 4. Point m on the train arm being the pivot point or axis may be considered as having zero velocity. Draw a line from m to e which intersects a continuation of ab at f. Hence af represents the velocity of the follower or train arm at radius ma.

Step 5. Since ab represents the velocity of the driver at radius ma,

$$N = \frac{ab}{af}$$

It will be noted that ab is shorter than af, hence N will be a fraction indicating that only a fraction of a turn of the driver will be required to rotate the follower a full turn.

Example: In Fig. B if $D = 32$; $D_2 = 20$; $D_3 = 12$ and $D_4 = 24$, find N by the analytical method.

Solution:

$$N = 1 - \frac{D_2}{D_4} \times \frac{D_3}{D} = 1 - \frac{20}{24} \times \frac{12}{32} = 1 - \frac{5}{16} = \frac{11}{16}$$

turn of the driver for each turn of the follower.

Two Planet Gears with Train Arm Acting as Driver and External Gear as Follower

PROBLEM 4: As shown in Fig. 4, two external spur planet gears mesh with two internal spur gears which are the fixed and follower members, respectively. The train arm acts as the driver. If the pitch diameters D, D_2, D_3 and D_4 of the gears are known, it is required to find N.

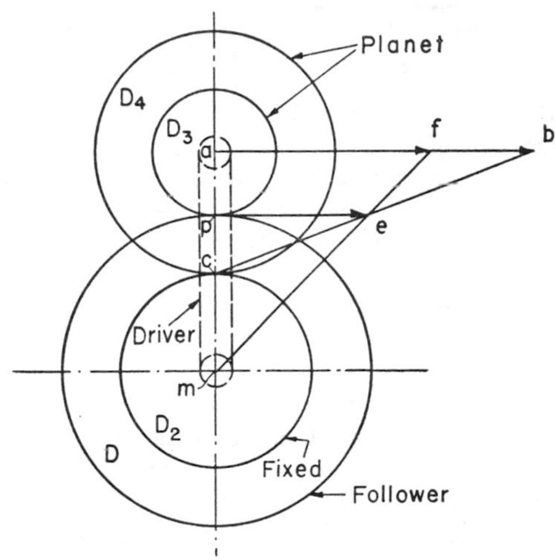

FIG. 4. Two Planet Gears with Train Arm Acting as Driver and External Gear as Follower

Formulas: $$N = \frac{D_4D}{D_4D - D_2D_3} \quad \text{(By Analytical Method)}$$

$$N = \frac{ab}{af} \quad \text{(By Graphical Method)}$$

Derivation of Formulas: *By the Analytical Method*

Step 1. Entire mechanism is locked together and rotated clockwise once about m.

Step 2. Fixed gear D_2 is rotated back one turn in counter-clockwise direction with train arm held in fixed position.

Step 3. $X = 0$ turns of driver during *Step* 2.

Step 4. $Y = \frac{D_2}{D_4} \times \frac{D_3}{D}$ turns of follower during *Step* 2.

Step 5. It can be seen that the follower rotates in a counter-clockwise or negative direction during *Step* 2, hence

$$N = \frac{1}{1 - Y}$$

or

$$N = \frac{1}{1 - \frac{D_2}{D_4} \times \frac{D_3}{D}} = \frac{1}{\frac{D_4 D - D_2 D_3}{D_4 D}} = \frac{D_4 D}{D_4 D - D_2 D_3}$$

By the Graphical Method (See Fig. 4)

Step 1. Draw *ab* of any convenient length from point *a* which is the center of the outermost bearing pin carried by the driver, which is the train arm. This represents the linear velocity of point *a* on the driver. Line *ab* also represents the linear velocity of point *a* on the planet gear which meshes with the fixed gear.

Step 2. Draw a line from point *c* on the planet gear which is in mesh with the fixed gear to point *b*.

Step 3. Draw a line from point *p*, which is the point of tangency between the pitch circles of planet gear and the follower, parallel to *ab* until it intersects *bc* at *e*. Line *pe* is then the velocity of the follower at radius *mp*.

Step 4. Draw a line from the axis *m* of the follower through point *e* until it intersects *ab* at *f*. Then *af* represents the linear velocity of the follower at radius *ma*.

Step 5. But *ab* represents the linear velocity of the driver at radius *ma*, hence

$$N = \frac{ab}{af}$$

Example: In Fig. 4 if $D = 40$; $D_2 = 24$; $D_3 = 16$ and $D_4 = 32$, find N by the analytical method.

Solution:

$$N = \frac{D_4 D}{D_4 D - D_2 D_3}$$

$$N = \frac{32 \times 40}{32 \times 40 - 24 \times 16} = \frac{1280}{896} = 1\frac{384}{896} = 1\tfrac{3}{7}$$ turns of the driver for each turn of follower.

Two Planet Gears with Train Arm Acting as Idler and Internal Gear as Driver

PROBLEM 5: In Fig. 5 the driver is an internal spur gear while the follower, the fixed gear and the two planetary gears are external spur gears. The train arm in this particular planetary gearing mechanism acts simply as an idler. If the pitch diameters D, D_2, D_3, D_4 and D_5 are known it is required to find N.

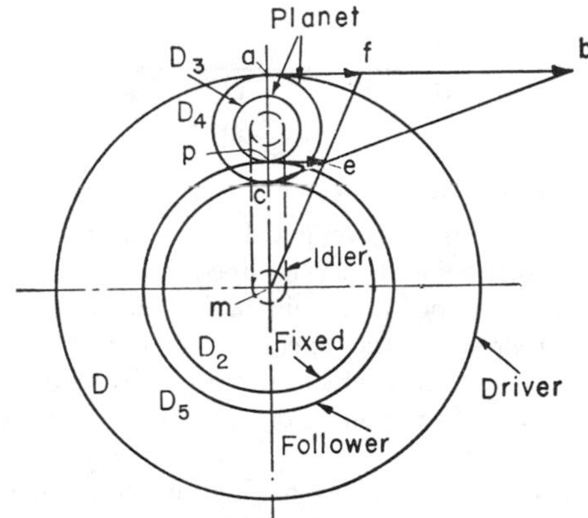

FIG. 5. Two Planet Gears with Train Arm Acting as Idler and Internal Gear as Driver

Formulas:

$$N = \frac{1 + \dfrac{D_2}{D}}{1 - \dfrac{D_2 \times D_3}{D_4 \times D}} \qquad \text{(By Analytical Method)}$$

$$N = \frac{ab}{af} \qquad \text{(By Graphical Method)}$$

Derivation of Formulas: *By the Analytical Method*

Step 1. Entire mechanism locked together and rotated clockwise once about m.

Step 2. Fixed gear D_2 rotated back one turn in counter-clockwise direction with train arm held in fixed position.

Step 3. $X = \frac{D_2}{D_4} \times \frac{D_4}{D} = \frac{D_2}{D}$ turns of driver in *Step* 2.

Step 4. $Y = \frac{D_2}{D_4} \times \frac{D_3}{D_5}$ turns of follower in *Step* 2.

Step 5. It can be seen from Fig. 5 that the driver is rotated in a clockwise or positive direction and the follower in a counter-clockwise or negative direction in *Step* 2, hence:

$$N = \frac{1 + X}{1 - Y} = \frac{1 + \frac{D_2}{D}}{1 - \frac{D_2 \times D_3}{D_4 \times D_5}}$$

By the Graphical Method

Step 1. Draw *ab* of any convenient length tangent to pitch circle of driver at intersection with center line of train arm. This line represents the linear velocity of point *a* on the pitch circle of the driver. It also represents the linear velocity of point *a* on the pitch circle of the planet gear which meshes with the driver.

Step 2. Draw a line from point *c* on the planet gear which meshes with the fixed gear to point *b*.

Step 3. Draw a line from point *p* which is the point of tangency of pitch circles of the planet gear and the follower parallel to *ab* until it intersects *cb* at *e*. The line *pe* then represents the velocity of the follower at radius *mp*.

Step 4. Draw a line from point *m*, which is the axis of the follower through point *e* until it intersects *ab* at *f*. Then *af* represents the velocity of the follower at radius *ma*.

Step 5. But *ab* represents the velocity of the driver at radius *ma*, hence $N = \frac{ab}{af}$ turns of the driver for each turn of the follower.

Example: In Fig. 5, if $D = 20$; $D_2 = 10$; $D_3 = 3$; $D_4 = 5$; and $D_5 = 12$, find *N*, by the analytical method.

Solution:

$$N = \frac{1 + \frac{10}{20}}{1 - \frac{10 \times 3}{5 \times 12}} = \frac{1\frac{1}{2}}{\frac{1}{2}} = 3 \text{ turns of driver for each turn of follower.}$$

Single Planet Gear with Train Arm Acting as Driver and Internal Gear as Secondary Driver

PROBLEM 6: Fig. 6 shows a compound drive planetary mechanism in which a secondary driver takes the place of a fixed gear. By changing the speed of rotation of this secondary driver, the ratio of driver and follower speeds can be varied until exactly the desired ratio is obtained. In this particular case the driver is the train arm and the planet gear and follower are both external spur gears. If the pitch diameters, or the numbers of teeth of gears D, D_2, and D_3 are known, and the ratio k of the secondary driver speed in revolutions per minute to the driver speed in revolutions per minute, it is required to find N, the number of turns of the driver for each turn of the follower.

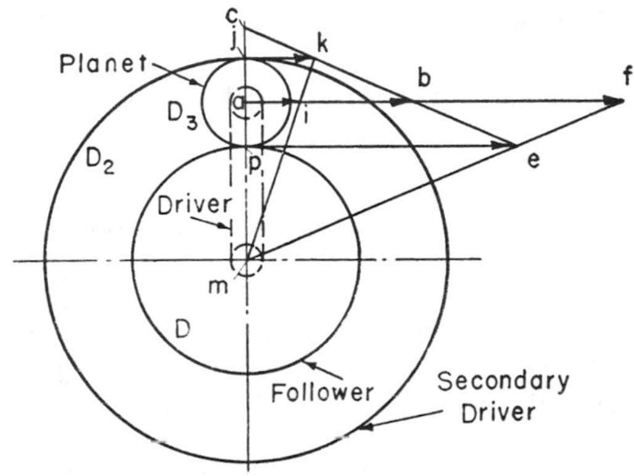

FIG. 6. Single Planet Gear with Train Arm Acting as Driver and Internal Gear as Secondary Driver

Formula: $$N = \frac{D}{D + (1 - k)D_2} \quad \text{(By Analytical Method)}$$

$$N = \frac{ab}{af} \quad \text{(By Graphical Method)}$$

Derivation of Formulas: *By the Analytical Method*

Step 1. The entire mechanism is locked together and rotated one turn in a clockwise direction about m.

Step 2. Since the gear D_2 rotates at k times the speed of the driver in terms of R.P.M., it normally should have made k turns in *Step* 1 (clockwise, if k is positive; counter-clockwise, if k is negative). Therefore to bring gear D_2 to the position it should be (comparable to turning the fixed gear in previous problems back one turn) it must be rotated $k - 1$ revolutions in this step. If $k - 1$ is negative, it signifies that gear D_2 should be rotated in a counter-clockwise direction, if positive, in a clockwise direction.

In this step, the mechanism is considered to be unlocked again so that the various parts are free to turn except for the train arm which is held in a fixed position.

Step 3. Since the driver is the train arm, it makes no turns during *Step* 2, so that $X = 0$.

Step 4. For each turn of the gear D_2, the follower will make $\frac{D_2}{D_3} \times \frac{D_3}{D}$ or $\frac{D_2}{D}$ turns. Hence, if gear D_2 is rotated $k - 1$ turns during *Step* 2, then the follower will make $(k - 1)\frac{D_2}{D}$ turns in the opposite direction so that $Y = -(k - 1)\frac{D_2}{D} = (1 - k)\frac{D_2}{D}$.

Step 5. The direction that the follower rotates during *Step* 2 will be determined by the sign of the expression $(1 - k)\frac{D_2}{D}$. Thus, in the expression for N a plus sign is placed before Y, but when the equivalent of Y is substituted, this sign will become plus or minus depending upon the value of k.

$$N = \frac{1}{1 + Y} = \frac{1}{1 + (1 - k)\frac{D_2}{D}} = \frac{1}{\frac{D + (1 - k)D_2}{D}}$$

or

$$N = \frac{D}{D + (1 - k)D_2}$$

By the Graphical Method

Step 1. Draw ab of any convenient length from point a, the center of the outermost bearing pin carried by the train arm, which is the driver. This line represents the linear velocity of point a on the driver and also of point a on planet gear D_3.

Step 2. If the gear D_2 is rotating at k times the speed of the driver, then k times ab or ai represents the linear velocity of point a on gear D_2.

Step 3. Draw a line from point m, which is the axis of gear D_2 and therefore the point of zero velocity for gear D_2, through point i.

Step 4. Draw a line from point j, which is the point of tangency of the secondary driver D_2 and the planet gear, parallel with ab until it intersects an extension of mi at point k. Then, jk is the linear velocity of point j on the secondary driver and on the planet gear.

Step 5. Draw a line through points k and b. Then draw a line parallel to ab from point p until it intersects an extension of kb at e. Line pe then represents the linear velocity of point p on the planet gear and also point p on the follower. It is, therefore, the linear velocity of the follower at radius mp.

Step 6. Draw a line from point m, which is the axis and hence point of zero velocity of the follower, through point e until it intersects an extension of ab at f. Line af represents, then, the linear velocity of the follower at radius ma.

Step 7. But ab represents the linear velocity of the driver at radius ma. Hence

$N = ab$ turn of the driver for each af turn of the follower.

Example 1: In Fig. 6 if $D = 20$, $D_2 = 36$ and $D_3 = 8$, and $k = +\frac{1}{3}$, what will be the value of N?

Solution: Using the analytical method:

$$N = \frac{D}{D + (1 - k)D_2}$$

$$N = \frac{20}{20 + (1 - \frac{1}{3})36} = \frac{20}{20 + 24}$$

$N = \frac{5}{11}$ turn of the driver for each turn of the follower.

Example 2: In Fig. 6 if $D = 20$, $D_2 = 36$ and $D_3 = 8$, and $k = -\frac{1}{3}$, what will be the value of N?

Solution: Using the analytical method:

$$N = \frac{20}{20 + (1 + \frac{1}{3})36} = \frac{20}{68}$$

$N = \frac{5}{17}$ turn of the driver for each turn of the follower.

Two Pairs of Gears of Different Circular Pitch

PROBLEM 7: In Fig. 7 is shown a diagrammatic sketch of a planetary gearing mechanism composed of two pairs of gears of different circular pitch. The pitch circles of all four gears have been laid out to the same scale and in proportion to their respective numbers of teeth T, T_2, T_3 and T_4. Because of this fact, one pair of pitch circles—that of the large planet gear and the driving gear—are not tangent to each other, but overlap. If correctly drawn, the pitch

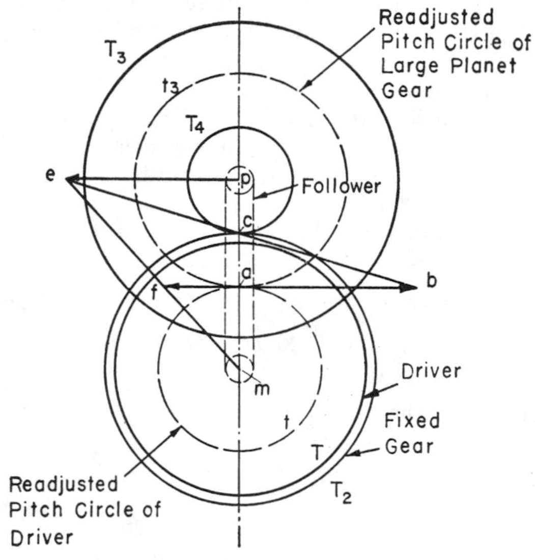

FIG. 7. Two Pairs of Gears of Different Circular Pitch

circles should be laid out in proportion to the pitch diameters of the gears, in which case the pitch circle of the large planet gear and the driver would be tangent to each other. A graphical solution can then be found.

Let it be assumed, however, that only the tooth numbers T, T_2, T_3 and T_4 of the four gears are known and it is required to find N, the number of turns of the driver for each turn of the follower by the analytical method and by the graphical method.

Formulas: $N = 1 - \frac{T_2}{T_4} \times \frac{T_3}{T}$ (By Analytical Method)

$N = -\frac{ab}{af}$ (By Graphical Method)

Analysis of Problem: If the analytical method of solution is used, only the number of teeth of the respective gears need be known. The fact that the two pairs of gears have different diametral pitches does not affect this method of solution.

In using the graphical method, however, it is necessary to redraw the pitch circles of the large planet gear and the driving gear (shown in Fig. 7 as T_3 and T respectively so that they are tangent to each other, t_3 and t, before the graphical method can proceed. If the pitch diameters or the diametral pitches of the respective gears were known, then this diagram could be drawn directly. If only the numbers of teeth are known, then a computation has to be made to determine the proper scale size of these pitch circles so that they have the ratio numbers of teeth, T and T_3, and yet be tangent to each other.

This is done by taking the distance as laid out between the centers of the other pair of gears, which are the small planet gear and the fixed gear, and dividing it into two parts proportional to the numbers of teeth in the large planet gear and the driver. These distances are, then, the respective radii of the large planet gear and the driver. When these re-adjusted pitch circles are drawn as shown in Fig. 7, the graphical solution proceeds as outlined in the general description of the method.

Derivation of Formulas: *By the Analytical Method*

Step 1. The entire mechanism is locked together and rotated one turn in a clockwise direction, then $N = 1$.

Step 2. The fixed gear is rotated back one turn in a counter-clockwise direction with the train arm held in a fixed position.

Step 3. $X = \frac{T_2}{T_4} \times \frac{T_3}{T}$ turns of driver during *Step* 2.

Step 4. $Y = 0$ turns of follower during *Step* 2.

Step 5. It can be seen from Fig. 7 that during *Step* 2 the driver will rotate in a counter-clockwise or negative direction, therefore

$$N = \frac{1 - X}{1} = 1 - \frac{T_2}{T_4} \times \frac{T_3}{T}$$

By the Graphical Method

Assuming that the diagram of this planetary mechanism has been drawn so that the re-adjusted pitch circles of the driver and the large planet gear are tangent to each other as shown in Fig. 7, the graphical method proceeds as follows:

Step 1. Draw *ab* of any convenient length from point *a* tangent to the pitch circle of the driver and at right angles to the center line of the train arm.

This represents the linear velocity of point *a* on the driver and on the large planet gear and also point *a* on the small planet gear, if its radius were increased from *pc* to *pa*.

Step 2. Draw a line from *c*, which is the point of tangency of the small planet gear and the fixed gear to *b*.

Step 3. Draw a line from point *p*, which is the center of the bearing pin, carried by the train arm, on which both planet gears rotate, parallel with *ab* until it intersects an extension of *bc* at *e*. Line *pe* then represents the linear velocity of the center of the planet gears. It also represents the linear velocity of point *p* on the train arm.

Step 4. Draw a line from point *m* which is the point of zero velocity for the train arm or follower to point *e*. It will be noted that *me* intersects an extension of *ab* at *f*. Line *af* represents the linear velocity of point *a* on the follower or, in other words, the velocity of the follower at radius *ma*.

Step 5. But *ab* represents the velocity of the driver at radius *ma*. It will be noted, however, that *af* and *ab* extend in opposite directions indicating that one is minus and the other is plus. Hence,

$$N = -\frac{ab}{af} \text{ turns of the driver for each turn of the driver}$$

This negative sign means that follower and driver will rotate in opposite directions.

Example: In Fig. 7 let $T = 30$; $T_2 = 32$; $T_3 = 36$ and $T_4 = 12$. It is required to find N.

Solution: Using the analytical method:

$$N = 1 - \frac{T_2}{T_4} \times \frac{T_3}{T} = 1 - \frac{32}{12} \times \frac{36}{30} = 1 - \frac{48}{15} = -2\tfrac{1}{5} \text{ turns of the}$$

driver for each turn of the follower.

Single Bevel Planet Gear as Idler

PROBLEM 8: In Fig. 8A is shown a planetary gearing mechanism consisting of three bevel gears, one of which is fixed, one of which acts as a planet gear, and one of which acts as the driver. The follower is the train arm. If the respective diameters D, D_2 and D_3 of the bevel gears are known, it is required to find the number of turns N of the driver for each turn of the follower.

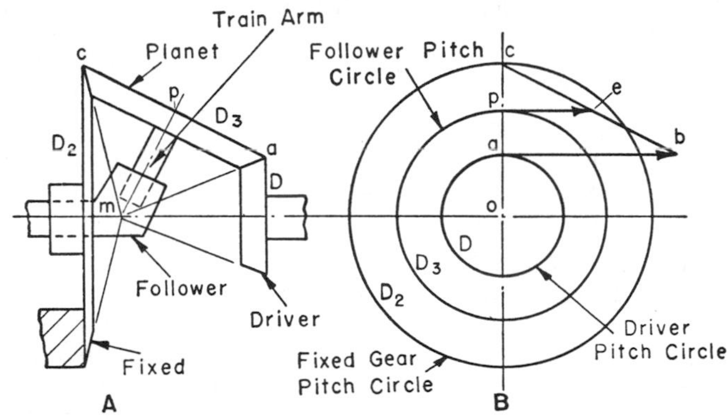

FIG. 8A. Single Bevel Planet Gear Acting as Idler. B. Diagram for Graphical Solution

Formulas: $N = 1 + \dfrac{D_2}{D}$ (Analytical Method)

$$N = \frac{ab \times mp}{ma \times pe} \quad \text{(Graphical Method)}$$

Analysis of Problem: Where the planet gear is an idler, the analytical method of solution is exactly the same as for the spur gear type of planetary mechanism. Where the planet gear acts as the follower, a special analysis is required and will be explained in a subsequent problem.

Derivation of Formulas: *By the Analytical Method*

Step 1. Entire mechanism is locked together and rotated one turn in a clockwise direction (looking at the mechanism from the right, as shown in Fig. 8B) about the axis of the driver.

Step 2. The train arm, which is the follower in this particular case, is now held motionless and the fixed gear is rotated back one turn in a counter-clockwise direction.

Step 3. $X = \frac{D_2}{D_3} \times \frac{D_3}{D} = \frac{D_2}{D}$ turns of the driver during *Step* 2.

Step 4. $Y = 0$ turns of the follower during *Step* 2.

Step 5. It can be seen from Fig. 8A that the driver will rotate in a clockwise direction during *Step* 2, hence

$$N = \frac{1 + X}{1}$$

or

$$N = \frac{1 + \frac{D_2}{D}}{1} = 1 + \frac{D_2}{D}$$

By Graphical Method

Step 1. Referring to Fig. 8B, draw *ab* from point *a* on the pitch circle of the driver and at right angles to the vertical center line to represent the linear velocity of this point. Line *ab* also represents the linear velocity of point *a* on the planet gear. Point *c* on the pitch circle of the planet gear which is its point of tangency with the fixed gear may be considered to have a momentary linear velocity of zero.

Step 2. Draw a line from *c* to *b*.

Step 3. Draw a line from *p* parallel to *ab* until it intersects *bc* at *e*. Then *pe* is the linear velocity of point *p* on the planet gear. But point *p* may also be considered to be the end point of the axis of the train arm which is the follower. Hence *pe* is the velocity of point *p* on the follower.

Step 4. The angular velocity of the follower is equal to the linear velocity of any point on the follower divided by its radius from the axis of rotation or $\frac{pe}{op}$.

Similarly the angular velocity of the driver is $\frac{ab}{oa}$

Step 5. The number of turns *N* of the driver for each turn of the follower is, then, equal to the angular velocity of the driver divided by the angular velocity of the follower or

$$N = \frac{ab}{oa} \div \frac{pe}{op}$$

or

$$N = \frac{ab \times op}{oa \times pe}$$

Example: In Fig. 8A let $D = 16$, $D_2 = 40$ and $D_3 = 27$. It is required to find N, using the analytical method.

Solution:

$$N = 1 + \frac{D_2}{D}$$

$$N = 1 + \frac{40}{16} = 3\tfrac{1}{2}$$ turns of driver for each turn of follower.

Two Bevel Planet Gears as Idlers

PROBLEM 9: In Fig. 9A is shown what is known as the Humpage bevel type of reducing gear. It consists of a fixed bevel gear, two bevel planet gears which are mounted together, a bevel gear driver and a bevel gear follower. If the pitch diameters D, D_1, D_2, D_3 and D_4 are given, it is required to find the number of turns N of the driver for each turn of the follower.

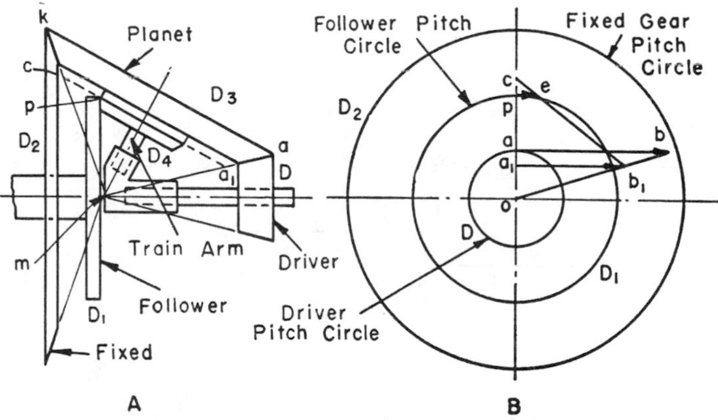

FIG. 9A. Two Bevel Planet Gears Acting as Idlers. B. Diagram for Graphical Solution

Formulas:

$$N = \frac{D + D_2}{D} \times \frac{D_3D_1}{D_3D_1 - D_2D_4} \quad \text{(By Analytical Method)}$$

$$N = \frac{ab \times op}{oa \times pe} \quad \text{(By Graphical Method)}$$

Derivation of Formulas: *By the Analytical Method*

Step 1. The entire mechanism is locked together and given one turn in a clockwise direction (looking toward the mechanism from the right side as shown in Fig. 9B) about the axis of the driver and the follower as shown in Fig. 9B.

Step 2. The various members of the mechanism are now considered free to turn, the train arm is held in a fixed position and the fixed gear is turned back one turn in a counter-clockwise direction.

Step 3. $X = \frac{D_2}{D_3} \times \frac{D_3}{D}$ turns of the driver during *Step* 2.

Step 4. $Y = \frac{D_2}{D_3} \times \frac{D_4}{D_1}$ turns of the follower during *Step* 2.

Step 5. It can be seen from Fig. 9A that during *Step* 2 the driver will rotate in a clockwise (looking towards the driver from the right-hand side) or positive direction and the follower will rotate in a counter-clockwise or negative direction, hence the number of turns N of the driver for each turn of the follower is

$$N = \frac{1 + X}{1 - Y}$$

or substituting the equivalents of X and Y:

$$N = \frac{1 + \frac{D_2}{D_3} \times \frac{D_3}{D}}{1 - \frac{D_2}{D_3} \times \frac{D_4}{D_1}} = \frac{1 + \frac{D_2}{D}}{1 - \frac{D_2 \times D_4}{D_3 \times D_1}}$$

$$N = \frac{\frac{D + D_2}{D}}{\frac{D_3D_1 - D_2D_4}{D_3D_1}} = \frac{D + D_2}{D} \times \frac{D_3D_1}{D_3D_1 - D_2D_4}$$

By the Graphical Method

Step 1. Draw a line of any convenient length from point *a* on the pitch circle of the driver and at right angles to the vertical center line as shown in Fig. 9B. Then *ab* represents the linear velocity of this point *a*.

Step 2. It will be noted that the pitch circle of the smaller planet wheel makes contact with the follower at point *p* and it is the velocity of this point which must be found if *N* is to be determined.

Let it be assumed that the larger and smaller planet gears, being fixed together, are part of a cone having an apex *m* and sides *km* and *am* as shown in Fig. 9A. Similarly let it be assumed that the driver is part of a cone also having its apex at *m* and one side *am* coincident with side *am* of the planet gear cone. Point *m* being the center of rotation of both cones may be considered as having zero velocity.

Step 3. In Fig. 9B draw a line from point *o* to point *b*. The velocity of any point along the edge *ma* of the driver cone in Fig. 9A will be represented by a line drawn from that point as shown in Fig. 9B parallel to *ab* until it intersects *ob*.

Step 4. In the cone *kma* the point a_1 is at the same radial distance from *m* as points *p* and *c*. It will be noted that point *c* may be considered as having zero velocity since it is on the edge *km* of the fixed gear.

Returning now to Fig. 9B let a line be drawn from point a_1 parallel to *ab* until it intersects *ob* at b_1. Then a_1b_1 represents the velocity of point a_1 on the driver cone. It also represents the velocity of point a_1 on the planet gear cone.

Step 5. Draw a line from point *c* in Fig. 9B to point b_1. Then a line drawn from any point on the planet gear which is located on the vertical center line in Fig. 9B, parallel to a_1b_1 until it intersects b_1c will represent the linear velocity of that point.

Step 6. Draw a line from point *p* in Fig. 9B parallel to a_1b_1 until it intersects b_1c at *e*. Then *pe* represents the linear velocity of point *p* on the planet gear. It also represents the linear velocity of point *p* on the follower.

Step 7. The angular velocity of the driver is equal to the linear velocity of point *a* divided by its radius of rotation or $\frac{ab}{oa}$. Similarly the angular velocity of the follower is equal to the linear velocity of point *p* divided by its radius of rotation or $\frac{pe}{op}$.

Step 8. The number of revolutions of the driver for each turn of the follower is equal to the angular velocity of the driver divided by the angular velocity of the follower or

$$N = \frac{ab}{oa} \div \frac{pe}{op} = \frac{ab \times op}{oa \times pe}$$

Example: In Fig. 9A let $D = 3$; $D_1 = 8$; $D_2 = 12$; $D_3 = 9$ and $D_4 = 4$. Find N, using the analytical method.

Solution:

$$N = \frac{D + D_2}{D} \times \frac{D_3 D_1}{D_3 D_1 - D_2 D_4}$$

$$N = \frac{3 + 12}{3} \times \frac{9 \times 8}{9 \times 8 - 12 \times 4}$$

$$N = \frac{15}{3} \times \frac{72}{24}$$

$N = 5 \times 3 = 15$ turns of driver for each turn of follower.

Bevel Planet Gear Acting as Follower

PROBLEM 10: In Fig. 10 is shown a bevel gear planetary mechanism in which the follower is the planet gear. If the pitch diameters D, D_2 and D_3 are given, it is required to find N, the number of turns of the driver for each turn of the follower.

Formulas:

$$N = \frac{D_3}{D_3 \cos \phi + D_2} \quad \text{(By Analytical Method)}$$

$$N = \frac{ac}{oa} \quad \text{(By Graphical Method)}$$

Analysis of Problem: Solution of this problem by the analytical method proceeds according to the steps previously described. The first time this type of problem (in which the follower is a bevel planet gear) is faced, there may be some doubt as to whether or not the follower rotates about its axis during *Step* 1 when the mechanism is locked together and rotated about the axis of the driver. If the axis of the follower were at right angles to the axis of the driver, it would not rotate about its own axis as the entire mechanism is rotated about the axis of the driver. If, on the other hand, the axis of the follower coincided with or was parallel to the axis of the driver it would rotate one turn as the entire mechanism was rotated one turn about the axis of the driver. For any intermediate position of the follower shaft with respect to the driver axis that is from 0° to 90°, the rotation of the follower about its axis for one turn of the entire mechanism about the axis of the driver will equal the cosine of the angle between the driver and follower shafts.

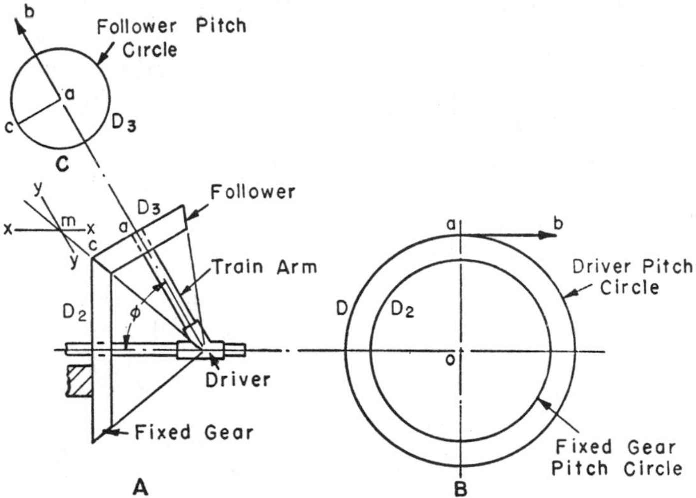

FIG. 10A. Bevel Planet Gear Acting as Follower. B and C. Diagrams for Graphical Solution

In making a graphical analysis of this problem it will be noted from Fig. 10A that point *a* on the follower is also a point on the driver. Hence it is only necessary to draw a line representing the linear velocity of this point on the pitch circle of the driver (Fig. 10B) and also at the center of the pitch circle of the follower (Fig. 10C) and then establish the respective angular velocities of driver and follower.

Derivation of Formulas: *By the Analytical Method*

Step 1. Entire mechanism is locked together and rotated once in a clockwise direction, looking at the fixed gear from the right. Thus the driver will be rotated one turn and the follower, cos ϕ turn.

Step 2. The various parts of the mechanism are now considered to be free to turn, with the exception of the train arm which is held fixed, and the fixed gear is rotated back one turn in a counterclockwise direction.

Step 3. $X = 0$ turns of driver during *Step* 2.

Step 4. $Y = \frac{D_2}{D_3}$ turns of follower during *Step* 2.

Step 5. It can be seen from Fig. 10A that the follower will be rotated in a clockwise direction during *Step* 2, hence

$$N = \frac{1}{\cos\phi + \frac{D_2}{D_3}} = \frac{1}{\frac{D_3 \cos\phi + D_2}{D_3}} = \frac{D_3}{D_3 \cos\phi + D_2}$$

By the Graphical Method

Before proceeding with the graphical solution of this problem, it is necessary to lay out the bevel gears correctly. First draw the axes of the two bevel gears at the given angle ϕ with each other. Next draw a line x–x parallel to the axis of one of the bevel gears (in this case the fixed gear, as shown in Fig. 10A) and at some scale distance from this axis to represent its pitch radius. Then draw a line y–y parallel to the axis of the other bevel gear (in this case the follower as shown in Fig. 10A) and at the same scale distance from the axis to represent its pitch radius. The point of intersection m of lines x–x and y–y is a point on the tangent line of the pitch cones of these two bevel gears. This tangent line is the line drawn through point m and the point of intersection of the axes of the two bevel gears. With this tangent line established, the gears are laid out at any convenient distance from the intersection of their axes and the graphical solution proceeds as follows:

Step 1. Draw a line of any convenient length from point a and tangent to the pitch circle of the driver as shown in Fig. 10B. Then, ab represents the linear velocity of this point on the driver.

Step 2. But point a is also the center of the follower, which is rotating momentarily about point c as a pivot point. Hence, draw ab at right angles to ac as shown in Fig. 10C. Then ab represents the linear velocity of point a on the follower.

Step 3. The angular velocity of the driver equals the linear velocity of any point on the driver divided by its radius of rotation, or $\frac{ab}{oa}$.

Step 4. The angular velocity of the follower equals the linear velocity of any point on the follower divided by its radius of rotation or $\frac{ab}{ac}$.

Step 5. The number of revolutions N of the driver for each revolution of the follower equals the angular velocity of the driver divided by the angular velocity of the follower or

$$N = \frac{ab}{oa} \div \frac{ab}{ac}$$

or

$$N = \frac{ab}{oa} \times \frac{ac}{ab} = \frac{ac}{oa}$$

Example: In Fig. 10 let $D_2 = 20$, $D_3 = 10$ and angle $\phi = 60$ degrees. Find N, using the analytical method.

Solution:

$$N = \frac{D_3}{D_3 \cos \phi + D_2} = \frac{10}{10 \times 0.5 + 20} = \frac{10}{25}$$

$N = 0.4$ revolution of driver for each revolution of follower

Eccentrically Mounted Gear Acts as Planet and Fixed Gear

PROBLEM 11: In Fig. 11 is shown a speed reducing planetary mechanism. An eccentric A carried by the driving shaft imparts motion to gear B which is prevented from rotating about its own axis by two rollers C that slide in slots in a fixed plate D. If the number of teeth T in gear B and T_2 in gear E are known, it is required to find N, the number of turns of the driver for each turn of the follower.

Analysis of Problem: The unusual feature of this planetary mechanism is that gear B acts as both a planet gear and a fixed gear. It acts as a planet gear because its center is movable, that is, its center rotates about the axis of the shaft. It acts as a fixed gear because it does not rotate about its own axis. The eccentric A acts as the train arm. Gear E is the follower.

Formula:

$$N = \frac{T_2}{T_2 - T}$$

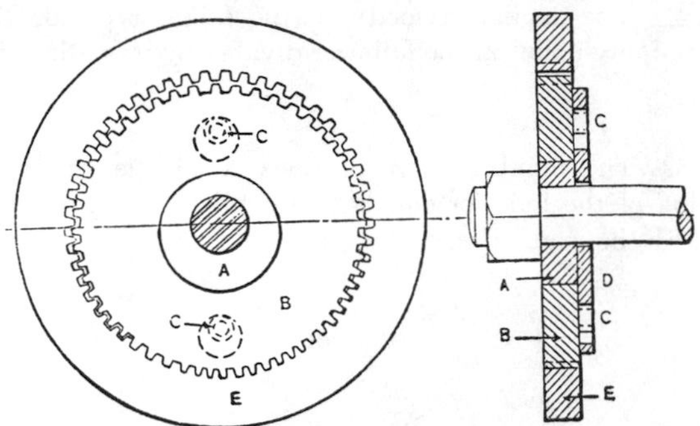

FIG. 11. Eccentrically Mounted Gear B Acts as Planet and Fixed Gear. Pins C Prevent its Rotation About its Own Axis

Solution: The analytical method will be used to solve this problem.

Step 1. The entire mechanism is locked together and rotated one turn in a clockwise direction, then $N = 1$.

Step 2. The fixed gear B is rotated back one turn in a counter-clockwise direction with the train arm (eccentric) held in a fixed position.

Step 3. $X = 0$ turns of driver during *Step* 2.

Step 4. $Y = \frac{T}{T_2}$ turns of follower during *Step* 2.

Step 5. It can be seen from Fig. 11 that the follower will rotate in a counter-clockwise or negative direction during *Step* 2, therefore

$$N = \frac{1}{1 - Y} = \frac{1}{1 - \frac{T}{T_2}} = \frac{1}{\frac{T_2 - T}{T_2}} = \frac{T_2}{T_2 - T}$$

Example: In Fig. 11 let $T = 54$ and $T_2 = 56$. Find N, using the analytical method.

Solution:
$$N = \frac{56}{56 - 54} = \frac{28}{1}$$

or 28 turns of driving shaft for each turn of follower gear.

Selecting Gears for a Specified Planetary Ratio

When the numbers of teeth in the gears of a planetary drive are known, it is relatively simple to calculate the ratio of the gearing from whatever formula is applicable. The converse problem, of finding suitable gears for a specified ratio, is not always so easy. The gears selected must not only produce the required ratio, they must also satisfy certain center-distance requirements. As an example, consider the planetary gear formula:

$$R = 1 - \frac{C}{Y} \times \frac{X}{B}$$

In this formula, R is the ratio of gearing desired, C and Y are the numbers of teeth on one pair of mating gears, and X and B are the numbers of teeth on the other pair of mating gears. Pair C - Y operates on the same center distance as pair X - B so that the sum of the numbers of teeth C+Y must be equal to the sum of the numbers of teeth X+B.

It is sometimes possible to meet the center distance requirements by making pair C - Y of a different diametral pitch than pair X - B. However, when either of the two following methods can be applied, a simpler solution using a single diametral pitch is obtained.

Solution A:

$$R = 1 - \frac{C}{Y} \times \frac{X}{B},$$

or

$$(1 - R) = \frac{C}{Y} \times \frac{X}{B} \qquad (1)$$

For this particular gear train, pairs $\frac{C}{Y}$ and $\frac{X}{B}$ operate at the same center distances and must satisfy the condition $X + Y = C + Y$ if the gears are to mesh properly. The easiest way to satisfy this condition is to make $C = X$ and $Y = B$. The problem is then reduced to finding the ratio of a single pair of gears.

$$(1 - R) = \frac{C}{Y} \times \frac{C}{Y}, \text{ since } \frac{C}{Y} = \frac{X}{B}$$

$$(1 - R) = \left(\frac{C}{Y}\right)^2$$

$$\log\ (1 - R) = 2 \log \frac{C}{Y} \qquad (2)$$

As an example showing how Equation (2) can be used to determine the gears required, assume that $R = 0.9497$.

Then

$$(1 - R) = 2 \log \frac{C}{Y}$$

$$(1 - 0.9497) = \log 0.0503 = 8.70157 - 10$$

$$2 \log \frac{C}{Y} = 8.70157 - 10$$

$$\log \frac{C}{Y} = \frac{8.70157 - 10}{2} = \frac{18.70157 - 20}{2} = 9.35078 - 10$$

On page 156 of Machinery's "*14,000 Gear Ratios*" the nearest logarithm of a gear ratio corresponding to 9.35078 – 10 is 9.3508274 –10, and the gear ratio is given as $\frac{24}{107}$. Checking this gear ratio in the formula

$$R = 1 - \frac{C}{Y} \times \frac{X}{B}$$

$$0.9497 = 1 - \frac{24 \times 24}{107 \times 107} = 0.94969$$

The error resulting from use of this pair of gears is:

0.94970

0.94969

0.00001

Electronic calculators and computers may be used in place of tables of logarithms and gear ratios. Starting with

$$(1 - R) = \left(\frac{C}{Y}\right)^2, \text{ and } R = 0.9497$$

$$\frac{C}{Y} = 0.224277 \approx \frac{22}{100}$$

Attempt to keep the numbers of teeth on the gears between 20 and 120 for ease of procurement or fabrication. Clearly, meshing gears with

22 and 100 teeth will provide a speed ratio that is too high (R = 0.9516). With a calculator or short computer program, a series of integer ratios can be computed quickly and the best approximation selected. By simply changing the numerator or denominator, or both, by adding or subtracting 1, 2, 3,4 ,5, etc. a series of trials can quickly be made. For example:

20/88 = 0.2273, high. 20/89 = 0.2247, off. 20/90 = 0.2222, low
32/101 = 0.2277, high. 23/102 = 0.22549, off. 23/103 = 0.2233, low
24/106 = 0.2264, high. 24/107 = 0.224229, ideal

Thus the same solution of $\frac{C}{Y} = \frac{24}{107}$ is accomplished by trial and error in a relatively short time.

Solution B:

Starting with Equation (1),

$$(1-R) = \frac{C}{Y} \times \frac{X}{B}$$

$$(1-0.9497) = 0.0503 = \frac{C}{Y} \times \frac{X}{B}$$

Again assuming $\frac{C}{Y} = \frac{X}{B}$, then $\frac{C}{Y} = \sqrt{0.0503} = 0.2242766$

From the table mentioned above, we locate the pair $\frac{24}{107}$ which corresponds to a ratio of 0.224299.

An alternative to using the pair $\frac{24}{107}$ for both $\frac{X}{B}$ and $\frac{C}{Y}$, a variation may be introduced by taking one pair, say $\frac{X}{B} = \frac{23}{108}$ and the other pair $\frac{C}{Y} = \frac{23}{108}$. These values are above and below the ratio $\frac{24}{107}$ and are neighboring values taken from the table mentioned above. Checking,

$$R = 1 - \frac{C}{Y} \times \frac{X}{B} = 1 - \frac{25}{106} \times \frac{23}{108} = 0.94977$$

giving an error of

$$\begin{array}{r} 0.94977 \\ \underline{0.94970} \\ 0.00007 \end{array}$$

22

Trial and Error Solutions — When and How They Are Used

Occasionally, mechanical problems crop up in which the solution either cannot be obtained directly, or else a direct solution proves to be too cumbersome or complicated. In solving such problems, trial and error methods are often employed. Usually the procedure consists in first establishing the least cumbersome or complicated equation possible in terms of the unknown factor. A succession of trial values are then substituted for the unknown factor until the equation is "satisfied," that is, both sides are numerically equal within the required limits of accuracy.

To be practicable, however, the substitution of trial values should not be carried forward in a "hit or miss" fashion, but should proceed along a definite course in which the answer is progressively determined to greater and greater accuracy. One way of learning how to do this is to follow through in step-by-step fashion, the trial and error solutions of several different problems such as are presented in this chapter.

It will be noted that in some instances where a considerable number of trial solutions have to be made, the setting up of a table to record the computations made for each trial lessens the chance for error and facilitates a comparison of the accuracy of each successive trial.

To anyone used to solving problems by formula and obtaining, directly, values that are "exact," the methods of "trial and error" may seem, at first, roundabout and inaccurate. The following problems are presented to show that the desired values can be obtained quite directly by "trial and error" methods and to the degree of accuracy required.

Radius of Tangent Arc

PROBLEM 1: To find radius x. Given an arc of radius r, as shown in Fig. 1, with two tangents to this arc at right angles to each other. It is required to find the radius x of another arc which, when placed outside of and tangent to the first arc, will have tangents parallel to those of the first arc which are respectively at given distances a and b apart.

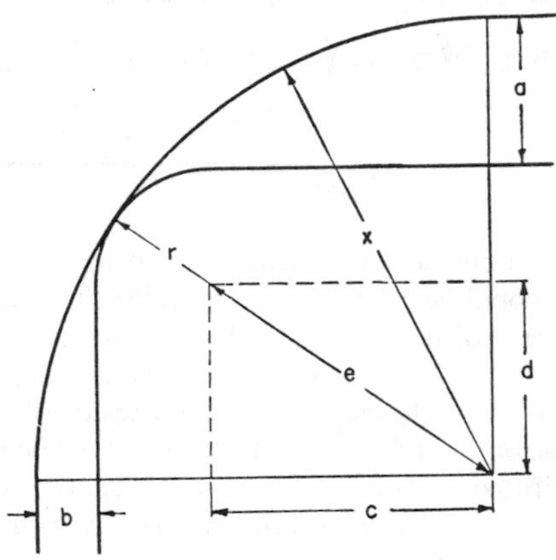

FIG. 1. To Find Radius x of Arc Tangent to Another Arc of Radius r, When Distances a and b are Known

Analysis of Problem: From the diagram it can be seen that three equations can be established for x. One in terms of e and r and then converted to d, c, and r; one in terms of d, r, and a; and one in terms of c, r, and b. From these equations the formula given below can be derived. It is possible to solve this formula for x by transposing r to the left-hand side of the equation and then squaring both sides, which will result in a quadratic equation. It will be found, however, that the quadratic equation is cumbersome to handle, while the trial and error insertion of values for x beginning with one that is determined by measuring a scale drawing, will at least be as rapid a procedure while offering fewer chances for arithmetical errors.

Formula:

$$x = r + \sqrt{(x - r - a)^2 + (x - r - b)^2}$$

(Solved by trial and error insertion of values for x.)

Derivation of Formula:

$$x = r + e \tag{1}$$

$$e^2 = c^2 + d^2 \tag{2a}$$

$$e = \sqrt{c^2 + d^2} \tag{2b}$$

$$x = r + \sqrt{c^2 + d^2} \tag{3}$$

$$x = d + r + a \tag{4a}$$

$$d = x - r - a \tag{4b}$$

$$x = c + r + b \tag{5a}$$

$$c = x - r - b \tag{5b}$$

Substituting the equivalents of d in Equation (4b) and c in Equation (5b) in Equation 3:

$$x = r + \sqrt{(x - r - b)^2 + (x - r - a)^2} \tag{6}$$

Example: In Fig. 1 let $r = 2.3$; $a = 2.9$; and $b = 1.15$. Find radius x.

Solution: Measurement of a scale drawing shows $x = 9$ approximately. Use this value for a first trial and insert in Equation (6) together with given values.

$$x = 2.3 + \sqrt{(9 - 2.3 - 1.15)^2 + (9 - 2.3 - 2.9)^2} \tag{6}$$

$$x = 2.3 + \sqrt{(5.55)^2 + (3.8)^2}$$

$$x = 2.3 + \sqrt{30.80 + 14.44} = 2.3 + \sqrt{45.24}$$

$$x = 2.3 + 6.73 = 9.03$$

The assumed value of x is too large. Try a smaller value, say 8.9.

$$x = 2.3 + \sqrt{(8.9 - 2.3 - 1.15)^2 + (8.9 - 2.3 - 2.9)^2}$$

$$x = 2.3 + \sqrt{29.70 + 12.96} = 2.3 + \sqrt{42.66}$$

$$x = 2.3 + 6.53 = 8.83$$

The value 8.9 is too small, try 8.95.

$$x = 2.3 + \sqrt{(8.95 - 2.3 - 1.15)^2 + (8.95 - 2.3 - 2.9)^2}$$

$$x = 2.3 + \sqrt{30.25 + 14.06} = 2.3 + \sqrt{44.31}$$

$$x = 2.3 + 6.66 = 8.96$$

Try 8.94.

$$x = 2.3 + \sqrt{(8.94 - 2.3 - 1.15)^2 + (8.94 - 2.3 - 2.9)^2}$$

$$x = 2.3 + \sqrt{30.14 + 13.99} = 2.3 + \sqrt{44.13}$$

$$x = 2.3 + 6.64 = 8.94$$

Therefore, the correct value of x to two decimal places is 8.94.

Note: If x were known and r were to be found, Equation (6) would be written

$$r = x - \sqrt{(x - r - b)^2 + (x - r - a)^2}$$

The trial-and-error solution would be similar to that outlined above except that successive trial values of r would be substituted until the equation was satisfied.

Base and Altitude of Triangle Circumscribing Known Rectangle

PROBLEM 2: In Fig. 2 it is required to find the base x and altitude y of a right-angled triangle having a hypotenuse of 12 inches in which a rectangle having a length of 4 inches and a width of 3 inches can be located as shown.

Analysis of Problem: In the right-angled triangle abc

$$ac = bc \csc A = 3 \csc A \qquad (1)$$

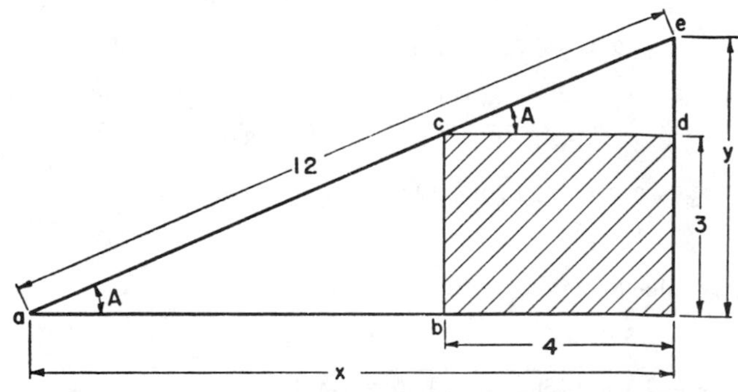

FIG. 2. To Find Base x and Altitude y of Right Angled Triangle, When Length of Hypotenuse and Length and Width of Inscribed Rectangle are Known

In the right-angled triangle *cde*

$$ce = cd \sec A = 4 \sec A \qquad (2)$$

But

$$ac + ce = 12 \qquad (3)$$

Hence

$$3 \csc A + 4 \sec A = 12 \qquad (4a)$$

Dividing both sides of Equation (4a) by 4

$$0.75 \csc A + \sec A = 3 \qquad (4b)$$

The value of angle A, and hence the values of sides x and y, can now be found by trial and error solution of this equation. A rough estimate is used as the first trial value.

Procedure: As a first trial value let $A = 25°$.

Then

$$0.75 \csc 25° + \sec 25° = ?$$

$$0.75 \times 2.3662 + 1.1034 = 2.8780$$

As a second trial value let $A = 30°$

Then

$$0.75 \csc 30° + \sec 30° = ?$$

$$0.75 \times 2.0000 + 1.1547 = 2.6547$$

Since we wish to find a value of A which will make the left-hand side of this equation equal to 3 and since when A is 30° a value farther away from 3 is obtained than when A is 25°, it may be concluded that the desired value of A is less than 25°.

As a third trial value let $A = 20°$

Then

$$0.75 \csc 20° + \sec 20° = ?$$

$$0.75 \times 2.9238 + 1.0642 = 3.2570$$

Hence it may be concluded that the desired value of A is greater than 20°.

Trial No.	A	sec X	csc x	0.75 csc x	sec x + 0.75 csc x
1	25°	1.1034	2.3662	1.7746	2.8780
2	30°	1.1547	2.0000	1.5000	2.6547
3	20°	1.0642	2.9238	2.1928	3.2570
4	22½°	1.0824	2.6131	1.9598	3.0422
5	23°	1.0864	2.5593	1.9195	3.0059
6	23°10′	1.0877	2.5419	1.9064	2.9941
7	23°5′	1.0870	2.5506	1.9130	3.0000

These first three trials and the successive ones are tabulated in the accompanying table. It will be noted that a fourth trial value half way between 20° and 25° is taken and a fifth trial value slightly greater than this. The sixth trial value is a little too large but dropping back 5 minutes to 23°5′ gives the desired result.

Therefore $x = 12 \cos A = 12 \cos 23°5'$

$$x = 12 \times 0.91993 = 11.039 \text{ inches}$$

$$y = 12 \sin A = 12 \sin 23°5'$$

$$y = 12 \times 0.39207 = 4.705 \text{ inches}$$

Length of Rectangle Inscribed Within Rectangle

PROBLEM 3: In Fig. 3 it is required to find the length Z of a rectangle having a width of 3 inches that can be inscribed in another rectangle having a length of 15 inches and a width of 10 inches. By "inscribed" is meant that only the four corners of smaller rectangle shall touch the perimeter of the larger rectangle.

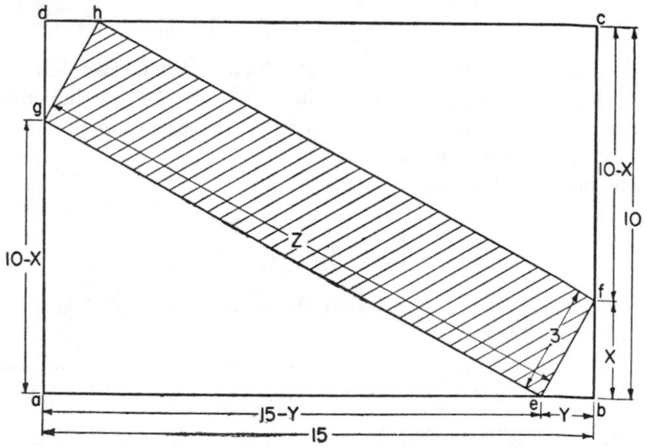

FIG. 3. To Find Length Z of Inscribed Rectangle of Known Width When Length and Width of Larger Rectangle are Known

Analysis of Problem: As can be seen from Fig. 3 there are three unknowns X, Y and Z. The procedure is to set up two equations, in both of which at least two of these unknowns appear. Values are substituted for one of the unknowns in one equation and it is then solved for the value of the other unknown. These two values are substituted in the second equation. It usually will be found, since they are first trial values, that they do not satisfy this second equation. A new value for one of the unknowns is now substituted in the first equation and it is solved for the second unknown. These "second trial values" are substituted in the second equation. A comparison between the second equation with the first trial values inserted and the second equation with the second trial values inserted will probably indicate about what trial value is to be selected for a third attempt to satisfy both first and second equations. The outline of the solution of this specific problem will make these steps clear.

Procedure: Three equations are employed:

(1) $X^2 + Y^2 = 9$ is used to determine the respective trial values of X and Y. This equation is based on the geometrical theorem that in any right angle triangle (in this case triangle *bef*) the sum of the squares of the sides equals the square of the hypotenuse. It can be seen from this equation that neither X nor Y can be greater than 3. Hence, let the first trial value for Y be taken as 1. As shown in the table on page **22–8**, X then equals 2.828.

The second equation is then used to determine if these values of X and Y will satisfy the requirements of the problem and, when the correct values of X and Y are found, this equation is also used to determine the value of Z. As can be seen from the accompanying diagram, Z is the unknown length to be determined.

(2) $\dfrac{Y}{10 - X} = \dfrac{X}{15 - Y} = \dfrac{3}{Z}$ is based on the geometrical theorem that in any two similar triangles, their respective sides are proportional to each other. Thus, triangle *bef* is similar to triangle *age* since their corresponding sides are at right angles to each other, and hence their corresponding angles are equal.

Now, substituting the first trial values of X and Y in this equation, we find that $\dfrac{X}{15 - Y} = 0.2020$ and $\dfrac{Y}{10 - X} = 0.1394$. Since Equation (2) indicates that these two values should be equal, it is evident that the assumed values of X and Y are incorrect.

Next, let $Y = 2$, then $X = 2.236$. With these assumed values, $\dfrac{X}{15 - Y} = 0.1720$ and $\dfrac{Y}{10 - X} = 0.2576$. It will be noted that where in the first instance $\frac{X}{15 - Y}$ was greater than $\frac{Y}{10 - X}$, in the second instance the reverse is true. Hence, it may be concluded that the assumed value for Y of 2 is too large and the correct value lies somewhere between 1 and 2.

Next, let $Y = 1.5$. Then $X = 2.598$ and $\frac{X}{15 - Y} = 0.1924$ and $\frac{Y}{10 - X} = 0.2026$. Since $\frac{X}{15 - Y}$ is less than $\frac{Y}{10 - X}$ this shows that the correct value of Y lies between 1 and 1.5. The successive approximations, using values for Y of 1.45, 1.44, 1.43 and 1.431 are shown in the accompanying table. The value of 1.431 for Y makes $\frac{X}{15 - Y}$ and $\frac{Y}{10 - X}$ equal to within 0.00002. Hence, for the purpose of this particular problem, the values of $Y = 1.431$ and $X = 2.6367$ will be assumed as correct. (A more exact determination of these values could be made if required.)

From Equation (2) it is noted that $\frac{3}{Z}$ is equal to both $\frac{X}{15 - Y}$ and $\frac{Y}{10 - X}$. To determine the value of Z, an average of 0.19432 and 0.19434 is taken, or 0.19433. Thus,

$$\frac{3}{Z} = 0.19433$$

$$Z = \frac{3}{0.19433} = 15.438$$

Y	Y^2	X^2 $(9 - Y^2)$	X	$\frac{X}{15 - Y}$	$\frac{Y}{10 - X}$
1	1	8	2.828	0.2020	0.1394
2	4	5	2.236	0.1720	0.2576
1.5	2.25	6.75	2.598	0.1924	0.2026
1.45	2.1025	6.8975	2.626	0.1938	0.1966
1.44	2.0736	6.9264	2.632	0.1941	0.1954
1.43	2.0449	6.9551	2.6372	0.1943	0.1942
1.431	2.0478	6.9522	2.6367	0.19432	0.19434

Now a third equation is set up to check this value of Z.

(3) $(15 - Y)(10 - X) + XY + 3Z = 150$ is based on the fact

that the sum of the areas of the two large triangles (which together comprise a rectangle with sides $15 - Y$ and $10 - X$) and the two small triangles (which together comprise a rectangle with sides X and Y) and the rectangle with sides 3 and Z equals the area of the entire large rectangle, the sides of which are 10 and 15.

Substituting the respective values of $X = 2.6367$, $Y = 1.431$ and $Z = 15.438$ in Equation (3), it will be found that these are correct.

$$(15 - Y)(10 - X) + XY + 3Z = 150$$

$$(15 - 1.431)(10 - 2.6367) + (2.6367 \times 1.431) + (3 \times 15.438)$$
$$= 99.913 + 3.773 + 46.314 = 150.000$$

Radius and Height of Arc of a Segment

PROBLEM 4: To find the radius R and height h of a given arc, Fig. 4, when only the length l of the arc and the length c of the chord subtending it are known.

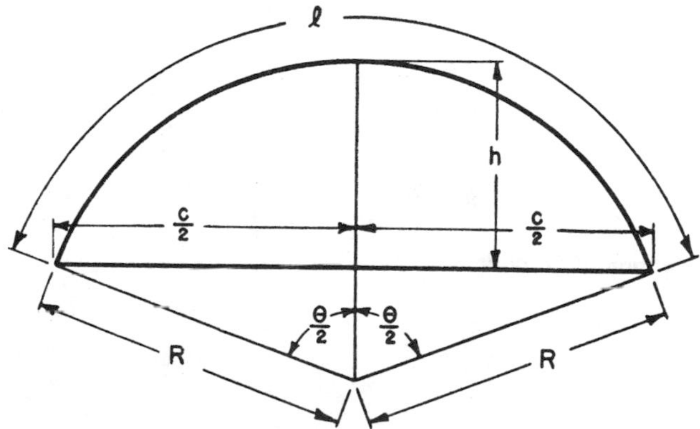

FIG. 4. To Find Radius R and Height h of Arc When Length of Arc l and Length of Chord c are Known

Analysis of Problem: The first step is to establish two equations involving the two known values l and c and two unknown values, which in this case, are R and Θ, where Θ is the angle subtended by the arc. Each of these equations is rearranged with R on one side and the remaining terms on the other. The two equivalents of R are now set equal to each other and rearranged to give Formula (1) shown below. Using this formula, Θ is found by trial-and-error substitution of values for $\hat{\Theta}$ in radians in the expression $\frac{\hat{\Theta}}{\sin\frac{\Theta}{2}}$ until the value of $\frac{2l}{c}$ is satisfied.

Since the length of an arc is equal to its radius times the angle (in radians) which it subtends, Formula (2) given below is established to find R in terms of l and $\hat{\Theta}$.

Referring to Fig. 4, it will be seen that h is equal to R minus one side of a right-angled triangle of which R is the hypotenuse and $\frac{C}{2}$ is the other side. Hence, Formula (3) is established for finding h in terms of R and c.

Formulas:

$$\frac{\hat{\Theta}}{\sin\frac{\Theta}{2}} = \frac{2l}{c} \text{ } (\hat{\Theta} \text{ is in radians}) \tag{1}$$

$$R = \frac{l}{\hat{\Theta}} \text{ } (\hat{\Theta} \text{ is in radians}) \tag{2}$$

$$h = R - \sqrt{R^2 - \frac{c^2}{4}} \tag{3}$$

Derivation of Formulas:

$$\sin\frac{\Theta}{2} = \frac{\frac{c}{2}}{R} \tag{1a}$$

$$R = \frac{\frac{c}{2}}{\sin\frac{\Theta}{2}} \tag{1b}$$

$$l = R\hat{\Theta} \text{ where } \hat{\Theta} \text{ is in radians} \tag{2a}$$

$$R = \frac{l}{\hat{\Theta}} \tag{2b}$$

Setting the equivalents of R in Equations (1b) and (2b) equal to each other:

$$\frac{\frac{c}{2}}{\sin\frac{\theta}{2}} = \frac{l}{\hat{\theta}} \tag{3a}$$

$$\frac{\hat{\theta}}{\sin\frac{\theta}{2}} = \frac{2l}{c} \tag{3b}$$

$$R^2 - \left(\frac{c}{2}\right)^2 = (R - h)^2 \tag{4a}$$

$$\sqrt{R^2 - \left(\frac{c}{2}\right)^2} = R - h \tag{4b}$$

$$h = R - \sqrt{R^2 - \frac{c^2}{4}} \tag{4c}$$

Example: Find the radius and height of an arc, the length of which is 7.286 inches if the length of the chord subtending it is 6.243 inches.

Solution: $l = 7.286 \qquad c = 6.243$

$$\frac{\hat{\theta}}{\sin\frac{\theta}{2}} = \frac{2 \times 7.286}{6.243} = 2.3341 \tag{3b}$$

Trial No.	θ (degrees)	$\hat{\theta}$ (radians)	$\sin\frac{\theta}{2}$	$\hat{\theta} \div \sin\frac{\theta}{2}$
1	45°	0.7854	0.38268	2.052
2	70°	1.2217	0.57358	2.130
3	120°	2.0944	0.86603	2.418
4	110°	1.9199	0.81915	2.344
5	108°	1.8850	0.80902	2.330
6	109°	1.9024	0.81411	2.3368
7	108°30′	1.89368	0.81157	2.3333
8	108°45′	1.89805	0.81284	2.3351
9	108°38′	1.89601	0.81225	2.3343
10	108°37′	1.89572	0.81216	2.3342
11	108°36½′	1.89557	0.81212	2.3341

Trial values of Θ will now be substituted in the expression $\dfrac{\hat{\Theta}}{\sin \frac{\Theta}{2}}$ until the value 2.3341 is obtained. As in most trial-and-error solutions, the answer is reached more quickly and surely if the assumed and computed values are arranged in tabular form as on page **22–11**. Therefore Θ is $108°36\frac{1}{2}'$ and $\hat{\Theta} = 1.89557$ radians

$$R = \frac{7.286}{1.89557} = 3.8437 \text{ inches} \tag{2b}$$

$$h = 3.8437 - \sqrt{3.8437^2 - \frac{6.243^2}{4}} \tag{4c}$$

$$h = 3.8437 - \sqrt{14.7740 - 9.7438}$$

$$h = 3.8437 - 2.2428 = 1.6009 \text{ inches}$$

Length and Sag of Catenary Cable

PROBLEM 5: A flexible steel cable having a weight of w pounds per foot of length is to be suspended between two points on the same level at a distance $2d$ feet apart as shown in Fig. 5. How much sag s in feet at the mid-point will occur for a given tension of T pounds at each end of the cable and how long l, in feet, will the cable be?

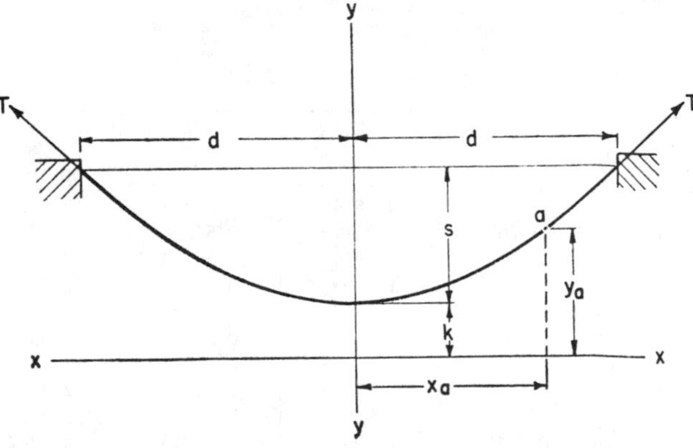

FIG. 5. To Find Sag s and Length l of Flexible Steel Cable of Unit Weight w for Given Tension T and Horizontal Distance $2d$

Analysis of Problem: If a flexible cable of uniform size and weight is freely suspended between two supports with the ends of the cable on the same level, it will assume the form of a catenary curve. The solution of this problem is, therefore, based on equations for such a curve.

In order to establish these equations, x- and y-reference axes are drawn as shown in Fig. 5. If the x-axis is located so that the vertical distance from it to the lowest point on the catenary is equal to the tension in the cable at that point divided by the weight of the cable in pounds per foot of length, then it can be shown that the vertical distance from the x-axis to any point on the catenary is equal to the tension at that point divided by the weight in pounds per foot. The y-axis is drawn through the mid-point of the catenary as shown.

Formulas:

$$\frac{T}{w} = k \cosh \frac{d}{k}$$

$$s = \frac{T}{w} - k$$

$$l = 2\sqrt{\left(\frac{T}{w}\right)^2 - k^2}$$

Note: — A catenary cable should not be confused with a parabolic cable. A parabolic cable, like a catenary cable, is a flexible cable suspended from two points; however, it carries a load, in addition to its own weight, which is uniformly distributed horizontally.

The cables of a suspension bridge are examples of parabolic cables since the weight of the roadbed, which is supported by the cables, is uniformly distributed horizontally. The formulas which apply to parabolic cables customarily do not take into account the weight of the cable itself since this is usually small compared to the uniformly distributed load. It is for this reason (the neglecting of the cable weight) that equations for parabolic cables can be established which do not require trial and error methods of solution and can be solved directly.

Derivation of Formulas: The location of any point a on the catenary in terms of the horizontal distance x_a from the y-axis and vertical distance y_a from the x-axis is expressed by the general equation for the catenary:

$$y_a = k \cosh \frac{x_a}{k} \tag{1}$$

where k is equal to the cable tension at the lowest point in the catenary divided by the weight per foot.

If a is the end point of the catenary, then y_a is equal to the tension at that point divided by the weight per foot or $\frac{T}{w}$ and x_a is equal to d. Making these substitutions in Equation (1):

$$\frac{T}{w} = k \cosh \frac{d}{k} \tag{2}$$

In this equation, T, w, and d are known. The value of k is found by trial-and-error substitution of values until one is found which satisfies the equation. *Cosh* stands for hyperbolic cosine and is found by referring to a table of hyperbolic functions. (It should be noted that the value of $\frac{d}{k}$ is expressed in radians and the entry values of tables of hyperbolic functions are in radians rather than in degrees.)

By referring to Fig. 5, it can be seen that the vertical distance from the x-axis to the end point of the cable is equal to the sag plus the vertical distance from the x-axis to the lowest point of the catenary or:

$$y \text{ (endpoint)} = \frac{T}{w} = s + k \tag{3a}$$

In this equation, T, w and k are known, so that it can be solved for s.

$$s = \frac{T}{w} - k \tag{3b}$$

It can be shown that the horizontal component of the stress at any point in the cable is a constant value equal to the weight per foot times k or

$$T_x = wk \tag{4}$$

and the vertical component varies with the linear distance r along the cable from the mid-point to the point in question and is equal to the product of this distance times the weight per foot or

$$T_y = wr \tag{5}$$

The vertical component of the stress or tension at the end of the cable will therefore be equal to one-half its length times the weight per foot. The total tension T at the end of the cable will be equal to the vector sum of the horizontal and vertical components which is equal to the square root of the sum of their squares or

$$T = \sqrt{T_y^2 + T_x^2} = \sqrt{\left(w\frac{l}{2}\right)^2 + (wk)^2} \tag{6}$$

Rearranging this equation to solve it for l

$$T^2 = w^2\left(\frac{l^2}{4} + k^2\right) \tag{7a}$$

$$l^2 = 4\left(\frac{T^2}{w^2} - k^2\right) \tag{7b}$$

$$l = 2\sqrt{\left(\frac{T}{w}\right)^2 - k^2} \tag{7c}$$

In this equation, T, w and k are known so that it can be solved for l, which is the only unknown.

Example: A steel cable weighing 2.34 pounds per foot of length is suspended between two points at the same level and 125 feet apart. If the tension at the points of support is to be 400 pounds, how much sag will there be and how long a cable is needed?

Solution: $w = 2.34$; $2d = 125$; $T = 400$.

$$\frac{400}{2.34} = k \cosh \frac{62.5}{k} \tag{2}$$

The equation will now be solved by substitution of trial values for k:

Let $k = 150$

$$\begin{aligned} 170.9 &= 150 \times \cosh \frac{62.5}{150} \\ &= 150 \times \cosh 0.416 \text{ radians} \\ &= 150 \times 1.0878 = 163.2 \end{aligned}$$

hence 150 is incorrect. Try 160:

$$\begin{aligned} 170.9 &= 160 \times \cosh \frac{62.5}{160} \\ &= 160 \times \cosh 0.391 \text{ radians} \\ &= 160 \times 1.0774 = 172.4 \end{aligned}$$

hence 160 is too high. Try 159

$$170.9 = 159 \times \cosh \frac{62.5}{159}$$
$$= 159 \times \cosh 0.393 \text{ radians}$$
$$= 159 \times 1.0782 = 171.4$$

Try a slightly smaller value than 159:

Let $$k = 158.5$$
$$170.9 = 158.5 \times \cosh \frac{62.5}{158.5}$$
$$= 158.5 \times \cosh 0.394 \text{ radians}$$
$$= 158.5 \times 1.0786 = 171.0$$

Let $$k = 158.4$$
$$170.9 = 158.4 \times \cosh \frac{62.5}{158.4}$$
$$= 158.4 \cosh 0.3946 \text{ radians}$$
$$= 158.4 \times 1.0789 = 170.9$$

therefore, $$k = 158.4$$

$$l = 2\sqrt{\left(\frac{400}{2.34}\right)^2 - 158.4^2} \qquad (7c)$$
$$l = 2\sqrt{29{,}207 - 25{,}091}$$
$$l = 2\sqrt{4116} = 2 \times 64.2 = 128.4 \text{ feet}$$

$$s = \frac{400}{2.34} - 158.4 = 170.9 - 158.4 = 12.5 \text{ feet} \qquad (3b)$$

Ratio of Geometric Series

PROBLEM 6: There are 47 terms in a geometrical series of which the first is 0.029. The sum of the 47 terms is 5.950. It is required to find the ratio for the series.

Analysis of Problem: This is a problem in which the first term a, the number of terms n, and the sum of the terms S are known and the ratio r is to be found.

$$r^n = \frac{Sr}{a} + \frac{a - S}{a}$$ (See page 2–29)

Since r^n appears on one side of this formula and r on the other, it requires a trial-and-error solution.

Procedure: The first step is to rearrange the formula slightly to facilitate the insertion of trial values.

$$r = \sqrt[n]{\frac{Sr}{a} + \frac{a - S}{a}}$$

Substituting the given numerical values: $a = 0.029$; $n = 47$ and $S = 5.950$:

$$r = \sqrt[47]{\frac{5.95r}{0.029} + \frac{0.029 - 5.950}{0.029}}$$

Simplifying: $r = \sqrt[47]{205.17r - 204.17}$

Since this is a geometric progression, r must have some value other than 1.

As a first trial value let $r = 2$

$$2 = \sqrt[47]{410.34 - 204.17} = \sqrt[47]{206.17}$$

$$\log 206.17 = 2.31423$$

$$\log \sqrt[47]{206.17} = 2.31423 \div 47 = 0.04924$$

$$\sqrt[47]{206.17} = 1.12 \text{ app.}$$

Since the equation does not balance, that is, we have 2 on the left-hand side and 1.12 on the right-hand side, a second trial is made.

Let $r = 3$, then

$$3 = \sqrt[47]{615.51 - 204.17} = \sqrt[47]{411.34}$$

$$\log 411.34 = 2.61420$$

$$\log \sqrt[47]{411.34} = 2.61420 \div 47 = 0.05562$$

$$\sqrt[47]{411.34} = 1.14 \text{ app.}$$

The unbalance or inequality is worse than in the first trial, hence for the third trial let r equal a value *smaller* than that used in the first trial.

Let $r = 1.05$, then

$$1.05 = \sqrt[47]{215.43 - 204.17} = \sqrt[47]{11.26}$$

$$\log 11.26 = 1.05154$$

$$\log \sqrt[47]{11.26} = 1.05154 \div 47 = 0.02237$$

$$\sqrt[47]{11.26} = 1.053$$

indicating that the value of r must be close to 1.05.

Let $r = 1.06$

$$1.06 = \sqrt[47]{217.48 - 204.17} = \sqrt[47]{13.31}$$

$$\log 13.31 = 1.12418$$

$$\log \sqrt[47]{13.31} = 1.12418 \div 47 = 0.02392$$

$$\sqrt[47]{13.31} = 1.057$$

It will be noted that the value on the right-hand side is now 0.003 *less* than that on the left and that when r was assumed to be 1.05 the value on the right-hand side was 0.003 more than that on the left-hand side. Thus, a trial value about halfway between 1.05 and 1.06 is indicated.

Let $r = 1.055$

$$1.055 = \sqrt[47]{216.45 - 204.17} = \sqrt[47]{12.28}$$

$$\log \sqrt[47]{12.28} = 1.08920 \div 47 = 0.02317$$

$$\sqrt[47]{12.28} = 1.055$$

indicating that this is the correct ratio to three decimal places.

Arrangement for Greatest Number of Balls in Cubical Box

PROBLEM 7: How many 1-inch balls can be placed in a box having inside dimensions of 12 inches by 12 inches by 12 inches?

Analysis of Problem: In this problem the solution is effected, not by trial and error substitution of values in a formula, but by a trial and error consideration of the various ways in which the balls can be arranged. The balls can be arranged in two types of layers. In one, there will be 12 balls on a side or 144 balls in the layer. In the other—which can be arranged only directly on top of the first type of layer by placing balls in the recesses formed by that layer, as shown in Figs. 6A and 6B—there will be 11 balls on a side or 121 balls in the layer. The problem is, then, to determine what combination of these two types of layers will permit the most balls to be placed in the box.

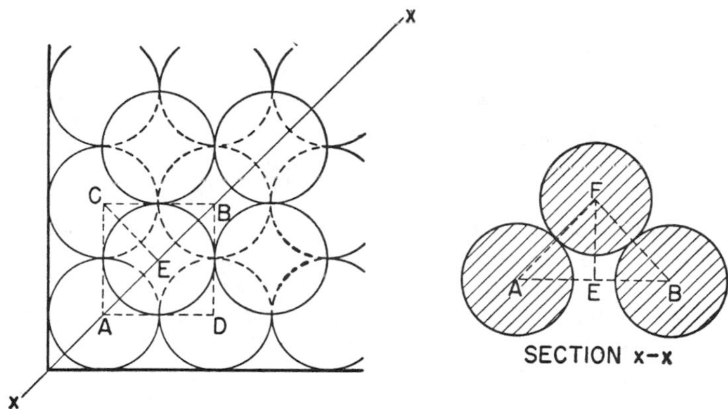

FIG. 6. To Find Greatest Number of One-Inch Balls that Can Be Fitted Into a Cubical Box. *A*. Plan View Showing Two Layers. *B*. Vertical Section Taken Through *x–x*

Twelve layers of the first type can be placed in the box since the height of twelve balls, one on top of the other would be 12 inches. With this arrangement, the total number of balls T_1 in the box would be:

$$T_1 = 12 \times 144 = 1728 \text{ balls}$$

To find the number of balls that could be placed in the box in alternate layers of 144 balls and 121 balls each, it is first necessary to compute the distance between the centers of adjoining layers. This distance is represented by the altitude EF of triangle ABF in Fig. 6B which is section *x-x* of Fig. 6A. In Fig. 6B, AB is the hypotenuse of a right-angled triangle ABC of which sides AC and BC are known. (*Note:* Angle ACB is a right angle since sides AC and CB are respectively parallel to two sides of the cubical box.)

Hence, AB can be found and, since CE is the altitude of an isosceles triangle, AE equals EB. In Fig. 6B with AE and AF known, EF can be found since AEF is a right-angled triangle. With EF known, the number of layers of 144 balls and 121 balls each that can be placed in the box is then computed.

Procedure:

$$AB = \sqrt{\overline{AC}^2 + \overline{CB}^2}$$
$$= \sqrt{1^2 + 1^2} = \sqrt{2} = 1.414$$

$$AE = \frac{AB}{2} = \frac{1.414}{2} = 0.707$$

$$FE = \sqrt{\overline{AF}^2 - \overline{AE}^2}$$
$$= \sqrt{1^2 - 0.707^2} = \sqrt{0.5} = 0.707$$

From the bottom of the box to the plane of centers of the bottom layer of balls is 0.5 inch. From the plane of centers of the bottom row of balls to the plane of centers of the top row of balls is equal to $(N - 1) \times 0.707$, where N is the total number of layers. From the plane of centers of the top layer to the top of the box cannot be less than 0.5 inch.

Hence, $$0.5 + (N - 1) \times 0.707 + 0.5 = 12 \text{ inches}$$

$$N - 1 = \frac{12 - 0.5 - 0.5}{0.707}$$

$$N = \frac{11}{0.707} + 1 = 16.56$$

Since N must be a whole number, 16 is taken as 17 would be too many. Therefore, there will be 8 layers of 144 balls arranged alternately with 8 layers of 121 balls. The total number of balls T_2 in these 16 layers would be:

$$T_2 = 8 \times 144 + 8 \times 121 = 2120$$

The height H from the bottom of the box to the top of the top layer is:

$$H = 0.5 + (16 - 1) \times 0.707 + 0.5$$
$$H = 0.5 + 10.6 + 0.5 = 11.6 \text{ inches}$$

The total height of the balls is thus 0.4 inch less than 12 inches.

The next point to determine is whether this excess height is great enough to permit the substitution of one layer of 144 balls (with the balls resting directly on top of these in the layer beneath) in place of a layer of 121 balls (where the balls rest in recesses formed by the balls in the layer beneath). If this were done, the over-all height H would be:

$$H = 0.5 + (15 - 1) \times 0.707 + 1.0 + 0.5$$

$$= 0.5 + 9.9 + 1.5 = 11.9 \text{ inches}$$

Hence, this arrangement of 9 layers of 144 balls each and 7 layers of 121 balls each can be used and the total number of balls T_3 will be:

$$T_3 = 9 \times 144 + 7 \times 121 = 2143 \text{ balls}$$

23

Errors—Their Occurrence and Systematic Handling

In referring to "errors" in connection with mechanical problems, it might be supposed that mistakes in the working out of such problems are what is meant. In this chapter, however, what is meant by errors are those inaccuracies which inevitably appear in a problem because of the degree of accuracy of the measurements being used, the limitations of the formula being applied, or the deliberate rounding off of figures used in the computations.

Errors in this sense, are normally present in the solution of all mechanical problems. In some cases they are of no importance and can be ignored. There are a number of problems, however, in which the recognition of such errors and the ability to estimate their magnitude is a real aid in helping to prevent misinterpretation of the answers obtained from the standpoint of their accuracy.

In this chapter, some of the sources of inaccuracy will be discussed with particular reference to measurements. What happens when inaccurate figures are added, subtracted, multiplied or divided, and the proper procedure for "rounding off" figures are also taken up. Finally, a method which helps to prevent errors in the location of the decimal point when figures of widely different magnitudes are multiplied together or are divided one by another, is described.

Sources of Inaccuracy. Those errors which inevitably appear in calculations, that is, those errors which are not due to mistakes in computation or to improper application of a formula, may arise either from the formula, itself, or from the data used in it. Thus, the formula may be an approximate one, such as those given for finding the area of a circular segment on page 16–3. This source of error is, of course, recognized by the computer, and, where necessary, the amount of error may be computed as shown by examples in Chapter 16.

Where the formula selected for use is an empirical one, it may be recognized that an error is introduced, but because no exact formula or convenient method of measurement is available for comparison, the amount of error can only be estimated.

Where certain constants such as π or e are used, it might be supposed that some inaccuracy will be introduced by their use. This inaccuracy may be ignored, however, since such constants can be expressed to greater accuracy than the other data employed in the computation. The ratio π, for example, may be expressed as 3.1416, 3.14159, 3.141593, etc., depending upon the accuracy of the other data employed.

Another source of error arises from the fact that the data used may be such as to make an exact result impossible. Under the section entitled, "Special Methods of Computation" examples are given of finding the areas of irregular plane figures. In those cases where the boundary of the figure is a curve for which the formula is not known, certain assumptions have to be made and a formula used which is based on these assumptions. In most cases of computation faced by the practical man, however, the source of error arises from data which are based upon measurements.

Measurements are Approximations. It may be said that all figures representing measurements or obtained from computations in which measured values are used, are approximations. The more accurate the measuring instrument, the more skillfully it is used, the more accurate will the resulting measurements usually be. But nevertheless they are still approximations and should be recognized as such. Thus, whenever measured values are to be used in mathematical computations they should be expressed in such a way as to indicate their accuracy. If a measurement is 5 inches to the nearest thousandth of an inch, it should be written as 5.000 inches not 5 inches. If a measurement is 3.9 inches to the nearest ten-thousandth of an inch, it should be written 3.9000 inches, and not 3.9 inches. In other words, zeros should be inserted, where required, to the number of decimal places which indicates the accuracy of a given figure.

Rounding off Figures. Where a figure represents a highly accurate measurement but is to be used in a computation where such an accurate value is not required, the figure may be *rounded off*. Thus, if a measurement were 5.2391 inches and a value to the nearest thousandth was satisfactory for the computation in question, it would be written 5.239; if only to the nearest hundredth, then 5.24; if only to the nearest tenth, then 5.2.

The following rules have been recommended as an American Standard. "In setting up rules for rounding off decimals there are three general cases that should be considered. They may be stated as follows:

"**Rule 1.** When the figure next beyond the last figure or place to be retained is less than 5, the figure in the last place retained should be kept unchanged.

Example: 1.2342 1.234 1.23 1.2

"**Rule 2.** When the figure next beyond the last figure or place to be retained is more than 5, the figure in the last place retained should be increased by 1.

Example: 1.6789 1.679 1.68 1.7

"**Rule 3.** When the figure next beyond the last figure or place to be retained is 5, and (*a*) there are no figures, or only zeros, beyond this 5, if the figure in the last place to be retained is odd, it should be increased by 1; if even, it should be kept unchanged; (*b*) if the 5 next beyond the figure in the last place to be retained is followed by any figures other than zero, the figure in the last place retained should be increased by 1, whether odd or even."

Example (*a*)		*Example* (*b*)	
1.35	1.4	1.3501	1.4
1.3500	1.4	1.3599	1.4
1.45	1.4	1.4501	1.5
1.4500	1.4	1.4599	1.5

It should be noted that when more than one digit is dropped by rounding off, the final value is obtained, not by a series of successive roundings in which one digit is dropped at a time, but by dropping all digits beyond the last digit to be retained and following the rules given above.

For example, when 0.5499 is rounded off to one decimal place it becomes 0.5 and not 0.6 as it would if successive roundings were applied.

Addition and Subtraction. The accuracy of a sum of figures cannot be greater than that of the least accurate of the figures being added. It may be much less. Take the following example:

$$\begin{array}{r} 6.34 \\ 5.192 \\ 3.1 \\ 8.9658 \\ \hline 23.5978 \end{array}$$

Presuming that each figure has been written down correctly to indicate its accuracy, the least accurate figure is 3.1. This figure is accurate only to the nearest tenth. Hence, the answer can be expected to be accurate to the nearest tenth and should be written 23.6. Of course, if the above figures had been written

$$\begin{array}{r} 6.3400 \\ 5.1920 \\ 3.1000 \\ 8.9658 \\ \hline 23.5978 \end{array}$$

indicating that each had been measured to the nearest ten-thousandth, then the answer would be written as 23.5978. It should be recognized, however, that the error in any sum of figures may be greater than the error in any one of them. Thus, in the above example, if each of the figures being added had an error of plus or minus 0.00005, then the resulting errors might be:

$$\begin{array}{r} +0.00005 \\ +0.00005 \\ +0.00005 \\ +0.00005 \\ \hline +0.00020 \end{array} \qquad \begin{array}{r} +0.00005 \\ -0.00005 \\ -0.00005 \\ +0.00005 \\ \hline 0.0000 \end{array} \qquad \begin{array}{r} -0.00005 \\ -0.00005 \\ -0.00005 \\ -0.00005 \\ \hline -0.00020 \end{array}$$

to give but three possible combinations. In the first case, the correct sum would be $23.5978 - 0.0002 = 23.5976$; in the second case, it would be $23.5978 + 0.0000 = 23.5978$; and in the third case, it would be $23.5978 + 0.0002 = 23.5980$. Thus, it can be seen that the sum of a group of figures measured to the nearest ten-thousandth may or may not be correct to the nearest ten-thousandth. It is frequently assumed, however, that in adding a long column of figures there is a more or less equal distribution of plus and minus errors which tend to balance each other out. This depends, of course, on the type of data which the figures represent. In some cases where a small number of figures of equal accuracy are added, the last place in their sum is rounded off as being inaccurate.

These same general considerations apply in the case of subtraction except, that since only two numbers are usually involved, there is less chance for a balancing out of errors, but also less chance for a large accumulated error.

Multiplication. The accuracy of the product of two figures may be much less than the accuracy of either due to the compounding of the error in each, thus:

$$\begin{array}{r} 12.31 \pm 0.005 \\ 1.05 \pm 0.005 \\ \hline 6155 \\ 12310 \\ \hline 12.9255 \end{array}$$

In the first place, the product should be expressed to an accuracy of one decimal place less than that of the least accurate of its factors. Hence, the above answer should be written 12.9. To indicate the possible error in the product of the above figures, let it be assumed that each figure was in error by +0.005 inch. Then the resulting error e in the product would be found by the equation $e = 12.31 \times 0.005 + 1.05 \times 0.005 + 0.005 \times 0.005$. Neglecting the last term, which is too small to materially affect the total error, $e = 0.0668$, which is considerably more than the error of either factor.

Division. It might be supposed that in division the error of the quotient would be less than that of either the dividend or the divisor, but such is not the case. As in multiplication, the error in the answer may be much greater than that in either of the factors.

General Rules. Where extremely accurate results are not required, the following rule may be used as a guide in addition and subtraction: Express the answer to as many decimal places as the least accurate figure being added.

In multiplication and division, express the answer to one less decimal place than the least accurate of the figures used. In other words, if an answer to three decimal places is desired in an operation requiring multiplication or division, the factors that are used should be accurate to at least four decimal places.

For extreme accuracy in computations, reference should be made to books covering the theory of errors, where equations indicating the probable error in various types of computations can be found.

Avoiding Errors in Locating the Decimal Point. Computations by longhand or by slide rule involving very large or very small numbers can often be carried out more easily and with less chance of error, particularly with regard to placement of the decimal point, if the powers of ten system of notation is used.

Powers of Ten Notation. In this system of notation every number is expressed by two factors, one of which is some integer from 1 to 9 followed by a decimal and the other is some power of 10.

Thus, 10,000 is expressed as 1.0000×10^4 and 10,463 as 1.0463×10^4. The number 43 is expressed 4.3×10 and 568 is expressed 5.68×10^2.

In the case of decimals, the number 0.0001 which as a fraction is $\frac{1}{10{,}000}$ is expressed as 1×10^{-4} and 0.0001463 is expressed as 1.463×10^{-4}. The decimal 0.498 is expressed as 4.98×10^{-1} and 0.03146 is expressed as 3.146×10^{-2}.

Rules for Converting any Number to Powers of Ten Notation. Any number can be converted to the powers of ten notation by means of one of two rules.

Rule 1. If the number is a whole number or a whole number and a decimal so that it has digits to the left of the decimal point, the decimal point is moved a sufficient number of places to the *left* to bring it to the immediate right of the first digit. With the decimal point shifted to this position, the number so written comprises the *first* factor when written in powers of ten notation.

The number of places that the decimal point is moved to the left, to bring it immediately to the right of the first digit is the *positive* index or power of 10 that comprises the *second* factor when written in powers of ten notation.

Thus, to write 4639 in this notation, the decimal point is moved three places to the left giving the two factors: 4.639×10^3. Similarly,

$$431.412 = 4.31412 \times 10^2$$

$$986388 = 9.86388 \times 10^5$$

$$7006 = 7.006 \times 10^3$$

Rule 2. If the number is a decimal, i.e., it has digits entirely to the right of the decimal point, then the decimal point is moved a sufficient number of places to the *right* to bring it immediately to the right of the first digit. With the decimal point shifted to this position, the number so written comprises the *first* factor when written in powers of ten notation.

The number of places that the decimal point is moved to the *right* to bring it immediately to the right of the first digit is the *negative* index or power of 10 that follows the number when written in powers of ten notation.

Thus, to bring the decimal point in 0.005721 to the immediate right of the first digit which is 5, it must be moved *three* places to the right, giving the two factors: 5.721×10^{-3}. Similarly,

$$0.469 = 4.69 \times 10^{-1}$$

$$0.0000516 = 5.16 \times 10^{-5}$$

Multiplying Numbers Written in Powers of Ten Notation. When multiplying two numbers written in the powers of ten notation together, the procedure is as follows:

1. Multiply the first factor of one number by the first factor of the other to obtain the first factor of the product.

2. Add the index of the second factor (which is some power of 10) of one number to the index of the second factor of the other number to obtain the index of the second factor (which is some power of 10) in the product. Thus,

$$(4.31 \times 10^{-2}) \times (9.0125 \times 10) = (4.31 \times 9.0125) \times 10^{-2+1} = 38.844 \times 10^{-1}$$

$$(5.986 \times 10^{4}) \times (4.375 \times 10^{3}) = (5.986 \times 4.375) \times 10^{4+3} = 26.189 \times 10^{7}$$

in each case rounding the first factor off to three decimal places.

When multiplying several numbers written in this notation together, the procedure is the same. All of the first factors are multiplied together to get the first factor of the product and all of the indices of the respective powers of ten are added together, taking into account their respective signs, to get the index of the second factor of the product. Thus, $(4.02 \times 10^{-3}) \times (3.987 \times 10) \times (4.863 \times 10^{5}) = (4.02 \times 3.987 \times 4.863) \times (10^{-3+1+5}) = 77.94 \times 10^{3}$ rounding off the first factor to two decimal places.

Dividing Numbers Written in Powers of Ten Notation. When dividing one number by another when both are written in this notation, the procedure is as follows:

1. Divide the first factor of the dividend by the first factor of the divisor to get the first factor of the quotient.

2. Subtract the index of the second factor of the divisor from the index of the second factor of the dividend, taking into account their respective signs, to get the index of the second factor of the quotient. Thus,

$$(4.31 \times 10^{-2}) \div (9.0125 \times 10) =$$
$$(4.31 \div 9.0125) \times (10^{-2-1}) = 0.4782 \times 10^{-3}$$

It can be seen, then, that where several numbers of different magnitudes are to be multiplied and divided this system of notation is helpful.

Example: Find the quotient of $\dfrac{250 \times 4698 \times 0.00039}{43678 \times 0.002 \times 0.0147}$

Solution: Changing all of these numbers to powers of ten notation and performing the operations indicated:

$$\frac{(2.5 \times 10^{2}) \times (4.698 \times 10^{3}) \times (3.9 \times 10^{-4})}{(4.3678 \times 10^{4}) \times (2 \times 10^{-3}) \times (1.47 \times 10^{-2})} =$$

$$\frac{(2.5 \times 4.698 \times 3.9)(10^{2+3-4})}{(4.3678 \times 2 \times 1.47)(10^{4-3-2})} = \frac{45.806 \times 10}{12.841 \times 10^{-1}}$$

$$= 3.5672 \times 10^{1-(-1)}$$

$$= 3.5672 \times 10^{2}$$

$$= 356.72$$

24

Working Formulas Based Upon Tests and Experience

The practical man is usually aware that there is a marked difference between mechanical problems involving only dimensions or quantities such as length, width, thickness, angularity, area, volume, or number of units, and those problems involving properties of material such as tensile strength, thermal expansion, elongation, impact strength, etc. He should also have clearly in mind just where the difference lies between these two types of problems and what this difference signifies in their solution.

The first group of problems, mentioned above, usually can be worked out by exact mathematical formulas which are called *rational* or *non-empirical* formulas, because they are arrived at by strict mathematical derivation from accepted mathematical concepts. Thus, the formula for the length of chord c in terms of height of arc h and radius of circle r:

$$c = 2\sqrt{h(2r - h)}$$

is a rational formula. Such errors as may occur in the results obtained by using rational formulas are due to inaccuracies in the data employed or in the mathematical operations performed.

The second group of problems mentioned above, many of which have to do with the designing of machine parts of structural members, usually are not solved by rational formulas, but by formulas based upon tests and experience and fitted to experimental data and working conditions. These formulas, which are called *empirical*, *semi-empirical*, or if the descriptive term may be applied, *working* formulas, may range from those which are widely known and used, as Rankine's formulas for columns; the Guest formula for shafts; and the Lewis formula for the power transmitting capacity of gearing, to some formula of extremely limited application established to fit a specific set of data or conditions. The results which are obtained by the use of empirical formulas vary widely in their degree of accuracy and, in some cases, may be no more than "educated guesses."

From the standpoint of practical application, the first step is to ascertain if the formula to be employed is empirical or not. If it is empirical, the next step is to determine, if possible, why it is empirical and what its limitations are in application, that is, under what range of conditions it may be applied and what degree of accuracy may be expected in the results obtained by its use.

The empirical formula may frequently be recognized by the presence of certain arbitrary constants which appear as multipliers, divisors, addends, or subtrahends. For example, in the Rankine or Gordon formula for columns or struts (in which p is the ultimate load in pounds per square inch; S is the ultimate compressive strength of the material in pounds per square inch; l is the length of the column or strut in inches; and r is the least radius of gyration in inches).

$$p = \frac{S}{1 + K\left(\frac{l}{r}\right)^2}$$

the factor K is an empirical constant established by tests for a given material and end condition.

In other cases, however, no arbitrary constant appears in the working formula. This may be because it is rigidly derived from a mathematical standpoint; yet it is not a rational formula, since the assumptions from which the derivation proceeds are not known to be absolutely correct. Many formulas for stress and strain are mathematically derived, but are based on assumptions that are not strictly true, such as the perfect elasticity and homogeneity of materials. In spite of these assumptions, such semi-empirical formulas are quite useful and practical in engineering design.

In some cases, a working formula may be spotted by the presence of a factor of safety which is inserted to take care of unknown conditions or properties. Thus, a formula to determine the safe stress to which a part may be subjected, may be based upon a static condition in which a fixed load of constant value is applied. The conditions of application may call for a load of rapidly varying magnitude. A factor of safety is introduced into the formula to make sure that the working stress obtained will be well within safe limits. Or, the material under consideration may not be homogeneous, so that its physical properties may vary markedly from piece to piece, or even for different sections of a given piece. In such cases, a standard formula is used together with a suitable factor of safety to cover this uncertainty about the properties of the non-homogeneous material.

A formula may sometimes be suspected to be empirical if it is not based upon a general theory, as in the case of certain formulas for the strength of flat plates. The desired strength values for each shape of plate are found by a special method which, in most cases, makes certain arbitrary assumptions about the dangerous section or the reaction of the supports. The resulting formulas, being of an approximate nature, are working formulas which indicate in a general way, the dimensions required for a given shape of plate, type of support and load. In applying them, suitable factors of safety should be used.

Once the formula has been recognized as empirical and its limits of application established, the person who is applying this formula to his problem should proceed accordingly. He should make certain that the conditions of the problem at hand warrant the use of the particular formula and, knowing the accuracy to be expected, he should not attempt to obtain or use data which are more precise than the results warrant. Finally, in interpreting the results, he should keep in mind the possible range of error and, if possible, test and weigh the results in the light of the intended application and on the basis of his best judgment and experience.

The problems which follow, illustrate the use of a number of different working formulas in the solution of mechanical problems.

Pressure on Lathe Tool Turning Cast Iron

PROBLEM 1: Find the average pressure on a lathe tool which is turning hard cast iron and is taking a chip ⅛-inch deep with a feed of 1/16-inch per revolution.

Analysis of Problem: The formula given below is an empirical one developed by F. W. Taylor for finding the average pressure on a lathe tool. The first step in solving this problem, after substituting the given values in the formula, is to change the fractional exponents to decimal values. The answer can then be readily obtained by the use of logarithms.

Formula: $P = CD^{14/15}F^{3/4}$

where P = average cutting pressure on tool, in pounds;
C = a constant (69,000 for hard cast iron);
D = depth of cut, in inches
F = feed per revolution, in inches

Solution: $P = 69{,}000 \times 0.125^{14/15} \times 0.0625^{3/4}$

Converting the exponents to decimals:

$$P = 69{,}000 \times 0.125^{0.93} \times 0.0625^{0.75}$$
$$\log P = \log 69{,}000 + \log 0.125^{0.93} + \log 0.0625^{0.75}$$
$$\log P = \log 69{,}000 + 0.93 \log 0.125 + 0.75 \log 0.0625$$
$$\log P = 4.83855 + 0.93 \times (9.09691 - 10) + 0.75 \times (8.79588 - 10)$$

Adding and subtracting 90 to and from the characteristic so that when multiplied by the index of the power, the minus part of the characteristic will be a whole number.

$$\begin{aligned}
\log 0.125 &= 9.09691 - 10 \\
&\quad +90 \quad - 90 \\
\log 0.125 &= 99.09691 - 100 \\
0.93 \log 0.125 &= 92.16013 - 93 = 9.16013 - 10
\end{aligned}$$

$$\begin{aligned}
\log 0.0625 &= 8.79588 - 10 \\
&\quad +90 \quad - 90 \\
\log 0.0625 &= 98.79588 - 100 \\
0.75 \log 0.0625 &= 74.09691 - 75 = 9.09691 - 10
\end{aligned}$$

Adding the logarithms of the three factors in the formula:

$$\begin{aligned}
\log 69{,}000 &= 4.83855 \\
0.93 \log 0.125 &= 9.16013 - 10 \\
0.75 \log 0.0625 &= 9.09691 - 10 \\
\log P &= 23.09559 - 20 = 3.09559 \\
P &= 1246, \text{ say } 1250 \text{ pounds}
\end{aligned}$$

Torque Required in Drilling High-Manganese Steel

PROBLEM 2: Find the torque required in drilling S.A.E. 1045 steel in annealed condition with a $\frac{1}{4}$-inch diameter high-speed steel twist drill and a feed of 0.006 inch per revolution.

Analysis of Problem: The formula given below is an empirical one developed at the University of Michigan as a result of tests in drilling high-manganese steel having a hardness of approximately 288 Brinell with cobalt high-speed steel twist drills. Since decimal powers appear in the formula, this problem is solved by using logarithms.

Formula: $T = kf^{0.78}d^{1.8}$

Where T is torque required, in pound-feet

k is a constant (1590 for annealed 1045 steel)

f is the feed, in inches per revolution

d is the drill diameter in inches

Solution:

$$T = 1590 \times 0.006^{0.78} \times 0.25^{1.8}$$
$$\log T = \log 1590 + \log 0.006^{0.78} + \log 0.25^{1.8}$$
$$\log T = \log 1590 + 0.78 \log 0.006 + 1.8 \log 0.25$$
$$\log T = 3.20140 + 0.78 \times (7.77815 - 10) + 1.8 \times (9.39794 - 10)$$

Adding and subtracting 90 to and from the characteristic of the log of 0.006 so that when it is multiplied by the index of the power, the minus part of the characteristic will be a whole number.

$$\begin{aligned} \log 0.006 &= 7.77815 - 10 \\ & +90 \quad\; - 90 \\ \log 0.006 &= 97.77815 - 100 \\ 0.78 \log 0.006 &= 76.26696 - 78 \end{aligned}$$

The characteristic of the log of 0.25 does not need to be altered for multiplying by the index of the power.

$$\begin{aligned} \log 0.25 &= 9.39794 - 10 \\ 1.8 \log 0.25 &= 16.91629 - 18 \end{aligned}$$

Adding the logarithms of the three factors in the formula:

$$\begin{aligned} \log 1590 &= 3.20140 \\ 0.78 \log 0.006 &= 76.26696 - 78 \\ 1.8 \log 0.25 &= 16.91629 - 18 \\ \log T &= 96.38465 - 96 \\ \log T &= 0.38465 \\ T &= 2.425 \text{ say } 2.4 \text{ pound-feet} \end{aligned}$$

Power Loss in Plain Bearing

PROBLEM 3: Find the power loss expressed in kilowatts for a plain bearing of length L in inches and diameter D in inches, if it is subjected to a bearing pressure P with relation to the projected area A (bearing length $\times$ diameter) and the journal rotates at surface velocity V in feet per minute.

Analysis of Problem: The formula shown below is empirical and gives the approximate power loss in kilowatts of a plain bearing. Although the power lost in the bearing is in the form of heat due to friction, it is sometimes convenient to express the loss in electrical units if, for example, the bearing is located in an electric motor. This enables a direct comparison of this loss with the power consumption of the motor. Due to the presence of fractional exponents, logarithms will be used in the computation.

Formula: $$\text{KW loss} = \frac{0.38AP^{0.4}V^{1.2}}{10^6}$$

Example: What is the power loss measured in kilowatts of a motor bearing that is $2\frac{1}{2}$ inches long and $1\frac{1}{4}$ inches in diameter, if the journal rotates at 1500 R.P.M. and the bearing is subjected to a pressure of 90 pounds per square inch of projected area?

$$L = 2.5 \text{ inches} \quad D = 1.25 \text{ inches} \quad N = 1500 \text{ R.P.M.}$$

$$P = 90 \text{ pounds per square inch}$$

$$\text{Projected area} = A = LD = 2.5 \times 1.25 = 3.125 \text{ sq. in.}$$

$$\text{Velocity} = V = \frac{\pi DN}{12} = \frac{3.1416 \times 1.25 \times 1500}{12}$$

$$= 490.9 \text{ ft. per min.}$$

$$\text{KW loss} = \frac{0.38 \times 3.125 \times 90^{0.4} \times 490.9^{1.2}}{10{,}000{,}000}$$

$$\begin{aligned}
\log 0.38 &= 9.57978 - 10 \\
\log 3.125 &= .49485 \\
0.4 \log 90 = 0.4 \times 1.95424 &= .78170 \\
1.2 \log 490.9 = 1.2 \times 2.69099 &= 3.22919 \\
\hline
& 14.08552 - 10 \\
-\log 10{,}000{,}000 &= -7.00000 \\
\hline
\log \text{KW loss} & 7.08552 - 10 \\
\text{loss} &= 0.00122 \text{ KW}
\end{aligned}$$

Size of Pulley Set Screw for Given Horsepower

PROBLEM 4: What diameter d cup-point set-screw is required to hold a pulley fixed to a shaft of diameter D rotating at n revolutions per minute and transmitting p horsepower?

Analysis of Problem: Experimental tests have shown that the torsional holding power t of cup-point set-screws can be expressed by the empirical formula $t = 1250Dd^{2.3}$. If this torsional holding power is set to equal the equivalent torque for the given horsepower and R.P.M. of the shaft, an expression for d in terms of p, D, and n can be derived.

Formula: $d = \sqrt[2.3]{\dfrac{50.4p}{D \times n}}$ inches

Derivation of Formula:

$$t = 1250 \times D \times d^{2.3} \tag{1}$$

Let T = torque to be transmitted by shaft

$$T = \frac{12 \times 33{,}000 \times p}{2\pi n} \text{ pound-inches} \tag{2a}$$

$$T = \frac{63{,}025p}{n} \text{ pound-inches} \tag{2b}$$

If t is to equal T, then

$$1250 \times D \times d^{2.3} = \frac{63{,}025p}{n} \tag{3a}$$

$$d^{2.3} = \frac{50.4 \times p}{D \times n} \tag{3b}$$

$$d = \sqrt[2.3]{\frac{50.4 \times p}{D \times n}} \tag{3c}$$

Example: What diameter cup-point set-screw will be required to hold a pulley fast to a 2-inch diameter shaft that is to transmit 25 horsepower at 1200 R.P.M.?

Solution: Since this problem involves the use of a fractional exponent, logarithms will be employed in its solution.

$$d = \sqrt[2.3]{\frac{50.4 \times 25}{2 \times 1200}} = \sqrt[2.3]{\frac{50.4 \times 25}{2400}}$$

$$\begin{aligned}
\log 50.4 &= 1.70243 \\
\log 25 &= 11.39794 - 10 \\
&\overline{13.10037 - 10} \\
-\log 2400 &= -3.38021 \\
\log d^{2.3} &= 9.72016 - 10
\end{aligned}$$

$$\log d = \frac{9.72016 - 10}{2.3} = \frac{22.72016 - 23}{2.3} = 9.87833 - 10$$

$$d = 0.756 \text{ inch}$$

A $\frac{3}{4}$-inch set-screw would be used.

Adiabatic Compression and Expulsion of Air

PROBLEM 5: Find the horsepower, H.P., required to adiabatically compress V cubic feet of air per second from an initial absolute pressure P_1 in pounds per square foot to a final absolute pressure P_2 in pounds per square foot and expel it from the cylinder.

Analysis of Problem: The *work* W in foot-pounds for the compression and expulsion of air, adiabatically compressed, that is without transmission of heat to or from it, is given by the equation:

$$W = 3.46 P_1 V_1 \left[\left(\frac{P_2}{P_1}\right)^{0.29} - 1\right]$$

where V_1 is the volume of air being compressed in cubic feet. If V, the cubic feet of air per second being compressed is substituted for V_1, then the equation will give the *power* required in foot-pounds per second. From this, the horsepower can be readily computed.

Since the formula for horsepower required involves the use of a decimal exponent, logarithms are used in the numerical solution as illustrated in the example given below.

Formula: $\text{H.P.} = 0.00629 P_1 V \left[\left(\frac{P_2}{P_1}\right)^{0.29} - 1\right]$

Derivation of Formula:

$$W = 3.46 P_1 V_1 \left[\left(\frac{P_2}{P_1}\right)^{0.29} - 1\right] \tag{1}$$

Substituting V for V_1 the formula gives the power F in foot-pounds per second:

$$F = 3.46P_1V\left[\left(\frac{P_2}{P_1}\right)^{0.29} - 1\right] \tag{2}$$

and

$$\text{H.P.} = \frac{3.46}{550}P_1V\left[\left(\frac{P_2}{P_1}\right)^{0.29} - 1\right] \tag{3a}$$

or

$$\text{H.P.} = 0.00629P_1V\left[\left(\frac{P_2}{P_1}\right)^{0.29} - 1\right] \tag{3b}$$

Example: Find the horsepower required to compress 10 cubic feet of air per second from 14.7 to 176.4 pounds per square inch absolute pressure and to expel the air from the cylinder. Frictional and other losses are to be disregarded.

Solution: $P_1 = 14.7 \times 144 = 2116.8$ pounds per square foot; $P_2 = 176.4 \times 144 = 25{,}401.6$ pounds per square foot; $V = 10$ cubic feet per second

$$\text{H.P.} = 0.00629 \times 2116.8 \times 10\left[\left(\frac{25{,}401.6}{2116.8}\right)^{0.29} - 1\right] \tag{3b}$$

$$\text{H.P.} = 133.15(12^{0.29} - 1)$$

To find the value of $12^{0.29}$ logarithms are used thus:

$$\log 12^{0.29} = 0.29 \log 12$$

$$0.29 \log 12 = 0.29 \times 1.07918 = 0.31296$$

0.31296 is the logarithm of 2.056 app.

hence $$12^{0.29} = 2.056$$

so that $$\text{H.P.} = 133.15\ (2.056 - 1)$$

$$\text{H.P.} = 141$$

Collapsing Pressure of Lap-Welded Bessemer Steel Tubes

PROBLEM 6: Find the collapsing pressure P and safe pressure P_s of a modern lap-welded Bessemer steel tube having outside diameter D and wall thickness t.

Analysis of Problem: Two empirical formulas for lap-welded Bessemer steel tubes are given below. The first is for values of P greater than 580 pounds per square inch and the second is for values of P less than 580 pounds per square inch. These formulas are substantially correct for all lengths of pipe greater than six diameters between transverse joints that tend to hold the pipe in circular form.

Formulas: $P = 86{,}670 \dfrac{t}{D} - 1386$ (1)

$$P = 50{,}210{,}000 \left(\frac{t}{D}\right)^3 \quad (2)$$

where t and D are in inches.

Example: Find the collapsing pressure and safe working pressure of a lap-welded Bessemer steel tube that has an outside diameter of 9 inches and a wall thickness of 0.217 inch.

Solution: Since it is not known whether Formula (1) or Formula (2) should be employed, P being unknown, try Formula (1) for collapsing pressures above 580 pounds per square inch.

$$P = 86{,}670 \times \frac{0.217}{9} - 1386$$

$$P = 2089 - 1386 = 703 \text{ pounds per square inch}$$

Using a safety factor of 5 for ordinary service, the safe working pressure P_s would be

$$P_s = \frac{P}{5} = \frac{703}{5} = 140 \text{ pounds per square inch, app.}$$

Use of Rankine's Formulas for Columns

PROBLEM 7: Find the maximum stress s and maximum force F that can be carried safely by a horizontal brace in a painter's scaffold. The brace is a steel pipe that is relatively free to rotate slightly at its ends, where it is attached to other pipes in the scaffold.

Analysis of Problem: Numerous empirical formulas exist for finding the maximum load to which a column may safely be subjected. These formulas vary for the material, end condition, and slenderness ratio (ratio of length of the column L in inches to the least radius of gyration r of the cross section). For columns having a slenderness ratio between 20 and 100, the Rankine, or Gordon Formula, given below, may be used.

Formulas: $s = \dfrac{S_u}{1 + K(L/r)^2}$ and $K = \dfrac{S_u}{(C \pi^2 E)}$

where S_u is the ultimate tensile stress, E is the modulus of elasticity of the material, and C is an empirical factor established by tests for the given end conditions.

Example: Find the maximum safe stress and force that a 2-inch diameter standard steel pipe can support as a horizontal brace in a painter's scaffold. The pipe is 7.5 feet long and is relatively free to rotate where it is bolted to the other pipes in the structure. The cross sectional area of the pipe is 1.07 square inches, and the radius of gyration is 0.787 inches. Use a factor of safety of 3, because personal safety is involved. The ultimate stress of structural steel is 70,000 psi and the yield stress is 36,000 lb per in^2.

The slenderness ratio is 90/0.787 = 114, so the Rankine formula given above, may be applied. $C = 1$ for the rotationally free end conditions, and $E = 30$ million lb per in^2 for steel.

$$K = \frac{70{,}000}{(1 \times \pi^2 \times 30{,}000{,}000)} = 0.0002364$$

The maximum stress must satisfy both of the following conditions.

$$\text{Maximum } s = \frac{70{,}000}{(1 + 0.0002364\,(114)^2)} \quad \text{and maximum } s = \leq S_y$$

therefore maximum s = 17,180 lb per in^2, which satisfies both conditions.

The safe stress is then 17,180/3 = 5720 lb per in^2, taking into account the factor of safety of 3. The safe force is found by multiplying the stress by the cross sectional area:

$$F = 5720(1.07 \text{ in}^2) = 6120 \text{ lb}.$$

The American Institute of Steel Construction suggests a stress of 11,390 lb per in^2, which has a built-in factor of safety of 1.92.

Limiting Load for Wear of Gears

PROBLEM 8: To find the limiting static load for wear L_w of two external spur gears which are in mesh, if the materials of which the gears are made; the pressure angle A of the gears; the pitch diameter of the pinion D_p in inches; the face width F in inches; the number of teeth in the gear N_g; and the number of teeth in the pinion N_p are known.

Analysis of Problem: In checking the horsepower capacity of spur gears, the beam strength of the teeth and the wear load limit are computed and the load-carrying capacity is based on the lower of the two computed values. In this problem, however, only the limiting load for wear will be considered.

The first formula given below is Buckingham's empirical formula for wear load limit of spur gears. It contains a load stress factor K which is computed by the second formula given below and a factor Q which is computed by the third formula given below.

Formulas:

$$L_w = D_p F K Q \tag{1}$$

where

$$K = \frac{S_c^2 \sin A}{1.4}\left(\frac{1}{E_1} + \frac{1}{E_2}\right) \tag{2}$$

$$Q = \frac{2N_g}{N_g + N_p} \text{ for external gears} \tag{3a}$$

$$Q = \frac{2N_g}{N_g - N_p} \text{ for internal gears} \tag{3b}$$

E_1 and E_2 = moduli of elasticity for material of each gear, respectively

S_c = surface endurance limit of gear materials in pounds per square inch

Example: Find the limiting static load for wear of an external steel spur gear mating with a steel pinion if both gears have a $14\frac{1}{2}$-degree pressure angle; the pitch diameter of the pinion is 6 inches; the face width is $3\frac{1}{8}$ inches; the number of teeth on the pinion is 24 and on the gear is 42. Assume that the steel in the gear has a Brinell hardness of 200 and that in the pinion 300, and the surface endurance limit of the steel is 90,000 pounds per square inch.

Solution: $A = 14\frac{1}{2}°$; $D_p = 6$; $F = 3.125$; $N_p = 24$; $N_g = 42$

$$Q = \frac{2 \times 42}{24 + 42} = \frac{84}{66} = 1.273 \tag{3a}$$

$$K = \frac{90{,}000^2 \times 0.25038}{1.4}\left(\frac{1}{30{,}000{,}000} + \frac{1}{30{,}000{,}000}\right) \tag{2}$$

$$K = \frac{8{,}100{,}000{,}000 \times 0.25038 \times 2}{1.4 \times 30{,}000{,}000} = 96.6$$

$$L_w = 6 \times 3.125 \times 96.6 \times 1.273 = 2305 \tag{1}$$

say 2300 pounds.

25

Miscellaneous Mechanical Problems

A number of problems not readily classifiable under previous headings have been brought together in this chapter. They relate to such mathematical elements as spirals, helices, volumes of spheres, cylinders and frustums of cones and to some phases of mechanical work not previously touched upon, such as the use of a vector diagram in finding the angle for setting a tool slide, figuring the expansion of metal parts due to increase in temperature and computing the overall production rate of a given piece where several operations on it at different rates are being performed.

Finding Length of a Spiral

PROBLEM 1: To determine the length of a closely rolled belt when the inside and outside diameters, thickness of the roll and number of turns are known.

Analysis of Problem: The problem is one of finding the length of a spiral. One method of doing this is to take the average length of one turn of the spiral and multiply it by the number of turns. The average length of one turn is found by adding the inside and outside diameters of the spiral, dividing by 2 and multiplying by π. This is, of course, an approximate method. It can be shown that the same result will be obtained, if the sum of the diameter of the opening or hole at the center of the spiral and the total thickness of the spiral is multiplied by π times the number of turns in the spiral. In the case of this specific problem, this method is somewhat more convenient, since only one measurement is required.

Formulas: Let L = length of belt; d = diameter of opening or hole at center; D = diameter of roll; t = thickness of roll; n = number of turns in roll; then

$$L = \left(\frac{D + d}{2}\right)\pi \times n$$

or

$$L = (d + t)\,\pi \times n$$

Derivation of Formula: The first formula is based on the assumption that a tightly wrapped spiral is approximated by a number of concentric circles, the number of which is equal to the number of turns in the spiral, the smallest circle of which has a diameter equal to the diameter of the hole or opening at the center of the spiral and the largest circle of which is equal to the outside "diameter" of the spiral.

For the sake of simplicity, the roll may be considered to be a ring of diameters d and D and thickness t. The average diameter is then $\frac{D + d}{2}$ and the average circumference $= \left(\frac{D + d}{2}\right)\pi$. The average circumference multiplied by the number of turns gives the length L.

$$L = \left(\frac{D + d}{2}\right)\pi \times n \qquad (1)$$

but

$$D = d + 2t \qquad (2)$$

hence

$$L = \frac{d + 2t + d}{2}\pi \times n = (d + t)\pi \times n \qquad (3)$$

Example: Given a roll of belting having an outside diameter of $37\frac{1}{4}$ inches, an inside diameter of $7\frac{5}{8}$ inches, a roll thickness of $14\frac{13}{16}$ inches, and $79\frac{1}{3}$ turns. Find the length of the belting.

Solution: $D = 37\frac{1}{4}$; $t = 14\frac{13}{16}$; $d = 7\frac{5}{8}$; $n = 79\frac{1}{3}$

$$L = \frac{37\frac{1}{4} + 7\frac{5}{8}}{2} \times 3.1416 \times 79\frac{1}{3} \qquad (1)$$

$$L = 22.438 \times 3.1416 \times 79.33$$

$$L = 5592.07 \text{ inches} = 466 \text{ feet approximately}$$

or, by the use of Formula (3):

$$L = (7\tfrac{5}{8} + 14\tfrac{13}{16})3.1416 \times 79\tfrac{1}{3} \qquad (3)$$

$$L = 22.438 \times 3.1416 \times 79.33$$

$$L = 5592.07 \text{ inches} = 466 \text{ feet approximately}$$

Diameter of Spiral Roll by Arithmetical Series

PROBLEM 2: Find the outside diameter D_o of a roll of material when the length L, thickness of the material t and diameter D_i of the hole in the center or the roll on which the material is wound, as shown in Fig. 1, are known. Variations due to stretching or softness of the material are to be ignored.

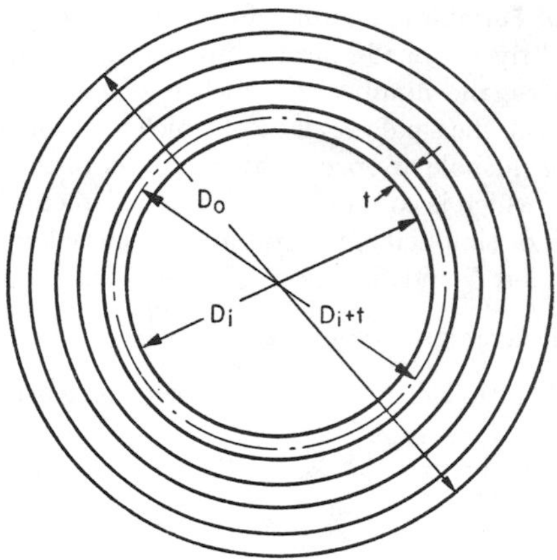

FIG. 1. To Find Outside Diameter D_o when Length L, Thickness t, and Diameter D_i are Known

Analysis of Problem: The length of the material may be considered to be the sum of the lengths, or rather mean circumferences of each successive layer. If it is assumed that each layer is circular and not a spiral, as is actually the case, the formula for the length of the material may be written.

$$L = (D_i + t)\pi + (D_i + 3t)\pi + (D_i + 5t)\pi + \cdots$$

This is, in fact, an arithmetical progression in which there is a common difference (d) equal to $2t\pi$. In this progression $(D_i + t)\pi$ equals the first term (a) and L equals the sum (S) of the number of terms in the progression (using the notation for arithmetical progressions given on page 2–28).

The problem is to determine the last term (l) in the progression, which equals the mean circumference of the last turn of the material.

When the three values (a), (d) and (S) are known in this arithmetical series, the value of the last term l can be found by the formula given on page 2–28.

$$l = -\frac{d}{2} + \frac{1}{2}\sqrt{8dS + (2a - d)^2}$$

After finding the value of (l), which in this case equals the mean circumference of the outside layer, the outside diameter D_o is then equal to $(l \div \pi) + t$.

Example: Find the outside diameter of a roll of belting which is 20 feet long, $\frac{1}{4}$ inch thick and is to be wound on a roll 1 foot in diameter.

Solution:

$$L = 20 \times 12 = 240 \text{ inches};\ t = 0.25 \text{ inch};\ D_i = 12 \text{ inches}$$

$$a = (D_i + t)\pi = (12 + 0.25)3.1416 = 38.48 \text{ inches}$$

$$d = 2t \times \pi = 0.5 \times 3.1416 = 1.57 \text{ inches}$$

$$S = L = 240 \text{ inches}$$

$$l = -\frac{1.57}{2} + \frac{1}{2}\sqrt{8 \times 1.57 \times 240 + (2 \times 38.48 - 1.57)^2}$$

$$l = -0.785 + 0.5\sqrt{3014.40 + 5683.65}$$

$$l = -0.785 + 0.5 \times 93.26$$

$$l = -0.785 + 46.63 = 45.845 \text{ inches}$$

$$D_o = \left(\frac{45.845}{3.1416} + 0.25\right) = 14.59 + 0.25 = 14.84, \text{ say 15 inches,}$$

since the outside diameter, in this case, must be a multiple of 0.25.

Finding Length of a Helix

PROBLEM 3: To find the length l of wire required to make an open-coil spring of mean diameter d, having n turns and a distance or pitch h between turns.

Analysis of Problem: An open-coil spring takes the form of a helix and the problem then becomes that of finding the length of a helix of the dimensions given. The formula shown below, gives the length of a helix in terms of diameter, pitch and number of turns.

Formula: $l = n\sqrt{\pi^2 d^2 + h^2}$

Example: Find the length of wire required to form an open-coil spring having 15 turns, a mean diameter of 3.20 inches and a pitch of 0.875 inch.

Solution:

$$n = 15;\ d = 3.2;\ h = 0.875$$
$$l = 15\sqrt{\pi^2 \times 3.2^2 + 0.875^2}$$
$$l = 15\sqrt{9.8696 \times 10.24 + 0.7656}$$
$$l = 15\sqrt{101.0647 + 0.7656}$$
$$l = 15\sqrt{101.8303}$$
$$l = 15 \times 10.09$$
$$l = 151.35 \text{ inches}$$

Locating Points on Periphery of Ellipse

PROBLEM 4: Given an ellipse of known major diameter $2b$ and known minor diameter $2a$ also the six distances y_1, y_2, y_3, y_4, y_5 and y_6 from the x-axis of six points on the periphery of the ellipse as shown in Fig. 2. It is required to find the distances x_1, x_2, x_3, x_4, x_5 and x_6 of these points from the y-axis.

Analysis of Problem: The solution of this problem is based on analytical geometry but does not require a background in this subject to use it. The equation for any point on the periphery of the ellipse and at distance x from the vertical or y-reference axis and at distance y from the horizontal or x-reference axis when the major diameter is along the y-axis and is equal to $2b$ and the minor diameter is along the x-axis and is equal to $2a$ is:

$$\frac{x^2}{a^2} + \frac{y^2}{b^2} = 1$$

Thus, for any point on the periphery of the ellipse of known major and minor diameters, if y is known, x can be found directly.

Formula:

$$\frac{x^2}{a^2} + \frac{y^2}{b^2} = 1$$

Example: The major diameter of an ellipse lies along the y-axis and is equal to 20.6 inches. The minor diameter lies along the x-axis and is equal to 10 inches. If the respective distances of six points on the periphery of the ellipse from the x-axis are: $y_1 = 1.72$; $y_2 = 3.43$; $y_3 = 5.15$; $y_4 = 6.87$; $y_5 = 8.58$; and $y_6 = 9.87$, what will be

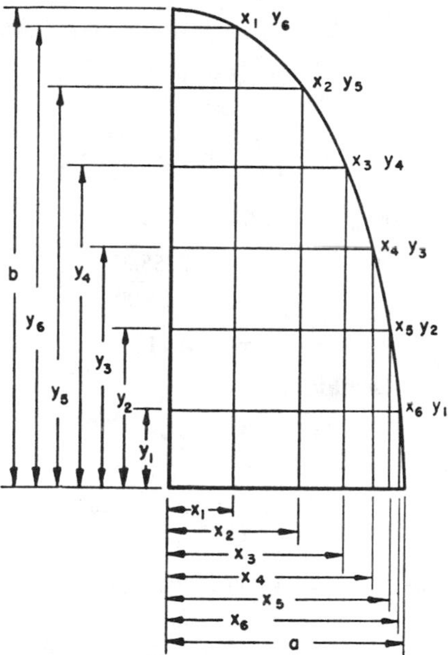

FIG. 2. To Find Distances x_1–x_6 when Major Diameter $2b$, Minor Diameter $2a$ and Distances y_1–y_6 are Known

the respective distances x_6, x_5, x_4, x_3, x_2, and x_1 of these points from the y-axis.

Solution: $a = \dfrac{10}{2} = 5$ and $b = \dfrac{20.6}{2} = 10.3$

For point x_6, y_1 (see Fig. 2)

$$\frac{x_6^2}{a^2} + \frac{y_1^2}{b^2} = 1$$

$$\frac{x_6^2}{5^2} + \frac{1.72^2}{10.3^2} = \frac{x_6^2}{25} + \frac{2.96}{106.09} = 1$$

$$106.09x_6^2 + 74.00 = 2652.25$$

$$x_6^2 = \frac{2652.25 - 74.00}{106.09} = 24.30$$

$$x_6 = 4.93$$

For point x_5, y_2 (see Fig. 2)

$$\frac{x_5^2}{a^2} + \frac{y_2^2}{b^2} = 1$$

$$\frac{x_5^2}{5^2} + \frac{3.43^2}{10.3^2} = \frac{x_5^2}{25} + \frac{11.76}{106.09} = 1$$

$$106.09x_5^2 + 294.00 = 2652.25$$

$$x_5^2 = \frac{2358.25}{106.09} = 22.23$$

$$x_5 = 4.71$$

For point x_4, y_3 (see Fig. 2)

$$\frac{x_4^2}{a^2} + \frac{y_3^2}{b^2} = 1$$

$$\frac{x_4^2}{5^2} + \frac{5.15^2}{10.3^2} = \frac{x_4^2}{25} + \frac{26.52}{106.09} = 1$$

$$106.09x_4^2 + 638.00 = 2652.25$$

$$x_4^2 = \frac{2014.25}{106.09} = 18.99$$

$$x_4 = 4.36$$

For point x_3, y_4 (see Fig. 2)

$$\frac{x_3^2}{a^2} + \frac{y_4^2}{b^2} = 1$$

$$\frac{x_3^2}{5^2} + \frac{6.87^2}{10.3^2} = \frac{x_3^2}{25} + \frac{47.20}{106.09} = 1$$

$$106.09x_3^2 + 1180.00 = 2652.25$$

$$x_3^2 = \frac{2652.25 - 1180.00}{106.09} = 13.88$$

$$x_4 = 3.73$$

For point x_2, y_5 (see Fig. 2)

$$\frac{x_2^2}{a^2} + \frac{y_5^2}{b^2} = 1$$

$$\frac{x_2^2}{5^2} + \frac{8.58^2}{10.3^2} = \frac{x_2^2}{25} + \frac{73.62}{106.09} = 1$$

$$106.09x_2^2 + 1840.50 = 2652.25$$

$$x_2^2 = \frac{2652.25 - 1840.50}{106.09}$$

$$x_2^2 = 7.65$$

$$x_2 = 2.77$$

For point x_1, y_6 (see Fig. 2)

$$\frac{x_1^2}{a^2} + \frac{y_6^2}{b^2} = 1$$

$$\frac{x_1^2}{5^2} + \frac{9.87^2}{10.3} = \frac{x_1^2}{25} + \frac{97.42}{106.09} = 1$$

$$106.09x_1^2 + 2435.50 = 2652.25$$

$$x_1^2 = \frac{2652.25 - 2435.50}{106.09}$$

$$x_1^2 = 2.04$$

$$x_1 = 1.43$$

Note: This method may be used to plot points for drawing an ellipse of known major and minor diameters. A sufficient number of y values are set down. These may be any series of values ranging from 0 to one-half the diameter which lies on the y-axis. The corresponding x values are then computed, the points plotted on cross-section paper and the ellipse periphery drawn through these points. Although only one-quarter of the ellipse is thus computed the points for the other three quarters may be established without additional computation. This is so because the ellipse is symmetrical and every point on its periphery in one quarter has a corresponding point in each of the other three quarters. Thus, the point $x = 3$, $y = 5$ in one quarter has the corresponding points $x = -3$, $y = 5$; $x = -3$, $y = -5$; and $x = 3$, $y = -5$ located respectively in the other three quarters. This is following the conventional notation where all points located to the left of the y-axis have minus x values and all points located below the x-axis have minus y values.

Diameter of Pulleys for Given Center Distance, Speed Ratio and Length of Belt

PROBLEM 5: In a pulley design for V-belts the speed ratio R, center distance C, and length of belt L are known, as shown in Fig. 3. It is required to find the diameters D and d of the large and small pulleys, respectively.

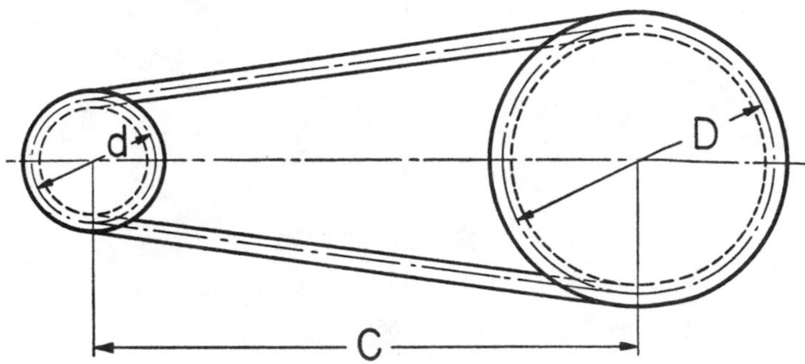

FIG. 3. To Find Diameters D and d when Speed Ratio R, Center Distance C and Belt Length L are Known

Analysis of Problem: One of the standard formulas for computing the length of belt L required for a pair of pulleys of diameters D and d and at a given center distance C apart is:

$$L = 2C + \frac{\pi}{2}(D + d) + \frac{(D - d)^2}{4C}$$

If the equivalent Rd is substituted for D in this equation, there will result a quadratic equation which can be solved for d. With d known, D can be found.

Formulas:

$$\frac{(R - 1)^2}{2\pi C(R + 1)} d^2 + d = \frac{2(L - 2C)}{\pi(R + 1)}$$

to be solved as a quadratic.

Derivation of Formula:

$$L = 2C + \frac{\pi}{2}(D + d) + \frac{(D - d)^2}{4C} \qquad (1)$$

but $$\frac{D}{d} = R \qquad \text{or} \qquad D = Rd \tag{2}$$

Substituting *Rd* for *D*

$$L = 2C + \frac{\pi}{2}(Rd + d) + \frac{(Rd - d)^2}{4C} \tag{2a}$$

or $$L = 2C + \frac{\pi d}{2}(R + 1) + \frac{d^2(R - 1)^2}{4C} \tag{2b}$$

Rearranging Equation (2b)

$$\frac{\pi d}{2}(R + 1) = L - 2C - \frac{d^2(R - 1)^2}{4C} \tag{2c}$$

hence $$d = \frac{2(L - 2C)}{\pi(R + 1)} - \frac{d^2(R - 1)^2}{2\pi C(R + 1)} \tag{2d}$$

This is a quadratic equation and can be written:

$$\frac{(R - 1)^2}{2\pi C(R + 1)}d^2 + d = \frac{2(L - 2C)}{\pi(R + 1)} \tag{2e}$$

The formula for solving a quadratic of the form $ad^2 + bd = c$ is

$$d = \frac{-b \pm \sqrt{b^2 + 4ac}}{2a} \tag{3}$$

In this case, if the numerical values of R, L and C in Equation (2e) are known, then the value of d can be found by substituting the coefficients of d^2 and d and the constant $\dfrac{2(L - 2C)}{\pi(R + 1)}$ in Equation (3).

Example: Let the length of a V-belt at the neutral axis be 61.12 inches, the center distance between the pulleys is to be 18.8 inches and the speed ratio is to be 3.35. Find the pitch diameters D and d of the two V-belt sheaves.

Solution: $L = 61.12$; $C = 18.8$; $R = 3.35$

$$\frac{(3.35 - 1)^2}{2\pi \times 18.8(3.35 + 1)}d^2 + d = \frac{2(61.12 - 2 \times 18.8)}{\pi(3.35 + 1)} \tag{2e}$$

$$\frac{5.5225}{513.84}d^2 + d = \frac{47.04}{13.666}$$

$$0.01075d^2 + d = 3.442$$

This is a quadratic equation in the form $ad^2 + bd = c$, in which $a = 0.01075$; $b = 1$ and $c = 3.442$,

hence
$$d = \frac{-1 \pm \sqrt{1 + 4 \times 0.01075 \times 3.442}}{2 \times 0.01075}$$

$$d = \frac{-1 \pm \sqrt{1.1480}}{0.0215} = \frac{-1 \pm 1.0714}{0.0215}$$

$$d = 3.321$$

or
$$d = -96.34$$

If the minus value is considered to be inapplicable, then:

$$D = Rd = 3.35 \times 3.321$$
$$D = 11.125 \text{ inches}$$

and
$$d = 3.321 \text{ inches}$$

Finding Tool-Slide Angle by Use of Vector Diagram

PROBLEM 6: To Find Angle *A*. This problem involves determining the angle *A* (Fig. 4) at which the tool-slide of a boring mill is set for obtaining a given half-angle of taper *D*, not within the angular setting range of the tool-slide, and obtained by using a known vertical feed *oa* (actually at angle *A* with the vertical) and a known horizontal feed *oc*.

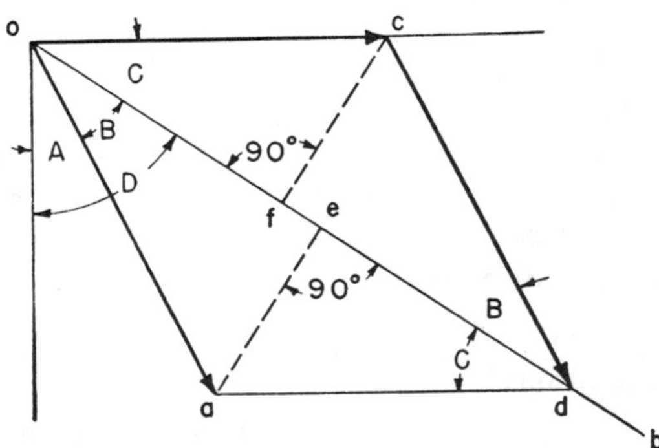

FIG. 4. To Find Angle *A* for Setting Tool-slide of Boring Mill when Angle *D*, Vertical Feed *oa* and Horizontal Feed *oc* are Known

Analysis of Problem: As can be seen from Fig. 4, this is a problem involving the resultant *od* of two vectors (quantities involving both magnitude and direction) *oa* and *oc*.

Graphical Solution: If the accuracy required were not too great, the problem could be worked out graphically. Vector *oc* is drawn in a horizontal direction and scaled to a length to equal the horizontal feed. A line *ob* is drawn at an angle *D* from the vertical and of indeterminate length to represent the resultant movement of the tool-slide. With point *c* as a center and a radius scaled to equal the vertical feed, an arc is struck which intercepts vector *ob* at point *d*. The angle which the vector *cd* makes with the vertical is the required angle *A*. This is shown in Fig. 4, as an angle between the vertical and vector *oa* which is drawn parallel to *cd* from point *o* and to the same length.

Trigonometrical Solution: In Fig. 4, *oc* represents the horizontal movement of the slide, *oa* the vertical movement, *od* the resultant movement at a given angle *D* with the vertical. From point *c* drop a perpendicular *cf* to *od* and similarly a perpendicular *ae* from point *a*. Then in right-angled triangle *ocf*, *cf* can be found, since *oc* and angle C $(90° - D)$ are known. But $ae = fc$ since they are corresponding sides of equal triangles *ofc* and *aed*. Since *oa* and *ae* are now known, angle *B* and hence angle *A* can be found.

Formulas:

$$\sin B = \frac{oc \sin (90° - D)}{oa}$$

$$A = D - B$$

Derivation of Formula:

$$fc = oc \sin C = oc \sin (90° - D) \quad (1)$$

$$\sin B = \frac{ae}{oa} \quad (2)$$

but

$$fc = ae \quad (3)$$

therefore

$$\sin B = \frac{oc \sin (90° - D)}{oa} \quad (4)$$

$$A = D - B \quad (5)$$

Example: Given $oc = 7.8$; $oa = 8.6$; and $D = 57°$. Find *A*.

Solution: $\sin B = \dfrac{7.8 \times \sin (90° - 57°)}{8.6}$ (4)

$$\sin B = \frac{7.8 \times 0.54464}{8.6} = 0.49398; \quad B = 29°36'$$

$$A = 57° - 29°36' = 27°24' \tag{5}$$

Computing Outside Radius of Hollow Sphere

PROBLEM 7: To compute the outside radius R of a hollow sphere when the thickness T of the shell and the ratio K of the volume of the shell to the total volume of the sphere is known.

Analysis of Problem: The radius of the hollow sphere enclosed by the shell, and hence its volume, can be expressed in terms of the outside radius R and the thickness T of the shell. Since the total volume of the sphere minus the volume of the hollow part is equal to the volume of the shell, an equation can be established in terms of R, T and K which can be solved for R.

Formula: $$R = \frac{T}{1 - \sqrt[3]{1 - K}}$$

Derivation of Formula:

Total volume of sphere $= \frac{4}{3}\pi R^3$ (1)

Volume of hollow part $= \frac{4}{3}\pi(R - T)^3$ (2)

Volume of shell $= K \times \frac{4}{3}\pi R^3$ (3)

hence $$K \times \tfrac{4}{3}\pi R^3 = \tfrac{4}{3}\pi R^3 - \tfrac{4}{3}\pi(R - T)^3 \tag{4a}$$

Dividing all terms by $\frac{4}{3}\pi$.

$$KR^3 = R^3 - (R - T)^3 \tag{4b}$$

or $$R^3 - KR^3 = (R - T)^3 \tag{4c}$$

taking the cube root of both sides of the equation

$$R\sqrt[3]{1 - K} = R - T \tag{4d}$$

or $$R - R\sqrt[3]{1 - K} = T \tag{4e}$$

$$R = \frac{T}{1 - \sqrt[3]{1 - K}} \tag{4f}$$

Example: Find the outside radius of a hollow sphere with a shell thickness of $2\frac{1}{4}$ inches, if the volume of the shell is 35 per cent of the total volume of the sphere.

Solution: $T = 2.25$ and $K = 0.35$

$$R = \frac{2.25}{1 - \sqrt[3]{1 - 0.35}} = \frac{2.25}{1 - 0.866} = 16.79 \text{ inches} \quad (4f)$$

Relation of Volume of Sphere to its Diameter

PROBLEM 8: If a steel ball of diameter x weighs a grains, what must the diameter y of a similar ball be to weigh b grains?

Analysis of Problem: The volumes of spheres are to one another as the cubes of their diameters. If they are of material having the same density, their weights are also proportional to the cubes of their diameters. Hence, if x and y represent the respective diameters of two spheres and a and b their given weights, then $x^3 : y^3 = a : b$.

Formula: $y^3 = \frac{b}{a} x^3$

Example: Let $x = 0.718$ inch; $y =$ required diameter; $a = 380$ grains; and $b = 383$ grains. Find y.

Solution:

$$y^3 = \frac{383 \times (0.718)^3}{380} = \frac{383 \times 0.370}{380}$$

$$y^3 = 0.3729; \quad y = 0.720 \text{ inch}$$

Diameter of Hole to Reduce Weight of Sphere Given Amount

PROBLEM 9: To determine diameter d in Fig. 5. A hole is to be drilled through a sphere of diameter D in order to reduce its weight W by a given amount w. What is the diameter d of the hole in terms of the diameter D of the sphere and weights W and w.

Analysis of Problem: Since it is presumed that the sphere is composed of homogeneous material and is therefore of uniform density throughout, the problem is essentially that of finding the *volume* of that part of the sphere removed by drilling the hole which will reduce

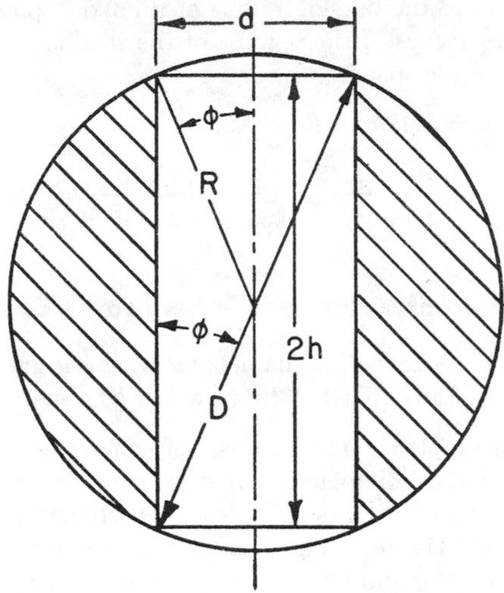

FIG. 5. To Determine Diameter d of Hole Drilled in Sphere of Diameter D To Reduce Weight W by Given Amount w

the volume of the sphere in the same proportion as the desired weight reduction.

As shown in Fig. 5, that part of the sphere removed by the drill may, for the sake of simplified calculation, be considered as made up of three sections: a cylinder of diameter d and length $2h$ and two spherical segments of height $R - h$ and chordal width d.

The volume of the cylinder and the two spherical segments is determined in terms of d, h and R, and is then set equal to that proportion of the total volume of the sphere by which the total weight of the sphere is to be reduced.

Formula: $d = D \sin \phi$; where $\phi = \text{arc} \cos \sqrt[3]{1 - \frac{w}{W}}$

Derivation of Formula: The height of each spherical segment is equal to $R - h$ and the radius of the base of each segment is $\frac{d}{2}$.

The volume V_{seg} of a spherical segment of height $R - h$ in a sphere of radius R is:

$$V_{seg} = \pi(R - h)^2\left(R - \frac{R - h}{3}\right) \tag{1a}$$

$$V_{seg} = \frac{\pi(R - h)^2\ (2R + h)}{3} \tag{1b}$$

The volume V_{cyl} of the cylinder which has a height of $2h$ and a diameter of d is:

$$V_{cyl} = \frac{\pi d^2}{4} \times 2h \tag{2a}$$

$$V_{cyl} = \frac{\pi d^2 h}{2} \tag{2b}$$

It can be seen from Fig. 5 that d, D and $2h$ form a right-angled triangle of which D is the hypotenuse, hence

$$d^2 = D^2 - 4h^2 = 4R^2 - 4h^2 \qquad \text{(3a and 3b)}$$

Then substituting for d^2 in Equation (2b):

$$V_{cyl} = \frac{4\pi h\ (R^2 - h^2)}{2} = 2\pi h\ (R^2 - h^2) \qquad \text{(4a and 4b)}$$

The volume V_{sph} of the sphere which has a radius R is:

$$V_{sph} = \tfrac{4}{3}\pi R^3 \tag{5}$$

If w is the weight of the sphere to be removed by drilling the hole of diameter d and W is the total weight of the sphere, the volume of the cylinder plus the two segments is equal to $\frac{w}{W}$ times the volume of the sphere, thus:

$$2\pi h(R^2 - h^2) + 2\,\frac{\pi(R - h)^2(2R + h)}{3} = \frac{w}{W}\left(\frac{4}{3}\,\pi R^3\right) \tag{6a}$$

Dividing both sides of Equation (6a) by 2π:

$$h(R^2 - h^2) + \frac{(R - h)^2(2R + h)}{3} = \frac{w}{W} \times \frac{2R^3}{3} \tag{6b}$$

Multiplying both sides of Equation (6b) by 3 and carrying out the multiplications of terms indicated:

$$3R^2h - 3h^3 + (R^2 - 2Rh + h^2)(2R + h) = \frac{w}{W} \times 2R^3 \tag{6c}$$

$$3R^2h - 3h^3 + 2R^3 - 4R^2h + 2Rh^2 + R^2h - 2Rh^2 + h^3 = \frac{w}{W} \times 2R^3 \quad (6d)$$

Combining and eliminating terms:

$$2R^3 - 2h^3 = 2R^3 \times \frac{w}{W} \quad (6e)$$

$$R^3 - h^3 = R^3 \times \frac{w}{W} \quad (6f)$$

Referring to Fig. 5,

$$h = R \cos \phi \quad (7)$$

Substituting for h in Equation (6f)

$$R^3 - R^3 \cos^3 \phi = R^3 \times \frac{w}{W} \quad (8a)$$

$$1 - \cos^3 \phi = \frac{w}{W} \quad (8b)$$

$$\cos^3 \phi = 1 - \frac{w}{W} \quad (8c)$$

$$\cos \phi = \sqrt[3]{1 - \frac{w}{W}} \quad (8d)$$

and

$$\phi = \text{arc cos} \sqrt[3]{1 - \frac{w}{W}} \quad (8e)$$

From Fig. 5, it can be seen that

$$d = D \sin \phi \quad (9)$$

Example: Find the diameter of a hole to be drilled through a sphere of 4.82-inch diameter to remove one-sixth of its weight.

Solution: $D = 4.82; \ \frac{w}{W} = \frac{1}{6}$

Then $\phi = \text{arc cos} \sqrt[3]{1 - \frac{1}{6}}$ (8d)

$\phi = \text{arc cos} \sqrt[3]{0.83333}$

$\phi = \text{arc cos } 0.94104 = 19°46.4'$

$d = 4.82 \sin 19°46.4'$ (9)

$d = 4.82 \times 0.33830 = 1.63$ inches

Finding Partial Volume of Horizontal Cylindrical Tank

PROBLEM 10: To find the volume V of water in a horizontal cylindrical tank, Fig. 6, of length l and radius r when the depth h of the water is known. The tank has flat ends.

Analysis of Problem: As can be seen from Fig. 6 the volume of water can be computed by first finding the area of its cross-section, which is in the form of a circular segment and then multiplying this by the length of the tank. The area of a circular segment can be found if the radius of its arc and the angle which intercepts this arc are known. In this case, the angle Θ can be found since it lies in right-angled triangle *mno* of which hypotenuse *mo* and side *no* are known.

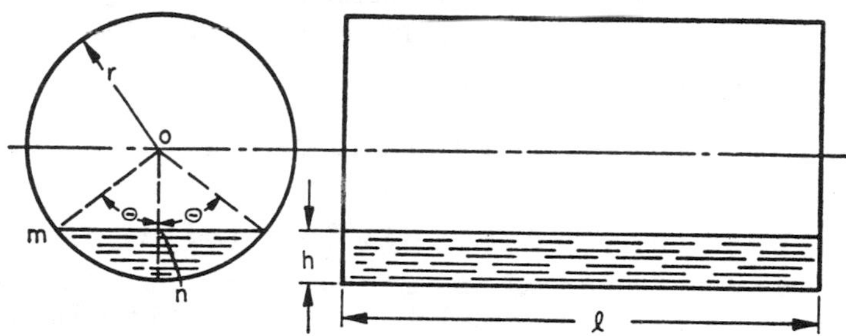

FIG. 6. To Find Volume V of Water in Cylindrical Tank of Length l and Radius r when Depth h is Known

Formula: $V = \frac{r^2}{2}(2\hat{\Theta} - \sin 2\Theta)l$, where $\hat{\Theta}$ is in radians

Derivation of Formula:

$$\cos \Theta = \frac{on}{om} = \frac{r - h}{r} \quad (1)$$

If A = area of the segment

$$A = \frac{r^2}{2}(2\hat{\Theta} - \sin 2\Theta), \text{ where } \hat{\Theta} \text{ is in radians} \quad (2)$$

but

$$V = A \times l \quad (3)$$

hence

$$V = \frac{r^2}{2}(2\hat{\Theta} - \sin 2\Theta)l \quad (4)$$

Example: Find the number of gallons of water in a cylindrical tank that is 10 feet long and 4 feet 6 inches in diameter when the depth of water in the tank is 15 inches.

Solution: $h = 15$ inches; $l = 10 \times 12 = 120$ inches; $r = \dfrac{4.5 \times 12}{2}$ $= 27$ inches

$$\cos\Theta = \frac{27 - 15}{27} = 0.4444$$

$$H = 63°37' = 1.1104 \text{ radians}$$

$$V = \frac{27^2}{2}(2 \times 1.1104 - \sin 127°14') \times 120$$

$$V = 364.5(2.2208 - 0.7962) \times 120$$

$$V = 364.5 \times 1.4246 \times 120 = 62{,}312 \text{ cubic inches}$$

Since 1 gallon = 231 cubic inches

$$V = \frac{62312}{231} = 269.7 \text{ or about 270 gallons}$$

Finding Partial Volume of a Conic Frustum

PROBLEM 11: Given a tank having the shape of a frustum of a cone as shown in cross-section through the center, Fig. 7, to determine the height x to which it would have to be filled with water if the required volume of water is $1/n$ of the volume of the tank.

Analysis of Problem: Extend the sides of the cone until they meet at the apex O. In this cross-sectional figure there are, then, three cones: one with a base equal to $2r$; one with a base equal to $2u$; and one with a base equal to $2R$. The volumes of these cones are to each other as the cubes of their base radii so that the volumes of the smallest and intermediate cones can be expressed in terms of the volume of the largest cone and the respective radii. Since the volume of water is equal to the volume of the largest cone minus the volume of the intermediate cone, and the volume of the tank is equal to the volume of the largest cone minus the volume of the smallest cone, an equation for the ratio of volume of water to volume of tank can be established in terms of the volume of the largest cone and the respective radii.

This equation can be set up so that u, the base radius of the intermediate cone is on one side. Another equation based on the similar triangles having $(u - r)$ and $(R - r)$ for a base, respectively, can be established in terms of u, r, R, h and x. This equation can also be set up so that u is on one side. The equivalents of u in each equation are then set equal to each other and the resulting equation is solved for x.

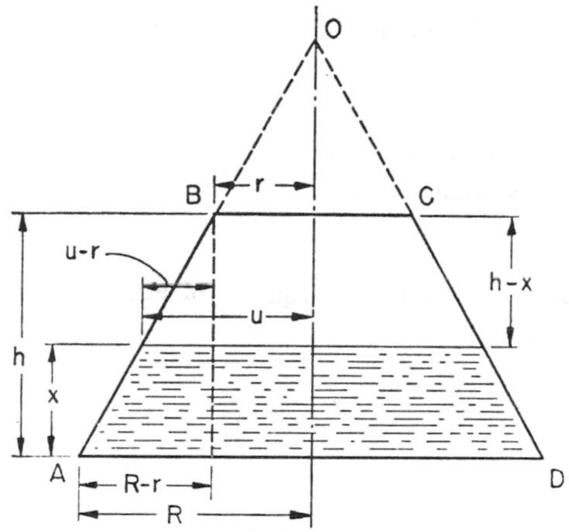

FIG. 7. To Determine Height x of Water, if Volume of Water is to be $\frac{1}{n}$ of Volume of Tank

Formula:
$$x = \frac{h\left(R - \sqrt{\frac{R^3\ (n-1) + r^3}{n}}\right)}{R - r}$$

Derivation of Formula: Let the volume of the three cones whose bases are circles having radii r, u and R be designated respectively by V_r, V_u and V_R.

Then
$$V_r : V_R = r^3 : R^3 \tag{1}$$

and
$$V_u : V_R = u^3 : R^3 \tag{2}$$

since corresponding volumes of cones having the same ratio of height to base radius vary as the cubes of their respective base radii.

Hence
$$V_r = \frac{r^3}{R^3}\ V_R \tag{3}$$

and
$$V_u = \frac{u^3}{R^3} V_R \tag{4}$$

$$\text{The volume of water} = V_R - V_u$$
$$\text{The volume of the tank} = V_R - V_r$$

Substituting for V_r and V_u their equivalents as given in Equations (3) and (4):

$$\frac{\text{Volume of water}}{\text{Volume of tank}} = \frac{V_R - \dfrac{u^3}{R^3} V_R}{V_R - \dfrac{r^3}{R^3} V_R} = \frac{1}{n} \tag{5}$$

Dividing numerator and denominator of Equation (5) by V_R:

$$\frac{\text{Volume of water}}{\text{Volume of tank}} = \frac{1 - \dfrac{u^3}{R^3}}{1 - \dfrac{r^3}{R^3}} = \frac{\dfrac{R^3 - u^3}{R^3}}{\dfrac{R^3 - r^3}{R^3}} = \frac{1}{n} \tag{6a}$$

$$\frac{\text{Volume of water}}{\text{Volume of tank}} = \frac{R^3 - u^3}{R^3 - r^3} = \frac{1}{n} \tag{6b}$$

$$nR^3 - nu^3 = R^3 - r^3 \tag{6c}$$

$$u^3 = \frac{R^3(n-1) + r^3}{n} \tag{6d}$$

$$u = \sqrt[3]{\frac{R^3(n-1) + r^3}{n}} \tag{6e}$$

Referring to Fig. 7 it will be seen that according to the geometrical theorem for similar triangles:

$$\frac{u - r}{R - r} = \frac{h - x}{h} \tag{7a}$$

$$hu - hr = Rh - Rx - hr + rx \tag{7b}$$

$$hu = Rh - x(R - r) \tag{7c}$$

$$u = R - \frac{x}{h}(R - r) \tag{7d}$$

Now, equating Equations (6e) and (7d) and solving for x:

$$R - \frac{x}{h}(R - r) = \sqrt[3]{\frac{R^3(n-1) + r^3}{n}} \quad (8a)$$

$$x = \frac{h\left(R - \sqrt[3]{\frac{R^3(n-1) + r^3}{n}}\right)}{R - r} \quad (8b)$$

Example: Given a tank 8 feet in diameter at the top, 10 feet in diameter at the bottom and 6 feet high. What depth of water would be required to fill the tank one-third full?

Solution: $R = 5$; $r = 4$; $h = 6$; $n = 3$

$$x = \frac{6\left(5 - \sqrt[3]{\frac{(5)^3(3-1) + (4)^3}{3}}\right)}{5 - 4} \quad (8b)$$

$$x = 30 - 6\sqrt[3]{\frac{250 + 64}{3}}$$

$$x = 30 - 6 \times 4.71 = 1.74 \text{ ft. or } 20.9 \text{ inches}$$

Temperature Increase Required to Enlarge Diameter of Steel Band Given Amount

PROBLEM 12: A steel band is to be slipped over a steel cylinder of slightly larger diameter. If the inside diameter of the band at room temperature T_1 is designated as d, and the diameter of the cylinder at the same temperature, plus twice the required clearance, is designated as D, to what temperature T_2 must the band be heated if it is to be slipped over the cylinder? Let the linear coefficient of temperature expansion for steel be designated as k.

Analysis of Problem: The difference between the circumference of a circle having a diameter equal to the diameter of the cylinder plus twice the clearance, and the circumference of the steel band at room temperature is the increase in circumference or "length" of the band that must be effected by heating. If k is the increase in "length" per inch of "length" per degree Fahrenheit rise in temperature, then k times the circumference of the band gives the increase in circumference of the band for each degree Fahrenheit rise in temperature. Knowing this and the required increase in "length," the required rise in temperature is readily found.

Formula: $$T_2 = T_1 + \frac{1}{k}\left(\frac{D}{d} - 1\right)$$

Derivation of Formula:

Let x = increase in circumference of band required

y = increase in temperature (deg. F) required to produce x

z = increase in circumference of band for each degree F rise in temperature

$$x = \pi D - \pi d \tag{1}$$

$$z = k \times \pi d \tag{2}$$

$$y = \frac{x}{z} = \frac{\pi D - \pi d}{k\pi d} = \frac{1}{k}\left(\frac{D}{d} - 1\right) \tag{3}$$

$$T_2 = T_1 + y = T_1 + \frac{1}{k}\left(\frac{D}{d} - 1\right) \tag{4}$$

Example: A band of 7.285-inch inside diameter is to be slipped over a cylinder of 7.3125-inch diameter. If the room temperature is 70 degrees F, to what temperature must the band be heated to permit it to be slipped over the cylinder? Assume that a clearance between band and cylinder of 0.002 inch is called for and the coefficient of linear expansion of steel to be 0.00000633 inch per inch per degree F.

Solution: $T_1 = 70$; $D = 7.3125 + 0.004 = 7.3165$; $d = 7.25$; $k = 0.00000633$.

$$T_2 = 70 + \frac{1}{0.00000633}\left(\frac{7.3165}{7.285} - 1\right)$$

$$T_2 = 70 + \frac{1{,}000{,}000}{6.33}(1.0043 - 1)$$

$$T_2 = 70 + 157{,}978 \times 0.0043$$

$$T_2 = 70 + 680 = 750 \text{ degrees F.}$$

PROBLEM 13: In the previous problem, after the steel band is slipped over the steel cylinder and is cooled to room temperature, what average tensile stress, s, is induced in the band, assuming no reduction in size of the cylinder. Assume modulus of elasticity of the steel to be 30,000,000.

Analysis of Problem: If the band were free to shrink as it cools, it would return to its former size. Since it is now on the cylinder, which is assumed to be absolutely rigid, this shrinkage is prevented and a tensile stress is induced in the band. The ratio of this stress to the unit decrease in circumference that would occur if the band were free to shrink is equal to the modulus of elasticity of the steel. Since the length of the band and the amount of shrinkage are known, the unit decrease in circumference, in this case, the decrease per inch of length, can be computed, and since the modulus of elasticity is known, the induced tensile stress can be computed.

Formula: $$s = E\left(1 - \frac{d}{D}\right)$$

Derivation of Formula:

Let x' = total decrease in circumference
w = unit decrease in circumference
s = unit tensile stress induced by shrinkage
D' = diameter of cylinder
E = modulus of elasticity

and x, and d as in the previous problem

$$x' = x = \pi D' - \pi d \qquad (1)$$

$$w = \frac{\pi D' - \pi d}{\pi D'} = 1 - \frac{d}{D'} \qquad (2)$$

$$s = E \times w \qquad (3)$$

$$s = E\left(1 - \frac{d}{D'}\right)$$

Example: If the band mentioned in the example in Problem 12 is cooled back to room temperature after being slipped over the cylinder, what tensile stress is induced in it?

Solution: $E = 30{,}000{,}000$; $d = 7.285$; $D' = 7.3125$

$$s = 30{,}000{,}000\left(1 - \frac{7.285}{7.3125}\right)$$

$s = 30{,}000{,}000 \times 0.00376 = 112{,}800$ pounds per square inch

Figuring Metal Expansion Due to Increase in Temperature

PROBLEM 14: How many degrees Fahrenheit t must the temperature of a steel ring having a tapered bore be increased so that it can be slipped with clearance c over the large end of a tapered steel seat on which it is to be shrink fitted. The large and small diameters D_1 and D_2 of the tapered seat are known, the allowance a for the shrinkage fit of the ring on the seat, and the thermal coefficient of expansion k of the ring material.

Analysis of Problem: The first step is to determine the diameter d of the small end of the ring bore. This is equal to the small diameter D_2 of the seat minus the product of this diameter and the shrinkage allowance a.

The next step is to determine how much the small diameter of the ring must be increased by heating to permit it to be slipped easily over the large end of the seat. The difference between the large diameter of the seat and the small diameter of the bore is the minimum increase in diameter of the ring bore that must be effected by heating. To this should be added some clearance c. The total increase in diameter required divided by the product of the diameter of the small end of the ring bore before heating and the coefficient of thermal expansion will give the temperature rise required.

Formula:

$$t = \frac{D_1 - D_2(1 - a) + c}{D_2(1 - a)k}$$

Derivation of Formula:

$$d = D_2 - D_2 \times a = D_2(1 - a) \tag{1}$$

Let x = minimum increase in small diameter of ring bore required

$$x = D_1 - d \tag{2}$$

Then total increase in small diameter of ring bore required is $x + c$ or

$$x + c = D_1 - d + c \tag{3}$$

Let t = increase in temperature in deg. F required to cause desired increase in diameter of bore.

$$t = \frac{D_1 - d + c}{d \times k} \tag{4}$$

Substituting the equivalent of d from Equation (1)

$$t = \frac{D_1 - D_2(1 - a) + c}{D_2(1 - a)k} \qquad (5)$$

Example: How many degrees F must the temperature of a steel ring having a tapered bore be increased, if the large diameter of the steel seat over which it is to be fitted is 15.750 inches, the small diameter of the seat is 15.715 inches, the allowance for shrinkage fit is to be 0.001 inch per inch diameter and the clearance between the small diameter of the bore of the heated ring and the large diameter of the seat is to be 0.03 inch. Assume a coefficient of thermal expansion for steel of 0.0000064 inch per inch.

Solution: $D_1 = 15.750$ inches; $D_2 = 15.715$ inches; $a = 0.001$ inch per inch; $c = 0.03$ inch; and $k = 0.0000064$ inch per inch.

$$t = \frac{15.750 - 15.715(1 - 0.001) + 0.03}{15.715(1 - 0.001)0.0000064} \qquad (5)$$

$$t = \frac{15.750 - 15.699 + 0.030}{15.699 \times 0.0000064}$$

Using the powers of ten system of notation explained in Chapter 23

$$t = \frac{8.1 \times 10 - 2}{1.0 \times 10 - 4} = 8.1 \times 10^2 = 810$$

The temperature of the ring should be increased by about 810 degrees Fahrenheit.

Combined Production Rate for Several Operations at Different Rates

PROBLEM 15: To determine the rate of production in terms of x pieces per hour of a part on which several operations are being performed at the rate of a, b, c, d pieces per hour respectively.

Analysis of Problem: Let it be assumed that the part has a drilling operation performed on it at the rate of a pieces per hour and a milling operation at the rate of b pieces per hour. It takes $\frac{1}{a}$ hour to drill a piece and $\frac{1}{b}$ hour to mill it. The total time to complete the piece would then be $\frac{1}{a} + \frac{1}{b} = \frac{a + b}{ab}$ hour.

Now let it be supposed that a third operation of grinding is performed at the rate of c pieces per hour. The time required to perform this operation on one piece would be $\frac{1}{c}$ hour and the total time required to complete the piece would now be

$$\frac{a+b}{ab}+\frac{1}{c}=\frac{ac+bc+ab}{abc} \text{ hour per piece}$$

Finally let it be assumed that a fourth operation of superfinishing is performed at the rate of d pieces per hour. The time required to complete this operation on one piece would be $\frac{1}{d}$ hour and the total time required to complete the piece would now be

$$\frac{ac+bc+ab}{abc}+\frac{1}{d}=\frac{acd+bcd+abd+abc}{abcd}$$

It can now be seen by inspection that the total time required to produce a piece upon which several operations are performed at different rates a, b, c, d, etc., pieces per hour is equal to the sum of all possible products of $(n-1)$ rates, if the total number of rates is n, divided by the product of all the rates.

When the total time to produce one piece is known the rate of production is equal to the reciprocal of this figure. For the case just analyzed, the rate of production is $\frac{abcd}{acd+bcd+abd+abc}$.

Formula: $$\frac{abcd}{acd+bcd+abd+abc} \quad \text{or} \quad \frac{abcd}{abc+bcd+acd+abd}$$

Example: Find the combined rate of production x of parts on which a drilling operation is performed at the rate of 225 pieces per hour, a milling operation at the rate of 140 pieces per hour, a grinding operation at the rate of 430 pieces per hour and a superfinishing operation at the rate of 370 pieces per hour, neglecting waiting time between operations.

Solution: $a = 225$; $b = 140$; $c = 430$; $d = 370$

$$x = \frac{225 \times 140 \times 430 \times 370}{225 \times 430 \times 370 + 140 \times 430 \times 370 + 225 \times 140 \times 370 + 225 \times 140 \times 430}$$

$$x = \frac{5{,}011{,}650{,}000}{35{,}797{,}500 + 22{,}274{,}000 + 11{,}655{,}000 + 13{,}545{,}000}.$$

$$x = \frac{5{,}011{,}650{,}000}{83{,}271{,}500} = 60 \text{ pieces per hour, approximately.}$$

Computing Series of Speeds in Geometric Progression

PROBLEM 16: A series of speeds to be taken from a set of change gears are to be in geometric progression and the lowest speed a the highest speed l and the number of steps n are known. Find the successive speeds.

Analysis of Problem: In order to find each step in the series of speeds it is necessary to determine the ratio r for this series. On page 2–29 is given a table of formulas for geometric series. In this table will be found a formula for the ratio r when the first term a, last term l, and number of terms n are known. With the ratio known, the second term or speed is found by multiplying the first or lowest speed by the ratio, the third is then found by multiplying the second by this ratio and so on.

Formula:

$$r = \sqrt[n-1]{\frac{l}{a}}$$

Example: Find the successive speeds to be delivered by a set of change gears, if the lowest speed is to be 120 R.P.M., the highest speed is to be 1650 R.P.M. and the number of steps is to be 6.

Solution: $a = 120;\ l = 1650;\ n = 6$

$$r = \sqrt[6-1]{\frac{1650}{120}} = \sqrt[5]{\frac{165}{12}} = \sqrt[5]{13.75}$$

This calculation is performed by the use of logarithms:

$$\log 13.75 = 1.13830$$

$$\log \sqrt[5]{13.75} = \frac{1.13830}{5} = 0.22766$$

$$\sqrt[5]{13.75} = 1.689 \text{ say } 1.69$$

Then 2nd speed $= 120 \times 1.69 = 202.80$, say 203 R.P.M.

3rd speed $= 203 \times 1.69 = 343.07$, say 343 R.P.M.

4th speed $= 343 \times 1.69 = 579.67$, say 580 R.P.M.
5th speed $= 580 \times 1.69 = 980.20$, say 980 R.P.M.

If the sixth or final speed is to be 1650 R.P.M., the ratio of the sixth speed to the fifth is:

$$\frac{1650}{980} = 1.684$$

or only slightly less than the required ratio of 1.69.

Solving Quadratic Equation having Multiple Terms

PROBLEM 17: Given two pairs of cone pulleys, as shown in Fig. 8, in which the diameters *d*, *e*, *f* and *g* of the steps of the smaller pulley are known as well as diameter *D* of the largest step of the larger pulley. The distance *S* between the shaft centers is also known. It is required to find the diameters *E*, *F* and *G* of the remaining steps on the larger pulley which will result in the belt tension being the same for all steps.

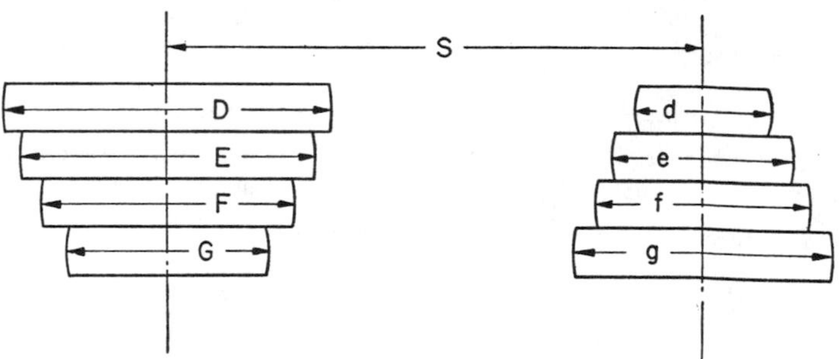

FIG. 8. To Find Pulley Diameters *E*, *F* and *G* when Pulley Diameters *D* and *d*, *e*, *f*, *g*, as well as Center Distance *S* are Known

Analysis of Problem: The belt tension will remain constant for each pair of steps of the pulleys if they are of such size that the proper length of belt for each pair of steps is the same. The first step is, then, to set up the formula for length of belt required to pass over steps *D* and *d*, the diameters of which are known. From this formula can be derived a general formula for finding the diameter of any unknown step on either pulley that will cause the belt tension to be the same.

Formulas: $E = \sqrt{(\pi S + D + d)^2 - 4(Dd + \pi Se)} - \pi S + e$

$$F = \sqrt{(\pi S + D + d)^2 - 4(Dd + \pi Sf)} - \pi S + f$$

$$G = \sqrt{(\pi S + D + d)^2 - 4(Dd + \pi Sg)} - \pi S + g$$

Derivation of Formulas: If L is the length of the belt passing over two pulleys having respective diameters D and d and at a center distance S apart, then, as was shown in Problem 5 of this chapter:

$$L = 2S + \frac{\pi}{2}(D + d) + \frac{(D - d)^2}{4S} \qquad (1)$$

If X is the diameter of any step on the larger pulley and x is the diameter of the corresponding step on the smaller pulley, then:

$$L = 2S + \frac{\pi}{2}(X + x) + \frac{(X - x)^2}{4S} \qquad (2)$$

Equating the right-hand sides of Equations (1) and (2):

$$2S + \frac{\pi}{2}(D + d) + \frac{(D - d)^2}{4S} = 2S + \frac{\pi}{2}(X + x) + \frac{(X - x)^2}{4S} \qquad (3a)$$

Clearing of fractions by multiplying each term by $4S$ and cancelling out the first term of each side of the equation:

$$2\pi S(D + d) + (D - d)^2 = 2\pi S(X + x) + (X - x)^2 \qquad (3b)$$

Subtracting $4\pi Sx$ from both sides of the equation:

$$2\pi S(D + d) + (D - d)^2 - 4\pi Sx = 2\pi S(X + x) - 4\pi Sx + (X - x)^2 \qquad (3c)$$

Combining the Sx terms on the right-hand side of the equation:

$$2\pi S(D + d) + (D - d)^2 - 4\pi Sx = 2\pi S(X - x) + (X - x)^2 \qquad (3d)$$

Rearranging in terms of descending powers of $(X - x)$ and placing all terms on the right-hand side of the equation:

$$0 = (X - x)^2 + 2\pi S(X - x) + 4\pi Sx - 2\pi S(D + d) - (D - d)^2 \qquad (3e)$$

This is a quadratic of the form of $(X - x)^2 + b(X - x) + c = 0$ in which $a = 1$; $b = 2\pi S$; and $c = 4\pi Sx - 2\pi S(D + d) - (D - d)^2$.

The formula (see page **2**–20) for solving a quadratic in this form is:

$$(X - x) = \frac{-b \pm \sqrt{b^2 - 4ac}}{2a} \tag{4}$$

Substituting the equivalents of a, b and c in Equation (4) and taking the plus sign before the radical:

$$(X - x) = \frac{-2\pi S + \sqrt{4\pi^2 S^2 - 16\pi Sx + 8\pi S(D + d) + 4(D - d)^2}}{2} \tag{5a}$$

$$(X - x) = -\pi S + \sqrt{\pi^2 S^2 - 4\pi Sx + 2\pi S(D + d) + (D - d)^2} \tag{5b}$$

That part of Equation (5b) under the radical can be rearranged for easier handling. Let it be considered by itself for the moment. First expand the term $(D - d)^2$:

$$\pi^2 S^2 - 4\pi Sx + 2\pi S(D + d) + (D^2 - 2Dd + d^2)$$

Next add $4Dd$ to the last term and establish a new term of $-4Dd$ to balance this addition:

$$\pi^2 S^2 - 4\pi Sx + 2\pi S(D + d) + (D^2 + 2Dd + d^2) - 4Dd$$

If the second term is now placed last and the last term re-expressed as $(D + d)^2$ is placed third:

$$\pi^2 S^2 + 2\pi S(D + d) + (D + d)^2 - 4Dd - 4\pi Sx$$

It will now be seen that the first three terms constitute a perfect square which may be written: $[\pi S + (D + d)]^2$ or $(\pi S + D + d)^2$ so that the entire expression under the radical in Equation (5b) becomes

$$X - x = -\pi S + \sqrt{(\pi S + D + d)^2 - 4(Dd + \pi Sx)} \tag{5c}$$

or

$$X = \sqrt{(\pi S + D + d)^2 - 4(Dd + \pi Sx)} - \pi S + x \tag{5d}$$

Now if X is E and $x = e$:

$$E = \sqrt{(\pi S + D + d)^2 - 4(Dd + \pi Se)} - \pi S + e \tag{6}$$

and if X is F and $x = f$:

$$F = \sqrt{(\pi S + D + d)^2 - 4(Dd + \pi Sf)} - \pi S + f \tag{7}$$

and if X is G and $x = g$:

$$G = \sqrt{(\pi S + D + d)^2 - 4(Dd + \pi Sg) - \pi S + g} \tag{8}$$

Example: The diameters of the steps of the smaller pulley are 5, 6.25, 8, and 10 inches, respectively, and the largest diameter of the larger pulley is 12.5 inches. The distance between shaft centers is 24 inches. Find the diameters of the other three steps of the larger pulley.

Solution: $d = 5$; $e = 6.25$; $f = 8$; $g = 10$; $D = 12.5$ and $S = 24$

$$E = \sqrt{(\pi \times 24 + 12.5 + 5)^2 - 4(12.5 \times 5 + \pi \times 24 \times 6.25)} - \pi \times 24 + 6.25$$

$$E = \sqrt{8630.40 - 2135.00} - 75.40 + 6.25$$

$$E = 80.59 - 75.40 + 6.25 = 11.44 \text{ inches}$$

$$F = \sqrt{(\pi \times 24 + 12.5 + 5)^2 - 4(12.5 \times 5 + \pi \times 24 \times 8)} - \pi \times 24 + 8$$

$$F = \sqrt{8630.40 - 2662.8} - 75.40 + 8$$

$$F = 77.25 - 75.40 + 8 = 9.85 \text{ inches}$$

$$G = \sqrt{(\pi \times 24 + 12.5 + 5)^2 - 4(12.5 \times 5 + \pi \times 24 \times 10)} - \pi \times 24 + 10$$

$$G = \sqrt{8630.40 - 3266.00} - 75.40 + 10$$

$$G = 73.24 - 75.40 + 10 = 7.84 \text{ inches}$$

Every Specification on a Drawing isn't Gospel

The drawing for a precision inspection fixture specified the locations of three holes by means of two dimensions and an angle, as shown in Fig. 9. In making this fixture, the toolmaker used the 1.7500- and 6.5625-

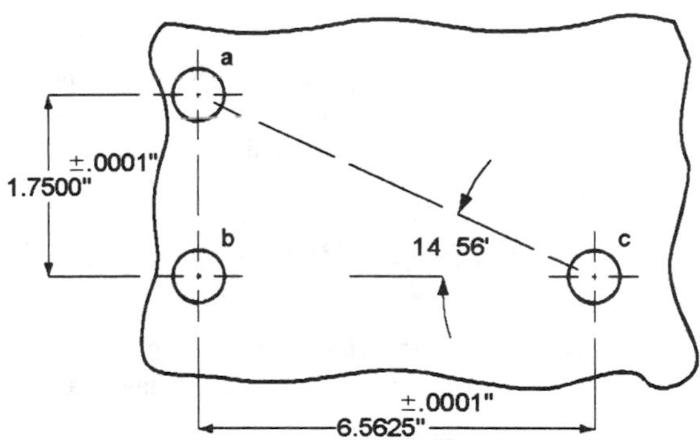

Fig. 9. Locations of holes (a) and (c) relative to hole (b) are sufficiently specified by the coordinate dimensions. Showing the angle is unnecessary and in this example, led to confusion.

inch dimensions shown as coordinates in boring holes *a* and *c* relative to hole *b*. As a final check on their locations, the distance between holes *a* and *c* was measured, using plugs inserted into the holes. The center distance *ac* was found to be 6.7918 inches, 0.0007 inches more than it should have been according to the following calculations:

$$ac = \frac{1.7500}{\sin 14° 56'} = \frac{1.7500}{0.25769} = 6.7911 \text{ inches}$$

Being aware of the accuracy of a precision jig-boring operation, the toolmaker naturally assumed that the value for sin 14° 56′ used in the computation was in error. He therefore recomputed *ac* using the cosine of 14° 56′: Because this calculated distance was within 0.0001 inch of the measured value, he concluded that the value for sin 14° 56′ previously used was given incorrectly in the trigonometric table from which it had been taken. However, the value was not in error. He therefore recomputed *ac* using the cosine of 14° 56′:

$$ac = \frac{6.5625}{\cos 14° 56'} = \frac{6.5625}{0.96623} = 6.7919 \text{ inches}$$

This calculated distance was within 0.0001 inch of the measured value, so he concluded that the value for sin 14° 56′ previously used was given incorrectly in the trigonometric table from which it was taken. However, the value was not in error.

Solution: The root of the trouble is the 14° 56′ angle shown on the drawing. This angle should have been 14° 55.9′ which is the angle corresponding to a right-angled triangle with sides measuring 1.7500 and 6.5625 inches long.:

$$\frac{1.7500}{6.5625} = 0.26667 = \tan 14° 55.9' \text{ by interpolation}$$

The angle need not have been shown on the drawing at all because the angle could have been determined easily by the Pythagorean Theorem:

$$ac = \sqrt{1.7500^2 + 6.5625^2} = 6.7918$$

Although the 14° 56′ angle on the drawing was incorrect, it should be noted that use of the cosine of this angle in the second computation produced an almost-correct result, whereas use of the sine value produced a considerable error. This result can be explained simply by reference to a trigonometric table, where it will be seen that, for small angles, the cosine of an angle changes at a slower rate than the sine of the angle. For example, the difference between sin 14° 55.9′ and sin 14° 56′ is only 0.00003, and the difference between cos 14° 55.9′ and cos 14° 56′ is only 0.00001.

The conclusion to be drawn from this observation is that, when calculations involving sine and cosine are to be made, it is preferable to use the cosine function for angles up to 45 degrees and the sine function for angles over 45 degrees.

26

Refresher Questions About Mathematics and Mechanics

In the course of correspondence with men working with mechanical problems, it has been found that certain terms used in mathematics, mechanics and strength of materials are frequently a source of misunderstanding and difficulty. A number of questions have been asked and answered in this chapter to aid in clarifying the proper use of these terms in connection with the solving of mechanical problems. A brief review of these questions should be helpful to all who are working in this field.

Due to the diverse nature of the questions, it has not been possible to arrange them in a completely logical order. However, questions relating to each other have been grouped together and, in general, it will be found that those in the first half of the chapter relate to mathematics and those in the last half to mechanics and strength of materials.

Question: What is the difference between a *supplementary* and a *complementary* angle?

Answer: Two angles which together equal 180 degrees are called supplementary angles and each is said to be the supplement of the other. Two angles which together equal 90 degrees are called complementary angles and each is said to be the complement of the other.

Question: What is a *radian*?

Answer: The radian is a unit of linear measure for angles and indicates the proportionate length of arc subtended by the angle. By proportionate length is meant the ratio of the length of the arc to the radius of the circle of which it is a part. The proportionate length of arc is used because if the *actual* length of an arc subtended by a given angle were used, it might be any value, depending upon the radius of the circle. The proportionate length, however, is always a constant value for a given angle. If a given angle subtended an arc that was just equal in length to its radius R, the angle would then be equal to 1 radian, since $R \div R = 1$. The circumference of a circle is equal to $2\pi R$ so that there are $2\pi R \div R$ or 2π radians in a complete circle. Since 2π radians equals 360 degrees, 1 degree = $2\pi \div 360 = 0.0174532925+$ radian.

Question: How is an angle expressed in degrees converted to radians?

Answer: Conversion tables giving the equivalents of degrees, minutes and seconds in terms of radians are given in most handbooks. When such a table is not readily available, the angle is first converted to degrees and decimal of a degree and is then multiplied by 0.0174532925+ expressed to as many significant figures as accuracy requires.

Example: Find the equivalent of 16 degrees, 34 minutes and 28 seconds in terms of radians.

Solution:

$$16°34'28'' = 16°2068'' = 16\,\frac{2068°}{3600} = 16.5744°$$

$$16.5744 \times 0.0174533 = 0.289278 \text{ radian}$$

Question: When are angles expressed in radians?

Answer: There are many instances when it is more suitable to express an angle in radians rather than in degrees. For example the angular velocity of a rotating body is expressed in angular measure and equals the angle through which any point other than the center of rotation turns in one second. This angle is usually expressed in radians.

Another instance is in the computation of the area of a sector of a circle. The area is equal to the product of one-half the length of the arc by the radius. Since the length of an arc of radius R subtending an angle A is equal to RA when A is measured in radians, then the area of the sector would be $\frac{1}{2}R^2A$.

Question: What is the difference between the sexigesimal and circular systems of angular measure?

Answer: In the sexigesimal system the circumference of a circle is divided into 360 equal parts called degrees; so that the angle which each part subtends is equal to a degree. Each degree is divided into

60 equal parts called minutes and each minute is divided into 60 equal parts called seconds.

In the circular system of measure the unit is the *radian.* If, instead of dividing the circumference of a circle into 360 equal parts, its radius is laid out around the circumference, the angle subtended by the arc equal in length to the radius is called a radian. There are 2π radians in a complete circle or, in other words, 2π radians equal 360 degrees.

Question: What is π and how is its value expressed?

Answer: Pi or π is the ratio of the circumference of any circle to its diameter. Pi is an incommensurable number, that is, it cannot be expressed exactly by any arithmetical number. Four approximate values are 22/7; 3.1416; 335/113; 3.14159265. Each of the three latter values is more accurate than any preceding it.

Question: What is the involute of a circle?

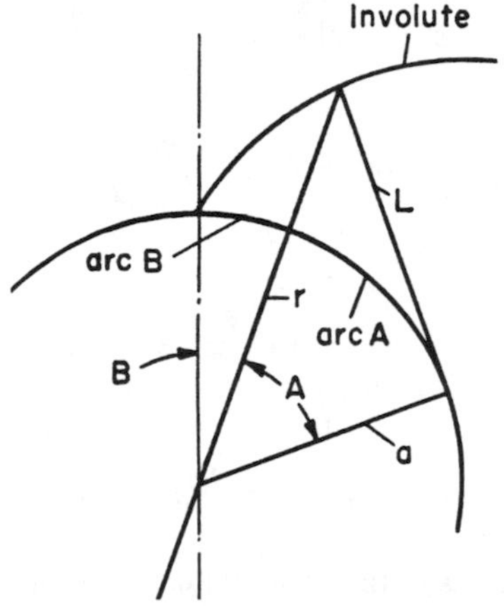

FIG. 1. Diagram Showing Relationship of Involute A to Tangent A and Arc A

Answer: The involute of a circle is the curve that is described by the end of a line L which is unwound from the circumference of the circle as shown in Fig. 1.

Question: What is the involute function of an angle?

Answer: In Fig. 1 the angle $(A + B)$ is subtended by that part of the base circle from which line L has been unwound. The involute of angle A is defined as equal to angle B in radians, and since

$$\text{angle } B \text{ in radians} = \frac{\text{arc } B}{a} = \frac{L - \text{arc } A}{a} = \frac{L}{a} - \frac{\text{arc } A}{a}$$

then, $$\text{inv } A = \frac{L}{a} - \frac{\text{arc } A}{a} = \tan A - \text{angle } A \text{ in radians}$$

Question: Where are involute functions used?

Answer: Involute functions are used frequently in computing certain dimensions of gears having teeth with an involute profile, as for example when measurements are taken over rolls placed between the gear teeth to check the arc tooth thickness of the gear teeth.

Question: When two sides a and b and the included angle C of an oblique-angled triangle are known and the third side c is to be found, the formula $c = \sqrt{a^2 + b^2 - 2ab \cos C}$ is commonly used. Can this formula be used regardless of whether the given angle C is greater or less than 90 degrees?

Answer: This formula can be used regardless of the size of the given angle C, since when it is less than 90 degrees, its cosine is positive and hence the expression $2ab \cos C$ remains negative and is *subtracted* from $a^2 + b^2$; and when the given angle is greater than 90 degrees, its cosine is negative and hence the expression $2ab \cos C$ becomes positive and is *added* to $a^2 + b^2$.

Example: Find the third side of an oblique-angled triangle if one side is 7, the second side is 9 and the angle between them is 30 degrees. If the angle between them is 150 degrees.

Solution: $a = 7;\ b = 9;\ C = 30°;\ \cos 30° = 0.86603$

$$c = \sqrt{49 + 81 - 2 \times 7 \times 9 \times 0.86603}$$

$$c = \sqrt{20.88} = 4.57$$

$a = 7;\ b = 9;\ C = 150°;\ \cos 150° = -0.86603$

$$c = \sqrt{49 + 81 + 2 \times 7 \times 9 \times 0.86603}$$

$$c = \sqrt{239.12} = 15.46$$

Question: What does the term *arc sin* mean?

Answer: The term arc sin is simply a "shorthand" way of writing "The angle for which the sine equals . . ." Thus arc sin $\frac{a}{b}$ means the angle for which the sine is $\frac{a}{b}$. Other similar designations are *arc* *cos*, *arc tan*, *arc cot*, *arc sec* and *arc cosec*. These designations are also written sin^{-1}, cos^{-1}, tan^{-1}, cot^{-1}, sec^{-1}, $cosec^{-1}$ and have correspondingly the same meaning. Thus, $\cos^{-1} 0.86603$ = arc cos 0.86603 = the angle for which the cosine is 0.86603 = 30 degrees.

Question: What are the *versed sine* and *coversed sine* and where are they used?

Answer: The versed sine of an angle, sometimes written *vers*, equals 1 minus its cosine and, similarly, the coversed sine of an angle, sometimes written *covers*, equals 1 minus its sine. These two trigonometric functions are shown in Fig. 2 which is the customary diagram for showing the more commonly used functions, such as sine, cosine, etc.

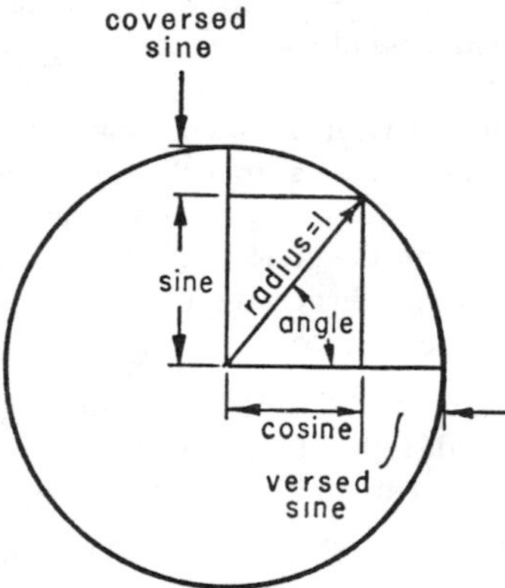

FIG. 2. Diagram Showing Versed Sine and Coversed Sine of an Angle

In certain types of problems, the dimension represented by either of these functions might be obtained for any given angle by multiplying the value found in tables of these functions by the known radius. As an illustration, suppose the height H of an arc as shown in Fig. 3 was required.

This height might be found by using the following formula, assuming that dimensions R and C are known:

$$H = R - \sqrt{R^2 - (C \div 2)^2}$$

A simpler and easier method would be to find the versed sine of angle a in a table and multiply it by the radius R.

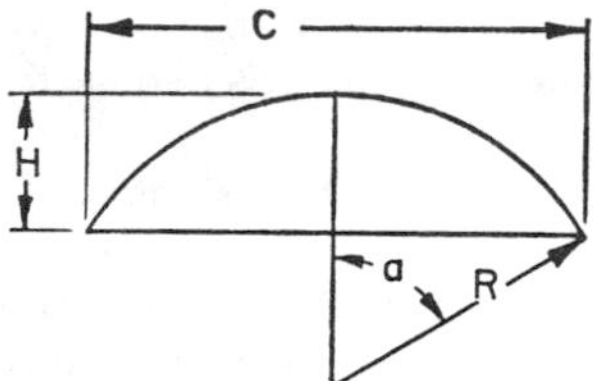

FIG. 3. Versed Sine of Angle *a* when Multiplied by Radius *R* Gives Height *H*

Question: What are *coordinates*?

Answer: In drawing a chart having two reference axes at right angles to each other, the location of any point on this chart may be designated by the respective distances of this point from the two axes when measured at right angles to them. These two distances are called the coordinates of the point. In plane analytic geometry, the horizontal reference axis is called the *x-axis* and the vertical axis the *y-axis*. Similarly, the coordinate measured from the *y-axis*, and parallel to the *x-axis* is called the *x*-coordinate or *abscissa* and the coordinate measured from the *x*-axis and parallel to the *y*-axis is called the *y*-coordinate or *ordinate*. In solid analytic geometry there are three reference axes, the third axis being at right angles to the *x*- and *y*-axes and is called the *z*-axis. Similarly, the coordinate measured from the plane of the *x*- and *y*-axes and parallel to the *z*-axis is called the *z*-coordinate.

Question: What are the differences between a *cycloid*, an *epicycloid* and a *hypocycloid*?

Answer: A cycloid is a geometrical curve which is traced by a point located on the periphery of a circle as the circle rolls along a straight line. The epicycloid is a curve traced by a point located on the periphery of a circle as it rolls on the outside of another circle. The hypocycloid is a curve traced by a point located on the periphery of a circle as it rolls along the inside of a larger circle.

Question: What are the differences between an *ellipse*, a *hyperbola* and a *parabola*?

Answer: An ellipse is a plane curve, any point on which has the *sum* of its distances from two fixed points (called the focii) equal to a constant.

A hyperbola is a plane curve, any point on which has the *difference* of its distances from two fixed points (called the focii) equal to a constant.

A parabola is a plane curve, any point on which has its distance from a fixed line (called the directrix) equal to its distance from a fixed point (called the focus).

Question: What is meant by the *generation* of a line, a surface or a solid?

Answer: The meaning of generation, as so used, is *the tracing out* of a line, a surface or a solid by the movement of a point, a line or a surface, respectively. Thus, the path of a moving point is a line; the path of a moving line is, in general, a surface; and the path of a moving surface is, in general, a solid.

Question: What is the difference between an *arithmetical mean* and a *geometric mean*?

Answer: The arithmetical mean of two numbers is equal to one-half their sum; the geometric mean is equal to the square root of their product.

Question: What is the difference between an *arithmetical average* and a *root-mean-square* average?

Answer: An arithmetical average of a number of measurements, for example, is equal to their sum divided by the number of measurements. Thus, the arithmetical average of 4, 6, 3 and 7 inches is

$$(4 + 6 + 3 + 7) \div 4 = 5 \text{ inches}$$

The root-mean-square average of a number of measurements, for example, is equal to the square root of the quotient obtained by squaring each measurement, adding the squares and dividing by the number of measurements. Thus, the root-mean-square average of 4, 6, 3 and 7 inches is:

$$\text{Root-mean-square average} = \sqrt{\frac{4^2 + 6^2 + 3^2 + 7^2}{4}} = 5.244$$

The root-mean-square average tends to give greater emphasis to the high numbers in the group being averaged, so that if there is considerable difference in the size of the numbers in the group, the root-mean-square average will result in an appreciably higher value than the arithmetical average.

An example of the use of the root-mean-square average, where it is desired to give emphasis to the higher values being averaged, is in indicating the degree of surface roughness by an average of a given number of surface deviations from a mean reference line. By using the root-mean-square average, greater weight is given to the larger surface deviations than to the smaller ones.

Question: What is meant by the statement that a number is expressed "to so many significant figures?"

Answer: The significant figures in any number are those which are not in doubt. In common practice they are taken to be those beginning with the first figure other than zero and ending with the last figure other than zero. In the number 436,090, for example, there are five significant figures, beginning with the 4 and ending with the 9. Similarly in the number 0.0040003 there are five significant figures beginning with the 4 and ending with the 3. In the first case, if the last zero were not in doubt, then it would be a significant figure.

When the statement is made that a number is accurate to five significant figures, it means that the error present is not any greater than the value expressed by 5 in the next place after the last significant figure. Thus 8931.2 is accurate within 5 in the hundredths place, or ± 0.05 and 4,327,100 would be taken as accurate to within 5 in the tens (not tenths) place or ± 50 unless it were made clear that the first, or first and second zeros were also significant, that is, not in doubt. The error would then be ± 5 or ± 0.5, respectively.

Question: What is the difference between an *arithmetical* and a *geometrical progression*?

Answer: An arithmetical progression is a series of numbers in which each term is obtained by adding or subtracting a given quantity to or from the next preceding term.

A geometrical progression is a series of numbers in which each succeeding term is produced by multiplying the preceding term by a given factor.

Question: What is the difference between a *permutation* and a *combination*?

Answer: A permutation of a group of things is any arrangement of any number of them in a definite order. If n is the number of things in the group and r is the number of them arranged in a definite order at one time, the total number of permutations P of n things, taken r at a time, is given by the formula:

$$_nP_r = n(n-1)(n-2)(n-3)\cdots(n-r+1)$$

or

$$_nP_r = \frac{n\,!}{(n-r)\,!}$$

A combination of a group of things is any number of them taken without regard to their order. If n is the number of things in the group and r is the number taken in a combination at one time, then the total number of combinations C of n things taken r at a time is given by the formula:

$$_nC_r = \frac{n(n-1)(n-2)(n-3)\cdots(n-r+1)}{r(r-1)(r-2)\cdots(1)} \quad \text{or} \quad _nC_r = \frac{_nP_r}{r!}$$

Thus it can be seen that the number of combinations of n things taken r at a time is equal to the number of permutations of n things arranged r at a time divided by factorial r.

Question: What does *interpolation* mean?

Answer: The noun interpolation comes from the verb "to interpolate" which means "to insert between." It is used commonly in the sense of interpolating or inserting a value between two values given in a table such as a table of logarithms or a table of trigonometrical functions.

For example, suppose it were required to find the value of sine 4 degrees, 42 minutes and 16 seconds. Reference to a table of natural trigonometrical functions shows that angles are given in it only to the nearest minute. It is necessary, therefore, to interpolate to find the desired value. Since 4 degrees, 42 minutes and 16 seconds lies between 4 degrees, 42 minutes and 4 degrees, 43 minutes, the sine of 4 degrees, 42 minutes and 16 seconds lies between the sine of 4 degrees, 42 minutes and the sine of 4 degrees, 43 minutes, both of which are given in the table.

There is a difference of 60 seconds between 4 degrees, 42 minutes and 4 degrees, 43 minutes. There is a difference of 16 seconds between the given angle (4 degrees, 42 minutes, 16 seconds) and 4 degrees, 42 minutes. Hence it may be said that the given angle is $\frac{16}{60}$ of the way from angle $4°42'$ to angle $4°43'$. It is assumed, without appreciable error, that the sine of the given angle also lies $\frac{16}{60}$ of the way from the sine of $4°42'$ to the sine of $4°43'$. The interpolation is made on this assumption, which is, in fact, that the sine of an angle varies directly as the numerical value of the angle in degrees, minutes and seconds. Actually it does not, but for small intervals of less than one minute and for sine values given to five significant figures, no appreciable error is introduced by interpolating on this basis.

Question: If a wooden block weighing 50 pounds were moved 10 feet along the surface of a smooth iron plate would the amount of work done be 50 × 10 or 500 foot-pounds?

Answer: The work done would be the product of the force required to move the block and the distance through which it was moved. The force required to move the block depends upon the frictional resistance between the block and the plate upon which it rests. This frictional resistance is expressed as a ratio of the horizontal force required to move the block divided by the vertical force exerted by the block on the plate. This ratio is called the *coefficient of friction.* It is called the coefficient of static friction for bodies at rest, coefficient of sliding friction where two surfaces are rubbing together and coefficient of rolling friction where one surface is rolling on another. Each of these three coefficients of friction vary with the nature of the surfaces in contact. If in the question referred to, the force required to initially move the block from rest were neglected and only the force required to keep the block in motion is considered, it would equal the weight of the block times the coefficient of sliding friction for dry wood on cast iron. If this coefficient is taken to be 0.49 then the work W is found by: $W = 50 \times 0.49 \times 10 = 245$ foot-pounds.

Question: How is the *coefficient of linear expansion used?*

Answer: The coefficient of linear expansion for a given material is the linear expansion per unit length it undergoes for each degree rise in temperature. The coefficient of linear expansion based on the Fahrenheit temperature scale is five-ninths of the coefficient based on the Centigrade scale. The coefficient of linear expansion for nickel, for example, is 0.000007 based on the Fahrenheit scale. A nickel rod when heated would increase in length by 0.000007 inch for each inch in length and degree temperature rise. Therefore, if it were 20 inches long and its temperature were raised from 70 degrees F to 150 degrees F, its length would increase by $20 \times 80 \times 0.000007 = 0.0112$ inch.

Question: What is the difference between *specific heat* and *latent heat?*

Answer: The specific heat of a given substance is the ratio of the amount of heat required to raise the temperature of the substance one degree on a given temperature scale to that required to raise an equal mass of water one degree on the same temperature scale. The specific heat of a substance is numerically equal to the number of calories required to raise the temperature of one gram of the substance one degree Centrigrade and also to the number of British Thermal Units required to raise the temperature of 1 pound of the substance one degree Fahrenheit.

Latent heat may signify *heat of fusion* or *heat of vaporization.* The heat of fusion of any substance may be defined either as the number of calories required to convert one gram of the solid at the melting point into liquid at the same temperature or as the number of British Thermal Units required to convert one pound of the solid at the melting point into liquid at the same temperature.

The heat of vaporization of any substance may be defined as the number of calories required to convert one gram of the liquid at a certain temperature into a vapor at the same temperature under a specified pressure or as the number of British Thermal Units required to convert one pound of the liquid into vapor at the same temperature under a specified pressure.

Question: What is the difference between the *weight* and *mass* of a body?

Answer: The term *weight* is commonly used in two different ways and because of this double usage, confusion sometimes arises when it is compared with the term *mass.*

In common usage, the term *weight* means the quantity of matter in a given body. As so used, it has the same meaning as *mass.* For example, the mass of a platinum cylinder of established size, which is preserved in the English Standards Office, Westminster, is equal to one pound. According to this usage, the *weight* of this standard reference cylinder is also equal to one pound. Also according to this usage, the density of a given substance is equal to its *weight or mass* per unit volume.

As used in science, and therefore in theoretical mechanics, which is of particular interest to those working with mechanical problems, the term *weight* denotes the force of gravity exerted on a given body. The term *mass* denotes the ratio of force acting on the body to the acceleration imparted to the body by this force. According to this usage, *weight* and *mass* have quite different meanings. They may or may not be numerically equal to each other depending upon the units employed.

Thus, if the English system of units is employed and weight is expressed in pounds, force in pounds and acceleration in feet per second per second, *mass* is equal to *weight* divided by acceleration due to gravity in feet per second per second — usually taken to be 32.16. Hence, when these units are used in an equation of theoretical mechanics involving *mass, weight* divided by acceleration due to gravity (commonly written as W/g) can be substituted for *mass* (commonly written as M).

If, however, force is expressed in poundals (32.16 poundals being equal to 1 pound, if 32.16 feet per second per second is the value taken for acceleration due to gravity) and, according to the English system, *weight* is expressed in pounds and acceleration in feet per second per second, *mass* is numerically equal to *weight* and they can be used interchangeably in the equations of theoretical mechanics involving such units. The poundal unit is used very rarely, however.

Similarly, if the metric system of units is employed and weight is measured in grams, force in grams and acceleration in centimeters per second per second, *mass* is equal to *weight* divided by 980.2, or whatever is taken to be the acceleration due to gravity in centimeters per second per second. Hence, when these units are used in the equations of theoretical mechanics involving *mass, weight* divided by acceleration due to gravity can be substituted for *mass*.

If, however, force is expressed in dynes (980.2 dynes being equal to 1 gram, if 980.2 centimeters per second per second is the value taken for the acceleration due to gravity) and, according to the metric system, *weight* is expressed in grams and acceleration is expressed in centimeters per second per second, *mass* is numerically equal to *weight* and they can be used interchangeably in equations of mechanics involving such units.

Question: How does the weight of a body vary with relation to its position below or above the surface of the earth?

Answer: The attractive force of gravity acting on a body is greatest at the surface of the earth; above the surface it decreases as the square of the distance to the center of the earth and below the surface it decreases as the distance to the center decreases.

Question: What is the distinction between the *density* and *specific gravity* of a substance?

Answer: The density of any solid, fluid or gaseous substance is the mass or weight of that substance per unit volume. The numerical value of the density of a substance will depend on the units employed.

Thus, density may be expressed in pounds per cubic inch, pounds per cubic foot, grams per cubic centimeter, etc. In engineering work, the density of a solid or liquid substance may be expressed in grams per cubic centimeter, without naming the units. When so expressed, the density will, for all practical purposes, be equal to the specific gravity.

The term specific gravity as applied to solids and liquids, indicates the ratio of the weight of a given volume of the substance to the same volume of water. Because the density of water varies slightly with the temperature, it is customary to make comparisons on the basis of water at a temperature of 4 degrees Centigrade (39.2 degrees F).

As applied to gases, specific gravity indicates the ratio of a given volume of the gas to an equal volume of air. The comparison is made at a temperature of 0 degrees Centigrade (32 degrees F) and a pressure of 760 millimeters of mercury, which is standard atmospheric pressure.

Question: Since the pressure exerted by a fluid increases with its depth, will a solid body that is denser than the liquid sink only to a certain level where it would be supported by the pressure exerted by the liquid?

Answer: No, the body would not stop sinking at a given level since the downward pressure exerted by the liquid on the top surface of the body plus the weight of the body is always greater than the upward pressure of the liquid on the bottom surface of the body. Thus, if the body were a cube located in the fluid with two faces in a horizontal position the total pressure exerted by the fluid on the top face and tending to push the cube down would be less than the total pressure exerted by the fluid on the bottom face and tending to push the cube up by an amount equal to the weight of the water displaced by the cube. But, the weight of the cube is greater than the weight of the water displaced by the cube, hence the total force causing the cube to sink is always greater than the force supporting it, so that the cube would continue to sink, even in the deepest part of the Atlantic or Pacific Ocean.

Question: Does the pressure exerted by a liquid on a vertical surface depend upon the distance that the liquid extends back from the surface in a horizontal direction? For example, does the pressure exerted by water at a given depth on a dam 100 feet long and 50 feet high depend upon whether the water extends 6 inches, 6 feet or 6 miles back from the face of the dam.

Answer: No, the distance that a fluid extends back of a vertical surface does not affect its pressure on that surface. The total pressure exerted on a vertical surface is dependent only on the area of the surface below the surface of the liquid and the vertical distance from the surface of the liquid to the center of gravity or, more properly speaking, the centroid of the surface below the liquid. In other words, the pressure exerted by water at a given depth on the dam would be the same regardless of whether it extended 6 inches, 6 feet, or 6 miles back from the face of the dam.

Question: Does the effect of a force on a body depend upon whether the body is at rest or in motion or whether the force acts alone or in combination with other forces?

Answer: The effect of a force in tending to produce motion of the body on which it acts is the same whether the body is at rest or in motion or whether the force acts alone or in combination with other forces. According to Newton's second law of motion: "The rate of change of momentum of a body is proportional to the force acting on the body and is in the direction of the force."

Question: If a single force were acting to change the state of motion of a body could it be said that only this single force is involved?

Answer: No, since according to Newton's third law "For every action there is an equal and opposite reaction." When a force acts on a body, the body offers a resistance equal and opposite to the force exerted on it.

Question: What are the three properties or characteristics of a force and how are they represented diagrammatically?

Answer: The three properties or characteristics of a force are its magnitude, position of its line of action, and its direction along this line of action, sometimes called *sense.* These three characteristics of a force are shown diagrammatically by an arrow called a *vector.* The length of the arrow is scaled to the magnitude of the force, the body or line of the arrow indicates the position of the line of action, and the head of the arrow indicates the direction of the force. When a diagram shows a force acting on a body, the vector is usually drawn so that one end of it is at the *point of application* of the force.

Question: What is a *couple*?

Answer: A couple is formed by the action of two equal, opposite and parallel forces which are not in the same straight line.

Question: Does the moment of a couple depend upon the location of the axis about which it acts?

Answer: The moment of a couple is equal to the product of one of the forces composing it and the perpendicular distance between the action-lines of the two forces and is independent of the location of the axis, providing this axis is at right angles to the plane of the couple.

Question: What are the three properties or characteristics of a couple?

Answer: The three properties or characteristics of a couple are the magnitude of its moment, the direction of its rotation, and the direction or slope of the plane in which it is located. A couple may be shown diagrammatically by two equal and parallel vectors which are drawn opposite to each other. In a scale drawing, the product of the length of one vector and the perpendicular distance between them indicates the moment of the couple. A couple may also be indicated diagrammatically by a single vector. In scale drawings, the length of this vector indicates the moment of the couple. It is drawn at right angles to the plane of the couple so that the direction of this plane is thus clearly indicated and the head of the vector arrow is so drawn that, when the plane of the couple is viewed in this direction, the couple will have a clockwise direction of rotation.

Question: What is meant by the statement that the forces acting on a body are in equilibrium?

Answer: When all the forces acting on a body are in equilibrium, it means that their combined effect does not change the state of motion of the body, that is, it either remains at rest or moves at a constant velocity in a straight line. When the forces acting on a body are *not* in equilibrium there results a change in velocity of the body, a change in direction of motion of the body, or a change of both velocity and direction of motion.

Question: What is meant by *torque*?

Answer: By torque is meant the turning moment which causes or tends to cause a given body to rotate. In the English or American system of units it is usually expressed as pound-feet or ounce-inches which represent the product of the force applied and the radial distance of its point of application from the axis of rotation. Thus a torque of 54 pound-feet might be the result of a force of 9 pounds applied 6 feet from the axis of rotation; a force of 27 pounds applied 2 feet from the axis of rotation; a force of 54 pounds applied 1 foot from the axis of rotation, etc.

Question: Is *inertia* a force?

Answer: Inertia is not a force. It is that property of a body which makes it tend to remain in its state of rest or motion as expressed by Newton's first law: *Every body persists in its state of rest or of uniform motion in a straight line, unless it be compelled by some force to change that state.*

Question: What is *momentum*?

Answer: The momentum of a body is the product of its mass and velocity. Thus, when a body is not in motion it has no momentum. Since the rate of change of momentum of a body is proportional to the force applied to it and since the mass of the body remains constant, it follows that the rate of change in momentum is solely due to the rate of change in velocity of the body. Hence, it may be said, that the rate of change of velocity, or in other words the acceleration, of a body is proportional to the force applied. Momentum may also be defined as that constant force which, resisting the movement of the given body, would bring it to rest in one second. If the force were expressed in pounds, momentum would then be expressed in pound-seconds.

Question: What is *angular velocity*?

Answer: The angular velocity of a rotating body is the angle through which any radius of the body turns in one second. It is usually expressed in radians.

Question: What is the *center of oscillation*?

Answer: The center of oscillation in a body oscillating about a horizontal axis that does not pass through its center of gravity is that point on a line drawn through the center of gravity and perpendicular to the axis, the motion of which is the same as if the whole mass of the body were concentrated at that point.

Question: What is the *radius of oscillation*?

Answer: The radius of oscillation of a body is the distance from the center of oscillation to the point of suspension.

Question: What is *centrifugal force*?

Answer: When a body rotates about an axis which does not pass through its center of mass, it exerts an outward radial force on the axis. This force is called centrifugal force.

Question: How is centrifugal force computed?

Answer: When a body of mass m rotates with constant *angular* velocity v, expressed in radians, about any axis not through the center of mass, it exerts a centrifugal force F on the axis equal to:

$$F = mv^2r.$$

Where r is the perpendicular distance from the axis of rotation to the center of mass.

Question: When a body rotates about an axis through its center of mass, is there any centrifugal force exerted on the body?

Answer: In such a case, the various particles of the body are each subjected to centrifugal force which tends to act radially outward on each particle. If the speed of rotation is high enough, the centrifugal forces acting on these particles will be greater than the cohesive forces holding them together and the body, which may be a grinding wheel, a flywheel, a turbine rotor, etc., will then fly apart. However, it should be noted that if the axis of rotation passes through the center of mass of the body, the resultant of all the centrifugal forces acting on the various particles of the body is equal to zero, or in other words, no centrifugal force is exerted on the axis.

Question: What is the *center of mass* of a body?

Answer: The center of mass of a body is that point which moves as if the whole body were concentrated at it, and to which all the forces acting on the body could be transferred with the directions unchanged.

Question: What is the *center of gravity* of a body?

Answer: The center of gravity of a body is the point at which the resultant of all the forces of gravity acting upon the respective particles of the body may be considered as acting. In other words, it is a point at which, if a single force of gravity were to act in place of all the other forces, and equal in intensity to their sum, the effect on the body would be the same as before.

Question: What is the *centroid* of a figure or line?

Answer: It is more accurate to refer to the *centroid* of a plane figure or line rather than to its *center of gravity* since it does not have weight. The centroid is sometimes called the *center of area* or *center of arc*. It may be defined as that point at which the whole area of the figure or length of line may be considered as being concentrated without affecting the moment of the figure or line with respect to any line in the plane.

Question: When can the *radius to the center of gravity* of a body be substituted for the *radius to its center of mass*, as in the case of computing centrifugal force?

Answer: For all practical purposes, the center of gravity of all bodies found on the earth's surface may be considered as coinciding with the center of mass.

Question: What is *moment of inertia*?

Answer: The term moment of inertia is applied to a rotating body and is a measure of its resistance to change in velocity of rotation (either acceleration or deceleration). The moment of inertia of a given body depends upon the mass of the body and location of the axis about which the body rotates.

Question: How is the moment of inertia of a body found?

Answer: If m_1, m_2, $\cdots$ are the respective masses of the particles constituting a body of total mass M and the respective perpendicular distances of these particles from the axis of rotation are r_1, $r_2 \cdots$, the moment of inertia I of the body is equal to:

$$I = (m_1 r_1^2 + m_2 r_2^2 \cdots)$$

Most handbooks have tables giving moments of inertia of various geometrically shaped bodies about specified axes.

Question: What is the *radius of gyration*?

Answer: The radius of gyration of a body about a given axis is that distance k from the axis at which its whole mass might suppose to be concentrated without changing its moment of inertia about that axis.

Question: How is the radius of gyration found?

Answer: If the moment of inertia I of a body is divided by its total mass M the result is actually the average or mean of the sum of $r_1^2 + r_2^2 + \cdots$ representing the squares of the distances of the respective particles from the axis of rotation. This result is designated as k^2 and k is called the radius of gyration.

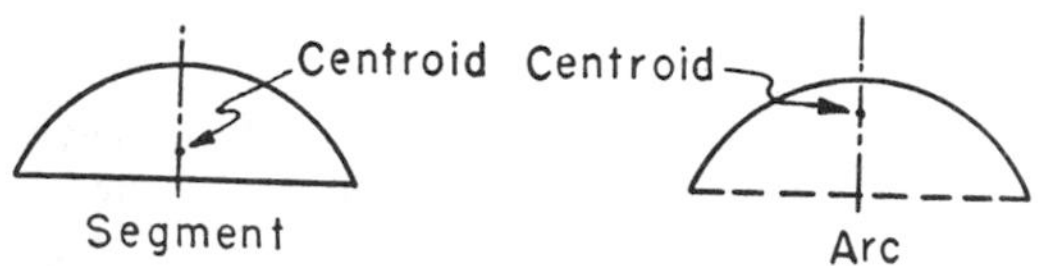

FIG. 4. Approximate Location of Centroid of a Circular Segment (Left) and of Circular Arc (Right)

Question: In Fig. 4 is shown the approximate location of the centroids (center of gravity) of a circular segment and of a circular arc of the same size as that in the segment. Why is there a difference in location of this point for each of these two figures?

Answer: In one case, the point represents the centroid of a curved line, and in the other, the centroid of an area. It can be readily seen that if this line had weight, its center of gravity would be closer to the arc than if this same weight were spread out over a segment subtended by an arc of the same size.

Question: What is *acceleration due to gravity*?

Answer: Acceleration due to gravity, which is commonly given the symbol g, is the rate of change of velocity in a freely falling body caused by the force of gravity, neglecting atmospheric resistance, as determined for a given point on the earth's surface. As expressed in the English system of units, it varies from 32.0878 feet per second per second at 0 degrees latitude and sea-level to 32.2577 feet per second per second at 90 degrees latitude and sea level. In problems of theoretical mechanics, the value taken is usually 32.16 feet per second per second, which is for a latitude of 41 degrees and sea level.

Question: Would a bullet or projectile fired vertically into the atmosphere have the same velocity at the time of striking the earth as it had upon leaving the muzzle of the gun?

Answer: Theoretically it should have the same velocity at the time it strikes the ground as when it leaves the muzzle of the gun. Actually, however, there are a number of factors which act to reduce the velocity at which the bullet or projectile strikes the ground. If every factor is considered, the problem becomes quite a complex one. The most important factor is probably the relation between velocity and atmospheric density, there being a very large air resistance at high velocities. Another factor of some importance is the position of the projectile when going up and when coming down. Due to its spinning action, the projectile maintains its pointed end foremost as it ascends. As the bullet descends, however, there would probably be more or less wobbling or turning over which would increase the effective or projected area against which the air resistance would be acting. Therefore the bullet would be moving more slowly when it hit the earth than when it left the muzzle of the gun.

Question: What is the difference between *work* and *power*?

Answer: These two terms are frequently confused. Work may be defined as the conversion of energy resulting from the movement of a force against a continuously acting resistance. Work is commonly measured in foot-pounds, which is the product of the moving force, in pounds, and the distance through which it moves, in feet.

Power, on the other hand, is the *rate* of doing work and represents the conversion of energy at a certain rate. A common unit of measurement is the horsepower which is equivalent to 33,000 foot-pounds per minute.

Question: How did the unit of measurement *horsepower* originate?

Answer: The horsepower unit was introduced by James Watt, the great improver of the steam engine, for the purpose of designating the power developed by the engine. It is said that he had ascertained by experiments that an average cart horse could develop 22,000 foot-pounds of work per minute, and being anxious to give good value to the purchasers of his engines, he added 50 per cent to this amount, thus obtaining the 33,000 foot-pounds per minute unit by which the power of steam and other engines has ever since been indicated.

Question: A horsepower-hour is said to be equal to 2545 British Thermal Units. How is this number determined?

Answer: If a machine having a capacity of one horsepower were to run for one hour, it would perform 1,980,000 foot-pounds of work. Since one B.T.U. is equal to 778 foot-pounds of work, one horsepower is equal to $1{,}980{,}000 \div 778 = 2545$ B.T.U., very nearly.

It should be noted that a horsepower-hour is not a unit of *power*, but a unit of *work*, and is equivalent to 1,980,000 foot-pounds. Similarly, 200 horsepower-hours may mean the work done by a 200 horsepower machine in one hour, or by a 25 horsepower machine in eight hours, etc., but in any case, it would equal $1{,}980{,}000 \times 200 = 396{,}000{,}000$ foot-pounds $= 2545 \times 200 = 509{,}000$ B.T.U.

Question: How does *mechanical efficiency* differ from *absolute efficiency*?

Answer: Mechanical efficiency is the ratio, expressed as a percentage, of the amount of energy which a machine transforms into useful work to the sum of this useful energy plus the energy which is lost as heat, due to friction. If E_m is mechanical efficiency; W is the useful energy; and F is the energy loss due to friction, then:

$$E_m = \frac{W}{W + F}$$

Absolute efficiency is the ratio, expressed as a percentage, of the amount of energy which a machine transforms into useful work to the total amount of energy it receives. In the case of a steam engine, an internal combustion engine or other prime mover, the total amount of energy supplied is much larger than the sum of the energy converted into useful work and the energy lost due to friction. In the case of the steam engine, there are heat losses due to radiation and steam condensation and in the case of an internal combustion engine, heat is dissipated either through the cooling water or directly to the atmosphere. All of these energy losses must be added to the energy converted into useful work and the energy loss due to friction to obtain the total energy delivered to the machine. In the case of some machines, such as electric motors, it is feasible to measure this energy input directly. If L is all of the energy loss, except that due to friction; T is total energy input; E_A is absolute efficiency; W is energy converted into useful work; and F is energy loss due to friction, then:

$$E_A = \frac{W}{T} = \frac{W}{W + F + L}$$

Ordinarily, the term efficiency as applied to engines and other classes of machinery, means mechanical efficiency. In the case of manufacturing machinery, the energy available at the driven or working end divided by the energy supplied to the initial driving shaft equals the mechanical efficiency.

Question: Can the working efficiencies of such machine elements as gears, bearings, belts, chains, etc. be determined?

Answer: The efficiencies of such working elements have been determined by various carefully conducted tests. However, the efficiency of a given machine element may vary over a fairly wide range due to differences in operating conditions, accuracy of construction, lubrication and other factors.

Under average conditions, the following efficiencies may serve as a general guide: plain bearings, 95 to 98 per cent; roller bearings and ball bearings, 98 to 99 per cent; spur gears with cut teeth and with bearings, 96 per cent; bevel gears, 95 per cent; belt drives, 96 to 98 per cent; roller chain drives, 95 to 97 per cent; high-class silent chain transmissions, 97 to 99 per cent.

Question: Do *allowable stress* and *working stress* have the same meaning?

Answer: Allowable stress and working stress both mean the maximum stress of the kind to which a given material can be subjected safely for a given condition of service. The difference between the allowable stress and the damaging stress is the margin of safety against damage or failure. By the expression "maximum stress of the kind" is meant whether the stress is tensional, compressive, torsional or some combination of these.

Question: What is meant by *damaging stress*?

Answer: Damaging stress is the minimum unit stress which will render a member unfit for service before the end of its normal life, taking into consideration the kind of stress, the material of which the member is made and the type of service to which it is subjected. Failure or damage may occur by rupture; fatigue cracking; excessive permanent deformation; too rapid a rate of creep (continuous increase in distortion); or excessive hardening due to cold working during service.

Question: What is the difference between the *apparent elastic limit*, *elastic limit* and *proportional limit* of a material?

Answer: The apparent elastic limit, also known as Johnson's apparent elastic limit, is the unit stress at which the rate of deformation of the material is 50 per cent greater than the initial rate of deformation. This value is used in the case of such materials as cast iron or soft copper where the ratio of stress to strain is not constant for any appreciable range.

Elastic limit may be defined as the least unit stress that will cause a set or permanent deformation in the material.

Proportional limit is the greatest unit stress to which a material can be subjected without deviating from a fixed ratio of stress to strain.

In the case of metals, experiments have shown that elastic limit and proportional limit are nearly equal to each other.

Question: What is the meaning of *ultimate strength*?

Answer: The ultimate strength of a material in tension, compression or shear, respectively, is the greatest unit tensile, compressive or shearing stress to which it can be subjected without breaking. This ultimate strength is equal to the breaking load divided by the original cross-section.

Question: What is the distinction between *yield strength* and *yield point*?

Answer: The yield strength of a material is the unit stress at which it exhibits a specified permanent deformation or set. The value of this set is often taken as 0.2 per cent (0.002 inch per inch) of the gage length.

The yield point, on the other hand, is the lowest unit stress at which unit strain begins to increase without increase in unit stress. Since only a few materials exhibit a true yield point, the term is often used synonymously for yield strength. The use of yield strength as a general property in preference to yield point is now accepted, since it can be determined more accurately.

Question: What is meant by the term *strain energy*?

Answer: Strain energy or elastic energy is the amount of mechanical energy stored up in a stressed material and is equal to the work done in producing the stress. In the case of a helical spring, compressed to a certain amount, for example, the strain energy in this spring would be equal to the work done in compressing the spring the given amount, neglecting the very small amount of energy lost as heat. When the force acting to produce the stress in the body is removed the strain energy is given up by the body.

Question: How is the *factor of safety* for a given material determined and applied?

Answer: In order to obtain the safe working stress, the ultimate strength of the material is divided by the factor of safety which is the product of several factors such as the ratio of the ultimate strength of the material to the elastic limit of the material; a factor representing the character of the stress, that is, whether it is caused by a static load, a load varying between zero and maximum or a load which alternately produces a tension and compression equal in amount; a factor representing the manner in which the load is applied, that is whether slowly, rapidly or rapidly with impact; and a factor of ignorance where accidental overload, unexpectedly severe service, unreliable or imperfect materials, or other unknown conditions must be guarded against. These component factors, which are multiplied together to obtain the factor of safety, may be found in various handbooks such as *Machinery's Handbook.*

If a part made of machine steel has an ultimate strength of 60,000 pounds per square inch and a factor of safety of 6 were used, the allowable or working stress would, then, be 10,000 pounds per square inch.

Question: What is the meaning of *fatigue failure*?

Answer: Fatigue failure is the rupture of a material due to repeated stresses which are less than that required to produce immediate rupture. Fatigue is most severe when the material is subjected to alternate compression and tension, as in alternate bending.

Question: What is the *endurance limit* of a material?

Answer: The endurance limit or fatigue strength of a material, when not otherwise specified, is usually taken to mean the maximum reverse stress, that is the maximum stress in alternate compression and tension, to which it can be subjected an indefinite number of times without rupture.

Question: What is the *neutral surface* and the *neutral axis* in a beam?

Answer: When a rod, bar or beam is subjected to bending, the fibers on the convex side or outside of the bend are placed in tension, whereas those on the concave side or inside of the bend are placed in compression. Somewhere between these two sides of the beam there is a layer of fibers which are subjected to neither tension nor compression. This layer of fibers constitutes what is called the *neutral surface.* The intersection of this surface with any cross-section of the rod, bar or beam is the neutral axis of that section, along which there is a line of fibers that are in neither tension nor compression.

Question: What is the meaning of the term *section modulus*?

Answer: Section modulus is a term applied to the cross-sectional area of a rod, bar or beam and equals the moment of inertia of this area with respect to the neutral axis, divided by the distance from the neutral axis to the extreme fiber of the cross-section. Section modulus values for different shapes of cross-sections are given in handbooks and are useful in finding the maximum allowable bending moment of the bar, rod or beam. This maximum allowable bending moment is equal to the product of the section modulus and the maximum allowable unit stress for the material of which the rod, bar or beam is made. Similarly, the tensile stress at any point along a rod, bar or beam subject to bending can be found by dividing the bending moment at this point by the section modulus.

Question: What is the *modulus of elasticity* of a material?

Answer: The modulus of elasticity is based on Hooke's Law which states that within the elastic limits of a material, deformation in tension or compression is proportional to the stress which produces it. This constant ratio of the stress in pounds per square inch which theoretically would be required to produce an elongation of 1 inch per inch of length is called Young's modulus, or the modulus of elasticity of the material. It is expressed in pounds per square inch.

Actually, due to such factors as lack of homogeneity in materials, Hooke's law does not strictly apply and there is some variation in the modulus of elasticity for a given material, particularly in the case of certain materials such as cast iron and concrete. However, for ordinary purposes, Hooke's law may be considered as holding true up to the proportional limit of a given material.

27

Mathematical Tables—Aids for Solving Mechanical Problems

Mathematical Signs and Abbreviations

In this table are given the signs and abbreviations that are commonly used in mathematical formulas and calculations.

The delta, differential, integral and limit signs, which are employed in the calculus, do not appear in the problems as worked out in this manual.

$+$	Plus (sign of addition)	π	Pi (3.1416)
$+$	Positive	Σ	Sigma (sign of summation)
$-$	Minus (sign of subtraction)	ω	Omega (angles measured in radians)
$-$	Negative		
$\pm$ ($\mp$)	Plus or minus (minus or plus)	g	Acceleration due to gravity (32.16 ft. per sec. per sec.)
$\times$	Multiplied by (multiplication sign)		
$\cdot$	Multiplied by (multiplication sign)	i (or j)	Imaginary quantity ($\sqrt{-1}$)
$\div$	Divided by (division sign)	sin	Sine
$:$	Divided by (division sign)	cos	Cosine
$:$	Is to (in proportion)	tan	Tangent
$=$	Equals	(tg)	Tangent
$\neq$	Is not equal to	(tang)	Tangent
$\equiv$	Is identical to	cot	Cotangent
$::$	Equals (in proportion)	(ctg)	Cotangent
$\cong$	Approximately equals	sec	Secant
$\approx$	Approximately equals	cosec	Cosecant
$>$	Greater than	versin	Versed sine
$<$	Less than	covers	Coversed sine
$\geqq$	Greater than or equal to	$\sin^{-1} a$	Arc the sine of which is a
$\leqq$	Less than or equal to	arc sin a	Arc the sine of which is a
$\rightarrow$	Approaches as a limit	$(\sin a)^{-1}$	Reciprocal of sin a ($1 \div \sin a$)
$\propto$	Varies directly as		
$\therefore$	Therefore	sinh x	Hyperbolic sine of x
$\sqrt{\ }$	Square root	cosh x	Hyperbolic cosine of x
$\sqrt[3]{\ }$	Cube root	Δ	Delta (increment of)
$\sqrt[4]{\ }$	4th root	δ	Delta (variation of)
$\sqrt[n]{\ }$	nth root	d	Differential (in calculus)
a^2	a squared (2d power of a)	$\int$	Integral (in calculus)
a^3	a cubed (3d power of a)		
a^4	4th power of a	$\int_b^a$	Integral between the limits a and b
a^n	nth power of a		
a^{-n}	$1 \div a^n$	!	$5! = 1 \times 2 \times 3 \times 4 \times 5$
		$\angle$	Angle
$\frac{1}{n}$	Reciprocal value of n	$\llcorner$	Right angle
		$\perp$	Perpendicular to
log	Logarithm	$\triangle$	Triangle
hyp. log	Hyperbolic, natural or Napierian logarithm	$\odot$	Circle
nat. log	Hyperbolic, natural or Napierian logarithm	$\square$	Parallelogram
$\log_e$	Hyperbolic, natural or Napierian logarithm	$\circ$	Degree (circular arc or temperature)
ln	Hyperbolic, natural or Napierian logarithm		
e	Base of hyp. logarithms (2.71828)	$'$	Minutes or feet
		$''$	Seconds or inches
lim.	Limit value (of an expression)	a'	a prime
∞	Infinity	a''	a double prime
α	Alpha — commonly used to denote angles	a_1	a sub one
β	Beta — commonly used to denote angles	a_2	a sub two
γ	Gamma — commonly used to denote angles	a_n	a sub n
θ	Theta — commonly used to denote angles	( )	Parentheses
ϕ	Phi	[ ]	Brackets
μ	Mu (coefficient of friction)	{ }	Braces

Commonly Used Constants

A *constant* is a value that does not change or is not a variable. Constants at one stage of a mathematical investigation may be variables at another stage. However, an *absolute constant*, such as π, has the same value under all circumstances.

Constant	Numerical Value	Logarithm	Constant	Numerical Value	Logarithm
π	3.141593	0.49715	Weight in pounds of:		
2π	6.283185	0.79818	Water column, 1″×1″×1 ft.	0.4335	$\bar{1}.63699$
$\pi \div 4$	0.785398	$\bar{1}.89509$	1 U.S. gallon of water, 39.1°F.	8.34	0.92117
π^2	9.869604	0.99430	1 cu. ft. of water, 39.1° F...	62.4245	1.79535
π^3	31.006277	1.49145	1 cu. in. of water, 39.1° F...	0.0361	$\bar{2}.55751$
$1 \div \pi$	0.318310	$\bar{1}.50285$	1 cu. ft. of air, 32° F., atmos-		
$1 \div \pi^2$	0.101321	$\bar{1}.00570$	pheric pressure	0.08073	$\bar{2}.90703$
$1 \div \pi^3$	0.032252	$\bar{2}.50855$	Volume in cu. ft. of:		
$\sqrt{\pi}$	1.772454	0.24858	1 pound of water, 39.1° F...	0.01602	$\bar{2}.20465$
$\sqrt[3]{\pi}$	1.464592	0.16572	1 pound of air, 32° F., atmos-		
g	32.16	1.50732	pheric pressure	12.387	1.09297
g^2	1034.266	3.01463	Volume in gallons of 1 pound		
$2g$	64.32	1.80835	of water, 39.1° F.........	0.1199	$\bar{1}.07883$
$1 \div 2g$	0.01555	$\bar{2}.19165$	Volume in cu. in. of 1 pound of		
$\sqrt{2g}$	8.01998	0.90417	water, 39.1° F.	27.70	1.44249
$1 \div \sqrt{g}$	0.17634	$\bar{1}.24635$	One cubic ft. in gallons	7.4805	0.87393
$\pi \div \sqrt{g}$	0.55399	$\bar{1}.74350$	Atmospheric pressure in		
e	2.71828	0.43429	pounds per sq. in........	14.696	1.16720

Useful Constants Multiplied and Divided by 1 to 10

Constant	Multiplied by:							
	2	3	4	5	6	7	8	9
0.7854	1.5708	2.3562	3.1416	3.9270	4.7124	5.4978	6.2832	7.0686
3.1416	6.2832	9.4248	12.566	15.708	18.850	21.991	25.133	28.274
14.7	29.4	44.1	58.8	73.5	88.2	102.9	117.6	132.3
32.16	64.32	96.48	128.64	160.80	192.96	225.12	257.28	289.44
64.32	128.64	192.96	257.28	321.60	385.92	450.24	514.56	578.88
144	288	432	576	720	864	1,008	1,152	1,296
778	1,556	2,334	3,112	3,890	4,668	5,446	6,224	7,002
1,728	3,456	5,184	6,912	8,640	10,368	12,096	13,824	15,552
33,000	66,000	99,000	132,000	165,000	198,000	231,000	264,000	297,000
Constant	**Divided by:**							
	2	**3**	**4**	**5**	**6**	**7**	**8**	**9**
0.7854	0.3927	0.2618	0.1964	0.1571	0.1309	0.1122	0.0982	0.0873
3.1416	1.5708	1.0472	0.7854	0.6283	0.5236	0.4488	0.3927	0.3490
14.7	7.350	4.900	3.625	2.940	2.450	2.100	1.838	1.633
32.16	16.080	10.720	8.040	6.432	5.360	4.594	4.020	3.573
64.32	32.160	21.440	16.080	12.864	10.720	9.189	8.040	7.147
144	72	48	36	28.800	24	20.571	18	16
778	389	259.33	194.50	155.60	129.67	111.14	97.25	86.44
1,728	864	576	432	345.60	288	246.86	216	192
33,000	16,500	11,000	8250	6600	5500	4714.3	4125	3666.7

Prime Numbers and Smallest Factors

1 **1199**

From to	0 100	100 200	200 300	300 400	400 500	500 600	600 700	700 800	800 900	900 1000	1000 1100	1100 1200
1	P	P	3	7	P	3	P	P	3	17	7	3
3	P	P	7	3	13	P	3	19	11	3	17	P
5	P	3	5	5	3	5	5	3	5	5	3	5
7	P	P	3	P	11	3	P	7	3	P	19	3
9	3	P	11	3	P	P	3	P	P	3	P	P
11	P	3	P	P	3	7	13	3	P	P	3	11
13	P	P	3	P	7	3	P	23	3	11	P	3
15	3	5	5	3	5	5	3	5	5	3	5	5
17	P	3	7	P	3	11	P	3	19	7	3	P
19	P	7	3	11	P	3	P	P	3	P	P	3
21	3	11	13	3	P	P	3	7	P	3	P	19
23	P	3	P	17	3	P	7	3	P	13	3	P
25	5	5	3	5	5	3	5	5	3	5	5	3
27	3	P	P	3	7	17	3	P	P	3	13	7
29	P	3	P	7	3	23	17	3	P	P	3	P
31	P	P	3	P	P	3	P	17	3	7	P	3
33	3	7	P	3	P	13	3	P	7	3	P	11
35	5	3	5	5	3	5	5	3	5	5	3	5
37	P	P	3	P	19	3	7	11	3	P	17	3
39	3	P	P	3	P	7	3	P	P	3	P	17
41	P	3	P	11	3	P	P	3	29	P	3	7
43	P	11	3	7	P	3	P	P	3	23	7	3
45	3	5	5	3	5	5	3	5	5	3	5	5
47	P	3	13	P	3	P	P	3	7	P	3	31
49	7	P	3	P	P	3	11	7	3	13	P	3
51	3	P	P	3	11	19	3	P	23	3	P	P
53	P	3	11	P	3	7	P	3	P	P	3	P
55	5	5	3	5	5	3	5	5	3	5	5	3
57	3	P	P	3	P	P	3	P	P	3	7	13
59	P	3	7	P	3	13	P	3	P	7	3	19
61	P	7	3	19	P	3	P	P	3	31	P	3
63	3	P	P	3	P	P	3	7	P	3	P	P
65	5	3	5	5	3	5	5	3	5	5	3	5
67	P	P	3	P	P	3	23	13	3	P	11	3
69	3	13	P	3	7	P	3	P	11	3	P	7
71	P	3	P	7	3	P	11	3	13	P	3	P
73	P	P	3	P	11	3	P	P	3	7	29	3
75	3	5	5	3	5	5	3	5	5	3	5	5
77	7	3	P	13	3	P	P	3	P	P	3	11
79	P	P	3	P	P	3	7	19	3	11	13	3
81	3	P	P	3	13	7	3	11	P	3	23	P
83	P	3	P	P	3	11	P	3	P	P	3	7
85	5	5	3	5	5	3	5	5	3	5	5	3
87	3	11	7	3	P	P	3	P	P	3	P	P
89	P	3	17	P	3	19	13	3	7	23	3	29
91	7	P	3	17	P	3	P	7	3	P	P	3
93	3	P	P	3	17	P	3	13	19	3	P	P
95	5	3	5	5	3	5	5	3	5	5	3	5
97	P	P	3	P	7	3	17	P	3	P	P	3
99	3	P	13	3	P	P	3	17	29	3	7	11

The way in which this table may be used to find all the prime factors of a given number is explained on page 1–2.

Prime Numbers and Smallest Factors

1201 **2399**

From to	1200 1300	1300 1400	1400 1500	1500 1600	1600 1700	1700 1800	1800 1900	1900 2000	2000 2100	2100 2200	2200 2300	2300 2400
1	P	P	3	19	P	3	P	P	3	11	31	3
3	3	P	23	3	7	13	3	11	P	3	P	7
5	5	3	5	5	3	5	5	3	5	5	3	5
7	17	P	3	11	P	3	13	P	3	7	P	3
9	3	7	P	3	P	P	3	23	7	3	47	P
11	7	3	17	P	3	29	P	3	P	P	3	P
13	P	13	3	17	P	3	7	P	3	P	P	3
15	3	5	5	3	5	5	3	5	5	3	5	5
17	P	3	13	37	3	17	23	3	P	29	3	7
19	23	P	3	7	P	3	17	19	3	13	7	3
21	3	P	7	3	P	P	3	17	43	3	P	11
23	P	3	P	P	3	P	P	3	7	11	3	23
25	5	5	3	5	5	3	5	5	3	5	5	3
27	3	P	P	3	P	11	3	41	P	3	17	13
29	P	3	P	11	3	7	31	3	P	P	3	17
31	P	11	3	P	7	3	P	P	3	P	23	3
33	3	31	P	3	23	P	3	P	19	3	7	P
35	5	3	5	5	3	5	5	3	5	5	3	5
37	P	7	3	29	P	3	11	13	3	P	P	3
39	3	13	P	3	11	37	3	7	P	3	P	P
41	17	3	11	23	3	P	7	3	13	P	3	P
43	11	17	3	P	31	3	19	29	3	P	P	3
45	3	5	5	3	5	5	3	5	5	3	5	5
47	29	3	P	7	3	P	P	3	23	19	3	P
49	P	19	3	P	17	3	43	P	3	7	13	3
51	3	7	P	3	13	17	3	P	7	3	P	P
53	7	3	P	P	3	P	17	3	P	P	3	13
55	5	5	3	5	5	3	5	5	3	5	5	3
57	3	23	31	3	P	7	3	19	11	3	37	P
59	P	3	P	P	3	P	11	3	29	17	3	7
61	13	P	3	7	11	3	P	37	3	P	7	3
63	3	29	7	3	P	41	3	13	P	3	31	17
65	5	3	5	5	3	5	5	3	5	5	3	5
67	7	P	3	P	P	3	P	7	3	11	P	3
69	3	37	13	3	P	29	3	11	P	3	P	23
71	31	3	P	P	3	7	P	3	19	13	3	P
73	19	P	3	11	7	3	P	P	3	41	P	3
75	3	5	5	3	5	5	3	5	5	3	5	5
77	P	3	7	19	3	P	P	3	31	7	3	P
79	P	7	3	P	23	3	P	P	3	P	43	3
81	3	P	P	3	41	13	3	7	P	3	P	P
83	P	3	P	P	3	P	7	3	P	37	3	P
85	5	5	3	5	5	3	5	5	3	5	5	3
87	3	19	P	3	7	P	3	P	P	3	P	7
89	P	3	P	7	3	P	P	3	P	11	3	P
91	P	13	3	37	19	3	31	11	3	7	29	3
93	3	7	P	3	P	11	3	P	7	3	P	P
95	5	3	5	5	3	5	5	3	5	5	3	5
97	P	11	3	P	P	3	7	P	3	13	P	3
99	3	P	P	3	P	7	3	P	P	3	11	P

The way in which this table may be used to find all the prime factors of a given number is explained on page 1–2.

Prime Numbers and Smallest Factors

2401 **3599**

From to	2400 2500	2500 2600	2600 2700	2700 2800	2800 2900	2900 3000	3000 3100	3100 3200	3200 3300	3300 3400	3400 3500	3500 3600
1	7	41	3	37	P	3	P	7	3	P	19	3
3	3	P	19	3	P	P	3	29	P	3	41	31
5	5	3	5	5	3	5	5	3	5	5	3	5
7	29	23	3	P	7	3	31	13	3	P	P	3
9	3	13	P	3	53	P	3	P	P	3	7	11
11	P	3	7	P	3	41	P	3	13	7	3	P
13	19	7	3	P	29	3	23	11	3	P	P	3
15	3	5	5	3	5	5	3	5	5	3	5	5
17	P	3	P	11	3	P	7	3	P	31	3	P
19	41	11	3	P	P	3	P	P	3	P	13	3
21	3	P	P	3	7	23	3	P	P	3	11	7
23	P	3	43	7	3	37	P	3	11	P	3	13
25	5	5	3	5	5	3	5	5	3	5	5	3
27	3	7	37	3	11	P	3	53	7	3	23	P
29	7	3	11	P	3	29	13	3	P	P	3	P
31	11	P	3	P	19	3	7	31	3	P	47	3
33	3	17	P	3	P	7	3	13	53	3	P	P
35	5	3	5	5	3	5	5	3	5	5	3	5
37	P	43	3	7	P	3	P	P	3	47	7	3
39	3	P	7	3	17	P	3	43	41	3	19	P
41	P	3	19	P	3	17	P	3	7	13	3	P
43	7	P	3	13	P	3	17	7	3	P	11	3
45	3	5	5	3	5	5	3	5	5	3	5	5
47	P	3	P	41	3	7	11	3	17	P	3	P
49	31	P	3	P	7	3	P	47	3	17	P	3
51	3	P	11	3	P	13	3	23	P	3	7	53
53	11	3	7	P	3	P	43	3	P	7	3	11
55	5	5	3	5	5	3	5	5	3	5	5	3
57	3	P	P	3	P	P	3	7	P	3	P	P
59	P	3	P	31	3	11	7	3	P	P	3	P
61	23	13	3	11	P	3	P	29	3	P	P	3
63	3	11	P	3	7	P	3	P	13	3	P	7
65	5	3	5	5	3	5	5	3	5	5	3	5
67	P	17	3	P	47	3	P	P	3	7	P	3
69	3	7	17	3	19	P	3	P	7	3	P	43
71	7	3	P	17	3	P	37	3	P	P	3	P
73	P	31	3	47	13	3	7	19	3	P	23	3
75	3	5	5	3	5	5	3	5	5	3	5	5
77	P	3	P	P	3	13	17	3	29	11	3	7
79	37	P	3	7	P	3	P	11	3	31	7	3
81	3	29	7	3	43	11	3	P	17	3	59	P
83	13	3	P	11	3	19	P	3	7	17	3	P
85	5	5	3	5	5	3	5	5	3	5	5	3
87	3	13	P	3	P	29	3	P	19	3	11	17
89	19	3	P	P	3	7	P	3	11	P	3	37
91	47	P	3	P	7	3	11	P	3	P	P	3
93	3	P	P	3	11	41	3	31	37	3	7	P
95	5	3	5	5	3	5	5	3	5	5	3	5
97	11	7	3	P	P	3	19	23	3	43	13	3
99	3	23	P	3	13	P	3	7	P	3	P	59

The way in which this table may be used to find all the prime factors of a given number is explained on page 1–2.

Prime Numbers and Smallest Factors

3601 **4799**

From to	3600 3700	3700 3800	3800 3900	3900 4000	4000 4100	4100 4200	4200 4300	4300 4400	4400 4500	4500 4600	4600 4700	4700 4800
1	13	P	3	47	P	3	P	11	3	7	43	3
3	3	7	P	3	P	11	3	13	7	3	P	P
5	5	3	5	5	3	5	5	3	5	5	3	5
7	P	11	3	P	P	3	7	59	3	P	17	3
9	3	P	13	3	19	7	3	31	P	3	11	17
11	23	3	37	P	3	P	P	3	11	13	3	7
13	P	47	3	7	P	3	11	19	3	P	7	3
15	3	5	5	3	5	5	3	5	5	3	5	5
17	P	3	11	P	3	23	P	3	7	P	3	53
19	7	P	3	P	P	3	P	7	3	P	31	3
21	3	61	P	3	P	13	3	29	P	3	P	P
23	P	3	P	P	3	7	41	3	P	P	3	P
25	5	5	3	5	5	3	5	5	3	5	5	3
27	3	P	43	3	P	P	3	P	19	3	7	29
29	19	3	7	P	3	P	P	3	43	7	3	P
31	P	7	3	P	29	3	P	61	3	23	11	3
33	3	P	P	3	37	P	3	7	11	3	41	P
35	5	3	5	5	3	5	5	3	5	5	3	5
37	P	37	3	31	11	3	19	P	3	13	P	3
39	3	P	11	3	7	P	3	P	23	3	P	7
41	11	3	23	7	3	41	P	3	P	19	3	11
43	P	19	3	P	13	3	P	43	3	7	P	3
45	3	5	5	3	5	5	3	5	5	3	5	5
47	7	3	P	P	3	11	31	3	P	P	3	47
49	41	23	3	11	P	3	7	P	3	P	P	3
51	3	11	P	3	P	7	3	19	P	3	P	P
53	13	3	P	59	3	P	P	3	61	29	3	7
55	5	5	3	5	5	3	5	5	3	5	5	3
57	3	13	7	3	P	P	3	P	P	3	P	67
59	P	3	17	37	3	P	P	3	7	47	3	P
61	7	P	3	17	31	3	P	7	3	P	59	3
63	3	53	P	3	17	23	3	P	P	3	P	11
65	5	3	5	5	3	5	5	3	5	5	3	5
67	19	P	3	P	7	3	17	11	3	P	13	3
69	3	P	53	3	13	11	3	17	41	3	7	19
71	P	3	7	11	3	43	P	3	17	7	3	13
73	P	7	3	29	P	3	P	P	3	17	P	3
75	3	5	5	3	5	5	3	5	5	3	5	5
77	P	3	P	41	3	P	7	3	11	23	3	17
79	13	P	3	23	P	3	11	29	3	19	P	3
81	3	19	P	3	7	37	3	13	P	3	31	7
83	29	3	11	7	3	47	P	3	P	P	3	P
85	5	5	3	5	5	3	5	5	3	5	5	3
87	3	7	13	3	61	53	3	41	7	3	43	P
89	7	3	P	P	3	59	P	3	67	13	3	P
91	P	17	3	13	P	3	7	P	3	P	P	3
93	3	P	17	3	P	7	3	23	P	3	13	P
95	5	3	5	5	3	5	5	3	5	5	3	5
97	P	P	3	7	17	3	P	P	3	P	7	3
99	3	29	7	3	P	13	3	53	11	3	37	P

The way in which this table may be used to find all the prime factors of a given number is explained on page 1–2.

Prime Numbers and Smallest Factors

4801 **5999**

From to	4800 4900	4900 5000	5000 5100	5100 5200	5200 5300	5300 5400	5400 5500	5500 5600	5600 5700	5700 5800	5800 5900	5900 6000
1	P	13	3	P	7	3	11	P	3	P	P	3
3	3	P	P	3	11	P	3	P	13	3	7	P
5	5	3	5	5	3	5	5	3	5	5	3	5
7	11	7	3	P	41	3	P	P	3	13	P	3
9	3	P	P	3	P	P	3	7	71	3	37	19
11	17	3	P	19	3	47	7	3	31	P	3	23
13	P	17	3	P	13	3	P	37	3	29	P	3
15	3	5	5	3	5	5	3	5	5	3	5	5
17	P	3	29	7	3	13	P	3	41	P	3	61
19	61	P	3	P	17	3	P	P	3	7	11	3
21	3	7	P	3	23	17	3	P	7	3	P	31
23	7	3	P	47	3	P	11	3	P	59	3	P
25	5	5	3	5	5	3	5	5	3	5	5	3
27	3	13	11	3	P	7	3	P	17	3	P	P
29	11	3	47	23	3	73	61	3	13	17	3	7
31	P	P	3	7	P	3	P	P	3	11	7	3
33	3	P	7	3	P	P	3	11	43	3	19	17
35	5	3	5	5	3	5	5	3	5	5	3	5
37	7	P	3	11	P	3	P	7	3	P	13	3
39	3	11	P	3	13	19	3	29	P	3	P	P
41	47	3	71	53	3	7	P	3	P	P	3	13
43	29	P	3	37	7	3	P	23	3	P	P	3
45	3	5	5	3	5	5	3	5	5	3	5	5
47	37	3	7	P	3	P	13	3	P	7	3	19
49	13	7	3	19	29	3	P	31	3	P	P	3
51	3	P	P	3	59	P	3	7	P	3	P	11
53	23	3	31	P	3	53	7	3	P	11	3	P
55	5	5	3	5	5	3	5	5	3	5	5	3
57	3	P	13	3	7	11	3	P	P	3	P	7
59	43	3	P	7	3	23	53	3	P	13	3	59
61	P	11	3	13	P	3	43	67	3	7	P	3
63	3	7	61	3	19	31	3	P	7	3	11	67
65	5	3	5	5	3	5	5	3	5	5	3	5
67	31	P	3	P	23	3	7	19	3	73	P	3
69	3	P	37	3	11	7	3	P	P	3	P	47
71	P	3	11	P	3	41	P	3	53	29	3	7
73	11	P	3	7	P	3	13	P	3	23	7	3
75	3	5	5	3	5	5	3	5	5	3	5	5
77	P	3	P	31	3	19	P	3	7	53	3	43
79	7	13	3	P	P	3	P	7	3	P	P	3
81	3	17	P	3	P	P	3	P	13	3	P	P
83	19	3	13	71	3	7	P	3	P	P	3	31
85	5	5	3	5	5	3	5	5	3	5	5	3
87	3	P	P	3	17	P	3	37	11	3	7	P
89	P	3	7	P	3	17	11	3	P	7	3	53
91	67	7	3	29	11	3	17	P	3	P	43	3
93	3	P	11	3	67	P	3	7	P	3	71	13
95	5	3	5	5	3	5	5	3	5	5	3	5
97	59	19	3	P	P	3	23	29	3	11	P	3
99	3	P	P	3	7	P	3	11	41	3	17	7

The way in which this table may be used to find all the prime factors of a given number is explained on page 1–2.

Prime Numbers and Smallest Factors

6001 **7199**

From to	6000 6100	6100 6200	6200 6300	6300 6400	6400 6500	6500 6600	6600 6700	6700 6800	6800 6900	6900 7000	7000 7100	7100 7200
1	17	P	3	P	37	3	7	P	3	67	P	3
3	3	17	P	3	19	7	3	P	P	3	47	P
5	5	3	5	5	3	5	5	3	5	5	3	5
7	P	31	3	7	43	3	P	19	3	P	7	3
9	3	41	7	3	13	23	3	P	11	3	43	P
11	P	3	P	P	3	17	11	3	7	P	3	13
13	7	P	3	59	11	3	17	7	3	31	P	3
15	3	5	5	3	5	5	3	5	5	3	5	5
17	11	3	P	P	3	7	13	3	17	P	3	11
19	13	29	3	71	7	3	P	P	3	11	P	3
21	3	P	P	3	P	P	3	11	19	3	7	P
23	19	3	7	P	3	11	37	3	P	7	3	17
25	5	5	3	5	5	3	5	5	3	5	5	3
27	3	11	13	3	P	61	3	7	P	3	P	P
29	P	3	P	P	3	P	7	3	P	13	3	P
31	37	P	3	13	59	3	19	53	3	29	79	3
33	3	P	23	3	7	47	3	P	P	3	13	7
35	5	3	5	5	3	5	5	3	5	5	3	5
37	P	17	3	P	41	3	P	P	3	7	31	3
39	3	7	17	3	47	13	3	23	7	3	P	11
41	7	3	79	17	3	31	29	3	P	11	3	37
43	P	P	3	P	17	3	7	11	3	53	P	3
45	3	5	5	3	5	5	3	5	5	3	5	5
47	P	3	P	11	3	P	17	3	41	P	3	7
49	23	11	3	7	P	3	61	17	3	P	7	3
51	3	P	7	3	P	P	3	43	13	3	11	P
53	P	3	13	P	3	P	P	3	7	17	3	23
55	5	5	3	5	5	3	5	5	3	5	5	3
57	3	47	P	3	11	79	3	29	P	3	P	17
59	73	3	11	P	3	7	P	3	19	P	3	P
61	11	61	3	P	7	3	P	P	3	P	23	3
63	3	P	P	3	23	P	3	P	P	3	7	13
65	5	3	5	5	3	5	5	3	5	5	3	5
67	P	7	3	P	29	3	59	67	3	P	37	3
69	3	31	P	3	P	P	3	7	P	3	P	67
71	13	3	P	23	3	P	7	3	P	P	3	71
73	P	P	3	P	P	3	P	13	3	19	11	3
75	3	5	5	3	5	5	3	5	5	3	5	5
77	59	3	P	7	3	P	11	3	13	P	3	P
79	P	37	3	P	11	3	P	P	3	7	P	3
81	3	7	11	3	P	P	3	P	7	3	73	43
83	7	3	61	13	3	29	41	3	P	P	3	11
85	5	5	3	5	5	3	5	5	3	5	5	3
87	3	23	P	3	13	7	3	11	71	3	19	P
89	P	3	19	P	3	11	P	3	83	29	3	7
91	P	41	3	7	P	3	P	P	3	P	7	3
93	3	11	7	3	43	19	3	P	61	3	41	P
95	5	3	5	5	3	5	5	3	5	5	3	5
97	7	P	3	P	73	3	37	7	3	P	47	3
99	3	P	P	3	67	P	3	13	P	3	31	23

The way in which this table may be used to find all the prime factors of a given number is explained on page 1–2.

Prime Numbers and Smallest Factors

7201 **8399**

From to	7200 7300	7300 7400	7400 7500	7500 7600	7600 7700	7700 7800	7800 7900	7900 8000	8000 8100	8100 8200	8200 8300	8300 8400
1	19	7	3	13	11	3	29	P	3	P	59	3
3	3	67	11	3	P	P	3	7	53	3	13	19
5	5	3	5	5	3	5	5	3	5	5	3	5
7	P	P	3	P	P	3	37	P	3	11	29	3
9	3	P	31	3	7	13	3	11	P	3	P	7
11	P	3	P	7	3	11	73	3	P	P	3	P
13	P	71	3	11	23	3	13	41	3	7	43	3
15	3	5	5	3	5	5	3	5	5	3	5	5
17	7	3	P	P	3	P	P	3	P	P	3	P
19	P	13	3	73	19	3	7	P	3	23	P	3
21	3	P	41	3	P	7	3	89	13	3	P	53
23	31	3	13	P	3	P	P	3	71	P	3	7
25	5	5	3	5	5	3	5	5	3	5	5	3
27	3	17	7	3	29	P	3	P	23	3	19	11
29	P	3	17	P	3	59	P	3	7	11	3	P
31	7	P	3	17	13	3	41	7	3	47	P	3
33	3	P	P	3	17	11	3	P	29	3	P	13
35	5	3	5	5	3	5	5	3	5	5	3	5
37	P	11	3	P	7	3	17	P	3	79	P	3
39	3	41	43	3	P	71	3	17	P	3	7	31
41	13	3	7	P	3	P	P	3	11	7	3	19
43	P	7	3	19	P	3	11	13	3	17	P	3
45	3	5	5	3	5	5	3	5	5	3	5	5
47	P	3	11	P	3	61	7	3	13	P	3	17
49	11	P	3	P	P	3	47	P	3	29	73	3
51	3	P	P	3	7	23	3	P	83	3	37	7
53	P	3	29	7	3	P	P	3	P	31	3	P
55	5	5	3	5	5	3	5	5	3	5	5	3
57	3	7	P	3	13	P	3	73	7	3	23	61
59	7	3	P	P	3	P	29	3	P	41	3	13
61	53	17	3	P	47	3	7	19	3	P	11	3
63	3	37	17	3	79	7	3	P	11	3	P	P
65	5	3	5	5	3	5	5	3	5	5	3	5
67	13	53	3	7	11	3	P	31	3	P	7	3
69	3	P	7	3	P	17	3	13	P	3	P	P
71	11	3	31	67	3	19	17	3	7	P	3	11
73	7	73	3	P	P	3	P	7	3	11	P	3
75	3	5	5	3	5	5	3	5	5	3	5	5
77	19	3	P	P	3	7	P	3	41	13	3	P
79	29	47	3	11	7	3	P	79	3	P	17	3
81	3	11	P	3	P	31	3	23	P	3	7	17
83	P	3	7	P	3	43	P	3	59	7	3	83
85	5	5	3	5	5	3	5	5	3	5	5	3
87	3	83	P	3	P	13	3	7	P	3	P	P
89	37	3	P	P	3	P	7	3	P	19	3	P
91	23	19	3	P	P	3	13	61	3	P	P	3
93	3	P	59	3	7	P	3	P	P	3	P	7
95	5	3	5	5	3	5	5	3	5	5	3	5
97	P	13	3	71	43	3	53	11	3	7	P	3
99	3	7	P	3	P	11	3	19	7	3	43	37

The way in which this table may be used to find all the prime factors of a given number is explained on page **1–2**.

Prime Numbers and Smallest Factors

8401 **9599**

From to	8400 8500	8500 8600	8600 8700	8700 8800	8800 8900	8900 9000	9000 9100	9100 9200	9200 9300	9300 9400	9400 9500	9500 9600
1	31	P	3	7	13	3	P	19	3	71	7	3
3	3	11	7	3	P	29	3	P	P	3	P	13
5	5	3	5	5	3	5	5	3	5	5	3	5
7	7	47	3	P	P	3	P	7	3	41	23	3
9	3	67	P	3	23	59	3	P	P	3	97	37
11	13	3	79	31	3	7	P	3	61	P	3	P
13	47	P	3	P	7	3	P	13	3	67	P	3
15	3	5	5	3	5	5	3	5	5	3	5	5
17	19	3	7	23	3	37	71	3	13	7	3	31
19	P	7	3	P	P	3	29	11	3	P	P	3
21	3	P	37	3	P	11	3	7	P	3	P	P
23	P	3	P	11	3	P	7	3	23	P	3	89
25	5	5	3	5	5	3	5	5	3	5	5	3
27	3	P	P	3	7	79	3	P	P	3	11	7
29	P	3	P	7	3	P	P	3	11	19	3	13
31	P	19	3	P	P	3	11	23	3	7	P	3
33	3	7	89	3	11	P	3	P	7	3	P	P
35	5	3	5	5	3	5	5	3	5	5	3	5
37	11	P	-3	P	P	3	7	P	3	P	P	3
39	3	P	53	3	P	7	3	13	P	3	P	P
41	23	3	P	P	3	P	P	3	P	P	3	7
43	P	P	3	7	37	3	P	41	3	P	7	3
45	3	5	5	3	5	5	3	5	5	3	5	5
47	P	3	P	P	3	23	83	3	7	13	3	P
49	7	83	3	13	P	3	P	7	3	P	11	3
51	3	17	41	3	53	P	3	P	11	3	13	P
53	79	3	17	P	3	7	11	3	19	47	3	41
55	5	5	3	5	5	3	5	5	3	5	5	3
57	3	43	11	3	17	13	3	P	P	3	7	19
59	11	3	7	19	3	17	P	3	47	7	3	11
61	P	7	3	P	P	3	13	P	3	11	P	3
63	3	P	P	3	P	P	3	7	59	3	P	73
65	5	3	5	5	3	5	5	3	5	5	3	5
67	P	13	3	11	P	3	P	89	3	17	P	3
69	3	11	P	3	7	P	3	53	13	3	17	7
71	43	3	13	7	3	P	47	3	73	P	3	17
73	37	P	3	31	19	3	43	P	3	7	P	3
75	3	5	5	3	5	5	3	5	5	3	5	5
77	7	3	P	67	3	47	29	3	P	P	3	61
79	61	23	3	P	13	3	7	67	3	83	P	3
81	3	P	P	3	83	7	3	P	P	3	19	11
83	17	3	19	P	3	13	31	3	P	11	3	7
85	5	5	3	5	5	3	5	5	3	5	5	3
87	3	31	7	3	P	11	3	P	37	3	53	P
89	13	3	P	11	3	89	61	3	7	41	3	43
91	7	11	3	59	17	3	P	7	3	P	P	3
93	3	13	P	3	P	17	3	29	P	3	11	53
95	5	3	5	5	3	5	5	3	5	5	3	5
97	29	P	3	19	7	3	11	17	3	P	P	3
99	3	P	P	3	11	P	3	P	17	3	7	29

The way in which this table may be used to find all the prime factors of a given number is explained on page 1–2.

Powers, Roots and Reciprocals

1 **50**

No.	Square	Cube	Sq. Root	Cube Root	Reciprocal	No.
1	1	1	1.00000	1.00000	1.0000000	1
2	4	8	1.41421	1.25992	0.5000000	2
3	9	27	1.73205	1.44225	0.3333333	3
4	16	64	2.00000	1.58740	0.2500000	4
5	25	125	2.23607	1.70998	0.2000000	5
6	36	216	2.44949	1.81712	0.1666667	6
7	49	343	2.64575	1.91293	0.1428571	7
8	64	512	2.82843	2.00000	0.1250000	8
9	81	729	3.00000	2.08008	0.1111111	9
10	100	1,000	3.16228	2.15443	0.1000000	10
11	121	1,331	3.31662	2.22398	0.0909091	11
12	144	1,728	3.46410	2.28943	0.0833333	12
13	169	2,197	3.60555	2.35133	0.0769231	13
14	196	2,744	3.74166	2.41014	0.0714286	14
15	225	3,375	3.87298	2.46621	0.0666667	15
16	256	4,096	4.00000	2.51984	0.0625000	16
17	289	4,913	4.12311	2.57128	0.0588235	17
18	324	5,832	4.24264	2.62074	0.0555556	18
19	361	6,859	4.35890	2.66840	0.0526316	19
20	400	8,000	4.47214	2.71442	0.0500000	20
21	441	9,261	4.58258	2.75892	0.0476190	21
22	484	10,648	4.69042	2.80204	0.0454545	22
23	529	12,167	4.79583	2.84387	0.0434783	23
24	576	13,824	4.89898	2.88450	0.0416667	24
25	625	15,625	5.00000	2.92402	0.0400000	25
26	676	17,576	5.09902	2.96250	0.0384615	26
27	729	19,683	5.19615	3.00000	0.0370370	27
28	784	21,952	5.29150	3.03659	0.0357143	28
29	841	24,389	5.38516	3.07232	0.0344828	29
30	900	27,000	5.47723	3.10723	0.0333333	30
31	961	29,791	5.56776	3.14138	0.0322581	31
32	1,024	32,768	5.65685	3.17480	0.0312500	32
33	1,089	35,937	5.74456	3.20753	0.0303030	33
34	1,156	39,304	5.83095	3.23961	0.0294118	34
35	1,225	42,875	5.91608	3.27107	0.0285714	35
36	1,296	46,656	6.00000	3.30193	0.0277778	36
37	1,369	50,653	6.08276	3.33222	0.0270270	37
38	1,444	54,872	6.16441	3.36198	0.0263158	38
39	1,521	59,319	6.24500	3.39121	0.0256410	39
40	1,600	64,000	6.32456	3.41995	0.0250000	40
41	1,681	68,921	6.40312	3.44822	0.0243902	41
42	1,764	74,088	6.48074	3.47603	0.0238095	42
43	1,849	79,507	6.55744	3.50340	0.0232558	43
44	1,936	85,184	6.63325	3.53035	0.0227273	44
45	2,025	91,125	6.70820	3.55689	0.0222222	45
46	2,116	97,336	6.78233	3.58305	0.0217391	46
47	2,209	103,823	6.85565	3.60883	0.0212766	47
48	2,304	110,592	6.92820	3.63424	0.0208333	48
49	2,401	117,649	7.00000	3.65931	0.0204082	49
50	2,500	125,000	7.07107	3.68403	0.0200000	50

Powers and roots of numbers are explained on pages **1–10** to **1–12**. Reciprocals are discussed on page **1–8**.

Natural Logarithms

3.51 **6.00**

No.	Log_e	No.	Log_e	No.	Log_e	No.	Log_e	No.	Log_e
3.51	1.2556	4.01	1.3888	4.51	1.5063	5.01	1.6114	5.51	1.7066
3.52	1.2585	4.02	1.3913	4.52	1.5085	5.02	1.6134	5.52	1.7084
3.53	1.2613	4.03	1.3938	4.53	1.5107	5.03	1.6154	5.53	1.7102
3.54	1.2641	4.04	1.3962	4.54	1.5129	5.04	1.6174	5.54	1.7120
3.55	1.2669	4.05	1.3987	4.55	1.5151	5.05	1.6194	5.55	1.7138
3.56	1.2698	4.06	1.4012	4.56	1.5173	5.06	1.6214	5.56	1.7156
3.57	1.2726	4.07	1.4036	4.57	1.5195	5.07	1.6233	5.57	1.7174
3.58	1.2754	4.08	1.4061	4.58	1.5217	5.08	1.6253	5.58	1.7192
3.59	1.2782	4.09	1.4085	4.59	1.5239	5.09	1.6273	5.59	1.7210
3.60	1.2809	4.10	1.4110	4.60	1.5261	5.10	1.6292	5.60	1.7228
3.61	1.2837	4.11	1.4134	4.61	1.5282	5.11	1.6312	5.61	1.7246
3.62	1.2865	4.12	1.4159	4.62	1.5304	5.12	1.6332	5.62	1.7263
3.63	1.2892	4.13	1.4183	4.63	1.5326	5.13	1.6351	5.63	1.7281
3.64	1.2920	4.14	1.4207	4.64	1.5347	5.14	1.6371	5.64	1.7299
3.65	1.2947	4.15	1.4231	4.65	1.5369	5.15	1.6390	5.65	1.7317
3.66	1.2975	4.16	1.4255	4.66	1.5390	5.16	1.6409	5.66	1.7334
3.67	1.3002	4.17	1.4279	4.67	1.5412	5.17	1.6429	5.67	1.7352
3.68	1.3029	4.18	1.4303	4.68	1.5433	5.18	1.6448	5.68	1.7370
3.69	1.3056	4.19	1.4327	4.69	1.5454	5.19	1.6467	5.69	1.7387
3.70	1.3083	4.20	1.4351	4.70	1.5476	5.20	1.6487	5.70	1.7405
3.71	1.3110	4.21	1.4375	4.71	1.5497	5.21	1.6506	5.71	1.7422
3.72	1.3137	4.22	1.4398	4.72	1.5518	5.22	1.6525	5.72	1.7440
3.73	1.3164	4.23	1.4422	4.73	1.5539	5.23	1.6544	5.73	1.7457
3.74	1.3191	4.24	1.4446	4.74	1.5560	5.24	1.6563	5.74	1.7475
3.75	1.3218	4.25	1.4469	4.75	1.5581	5.25	1.6582	5.75	1.7492
3.76	1.3244	4.26	1.4493	4.76	1.5602	5.26	1.6601	5.76	1.7509
3.77	1.3271	4.27	1.4516	4.77	1.5623	5.27	1.6620	5.77	1.7527
3.78	1.3297	4.28	1.4540	4.78	1.5644	5.28	1.6639	5.78	1.7544
3.79	1.3324	4.29	1.4563	4.79	1.5665	5.29	1.6658	5.79	1.7561
3.80	1.3350	4.30	1.4586	4.80	1.5686	5.30	1.6677	5.80	1.7579
3.81	1.3376	4.31	1.4609	4.81	1.5707	5.31	1.6696	5.81	1.7596
3.82	1.3403	4.32	1.4633	4.82	1.5728	5.32	1.6715	5.82	1.7613
3.83	1.3429	4.33	1.4656	4.83	1.5748	5.33	1.6734	5.83	1.7630
3.84	1.3455	4.34	1.4679	4.84	1.5769	5.34	1.6752	5.84	1.7647
3.85	1.3481	4.35	1.4702	4.85	1.5790	5.35	1.6771	5.85	1.7664
3.86	1.3507	4.36	1.4725	4.86	1.5810	5.36	1.6790	5.86	1.7681
3.87	1.3533	4.37	1.4748	4.87	1.5831	5.37	1.6808	5.87	1.7699
3.88	1.3558	4.38	1.4770	4.88	1.5851	5.38	1.6827	5.88	1.7716
3.89	1.3584	4.39	1.4793	4.89	1.5872	5.39	1.6845	5.89	1.7733
3.90	1.3610	4.40	1.4816	4.90	1.5892	5.40	1.6864	5.90	1.7750
3.91	1.3635	4.41	1.4839	4.91	1.5913	5.41	1.6882	5.91	1.7766
3.92	1.3661	4.42	1.4861	4.92	1.5933	5.42	1.6901	5.92	1.7783
3.93	1.3686	4.43	1.4884	4.93	1.5953	5.43	1.6919	5.93	1.7800
3.94	1.3712	4.44	1.4907	4.94	1.5974	5.44	1.6938	5.94	1.7817
3.95	1.3737	4.45	1.4929	4.95	1.5994	5.45	1.6956	5.95	1.7834
3.96	1.3762	4.46	1.4951	4.96	1.6014	5.46	1.6974	5.96	1.7851
3.97	1.3788	4.47	1.4974	4.97	1.6034	5.47	1.6993	5.97	1.7867
3.98	1.3813	4.48	1.4996	4.98	1.6054	5.48	1.7011	5.98	1.7884
3.99	1.3838	4.49	1.5019	4.99	1.6074	5.49	1.7029	5.99	1.7901
4.00	1.3863	4.50	1.5041	5.00	1.6094	5.50	1.7047	6.00	1.7918

For a discussion of natural logarithms and their application, see Chapter 5.

Natural Logarithms

6.01 **8.50**

No.	$\log_e$	No.	$\log_e$	No.	$\log_e$	No.	$\log_e$	No.	$\log_e$
6.01	1.7934	6.51	1.8733	7.01	1.9473	7.51	2.0162	8.01	2.0807
6.02	1.7951	6.52	1.8749	7.02	1.9488	7.52	2.0176	8.02	2.0819
6.03	1.7967	6.53	1.8764	7.03	1.9502	7.53	2.0189	8.03	2.0832
6.04	1.7984	6.54	1.8779	7.04	1.9516	7.54	2.0202	8.04	2.0844
6.05	1.8001	6.55	1.8795	7.05	1.9530	7.55	2.0215	8.05	2.0857
6.06	1.8017	6.56	1.8810	7.06	1.9544	7.56	2.0229	8.06	2.0869
6.07	1.8034	6.57	1.8825	7.07	1.9559	7.57	2.0242	8.07	2.0882
6.08	1.8050	6.58	1.8840	7.08	1.9573	7.58	2.0255	8.08	2.0894
6.09	1.8066	6.59	1.8856	7.09	1.9587	7.59	2.0268	8.09	2.0906
6.10	1.8083	6.60	1.8871	7.10	1.9601	7.60	2.0281	8.10	2.0919
6.11	1.8099	6.61	1.8886	7.11	1.9615	7.61	2.0295	8.11	2.0931
6.12	1.8116	6.62	1.8901	7.12	1.9629	7.62	2.0308	8.12	2.0943
6.13	1.8132	6.63	1.8916	7.13	1.9643	7.63	2.0321	8.13	2.0956
6.14	1.8148	6.64	1.8931	7.14	1.9657	7.64	2.0334	8.14	2.0968
6.15	1.8165	6.65	1.8946	7.15	1.9671	7.65	2.0347	8.15	2.0980
6.16	1.8181	6.66	1.8961	7.16	1.9685	7.66	2.0360	8.16	2.0992
6.17	1.8197	6.67	1.8976	7.17	1.9699	7.67	2.0373	8.17	2.1005
6.18	1.8213	6.68	1.8991	7.18	1.9713	7.68	2.0386	8.18	2.1017
6.19	1.8229	6.69	1.9006	7.19	1.9727	7.69	2.0399	8.19	2.1029
6.20	1.8245	6.70	1.9021	7.20	1.9741	7.70	2.0412	8.20	2.1041
6.21	1.8262	6.71	1.9036	7.21	1.9755	7.71	2.0425	8.21	2.1054
6.22	1.8278	6.72	1.9051	7.22	1.9769	7.72	2.0438	8.22	2.1066
6.23	1.8294	6.73	1.9066	7.23	1.9782	7.73	2.0451	8.23	2.1078
6.24	1.8310	6.74	1.9081	7.24	1.9796	7.74	2.0464	8.24	2.1090
6.25	1.8326	6.75	1.9095	7.25	1.9810	7.75	2.0477	8.25	2.1102
6.26	1.8342	6.76	1.9110	7.26	1.9824	7.76	2.0490	8.26	2.1114
6.27	1.8358	6.77	1.9125	7.27	1.9838	7.77	2.0503	8.27	2.1126
6.28	1.8374	6.78	1.9140	7.28	1.9851	7.78	2.0516	8.28	2.1138
6.29	1.8390	6.79	1.9155	7.29	1.9865	7.79	2.0528	8.29	2.1150
6.30	1.8405	6.80	1.9169	7.30	1.9879	7.80	2.0541	8.30	2.1163
6.31	1.8421	6.81	1.9184	7.31	1.9892	7.81	2.0554	8.31	2.1175
6.32	1.8437	6.82	1.9199	7.32	1.9906	7.82	2.0567	8.32	2.1187
6.33	1.8453	6.83	1.9213	7.33	1.9920	7.83	2.0580	8.33	2.1199
6.34	1.8469	6.84	1.9228	7.34	1.9933	7.84	2.0592	8.34	2.1211
6.35	1.8485	6.85	1.9242	7.35	1.9947	7.85	2.0605	8.35	2.1223
6.36	1.8500	6.86	1.9257	7.36	1.9961	7.86	2.0618	8.36	2.1235
6.37	1.8516	6.87	1.9272	7.37	1.9974	7.87	2.0631	8.37	2.1247
6.38	1.8532	6.88	1.9286	7.38	1.9988	7.88	2.0643	8.38	2.1258
6.39	1.8547	6.89	1.9301	7.39	2.0001	7.89	2.0656	8.39	2.1270
6.40	1.8563	6.90	1.9315	7.40	2.0015	7.90	2.0669	8.40	2.1282
6.41	1.8579	6.91	1.9330	7.41	2.0028	7.91	2.0681	8.41	2.1294
6.42	1.8594	6.92	1.9344	7.42	2.0041	7.92	2.0694	8.42	2.1306
6.43	1.8610	6.93	1.9359	7.43	2.0055	7.93	2.0707	8.43	2.1318
6.44	1.8625	6.94	1.9373	7.44	2.0069	7.94	2.0719	8.44	2.1330
6.45	1.8641	6.95	1.9387	7.45	2.0082	7.95	2.0732	8.45	2.1342
6.46	1.8656	6.96	1.9402	7.46	2.0096	7.96	2.0744	8.46	2.1353
6.47	1.8672	6.97	1.9416	7.47	2.0109	7.97	2.0757	8.47	2.1365
6.48	1.8687	6.98	1.9430	7.48	2.0122	7.98	2.0769	8.48	2.1377
6.49	1.8703	6.99	1.9445	7.49	2.0136	7.99	2.0782	8.49	2.1389
6.50	1.8718	7.00	1.9459	7.50	2.0149	8.00	2.0794	8.50	2.1401

For a discussion of natural logarithms and their application, see Chapter 5.

Tables of common logarithms are given on pages 27–104 to 27–121.

Natural Logarithms

No.	Log$_e$	No.	Log$_e$	No.	Log$_e$	No.	Log$_e$	No.	Log$_e$
8.51	2.1412	9.01	2.1983	9.51	2.2523	10.25	2.3273	41	3.7136
8.52	2.1424	9.02	2.1994	9.52	2.2534	10.50	2.3514	42	3.7377
8.53	2.1436	9.03	2.2006	9.53	2.2544	10.75	2.3749	43	3.7612
8.54	2.1448	9.04	2.2017	9.54	2.2555	11.00	2.3979	44	3.7842
8.55	2.1459	9.05	2.2028	9.55	2.2565	11.25	2.4204	45	3.8067
8.56	2.1471	9.06	2.2039	9.56	2.2576	11.50	2.4423	46	3.8286
8.57	2.1483	9.07	2.2050	9.57	2.2586	11.75	2.4638	47	3.8501
8.58	2.1494	9.08	2.2061	9.58	2.2597	12.00	2.4849	48	3.8712
8.59	2.1506	9.09	2.2072	9.59	2.2607	12.25	2.5055	49	3.8918
8.60	2.1518	9.10	2.2083	9.60	2.2618	12.50	2.5257	50	3.9120
8.61	2.1529	9.11	2.2094	9.61	2.2628	12.75	2.5455	51	3.9318
8.62	2.1541	9.12	2.2105	9.62	2.2638	13.00	2.5649	52	3.9512
8.63	2.1552	9.13	2.2116	9.63	2.2649	13.25	2.5840	53	3.9703
8.64	2.1564	9.14	2.2127	9.64	2.2659	13.50	2.6027	54	3.9890
8.65	2.1576	9.15	2.2138	9.65	2.2670	13.75	2.6210	55	4.0073
8.66	2.1587	9.16	2.2148	9.66	2.2680	14.00	2.6391	56	4.0254
8.67	2.1599	9.17	2.2159	9.67	2.2690	14.25	2.6568	57	4.0431
8.68	2.1610	9.18	2.2170	9.68	2.2701	14.50	2.6741	58	4.0604
8.69	2.1622	9.19	2.2181	9.69	2.2711	14.75	2.6912	59	4.0775
8.70	2.1633	9.20	2.2192	9.70	2.2721	15.00	2.7081	60	4.0943
8.71	2.1645	9.21	2.2203	9.71	2.2732	15.50	2.7408	61	4.1109
8.72	2.1656	9.22	2.2214	9.72	2.2742	16.00	2.7726	62	4.1271
8.73	2.1668	9.23	2.2225	9.73	2.2752	16.50	2.8034	63	4.1431
8.74	2.1679	9.24	2.2235	9.74	2.2762	17.00	2.8332	64	4.1589
8.75	2.1691	9.25	2.2246	9.75	2.2773	17.50	2.8622	65	4.1744
8.76	2.1702	9.26	2.2257	9.76	2.2783	18.00	2.8904	66	4.1897
8.77	2.1713	9.27	2.2268	9.77	2.2793	18.50	2.9178	67	4.2047
8.78	2.1725	9.28	2.2279	9.78	2.2803	19.00	2.9444	68	4.2195
8.79	2.1736	9.29	2.2289	9.79	2.2814	19.50	2.9704	69	4.2341
8.80	2.1748	9.30	2.2300	9.80	2.2824	20.00	2.9957	70	4.2485
8.81	2.1759	9.31	2.2311	9.81	2.2834	21	3.0445	71	4.2627
8.82	2.1770	9.32	2.2322	9.82	2.2844	22	3.0910	72	4.2767
8.83	2.1782	9.33	2.2332	9.83	2.2854	23	3.1355	73	4.2905
8.84	2.1793	9.34	2.2343	9.84	2.2865	24	3.1781	74	4.3041
8.85	2.1804	9.35	2.2354	9.85	2.2875	25	3.2189	75	4.3175
8.86	2.1815	9.36	2.2364	9.86	2.2885	26	3.2581	76	4.3307
8.87	2.1827	9.37	2.2375	9.87	2.2895	27	3.2958	77	4.3438
8.88	2.1838	9.38	2.2386	9.88	2.2905	28	3.3322	78	4.3567
8.89	2.1849	9.39	2.2396	9.89	2.2915	29	3.3673	79	4.3694
8.90	2.1861	9.40	2.2407	9.90	2.2925	30	3.4012	80	4.3820
8.91	2.1872	9.41	2.2418	9.91	2.2935	31	3.4340	82	4.4067
8.92	2.1883	9.42	2.2428	9.92	2.2946	32	3.4657	84	4.4308
8.93	2.1894	9.43	2.2439	9.93	2.2956	33	3.4965	86	4.4543
8.94	2.1905	9.44	2.2450	9.94	2.2966	34	3.5264	88	4.4773
8.95	2.1917	9.45	2.2460	9.95	2.2976	35	3.5553	90	4.4998
8.96	2.1928	9.46	2.2471	9.96	2.2986	36	3.5835	92	4.5218
8.97	2.1939	9.47	2.2481	9.97	2.2996	37	3.6109	94	4.5433
8.98	2.1950	9.48	2.2492	9.98	2.3006	38	3.6376	96	4.5643
8.99	2.1961	9.49	2.2502	9.99	2.3016	39	3.6636	98	4.5850
9.00	2.1972	9.50	2.2513	10.00	2.3026	40	3.6889	100	4.6052

For a discussion of natural logarithms and their application, see Chapter 5.

Areas and Dimensions of Plane Figures

In the following tables are given the areas of plane figures, together with other formulas relating to their dimensions and properties; the surfaces of solids; and the volumes of solids. The notation used in the formulas is, as far as possible, given in the illustration accompanying them; where this has not been possible, it is given at the beginning of each set of formulas.

1. Square — Area; length of side and diagonal.
2. Rectangle — Area; length of sides and diagonal.
3. Parallelogram — Altitude and length of base.
4. Right-angled Triangle — Area; length of sides and hypotenuse.
5. Acute-angled Triangle — Area.

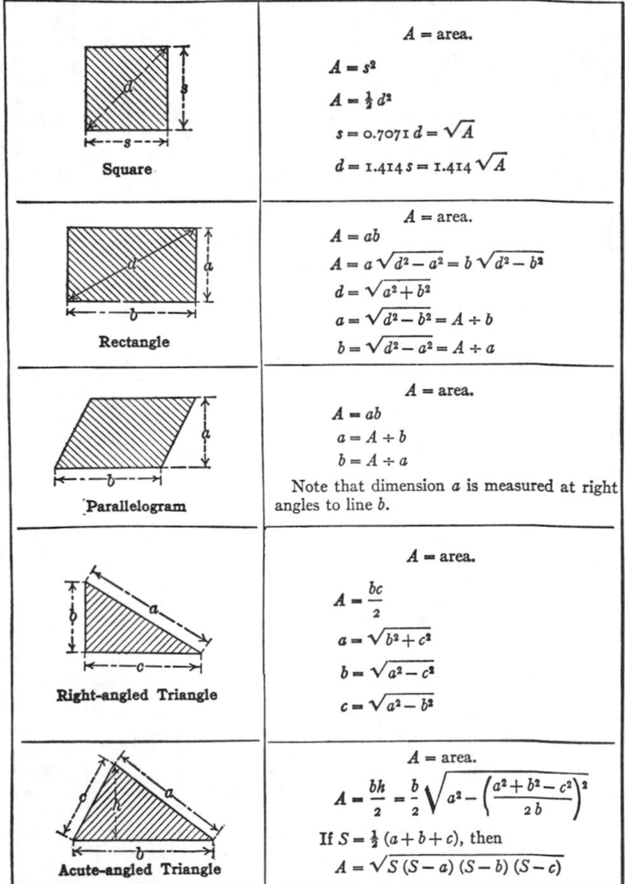

Square	A = area. $A = s^2$ $A = \frac{1}{2} d^2$ $s = 0.7071\, d = \sqrt{A}$ $d = 1.414\, s = 1.414 \sqrt{A}$
Rectangle	A = area. $A = ab$ $A = a\sqrt{d^2 - a^2} = b\sqrt{d^2 - b^2}$ $d = \sqrt{a^2 + b^2}$ $a = \sqrt{d^2 - b^2} = A \div b$ $b = \sqrt{d^2 - a^2} = A \div a$
Parallelogram	A = area. $A = ab$ $a = A \div b$ $b = A \div a$ Note that dimension a is measured at right angles to line b.
Right-angled Triangle	A = area. $A = \frac{bc}{2}$ $a = \sqrt{b^2 + c^2}$ $b = \sqrt{a^2 - c^2}$ $c = \sqrt{a^2 - b^2}$
Acute-angled Triangle	A = area. $A = \frac{bh}{2} = \frac{b}{2}\sqrt{a^2 - \left(\frac{a^2 + b^2 - c^2}{2b}\right)^2}$ If $S = \frac{1}{2}(a + b + c)$, then $A = \sqrt{S(S-a)(S-b)(S-c)}$

Examples Showing Use of Formulas

Below are given a number of examples showing the use of the formulas on the opposite page. Each section of the page corresponds to the opposite section on the previous page, and the illustration on that page should be referred to. The notation used in the illustrations is also used in the examples given.

1. Square — Area; length of side and diagonal.
2. Rectangle — Area; length of sides and diagonal.
3. Parallelogram — Area and length of base.
4. Right-angled Triangle — Area and length of sides.
5. Acute-angled Triangle — Area.

Square. — Assume that the side s of a square is 15 inches. Find the area and the length of the diagonal.

$$\text{Area} = A = s^2 = 15^2 = 225 \text{ square inches.}$$

$$\text{Diagonal} = d = 1.414\, s = 1.414 \times 15 = 21.21 \text{ inches.}$$

The area of a square is 625 square inches. Find the length of the side s and the diagonal d.

$$s = \sqrt{A} = \sqrt{625} = 25 \text{ inches.}$$

$$d = 1.414 \sqrt{A} = 1.414 \times 25 = 35.35 \text{ inches.}$$

Rectangle. — The side a of a rectangle is 12 inches, and the area 70.5 square inches. Find the length of the side b, and the diagonal d.

$$b = A \div a = 70.5 \div 12 = 5.875 \text{ inches.}$$

$$d = \sqrt{a^2 + b^2} = \sqrt{12^2 + 5.875^2} = \sqrt{178.516} = 13.361 \text{ inches.}$$

The sides of a rectangle are 30.5 and 11 inches long. Find the area.

$$\text{Area} = a \times b = 30.5 \times 11 = 335.5 \text{ square inches.}$$

Parallelogram. — The base b of a parallelogram is 16 feet. The height a is 5.5 feet. Find the area.

$$\text{Area} = A = a \times b = 5.5 \times 16 = 88 \text{ square feet.}$$

The area of a parallelogram is 12 square inches. The height is 1.5 inch. Find the length of the base b.

$$b = A \div a = 12 \div 1.5 = 8 \text{ inches.}$$

Right-angled Triangle. — The sides b and c in a right-angled triangle are 6 and 8 inches. Find side a and the area.

$$a = \sqrt{b^2 + c^2} = \sqrt{6^2 + 8^2} = \sqrt{36 + 64} = \sqrt{100} = 10 \text{ inches.}$$

$$A = \frac{b \times c}{2} = \frac{6 \times 8}{2} = \frac{48}{2} = 24 \text{ square inches.}$$

If $a = 10$ and $b = 6$, had been known, but not c, the latter would have been found as follows:

$$c = \sqrt{a^2 - b^2} = \sqrt{10^2 - 6^2} = \sqrt{100 - 36} = \sqrt{64} = 8 \text{ inches.}$$

Acute-angled Triangle. — If $a = 10$, $b = 9$, and $c = 8$ inches, what is the area of the triangle?

$$A = \frac{b}{2}\sqrt{a^2 - \left(\frac{a^2 + b^2 - c^2}{2\,b}\right)^2} = \frac{9}{2}\sqrt{10^2 - \left(\frac{10^2 + 9^2 - 8^2}{2 \times 9}\right)^2} = 4.5\sqrt{100 - \left(\frac{117}{18}\right)^2}$$

$$= 4.5\sqrt{100 - 42.25} = 4.5\sqrt{57.75} = 4.5 \times 7.60 = 34.20 \text{ square inches.}$$

Areas and Dimensions of Plane Figures

1. Obtuse-angled Triangle — Area.
2. Trapezoid — Area.
3. Trapezium — Area.
4. Regular Hexagon — Area and radius of circles.
5. Regular Octagon — Area and radius of circles.
6. Regular Polygon — Area; sizes of angles; radius of circles.

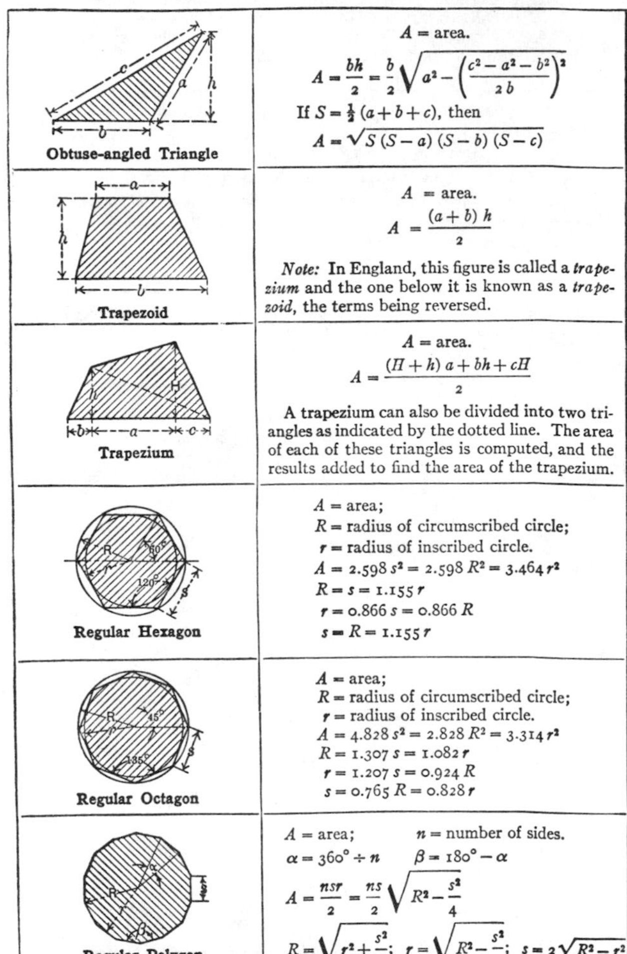

Obtuse-angled Triangle	A = area. $$A=\frac{bh}{2}=\frac{b}{2}\sqrt{a^2-\left(\frac{c^2-a^2-b^2}{2b}\right)^2}$$ If $S=\frac{1}{2}(a+b+c)$, then $$A=\sqrt{S(S-a)(S-b)(S-c)}$$
Trapezoid	A = area. $$A=\frac{(a+b)\,h}{2}$$ *Note:* In England, this figure is called a ***trapezium*** and the one below it is known as a ***trapezoid,*** the terms being reversed.
Trapezium	A = area. $$A=\frac{(H+h)\,a+bh+cH}{2}$$ A trapezium can also be divided into two triangles as indicated by the dotted line. The area of each of these triangles is computed, and the results added to find the area of the trapezium.
Regular Hexagon	A = area; R = radius of circumscribed circle; r = radius of inscribed circle. $A = 2.598\,s^2 = 2.598\,R^2 = 3.464\,r^2$ $R = s = 1.155\,r$ $r = 0.866\,s = 0.866\,R$ $s = R = 1.155\,r$
Regular Octagon	A = area; R = radius of circumscribed circle; r = radius of inscribed circle. $A = 4.828\,s^2 = 2.828\,R^2 = 3.314\,r^2$ $R = 1.307\,s = 1.082\,r$ $r = 1.207\,s = 0.924\,R$ $s = 0.765\,R = 0.828\,r$
Regular Polygon	A = area; n = number of sides. $\alpha = 360° \div n$ $\beta = 180° - \alpha$ $$A=\frac{nsr}{2}=\frac{ns}{2}\sqrt{R^2-\frac{s^2}{4}}$$ $R=\sqrt{r^2+\frac{s^2}{4}}$; $r=\sqrt{R^2-\frac{s^2}{4}}$; $s=2\sqrt{R^2-r^2}$

Examples Showing Use of Formulas

1. Obtuse-angled Triangle — Area.
2. Trapezoid — Area.
3. Trapezium — Area.
4. Regular Hexagon — Area and radius of inscribed circle.
5. Regular Octagon — Area and length of side.
6. Regular Polygon — Area; size of angles; radius of circles.

Obtuse-angled Triangle. — The side $a = 5$, side $b = 4$, and side $c = 8$ inches. Find the area.

$$S = \tfrac{1}{2}(a+b+c) = \tfrac{1}{2}(5+4+8) = \tfrac{1}{2} \times 17 = 8.5$$

$$A = \sqrt{S(S-a)(S-b)(S-c)} = \sqrt{8.5(8.5-5)(8.5-4)(8.5-8)}$$

$$= \sqrt{8.5 \times 3.5 \times 4.5 \times 0.5} = \sqrt{66.937} = 8.18 \text{ square inches.}$$

Trapezoid. — Side $a = 23$ feet, side $b = 32$ feet, and height $h = 12$ feet. Find the area.

$$A = \frac{(a+b)h}{2} = \frac{(23+32)12}{2} = \frac{55 \times 12}{2} = \frac{660}{2} = 330 \text{ square feet.}$$

Trapezium. — Let $a = 10$, $b = 2$, $c = 3$, $h = 8$, and $H = 12$ inches. Find the area.

$$A = \frac{(H+h)a + bh + cH}{2} = \frac{(12+8)10 + 2 \times 8 + 3 \times 12}{2}$$

$$= \frac{20 \times 10 + 16 + 36}{2} = \frac{252}{2} = 126 \text{ square inches.}$$

Regular Hexagon. — The side s of a regular hexagon is 4 inches. Find the area and the radius r of the inscribed circle.

$$A = 2.598\,s^2 = 2.598 \times 4^2 = 2.598 \times 16 = 41.568 \text{ square inches.}$$

$$r = 0.866\,s = 0.866 \times 4 = 3.464 \text{ inches.}$$

What is the length of the side of a hexagon that is described about a circle of 5 inches radius? — Here $r = 5$. Hence,

$$s = 1.155\,r = 1.155 \times 5 = 5.775 \text{ inches.}$$

Regular Octagon. — Find the area and the length of the side of an octagon that is inscribed in a circle of 12 inches diameter.

Diameter of circumscribed circle = 12 inches; hence, $R = 6$ inches.

$$A = 2.828\,R^2 = 2.828 \times 6^2 = 2.828 \times 36 = 101.81 \text{ square inches.}$$

$$s = 0.765\,R = 0.765 \times 6 = 4.590 \text{ inches.}$$

Regular Polygon. — Find the area of a polygon having 12 sides, inscribed in a circle of 8 inches radius. The length of the side s is 4.141 inches.

$$A = \frac{ns}{2}\sqrt{R^2 - \frac{s^2}{4}} = \frac{12 \times 4.141}{2}\sqrt{8^2 - \frac{4.141^2}{4}} = 24.846\sqrt{59.713}$$

$$= 24.846 \times 7.727 = 191.98 \text{ square inches.}$$

Areas and Dimensions of Plane Figures

1. Circle — Area; circumference; radius; diameter; arc.
2. Circular Sector — Area; length of arc; angle; radius.
3. Circular Segment — Area; chord; radius; arc; angle; height.
4. Circular Ring — Area.
5. Circular Ring Sector — Area.
6. Spandrel or Fillet — Area.

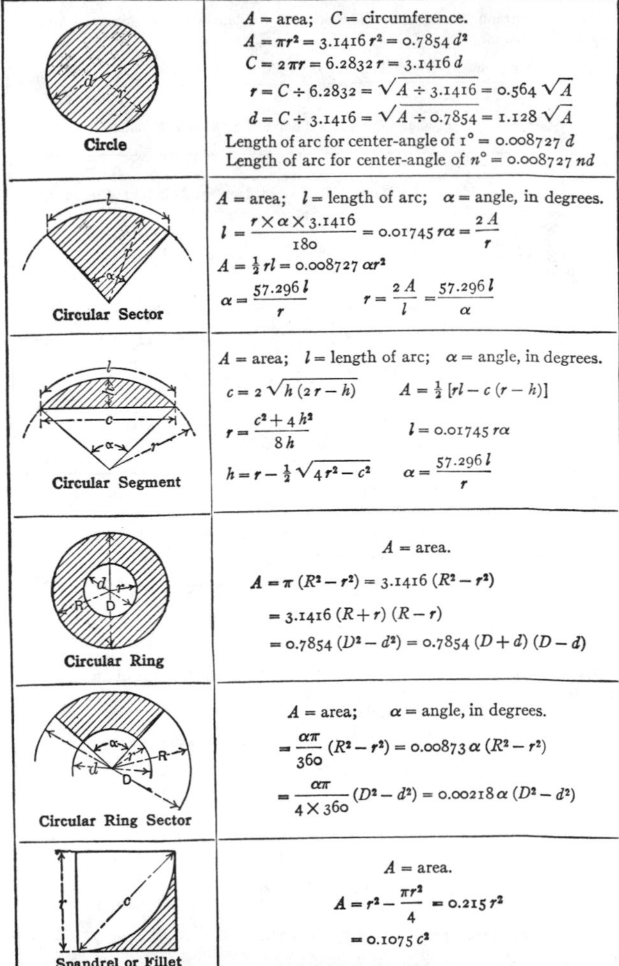

Circle	A = area; C = circumference. $A = \pi r^2 = 3.1416\, r^2 = 0.7854\, d^2$ $C = 2\pi r = 6.2832\, r = 3.1416\, d$ $r = C \div 6.2832 = \sqrt{A \div 3.1416} = 0.564 \sqrt{A}$ $d = C \div 3.1416 = \sqrt{A \div 0.7854} = 1.128 \sqrt{A}$ Length of arc for center-angle of $1° = 0.008727\, d$ Length of arc for center-angle of $n° = 0.008727\, nd$
Circular Sector	A = area; l = length of arc; α = angle, in degrees. $l = \frac{r \times \alpha \times 3.1416}{180} = 0.01745\, r\alpha = \frac{2A}{r}$ $A = \frac{1}{2} rl = 0.008727\, \alpha r^2$ $\alpha = \frac{57.296\, l}{r}$ $r = \frac{2A}{l} = \frac{57.296\, l}{\alpha}$
Circular Segment	A = area; l = length of arc; α = angle, in degrees. $c = 2\sqrt{h(2r - h)}$ $A = \frac{1}{2}[rl - c(r - h)]$ $r = \frac{c^2 + 4h^2}{8h}$ $l = 0.01745\, r\alpha$ $h = r - \frac{1}{2}\sqrt{4r^2 - c^2}$ $\alpha = \frac{57.296\, l}{r}$
Circular Ring	A = area. $A = \pi(R^2 - r^2) = 3.1416\,(R^2 - r^2)$ $= 3.1416\,(R + r)(R - r)$ $= 0.7854\,(D^2 - d^2) = 0.7854\,(D + d)(D - d)$
Circular Ring Sector	A = area; α = angle, in degrees. $= \frac{\alpha\pi}{360}(R^2 - r^2) = 0.00873\, \alpha\, (R^2 - r^2)$ $= \frac{\alpha\pi}{4 \times 360}(D^2 - d^2) = 0.00218\, \alpha\, (D^2 - d^2)$
Spandrel or Fillet	A = area. $A = r^2 - \frac{\pi r^2}{4} = 0.215\, r^2$ $= 0.1075\, c^2$

Examples Showing Use of Formulas

1. Circle — Area; circumference; diameter.
2. Circular Sector — Area; length of arc.
3. Circular Segment — Radius; chord.
4. Circular Ring — Area.
5. Circular Ring Sector — Area.
6. Spandrel or Fillet — Area.

Circle. — Find the area A and circumference C of a circle with a diameter of 2¾ inches.

$A = 0.7854\, d^2 = 0.7854 \times 2.75^2 = 0.7854 \times 2.75 \times 2.75 = 5.9396$ square inches.

$C = 3.1416\, d = 3.1416 \times 2.75 = 8.6394$ inches.

The area of a circle is 16.8 square inches. Find its diameter.

$$d = 1.128\sqrt{A} = 1.128\sqrt{16.8} = 1.128 \times 4.099 = 4.624 \text{ inches.}$$

Circular Sector. — The radius of a circle is 1½ inch, and angle α of a sector of the circle is 60 degrees. Find the area of the sector and the length of arc l.

$A = 0.008727\, \alpha r^2 = 0.008727 \times 60 \times 1.5^2 = 0.5236 \times 1.5 \times 1.5 = 1.178$ sq. inch.

$l = 0.01745\, r\alpha = 0.01745 \times 1.5 \times 60 = 1.5705$ inch.

Circular Segment. — The radius r of a circular segment is 60 inches and the height h is 8 inches. Find the length of the chord c.

$c = 2\sqrt{h\,(2r - h)} = 2\sqrt{8 \times (2 \times 60 - 8)} = 2\sqrt{896} = 2 \times 29.93 = 59.86$ inches.

If $c = 16$, and $h = 6$ inches, what is the radius of the circle of which the segment is a part?

$$r = \frac{c^2 + 4h^2}{8h} = \frac{16^2 + 4 \times 6^2}{8 \times 6} = \frac{256 + 144}{48} = \frac{400}{48} = 8\tfrac{1}{3} \text{ inches.}$$

Circular Ring. — Let the outside diameter $D = 12$ inches and the inside diameter $d = 8$ inches. Find area of ring.

$A = 0.7854\,(D^2 - d^2) = 0.7854\,(12^2 - 8^2) = 0.7854\,(144 - 64) = 0.7854 \times 80$
$= 62.83$ square inches.

By the alternative formula:

$A = 0.7854\,(D + d)\,(D - d) = 0.7854\,(12 + 8)\,(12 - 8) = 0.7854 \times 20 \times 4$
$= 62.83$ square inches.

Circular Ring Sector. — Find the area, if the outside radius $R = 5$ inches, the inside radius $r = 2$ inches, and $\alpha = 72$ degrees.

$$A = 0.00873\,\alpha\,(R^2 - r^2) = 0.00873 \times 72\,(5^2 - 2^2)$$
$$= 0.6286\,(25 - 4) = 0.6286 \times 21 = 13.2 \text{ square inches.}$$

Spandrel or Fillet. — Find the area of a spandrel, the radius of which is 0.7 inch.

$$A = 0.215\, r^2 = 0.215 \times 0.7^2 = 0.215 \times 0.7 \times 0.7 = 0.105 \text{ square inch.}$$

If chord c were given as 2.2 inches, what would be the area?

$$A = 0.1075\, c^2 = 0.1075 \times 2.2^2 = 0.1075 \times 4.84 = 0.520 \text{ square inch.}$$

Areas and Dimensions of Plane Figures

1. Ellipse — Area; circumference.
2. Hyperbola — Area.
3. Parabola — Length of arc.
4. Parabola — Area.
5. Segment of Parabola — Area.
6. Cycloid — Area; length of cycloid.

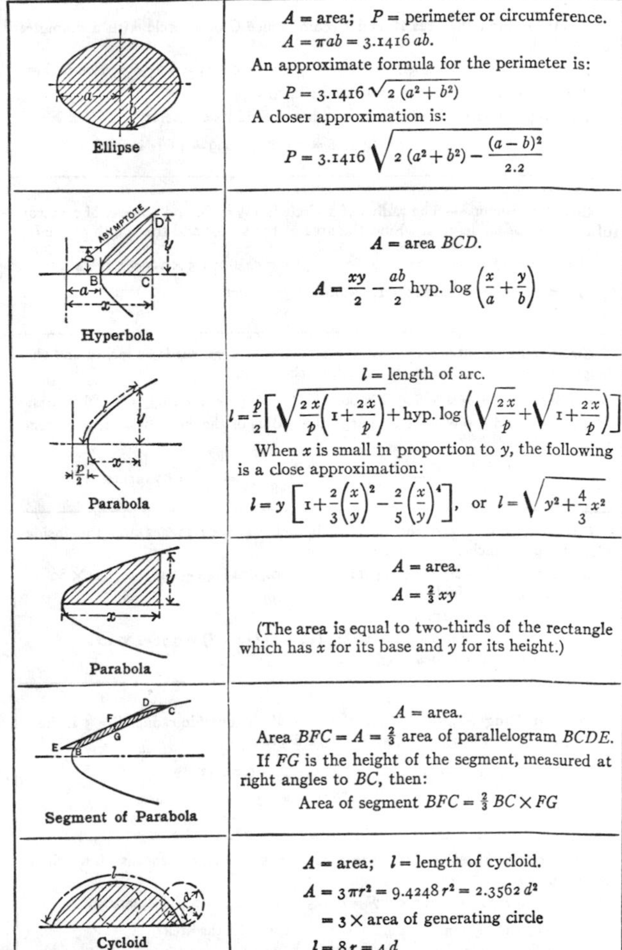

Ellipse	A = area; P = perimeter or circumference. $A = \pi ab = 3.1416\,ab.$ An approximate formula for the perimeter is: $P = 3.1416\sqrt{2\,(a^2+b^2)}$ A closer approximation is: $P = 3.1416\sqrt{2\,(a^2+b^2) - \dfrac{(a-b)^2}{2.2}}$
Hyperbola	A = area BCD. $A = \dfrac{xy}{2} - \dfrac{ab}{2}\ \text{hyp. log}\left(\dfrac{x}{a}+\dfrac{y}{b}\right)$
Parabola	l = length of arc. $l = \dfrac{p}{2}\left[\sqrt{\dfrac{2x}{p}\left(1+\dfrac{2x}{p}\right)} + \text{hyp. log}\left(\sqrt{\dfrac{2x}{p}}+\sqrt{1+\dfrac{2x}{p}}\right)\right]$ When x is small in proportion to y, the following is a close approximation: $l = y\left[1+\dfrac{2}{3}\left(\dfrac{x}{y}\right)^2 - \dfrac{2}{5}\left(\dfrac{x}{y}\right)^4\right]$, or $l = \sqrt{y^2+\dfrac{4}{3}x^2}$
Parabola	A = area. $A = \frac{2}{3}xy$ (The area is equal to two-thirds of the rectangle which has x for its base and y for its height.)
Segment of Parabola	A = area. Area $BFC = A = \frac{2}{3}$ area of parallelogram $BCDE$. If FG is the height of the segment, measured at right angles to BC, then: Area of segment $BFC = \frac{2}{3}\,BC \times FG$
Cycloid	A = area; l = length of cycloid. $A = 3\pi r^2 = 9.4248\,r^2 = 2.3562\,d^2$ $= 3 \times$ area of generating circle $l = 8r = 4d$

Examples Showing Use of Formulas

1. Ellipse — Area; circumference.
2. Hyperbola — Area.
3. Parabola — Length of arc.
4. Parabola — Area.
5. Segment of Parabola — Area.
6. Cycloid — Area; length of cycloid.

Ellipse. — The larger or major axis is 8 inches. The smaller or minor axis is 6 inches. Find the area and the approximate circumference. Here, then, $a = 4$, and $b = 3$.

$$A = 3.1416\,ab = 3.1416 \times 4 \times 3 = 37.699 \text{ square inches.}$$

$$P = 3.1416\sqrt{2\,(a^2 + b^2)} = 3.1416 \times \sqrt{2\,(4^2 + 3^2)} = 3.1416 \times \sqrt{2 \times 25}$$
$$= 3.1416\sqrt{50} = 3.1416 \times 7.071 = 22.214 \text{ inches.}$$

Hyperbola. — The half-axes a and b are 3 and 2 inches, respectively. Find area shown shaded in illustration for $x = 8$ and $y = 5$.

Inserting the known values in the formula:

$$A = \frac{8 \times 5}{2} - \frac{3 \times 2}{2} \times \text{hyp. log}\left(\frac{8}{3} + \frac{5}{2}\right) = 20 - 3 \times \text{hyp. log } 5.167$$
$$= 20 - 3 \times 1.6423 = 20 - 4.927 = 15.073 \text{ square inches.}$$

Parabola. — If $x = 2$ and $y = 24$ feet, what is the approximate length l of the parabolic curve?

$$l = y\left[1 + \frac{2}{3}\left(\frac{x}{y}\right)^2 - \frac{2}{5}\left(\frac{x}{y}\right)^4\right] = 24\left[1 + \frac{2}{3}\left(\frac{2}{24}\right)^2 - \frac{2}{5}\left(\frac{2}{24}\right)^4\right]$$
$$= 24\left[1 + \frac{2}{3} \times \frac{1}{144} - \frac{2}{5} \times \frac{1}{20{,}736}\right] = 24 \times 1.0046 = 24.11 \text{ feet.}$$

Parabola. — Let the dimension x in the illustration be 15 inches, and y, 9 inches. Find the area of the shaded portion of the parabola.

$$A = \tfrac{2}{3} \times xy = \tfrac{2}{3} \times 15 \times 9 = 10 \times 9 = 90 \text{ square inches.}$$

Segment of Parabola. — The length of the chord BC = 19.5 inches. The distance between lines BC and DE, measured at right angles to BC, is 2.25 inches. This is the height of the segment. Find the area.

$$\text{Area} = A = \tfrac{2}{3}\,BC \times FG = \tfrac{2}{3} \times 19.5 \times 2.25 = 29.25 \text{ square inches.}$$

Cycloid. — The diameter of the generating circle of a cycloid is 6 inches. Find the length l of the cycloidal curve, and the area enclosed between the curve and the base line.

$$l = 4\,d = 4 \times 6 = 24 \text{ inches.}$$
$$A = 2.3562\,d^2 = 2.3562 \times 6^2 = 2.3562 \times 36 = 84.82 \text{ square inches.}$$

Volumes and Dimensions of Solid Figures

1. Cube — Volume; length of side.
2. Square Prism — Volume; length of sides.
3. Prism — Volume; area of end surface.
4. Pyramid — Volume.
5. Frustum of Pyramid — Volume.
6. Wedge — Volume.

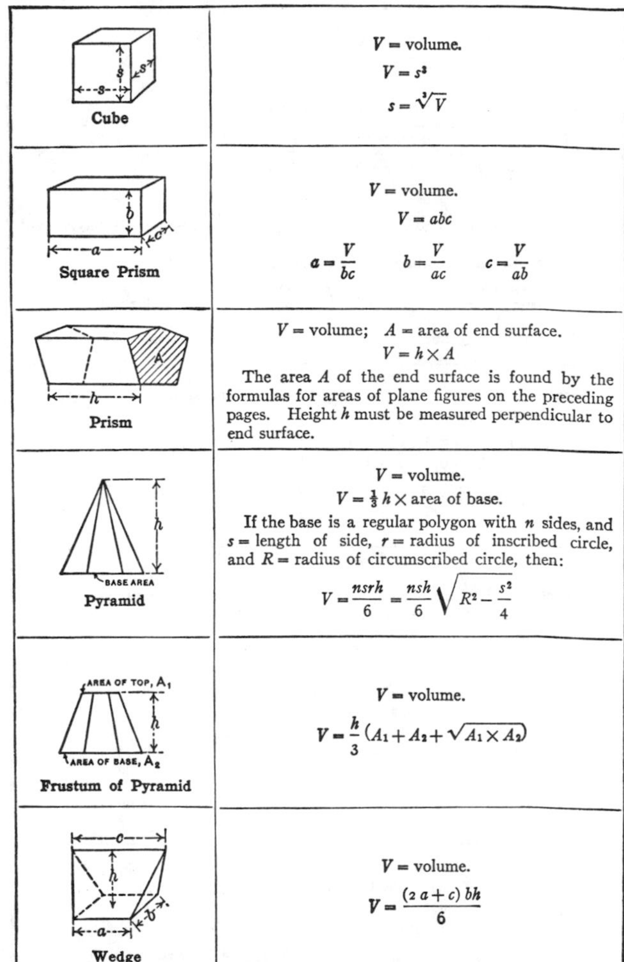

Figure	Formulas
Cube	V = volume. $V = s^3$ $s = \sqrt[3]{V}$
Square Prism	V = volume. $V = abc$ $a = \frac{V}{bc}$ $b = \frac{V}{ac}$ $c = \frac{V}{ab}$
Prism	V = volume; A = area of end surface. $V = h \times A$ The area A of the end surface is found by the formulas for areas of plane figures on the preceding pages. Height h must be measured perpendicular to end surface.
Pyramid	V = volume. $V = \frac{1}{3} h \times$ area of base. If the base is a regular polygon with n sides, and s = length of side, r = radius of inscribed circle, and R = radius of circumscribed circle, then: $V = \frac{nsrh}{6} = \frac{nsh}{6}\sqrt{R^2 - \frac{s^2}{4}}$
Frustum of Pyramid	V = volume. $V = \frac{h}{3}(A_1 + A_2 + \sqrt{A_1 \times A_2})$
Wedge	V = volume. $V = \frac{(2a + c)bh}{6}$

Examples Showing Use of Formulas

1. Cube — Volume; length of side.
2. Square Prism — Volume; length of side.
3. Prism — Volume; area of end surface.
4. Pyramid — Volume.
5. Frustum of Pyramid — Volume.
6. Wedge — Volume.

Cube. — The side of a cube equals 9.5 inches. Find its volume.

$$\text{Volume} = V = s^3 = 9.5^3 = 9.5 \times 9.5 \times 9.5 = 857.375 \text{ cubic inches.}$$

The volume of a cube is 231 cubic inches. What is the length of the side?

$$s = \sqrt[3]{V} = \sqrt[3]{231} = 6.136 \text{ inches.}$$

Square Prism. — In a square prism, $a = 6$, $b = 5$, $c = 4$. Find the volume.

$$V = a \times b \times c = 6 \times 5 \times 4 = 120 \text{ cubic inches.}$$

How high should a box be made to contain 25 cubic feet, if it is 4 feet long and $2\frac{1}{2}$ feet wide? Here, $a = 4$, $c = 2.5$, and $V = 25$. Then,

$$b = \text{depth} = \frac{V}{ac} = \frac{25}{4 \times 2.5} = \frac{25}{10} = 2.5 \text{ feet.}$$

Prism. — A prism having for its base a regular hexagon with a side s of 3 inches, is 10 inches high. Find the volume.

$$\text{Area of hexagon} = A = 2.598\,s^2 = 2.598 \times 9 = 23.382 \text{ square inches.}$$

$$\text{Volume of prism} = h \times A = 10 \times 23.382 = 233.82 \text{ cubic inches.}$$

Pyramid. —A pyramid, having a height of 9 feet, has a base formed by a rectangle, the sides of which are 2 and 3 feet, respectively. Find the volume.

$$\text{Area of base} = 2 \times 3 = 6 \text{ square feet;} \quad h = 9 \text{ feet.}$$

$$\text{Volume} = V = \tfrac{1}{3} h \times \text{area of base} = \tfrac{1}{3} \times 9 \times 6 = 18 \text{ cubic feet.}$$

Frustum of Pyramid. — The pyramid in the previous example is cut off $4\frac{1}{2}$ feet from the base, the upper part being removed. The sides of the rectangle forming the top surface of the frustum are, then, 1 and $1\frac{1}{2}$ foot long, respectively. Find the volume of the frustum.

$$\text{Area of top} = A_1 = 1 \times 1\tfrac{1}{2} = 1\tfrac{1}{2} \text{ sq. ft.} \quad \text{Area of base} = A_2 = 2 \times 3 = 6 \text{ sq. ft.}$$

$$V = \frac{4.5}{3}\left(1.5 + 6 + \sqrt{1.5 \times 6}\right) = 1.5\left(7.5 + \sqrt{9}\right) = 1.5 \times 10.5 = 15.75 \text{ cubic feet.}$$

Wedge. — Let $a = 4$ inches, $b = 3$ inches, and $c = 5$ inches. The height $h = 4.5$ inches. Find the volume.

$$V = \frac{(2a + c)\,bh}{6} = \frac{(2 \times 4 + 5) \times 3 \times 4.5}{6} = \frac{(8 + 5) \times 13.5}{6} = \frac{13 \times 13.5}{6}$$

$$= \frac{175.5}{6} = 29.25 \text{ cubic inches.}$$

Volumes and Dimensions of Solid Figures

1. Cylinder — Volume; area of surfaces.
2. Portion of Cylinder — Volume; surface area.
3. Portion of Cylinder — Volume; surface area.
4. Hollow Cylinder — Volume.
5. Cone — Volume; area of surface.
6. Frustum of Cone — Volume; surface area.

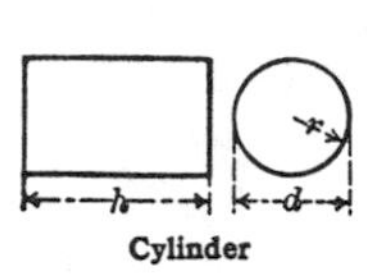

Cylinder

V = volume; S = area of cylindrical surface.

$$V = 3.1416\,r^2h = 0.7854\,d^2h$$

$$S = 6.2832\,rh = 3.1416\,dh$$

Total area A of cylindrical surface and end surfaces:

$$A = 6.2832\,r\,(r+h) = 3.1416\,d\,(\tfrac{1}{2}\,d+h)$$

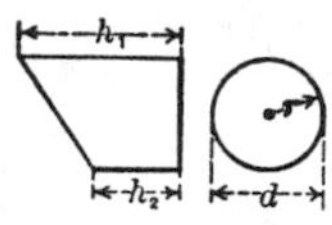

Portion of Cylinder

V = volume; S = area of cylindrical surface.

$$V = 1.5708\,r^2\,(h_1+h_2) = 0.3927\,d^2\,(h_1+h_2)$$

$$S = 3.1416\,r\,(h_1+h_2) = 1.5708\,d\,(h_1+h_2)$$

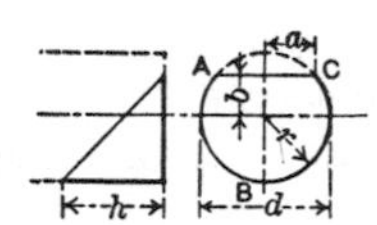

Portion of Cylinder

V = volume; S = area of cylindrical surface.

$$V = \left(\frac{2}{3}a^3 \pm b \times \text{area } ABC\right)\frac{h}{r \pm b}$$

$$S = (ad \pm b \times \text{length of arc } ABC)\,\frac{h}{r \pm b}$$

Use + when base area is larger, and − when base area is less than one-half the base circle.

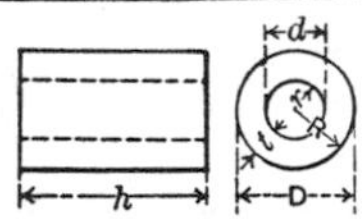

Hollow Cylinder

V = volume.

$$V = 3.1416\,h\,(R^2-r^2) = 0.7854\,h\,(D^2-d^2)$$

$$= 3.1416\,ht\,(2\,R-t) = 3.1416\,ht\,(D-t)$$

$$= 3.1416\,ht\,(2\,r+t) = 3.1416\,ht\,(d+t)$$

$$= 3.1416\,ht\,(R+r) = 1.5708\,ht\,(D+d)$$

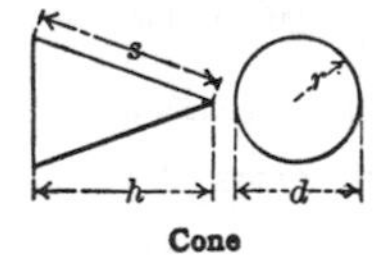

Cone

V = volume; A = area of conical surface.

$$V = \frac{3.1416\,r^2h}{3} = 1.0472\,r^2h = 0.2618\,d^2h$$

$$A = 3.1416\,r\,\sqrt{r^2+h^2} = 3.1416\,rs = 1.5708\,ds$$

$$s = \sqrt{r^2+h^2} = \sqrt{\frac{d^2}{4}+h^2}$$

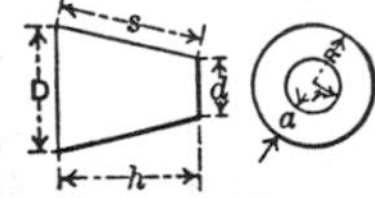

Frustum of Cone

V = volume; A = area of conical surface.

$$V = 1.0472\,h\,(R^2+Rr+r^2) = 0.2618\,h\,(D^2+Dd+d^2)$$

$$A = 3.1416\,s\,(R+r) = 1.5708\,s\,(D+d)$$

$$a = R-r \qquad s = \sqrt{a^2+h^2} = \sqrt{(R-r)^2+h^2}$$

Examples Showing Use of Formulas

1. Cylinder — Volume; area of surface.
2. Portion of Cylinder — Volume; surface area.
3. Portion of Cylinder — Volume.
4. Hollow Cylinder — Volume.
5. Cone — Volume; area of surface.
6. Frustum of Cone — Volume.

Cylinder. — The diameter of a cylinder is $2\frac{1}{2}$ inches. The length or height is 20 inches. Find the volume, and the area of the cylindrical surface S.

$$V = 0.7854\, d^2h = 0.7854 \times 2\tfrac{1}{2}^2 \times 20 = 0.7854 \times 6.25 \times 20 = 98.17 \text{ cubic inches.}$$

$$S = 3.1416\, dh = 3.1416 \times 2\tfrac{1}{2} \times 20 = 157.08 \text{ square inches.}$$

Portion of Cylinder. — A cylinder 5 inches in diameter, is cut off at an angle, as shown in the illustration. Dimension $h_1 = 6$, and $h_2 = 4$ inches. Find the volume and the area S of the cylindrical surface.

$$V = 0.3927\, d^2 (h_1 + h_2) = 0.3927 \times 5^2 \times (6 + 4) = 0.3927 \times 25 \times 10 = 98.175 \text{ cubic inches.}$$

$$S = 1.5708\, d\, (h_1 + h_2) = 1.5708 \times 5 \times 10 = 78.54 \text{ square inches.}$$

Portion of Cylinder. — Find the volume of a cylinder so cut off that line AC passes through the center of the base circle — that is, the base area is a half-circle. The diameter of the cylinder = 5 inches, and height h = 2 inches.

In this case $a = 2.5$; $b = 0$; area $ABC = \frac{1}{2} \times 0.7854 \times 5^2 = 9.82$; $r = 2.5$.

$$V = \left(\frac{2}{3} \times 2.5^3 + 0 \times 9.82\right) \frac{2}{2.5 + 0} = \frac{2}{3} \times 15.625 \times 0.8 = 8.33 \text{ cubic inches.}$$

Hollow Cylinder. — A cylindrical shell, 28 inches high, is 36 inches in outside diameter, and 4 inches thick. Find its volume.

$$V = 3.1416\, ht\, (D - t) = 3.1416 \times 28 \times 4\, (36 - 4) = 3.1416 \times 28 \times 4 \times 32 = 11{,}259.5 \text{ cubic inches.}$$

Cone. — Find the volume and area of conical surface of a cone, the base of which is a circle of 6 inches diameter, and the height of which is 4 inches.

$$V = 0.2618\, d^2h = 0.2618 \times 6^2 \times 4 = 0.2618 \times 36 \times 4 = 37.7 \text{ cubic inches.}$$

$$A = 3.1416\, r\sqrt{r^2 + h^2} = 3.1416 \times 3 \times \sqrt{3^2 + 4^2} = 9.4248 \times \sqrt{25} = 47.124 \text{ square inches.}$$

Frustum of Cone. — Find the volume of a frustum of a cone of the following dimensions: $D = 8$ inches; $d = 4$ inches; $h = 5$ inches.

$$V = 0.2618 \times 5\, (8^2 + 8 \times 4 + 4^2) = 0.2618 \times 5\, (64 + 32 + 16) = 0.2618 \times 5 \times 112 = 146.61 \text{ cubic inches.}$$

Volumes and Dimensions of Solid Figures

1. Sphere — Volume; area of surface.
2. Spherical Sector — Volume; total surface area.
3. Spherical Segment — Volume; surface area.
4. Spherical Zone — Volume; surface area.
5. Spherical Wedge — Volume; surface area.
6. Hollow Sphere — Volume.

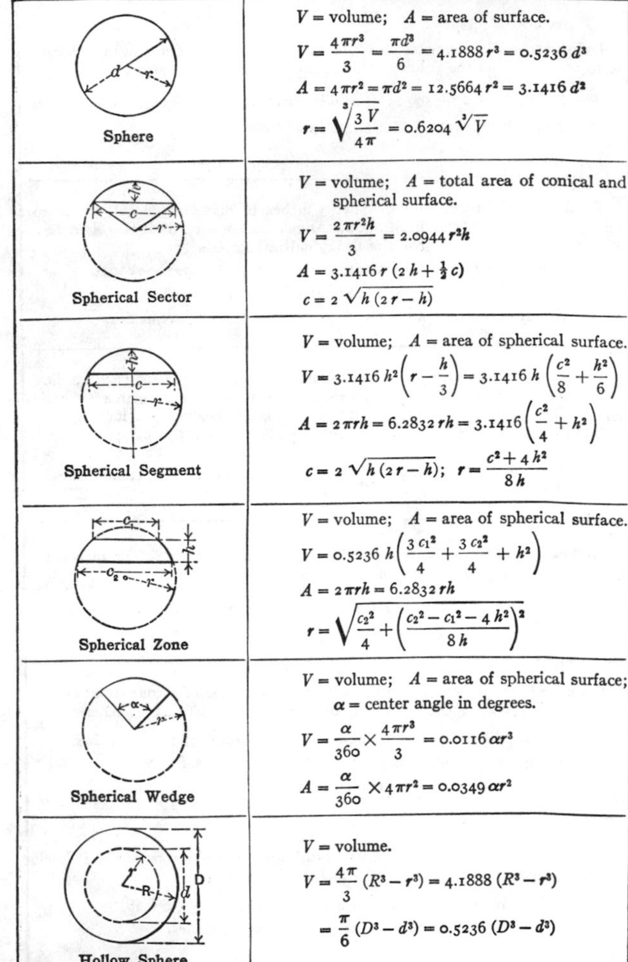

Figure	Formulas
Sphere	V = volume; A = area of surface. $V = \frac{4\pi r^3}{3} = \frac{\pi d^3}{6} = 4.1888\, r^3 = 0.5236\, d^3$ $A = 4\pi r^2 = \pi d^2 = 12.5664\, r^2 = 3.1416\, d^2$ $r = \sqrt[3]{\frac{3V}{4\pi}} = 0.6204 \sqrt[3]{V}$
Spherical Sector	V = volume; A = total area of conical and spherical surface. $V = \frac{2\pi r^2 h}{3} = 2.0944\, r^2 h$ $A = 3.1416\, r\,(2h + \frac{1}{2}c)$ $c = 2\sqrt{h\,(2r - h)}$
Spherical Segment	V = volume; A = area of spherical surface. $V = 3.1416\, h^2 \left(r - \frac{h}{3}\right) = 3.1416\, h \left(\frac{c^2}{8} + \frac{h^2}{6}\right)$ $A = 2\pi r h = 6.2832\, rh = 3.1416 \left(\frac{c^2}{4} + h^2\right)$ $c = 2\sqrt{h\,(2r - h)}$; $r = \frac{c^2 + 4h^2}{8h}$
Spherical Zone	V = volume; A = area of spherical surface. $V = 0.5236\, h \left(\frac{3c_1^2}{4} + \frac{3c_2^2}{4} + h^2\right)$ $A = 2\pi r h = 6.2832\, rh$ $r = \sqrt{\frac{c_2^2}{4} + \left(\frac{c_2^2 - c_1^2 - 4h^2}{8h}\right)^2}$
Spherical Wedge	V = volume; A = area of spherical surface; α = center angle in degrees. $V = \frac{\alpha}{360} \times \frac{4\pi r^3}{3} = 0.0116\, \alpha r^3$ $A = \frac{\alpha}{360} \times 4\pi r^2 = 0.0349\, \alpha r^2$
Hollow Sphere	V = volume. $V = \frac{4\pi}{3}(R^3 - r^3) = 4.1888\,(R^3 - r^3)$ $= \frac{\pi}{6}(D^3 - d^3) = 0.5236\,(D^3 - d^3)$

Examples Showing Use of Formulas

1. Sphere — Volume; area of surface.
2. Spherical Sector — Volume; length of chord.
3. Spherical Segment — Volume; radius.
4. Spherical Zone — Volume.
5. Spherical Wedge — Volume; surface area.
6. Hollow Sphere — Volume.

Sphere. — Find the volume and surface of a sphere 6.5 inches in diameter.

$V = 0.5236\,d^3 = 0.5236 \times 6.5^3 = 0.5236 \times 6.5 \times 6.5 \times 6.5 = 143.79$ cubic inches.

$A = 3.1416\,d^2 = 3.1416 \times 6.5^2 = 3.1416 \times 6.5 \times 6.5 = 132.73$ square inches.

The volume of a sphere is 64 cubic inches. Find its radius.

$r = 0.6204\sqrt[3]{64} = 0.6204 \times 4 = 2.4816$ inches.

Spherical Sector. — Find the volume of a sector of a sphere 6 inches in diameter, the height h of the sector being 1.5 inch. Also find length of chord c. — Here $r = 3$, and $h = 1.5$.

$V = 2.0944\,r^2h = 2.0944 \times 3^2 \times 1.5 = 2.0944 \times 9 \times 1.5 = 28.27$ cubic inches.

$c = 2\sqrt{h(2r - h)} = 2\sqrt{1.5(2 \times 3 - 1.5)} = 2\sqrt{6.75} = 2 \times 2.598$
$= 5.196$ inches.

Spherical Segment. — A segment of a sphere has the following dimensions: $h = 2$ inches; $c = 5$ inches. Find the volume V and the radius of the sphere of which the segment is a part.

$$V = 3.1416 \times 2 \times \left(\frac{5^2}{8} + \frac{2^2}{6}\right) = 6.2832 \times \left(\frac{25}{8} + \frac{4}{6}\right) = 6.2832 \times 3.792$$
$$= 23.825 \text{ cubic inches.}$$

$$r = \frac{5^2 + 4 \times 2^2}{8 \times 2} = \frac{25 + 16}{16} = \frac{41}{16} = 2\tfrac{9}{16} \text{ inches.}$$

Spherical Zone. — In a spherical zone, let $c_1 = 3$; $c_2 = 4$; and $h = 1.5$ inch. Find the volume.

$$V = 0.5236 \times 1.5 \times \left(\frac{3 \times 3^2}{4} + \frac{3 \times 4^2}{4} + 1.5^2\right) = 0.5236 \times 1.5 \times \left(\frac{27}{4} + \frac{48}{4} + 2.25\right)$$
$$= 0.5236 \times 1.5 \times 21 = 16.493 \text{ cubic inches.}$$

Spherical Wedge. — Find the area of the spherical surface and the volume of a wedge of a sphere. The diameter of the sphere is 4 inches, and the center angle α is 45 degrees.

$V = 0.0116 \times 45 \times 2^3 = 0.0116 \times 45 \times 8 = 4.176$ cubic inches.

$A = 0.0349 \times 45 \times 2^2 = 0.0349 \times 45 \times 4 = 6.282$ square inches.

Hollow Sphere. — Find the volume of a hollow sphere, 8 inches in outside diameter, with a thickness of material of 1.5 inch.
Here $R = 4$; $r = 4 - 1.5 = 2.5$.

$V = 4.1888\,(4^3 - 2.5^3) = 4.1888\,(64 - 15.625) = 4.1888 \times 48.375$
$= 202.63$ cubic inches.

Volumes and Dimensions of Solid Figures

1. Ellipsoid — Volume.
2. Paraboloid — Volume; area of surface.
3. Paraboloidal Segment — Volume.
4. Torus — Volume; area of surface.
5. Barrel — Approximate volume.
6. Cone — Paraboloid — Sphere — Cylinder.

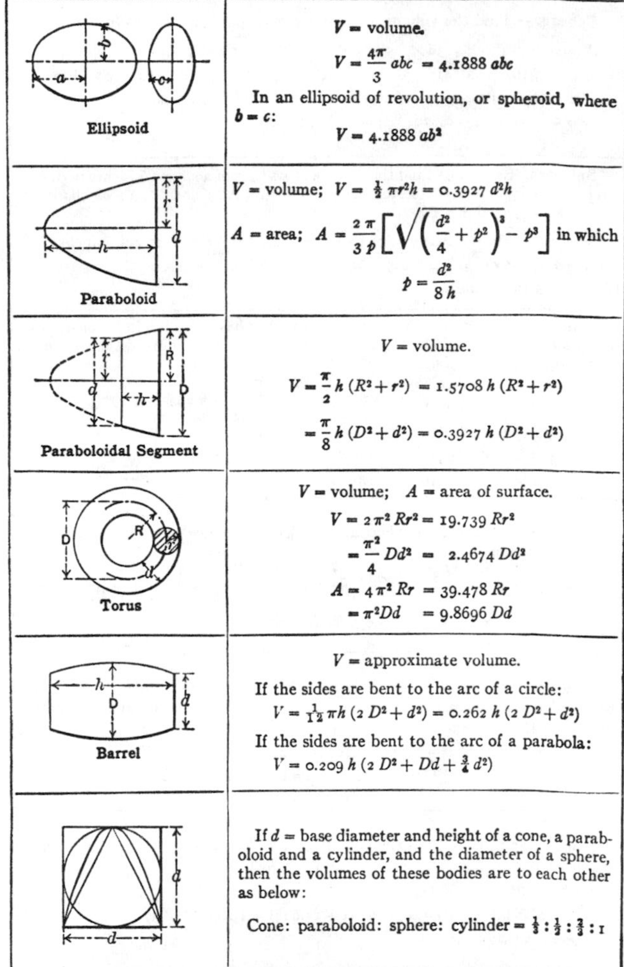

Figure	Formulas
Ellipsoid	V = volume. $V = \frac{4\pi}{3} abc = 4.1888\ abc$. In an ellipsoid of revolution, or spheroid, where $b = c$: $V = 4.1888\ ab^2$
Paraboloid	V = volume; $V = \frac{1}{2}\pi r^2 h = 0.3927\ d^2 h$. A = area; $A = \frac{2\pi}{3p}\left[\sqrt{\left(\frac{d^2}{4} + p^2\right)^3} - p^3\right]$ in which $p = \frac{d^2}{8h}$
Paraboloidal Segment	V = volume. $V = \frac{\pi}{2} h(R^2 + r^2) = 1.5708\ h(R^2 + r^2)$ $= \frac{\pi}{8} h(D^2 + d^2) = 0.3927\ h(D^2 + d^2)$
Torus	V = volume; A = area of surface. $V = 2\pi^2 Rr^2 = 19.739\ Rr^2$ $= \frac{\pi^2}{4} Dd^2 = 2.4674\ Dd^2$; $A = 4\pi^2 Rr = 39.478\ Rr$ $= \pi^2 Dd = 9.8696\ Dd$
Barrel	V = approximate volume. If the sides are bent to the arc of a circle: $V = \frac{1}{12}\pi h(2D^2 + d^2) = 0.262\ h(2D^2 + d^2)$. If the sides are bent to the arc of a parabola: $V = 0.209\ h(2D^2 + Dd + \frac{3}{4}d^2)$
	If d = base diameter and height of a cone, a paraboloid and a cylinder, and the diameter of a sphere, then the volumes of these bodies are to each other as below: Cone: paraboloid: sphere: cylinder $= \frac{1}{3} : \frac{1}{2} : \frac{2}{3} : 1$

Examples Showing Use of Formulas

1. Ellipsoid — Volume.
2. Paraboloid — Volume.
3. Paraboloidal Segment — Volume.
4. Torus — Volume; area of surface.
5. Barrel — Approximate volume.
6. Cone — Paraboloid — Sphere — Cylinder.

Ellipsoid or Spheroid.—Find the volume and area of surface of a spheroid in which $a = 5$, and $b = c = 1.5$ inch.

$$V = 4.1888 \times 5 \times 1.5^2 = 4.1888 \times 5 \times 2.25 = 47.124 \text{ cubic inches.}$$

NOTE: When the axis of revolution is the major axis, the solid is called a *prolate* spheroid; when it is the minor axis, an *oblate* spheroid.

Paraboloid. — Find the volume of a paraboloid in which $h = 12$ and $d = 5$ inches.

$$V = 0.3927\, d^2h = 0.3927 \times 5^2 \times 12 = 0.3927 \times 25 \times 12$$
$$= 0.3927 \times 300 = 117.81 \text{ cubic inches.}$$

Segment of Paraboloid. — Find the volume of a segment of a paraboloid in which $D = 5$ inches, $d = 3$ inches, and $h = 6$ inches.

$$V = 0.3927\, h\, (D^2 + d^2) = 0.3927 \times 6 \times (5^2 + 3^2) = 0.3927 \times 6 \times (25 + 9)$$
$$= 0.3927 \times 6 \times 34 = 80.11 \text{ cubic inches.}$$

Torus. — Find the volume and area of surface of a torus in which $d = 1.5$ and $D = 5$ inches.

$$V = 2.4674 \times 5 \times 1.5^2 = 2.4674 \times 5 \times 2.25 = 27.76 \text{ cubic inches.}$$
$$A = 9.8696 \times 5 \times 1.5 = 74.022 \text{ square inches}$$

Barrel. — Find the approximate contents of a barrel, the inside dimensions of which are $D = 24$ inches; $d = 20$ inches; $h = 48$ inches.

$$V = 0.262\, h\, (2\, D^2 + d^2) = 0.262 \times 48 \times (2 \times 24^2 + 20^2) = 0.262 \times 48$$
$$\times (1152 + 400) = 0.262 \times 48 \times 1552 = 19{,}518 \text{ cubic inches.}$$

Assume, as an example, that the diameter of the base of a cone, paraboloid and cylinder is 2 inches, that the height is 2 inches, and that the diameter of a sphere is 2 inches. Then the volumes, written in formula-form, are as below:

$$\underset{\text{Cone}}{\frac{3.1416 \times 2^2 \times 2}{12}} : \underset{\text{Paraboloid}}{\frac{3.1416 \times 2^2 \times 2}{8}} : \underset{\text{Sphere}}{\frac{3.1416 \times 2^3}{6}} : \underset{\text{Cylinder}}{\frac{3.1416 \times 2^2 \times 2}{4}} = \frac{1}{3} : \frac{1}{2} : \frac{2}{3} : 1$$

Conversion Tables for Radians and Degrees

Degrees Expressed in Radians

Deg.	Rad.	Deg.	Rad.	Deg.	Rad.	Deg.	Rad.	Deg.	Rad.	Deg.	Rad.
1	0.0175	31	0.5411	61	1.0647	91	1.5882	121	2.1118	151	2.6354
2	0.0349	32	0.5585	62	1.0821	92	1.6057	122	2.1293	152	2.6529
3	0.0524	33	0.5760	63	1.0996	93	1.6232	123	2.1468	153	2.6704
4	0.0698	34	0.5934	64	1.1170	94	1.6406	124	2.1642	154	2.6878
5	0.0873	35	0.6109	65	1.1345	95	1.6581	125	2.1817	155	2.7053
6	0.1047	36	0.6283	66	1.1519	96	1.6755	126	2.1991	156	2.7227
7	0.1222	37	0.6458	67	1.1694	97	1.6930	127	2.2166	157	2.7402
8	0.1396	38	0.6632	68	1.1868	98	1.7104	128	2.2340	158	2.7576
9	0.1571	39	0.6807	69	1.2043	99	1.7279	129	2.2515	159	2.7751
10	0.1745	40	0.6981	70	1.2217	100	1.7453	130	2.2689	160	2.7925
11	0.1920	41	0.7156	71	1.2392	101	1.7628	131	2.2864	161	2.8100
12	0.2094	42	0.7330	72	1.2566	102	1.7802	132	2.3038	162	2.8274
13	0.2269	43	0.7505	73	1.2741	103	1.7977	133	2.3213	163	2.8449
14	0.2443	44	0.7679	74	1.2915	104	1.8151	134	2.3387	164	2.8623
15	0.2618	45	0.7854	75	1.3090	105	1.8326	135	2.3562	165	2.8798
16	0.2793	46	0.8029	76	1.3265	106	1.8500	136	2.3736	166	2.8972
17	0.2967	47	0.8203	77	1.3439	107	1.8675	137	2.3911	167	2.9147
18	0.3142	48	0.8378	78	1.3614	108	1.8850	138	2.4086	168	2.9322
19	0.3316	49	0.8552	79	1.3788	109	1.9024	139	2.4260	169	2.9496
20	0.3491	50	0.8727	80	1.3963	110	1.9199	140	2.4435	170	2.9671
21	0.3665	51	0.8901	81	1.4137	111	1.9373	141	2.4609	171	2.9845
22	0.3840	52	0.9076	82	1.4312	112	1.9548	142	2.4784	172	3.0020
23	0.4014	53	0.9250	83	1.4486	113	1.9722	143	2.4958	173	3.0194
24	0.4189	54	0.9425	84	1.4661	114	1.9897	144	2.5133	174	3.0369
25	0.4363	55	0.9599	85	1.4835	115	2.0071	145	2.5307	175	3.0543
26	0.4538	56	0.9774	86	1.5010	116	2.0246	146	2.5482	176	3.0718
27	0.4712	57	0.9948	87	1.5184	117	2.0420	147	2.5656	177	3.0892
28	0.4887	58	1.0123	88	1.5359	118	2.0595	148	2.5831	178	3.1067
29	0.5061	59	1.0297	89	1.5533	119	2.0769	149	2.6005	179	3.1241
30	0.5236	60	1.0472	90	1.5708	120	2.0944	150	2.6180	180	3.1416

Minutes Expressed in Radians

Min.	Rad.	Min.	Rad.	Min.	Rad.	Min.	Rad.	Min.	Rad.	Min.	Rad.
1	0.0003	11	0.0032	21	0.0061	31	0.0090	41	0.0119	51	0.0148
2	0.0006	12	0.0035	22	0.0064	32	0.0093	42	0.0122	52	0.0151
3	0.0009	13	0.0038	23	0.0067	33	0.0096	43	0.0125	53	0.0154
4	0.0012	14	0.0041	24	0.0070	34	0.0099	44	0.0128	54	0.0157
5	0.0015	15	0.0044	25	0.0073	35	0.0102	45	0.0131	55	0.0160
6	0.0017	16	0.0047	26	0.0076	36	0.0105	46	0.0134	56	0.0163
7	0.0020	17	0.0049	27	0.0079	37	0.0108	47	0.0137	57	0.0166
8	0.0023	18	0.0052	28	0.0081	38	0.0111	48	0.0140	58	0.0169
9	0.0026	19	0.0055	29	0.0084	39	0.0113	49	0.0143	59	0.0172
10	0.0029	20	0.0058	30	0.0087	40	0.0116	50	0.0145	60	0.0175

Radians Expressed in Degrees, Minutes and Seconds

Rad.	Angle	Rad.	Angle	Rad.	Angle	Rad.	Angle	Rad.	Angle
0.001	0° 3′ 26″	0.008	0° 27′ 30″	0.06	3° 26′ 16″	0.4	22° 55′ 6″	2.0	114° 35′ 30″
0.002	0 6 53	0.009	0 30 56	0.07	4 0 39	0.5	28 38 52	3.0	171 53 14
0.003	0 10 19	0.01	0 34 23	0.08	4 35 1	0.6	34 22 39	4.0	229 10 59
0.004	0 13 45	0.02	1 8 45	0.09	5 9 24	0.7	40 6 25	5.0	286 28 44
0.005	0 17 11	0.03	1 43 8	0.1	5 43 46	0.8	45 50 12	6.0	343 46 29
0.006	0 20 38	0.04	2 17 31	0.2	11 27 33	0.9	51 33 58	7.0	401 4 14
0.007	0 24 4	0.05	2 51 53	0.3	17 11 19	1.0	57 17 45	8.0	458 21 58

For more exact calculations, the following equivalents may be used:

1 degree = 0.0174533 radian
1 radian = 57.295780 degrees

English System — Weights and Measures

Measures of Length

1 mile = 1760 yards = 5280 feet.
1 yard = 3 feet = 36 inches. 1 foot = 12 inches.
1 mil = 0.001 inch. 1 fathom = 2 yards = 6 feet.
1 rod = 5.5 yards = 16.5 feet. 1 hand = 4 inches. 1 span = 9 inches.
1 micro-inch = one millionth inch or 0.000001 inch. (1 micron = one millionth meter = 0.00003937 inch.)

Surveyor's Measure

1 mile = 8 furlongs = 80 chains.
1 furlong = 10 chains = 220 yards.
1 chain = 4 rods = 22 yards = 66 feet = 100 links.
1 link = 7.92 inches.

Nautical Measure

1 league = 3 nautical miles.
1 nautical mile = 6080.2 feet = 1.1516 statute mile. (The *knot*, which is a nautical unit of speed, is equivalent to a speed 1 nautical mile per hour.)
One degree at the equator = 60 nautical miles = 69.096 statute miles. 360 degrees = 21,600 nautical miles = 24,874.5 statute miles = circumference at equator.

Square Measure

1 square mile = 640 acres = 6400 square chains.
1 acre = 10 square chains = 4840 square yards = 43,560 square feet.
1 square chain = 16 square rods = 484 square yards = 4356 square feet.
1 square rod = 30.25 square yards = 272.25 square feet = 625 square links.
1 square yard = 9 square feet.
1 square foot = 144 square inches.
An acre is equal to a square, the side of which is 208.7 feet.

Cubic Measure

1 cubic yard = 27 cubic feet.
1 cubic foot = 1728 cubic inches.
The following measures are also used for wood and masonry:
1 cord of wood = 4 × 4 × 8 feet = 128 cubic feet.
1 perch of masonry = $16\frac{1}{2} \times 1\frac{1}{2} \times 1$ foot = $24\frac{3}{4}$ cubic feet.

Dry Measure

1 bushel (U. S. or Winchester struck bushel) = 1.2445 cubic foot = 2150.42 cubic inches.
1 bushel = 4 pecks = 32 quarts = 64 pints.
1 peck = 8 quarts = 16 pints.
1 quart = 2 pints.
1 heaped bushel = $1\frac{1}{4}$ struck bushel.
1 cubic foot = 0.8036 struck bushel.
1 British Imperial bushel = 8 Imperial gallons = 1.2837 cubic foot = 2218.19 cubic inches.

Liquid Measure

1 U. S. gallon = 0.1337 cubic foot = 231 cubic inches = 4 quarts = 8 pints.
1 quart = 2 pints = 8 gills.
1 pint = 4 gills.
1 British Imperial gallon = 1.2009 U. S. gallon = 277.42 cubic inches.
1 cubic foot = 7.48 U. S. gallons.

Avoirdupois or Commercial Weight

1 gross or long ton = 2240 pounds.
1 net or short ton = 2000 pounds.
1 pound = 16 ounces = 7000 grains.
1 ounce = 16 drachms = 437.5 grains.

The following measures for weight are now seldom used in the United States:

1 hundred-weight = 4 quarters = 112 pounds (1 gross or long ton = 20 hundred-weights); 1 quarter = 28 pounds; 1 stone = 14 pounds; 1 quintal = 100 pounds.

Troy Weight, used for Weighing Gold and Silver

1 pound = 12 ounces = 5760 grains.
1 ounce = 20 pennyweights = 480 grains.
1 pennyweight = 24 grains.
1 carat (used in weighing diamonds) = 3.086 grains.
1 grain Troy = 1 grain avoirdupois = 1 grain apothecaries' weight.

Apothecaries' Weight

1 pound = 12 ounces = 5760 grains.
1 ounce = 8 drachms = 480 grains.
1 drachm = 3 scruples = 60 grains.
1 scruple = 20 grains.

Measures of Pressure

1 pound per square inch = 144 pounds per square foot = 0.068 atmosphere = 2.042 inches of mercury at 62 degrees F. = 27.7 inches of water at 62 degrees F. = 2.31 feet of water at 62 degrees F.

1 atmosphere = 30 inches of mercury at 62 degrees F. = 14.7 pounds per square inch = 2116.3 pounds per square foot = 33.95 feet of water at 62 degrees F.

1 foot of water at 62 degrees F. = 62.355 pounds per square foot = 0.433 pound per square inch.

1 inch of mercury at 62 degrees F. = 1.132 foot of water = 13.58 inches of water = 0.491 pound per square inch.

Decimal Equivalents of Fractions of an Inch

Fraction	Decimal	Fraction	Decimal	Fraction	Decimal
1/64	0.015 625	11/32	0.343 75	43/64	0.671 875
1/32	0.031 25	23/64	0.359 375	11/16	0.687 5
3/64	0.046 875	3/8	0.375	45/64	0.703 125
1/16	0.062 5	25/64	0.390 625	23/32	0.718 75
5/64	0.078 125	13/32	0.406 25	47/64	0.734 375
3/32	0.093 75	27/64	0.421 875	3/4	0.750
7/64	0.109 375	7/16	0.437 5	49/64	0.765 625
1/8	0.125	29/64	0.453 125	25/32	0.781 25
9/64	0.140 625	15/32	0.468 75	51/64	0.796 875
5/32	0.156 25	31/64	0.484 375	13/16	0.812 5
11/64	0.171 875	1/2	0.500	53/64	0.828 125
3/16	0.187 5	33/64	0.515 625	27/32	0.843 75
13/64	0.203 125	17/32	0.531 25	55/64	0.859 375
7/32	0.218 75	35/64	0.546 875	7/8	0.875
15/64	0.234 375	9/16	0.562 5	57/64	0.890 625
1/4	0.250	37/64	0.578 125	29/32	0.906 25
17/64	0.265 625	19/32	0.593 75	59/64	0.921 875
9/32	0.281 25	39/64	0.609 375	15/16	0.937 5
19/64	0.296 875	5/8	0.625	61/64	0.953 125
5/16	0.312 5	41/64	0.640 625	31/32	0.968 75
21/64	0.328 125	21/32	0.656 25	63/64	0.984 375

Metric System Weights and Measures

In the metric system of measurements, the principal unit for length is the meter; the principal unit for capacity, the liter; and the principal unit for weight, the gram. The following prefixes are used for sub-divisions and multiples: milli = 1/1000; centi = 1/100; deci = 1/10; deca = 10; hecto = 100; kilo = 1000. In abbreviations, the sub-divisions are frequently used with a small letter and the multiples with a capital letter, although this practice is not universally followed everywhere where the metric system is used.

All the multiples and sub-divisions are not used commercially. Those ordinarily used for length are kilometer, meter, centimeter and millimeter; for capacity, square meter, square centimeter and square millimeter; for cubic measures, cubic meter, cubic decimeter (liter), cubic centimeter, and cubic millimeter. The most commonly used weights are the kilogram and gram. The metric system was legalized in the United States by an Act of Congress in 1866.

Measures of Length

1000 microns (μ)	=	1 millimeter (mm.).
10 millimeters	=	1 centimeter (cm.).
10 centimeters	=	1 decimeter (dm.).
10 decimeters	=	1 meter (m.).
1000 meters	=	1 kilometer (Km.).

Square Measure

100 square millimeters (mm.^{2})	=	1 square centimeter (cm.^{2}).
100 square centimeters	=	1 square decimeter (dm.^{2}).
100 square decimeters	=	1 square meter (m.^{2}).
1,000,000 square meters	=	1 square kilometer (Km.^{2}).

Surveyor's Square Measure

100 square meters (m.^{2})	=	1 are (ar.).
100 ares	=	1 hectare (har.).
100 hectares	=	1 square kilometer (Km.^{2}).

Cubic Measure

1000 cubic millimeters (mm.^{3})	=	1 cubic centimeter (cm.^{3}).
1000 cubic centimeters	=	1 cubic decimeter (dm.^{3}).
1000 cubic decimeters	=	1 cubic meter (m.^{3}).

Dry and Liquid Measure

10 milliliters (ml.)	=	1 centiliter (cl.).
10 centiliters	=	1 deciliter (dl.).
10 deciliters	=	1 liter (l.).
100 liters	=	1 hectoliter (Hl.).

1 liter = 1 cubic decimeter = the volume of 1 kilogram of pure water at a temperature of 39.2 degrees F.

Measures of Weight

1000 micrograms (μg.)	=	1 milligram (mg.).
10 milligrams	=	1 centigram (cg.).
10 centigrams	=	1 decigram (dg.).
10 decigrams	=	1 gram (g.).
10 grams	=	1 decagram (Dg.).
10 decagrams	=	1 hectogram (Hg.).
10 hectograms	=	1 kilogram (Kg.).
1000 kilograms	=	1 (metric) ton (T.).

Metric and English Conversion Table

Linear Measure

1 kilometer = 0.6214 mile.
1 meter = 39.37 inches. / 3.2808 feet. / 1.0936 yard.
1 centimeter = 0.3937 inch.
1 millimeter = 0.03937 inch.

1 mile = 1.609 kilometer.
1 yard = 0.9144 meter.
1 foot = 0.3048 meter.
1 foot = 304.8 millimeters.
1 inch = 2.54 centimeters.
1 inch = 25.4 millimeters.

1 micron = 0.00003937 inch.

Square Measure

1 square kilometer = 0.3861 square mile = 247.1 acres.
1 hectare = 2.471 acre = 107,640 square feet.
1 are = 0.0247 acre = 1076.4 square feet.
1 square meter = 10.764 square feet = 1.196 square yard.
1 square centimeter = 0.155 square inch.
1 square millimeter = 0.00155 square inch.

---

1 square mile = 2.5899 square kilometers.
1 acre = 0.4047 hectare = 40.47 ares.
1 square yard = 0.836 square meter.
1 square foot = 0.0929 square meter = 939 square centimeters.
1 square inch = 6.452 square centimeters = 645.2 square millimeters.

Cubic Measure

1 cubic meter = 35.314 cubic feet = 1.308 cubic yard.
1 cubic meter = 264.2 U. S. gallons.
1 cubic centimeter = 0.061 cubic inch.
1 liter (cubic decimeter) = 0.0353 cubic foot = 61.023 cubic inches.
1 liter = 0.2642 U. S. gallon = 1.0567 U. S. quart.

---

1 cubic yard = 0.7645 cubic meter.
1 cubic foot = 0.02832 cubic meter = 28.317 liters.
1 cubic inch = 16.38716 cubic centimeters.
1 U. S. gallon = 3.785 liters.
1 U. S. quart = 0.946 liter.

Mass

1 pound = 453.6 grams = 0.4536 kilograms
1 pound = 16 ounces avoirdupois = 7000 grains
1 pound troy = 373.2 grams = 0.3732 kilograins
1 pound troy = 12 ounces troy = 5760 grains
1 metric ton = 1000 kilograms

Weight

1 pound = 4.448 Newtons
1 Newton = 0.2248 pounds

1 short ton = 2000 pounds
1 long ton = 2240 pounds

Miscellaneous

1 pound per square inch = 6895 Pascals
1 Pascal = 0.0001450 pounds per inch^2 = 1 Newton per meter^2
1 Newton-meter = 0.7376 foot-pounds
1 foot-pound = 1.356 Newton-meters
1 calorie (kilogram calorie) = 3.968 B.T.U. (British Thermal Unit)
1 B.T.U. (British Thermal Unit) = 0.2520 calorie (kilogram calorie)

↓ 0° or 180° **Trigonometric and Involute Functions** 179° or 359° ↓

Minutes	Sine	Cosine	Tangent	Cotangent	Secant	Cosecant	Involute 0°–1°	Read Up	Minutes
0	0.000000	1.000000	0.000000	Infinite	1.000000	Infinite	0.0000000	Infinite	60
1	0.000291	1.000000	0.000291	3437.75	1.000000	3437.75	0.0000000	3436.176	59
2	0.000582	1.000000	0.000582	1718.87	1.000000	1718.87	0.0000000	1717.303	58
3	0.000873	1.000000	0.000873	1145.92	1.000000	1145.92	0.0000000	1144.345	57
4	0.001164	0.999999	0.001164	859.436	1.000001	859.437	0.0000000	857.8667	56
5	0.001454	0.999999	0.001454	687.549	1.000001	687.550	0.0000000	685.9795	55
6	0.001745	0.999998	0.001745	572.957	1.000002	572.958	0.0000000	571.3882	54
7	0.002036	0.999998	0.002036	491.106	1.000002	491.107	0.0000000	489.5372	53
8	0.002327	0.999997	0.002327	429.718	1.000003	429.719	0.0000000	428.1491	52
9	0.002618	0.999997	0.002618	381.971	1.000003	381.972	0.0000000	380.4028	51
10	0.002909	0.999996	0.002909	343.774	1.000004	343.775	0.0000000	342.2058	50
11	0.003200	0.999995	0.003200	312.521	1.000005	312.523	0.0000000	310.9538	49
12	0.003491	0.999994	0.003491	286.478	1.000006	286.479	0.0000000	284.9104	48
13	0.003782	0.999993	0.003782	264.441	1.000007	264.443	0.0000000	262.8738	47
14	0.004072	0.999992	0.004072	245.552	1.000008	245.554	0.0000000	243.9853	46
15	0.004363	0.999990	0.004363	229.182	1.000010	229.184	0.0000000	227.6152	45
16	0.004654	0.999989	0.004654	214.858	1.000011	214.860	0.0000000	213.2915	44
17	0.004945	0.999988	0.004945	202.219	1.000012	202.221	0.0000000	200.6529	43
18	0.005236	0.999986	0.005236	190.984	1.000014	190.987	0.0000000	189.4186	42
19	0.005527	0.999985	0.005527	180.932	1.000015	180.935	0.0000001	179.3669	41
20	0.005818	0.999983	0.005818	171.885	1.000017	171.888	0.0000001	170.3204	40
21	0.006109	0.999981	0.006109	163.700	1.000019	163.703	0.0000001	162.1355	39
22	0.006399	0.999980	0.006400	156.259	1.000020	156.262	0.0000001	154.6947	38
23	0.006690	0.999978	0.006691	149.465	1.000022	149.468	0.0000001	147.9009	37
24	0.006981	0.999976	0.006981	143.237	1.000024	143.241	0.0000001	141.6733	36
25	0.007272	0.999974	0.007272	137.507	1.000026	137.511	0.0000001	135.9439	35
26	0.007563	0.999971	0.007563	132.219	1.000029	132.222	0.0000001	130.6553	34
27	0.007854	0.999969	0.007854	127.321	1.000031	127.325	0.0000002	125.7584	33
28	0.008145	0.999967	0.008145	122.774	1.000033	122.778	0.0000002	121.2113	32
29	0.008436	0.999964	0.008436	118.540	1.000036	118.544	0.0000002	116.9778	31
30	0.008727	0.999962	0.008727	114.589	1.000038	114.593	0.0000002	113.0266	30
31	0.009017	0.999959	0.009018	110.892	1.000041	110.897	0.0000002	109.3303	29
32	0.009308	0.999957	0.009309	107.426	1.000043	107.431	0.0000003	105.8650	28
33	0.009599	0.999954	0.009600	104.171	1.000046	104.176	0.0000003	102.6097	27
34	0.009890	0.999951	0.009891	101.107	1.000049	101.112	0.0000003	99.54600	26
35	0.010181	0.999948	0.010181	98.2179	1.000052	98.2230	0.0000004	96.65733	25
36	0.010472	0.999945	0.010472	95.4895	1.000055	95.4947	0.0000004	93.92915	24
37	0.010763	0.999942	0.010763	92.9085	1.000058	92.9139	0.0000004	91.34845	23
38	0.011054	0.999939	0.011054	90.4633	1.000061	90.4689	0.0000005	88.90359	22
39	0.011344	0.999936	0.011345	88.1436	1.000064	88.1492	0.0000005	86.58412	21
40	0.011635	0.999932	0.011636	85.9398	1.000068	85.9456	0.0000005	84.38063	20
41	0.011926	0.999929	0.011927	83.8435	1.000071	83.8495	0.0000006	82.28464	19
42	0.012217	0.999925	0.012218	81.8470	1.000075	81.8531	0.0000006	80.28846	18
43	0.012508	0.999922	0.012509	79.9434	1.000078	79.9497	0.0000007	78.38514	17
44	0.012799	0.999918	0.012800	78.1263	1.000082	78.1327	0.0000007	76.56834	16
45	0.013090	0.999914	0.013091	76.3900	1.000086	76.3966	0.0000007	74.83230	15
46	0.013380	0.999910	0.013382	74.7292	1.000090	74.7359	0.0000008	73.17175	14
47	0.013671	0.999907	0.013673	73.1390	1.000093	73.1458	0.0000009	71.58187	13
48	0.013962	0.999903	0.013964	71.6151	1.000097	71.6221	0.0000009	70.05824	12
49	0.014253	0.999898	0.014254	70.1533	1.000102	70.1605	0.0000010	68.59680	11
50	0.014544	0.999894	0.014545	68.7501	1.000106	68.7574	0.0000010	67.19384	10
51	0.014835	0.999890	0.014836	67.4019	1.000110	67.4093	0.0000011	65.84589	9
52	0.015126	0.999886	0.015127	66.1055	1.000114	66.1130	0.0000012	64.54980	8
53	0.015416	0.999881	0.015418	64.8580	1.000119	64.8657	0.0000012	63.30263	7
54	0.015707	0.999877	0.015709	63.6567	1.000123	63.6646	0.0000013	62.10165	6
55	0.015998	0.999872	0.016000	62.4992	1.000128	62.5072	0.0000014	60.94436	5
56	0.016289	0.999867	0.016291	61.3829	1.000133	61.3911	0.0000014	59.82840	4
57	0.016580	0.999863	0.016582	60.3058	1.000137	60.3141	0.0000015	58.75160	3
58	0.016871	0.999858	0.016873	59.2659	1.000142	59.2743	0.0000016	57.71195	2
59	0.017162	0.999853	0.017164	58.2612	1.000147	58.2698	0.0000017	56.70754	1
60	0.017452	0.999848	0.017455	57.2900	1.000152	57.2987	0.0000018	55.73662	0
Minutes	Cosine	Sine	Cotangent	Tangent	Cosecant	Secant	Read Down	89°–90° Involute	Minutes

↑ 90° or 270° 89° or 269° ↑

↓ 1° or 181° Trigonometric and Involute Functions 178° or 358° ↓

Minutes	Sine	Cosine	Tangent	Cotangent	Secant	Cosecant	Involute 1°–2°	Read Up	Minutes
0	0.017452	0.999848	0.017455	57.2900	1.000152	57.2987	0.0000018	55.73662	60
1	0.017743	0.999843	0.017746	56.3506	1.000157	56.3595	0.0000019	54.79754	59
2	0.018034	0.999837	0.018037	55.4415	1.000163	55.4505	0.0000020	53.88876	58
3	0.018325	0.999832	0.018328	54.5613	1.000168	54.5705	0.0000021	53.00883	57
4	0.018616	0.999827	0.018619	53.7086	1.000173	53.7179	0.0000022	52.15641	56
5	0.018907	0.999821	0.018910	52.8821	1.000179	52.8916	0.0000023	51.33022	55
6	0.019197	0.999816	0.019201	52.0807	1.000184	52.0903	0.0000024	50.52907	54
7	0.019488	0.999810	0.019492	51.3032	1.000190	51.3129	0.0000025	49.75185	53
8	0.019779	0.999804	0.019783	50.5485	1.000196	50.5584	0.0000026	48.99749	52
9	0.020070	0.999799	0.020074	49.8157	1.000201	49.8258	0.0000027	48.26500	51
10	0.020361	0.999793	0.020365	49.1039	1.000207	49.1141	0.0000028	47.55345	50
11	0.020652	0.999787	0.020656	48.4121	1.000213	48.4224	0.0000029	46.86194	49
12	0.020942	0.999781	0.020947	47.7395	1.000219	47.7500	0.0000031	46.18965	48
13	0.021233	0.999775	0.021238	47.0853	1.000226	47.0960	0.0000032	45.53578	47
14	0.021524	0.999768	0.021529	46.4489	1.000232	46.4596	0.0000033	44.89959	46
15	0.021815	0.999762	0.021820	45.8294	1.000238	45.8403	0.0000035	44.28037	45
16	0.022106	0.999756	0.022111	45.2261	1.000244	45.2372	0.0000036	43.67745	44
17	0.022397	0.999749	0.022402	44.6386	1.000251	44.6498	0.0000037	43.09020	43
18	0.022687	0.999743	0.022693	44.0661	1.000257	44.0775	0.0000039	42.51801	42
19	0.022978	0.999736	0.022984	43.5081	1.000264	43.5196	0.0000040	41.96031	41
20	0.023269	0.999729	0.023275	42.9641	1.000271	42.9757	0.0000042	41.41655	40
21	0.023560	0.999722	0.023566	42.4335	1.000278	42.4452	0.0000044	40.88623	39
22	0.023851	0.999716	0.023857	41.9158	1.000285	41.9277	0.0000045	40.36885	38
23	0.024141	0.999709	0.024148	41.4106	1.000292	41.4227	0.0000047	39.86393	37
24	0.024432	0.999701	0.024439	40.9174	1.000299	40.9296	0.0000049	39.37105	36
25	0.024723	0.999694	0.024731	40.4358	1.000306	40.4482	0.0000050	38.88977	35
26	0.025014	0.999687	0.025022	39.9655	1.000313	39.9780	0.0000052	38.41968	34
27	0.025305	0.999680	0.025313	39.5059	1.000320	39.5185	0.0000054	37.96041	33
28	0.025595	0.999672	0.025604	39.0568	1.000328	39.0696	0.0000056	37.51157	32
29	0.025886	0.999665	0.025895	38.6177	1.000335	38.6307	0.0000058	37.07283	31
30	0.026177	0.999657	0.026186	38.1885	1.000343	38.2016	0.0000060	36.64384	30
31	0.026468	0.999650	0.026477	37.7686	1.000350	37.7818	0.0000062	36.22429	29
32	0.026759	0.999642	0.026768	37.3579	1.000358	37.3713	0.0000064	35.81386	28
33	0.027049	0.999634	0.027059	36.9560	1.000366	36.9695	0.0000066	35.41226	27
34	0.027340	0.999626	0.027350	36.5627	1.000374	36.5763	0.0000068	35.01921	26
35	0.027631	0.999618	0.027641	36.1776	1.000382	36.1914	0.0000070	34.63443	25
36	0.027922	0.999610	0.027933	35.8006	1.000390	35.8145	0.0000073	34.25768	24
37	0.028212	0.999602	0.028224	35.4313	1.000398	35.4454	0.0000075	33.88870	23
38	0.028503	0.999594	0.028515	35.0695	1.000406	35.0838	0.0000077	33.52726	22
39	0.028794	0.999585	0.028806	34.7151	1.000415	34.7295	0.0000080	33.17312	21
40	0.029085	0.999577	0.029097	34.3678	1.000423	34.3823	0.0000082	32.82606	20
41	0.029375	0.999568	0.029388	34.0273	1.000432	34.0420	0.0000085	32.48589	19
42	0.029666	0.999560	0.029679	33.6935	1.000440	33.7083	0.0000087	32.15238	18
43	0.029957	0.999551	0.029970	33.3662	1.000449	33.3812	0.0000090	31.82536	17
44	0.030248	0.999542	0.030262	33.0452	1.000458	33.0603	0.0000092	31.50463	16
45	0.030539	0.999534	0.030553	32.7303	1.000467	32.7455	0.0000095	31.19001	15
46	0.030829	0.999525	0.030844	32.4213	1.000476	32.4367	0.0000098	30.88133	14
47	0.031120	0.999516	0.031135	32.1181	1.000485	32.1337	0.0000101	30.57843	13
48	0.031411	0.999507	0.031426	31.8205	1.000494	31.8362	0.0000103	30.28114	12
49	0.031702	0.999497	0.031717	31.5284	1.000503	31.5442	0.0000106	29.98930	11
50	0.031992	0.999488	0.032009	31.2416	1.000512	31.2576	0.0000109	29.70278	10
51	0.032283	0.999479	0.032300	30.9599	1.000522	30.9761	0.0000112	29.42142	9
52	0.032574	0.999469	0.032591	30.6833	1.000531	30.6996	0.0000115	29.14509	8
53	0.032864	0.999460	0.032882	30.4116	1.000540	30.4280	0.0000118	28.87365	7
54	0.033155	0.999450	0.033173	30.1446	1.000550	30.1612	0.0000122	28.60698	6
55	0.033446	0.999441	0.033465	29.8823	1.000560	29.8990	0.0000125	28.34495	5
56	0.033737	0.999431	0.033756	29.6245	1.000570	29.6414	0.0000128	28.08745	4
57	0.034027	0.999421	0.034047	29.3711	1.000579	29.3881	0.0000131	27.83434	3
58	0.034318	0.999411	0.034338	29.1220	1.000589	29.1392	0.0000135	27.58553	2
59	0.034609	0.999401	0.034630	28.8771	1.000599	28.8944	0.0000138	27.34091	1
60	0.034899	0.999391	0.034921	28.6363	1.000610	28.6537	0.0000142	27.10036	0
Minutes	Cosine	Sine	Cotangent	Tangent	Cosecant	Secant	Read Down	88°–89° Involute	Minutes

↑ 91° or 271° 88° or 268° ↑

↓ 2° or 182° Trigonometric and Involute Functions 177° or 357° ↓

Minutes	Sine	Cosine	Tangent	Cotangent	Secant	Cosecant	Involute 2°–3°	Read Up	Minutes
0	0.034899	0.999391	0.034921	28.6363	1.000610	28.6537	0.0000142	27.10036	60
1	0.035190	0.999381	0.035212	28.3994	1.000620	28.4170	0.0000145	26.86380	59
2	0.035481	0.999370	0.035503	28.1664	1.000630	28.1842	0.0000149	26.63111	58
3	0.035772	0.999360	0.035795	27.9372	1.000640	27.9551	0.0000153	26.40222	57
4	0.036062	0.999350	0.036086	27.7117	1.000651	27.7298	0.0000157	26.17701	56
5	0.036353	0.999339	0.036377	27.4899	1.000661	27.5080	0.0000160	25.95542	55
6	0.036644	0.999328	0.036668	27.2715	1.000672	27.2898	0.0000164	25.73734	54
7	0.036934	0.999318	0.036960	27.0566	1.000683	27.0750	0.0000168	25.52270	53
8	0.037225	0.999307	0.037251	26.8450	1.000694	26.8636	0.0000172	25.31142	52
9	0.037516	0.999296	0.037542	26.6367	1.000704	26.6555	0.0000176	25.10342	51
10	0.037806	0.999285	0.037834	26.4316	1.000715	26.4505	0.0000180	24.89862	50
11	0.038097	0.999274	0.038125	26.2296	1.000726	26.2487	0.0000185	24.69695	49
12	0.038388	0.999263	0.038416	26.0307	1.000738	26.0499	0.0000189	24.49834	48
13	0.038678	0.999252	0.038707	25.8348	1.000749	25.8542	0.0000193	24.30271	47
14	0.038969	0.999240	0.038999	25.6418	1.000760	25.6613	0.0000198	24.11002	46
15	0.039260	0.999229	0.039290	25.4517	1.000772	25.4713	0.0000202	23.92017	45
16	0.039550	0.999218	0.039581	25.2644	1.000783	25.2841	0.0000207	23.73313	44
17	0.039841	0.999206	0.039873	25.0798	1.000795	25.0997	0.0000211	23.54881	43
18	0.040132	0.999194	0.040164	24.8978	1.000806	24.9179	0.0000216	23.36717	42
19	0.040422	0.999183	0.040456	24.7185	1.000818	24.7387	0.0000220	23.18815	41
20	0.040713	0.999171	0.040747	24.5418	1.000830	24.5621	0.0000225	23.01169	40
21	0.041004	0.999159	0.041038	24.3675	1.000842	24.3880	0.0000230	22.83773	39
22	0.041294	0.999147	0.041330	24.1957	1.000854	24.2164	0.0000235	22.66622	38
23	0.041585	0.999135	0.041621	24.0263	1.000866	24.0471	0.0000240	22.49712	37
24	0.041876	0.999123	0.041912	23.8593	1.000878	23.8802	0.0000245	22.33037	36
25	0.042166	0.999111	0.042204	23.6945	1.000890	23.7156	0.0000250	22.16592	35
26	0.042457	0.999098	0.042495	23.5321	1.000903	23.5533	0.0000256	22.00373	34
27	0.042748	0.999086	0.042787	23.3718	1.000915	23.3932	0.0000261	21.84374	33
28	0.043038	0.999073	0.043078	23.2137	1.000927	23.2352	0.0000266	21.68592	32
29	0.043329	0.999061	0.043370	23.0577	1.000940	23.0794	0.0000272	21.53022	31
30	0.043619	0.999048	0.043661	22.9038	1.000953	22.9256	0.0000277	21.37660	30
31	0.043910	0.999035	0.043952	22.7519	1.000965	22.7739	0.0000283	21.22502	29
32	0.044201	0.999023	0.044244	22.6020	1.000978	22.6241	0.0000288	21.07543	28
33	0.044491	0.999010	0.044535	22.4541	1.000991	22.4764	0.0000294	20.92781	27
34	0.044782	0.998997	0.044827	22.3081	1.001004	22.3305	0.0000300	20.78210	26
35	0.045072	0.998984	0.045118	22.1640	1.001017	22.1865	0.0000306	20.63827	25
36	0.045363	0.998971	0.045410	22.0217	1.001030	22.0444	0.0000312	20.49629	24
37	0.045654	0.998957	0.045701	21.8813	1.001044	21.9041	0.0000318	20.35612	23
38	0.045944	0.998944	0.045993	21.7426	1.001057	21.7656	0.0000324	20.21773	22
39	0.046235	0.998931	0.046284	21.6056	1.001071	21.6288	0.0000330	20.08108	21
40	0.046525	0.998917	0.046576	21.4704	1.001084	21.4937	0.0000336	19.94615	20
41	0.046816	0.998904	0.046867	21.3369	1.001098	21.3603	0.0000343	19.81289	19
42	0.047106	0.998890	0.047159	21.2049	1.001111	21.2285	0.0000349	19.68128	18
43	0.047397	0.998876	0.047450	21.0747	1.001125	21.0984	0.0000356	19.55128	17
44	0.047688	0.998862	0.047742	20.9460	1.001139	20.9698	0.0000362	19.42288	16
45	0.047978	0.998848	0.048033	20.8188	1.001153	20.8428	0.0000369	19.29603	15
46	0.048269	0.998834	0.048325	20.6932	1.001167	20.7174	0.0000376	19.17071	14
47	0.048559	0.998820	0.048617	20.5691	1.001181	20.5934	0.0000382	19.04690	13
48	0.048850	0.998806	0.048908	20.4465	1.001195	20.4709	0.0000389	18.92456	12
49	0.049140	0.998792	0.049200	20.3253	1.001210	20.3499	0.0000396	18.80367	11
50	0.049431	0.998778	0.049491	20.2056	1.001224	20.2303	0.0000403	18.68421	10
51	0.049721	0.998763	0.049783	20.0872	1.001238	20.1121	0.0000411	18.56614	9
52	0.050012	0.998749	0.050075	19.9702	1.001253	19.9952	0.0000418	18.44946	8
53	0.050302	0.998734	0.050366	19.8546	1.001268	19.8798	0.0000425	18.33412	7
54	0.050593	0.998719	0.050658	19.7403	1.001282	19.7656	0.0000433	18.22011	6
55	0.050883	0.998705	0.050949	19.6273	1.001297	19.6528	0.0000440	18.10740	5
56	0.051174	0.998690	0.051241	19.5156	1.001312	19.5412	0.0000448	17.99598	4
57	0.051464	0.998675	0.051533	19.4051	1.001327	19.4309	0.0000455	17.88582	3
58	0.051755	0.998660	0.051824	19.2959	1.001342	19.3218	0.0000463	17.77690	2
59	0.052045	0.998645	0.052116	19.1879	1.001357	19.2140	0.0000471	17.66920	1
60	0.052336	0.998630	0.052408	19.0811	1.001372	19.1073	0.0000479	17.56270	0
Minutes	Cosine	Sine	Cotangent	Tangent	Cosecant	Secant	Read Down	87°–88° Involute	Minutes

↑ 92° or 272° 87° or 267° ↑

↓ 3° or 183° Trigonometric and Involute Functions 176° or 356° ↓

Minutes	Sine	Cosine	Tangent	Cotangent	Secant	Cosecant	Involute 3°–4°	Read Up	Minutes
0	0.052336	0.998630	0.052408	19.0811	1.001372	19.1073	0.0000479	17.56270	60
1	0.052626	0.998614	0.052699	18.9755	1.001388	19.0019	0.0000487	17.45738	59
2	0.052917	0.998599	0.052991	18.8711	1.001403	18.8975	0.0000495	17.35321	58
3	0.053207	0.998583	0.053283	18.7678	1.001419	18.7944	0.0000503	17.25019	57
4	0.053498	0.998568	0.053575	18.6656	1.001434	18.6923	0.0000512	17.14829	56
5	0.053788	0.998552	0.053866	18.5645	1.001450	18.5914	0.0000520	17.04749	55
6	0.054079	0.998537	0.054158	18.4645	1.001465	18.4915	0.0000529	16.94778	54
7	0.054369	0.998521	0.054450	18.3655	1.001481	18.3927	0.0000537	16.84914	53
8	0.054660	0.998505	0.054742	18.2677	1.001497	18.2950	0.0000546	16.75155	52
9	0.054950	0.998489	0.055033	18.1708	1.001513	18.1983	0.0000555	16.65499	51
10	0.055241	0.998473	0.055325	18.0750	1.001529	18.1026	0.0000563	16.55945	50
11	0.055531	0.998457	0.055617	17.9802	1.001545	18.0079	0.0000572	16.46491	49
12	0.055822	0.998441	0.055909	17.8863	1.001562	17.9142	0.0000581	16.37136	48
13	0.056112	0.998424	0.056200	17.7934	1.001578	17.8215	0.0000591	16.27879	47
14	0.056402	0.998408	0.056492	17.7015	1.001594	17.7298	0.0000600	16.18717	46
15	0.056693	0.998392	0.056784	17.6106	1.001611	17.6389	0.0000609	16.09649	45
16	0.056983	0.998375	0.057076	17.5205	1.001628	17.5490	0.0000619	16.00673	44
17	0.057274	0.998359	0.057368	17.4314	1.001644	17.4600	0.0000628	15.91789	43
18	0.057564	0.998342	0.057660	17.3432	1.001661	17.3720	0.0000638	15.82995	42
19	0.057854	0.998325	0.057951	17.2558	1.001678	17.2848	0.0000647	15.74290	41
20	0.058145	0.998308	0.058243	17.1693	1.001695	17.1984	0.0000657	15.65672	40
21	0.058435	0.998291	0.058535	17.0837	1.001712	17.1130	0.0000667	15.57140	39
22	0.058726	0.998274	0.058827	16.9990	1.001729	17.0283	0.0000677	15.48692	38
23	0.059016	0.998257	0.059119	16.9150	1.001746	16.9446	0.0000687	15.40328	37
24	0.059306	0.998240	0.059411	16.8319	1.001763	16.8616	0.0000698	15.32046	36
25	0.059597	0.998223	0.059703	16.7496	1.001781	16.7794	0.0000708	15.23845	35
26	0.059887	0.998205	0.059995	16.6681	1.001798	16.6981	0.0000718	15.15724	34
27	0.060177	0.998188	0.060287	16.5874	1.001816	16.6175	0.0000729	15.07681	33
28	0.060468	0.998170	0.060579	16.5075	1.001833	16.5377	0.0000739	14.99716	32
29	0.060758	0.998153	0.060871	16.4283	1.001851	16.4587	0.0000750	14.91828	31
30	0.061049	0.998135	0.061163	16.3499	1.001869	16.3804	0.0000761	14.84015	30
31	0.061339	0.998117	0.061455	16.2722	1.001887	16.3029	0.0000772	14.76276	29
32	0.061629	0.998099	0.061747	16.1952	1.001905	16.2261	0.0000783	14.68610	28
33	0.061920	0.998081	0.062039	16.1190	1.001923	16.1500	0.0000794	14.61016	27
34	0.062210	0.998063	0.062331	16.0435	1.001941	16.0746	0.0000805	14.53494	26
35	0.062500	0.998045	0.062623	15.9687	1.001959	15.9999	0.0000817	14.46041	25
36	0.062791	0.998027	0.062915	15.8945	1.001977	15.9260	0.0000828	14.38658	24
37	0.063081	0.998008	0.063207	15.8211	1.001996	15.8527	0.0000840	14.31343	23
38	0.063371	0.997990	0.063499	15.7483	1.002014	15.7801	0.0000851	14.24095	22
39	0.063661	0.997972	0.063791	15.6762	1.002033	15.7081	0.0000863	14.16914	21
40	0.063952	0.997953	0.064083	15.6048	1.002051	15.6368	0.0000875	14.09798	20
41	0.064242	0.997934	0.064375	15.5340	1.002070	15.5661	0.0000887	14.02747	19
42	0.064532	0.997916	0.064667	15.4638	1.002089	15.4961	0.0000899	13.95759	18
43	0.064823	0.997897	0.064959	15.3943	1.002108	15.4267	0.0000911	13.88835	17
44	0.065113	0.997878	0.065251	15.3254	1.002127	15.3579	0.0000924	13.81972	16
45	0.065403	0.997859	0.065543	15.2571	1.002146	15.2898	0.0000936	13.75171	15
46	0.065693	0.997840	0.065836	15.1893	1.002165	15.2222	0.0000949	13.68429	14
47	0.065984	0.997821	0.066128	15.1222	1.002184	15.1553	0.0000961	13.61748	13
48	0.066274	0.997801	0.066420	15.0557	1.002203	15.0889	0.0000974	13.55125	12
49	0.066564	0.997782	0.066712	14.9898	1.002223	15.0231	0.0000987	13.48560	11
50	0.066854	0.997763	0.067004	14.9244	1.002242	14.9579	0.0001000	13.42052	10
51	0.067145	0.997743	0.067296	14.8596	1.002262	14.8932	0.0001013	13.35601	9
52	0.067435	0.997724	0.067589	14.7954	1.002282	14.8291	0.0001026	13.29206	8
53	0.067725	0.997704	0.067881	14.7317	1.002301	14.7656	0.0001040	13.22866	7
54	0.068015	0.997684	0.068173	14.6685	1.002321	14.7026	0.0001053	13.16580	6
55	0.068306	0.997664	0.068465	14.6059	1.002341	14.6401	0.0001067	13.10348	5
56	0.068596	0.997645	0.068758	14.5438	1.002361	14.5782	0.0001080	13.04169	4
57	0.068886	0.997625	0.069050	14.4823	1.002381	14.5168	0.0001094	12.98042	3
58	0.069176	0.997604	0.069342	14.4212	1.002401	14.4559	0.0001108	12.91966	2
59	0.069466	0.997584	0.069635	14.3607	1.002422	14.3955	0.0001122	12.85942	1
60	0.069756	0.997564	0.069927	14.3007	1.002442	14.3356	0.0001136	12.79968	0
Minutes	Cosine	Sine	Cotangent	Tangent	Cosecant	Secant	Read Down	86°–87° Involute	Minutes

↑ 93° or 273° 86° or 266° ↑

↓ 4° or 184° Trigonometric and Involute Functions 175° or 355° ↓

Minutes	Sine	Cosine	Tangent	Cotangent	Secant	Cosecant	Involute 4°–5°	Read Up	Minutes
0	0.069756	0.997564	0.069927	14.3007	1.002442	14.3356	0.0001136	12.79968	60
1	0.070047	0.997544	0.070219	14.2411	1.002462	14.2762	0.0001151	12.74044	59
2	0.070337	0.997523	0.070511	14.1821	1.002483	14.2173	0.0001165	12.68169	58
3	0.070627	0.997503	0.070804	14.1235	1.002503	14.1589	0.0001180	12.62343	57
4	0.070917	0.997482	0.071096	14.0655	1.002524	14.1010	0.0001194	12.56564	56
5	0.071207	0.997462	0.071389	14.0079	1.002545	14.0435	0.0001209	12.50833	55
6	0.071497	0.997441	0.071681	13.9507	1.002566	13.9865	0.0001224	12.45148	54
7	0.071788	0.997420	0.071973	13.8940	1.002587	13.9300	0.0001239	12.39510	53
8	0.072078	0.997399	0.072266	13.8378	1.002608	13.8739	0.0001254	12.33917	52
9	0.072368	0.997378	0.072558	13.7821	1.002629	13.8183	0.0001269	12.28369	51
10	0.072658	0.997357	0.072851	13.7267	1.002650	13.7631	0.0001285	12.22866	50
11	0.072948	0.997336	0.073143	13.6719	1.002671	13.7084	0.0001300	12.17407	49
12	0.073238	0.997314	0.073435	13.6174	1.002693	13.6541	0.0001316	12.11992	48
13	0.073528	0.997293	0.073728	13.5634	1.002714	13.6002	0.0001332	12.06619	47
14	0.073818	0.997272	0.074020	13.5098	1.002736	13.5468	0.0001347	12.01289	46
15	0.074108	0.997250	0.074313	13.4566	1.002757	13.4937	0.0001363	11.96001	45
16	0.074399	0.997229	0.074605	13.4039	1.002779	13.4411	0.0001380	11.90754	44
17	0.074689	0.997207	0.074898	13.3515	1.002801	13.3889	0.0001396	11.85548	43
18	0.074979	0.997185	0.075190	13.2996	1.002823	13.3371	0.0001412	11.80383	42
19	0.075269	0.997163	0.075483	13.2480	1.002845	13.2857	0.0001429	11.75257	41
20	0.075559	0.997141	0.075775	13.1969	1.002867	13.2347	0.0001445	11.70172	40
21	0.075849	0.997119	0.076068	13.1461	1.002889	13.1841	0.0001462	11.65125	39
22	0.076139	0.997097	0.076361	13.0958	1.002911	13.1339	0.0001479	11.60117	38
23	0.076429	0.997075	0.076653	13.0458	1.002934	13.0840	0.0001496	11.55148	37
24	0.076719	0.997053	0.076946	12.9962	1.002956	13.0346	0.0001513	11.50216	36
25	0.077009	0.997030	0.077238	12.9469	1.002978	12.9855	0.0001530	11.45321	35
26	0.077299	0.997008	0.077531	12.8981	1.003001	12.9368	0.0001548	11.40464	34
27	0.077589	0.996985	0.077824	12.8496	1.003024	12.8884	0.0001565	11.35643	33
28	0.077879	0.996963	0.078116	12.8014	1.003046	12.8404	0.0001583	11.30858	32
29	0.078169	0.996940	0.078409	12.7536	1.003069	12.7928	0.0001601	11.26109	31
30	0.078459	0.996917	0.078702	12.7062	1.003092	12.7455	0.0001619	11.21395	30
31	0.078749	0.996894	0.078994	12.6591	1.003115	12.6986	0.0001637	11.16716	29
32	0.079039	0.996872	0.079287	12.6124	1.003138	12.6520	0.0001655	11.12072	28
33	0.079329	0.996848	0.079580	12.5660	1.003161	12.6057	0.0001674	11.07461	27
34	0.079619	0.996825	0.079873	12.5199	1.003185	12.5598	0.0001692	11.02885	26
35	0.079909	0.996802	0.080165	12.4742	1.003208	12.5142	0.0001711	10.98342	25
36	0.080199	0.996779	0.080458	12.4288	1.003232	12.4690	0.0001729	10.93832	24
37	0.080489	0.996756	0.080751	12.3838	1.003255	12.4241	0.0001748	10.89355	23
38	0.080779	0.996732	0.081044	12.3390	1.003279	12.3795	0.0001767	10.84910	22
39	0.081069	0.996709	0.081336	12.2946	1.003302	12.3352	0.0001787	10.80497	21
40	0.081359	0.996685	0.081629	12.2505	1.003326	12.2913	0.0001806	10.76116	20
41	0.081649	0.996661	0.081922	12.2067	1.003350	12.2476	0.0001825	10.71766	19
42	0.081939	0.996637	0.082215	12.1632	1.003374	12.2043	0.0001845	10.67447	18
43	0.082228	0.996614	0.082508	12.1201	1.003398	12.1612	0.0001865	10.63159	17
44	0.082518	0.996590	0.082801	12.0772	1.003422	12.1185	0.0001885	10.58901	16
45	0.082808	0.996566	0.083094	12.0346	1.003446	12.0761	0.0001905	10.54673	15
46	0.083098	0.996541	0.083386	11.9923	1.003471	12.0340	0.0001925	10.50475	14
47	0.083388	0.996517	0.083679	11.9504	1.003495	11.9921	0.0001945	10.46306	13
48	0.083678	0.996493	0.083972	11.9087	1.003519	11.9506	0.0001965	10.42166	12
49	0.083968	0.996468	0.084265	11.8673	1.003544	11.9093	0.0001986	10.38055	11
50	0.084258	0.996444	0.084558	11.8262	1.003569	11.8684	0.0002007	10.33973	10
51	0.084547	0.996419	0.084851	11.7853	1.003593	11.8277	0.0002028	10.29919	9
52	0.084837	0.996395	0.085144	11.7448	1.003618	11.7873	0.0002049	10.25892	8
53	0.085127	0.996370	0.085437	11.7045	1.003643	11.7471	0.0002070	10.21893	7
54	0.085417	0.996345	0.085730	11.6645	1.003668	11.7073	0.0002091	10.17922	6
55	0.085707	0.996320	0.086023	11.6248	1.003693	11.6677	0.0002113	10.13978	5
56	0.085997	0.996295	0.086316	11.5853	1.003718	11.6284	0.0002134	10.10060	4
57	0.086286	0.996270	0.086609	11.5461	1.003744	11.5893	0.0002156	10.06169	3
58	0.086576	0.996245	0.086902	11.5072	1.003769	11.5505	0.0002178	10.02304	2
59	0.086866	0.996220	0.087196	11.4685	1.003794	11.5120	0.0002200	9.9846536	1
60	0.087156	0.996195	0.087489	11.4301	1.003820	11.4737	0.0002222	9.9465224	0
Minutes	Cosine	Sine	Cotangent	Tangent	Cosecant	Secant	Read Down	85°–86° Involute	Minutes

↑ 94° or 274° 85° or 265° ↑

↓ 5° or 185° Trigonometric and Involute Functions 174° or 354° ↓

Minutes	Sine	Cosine	Tangent	Cotangent	Secant	Cosecant	Involute 5°–6°	Read Up	Minutes
0	0.087156	0.996195	0.087489	11.4301	1.003820	11.4737	0.0002222	9.9465224	60
1	0.087446	0.996169	0.087782	11.3919	1.003845	11.4357	0.0002244	9.9086459	59
2	0.087735	0.996144	0.088075	11.3540	1.003871	11.3979	0.0002267	9.8710215	58
3	0.088025	0.996118	0.088368	11.3163	1.003897	11.3604	0.0002289	9.8336468	57
4	0.088315	0.996093	0.088661	11.2789	1.003923	11.3231	0.0002312	9.7965192	56
5	0.088605	0.996067	0.088954	11.2417	1.003949	11.2861	0.0002335	9.7596363	55
6	0.088894	0.996041	0.089248	11.2048	1.003975	11.2493	0.0002358	9.7229958	54
7	0.089184	0.996015	0.089541	11.1681	1.004001	11.2128	0.0002382	9.6865952	53
8	0.089474	0.995989	0.089834	11.1316	1.004027	11.1765	0.0002405	9.6504322	52
9	0.089763	0.995963	0.090127	11.0954	1.004053	11.1404	0.0002429	9.6145046	51
10	0.090053	0.995937	0.090421	11.0594	1.004080	11.1045	0.0002452	9.5788100	50
11	0.090343	0.995911	0.090714	11.0237	1.004106	11.0689	0.0002476	9.5433462	49
12	0.090633	0.995884	0.091007	10.9882	1.004133	11.0336	0.0002500	9.5081109	48
13	0.090922	0.995858	0.091300	10.9529	1.004159	10.9984	0.0002524	9.4731021	47
14	0.091212	0.995832	0.091594	10.9178	1.004186	10.9635	0.0002549	9.4383174	46
15	0.091502	0.995805	0.091887	10.8829	1.004213	10.9288	0.0002573	9.4037549	45
16	0.091791	0.995778	0.092180	10.8483	1.004240	10.8943	0.0002598	9.3694123	44
17	0.092081	0.995752	0.092474	10.8139	1.004267	10.8600	0.0002622	9.3352876	43
18	0.092371	0.995725	0.092767	10.7797	1.004294	10.8260	0.0002647	9.3013788	42
19	0.092660	0.995698	0.093061	10.7457	1.004321	10.7921	0.0002673	9.2676838	41
20	0.092950	0.995671	0.093354	10.7119	1.004348	10.7585	0.0002698	9.2342005	40
21	0.093239	0.995644	0.093647	10.6783	1.004375	10.7251	0.0002723	9.2009271	39
22	0.093529	0.995617	0.093941	10.6450	1.004403	10.6919	0.0002749	9.1678616	38
23	0.093819	0.995589	0.094234	10.6118	1.004430	10.6589	0.0002775	9.1350020	37
24	0.094108	0.995562	0.094528	10.5789	1.004458	10.6261	0.0002801	9.1023464	36
25	0.094398	0.995535	0.094821	10.5462	1.004485	10.5935	0.0002827	9.0698930	35
26	0.094687	0.995507	0.095115	10.5136	1.004513	10.5611	0.0002853	9.0376399	34
27	0.094977	0.995479	0.095408	10.4813	1.004541	10.5289	0.0002879	9.0055852	33
28	0.095267	0.995452	0.095702	10.4491	1.004569	10.4969	0.0002906	8.9737272	32
29	0.095556	0.995424	0.095995	10.4172	1.004597	10.4650	0.0002933	8.9420640	31
30	0.095846	0.995396	0.096289	10.3854	1.004625	10.4334	0.0002959	8.9105939	30
31	0.096135	0.995368	0.096583	10.3538	1.004653	10.4020	0.0002986	8.8793151	29
32	0.096425	0.995340	0.096876	10.3224	1.004682	10.3708	0.0003014	8.8482258	28
33	0.096714	0.995312	0.097170	10.2913	1.004710	10.3397	0.0003041	8.8173245	27
34	0.097004	0.995284	0.097464	10.2602	1.004738	10.3089	0.0003069	8.7866094	26
35	0.097293	0.995256	0.097757	10.2294	1.004767	10.2782	0.0003096	8.7560788	25
36	0.097583	0.995227	0.098051	10.1988	1.004795	10.2477	0.0003124	8.7257311	24
37	0.097872	0.995199	0.098345	10.1683	1.004824	10.2174	0.0003152	8.6955646	23
38	0.098162	0.995170	0.098638	10.1381	1.004853	10.1873	0.0003180	8.6655778	22
39	0.098451	0.995142	0.098932	10.1080	1.004882	10.1573	0.0003209	8.6357690	21
40	0.098741	0.995113	0.099226	10.0780	1.004911	10.1275	0.0003237	8.6061367	20
41	0.099030	0.995084	0.099519	10.0483	1.004940	10.0979	0.0003266	8.5766794	19
42	0.099320	0.995056	0.099813	10.0187	1.004969	10.0685	0.0003295	8.5473954	18
43	0.099609	0.995027	0.100107	9.989305	1.004998	10.0392	0.0003324	8.5182834	17
44	0.099899	0.994998	0.100401	9.960072	1.005028	10.0101	0.0003353	8.4893417	16
45	0.100188	0.994969	0.100695	9.931009	1.005057	9.981229	0.0003383	8.4605689	15
46	0.100477	0.994939	0.100989	9.902113	1.005086	9.952479	0.0003412	8.4319635	14
47	0.100767	0.994910	0.101282	9.873382	1.005116	9.923894	0.0003442	8.4035241	13
48	0.101056	0.994881	0.101576	9.844817	1.005146	9.895474	0.0003472	8.3752493	12
49	0.101346	0.994851	0.101870	9.816414	1.005175	9.867218	0.0003502	8.3471377	11
50	0.101635	0.994822	0.102164	9.788173	1.005205	9.839123	0.0003532	8.3191877	10
51	0.101924	0.994792	0.102458	9.760093	1.005235	9.811188	0.0003563	8.2913982	9
52	0.102214	0.994762	0.102752	9.732171	1.005265	9.783412	0.0003593	8.2637676	8
53	0.102503	0.994733	0.103046	9.704407	1.005295	9.755794	0.0003624	8.2362947	7
54	0.102793	0.994703	0.103340	9.676800	1.005325	9.728333	0.0003655	8.2089781	6
55	0.103082	0.994673	0.103634	9.649347	1.005356	9.701026	0.0003686	8.1818164	5
56	0.103371	0.994643	0.103928	9.622049	1.005386	9.673873	0.0003718	8.1548085	4
57	0.103661	0.994613	0.104222	9.594902	1.005416	9.646872	0.0003749	8.1279529	3
58	0.103950	0.994583	0.104516	9.567907	1.005447	9.620023	0.0003781	8.1012485	2
59	0.104239	0.994552	0.104810	9.541061	1.005478	9.593323	0.0003813	8.0746939	1
60	0.104528	0.994522	0.105104	9.514364	1.005508	9.566772	0.0003845	8.0482879	0
Minutes	Cosine	Sine	Cotangent	Tangent	Cosecant	Secant	Read Down	84°–85° Involute	Minutes

↑ 95° or 275° 84° or 264° ↑

↓ 6° or 186° Trigonometric and Involute Functions 173° or 353° ↓

Minutes	Sine	Cosine	Tangent	Cotangent	Secant	Cosecant	Involute 6°–7°	Read Up	Minutes
0	0.104528	0.994522	0.105104	9.514364	1.005508	9.566772	0.0003845	8.0482879	60
1	0.104818	0.994491	0.105398	9.487815	1.005539	9.540369	0.0003877	8.0220292	59
2	0.105107	0.994461	0.105692	9.461412	1.005570	9.514111	0.0003909	7.9959168	58
3	0.105396	0.994430	0.105987	9.435153	1.005601	9.487998	0.0003942	7.9699492	57
4	0.105686	0.994400	0.106281	9.409038	1.005632	9.462030	0.0003975	7.9441254	56
5	0.105975	0.994369	0.106575	9.383066	1.005663	9.436203	0.0004008	7.9184441	55
6	0.106264	0.994338	0.106869	9.357236	1.005694	9.410518	0.0004041	7.8929043	54
7	0.106553	0.994307	0.107163	9.331545	1.005726	9.384974	0.0004074	7.8675047	53
8	0.106843	0.994276	0.107458	9.305994	1.005757	9.359568	0.0004108	7.8422441	52
9	0.107132	0.994245	0.107752	9.280580	1.005788	9.334301	0.0004141	7.8171216	51
10	0.107421	0.994214	0.108046	9.255304	1.005820	9.309170	0.0004175	7.7921359	50
11	0.107710	0.994182	0.108340	9.230163	1.005852	9.284175	0.0004209	7.7672859	49
12	0.107999	0.994151	0.108635	9.205156	1.005883	9.259314	0.0004244	7.7425705	48
13	0.108289	0.994120	0.108929	9.180284	1.005915	9.234588	0.0004278	7.7179887	47
14	0.108578	0.994088	0.109223	9.155544	1.005947	9.209993	0.0004313	7.6935394	46
15	0.108867	0.994056	0.109518	9.130935	1.005979	9.185531	0.0004347	7.6692216	45
16	0.109156	0.994025	0.109812	9.106456	1.006011	9.161198	0.0004382	7.6450341	44
17	0.109445	0.993993	0.110107	9.082107	1.006043	9.136995	0.0004417	7.6209759	43
18	0.109734	0.993961	0.110401	9.057887	1.006076	9.112920	0.0004453	7.5970461	42
19	0.110023	0.993929	0.110695	9.033793	1.006108	9.088972	0.0004488	7.5732436	41
20	0.110313	0.993897	0.110990	9.009826	1.006141	9.065151	0.0004524	7.5495673	40
21	0.110602	0.993865	0.111284	8.985984	1.006173	9.041455	0.0004560	7.5260164	39
22	0.110891	0.993833	0.111579	8.962267	1.006206	9.017884	0.0004596	7.5025898	38
23	0.111180	0.993800	0.111873	8.938673	1.006238	8.994435	0.0004632	7.4792865	37
24	0.111469	0.993768	0.112168	8.915201	1.006271	8.971110	0.0004669	7.4561056	36
25	0.111758	0.993735	0.112463	8.891850	1.006304	8.947905	0.0004706	7.4330461	35
26	0.112047	0.993703	0.112757	8.868621	1.006337	8.924821	0.0004743	7.4101071	34
27	0.112336	0.993670	0.113052	8.845510	1.006370	8.901857	0.0004780	7.3872877	33
28	0.112625	0.993638	0.113346	8.822519	1.006403	8.879011	0.0004817	7.3645869	32
29	0.112914	0.993605	0.113641	8.799645	1.006436	8.856283	0.0004854	7.3420037	31
30	0.113203	0.993572	0.113936	8.776887	1.006470	8.833671	0.0004892	7.3195374	30
31	0.113492	0.993539	0.114230	8.754246	1.006503	8.811176	0.0004930	7.2971870	29
32	0.113781	0.993506	0.114525	8.731720	1.006537	8.788796	0.0004968	7.2749516	28
33	0.114070	0.993473	0.114820	8.709308	1.006570	8.766530	0.0005006	7.2528304	27
34	0.114359	0.993439	0.115114	8.687009	1.006604	8.744377	0.0005045	7.2308224	26
35	0.114648	0.993406	0.115409	8.664822	1.006638	8.722336	0.0005083	7.2089269	25
36	0.114937	0.993373	0.115704	8.642747	1.006671	8.700407	0.0005122	7.1871429	24
37	0.115226	0.993339	0.115999	8.620783	1.006705	8.678589	0.0005161	7.1654696	23
38	0.115515	0.993306	0.116294	8.598929	1.006739	8.656881	0.0005200	7.1439062	22
39	0.115804	0.993272	0.116588	8.577184	1.006773	8.635281	0.0005240	7.1224518	21
40	0.116093	0.993238	0.116883	8.555547	1.006808	8.613790	0.0005280	7.1011057	20
41	0.116382	0.993205	0.117178	8.534017	1.006842	8.592407	0.0005319	7.0798671	19
42	0.116671	0.993171	0.117473	8.512594	1.006876	8.571130	0.0005359	7.0587350	18
43	0.116960	0.993137	0.117768	8.491277	1.006911	8.549958	0.0005400	7.0377088	17
44	0.117249	0.993103	0.118063	8.470065	1.006945	8.528892	0.0005440	7.0167876	16
45	0.117537	0.993068	0.118358	8.448957	1.006980	8.507930	0.0005481	6.9959707	15
46	0.117826	0.993034	0.118653	8.427953	1.007015	8.487072	0.0005522	6.9752573	14
47	0.118115	0.993000	0.118948	8.407052	1.007049	8.466316	0.0005563	6.9546467	13
48	0.118404	0.992966	0.119243	8.386252	1.007084	8.445663	0.0005604	6.9341380	12
49	0.118693	0.992931	0.119538	8.365554	1.007119	8.425111	0.0005645	6.9137305	11
50	0.118982	0.992896	0.119833	8.344956	1.007154	8.404659	0.0005687	6.8934236	10
51	0.119270	0.992862	0.120128	8.324458	1.007190	8.384306	0.0005729	6.8732164	9
52	0.119559	0.992827	0.120423	8.304059	1.007225	8.364053	0.0005771	6.8531082	8
53	0.119848	0.992792	0.120718	8.283758	1.007260	8.343899	0.0005813	6.8330984	7
54	0.120137	0.992757	0.121013	8.263555	1.007295	8.323841	0.0005856	6.8131861	6
55	0.120426	0.992722	0.121308	8.243448	1.007331	8.303881	0.0005898	6.7933708	5
56	0.120714	0.992687	0.121604	8.223438	1.007367	8.284017	0.0005941	6.7736516	4
57	0.121003	0.992652	0.121899	8.203524	1.007402	8.264249	0.0005985	6.7540279	3
58	0.121292	0.992617	0.122194	8.183704	1.007438	8.244575	0.0006028	6.7344991	2
59	0.121581	0.992582	0.122489	8.163979	1.007474	8.224995	0.0006071	6.7150644	1
60	0.121869	0.992546	0.122785	8.144346	1.007510	8.205509	0.0006115	6.6957231	0
Minutes	Cosine	Sine	Cotangent	Tangent	Cosecant	Secant	Read Down	83°–84° Involute	Minutes

↑ 96° or 276° 83° or 263° ↑

↓ 7° or 187° Trigonometric and Involute Functions 172° or 352° ↓

Minutes	Sine	Cosine	Tangent	Cotangent	Secant	Cosecant	Involute 7°–8°	Read Up	Minutes
0	0.121869	0.992546	0.122785	8.144346	1.007510	8.205509	0.0006115	6.6957231	60
1	0.122158	0.992511	0.123080	8.124807	1.007546	8.186116	0.0006159	6.6764747	59
2	0.122447	0.992475	0.123375	8.105360	1.007582	8.166815	0.0006203	6.6573184	58
3	0.122735	0.992439	0.123670	8.086004	1.007618	8.147605	0.0006248	6.6382536	57
4	0.123024	0.992404	0.123966	8.066739	1.007654	8.128486	0.0006292	6.6192796	56
5	0.123313	0.992368	0.124261	8.047565	1.007691	8.109457	0.0006337	6.6003959	55
6	0.123601	0.992332	0.124557	8.028480	1.007727	8.090518	0.0006382	6.5816017	54
7	0.123890	0.992296	0.124852	8.009483	1.007764	8.071668	0.0006427	6.5628964	53
8	0.124179	0.992260	0.125147	7.990576	1.007801	8.052906	0.0006473	6.5442795	52
9	0.124467	0.992224	0.125443	7.971755	1.007837	8.034232	0.0006518	6.5257502	51
10	0.124756	0.992187	0.125738	7.953022	1.007874	8.015645	0.0006564	6.5073080	50
11	0.125045	0.992151	0.126034	7.934376	1.007911	7.997144	0.0006610	6.4889523	49
12	0.125333	0.992115	0.126329	7.915815	1.007948	7.978730	0.0006657	6.4706825	48
13	0.125622	0.992078	0.126625	7.897340	1.007985	7.960400	0.0006703	6.4524979	47
14	0.125910	0.992042	0.126920	7.878949	1.008022	7.942156	0.0006750	6.4343981	46
15	0.126199	0.992005	0.127216	7.860642	1.008059	7.923995	0.0006797	6.4163823	45
16	0.126488	0.991968	0.127512	7.842419	1.008097	7.905918	0.0006844	6.3984501	44
17	0.126776	0.991931	0.127807	7.824279	1.008134	7.887924	0.0006892	6.3806008	43
18	0.127065	0.991894	0.128103	7.806221	1.008172	7.870012	0.0006939	6.3628339	42
19	0.127353	0.991857	0.128399	7.788245	1.008209	7.852182	0.0006987	6.3451489	41
20	0.127642	0.991820	0.128694	7.770351	1.008247	7.834433	0.0007035	6.3275451	40
21	0.127930	0.991783	0.128990	7.752537	1.008285	7.816766	0.0007083	6.3100220	39
22	0.128219	0.991746	0.129286	7.734803	1.008323	7.799178	0.0007132	6.2925791	38
23	0.128507	0.991709	0.129582	7.717149	1.008361	7.781670	0.0007181	6.2752158	37
24	0.128796	0.991671	0.129877	7.699574	1.008399	7.764241	0.0007230	6.2579315	36
25	0.129084	0.991634	0.130173	7.682077	1.008437	7.746890	0.0007279	6.2407259	35
26	0.129373	0.991596	0.130469	7.664658	1.008475	7.729618	0.0007328	6.2235982	34
27	0.129661	0.991558	0.130765	7.647317	1.008513	7.712423	0.0007378	6.2065481	33
28	0.129949	0.991521	0.131061	7.630053	1.008552	7.695305	0.0007428	6.1895749	32
29	0.130238	0.991483	0.131357	7.612866	1.008590	7.678263	0.0007478	6.1726782	31
30	0.130526	0.991445	0.131652	7.595754	1.008629	7.661298	0.0007528	6.1558575	30
31	0.130815	0.991407	0.131948	7.578718	1.008668	7.644407	0.0007579	6.1391122	29
32	0.131103	0.991369	0.132244	7.561757	1.008706	7.627592	0.0007629	6.1224418	28
33	0.131391	0.991331	0.132540	7.544870	1.008745	7.610852	0.0007680	6.1058460	27
34	0.131680	0.991292	0.132836	7.528057	1.008784	7.594185	0.0007732	6.0893240	26
35	0.131968	0.991254	0.133132	7.511318	1.008823	7.577592	0.0007783	6.0728756	25
36	0.132256	0.991216	0.133428	7.494651	1.008862	7.561071	0.0007835	6.0565001	24
37	0.132545	0.991177	0.133725	7.478058	1.008902	7.544624	0.0007887	6.0401971	23
38	0.132833	0.991138	0.134021	7.461536	1.008941	7.528248	0.0007939	6.0239662	22
39	0.133121	0.991100	0.134317	7.445086	1.008980	7.511944	0.0007991	6.0078069	21
40	0.133410	0.991061	0.134613	7.428706	1.009020	7.495711	0.0008044	5.9917186	20
41	0.133698	0.991022	0.134909	7.412398	1.009059	7.479548	0.0008096	5.9757010	19
42	0.133986	0.990983	0.135205	7.396160	1.009099	7.463456	0.0008150	5.9597535	18
43	0.134274	0.990944	0.135502	7.379991	1.009139	7.447433	0.0008203	5.9438758	17
44	0.134563	0.990905	0.135798	7.363892	1.009178	7.431480	0.0008256	5.9280674	16
45	0.134851	0.990866	0.136094	7.347861	1.009218	7.415596	0.0008310	5.9123277	15
46	0.135139	0.990827	0.136390	7.331899	1.009258	7.399780	0.0008364	5.8966565	14
47	0.135427	0.990787	0.136687	7.316005	1.009298	7.384032	0.0008418	5.8810532	13
48	0.135716	0.990748	0.136983	7.300178	1.009339	7.368351	0.0008473	5.8655174	12
49	0.136004	0.990708	0.137279	7.284418	1.009379	7.352738	0.0008527	5.8500487	11
50	0.136292	0.990669	0.137576	7.268725	1.009419	7.337191	0.0008582	5.8346466	10
51	0.136580	0.990629	0.137872	7.253099	1.009460	7.321710	0.0008638	5.8193107	9
52	0.136868	0.990589	0.138169	7.237538	1.009500	7.306295	0.0008693	5.8040407	8
53	0.137156	0.990549	0.138465	7.222042	1.009541	7.290946	0.0008749	5.7888360	7
54	0.137445	0.990509	0.138761	7.206612	1.009581	7.275662	0.0008805	5.7736963	6
55	0.137733	0.990469	0.139058	7.191246	1.009622	7.260442	0.0008861	5.7586212	5
56	0.138021	0.990429	0.139354	7.175944	1.009663	7.245286	0.0008917	5.7436102	4
57	0.138309	0.990389	0.139651	7.160706	1.009704	7.230194	0.0008974	5.7286629	3
58	0.138597	0.990349	0.139948	7.145531	1.009745	7.215165	0.0009031	5.7137791	2
59	0.138885	0.990309	0.140244	7.130419	1.009786	7.200200	0.0009088	5.6989581	1
60	0.139173	0.990268	0.140541	7.115370	1.009828	7.185297	0.0009145	5.6841997	0
Minutes	Cosine	Sine	Cotangent	Tangent	Cosecant	Secant	Read Down	82°–83° Involute	Minutes

↑ 97° or 277° 82° or 262° ↑

↓ 8° or 188° **Trigonometric and Involute Functions** 171° or 351° ↓

Minutes	Sine	Cosine	Tangent	Cotangent	Secant	Cosecant	Involute 8°–9°	Read Up	Minutes
0	0.139173	0.990268	0.140541	7.115370	1.009828	7.185297	0.0009145	5.6841997	60
1	0.139461	0.990228	0.140837	7.100383	1.009869	7.170456	0.0009203	5.6695035	59
2	0.139749	0.990187	0.141134	7.085457	1.009910	7.155676	0.0009260	5.6548691	58
3	0.140037	0.990146	0.141431	7.070593	1.009952	7.140959	0.0009318	5.6402961	57
4	0.140325	0.990105	0.141728	7.055790	1.009993	7.126302	0.0009377	5.6257841	56
5	0.140613	0.990065	0.142024	7.041048	1.010035	7.111706	0.0009435	5.6113327	55
6	0.140901	0.990024	0.142321	7.026366	1.010077	7.097170	0.0009494	5.5969416	54
7	0.141189	0.989983	0.142618	7.011744	1.010119	7.082694	0.0009553	5.5826104	53
8	0.141477	0.989942	0.142915	6.997182	1.010161	7.068278	0.0009612	5.5683387	52
9	0.141765	0.989900	0.143212	6.982678	1.010203	7.053920	0.0009672	5.5541261	51
10	0.142053	0.989859	0.143508	6.968234	1.010245	7.039622	0.0009732	5.5399724	50
11	0.142341	0.989818	0.143805	6.953847	1.010287	7.025382	0.0009792	5.5258771	49
12	0.142629	0.989776	0.144102	6.939519	1.010329	7.011200	0.0009852	5.5118399	48
13	0.142917	0.989735	0.144399	6.925249	1.010372	6.997076	0.0009913	5.4978604	47
14	0.143205	0.989693	0.144696	6.911036	1.010414	6.983009	0.0009973	5.4839383	46
15	0.143493	0.989651	0.144993	6.896880	1.010457	6.968999	0.0010034	5.4700733	45
16	0.143780	0.989610	0.145290	6.882781	1.010499	6.955046	0.0010096	5.4562649	44
17	0.144068	0.989568	0.145587	6.868738	1.010542	6.941150	0.0010157	5.4425129	43
18	0.144356	0.989526	0.145884	6.854751	1.010585	6.927309	0.0010219	5.4288168	42
19	0.144644	0.989484	0.146181	6.840820	1.010628	6.913524	0.0010281	5.4151765	41
20	0.144932	0.989442	0.146478	6.826944	1.010671	6.899794	0.0010343	5.4015914	40
21	0.145220	0.989399	0.146776	6.813123	1.010714	6.886119	0.0010406	5.3880614	39
22	0.145507	0.989357	0.147073	6.799357	1.010757	6.872499	0.0010469	5.3745861	38
23	0.145795	0.989315	0.147370	6.785645	1.010801	6.858934	0.0010532	5.3611651	37
24	0.146083	0.989272	0.147667	6.771987	1.010844	6.845422	0.0010595	5.3477981	36
25	0.146371	0.989230	0.147964	6.758383	1.010887	6.831964	0.0010659	5.3344848	35
26	0.146659	0.989187	0.148262	6.744832	1.010931	6.818560	0.0010722	5.3212249	34
27	0.146946	0.989144	0.148559	6.731334	1.010975	6.805208	0.0010786	5.3080181	33
28	0.147234	0.989102	0.148856	6.717889	1.011018	6.791909	0.0010851	5.2948640	32
29	0.147522	0.989059	0.149154	6.704497	1.011062	6.778663	0.0010915	5.2817624	31
30	0.147809	0.989016	0.149451	6.691156	1.011106	6.765469	0.0010980	5.2687129	30
31	0.148097	0.988973	0.149748	6.677868	1.011150	6.752327	0.0011045	5.2557152	29
32	0.148385	0.988930	0.150046	6.664631	1.011194	6.739236	0.0011111	5.2427691	28
33	0.148672	0.988886	0.150343	6.651445	1.011238	6.726196	0.0011176	5.2298742	27
34	0.148960	0.988843	0.150641	6.638310	1.011283	6.713208	0.0011242	5.2170302	26
35	0.149248	0.988800	0.150938	6.625226	1.011327	6.700270	0.0011308	5.2042369	25
36	0.149535	0.988756	0.151236	6.612192	1.011371	6.687382	0.0011375	5.1914939	24
37	0.149823	0.988713	0.151533	6.599208	1.011416	6.674545	0.0011441	5.1788009	23
38	0.150111	0.988669	0.151831	6.586274	1.011461	6.661757	0.0011508	5.1661577	22
39	0.150398	0.988626	0.152129	6.573389	1.011505	6.649018	0.0011575	5.1535639	21
40	0.150686	0.988582	0.152426	6.560554	1.011550	6.636329	0.0011643	5.1410193	20
41	0.150973	0.988538	0.152724	6.547767	1.011595	6.623689	0.0011711	5.1285236	19
42	0.151261	0.988494	0.153022	6.535029	1.011640	6.611097	0.0011779	5.1160766	18
43	0.151548	0.988450	0.153319	6.522340	1.011685	6.598554	0.0011847	5.1036779	17
44	0.151836	0.988406	0.153617	6.509698	1.011730	6.586059	0.0011915	5.0913272	16
45	0.152123	0.988362	0.153915	6.497104	1.011776	6.573611	0.0011984	5.0790243	15
46	0.152411	0.988317	0.154213	6.484558	1.011821	6.561211	0.0012053	5.0667689	14
47	0.152698	0.988273	0.154510	6.472059	1.011866	6.548859	0.0012122	5.0545608	13
48	0.152986	0.988228	0.154808	6.459607	1.011912	6.536553	0.0012192	5.0423997	12
49	0.153273	0.988184	0.155106	6.447202	1.011957	6.524294	0.0012262	5.0302852	11
50	0.153561	0.988139	0.155404	6.434843	1.012003	6.512081	0.0012332	5.0182172	10
51	0.153848	0.988094	0.155702	6.422530	1.012049	6.499915	0.0012402	5.0061954	9
52	0.154136	0.988050	0.156000	6.410263	1.012095	6.487794	0.0012473	4.9942195	8
53	0.154423	0.988005	0.156298	6.398042	1.012141	6.475720	0.0012544	4.9822893	7
54	0.154710	0.987960	0.156596	6.385866	1.012187	6.463690	0.0012615	4.9704044	6
55	0.154998	0.987915	0.156894	6.373736	1.012233	6.451706	0.0012687	4.9585647	5
56	0.155285	0.987870	0.157192	6.361650	1.012279	6.439767	0.0012758	4.9467700	4
57	0.155572	0.987824	0.157490	6.349609	1.012326	6.427872	0.0012830	4.9350198	3
58	0.155860	0.987779	0.157788	6.337613	1.012372	6.416022	0.0012903	4.9233141	2
59	0.156147	0.987734	0.158086	6.325660	1.012419	6.404215	0.0012975	4.9116525	1
60	0.156434	0.987688	0.158384	6.313752	1.012465	6.392453	0.0013048	4.9000348	0
Minutes	Cosine	Sine	Cotangent	Tangent	Cosecant	Secant	Read Down	81°–82° Involute	Minutes

↑ 98° or 278° 81° or 261° ↑

↓ 9° or 189° Trigonometric and Involute Functions 170° or 350° ↓

Minutes	Sine	Cosine	Tangent	Cotangent	Secant	Cosecant	Involute 9°–10°	Read Up	Minutes
0	0.156434	0.987688	0.158384	6.313752	1.012465	6.392453	0.0013048	4.9000348	60
1	0.156722	0.987643	0.158683	6.301887	1.012512	6.380735	0.0013121	4.8884608	59
2	0.157009	0.987597	0.158981	6.290065	1.012559	6.369060	0.0013195	4.8769302	58
3	0.157296	0.987551	0.159279	6.278287	1.012605	6.357428	0.0013268	4.8654428	57
4	0.157584	0.987506	0.159577	6.266551	1.012652	6.345839	0.0013342	4.8539983	56
5	0.157871	0.987460	0.159876	6.254859	1.012699	6.334292	0.0013416	4.8425965	55
6	0.158158	0.987414	0.160174	6.243209	1.012747	6.322788	0.0013491	4.8312372	54
7	0.158445	0.987368	0.160472	6.231601	1.012794	6.311327	0.0013566	4.8199202	53
8	0.158732	0.987322	0.160771	6.220035	1.012841	6.299907	0.0013641	4.8086451	52
9	0.159020	0.987275	0.161069	6.208511	1.012889	6.288530	0.0013716	4.7974119	51
10	0.159307	0.987229	0.161368	6.197028	1.012936	6.277193	0.0013792	4.7862201	50
11	0.159594	0.987183	0.161666	6.185587	1.012984	6.265898	0.0013868	4.7750697	49
12	0.159881	0.987136	0.161965	6.174186	1.013031	6.254645	0.0013944	4.7639604	48
13	0.160168	0.987090	0.162263	6.162827	1.013079	6.243432	0.0014020	4.7528920	47
14	0.160455	0.987043	0.162562	6.151508	1.013127	6.232259	0.0014097	4.7418642	46
15	0.160743	0.986996	0.162860	6.140230	1.013175	6.221128	0.0014174	4.7308769	45
16	0.161030	0.986950	0.163159	6.128992	1.013223	6.210036	0.0014251	4.7199298	44
17	0.161317	0.986903	0.163458	6.117794	1.013271	6.198984	0.0014329	4.7090227	43
18	0.161604	0.986856	0.163756	6.106636	1.013319	6.187972	0.0014407	4.6981553	42
19	0.161891	0.986809	0.164055	6.095517	1.013368	6.177000	0.0014485	4.6873276	41
20	0.162178	0.986762	0.164354	6.084438	1.013416	6.166067	0.0014563	4.6765392	40
21	0.162465	0.986714	0.164652	6.073398	1.013465	6.155174	0.0014642	4.6657899	39
22	0.162752	0.986667	0.164951	6.062397	1.013513	6.144319	0.0014721	4.6550796	38
23	0.163039	0.986620	0.165250	6.051434	1.013562	6.133503	0.0014800	4.6444080	37
24	0.163326	0.986572	0.165549	6.040510	1.013611	6.122725	0.0014880	4.6337750	36
25	0.163613	0.986525	0.165848	6.029625	1.013659	6.111986	0.0014960	4.6231802	35
26	0.163900	0.986477	0.166147	6.018777	1.013708	6.101285	0.0015040	4.6126236	34
27	0.164187	0.986429	0.166446	6.007968	1.013757	6.090622	0.0015120	4.6021049	33
28	0.164474	0.986381	0.166745	5.997196	1.013807	6.079996	0.0015201	4.5916239	32
29	0.164761	0.986334	0.167044	5.986461	1.013856	6.069409	0.0015282	4.5811805	31
30	0.165048	0.986286	0.167343	5.975764	1.013905	6.058858	0.0015363	4.5707743	30
31	0.165334	0.986238	0.167642	5.965104	1.013954	6.048345	0.0015445	4.5604053	29
32	0.165621	0.986189	0.167941	5.954481	1.014004	6.037868	0.0015527	4.5500732	28
33	0.165908	0.986141	0.168240	5.943895	1.014054	6.027428	0.0015609	4.5397779	27
34	0.166195	0.986093	0.168539	5.933346	1.014103	6.017025	0.0015691	4.5295190	26
35	0.166482	0.986045	0.168838	5.922832	1.014153	6.006658	0.0015774	4.5192966	25
36	0.166769	0.985996	0.169137	5.912355	1.014203	5.996327	0.0015857	4.5091103	24
37	0.167056	0.985947	0.169437	5.901914	1.014253	5.986033	0.0015941	4.4989600	23
38	0.167342	0.985899	0.169736	5.891508	1.014303	5.975774	0.0016024	4.4888455	22
39	0.167629	0.985850	0.170035	5.881139	1.014353	5.965550	0.0016108	4.4787665	21
40	0.167916	0.985801	0.170334	5.870804	1.014403	5.955362	0.0016193	4.4687230	20
41	0.168203	0.985752	0.170634	5.860505	1.014453	5.945210	0.0016277	4.4587148	19
42	0.168489	0.985703	0.170933	5.850241	1.014504	5.935092	0.0016362	4.4487416	18
43	0.168776	0.985654	0.171233	5.840012	1.014554	5.925009	0.0016447	4.4388032	17
44	0.169063	0.985605	0.171532	5.829817	1.014605	5.914961	0.0016533	4.4288996	16
45	0.169350	0.985556	0.171831	5.819657	1.014656	5.904948	0.0016618	4.4190305	15
46	0.169636	0.985507	0.172131	5.809532	1.014706	5.894969	0.0016704	4.4091957	14
47	0.169923	0.985457	0.172430	5.799440	1.014757	5.885024	0.0016791	4.3993951	13
48	0.170209	0.985408	0.172730	5.789383	1.014808	5.875113	0.0016877	4.3896285	12
49	0.170496	0.985358	0.173030	5.779359	1.014859	5.865236	0.0016964	4.3798957	11
50	0.170783	0.985309	0.173329	5.769369	1.014910	5.855392	0.0017051	4.3701965	10
51	0.171069	0.985259	0.173629	5.759412	1.014962	5.845582	0.0017139	4.3605308	9
52	0.171356	0.985209	0.173929	5.749489	1.015013	5.835805	0.0017227	4.3508984	8
53	0.171643	0.985159	0.174228	5.739599	1.015064	5.826062	0.0017315	4.3412992	7
54	0.171929	0.985109	0.174528	5.729742	1.015116	5.816351	0.0017403	4.3317329	6
55	0.172216	0.985059	0.174828	5.719917	1.015167	5.806673	0.0017492	4.3221994	5
56	0.172502	0.985009	0.175127	5.710126	1.015219	5.797028	0.0017581	4.3126986	4
57	0.172789	0.984959	0.175427	5.700366	1.015271	5.787415	0.0017671	4.3032303	3
58	0.173075	0.984909	0.175727	5.690639	1.015323	5.777835	0.0017760	4.2937942	2
59	0.173362	0.984858	0.176027	5.680945	1.015375	5.768287	0.0017850	4.2843903	1
60	0.173648	0.984808	0.176327	5.671282	1.015427	5.758770	0.0017941	4.2750184	0
Minutes	Cosine	Sine	Cotangent	Tangent	Cosecant	Secant	Read Down	80°–81° Involute	Minutes

↑ 99° or 279° 80° or 260° ↑

↓ 10° or 190° Trigonometric and Involute Functions 169° or 349° ↓

Minutes	Sine	Cosine	Tangent	Cotangent	Secant	Cosecant	Involute 10°–11°	Read Up	Minutes
0	0.173648	0.984808	0.176327	5.671282	1.015427	5.758770	0.0017941	4.2750184	60
1	0.173935	0.984757	0.176627	5.661651	1.015479	5.749286	0.0018031	4.2656783	59
2	0.174221	0.984707	0.176927	5.652052	1.015531	5.739833	0.0018122	4.2563699	58
3	0.174508	0.984656	0.177227	5.642484	1.015583	5.730412	0.0018213	4.2470930	57
4	0.174794	0.984605	0.177527	5.632947	1.015636	5.721022	0.0018305	4.2378475	56
5	0.175080	0.984554	0.177827	5.623442	1.015688	5.711664	0.0018397	4.2286332	55
6	0.175367	0.984503	0.178127	5.613968	1.015741	5.702336	0.0018489	4.2194499	54
7	0.175653	0.984452	0.178427	5.604525	1.015793	5.693039	0.0018581	4.2102975	53
8	0.175939	0.984401	0.178727	5.595112	1.015846	5.683773	0.0018674	4.2011758	52
9	0.176226	0.984350	0.179028	5.585730	1.015899	5.674538	0.0018767	4.1920848	51
10	0.176512	0.984298	0.179328	5.576379	1.015952	5.665333	0.0018860	4.1830241	50
11	0.176798	0.984247	0.179628	5.567057	1.016005	5.656158	0.0018954	4.1739938	49
12	0.177085	0.984196	0.179928	5.557766	1.016058	5.647014	0.0019048	4.1649936	48
13	0.177371	0.984144	0.180229	5.548505	1.016111	5.637899	0.0019142	4.1560234	47
14	0.177657	0.984092	0.180529	5.539274	1.016165	5.628815	0.0019237	4.1470830	46
15	0.177944	0.984041	0.180829	5.530072	1.016218	5.619760	0.0019332	4.1381724	45
16	0.178230	0.983989	0.181130	5.520900	1.016272	5.610735	0.0019427	4.1292913	44
17	0.178516	0.983937	0.181430	5.511758	1.016325	5.601739	0.0019523	4.1204396	43
18	0.178802	0.983885	0.181731	5.502645	1.016379	5.592772	0.0019619	4.1116172	42
19	0.179088	0.983833	0.182031	5.493560	1.016433	5.583834	0.0019715	4.1028239	41
20	0.179375	0.983781	0.182332	5.484505	1.016487	5.574926	0.0019812	4.0940596	40
21	0.179661	0.983729	0.182632	5.475479	1.016541	5.566046	0.0019909	4.0853241	39
22	0.179947	0.983676	0.182933	5.466481	1.016595	5.557195	0.0020006	4.0766173	38
23	0.180233	0.983624	0.183234	5.457512	1.016649	5.548373	0.0020103	4.0679392	37
24	0.180519	0.983571	0.183534	5.448572	1.016703	5.539579	0.0020201	4.0592894	36
25	0.180805	0.983519	0.183835	5.439659	1.016757	5.530813	0.0020299	4.0506680	35
26	0.181091	0.983466	0.184136	5.430775	1.016812	5.522075	0.0020398	4.0420747	34
27	0.181377	0.983414	0.184437	5.421919	1.016866	5.513366	0.0020496	4.0335094	33
28	0.181663	0.983361	0.184737	5.413091	1.016921	5.504684	0.0020596	4.0249720	32
29	0.181950	0.983308	0.185038	5.404290	1.016975	5.496030	0.0020695	4.0164624	31
30	0.182236	0.983255	0.185339	5.395517	1.017030	5.487404	0.0020795	4.0079804	30
31	0.182522	0.983202	0.185640	5.386772	1.017085	5.478806	0.0020895	3.9995259	29
32	0.182808	0.983149	0.185941	5.378054	1.017140	5.470234	0.0020995	3.9910988	28
33	0.183094	0.983096	0.186242	5.369363	1.017195	5.461690	0.0021096	3.9826989	27
34	0.183379	0.983042	0.186543	5.360699	1.017250	5.453173	0.0021197	3.9743261	26
35	0.183665	0.982989	0.186844	5.352063	1.017306	5.444683	0.0021298	3.9659803	25
36	0.183951	0.982935	0.187145	5.343453	1.017361	5.436220	0.0021400	3.9576613	24
37	0.184237	0.982882	0.187446	5.334870	1.017416	5.427784	0.0021502	3.9493691	23
38	0.184523	0.982828	0.187747	5.326313	1.017472	5.419374	0.0021605	3.9411034	22
39	0.184809	0.982774	0.188048	5.317783	1.017527	5.410990	0.0021707	3.9328643	21
40	0.185095	0.982721	0.188349	5.309279	1.017583	5.402633	0.0021810	3.9246514	20
41	0.185381	0.982667	0.188651	5.300802	1.017639	5.394303	0.0021914	3.9164648	19
42	0.185667	0.982613	0.188952	5.292350	1.017695	5.385998	0.0022017	3.9083044	18
43	0.185952	0.982559	0.189253	5.283925	1.017751	5.377719	0.0022121	3.9001698	17
44	0.186238	0.982505	0.189555	5.275526	1.017807	5.369466	0.0022226	3.8920612	16
45	0.186524	0.982450	0.189856	5.267152	1.017863	5.361239	0.0022330	3.8839783	15
46	0.186810	0.982396	0.190157	5.258804	1.017919	5.353038	0.0022435	3.8759210	14
47	0.187096	0.982342	0.190459	5.250481	1.017976	5.344862	0.0022541	3.8678892	13
48	0.187381	0.982287	0.190760	5.242184	1.018032	5.336711	0.0022646	3.8598828	12
49	0.187667	0.982233	0.191062	5.233912	1.018089	5.328586	0.0022752	3.8519017	11
50	0.187953	0.982178	0.191363	5.225665	1.018145	5.320486	0.0022859	3.8439457	10
51	0.188238	0.982123	0.191665	5.217443	1.018202	5.312411	0.0022965	3.8360147	9
52	0.188524	0.982069	0.191966	5.209246	1.018259	5.304361	0.0023073	3.8281087	8
53	0.188810	0.982014	0.192268	5.201074	1.018316	5.296335	0.0023180	3.8202275	7
54	0.189095	0.981959	0.192570	5.192926	1.018373	5.288335	0.0023288	3.8123709	6
55	0.189381	0.981904	0.192871	5.184804	1.018430	5.280359	0.0023396	3.8045390	5
56	0.189667	0.981849	0.193173	5.176705	1.018487	5.272407	0.0023504	3.7967315	4
57	0.189952	0.981793	0.193475	5.168631	1.018544	5.264480	0.0023613	3.7889483	3
58	0.190238	0.981738	0.193777	5.160581	1.018602	5.256577	0.0023722	3.7811894	2
59	0.190523	0.981683	0.194078	5.152556	1.018659	5.248698	0.0023831	3.7734547	1
60	0.190809	0.981627	0.194380	5.144554	1.018717	5.240843	0.0023941	3.7657439	0
Minutes	Cosine	Sine	Cotangent	Tangent	Cosecant	Secant	Read Down	79°–80° Involute	Minutes

↑ 100° or 280° 79° or 259° ↑

↓ 11° or 191° Trigonometric and Involute Functions 168° or 348° ↓

Minutes	Sine	Cosine	Tangent	Cotangent	Secant	Cosecant	Involute 11°–12°	Read Up	Minutes
0	0.190809	0.981627	0.194380	5.144554	1.018717	5.240843	0.0023941	3.7657439	60
1	0.191095	0.981572	0.194682	5.136576	1.018774	5.233012	0.0024051	3.7580571	59
2	0.191380	0.981516	0.194984	5.128622	1.018832	5.225205	0.0024161	3.7503940	58
3	0.191666	0.981460	0.195286	5.120692	1.018890	5.217422	0.0024272	3.7427547	57
4	0.191951	0.981405	0.195588	5.112786	1.018948	5.209662	0.0024383	3.7351390	56
5	0.192237	0.981349	0.195890	5.104902	1.019006	5.201925	0.0024495	3.7275467	55
6	0.192522	0.981293	0.196192	5.097043	1.019064	5.194212	0.0024607	3.7199778	54
7	0.192807	0.981237	0.196494	5.089206	1.019122	5.186523	0.0024719	3.7124322	53
8	0.193093	0.981180	0.196796	5.081393	1.019180	5.178856	0.0024831	3.7049098	52
9	0.193378	0.981124	0.197099	5.073602	1.019239	5.171213	0.0024944	3.6974104	51
10	0.193664	0.981068	0.197401	5.065835	1.019297	5.163592	0.0025057	3.6899340	50
11	0.193949	0.981012	0.197703	5.058091	1.019356	5.155995	0.0025171	3.6824804	49
12	0.194234	0.980955	0.198005	5.050369	1.019415	5.148420	0.0025285	3.6750496	48
13	0.194520	0.980899	0.198308	5.042670	1.019473	5.140868	0.0025399	3.6676414	47
14	0.194805	0.980842	0.198610	5.034994	1.019532	5.133338	0.0025513	3.6602558	46
15	0.195090	0.980785	0.198912	5.027339	1.019591	5.125831	0.0025628	3.6528927	45
16	0.195376	0.980728	0.199215	5.019708	1.019650	5.118346	0.0025744	3.6455519	44
17	0.195661	0.980672	0.199517	5.012098	1.019709	5.110884	0.0025859	3.6382334	43
18	0.195946	0.980615	0.199820	5.004511	1.019769	5.103443	0.0025975	3.6309370	42
19	0.196231	0.980558	0.200122	4.996946	1.019828	5.096025	0.0026091	3.6236627	41
20	0.196517	0.980500	0.200425	4.989403	1.019887	5.088628	0.0026208	3.6164103	40
21	0.196802	0.980443	0.200727	4.981881	1.019947	5.081254	0.0026325	3.6091798	39
22	0.197087	0.980386	0.201030	4.974382	1.020006	5.073901	0.0026443	3.6019711	38
23	0.197372	0.980329	0.201333	4.966904	1.020066	5.066570	0.0026560	3.5947840	37
24	0.197657	0.980271	0.201635	4.959447	1.020126	5.059261	0.0026678	3.5876186	36
25	0.197942	0.980214	0.201938	4.952012	1.020186	5.051973	0.0026797	3.5804746	35
26	0.198228	0.980156	0.202241	4.944599	1.020246	5.044706	0.0026916	3.5733520	34
27	0.198513	0.980098	0.202544	4.937207	1.020306	5.037461	0.0027035	3.5662507	33
28	0.198798	0.980041	0.202847	4.929836	1.020366	5.030237	0.0027154	3.5591705	32
29	0.199083	0.979983	0.203149	4.922486	1.020426	5.023034	0.0027274	3.5521115	31
30	0.199368	0.979925	0.203452	4.915157	1.020487	5.015852	0.0027394	3.5450736	30
31	0.199653	0.979867	0.203755	4.907849	1.020547	5.008691	0.0027515	3.5380565	29
32	0.199938	0.979809	0.204058	4.900562	1.020608	5.001551	0.0027636	3.5310603	28
33	0.200223	0.979750	0.204361	4.893296	1.020668	4.994431	0.0027757	3.5240848	27
34	0.200508	0.979692	0.204664	4.886050	1.020729	4.987332	0.0027879	3.5171300	26
35	0.200793	0.979634	0.204967	4.878825	1.020790	4.980254	0.0028001	3.5101958	25
36	0.201078	0.979575	0.205271	4.871620	1.020851	4.973196	0.0028123	3.5032820	24
37	0.201363	0.979517	0.205574	4.864436	1.020912	4.966159	0.0028246	3.4963886	23
38	0.201648	0.979458	0.205877	4.857272	1.020973	4.959142	0.0028369	3.4895156	22
39	0.201933	0.979399	0.206180	4.850128	1.021034	4.952145	0.0028493	3.4826627	21
40	0.202218	0.979341	0.206483	4.843005	1.021095	4.945169	0.0028616	3.4758300	20
41	0.202502	0.979282	0.206787	4.835901	1.021157	4.938212	0.0028741	3.4690173	19
42	0.202787	0.979223	0.207090	4.828817	1.021218	4.931275	0.0028865	3.4622245	18
43	0.203072	0.979164	0.207393	4.821754	1.021280	4.924359	0.0028990	3.4554517	17
44	0.203357	0.979105	0.207697	4.814710	1.021341	4.917462	0.0029115	3.4486986	16
45	0.203642	0.979045	0.208000	4.807685	1.021403	4.910584	0.0029241	3.4419653	15
46	0.203927	0.978986	0.208304	4.800681	1.021465	4.903727	0.0029367	3.4352515	14
47	0.204211	0.978927	0.208607	4.793696	1.021527	4.896889	0.0029494	3.4285573	13
48	0.204496	0.978867	0.208911	4.786730	1.021589	4.890070	0.0029620	3.4218825	12
49	0.204781	0.978808	0.209214	4.779784	1.021651	4.883271	0.0029747	3.4152272	11
50	0.205065	0.978748	0.209518	4.772857	1.021713	4.876491	0.0029875	3.4085911	10
51	0.205350	0.978689	0.209822	4.765949	1.021776	4.869730	0.0030003	3.4019742	9
52	0.205635	0.978629	0.210126	4.759060	1.021838	4.862988	0.0030131	3.3953764	8
53	0.205920	0.978569	0.210429	4.752191	1.021900	4.856266	0.0030260	3.3887977	7
54	0.206204	0.978509	0.210733	4.745340	1.021963	4.849562	0.0030389	3.3822379	6
55	0.206489	0.978449	0.211037	4.738508	1.022026	4.842877	0.0030518	3.3756971	5
56	0.206773	0.978389	0.211341	4.731695	1.022089	4.836211	0.0030648	3.3691750	4
57	0.207058	0.978329	0.211645	4.724901	1.022151	4.829564	0.0030778	3.3626717	3
58	0.207343	0.978268	0.211949	4.718126	1.022214	4.822936	0.0030908	3.3561870	2
59	0.207627	0.978208	0.212253	4.711369	1.022277	4.816326	0.0031039	3.3497209	1
60	0.207912	0.978148	0.212557	4.704630	1.022341	4.809734	0.0031171	3.3432733	0
Minutes	Cosine	Sine	Cotangent	Tangent	Cosecant	Secant	Read Down	78°–79° Involute	Minutes

↑ 101° or 281° 78° or 258° ↑

↓ 12° or 192° Trigonometric and Involute Functions 167° or 347° ↓

Minutes	Sine	Cosine	Tangent	Cotangent	Secant	Cosecant	Involute 12°–13°	Read Up	Minutes
0	0.207912	0.978148	0.212557	4.704630	1.022341	4.809734	0.0031171	3.3432733	60
1	0.208196	0.978087	0.212861	4.697910	1.022404	4.803161	0.0031302	3.3368441	59
2	0.208481	0.978026	0.213165	4.691208	1.022467	4.796607	0.0031434	3.3304333	58
3	0.208765	0.977966	0.213469	4.684525	1.022531	4.790070	0.0031566	3.3240407	57
4	0.209050	0.977905	0.213773	4.677860	1.022594	4.783552	0.0031699	3.3176663	56
5	0.209334	0.977844	0.214077	4.671212	1.022658	4.777052	0.0031832	3.3113100	55
6	0.209619	0.977783	0.214381	4.664583	1.022722	4.770570	0.0031966	3.3049718	54
7	0.209903	0.977722	0.214686	4.657972	1.022785	4.764106	0.0032100	3.2986515	53
8	0.210187	0.977661	0.214990	4.651379	1.022849	4.757660	0.0032234	3.2923491	52
9	0.210472	0.977600	0.215294	4.644803	1.022913	4.751231	0.0032369	3.2860645	51
10	0.210756	0.977539	0.215599	4.638246	1.022977	4.744821	0.0032504	3.2797977	50
11	0.211040	0.977477	0.215903	4.631706	1.023042	4.738428	0.0032639	3.2735486	49
12	0.211325	0.977416	0.216208	4.625183	1.023106	4.732052	0.0032775	3.2673170	48
13	0.211609	0.977354	0.216512	4.618678	1.023170	4.725695	0.0032911	3.2611030	47
14	0.211893	0.977293	0.216817	4.612191	1.023235	4.719354	0.0033048	3.2549064	46
15	0.212178	0.977231	0.217121	4.605721	1.023299	4.713031	0.0033185	3.2487273	45
16	0.212462	0.977169	0.217426	4.599268	1.023364	4.706726	0.0033322	3.2425654	44
17	0.212746	0.977108	0.217731	4.592832	1.023429	4.700437	0.0033460	3.2364208	43
18	0.213030	0.977046	0.218035	4.586414	1.023494	4.694166	0.0033598	3.2302933	42
19	0.213315	0.976984	0.218340	4.580013	1.023559	4.687912	0.0033736	3.2241830	41
20	0.213599	0.976921	0.218645	4.573629	1.023624	4.681675	0.0033875	3.2180896	40
21	0.213883	0.976859	0.218950	4.567261	1.023689	4.675455	0.0034014	3.2120133	39
22	0.214167	0.976797	0.219254	4.560911	1.023754	4.669252	0.0034154	3.2059538	38
23	0.214451	0.976735	0.219559	4.554578	1.023819	4.663065	0.0034294	3.1999112	37
24	0.214735	0.976672	0.219864	4.548261	1.023885	4.656896	0.0034434	3.1938853	36
25	0.215019	0.976610	0.220169	4.541961	1.023950	4.650743	0.0034575	3.1878762	35
26	0.215303	0.976547	0.220474	4.535677	1.024016	4.644606	0.0034716	3.1818836	34
27	0.215588	0.976485	0.220779	4.529410	1.024082	4.638487	0.0034858	3.1759076	33
28	0.215872	0.976422	0.221084	4.523160	1.024148	4.632384	0.0035000	3.1699481	32
29	0.216156	0.976359	0.221389	4.516926	1.024214	4.626297	0.0035142	3.1640050	31
30	0.216440	0.976296	0.221695	4.510709	1.024280	4.620226	0.0035285	3.1580783	30
31	0.216724	0.976233	0.222000	4.504507	1.024346	4.614172	0.0035428	3.1521679	29
32	0.217008	0.976170	0.222305	4.498322	1.024412	4.608134	0.0035572	3.1462737	28
33	0.217292	0.976107	0.222610	4.492153	1.024478	4.602113	0.0035716	3.1403957	27
34	0.217575	0.976044	0.222916	4.486000	1.024544	4.596107	0.0035860	3.1345338	26
35	0.217859	0.975980	0.223221	4.479864	1.024611	4.590117	0.0036005	3.1286879	25
36	0.218143	0.975917	0.223526	4.473743	1.024678	4.584144	0.0036150	3.1228580	24
37	0.218427	0.975853	0.223832	4.467638	1.024744	4.578186	0.0036296	3.1170440	23
38	0.218711	0.975790	0.224137	4.461549	1.024811	4.572244	0.0036441	3.1112458	22
39	0.218995	0.975726	0.224443	4.455476	1.024878	4.566318	0.0036588	3.1054635	21
40	0.219279	0.975662	0.224748	4.449418	1.024945	4.560408	0.0036735	3.0996968	20
41	0.219562	0.975598	0.225054	4.443376	1.025012	4.554513	0.0036882	3.0939458	19
42	0.219846	0.975535	0.225360	4.437350	1.025079	4.548634	0.0037029	3.0882104	18
43	0.220130	0.975471	0.225665	4.431339	1.025146	4.542771	0.0037177	3.0824906	17
44	0.220414	0.975406	0.225971	4.425344	1.025214	4.536923	0.0037325	3.0767862	16
45	0.220697	0.975342	0.226277	4.419364	1.025281	4.531090	0.0037474	3.0710972	15
46	0.220981	0.975278	0.226583	4.413400	1.025349	4.525273	0.0037623	3.0654236	14
47	0.221265	0.975214	0.226889	4.407450	1.025416	4.519471	0.0037773	3.0597653	13
48	0.221548	0.975149	0.227194	4.401516	1.025484	4.513684	0.0037923	3.0541223	12
49	0.221832	0.975085	0.227500	4.395598	1.025552	4.507913	0.0038073	3.0484944	11
50	0.222116	0.975020	0.227806	4.389694	1.025620	4.502157	0.0038224	3.0428816	10
51	0.222399	0.974956	0.228112	4.383805	1.025688	4.496415	0.0038375	3.0372838	9
52	0.222683	0.974891	0.228418	4.377932	1.025756	4.490689	0.0038527	3.0317011	8
53	0.222967	0.974826	0.228724	4.372073	1.025824	4.484977	0.0038679	3.0261333	7
54	0.223250	0.974761	0.229031	4.366229	1.025892	4.479281	0.0038831	3.0205804	6
55	0.223534	0.974696	0.229337	4.360400	1.025961	4.473599	0.0038984	3.0150424	5
56	0.223817	0.974631	0.229643	4.354586	1.026029	4.467932	0.0039137	3.0095190	4
57	0.224101	0.974566	0.229949	4.348787	1.026098	4.462280	0.0039291	3.0040104	3
58	0.224384	0.974501	0.230255	4.343002	1.026166	4.456643	0.0039445	2.9985165	2
59	0.224668	0.974435	0.230562	4.337232	1.026235	4.451020	0.0039599	2.9930372	1
60	0.224951	0.974370	0.230868	4.331476	1.026304	4.445411	0.0039754	2.9875724	0
Minutes	Cosine	Sine	Cotangent	Tangent	Cosecant	Secant	Read Down	77°–78° Involute	Minutes

↑ 102° or 282° 77° or 257° ↑

↓ 13° or 193° Trigonometric and Involute Functions 166° or 346° ↓

Minutes	Sine	Cosine	Tangent	Cotangent	Secant	Cosecant	Involute 13°–14°	Read Up	Minutes
0	0.224951	0.974370	0.230868	4.331476	1.026304	4.445411	0.0039754	2.9875724	60
1	0.225234	0.974305	0.231175	4.325735	1.026373	4.439818	0.0039909	2.9821220	59
2	0.225518	0.974239	0.231481	4.320008	1.026442	4.434238	0.0040065	2.9766861	58
3	0.225801	0.974173	0.231788	4.314295	1.026511	4.428673	0.0040221	2.9712646	57
4	0.226085	0.974108	0.232094	4.308597	1.026581	4.423122	0.0040377	2.9658574	56
5	0.226368	0.974042	0.232401	4.302914	1.026650	4.417586	0.0040534	2.9604645	55
6	0.226651	0.973976	0.232707	4.297244	1.026719	4.412064	0.0040692	2.9550858	54
7	0.226935	0.973910	0.233014	4.291589	1.026789	4.406556	0.0040849	2.9497212	53
8	0.227218	0.973844	0.233321	4.285947	1.026859	4.401062	0.0041007	2.9443708	52
9	0.227501	0.973778	0.233627	4.280320	1.026928	4.395582	0.0041166	2.9390344	51
10	0.227784	0.973712	0.233934	4.274707	1.026998	4.390116	0.0041325	2.9337119	50
11	0.228068	0.973645	0.234241	4.269107	1.027068	4.384664	0.0041484	2.9284035	49
12	0.228351	0.973579	0.234548	4.263522	1.027138	4.379226	0.0041644	2.9231089	48
13	0.228634	0.973512	0.234855	4.257950	1.027208	4.373801	0.0041804	2.9178281	47
14	0.228917	0.973446	0.235162	4.252392	1.027278	4.368391	0.0041965	2.9125612	46
15	0.229200	0.973379	0.235469	4.246848	1.027349	4.362994	0.0042126	2.9073080	45
16	0.229484	0.973313	0.235776	4.241318	1.027419	4.357611	0.0042288	2.9020684	44
17	0.229767	0.973246	0.236083	4.235801	1.027490	4.352242	0.0042450	2.8968425	43
18	0.230050	0.973179	0.236390	4.230298	1.027560	4.346886	0.0042612	2.8916302	42
19	0.230333	0.973112	0.236697	4.224808	1.027631	4.341544	0.0042775	2.8864313	41
20	0.230616	0.973045	0.237004	4.219332	1.027702	4.336215	0.0042938	2.8812460	40
21	0.230899	0.972978	0.237312	4.213869	1.027773	4.330900	0.0043101	2.8760741	39
22	0.231182	0.972911	0.237619	4.208420	1.027844	4.325598	0.0043266	2.8709156	38
23	0.231465	0.972843	0.237926	4.202983	1.027915	4.320309	0.0043430	2.8657704	37
24	0.231748	0.972776	0.238234	4.197561	1.027986	4.315034	0.0043595	2.8606384	36
25	0.232031	0.972708	0.238541	4.192151	1.028057	4.309772	0.0043760	2.8555197	35
26	0.232314	0.972641	0.238848	4.186755	1.028129	4.304523	0.0043926	2.8504142	34
27	0.232597	0.972573	0.239156	4.181371	1.028200	4.299287	0.0044092	2.8453218	33
28	0.232880	0.972506	0.239464	4.176001	1.028272	4.294064	0.0044259	2.8402425	32
29	0.233163	0.972438	0.239771	4.170644	1.028343	4.288854	0.0044426	2.8351762	31
30	0.233445	0.972370	0.240079	4.165300	1.028415	4.283658	0.0044593	2.8301229	30
31	0.233728	0.972302	0.240386	4.159969	1.028487	4.278474	0.0044761	2.8250825	29
32	0.234011	0.972234	0.240694	4.154650	1.028559	4.273303	0.0044929	2.8200550	28
33	0.234294	0.972166	0.241002	4.149345	1.028631	4.268145	0.0045098	2.8150404	27
34	0.234577	0.972098	0.241310	4.144052	1.028703	4.263000	0.0045267	2.8100385	26
35	0.234859	0.972029	0.241618	4.138772	1.028776	4.257867	0.0045437	2.8050494	25
36	0.235142	0.971961	0.241925	4.133505	1.028848	4.252747	0.0045607	2.8000730	24
37	0.235425	0.971893	0.242233	4.128250	1.028920	4.247640	0.0045777	2.7951093	23
38	0.235708	0.971824	0.242541	4.123008	1.028993	4.242546	0.0045948	2.7901581	22
39	0.235990	0.971755	0.242849	4.117778	1.029066	4.237464	0.0046120	2.7852195	21
40	0.236273	0.971687	0.243157	4.112561	1.029138	4.232394	0.0046291	2.7802934	20
41	0.236556	0.971618	0.243466	4.107357	1.029211	4.227337	0.0046464	2.7753798	19
42	0.236838	0.971549	0.243774	4.102165	1.029284	4.222293	0.0046636	2.7704786	18
43	0.237121	0.971480	0.244082	4.096985	1.029357	4.217261	0.0046809	2.7655898	17
44	0.237403	0.971411	0.244390	4.091818	1.029430	4.212241	0.0046983	2.7607133	16
45	0.237686	0.971342	0.244698	4.086663	1.029503	4.207233	0.0047157	2.7558491	15
46	0.237968	0.971273	0.245007	4.081520	1.029577	4.202238	0.0047331	2.7509972	14
47	0.238251	0.971204	0.245315	4.076389	1.029650	4.197255	0.0047506	2.7461574	13
48	0.238533	0.971134	0.245624	4.071271	1.029724	4.192284	0.0047681	2.7413298	12
49	0.238816	0.971065	0.245932	4.066164	1.029797	4.187325	0.0047857	2.7365143	11
50	0.239098	0.970995	0.246241	4.061070	1.029871	4.182378	0.0048033	2.7317109	10
51	0.239381	0.970926	0.246549	4.055988	1.029945	4.177444	0.0048210	2.7269195	9
52	0.239663	0.970856	0.246858	4.050917	1.030019	4.172521	0.0048387	2.7221401	8
53	0.239946	0.970786	0.247166	4.045859	1.030093	4.167610	0.0048564	2.7173726	7
54	0.240228	0.970716	0.247475	4.040813	1.030167	4.162711	0.0048742	2.7126170	6
55	0.240510	0.970647	0.247784	4.035778	1.030241	4.157824	0.0048921	2.7078732	5
56	0.240793	0.970577	0.248092	4.030755	1.030315	4.152949	0.0049099	2.7031413	4
57	0.241075	0.970506	0.248401	4.025744	1.030390	4.148086	0.0049279	2.6984211	3
58	0.241357	0.970436	0.248710	4.020745	1.030464	4.143234	0.0049458	2.6937126	2
59	0.241640	0.970366	0.249019	4.015757	1.030539	4.138394	0.0049638	2.6890158	1
60	0.241922	0.970296	0.249328	4.010781	1.030614	4.133565	0.0049819	2.6843307	0
Minutes	Cosine	Sine	Cotangent	Tangent	Cosecant	Secant	Read Down	76°–77° Involute	Minutes

↑ 103° or 283° 76° or 256° ↑

↓ **14° or 194°** **Trigonometric and Involute Functions** **165° or 345°** ↓

Minutes	Sine	Cosine	Tangent	Cotangent	Secant	Cosecant	Involute 14°–15°	Read Up	Minutes
0	0.241922	0.970296	0.249328	4.010781	1.030614	4.133565	0.0049819	2.6843307	60
1	0.242204	0.970225	0.249637	4.005817	1.030688	4.128749	0.0050000	2.6796572	59
2	0.242486	0.970155	0.249946	4.000864	1.030763	4.123943	0.0050182	2.6749952	58
3	0.242769	0.970084	0.250255	3.995922	1.030838	4.119150	0.0050364	2.6703447	57
4	0.243051	0.970014	0.250564	3.990992	1.030913	4.114368	0.0050546	2.6657057	56
5	0.243333	0.969943	0.250873	3.986074	1.030989	4.109597	0.0050729	2.6610781	55
6	0.243615	0.969872	0.251183	3.981167	1.031064	4.104837	0.0050912	2.6564620	54
7	0.243897	0.969801	0.251492	3.976271	1.031139	4.100089	0.0051096	2.6518572	53
8	0.244179	0.969730	0.251801	3.971387	1.031215	4.095353	0.0051280	2.6472636	52
9	0.244461	0.969659	0.252111	3.966514	1.031290	4.090627	0.0051465	2.6426814	51
10	0.244743	0.969588	0.252420	3.961652	1.031366	4.085913	0.0051650	2.6381104	50
11	0.245025	0.969517	0.252729	3.956801	1.031442	4.081210	0.0051835	2.6335506	49
12	0.245307	0.969445	0.253039	3.951962	1.031518	4.076518	0.0052021	2.6290019	48
13	0.245589	0.969374	0.253348	3.947133	1.031594	4.071837	0.0052208	2.6244644	47
14	0.245871	0.969302	0.253658	3.942316	1.031670	4.067168	0.0052395	2.6199379	46
15	0.246153	0.969231	0.253968	3.937509	1.031746	4.062509	0.0052582	2.6154225	45
16	0.246435	0.969159	0.254277	3.932714	1.031822	4.057862	0.0052770	2.6109181	44
17	0.246717	0.969088	0.254587	3.927930	1.031899	4.053225	0.0052958	2.6064246	43
18	0.246999	0.969016	0.254897	3.923156	1.031975	4.048599	0.0053147	2.6019421	42
19	0.247281	0.968944	0.255207	3.918394	1.032052	4.043984	0.0053336	2.5974704	41
20	0.247563	0.968872	0.255516	3.913642	1.032128	4.039380	0.0053526	2.5930096	40
21	0.247845	0.968800	0.255826	3.908901	1.032205	4.034787	0.0053716	2.5885595	39
22	0.248126	0.968728	0.256136	3.904171	1.032282	4.030205	0.0053907	2.5841203	38
23	0.248408	0.968655	0.256446	3.899452	1.032359	4.025633	0.0054098	2.5796918	37
24	0.248690	0.968583	0.256756	3.894743	1.032436	4.021072	0.0054289	2.5752739	36
25	0.248972	0.968511	0.257066	3.890045	1.032513	4.016522	0.0054481	2.5708668	35
26	0.249253	0.968438	0.257377	3.885357	1.032590	4.011982	0.0054674	2.5664702	34
27	0.249535	0.968366	0.257687	3.880681	1.032668	4.007453	0.0054867	2.5620843	33
28	0.249817	0.968293	0.257997	3.876014	1.032745	4.002935	0.0055060	2.5577088	32
29	0.250098	0.968220	0.258307	3.871358	1.032823	3.998427	0.0055254	2.5533439	31
30	0.250380	0.968148	0.258618	3.866713	1.032900	3.993929	0.0055448	2.5489895	30
31	0.250662	0.968075	0.258928	3.862078	1.032978	3.989442	0.0055643	2.5446455	29
32	0.250943	0.968002	0.259238	3.857454	1.033056	3.984965	0.0055838	2.5403119	28
33	0.251225	0.967929	0.259549	3.852840	1.033134	3.980499	0.0056034	2.5359887	27
34	0.251506	0.967856	0.259859	3.848236	1.033212	3.976043	0.0056230	2.5316758	26
35	0.251788	0.967782	0.260170	3.843642	1.033290	3.971597	0.0056427	2.5273732	25
36	0.252069	0.967709	0.260480	3.839059	1.033368	3.967162	0.0056624	2.5230809	24
37	0.252351	0.967636	0.260791	3.834486	1.033447	3.962737	0.0056822	2.5187988	23
38	0.252632	0.967562	0.261102	3.829923	1.033525	3.958322	0.0057020	2.5145268	22
39	0.252914	0.967489	0.261413	3.825371	1.033604	3.953917	0.0057218	2.5102651	21
40	0.253195	0.967415	0.261723	3.820828	1.033682	3.949522	0.0057417	2.5060134	20
41	0.253477	0.967342	0.262034	3.816296	1.033761	3.945138	0.0057617	2.5017719	19
42	0.253758	0.967268	0.262345	3.811773	1.033840	3.940763	0.0057817	2.4975404	18
43	0.254039	0.967194	0.262656	3.807261	1.033919	3.936399	0.0058017	2.4933189	17
44	0.254321	0.967120	0.262967	3.802759	1.033998	3.932044	0.0058218	2.4891074	16
45	0.254602	0.967046	0.263278	3.798266	1.034077	3.927700	0.0058420	2.4849058	15
46	0.254883	0.966972	0.263589	3.793784	1.034156	3.923365	0.0058622	2.4807142	14
47	0.255165	0.966898	0.263900	3.789311	1.034236	3.919040	0.0058824	2.4765324	13
48	0.255446	0.966823	0.264211	3.784848	1.034315	3.914725	0.0059027	2.4723605	12
49	0.255727	0.966749	0.264523	3.780395	1.034395	3.910420	0.0059230	2.4681984	11
50	0.256008	0.966675	0.264834	3.775952	1.034474	3.906125	0.0059434	2.4640461	10
51	0.256289	0.966600	0.265145	3.771518	1.034554	3.901840	0.0059638	2.4599035	9
52	0.256571	0.966526	0.265457	3.767095	1.034634	3.897564	0.0059843	2.4557707	8
53	0.256852	0.966451	0.265768	3.762681	1.034714	3.893298	0.0060048	2.4516475	7
54	0.257133	0.966376	0.266079	3.758276	1.034794	3.889041	0.0060254	2.4475340	6
55	0.257414	0.966301	0.266391	3.753882	1.034874	3.884794	0.0060460	2.4434301	5
56	0.257695	0.966226	0.266702	3.749496	1.034954	3.880557	0.0060667	2.4393358	4
57	0.257976	0.966151	0.267014	3.745121	1.035035	3.876329	0.0060874	2.4352511	3
58	0.258257	0.966076	0.267326	3.740755	1.035115	3.872111	0.0061081	2.4311759	2
59	0.258538	0.966001	0.267637	3.736398	1.035196	3.867903	0.0061289	2.4271101	1
60	0.258819	0.965926	0.267949	3.732051	1.035276	3.863703	0.0061498	2.4230539	0
Minutes	Cosine	Sine	Cotangent	Tangent	Cosecant	Secant	Read Down	75°–76° Involute	Minutes

↑ **104° or 284°** **75° or 255°** ↑

↓ 15° or 195° Trigonometric and Involute Functions 164° or 344° ↓

Minutes	Sine	Cosine	Tangent	Cotangent	Secant	Cosecant	Involute 15°–16°	Read Up	Minutes
0	0.258819	0.965926	0.267949	3.732051	1.035276	3.863703	0.0061498	2.4230539	60
1	0.259100	0.965850	0.268261	3.727713	1.035357	3.859514	0.0061707	2.4190070	59
2	0.259381	0.965775	0.268573	3.723385	1.035438	3.855333	0.0061917	2.4149696	58
3	0.259662	0.965700	0.268885	3.719066	1.035519	3.851162	0.0062127	2.4109415	57
4	0.259943	0.965624	0.269197	3.714756	1.035600	3.847001	0.0062337	2.4069228	56
5	0.260224	0.965548	0.269509	3.710456	1.035681	3.842848	0.0062548	2.4029133	55
6	0.260505	0.965473	0.269821	3.706165	1.035762	3.838705	0.0062760	2.3989132	54
7	0.260785	0.965397	0.270133	3.701883	1.035843	3.834571	0.0062972	2.3949222	53
8	0.261066	0.965321	0.270445	3.697610	1.035925	3.830447	0.0063184	2.3909405	52
9	0.261347	0.965245	0.270757	3.693347	1.036006	3.826331	0.0063397	2.3869680	51
10	0.261628	0.965169	0.271069	3.689093	1.036088	3.822225	0.0063611	2.3830046	50
11	0.261908	0.965093	0.271382	3.684848	1.036170	3.818128	0.0063825	2.3790503	49
12	0.262189	0.965016	0.271694	3.680611	1.036252	3.814040	0.0064039	2.3751052	48
13	0.262470	0.964940	0.272006	3.676384	1.036334	3.809961	0.0064254	2.3711691	47
14	0.262751	0.964864	0.272319	3.672166	1.036416	3.805891	0.0064470	2.3672420	46
15	0.263031	0.964787	0.272631	3.667958	1.036498	3.801830	0.0064686	2.3633239	45
16	0.263312	0.964711	0.272944	3.663758	1.036580	3.797778	0.0064902	2.3594148	44
17	0.263592	0.964634	0.273256	3.659566	1.036662	3.793735	0.0065119	2.3555147	43
18	0.263873	0.964557	0.273569	3.655384	1.036745	3.789701	0.0065337	2.3516234	42
19	0.264154	0.964481	0.273882	3.651211	1.036827	3.785676	0.0065555	2.3477410	41
20	0.264434	0.964404	0.274194	3.647047	1.036910	3.781660	0.0065773	2.3438675	40
21	0.264715	0.964327	0.274507	3.642891	1.036993	3.777652	0.0065992	2.3400029	39
22	0.264995	0.964250	0.274820	3.638744	1.037076	3.773653	0.0066211	2.3361470	38
23	0.265276	0.964173	0.275133	3.634606	1.037159	3.769664	0.0066431	2.3322999	37
24	0.265556	0.964095	0.275446	3.630477	1.037242	3.765682	0.0066652	2.3284615	36
25	0.265837	0.964018	0.275759	3.626357	1.037325	3.761710	0.0066873	2.3246318	35
26	0.266117	0.963941	0.276072	3.622245	1.037408	3.757746	0.0067094	2.3208108	34
27	0.266397	0.963863	0.276385	3.618141	1.037492	3.753791	0.0067316	2.3169985	33
28	0.266678	0.963786	0.276698	3.614047	1.037575	3.749845	0.0067539	2.3131948	32
29	0.266958	0.963708	0.277011	3.609961	1.037659	3.745907	0.0067762	2.3093997	31
30	0.267238	0.963630	0.277325	3.605884	1.037742	3.741978	0.0067985	2.3056132	30
31	0.267519	0.963553	0.277638	3.601815	1.037826	3.738057	0.0068209	2.3018352	29
32	0.267799	0.963475	0.277951	3.597754	1.037910	3.734145	0.0068434	2.2980658	28
33	0.268079	0.963397	0.278265	3.593702	1.037994	3.730241	0.0068659	2.2943048	27
34	0.268359	0.963319	0.278578	3.589659	1.038078	3.726346	0.0068884	2.2905523	26
35	0.268640	0.963241	0.278891	3.585624	1.038162	3.722459	0.0069110	2.2868082	25
36	0.268920	0.963163	0.279205	3.581598	1.038246	3.718580	0.0069337	2.2830726	24
37	0.269200	0.963084	0.279519	3.577579	1.038331	3.714711	0.0069564	2.2793453	23
38	0.269480	0.963006	0.279832	3.573570	1.038415	3.710849	0.0069791	2.2756264	22
39	0.269760	0.962928	0.280146	3.569568	1.038500	3.706996	0.0070019	2.2719158	21
40	0.270040	0.962849	0.280460	3.565575	1.038584	3.703151	0.0070248	2.2682135	20
41	0.270320	0.962770	0.280773	3.561590	1.038669	3.699314	0.0070477	2.2645194	19
42	0.270600	0.962692	0.281087	3.557613	1.038754	3.695485	0.0070706	2.2608337	18
43	0.270880	0.962613	0.281401	3.553645	1.038839	3.691665	0.0070936	2.2571561	17
44	0.271160	0.962534	0.281715	3.549685	1.038924	3.687853	0.0071167	2.2534868	16
45	0.271440	0.962455	0.282029	3.545733	1.039009	3.684049	0.0071398	2.2498256	15
46	0.271720	0.962376	0.282343	3.541789	1.039095	3.680254	0.0071630	2.2461725	14
47	0.272000	0.962297	0.282657	3.537853	1.039180	3.676466	0.0071862	2.2425276	13
48	0.272280	0.962218	0.282971	3.533925	1.039266	3.672687	0.0072095	2.2388908	12
49	0.272560	0.962139	0.283286	3.530005	1.039351	3.668915	0.0072328	2.2352620	11
50	0.272840	0.962059	0.283600	3.526094	1.039437	3.665152	0.0072561	2.2316413	10
51	0.273120	0.961980	0.283914	3.522190	1.039523	3.661396	0.0072796	2.2280286	9
52	0.273400	0.961901	0.284229	3.518295	1.039609	3.657649	0.0073030	2.2244239	8
53	0.273679	0.961821	0.284543	3.514407	1.039695	3.653910	0.0073266	2.2208271	7
54	0.273959	0.961741	0.284857	3.510527	1.039781	3.650178	0.0073501	2.2172383	6
55	0.274239	0.961662	0.285172	3.506655	1.039867	3.646455	0.0073738	2.2136574	5
56	0.274519	0.961582	0.285487	3.502792	1.039953	3.642739	0.0073975	2.2100844	4
57	0.274798	0.961502	0.285801	3.498936	1.040040	3.639031	0.0074212	2.2065193	3
58	0.275078	0.961422	0.286116	3.495087	1.040126	3.635332	0.0074450	2.2029620	2
59	0.275358	0.961342	0.286431	3.491247	1.040213	3.631640	0.0074688	2.1994125	1
60	0.275637	0.961262	0.286745	3.487414	1.040299	3.627955	0.0074927	2.1958708	0
Minutes	Cosine	Sine	Cotangent	Tangent	Cosecant	Secant	Read Down	74°–75° Involute	Minutes

↑ 105° or 285° 74° or 254° ↑

↓ 16° or 196° Trigonometric and Involute Functions 163° or 343° ↓

Minutes	Sine	Cosine	Tangent	Cotangent	Secant	Cosecant	Involute 16°–17°	Read Up	Minutes
0	0.275637	0.961262	0.286745	3.487414	1.040299	3.627955	0.0074927	2.1958708	60
1	0.275917	0.961181	0.287060	3.483590	1.040386	3.624279	0.0075166	2.1923369	59
2	0.276197	0.961101	0.287375	3.479773	1.040473	3.620610	0.0075406	2.1888107	58
3	0.276476	0.961021	0.287690	3.475963	1.040560	3.616949	0.0075647	2.1852922	57
4	0.276756	0.960940	0.288005	3.472162	1.040647	3.613296	0.0075888	2.1817815	56
5	0.277035	0.960860	0.288320	3.468368	1.040735	3.609650	0.0076130	2.1782784	55
6	0.277315	0.960779	0.288635	3.464581	1.040822	3.606012	0.0076372	2.1747830	54
7	0.277594	0.960698	0.288950	3.460803	1.040909	3.602382	0.0076614	2.1712951	53
8	0.277874	0.960618	0.289266	3.457031	1.040997	3.598759	0.0076857	2.1678149	52
9	0.278153	0.960537	0.289581	3.453268	1.041085	3.595144	0.0077101	2.1643423	51
10	0.278432	0.960456	0.289896	3.449512	1.041172	3.591536	0.0077345	2.1608772	50
11	0.278712	0.960375	0.290211	3.445764	1.041260	3.587936	0.0077590	2.1574196	49
12	0.278991	0.960294	0.290527	3.442023	1.041348	3.584344	0.0077835	2.1539696	48
13	0.279270	0.960212	0.290842	3.438289	1.041436	3.580759	0.0078081	2.1505270	47
14	0.279550	0.960131	0.291158	3.434563	1.041524	3.577181	0.0078327	2.1470919	46
15	0.279829	0.960050	0.291473	3.430845	1.041613	3.573611	0.0078574	2.1436643	45
16	0.280108	0.959968	0.291789	3.427133	1.041701	3.570048	0.0078822	2.1402440	44
17	0.280388	0.959887	0.292105	3.423430	1.041789	3.566493	0.0079069	2.1368311	43
18	0.280667	0.959805	0.292420	3.419733	1.041878	3.562945	0.0079318	2.1334256	42
19	0.280946	0.959724	0.292736	3.416044	1.041967	3.559404	0.0079567	2.1300275	41
20	0.281225	0.959642	0.293052	3.412363	1.042055	3.555871	0.0079817	2.1266367	40
21	0.281504	0.959560	0.293368	3.408688	1.042144	3.552345	0.0080067	2.1232532	39
22	0.281783	0.959478	0.293684	3.405021	1.042233	3.548826	0.0080317	2.1198769	38
23	0.282062	0.959396	0.294000	3.401361	1.042322	3.545315	0.0080568	2.1165079	37
24	0.282341	0.959314	0.294316	3.397709	1.042412	3.541811	0.0080820	2.1131462	36
25	0.282620	0.959232	0.294632	3.394063	1.042501	3.538314	0.0081072	2.1097917	35
26	0.282900	0.959150	0.294948	3.390425	1.042590	3.534824	0.0081325	2.1064443	34
27	0.283179	0.959067	0.295265	3.386794	1.042680	3.531341	0.0081578	2.1031041	33
28	0.283457	0.958985	0.295581	3.383170	1.042769	3.527866	0.0081832	2.0997711	32
29	0.283736	0.958902	0.295897	3.379553	1.042859	3.524398	0.0082087	2.0964452	31
30	0.284015	0.958820	0.296213	3.375943	1.042949	3.520937	0.0082342	2.0931264	30
31	0.284294	0.958737	0.296530	3.372341	1.043039	3.517482	0.0082597	2.0898147	29
32	0.284573	0.958654	0.296846	3.368745	1.043129	3.514035	0.0082853	2.0865101	28
33	0.284852	0.958572	0.297163	3.365157	1.043219	3.510595	0.0083110	2.0832124	27
34	0.285131	0.958489	0.297480	3.361575	1.043309	3.507162	0.0083367	2.0799219	26
35	0.285410	0.958406	0.297796	3.358001	1.043400	3.503737	0.0083625	2.0766383	25
36	0.285688	0.958323	0.298113	3.354433	1.043490	3.500318	0.0083883	2.0733616	24
37	0.285967	0.958239	0.298430	3.350873	1.043581	3.496906	0.0084142	2.0700920	23
38	0.286246	0.958156	0.298747	3.347319	1.043671	3.493500	0.0084401	2.0668292	22
39	0.286525	0.958073	0.299063	3.343772	1.043762	3.490102	0.0084661	2.0635734	21
40	0.286803	0.957990	0.299380	3.340233	1.043853	3.486711	0.0084921	2.0603245	20
41	0.287082	0.957906	0.299697	3.336700	1.043944	3.483327	0.0085182	2.0570824	19
42	0.287361	0.957822	0.300014	3.333174	1.044035	3.479949	0.0085444	2.0538472	18
43	0.287639	0.957739	0.300331	3.329654	1.044126	3.476578	0.0085706	2.0506189	17
44	0.287918	0.957655	0.300649	3.326142	1.044217	3.473215	0.0085969	2.0473973	16
45	0.288196	0.957571	0.300966	3.322636	1.044309	3.469858	0.0086232	2.0441825	15
46	0.288475	0.957487	0.301283	3.319137	1.044400	3.466507	0.0086496	2.0409746	14
47	0.288753	0.957404	0.301600	3.315645	1.044492	3.463164	0.0086760	2.0377733	13
48	0.289032	0.957319	0.301918	3.312160	1.044583	3.459827	0.0087025	2.0345788	12
49	0.289310	0.957235	0.302235	3.308681	1.044675	3.456497	0.0087290	2.0313910	11
50	0.289589	0.957151	0.302553	3.305209	1.044767	3.453173	0.0087556	2.0282099	10
51	0.289867	0.957067	0.302870	3.301744	1.044859	3.449857	0.0087823	2.0250354	9
52	0.290145	0.956983	0.303188	3.298285	1.044951	3.446547	0.0088090	2.0218676	8
53	0.290424	0.956898	0.303506	3.294833	1.045043	3.443243	0.0088358	2.0187064	7
54	0.290702	0.956814	0.303823	3.291388	1.045136	3.439947	0.0088626	2.0155519	6
55	0.290981	0.956729	0.304141	3.287949	1.045228	3.436656	0.0088895	2.0124039	5
56	0.291259	0.956644	0.304459	3.284516	1.045321	3.433373	0.0089164	2.0092625	4
57	0.291537	0.956560	0.304777	3.281091	1.045413	3.430096	0.0089434	2.0061277	3
58	0.291815	0.956475	0.305095	3.277671	1.045506	3.426825	0.0089704	2.0029994	2
59	0.292094	0.956390	0.305413	3.274259	1.045599	3.423561	0.0089975	1.9998776	1
60	0.292372	0.956305	0.305731	3.270853	1.045692	3.420304	0.0090247	1.9967623	0
Minutes	Cosine	Sine	Cotangent	Tangent	Cosecant	Secant	Read Down	73°–74° Involute	Minutes

↑ 106° or 286° 73° or 253° ↑

↓ 17° or 197° Trigonometric and Involute Functions 162° or 342° ↓

Minutes	Sine	Cosine	Tangent	Cotangent	Secant	Cosecant	Involute 17°–18°	Read Up	Minutes
0	0.292372	0.956305	0.305731	3.270853	1.045692	3.420304	0.0090247	1.9967623	60
1	0.292650	0.956220	0.306049	3.267453	1.045785	3.417053	0.0090519	1.9936534	59
2	0.292928	0.956134	0.306367	3.264060	1.045878	3.413808	0.0090792	1.9905511	58
3	0.293206	0.956049	0.306685	3.260673	1.045971	3.410570	0.0091065	1.9874551	57
4	0.293484	0.955964	0.307003	3.257292	1.046065	3.407338	0.0091339	1.9843656	56
5	0.293762	0.955879	0.307322	3.253918	1.046158	3.404113	0.0091614	1.9812825	55
6	0.294040	0.955793	0.307640	3.250551	1.046252	3.400894	0.0091889	1.9782058	54
7	0.294318	0.955707	0.307959	3.247190	1.046345	3.397682	0.0092164	1.9751354	53
8	0.294596	0.955622	0.308277	3.243835	1.046439	3.394475	0.0092440	1.9720714	52
9	0.294874	0.955536	0.308596	3.240486	1.046533	3.391276	0.0092717	1.9690137	51
10	0.295152	0.955450	0.308914	3.237144	1.046627	3.388082	0.0092994	1.9659623	50
11	0.295430	0.955364	0.309233	3.233808	1.046721	3.384895	0.0093272	1.9629172	49
12	0.295708	0.955278	0.309552	3.230478	1.046815	3.381714	0.0093551	1.9598783	48
13	0.295986	0.955192	0.309870	3.227155	1.046910	3.378539	0.0093830	1.9568458	47
14	0.296264	0.955106	0.310189	3.223837	1.047004	3.375371	0.0094109	1.9538194	46
15	0.296542	0.955020	0.310508	3.220526	1.047099	3.372208	0.0094390	1.9507993	45
16	0.296819	0.954934	0.310827	3.217221	1.047193	3.369052	0.0094670	1.9477853	44
17	0.297097	0.954847	0.311146	3.213923	1.047288	3.365903	0.0094952	1.9447776	43
18	0.297375	0.954761	0.311465	3.210630	1.047383	3.362759	0.0095234	1.9417760	42
19	0.297653	0.954674	0.311784	3.207344	1.047478	3.359621	0.0095516	1.9387805	41
20	0.297930	0.954588	0.312104	3.204064	1.047573	3.356490	0.0095799	1.9357912	40
21	0.298208	0.954501	0.312423	3.200790	1.047668	3.353365	0.0096083	1.9328080	39
22	0.298486	0.954414	0.312742	3.197522	1.047763	3.350246	0.0096367	1.9298309	38
23	0.298763	0.954327	0.313062	3.194260	1.047859	3.347132	0.0096652	1.9268598	37
24	0.299041	0.954240	0.313381	3.191004	1.047954	3.344025	0.0096937	1.9238948	36
25	0.299318	0.954153	0.313700	3.187754	1.048050	3.340924	0.0097223	1.9209359	35
26	0.299596	0.954066	0.314020	3.184510	1.048145	3.337829	0.0097510	1.9179830	34
27	0.299873	0.953979	0.314340	3.181272	1.048241	3.334740	0.0097797	1.9150360	33
28	0.300151	0.953892	0.314659	3.178041	1.048337	3.331658	0.0098085	1.9120951	32
29	0.300428	0.953804	0.314979	3.174815	1.048433	3.328581	0.0098373	1.9091601	31
30	0.300706	0.953717	0.315299	3.171595	1.048529	3.325510	0.0098662	1.9062311	30
31	0.300983	0.953629	0.315619	3.168381	1.048625	3.322444	0.0098951	1.9033080	29
32	0.301261	0.953542	0.315939	3.165173	1.048722	3.319385	0.0099241	1.9003908	28
33	0.301538	0.953454	0.316258	3.161971	1.048818	3.316332	0.0099532	1.8974796	27
34	0.301815	0.953366	0.316578	3.158774	1.048915	3.313285	0.0099823	1.8945742	26
35	0.302093	0.953279	0.316899	3.155584	1.049011	3.310243	0.0100115	1.8916747	25
36	0.302370	0.953191	0.317219	3.152399	1.049108	3.307208	0.0100407	1.8887810	24
37	0.302647	0.953103	0.317539	3.149221	1.049205	3.304178	0.0100700	1.8858932	23
38	0.302924	0.953015	0.317859	3.146048	1.049302	3.301154	0.0100994	1.8830112	22
39	0.303202	0.952926	0.318179	3.142881	1.049399	3.298136	0.0101288	1.8801350	21
40	0.303479	0.952838	0.318500	3.139719	1.049496	3.295123	0.0101583	1.8772646	20
41	0.303756	0.952750	0.318820	3.136564	1.049593	3.292117	0.0101878	1.8743999	19
42	0.304033	0.952661	0.319141	3.133414	1.049691	3.289116	0.0102174	1.8715411	18
43	0.304310	0.952573	0.319461	3.130270	1.049788	3.286121	0.0102471	1.8686879	17
44	0.304587	0.952484	0.319782	3.127132	1.049886	3.283132	0.0102768	1.8658405	16
45	0.304864	0.952396	0.320103	3.123999	1.049984	3.280148	0.0103066	1.8629987	15
46	0.305141	0.952307	0.320423	3.120872	1.050081	3.277170	0.0103364	1.8601627	14
47	0.305418	0.952218	0.320744	3.117751	1.050179	3.274198	0.0103663	1.8573323	13
48	0.305695	0.952129	0.321065	3.114635	1.050277	3.271231	0.0103963	1.8545076	12
49	0.305972	0.952040	0.321386	3.111525	1.050376	3.268270	0.0104263	1.8516885	11
50	0.306249	0.951951	0.321707	3.108421	1.050474	3.265315	0.0104564	1.8488751	10
51	0.306526	0.951862	0.322028	3.105322	1.050572	3.262365	0.0104865	1.8460672	9
52	0.306803	0.951773	0.322349	3.102229	1.050671	3.259421	0.0105167	1.8432650	8
53	0.307080	0.951684	0.322670	3.099142	1.050769	3.256483	0.0105469	1.8404683	7
54	0.307357	0.951594	0.322991	3.096060	1.050868	3.253550	0.0105773	1.8376772	6
55	0.307633	0.951505	0.323312	3.092983	1.050967	3.250622	0.0106076	1.8348916	5
56	0.307910	0.951415	0.323634	3.089912	1.051066	3.247700	0.0106381	1.8321116	4
57	0.308187	0.951326	0.323955	3.086847	1.051165	3.244784	0.0106686	1.8293371	3
58	0.308464	0.951236	0.324277	3.083787	1.051264	3.241873	0.0106991	1.8265681	2
59	0.308740	0.951146	0.324598	3.080732	1.051363	3.238968	0.0107298	1.8238045	1
60	0.309017	0.951057	0.324920	3.077684	1.051462	3.236068	0.0107604	1.8210465	0
Minutes	Cosine	Sine	Cotangent	Tangent	Cosecant	Secant	Read Down	72°–73° Involute	Minutes

↑ 107° or 287° 72° or 252° ↑

↓ 18° or 198° Trigonometric and Involute Functions 161° or 341° ↓

Minutes	Sine	Cosine	Tangent	Cotangent	Secant	Cosecant	Involute 18°–19°	Read Up	Minutes
0	0.309017	0.951057	0.324920	3.077684	1.051462	3.236068	0.0107604	1.8210465	60
1	0.309294	0.950967	0.325241	3.074640	1.051562	3.233174	0.0107912	1.8182939	59
2	0.309570	0.950877	0.325563	3.071602	1.051661	3.230285	0.0108220	1.8155467	58
3	0.309847	0.950786	0.325885	3.068569	1.051761	3.227401	0.0108528	1.8128050	57
4	0.310123	0.950696	0.326207	3.065542	1.051861	3.224523	0.0108838	1.8100686	56
5	0.310400	0.950606	0.326528	3.062520	1.051960	3.221650	0.0109147	1.8073377	55
6	0.310676	0.950516	0.326850	3.059504	1.052060	3.218783	0.0109458	1.8046121	54
7	0.310953	0.950425	0.327172	3.056493	1.052161	3.215921	0.0109769	1.8018919	53
8	0.311229	0.950335	0.327494	3.053487	1.052261	3.213064	0.0110081	1.7991771	52
9	0.311506	0.950244	0.327817	3.050487	1.052361	3.210213	0.0110393	1.7964676	51
10	0.311782	0.950154	0.328139	3.047492	1.052461	3.207367	0.0110706	1.7937634	50
11	0.312059	0.950063	0.328461	3.044502	1.052562	3.204527	0.0111019	1.7910645	49
12	0.312335	0.949972	0.328783	3.041517	1.052663	3.201691	0.0111333	1.7883709	48
13	0.312611	0.949881	0.329106	3.038538	1.052763	3.198861	0.0111648	1.7856826	47
14	0.312888	0.949790	0.329428	3.035564	1.052864	3.196037	0.0111964	1.7829995	46
15	0.313164	0.949699	0.329751	3.032595	1.052965	3.193217	0.0112280	1.7803217	45
16	0.313440	0.949608	0.330073	3.029632	1.053066	3.190403	0.0112596	1.7776491	44
17	0.313716	0.949517	0.330396	3.026674	1.053167	3.187594	0.0112913	1.7749817	43
18	0.313992	0.949425	0.330718	3.023721	1.053269	3.184790	0.0113231	1.7723196	42
19	0.314269	0.949334	0.331041	3.020773	1.053370	3.181991	0.0113550	1.7696626	41
20	0.314545	0.949243	0.331364	3.017830	1.053471	3.179198	0.0113869	1.7670108	40
21	0.314821	0.949151	0.331687	3.014893	1.053573	3.176410	0.0114189	1.7643642	39
22	0.315097	0.949059	0.332010	3.011960	1.053675	3.173626	0.0114509	1.7617227	38
23	0.315373	0.948968	0.332333	3.009033	1.053777	3.170848	0.0114830	1.7590864	37
24	0.315649	0.948876	0.332656	3.006111	1.053878	3.168076	0.0115151	1.7564552	36
25	0.315925	0.948784	0.332979	3.003194	1.053981	3.165308	0.0115474	1.7538290	35
26	0.316201	0.948692	0.333302	3.000282	1.054083	3.162545	0.0115796	1.7512080	34
27	0.316477	0.948600	0.333625	2.997375	1.054185	3.159788	0.0116120	1.7485921	33
28	0.316753	0.948508	0.333949	2.994473	1.054287	3.157035	0.0116444	1.7459812	32
29	0.317029	0.948416	0.334272	2.991577	1.054390	3.154288	0.0116769	1.7433753	31
30	0.317305	0.948324	0.334595	2.988685	1.054492	3.151545	0.0117094	1.7407745	30
31	0.317580	0.948231	0.334919	2.985798	1.054595	3.148808	0.0117420	1.7381788	29
32	0.317856	0.948139	0.335242	2.982917	1.054698	3.146076	0.0117747	1.7355880	28
33	0.318132	0.948046	0.335566	2.980040	1.054801	3.143348	0.0118074	1.7330022	27
34	0.318408	0.947954	0.335890	2.977168	1.054904	3.140626	0.0118402	1.7304215	26
35	0.318684	0.947861	0.336213	2.974302	1.055007	3.137909	0.0118730	1.7278456	25
36	0.318959	0.947768	0.336537	2.971440	1.055110	3.135196	0.0119059	1.7252748	24
37	0.319235	0.947676	0.336861	2.968583	1.055213	3.132489	0.0119389	1.7227089	23
38	0.319511	0.947583	0.337185	2.965731	1.055317	3.129786	0.0119720	1.7201479	22
39	0.319786	0.947490	0.337509	2.962884	1.055420	3.127089	0.0120051	1.7175918	21
40	0.320062	0.947397	0.337833	2.960042	1.055524	3.124396	0.0120382	1.7150407	20
41	0.320337	0.947304	0.338157	2.957205	1.055628	3.121708	0.0120715	1.7124944	19
42	0.320613	0.947210	0.338481	2.954373	1.055732	3.119025	0.0121048	1.7099530	18
43	0.320889	0.947117	0.338806	2.951545	1.055836	3.116347	0.0121381	1.7074164	17
44	0.321164	0.947024	0.339130	2.948723	1.055940	3.113674	0.0121715	1.7048848	16
45	0.321439	0.946930	0.339454	2.945905	1.056044	3.111006	0.0122050	1.7023579	15
46	0.321715	0.946837	0.339779	2.943092	1.056148	3.108342	0.0122386	1.6998359	14
47	0.321990	0.946743	0.340103	2.940284	1.056253	3.105683	0.0122722	1.6973187	13
48	0.322266	0.946649	0.340428	2.937481	1.056357	3.103030	0.0123059	1.6948063	12
49	0.322541	0.946555	0.340752	2.934682	1.056462	3.100381	0.0123396	1.6922986	11
50	0.322816	0.946462	0.341077	2.931888	1.056567	3.097736	0.0123734	1.6897958	10
51	0.323092	0.946368	0.341402	2.929099	1.056672	3.095097	0.0124073	1.6872977	9
52	0.323367	0.946274	0.341727	2.926315	1.056777	3.092462	0.0124412	1.6848044	8
53	0.323642	0.946180	0.342052	2.923536	1.056882	3.089832	0.0124752	1.6823158	7
54	0.323917	0.946085	0.342377	2.920761	1.056987	3.087207	0.0125093	1.6798319	6
55	0.324193	0.945991	0.342702	2.917991	1.057092	3.084586	0.0125434	1.6773527	5
56	0.324468	0.945897	0.343027	2.915226	1.057198	3.081970	0.0125776	1.6748783	4
57	0.324743	0.945802	0.343352	2.912465	1.057303	3.079359	0.0126119	1.6724085	3
58	0.325018	0.945708	0.343677	2.909709	1.057409	3.076752	0.0126462	1.6699434	2
59	0.325293	0.945613	0.344002	2.906958	1.057515	3.074151	0.0126806	1.6674829	1
60	0.325568	0.945519	0.344328	2.904211	1.057621	3.071553	0.0127151	1.6650271	0
Minutes	Cosine	Sine	Cotangent	Tangent	Cosecant	Secant	Read Down	71°–72° Involute	Minutes

↑ 108° or 288° 71° or 251° ↑

↓ **19° or 199°** **Trigonometric and Involute Functions** **160° or 340°** ↓

Minutes	Sine	Cosine	Tangent	Cotangent	Secant	Cosecant	Involute 19°–20°	Read Up	Minutes
0	0.325568	0.945519	0.344328	2.904211	1.057621	3.071553	0.0127151	1.6650271	60
1	0.325843	0.945424	0.344653	2.901469	1.057727	3.068961	0.0127496	1.6625759	59
2	0.326118	0.945329	0.344978	2.898731	1.057833	3.066373	0.0127842	1.6601294	58
3	0.326393	0.945234	0.345304	2.895999	1.057939	3.063790	0.0128188	1.6576875	57
4	0.326668	0.945139	0.345630	2.893270	1.058045	3.061211	0.0128535	1.6552502	56
5	0.326943	0.945044	0.345955	2.890547	1.058152	3.058637	0.0128883	1.6528174	55
6	0.327218	0.944949	0.346281	2.887828	1.058258	3.056068	0.0129232	1.6503893	54
7	0.327493	0.944854	0.346607	2.885113	1.058365	3.053503	0.0129581	1.6479657	53
8	0.327768	0.944758	0.346933	2.882403	1.058472	3.050942	0.0129931	1.6455466	52
9	0.328042	0.944663	0.347259	2.879698	1.058579	3.048386	0.0130281	1.6431321	51
10	0.328317	0.944568	0.347585	2.876997	1.058686	3.045835	0.0130632	1.6407221	50
11	0.328592	0.944472	0.347911	2.874301	1.058793	3.043288	0.0130984	1.6383167	49
12	0.328867	0.944376	0.348237	2.871609	1.058900	3.040746	0.0131336	1.6359157	48
13	0.329141	0.944281	0.348563	2.868921	1.059007	3.038208	0.0131689	1.6335193	47
14	0.329416	0.944185	0.348889	2.866239	1.059115	3.035675	0.0132043	1.6311273	46
15	0.329691	0.944089	0.349216	2.863560	1.059222	3.033146	0.0132398	1.6287398	45
16	0.329965	0.943993	0.349542	2.860886	1.059330	3.030622	0.0132753	1.6263567	44
17	0.330240	0.943897	0.349868	2.858217	1.059438	3.028102	0.0133108	1.6239781	43
18	0.330514	0.943801	0.350195	2.855552	1.059545	3.025587	0.0133465	1.6216040	42
19	0.330789	0.943705	0.350522	2.852891	1.059653	3.023076	0.0133822	1.6192342	41
20	0.331063	0.943609	0.350848	2.850235	1.059762	3.020569	0.0134180	1.6168689	40
21	0.331338	0.943512	0.351175	2.847583	1.059870	3.018067	0.0134538	1.6145080	39
22	0.331612	0.943416	0.351502	2.844936	1.059978	3.015569	0.0134897	1.6121514	38
23	0.331887	0.943319	0.351829	2.842293	1.060087	3.013076	0.0135257	1.6097993	37
24	0.332161	0.943223	0.352156	2.839654	1.060195	3.010587	0.0135617	1.6074515	36
25	0.332435	0.943126	0.352483	2.837020	1.060304	3.008102	0.0135978	1.6051080	35
26	0.332710	0.943029	0.352810	2.834390	1.060412	3.005622	0.0136340	1.6027689	34
27	0.332984	0.942932	0.353137	2.831764	1.060521	3.003146	0.0136702	1.6004342	33
28	0.333258	0.942836	0.353464	2.829143	1.060630	3.000675	0.0137065	1.5981037	32
29	0.333533	0.942739	0.353791	2.826526	1.060739	2.998207	0.0137429	1.5957776	31
30	0.333807	0.942641	0.354119	2.823913	1.060849	2.995744	0.0137794	1.5934558	30
31	0.334081	0.942544	0.354446	2.821304	1.060958	2.993286	0.0138159	1.5911382	29
32	0.334355	0.942447	0.354773	2.818700	1.061067	2.990831	0.0138525	1.5888250	28
33	0.334629	0.942350	0.355101	2.816100	1.061177	2.988381	0.0138891	1.5865160	27
34	0.334903	0.942252	0.355429	2.813505	1.061287	2.985935	0.0139258	1.5842112	26
35	0.335178	0.942155	0.355756	2.810913	1.061396	2.983494	0.0139626	1.5819107	25
36	0.335452	0.942057	0.356084	2.808326	1.061506	2.981056	0.0139994	1.5796145	24
37	0.335726	0.941960	0.356412	2.805743	1.061616	2.978623	0.0140364	1.5773224	23
38	0.336000	0.941862	0.356740	2.803165	1.061727	2.976194	0.0140734	1.5750346	22
39	0.336274	0.941764	0.357068	2.800590	1.061837	2.973769	0.0141104	1.5727510	21
40	0.336547	0.941666	0.357396	2.798020	1.061947	2.971349	0.0141475	1.5704716	20
41	0.336821	0.941569	0.357724	2.795454	1.062058	2.968933	0.0141847	1.5681963	19
42	0.337095	0.941471	0.358052	2.792892	1.062168	2.966521	0.0142220	1.5659252	18
43	0.337369	0.941372	0.358380	2.790334	1.062279	2.964113	0.0142593	1.5636583	17
44	0.337643	0.941274	0.358708	2.787780	1.062390	2.961709	0.0142967	1.5613955	16
45	0.337917	0.941176	0.359037	2.785231	1.062501	2.959309	0.0143342	1.5591369	15
46	0.338190	0.941078	0.359365	2.782685	1.062612	2.956914	0.0143717	1.5568824	14
47	0.338464	0.940979	0.359694	2.780144	1.062723	2.954522	0.0144093	1.5546320	13
48	0.338738	0.940881	0.360022	2.777607	1.062834	2.952135	0.0144470	1.5523857	12
49	0.339012	0.940782	0.360351	2.775074	1.062945	2.949752	0.0144847	1.5501435	11
50	0.339285	0.940684	0.360679	2.772545	1.063057	2.947372	0.0145225	1.5479054	10
51	0.339559	0.940585	0.361008	2.770020	1.063168	2.944997	0.0145604	1.5456714	9
52	0.339832	0.940486	0.361337	2.767499	1.063280	2.942627	0.0145983	1.5434415	8
53	0.340106	0.940387	0.361666	2.764982	1.063392	2.940260	0.0146363	1.5412156	7
54	0.340380	0.940288	0.361995	2.762470	1.063504	2.937897	0.0146744	1.5389937	6
55	0.340653	0.940189	0.362324	2.759961	1.063616	2.935538	0.0147126	1.5367759	5
56	0.340927	0.940090	0.362653	2.757456	1.063728	2.933183	0.0147508	1.5345621	4
57	0.341200	0.939991	0.362982	2.754955	1.063840	2.930833	0.0147891	1.5323523	3
58	0.341473	0.939891	0.363312	2.752459	1.063953	2.928486	0.0148275	1.5301465	2
59	0.341747	0.939792	0.363641	2.749966	1.064065	2.926143	0.0148659	1.5279447	1
60	0.342020	0.939693	0.363970	2.747477	1.064178	2.923804	0.0149044	1.5257469	0
Minutes	Cosine	Sine	Cotangent	Tangent	Cosecant	Secant	Read Down	70°–71° Involute	Minutes

↑ **109° or 289°** **70° or 250°** ↑

↓ **20° or 200°** **Trigonometric and Involute Functions** **159° or 339°** ↓

Minutes	Sine	Cosine	Tangent	Cotangent	Secant	Cosecant	Involute 20°–21°	Read Up	Minutes
0	0.342020	0.939693	0.363970	2.747477	1.064178	2.923804	0.0149044	1.5257469	60
1	0.342293	0.939593	0.364300	2.744993	1.064290	2.921470	0.0149430	1.5235531	59
2	0.342567	0.939493	0.364629	2.742512	1.064403	2.919139	0.0149816	1.5213633	58
3	0.342840	0.939394	0.364959	2.740035	1.064516	2.916812	0.0150203	1.5191774	57
4	0.343113	0.939294	0.365288	2.737562	1.064629	2.914489	0.0150591	1.5169954	56
5	0.343387	0.939194	0.365618	2.735093	1.064743	2.912170	0.0150979	1.5148174	55
6	0.343660	0.939094	0.365948	2.732628	1.064856	2.909855	0.0151369	1.5126433	54
7	0.343933	0.938994	0.366278	2.730167	1.064969	2.907544	0.0151758	1.5104731	53
8	0.344206	0.938894	0.366608	2.727710	1.065083	2.905237	0.0152149	1.5083068	52
9	0.344479	0.938794	0.366938	2.725257	1.065196	2.902934	0.0152540	1.5061444	51
10	0.344752	0.938694	0.367268	2.722808	1.065310	2.900635	0.0152932	1.5039860	50
11	0.345025	0.938593	0.367598	2.720362	1.065424	2.898339	0.0153325	1.5018313	49
12	0.345298	0.938493	0.367928	2.717920	1.065538	2.896048	0.0153719	1.4996806	48
13	0.345571	0.938393	0.368259	2.715483	1.065652	2.893760	0.0154113	1.4975337	47
14	0.345844	0.938292	0.368589	2.713049	1.065766	2.891476	0.0154507	1.4953907	46
15	0.346117	0.938191	0.368919	2.710619	1.065881	2.889196	0.0154903	1.4932515	45
16	0.346390	0.938091	0.369250	2.708192	1.065995	2.886920	0.0155299	1.4911161	44
17	0.346663	0.937990	0.369581	2.705770	1.066110	2.884647	0.0155696	1.4889845	43
18	0.346936	0.937889	0.369911	2.703351	1.066224	2.882379	0.0156094	1.4868568	42
19	0.347208	0.937788	0.370242	2.700936	1.066339	2.880114	0.0156492	1.4847328	41
20	0.347481	0.937687	0.370573	2.698525	1.066454	2.877853	0.0156891	1.4826127	40
21	0.347754	0.937586	0.370904	2.696118	1.066569	2.875596	0.0157291	1.4804963	39
22	0.348027	0.937485	0.371235	2.693715	1.066684	2.873343	0.0157692	1.4783837	38
23	0.348299	0.937383	0.371566	2.691315	1.066799	2.871093	0.0158093	1.4762749	37
24	0.348572	0.937282	0.371897	2.688919	1.066915	2.868847	0.0158495	1.4741698	36
25	0.348845	0.937181	0.372228	2.686527	1.067030	2.866605	0.0158898	1.4720685	35
26	0.349117	0.937079	0.372559	2.684138	1.067146	2.864367	0.0159301	1.4699709	34
27	0.349390	0.936977	0.372890	2.681754	1.067262	2.862132	0.0159705	1.4678770	33
28	0.349662	0.936876	0.373222	2.679372	1.067377	2.859902	0.0160110	1.4657869	32
29	0.349935	0.936774	0.373553	2.676995	1.067493	2.857674	0.0160516	1.4637004	31
30	0.350207	0.936672	0.373885	2.674621	1.067609	2.855451	0.0160922	1.4616177	30
31	0.350480	0.936570	0.374216	2.672252	1.067726	2.853231	0.0161329	1.4595386	29
32	0.350752	0.936468	0.374548	2.669885	1.067842	2.851015	0.0161737	1.4574632	28
33	0.351025	0.936366	0.374880	2.667523	1.067958	2.848803	0.0162145	1.4553915	27
34	0.351297	0.936264	0.375211	2.665164	1.068075	2.846594	0.0162554	1.4533235	26
35	0.351569	0.936162	0.375543	2.662809	1.068191	2.844389	0.0162964	1.4512591	25
36	0.351842	0.936060	0.375875	2.660457	1.068308	2.842188	0.0163375	1.4491984	24
37	0.352114	0.935957	0.376207	2.658109	1.068425	2.839990	0.0163786	1.4471413	23
38	0.352386	0.935855	0.376539	2.655765	1.068542	2.837796	0.0164198	1.4450878	22
39	0.352658	0.935752	0.376872	2.653424	1.068659	2.835605	0.0164611	1.4430380	21
40	0.352931	0.935650	0.377204	2.651087	1.068776	2.833419	0.0165024	1.4409917	20
41	0.353203	0.935547	0.377536	2.648753	1.068894	2.831235	0.0165439	1.4389491	19
42	0.353475	0.935444	0.377869	2.646423	1.069011	2.829056	0.0165854	1.4369100	18
43	0.353747	0.935341	0.378201	2.644097	1.069129	2.826880	0.0166269	1.4348746	17
44	0.354019	0.935238	0.378534	2.641774	1.069246	2.824707	0.0166686	1.4328427	16
45	0.354291	0.935135	0.378866	2.639455	1.069364	2.822538	0.0167103	1.4308144	15
46	0.354563	0.935032	0.379199	2.637139	1.069482	2.820373	0.0167521	1.4287896	14
47	0.354835	0.934929	0.379532	2.634827	1.069600	2.818211	0.0167939	1.4267684	13
48	0.355107	0.934826	0.379864	2.632519	1.069718	2.816053	0.0168359	1.4247507	12
49	0.355379	0.934722	0.380197	2.630214	1.069836	2.813898	0.0168779	1.4227366	11
50	0.355651	0.934619	0.380530	2.627912	1.069955	2.811747	0.0169200	1.4207260	10
51	0.355923	0.934515	0.380863	2.625614	1.070073	2.809599	0.0169621	1.4187189	9
52	0.356194	0.934412	0.381196	2.623320	1.070192	2.807455	0.0170044	1.4167153	8
53	0.356466	0.934308	0.381530	2.621029	1.070311	2.805315	0.0170467	1.4147152	7
54	0.356738	0.934204	0.381863	2.618741	1.070429	2.803178	0.0170891	1.4127186	6
55	0.357010	0.934101	0.382196	2.616457	1.070548	2.801044	0.0171315	1.4107255	5
56	0.357281	0.933997	0.382530	2.614177	1.070668	2.798914	0.0171740	1.4087359	4
57	0.357553	0.933893	0.382863	2.611900	1.070787	2.796787	0.0172166	1.4067497	3
58	0.357825	0.933789	0.383197	2.609626	1.070906	2.794664	0.0172593	1.4047670	2
59	0.358096	0.933685	0.383530	2.607356	1.071025	2.792544	0.0173021	1.4027877	1
60	0.358368	0.933580	0.383864	2.605089	1.071145	2.790428	0.0173449	1.4008119	0
Minutes	Cosine	Sine	Cotangent	Tangent	Cosecant	Secant	Read Down	69°–70° Involute	Minutes

↑ **110° or 290°** **69° or 249°** ↑

↓ 21° or 201° Trigonometric and Involute Functions 158° or 338° ↓

Minutes	Sine	Cosine	Tangent	Cotangent	Secant	Cosecant	Involute 21°–22°	Read Up	Minutes
0	0.358368	0.933580	0.383864	2.605089	1.071145	2.790428	0.0173449	1.4008119	60
1	0.358640	0.933476	0.384198	2.602826	1.071265	2.788315	0.0173878	1.3988395	59
2	0.358911	0.933372	0.384532	2.600566	1.071384	2.786206	0.0174308	1.3968705	58
3	0.359183	0.933267	0.384866	2.598309	1.071504	2.784100	0.0174738	1.3949050	57
4	0.359454	0.933163	0.385200	2.596056	1.071624	2.781997	0.0175169	1.3929428	56
5	0.359725	0.933058	0.385534	2.593807	1.071744	2.779898	0.0175601	1.3909841	55
6	0.359997	0.932954	0.385868	2.591561	1.071865	2.777802	0.0176034	1.3890287	54
7	0.360268	0.932849	0.386202	2.589318	1.071985	2.775710	0.0176468	1.3870768	53
8	0.360540	0.932744	0.386536	2.587078	1.072106	2.773621	0.0176902	1.3851282	52
9	0.360811	0.932639	0.386871	2.584842	1.072226	2.771535	0.0177337	1.3831829	51
10	0.361082	0.932534	0.387205	2.582609	1.072347	2.769453	0.0177773	1.3812411	50
11	0.361353	0.932429	0.387540	2.580380	1.072468	2.767374	0.0178209	1.3793026	49
12	0.361625	0.932324	0.387874	2.578154	1.072589	2.765299	0.0178646	1.3773674	48
13	0.361896	0.932219	0.388209	2.575931	1.072710	2.763227	0.0179084	1.3754356	47
14	0.362167	0.932113	0.388544	2.573712	1.072831	2.761158	0.0179523	1.3735071	46
15	0.362438	0.932008	0.388879	2.571496	1.072952	2.759092	0.0179963	1.3715819	45
16	0.362709	0.931902	0.389214	2.569283	1.073074	2.757030	0.0180403	1.3696600	44
17	0.362980	0.931797	0.389549	2.567074	1.073195	2.754971	0.0180844	1.3677414	43
18	0.363251	0.931691	0.389884	2.564867	1.073317	2.752916	0.0181286	1.3658262	42
19	0.363522	0.931586	0.390219	2.562665	1.073439	2.750863	0.0181728	1.3639142	41
20	0.363793	0.931480	0.390554	2.560465	1.073561	2.748814	0.0182172	1.3620055	40
21	0.364064	0.931374	0.390889	2.558269	1.073683	2.746769	0.0182616	1.3601001	39
22	0.364335	0.931268	0.391225	2.556076	1.073805	2.744726	0.0183061	1.3581979	38
23	0.364606	0.931162	0.391560	2.553886	1.073927	2.742687	0.0183506	1.3562990	37
24	0.364877	0.931056	0.391896	2.551699	1.074049	2.740651	0.0183953	1.3544034	36
25	0.365148	0.930950	0.392231	2.549516	1.074172	2.738619	0.0184400	1.3525110	35
26	0.365418	0.930843	0.392567	2.547336	1.074295	2.736589	0.0184848	1.3506218	34
27	0.365689	0.930737	0.392903	2.545159	1.074417	2.734563	0.0185296	1.3487359	33
28	0.365960	0.930631	0.393239	2.542985	1.074540	2.732540	0.0185746	1.3468532	32
29	0.366231	0.930524	0.393574	2.540815	1.074663	2.730520	0.0186196	1.3449737	31
30	0.366501	0.930418	0.393910	2.538648	1.074786	2.728504	0.0186647	1.3430974	30
31	0.366772	0.930311	0.394247	2.536484	1.074909	2.726491	0.0187099	1.3412243	29
32	0.367042	0.930204	0.394583	2.534323	1.075033	2.724480	0.0187551	1.3393544	28
33	0.367313	0.930097	0.394919	2.532165	1.075156	2.722474	0.0188004	1.3374876	27
34	0.367584	0.929990	0.395255	2.530011	1.075280	2.720470	0.0188458	1.3356241	26
35	0.367854	0.929884	0.395592	2.527860	1.075403	2.718469	0.0188913	1.3337637	25
36	0.368125	0.929776	0.395928	2.525712	1.075527	2.716472	0.0189369	1.3319065	24
37	0.368395	0.929669	0.396265	2.523567	1.075651	2.714478	0.0189825	1.3300524	23
38	0.368665	0.929562	0.396601	2.521425	1.075775	2.712487	0.0190282	1.3282015	22
39	0.368936	0.929455	0.396938	2.519286	1.075899	2.710499	0.0190740	1.3263537	21
40	0.369206	0.929348	0.397275	2.517151	1.076024	2.708514	0.0191199	1.3245091	20
41	0.369476	0.929240	0.397611	2.515018	1.076148	2.706532	0.0191659	1.3226676	19
42	0.369747	0.929133	0.397948	2.512889	1.076273	2.704554	0.0192119	1.3208292	18
43	0.370017	0.929025	0.398285	2.510763	1.076397	2.702578	0.0192580	1.3189939	17
44	0.370287	0.928917	0.398622	2.508640	1.076522	2.700606	0.0193042	1.3171617	16
45	0.370557	0.928810	0.398960	2.506520	1.076647	2.698637	0.0193504	1.3153326	15
46	0.370828	0.928702	0.399297	2.504403	1.076772	2.696671	0.0193968	1.3135066	14
47	0.371098	0.928594	0.399634	2.502289	1.076897	2.694708	0.0194432	1.3116837	13
48	0.371368	0.928486	0.399971	2.500178	1.077022	2.692748	0.0194897	1.3098638	12
49	0.371638	0.928378	0.400309	2.498071	1.077148	2.690791	0.0195363	1.3080470	11
50	0.371908	0.928270	0.400646	2.495966	1.077273	2.688837	0.0195829	1.3062333	10
51	0.372178	0.928161	0.400984	2.493865	1.077399	2.686887	0.0196296	1.3044227	9
52	0.372448	0.928053	0.401322	2.491766	1.077525	2.684939	0.0196765	1.3026150	8
53	0.372718	0.927945	0.401660	2.489671	1.077650	2.682995	0.0197233	1.3008105	7
54	0.372988	0.927836	0.401997	2.487578	1.077776	2.681053	0.0197703	1.2990089	6
55	0.373258	0.927728	0.402335	2.485489	1.077902	2.679114	0.0198174	1.2972104	5
56	0.373528	0.927619	0.402673	2.483402	1.078029	2.677179	0.0198645	1.2954149	4
57	0.373797	0.927510	0.403011	2.481319	1.078155	2.675247	0.0199117	1.2936224	3
58	0.374067	0.927402	0.403350	2.479239	1.078281	2.673317	0.0199590	1.2918329	2
59	0.374337	0.927293	0.403688	2.477161	1.078408	2.671391	0.0200063	1.2900465	1
60	0.374607	0.927184	0.404026	2.475087	1.078535	2.669467	0.0200538	1.2882630	0
Minutes	Cosine	Sine	Cotangent	Tangent	Cosecant	Secant	Read Down	68°–69° Involute	Minutes

↑ 111° or 291° 68° or 248° ↑

↓ 22° or 202° Trigonometric and Involute Functions 157° or 337° ↓

Minutes	Sine	Cosine	Tangent	Cotangent	Secant	Cosecant	Involute 22°–23°	Read Up	Minutes
0	0.374607	0.927184	0.404026	2.475087	1.078535	2.669467	0.0200538	1.2882630	60
1	0.374876	0.927075	0.404365	2.473015	1.078662	2.667547	0.0201013	1.2864825	59
2	0.375146	0.926966	0.404703	2.470947	1.078788	2.665629	0.0201489	1.2847049	58
3	0.375416	0.926857	0.405042	2.468882	1.078916	2.663715	0.0201966	1.2829304	57
4	0.375685	0.926747	0.405380	2.466819	1.079043	2.661803	0.0202444	1.2811588	56
5	0.375955	0.926638	0.405719	2.464760	1.079170	2.659895	0.0202922	1.2793901	55
6	0.376224	0.926529	0.406058	2.462703	1.079297	2.657989	0.0203401	1.2776245	54
7	0.376494	0.926419	0.406397	2.460649	1.079425	2.656086	0.0203881	1.2758617	53
8	0.376763	0.926310	0.406736	2.458599	1.079553	2.654187	0.0204362	1.2741019	52
9	0.377033	0.926200	0.407075	2.456551	1.079680	2.652290	0.0204844	1.2723451	51
10	0.377302	0.926090	0.407414	2.454506	1.079808	2.650396	0.0205326	1.2705911	50
11	0.377571	0.925980	0.407753	2.452464	1.079936	2.648505	0.0205809	1.2688401	49
12	0.377841	0.925871	0.408092	2.450425	1.080065	2.646617	0.0206293	1.2670920	48
13	0.378110	0.925761	0.408432	2.448389	1.080193	2.644732	0.0206778	1.2653468	47
14	0.378379	0.925651	0.408771	2.446356	1.080321	2.642850	0.0207264	1.2636044	46
15	0.378649	0.925541	0.409111	2.444326	1.080450	2.640971	0.0207750	1.2618650	45
16	0.378918	0.925430	0.409450	2.442298	1.080578	2.639095	0.0208238	1.2601285	44
17	0.379187	0.925320	0.409790	2.440274	1.080707	2.637221	0.0208726	1.2583948	43
18	0.379456	0.925210	0.410130	2.438252	1.080836	2.635351	0.0209215	1.2566640	42
19	0.379725	0.925099	0.410470	2.436233	1.080965	2.633483	0.0209704	1.2549361	41
20	0.379994	0.924989	0.410810	2.434217	1.081094	2.631618	0.0210195	1.2532111	40
21	0.380263	0.924878	0.411150	2.432204	1.081223	2.629756	0.0210686	1.2514889	39
22	0.380532	0.924768	0.411490	2.430194	1.081353	2.627897	0.0211178	1.2497695	38
23	0.380801	0.924657	0.411830	2.428186	1.081482	2.626041	0.0211671	1.2480530	37
24	0.381070	0.924546	0.412170	2.426182	1.081612	2.624187	0.0212165	1.2463393	36
25	0.381339	0.924435	0.412511	2.424180	1.081742	2.622337	0.0212660	1.2446284	35
26	0.381608	0.924324	0.412851	2.422181	1.081872	2.620489	0.0213155	1.2429204	34
27	0.381877	0.924213	0.413192	2.420185	1.082002	2.618644	0.0213651	1.2412152	33
28	0.382146	0.924102	0.413532	2.418192	1.082132	2.616802	0.0214148	1.2395127	32
29	0.382415	0.923991	0.413873	2.416201	1.082262	2.614962	0.0214646	1.2378131	31
30	0.382683	0.923880	0.414214	2.414214	1.082392	2.613126	0.0215145	1.2361163	30
31	0.382952	0.923768	0.414554	2.412229	1.082523	2.611292	0.0215644	1.2344223	29
32	0.383221	0.923657	0.414895	2.410247	1.082653	2.609461	0.0216145	1.2327310	28
33	0.383490	0.923545	0.415236	2.408267	1.082784	2.607633	0.0216646	1.2310426	27
34	0.383758	0.923434	0.415577	2.406291	1.082915	2.605808	0.0217148	1.2293569	26
35	0.384027	0.923322	0.415919	2.404317	1.083046	2.603985	0.0217651	1.2276740	25
36	0.384295	0.923210	0.416260	2.402346	1.083177	2.602165	0.0218154	1.2259938	24
37	0.384564	0.923098	0.416601	2.400377	1.083308	2.600348	0.0218659	1.2243164	23
38	0.384832	0.922986	0.416943	2.398412	1.083439	2.598534	0.0219164	1.2226417	22
39	0.385101	0.922875	0.417284	2.396449	1.083571	2.596723	0.0219670	1.2209698	21
40	0.385369	0.922762	0.417626	2.394489	1.083703	2.594914	0.0220177	1.2193006	20
41	0.385638	0.922650	0.417967	2.392532	1.083834	2.593108	0.0220685	1.2176341	19
42	0.385906	0.922538	0.418309	2.390577	1.083966	2.591304	0.0221193	1.2159704	18
43	0.386174	0.922426	0.418651	2.388625	1.084098	2.589504	0.0221703	1.2143093	17
44	0.386443	0.922313	0.418993	2.386676	1.084230	2.587706	0.0222213	1.2126510	16
45	0.386711	0.922201	0.419335	2.384729	1.084362	2.585911	0.0222724	1.2109954	15
46	0.386979	0.922088	0.419677	2.382786	1.084495	2.584118	0.0223236	1.2093425	14
47	0.387247	0.921976	0.420019	2.380844	1.084627	2.582328	0.0223749	1.2076923	13
48	0.387516	0.921863	0.420361	2.378906	1.084760	2.580541	0.0224262	1.2060447	12
49	0.387784	0.921750	0.420704	2.376970	1.084892	2.578757	0.0224777	1.2043999	11
50	0.388052	0.921638	0.421046	2.375037	1.085025	2.576975	0.0225292	1.2027577	10
51	0.388320	0.921525	0.421389	2.373107	1.085158	2.575196	0.0225808	1.2011182	9
52	0.388588	0.921412	0.421731	2.371179	1.085291	2.573420	0.0226325	1.1994814	8
53	0.388856	0.921299	0.422074	2.369254	1.085424	2.571646	0.0226843	1.1978472	7
54	0.389124	0.921185	0.422417	2.367332	1.085558	2.569875	0.0227361	1.1962156	6
55	0.389392	0.921072	0.422759	2.365412	1.085691	2.568107	0.0227881	1.1945867	5
56	0.389660	0.920959	0.423102	2.363495	1.085825	2.566341	0.0228401	1.1929605	4
57	0.389928	0.920845	0.423445	2.361580	1.085959	2.564578	0.0228922	1.1913369	3
58	0.390196	0.920732	0.423788	2.359668	1.086092	2.562818	0.0229444	1.1897159	2
59	0.390463	0.920618	0.424132	2.357759	1.086226	2.561060	0.0229967	1.1880975	1
60	0.390731	0.920505	0.424475	2.355852	1.086360	2.559305	0.0230491	1.1864818	0
Minutes	Cosine	Sine	Cotangent	Tangent	Cosecant	Secant	Read Down	67°–68° Involute	Minutes

↑ 112° or 292° 67° or 247° ↑

↓ 23° or 203° Trigonometric and Involute Functions 156° or 336° ↓

Minutes	Sine	Cosine	Tangent	Cotangent	Secant	Cosecant	Involute 23°–24°	Read Up	Minutes
0	0.390731	0.920505	0.424475	2.355852	1.086360	2.559305	0.0230491	1.1864818	60
1	0.390999	0.920391	0.424818	2.353948	1.086495	2.557552	0.0231015	1.1848686	59
2	0.391267	0.920277	0.425162	2.352047	1.086629	2.555802	0.0231541	1.1832581	58
3	0.391534	0.920164	0.425505	2.350148	1.086763	2.554055	0.0232067	1.1816502	57
4	0.391802	0.920050	0.425849	2.348252	1.086898	2.552310	0.0232594	1.1800448	56
5	0.392070	0.919936	0.426192	2.346358	1.087033	2.550568	0.0233122	1.1784421	55
6	0.392337	0.919821	0.426536	2.344467	1.087167	2.548828	0.0233651	1.1768419	54
7	0.392605	0.919707	0.426880	2.342579	1.087302	2.547091	0.0234181	1.1752443	53
8	0.392872	0.919593	0.427224	2.340693	1.087437	2.545357	0.0234711	1.1736493	52
9	0.393140	0.919479	0.427568	2.338809	1.087573	2.543625	0.0235242	1.1720569	51
10	0.393407	0.919364	0.427912	2.336929	1.087708	2.541896	0.0235775	1.1704670	50
11	0.393675	0.919250	0.428256	2.335050	1.087843	2.540169	0.0236308	1.1688797	49
12	0.393942	0.919135	0.428601	2.333175	1.087979	2.538445	0.0236842	1.1672949	48
13	0.394209	0.919021	0.428945	2.331302	1.088115	2.536724	0.0237376	1.1657126	47
14	0.394477	0.918906	0.429289	2.329431	1.088251	2.535005	0.0237912	1.1641329	46
15	0.394744	0.918791	0.429634	2.327563	1.088387	2.533288	0.0238449	1.1625558	45
16	0.395011	0.918676	0.429979	2.325698	1.088523	2.531574	0.0238986	1.1609811	44
17	0.395278	0.918561	0.430323	2.323835	1.088659	2.529863	0.0239524	1.1594090	43
18	0.395546	0.918446	0.430668	2.321974	1.088795	2.528154	0.0240063	1.1578394	42
19	0.395813	0.918331	0.431013	2.320116	1.088932	2.526448	0.0240603	1.1562723	41
20	0.396080	0.918216	0.431358	2.318261	1.089068	2.524744	0.0241144	1.1547077	40
21	0.396347	0.918101	0.431703	2.316408	1.089205	2.523043	0.0241686	1.1531457	39
22	0.396614	0.917986	0.432048	2.314557	1.089342	2.521344	0.0242228	1.1515861	38
23	0.396881	0.917870	0.432393	2.312709	1.089479	2.519648	0.0242772	1.1500290	37
24	0.397148	0.917755	0.432739	2.310864	1.089616	2.517954	0.0243316	1.1484744	36
25	0.397415	0.917639	0.433084	2.309021	1.089753	2.516262	0.0243861	1.1469222	35
26	0.397682	0.917523	0.433430	2.307180	1.089890	2.514574	0.0244407	1.1453726	34
27	0.397949	0.917408	0.433775	2.305342	1.090028	2.512887	0.0244954	1.1438254	33
28	0.398215	0.917292	0.434121	2.303506	1.090166	2.511203	0.0245502	1.1422807	32
29	0.398482	0.917176	0.434467	2.301673	1.090303	2.509522	0.0246050	1.1407384	31
30	0.398749	0.917060	0.434812	2.299843	1.090441	2.507843	0.0246600	1.1391986	30
31	0.399016	0.916944	0.435158	2.298014	1.090579	2.506166	0.0247150	1.1376612	29
32	0.399283	0.916828	0.435504	2.296188	1.090717	2.504492	0.0247702	1.1361263	28
33	0.399549	0.916712	0.435850	2.294365	1.090855	2.502821	0.0248254	1.1345938	27
34	0.399816	0.916595	0.436197	2.292544	1.090994	2.501151	0.0248807	1.1330638	26
35	0.400082	0.916479	0.436543	2.290726	1.091132	2.499485	0.0249361	1.1315361	25
36	0.400349	0.916363	0.436889	2.288910	1.091271	2.497820	0.0249916	1.1300109	24
37	0.400616	0.916246	0.437236	2.287096	1.091410	2.496159	0.0250471	1.1284882	23
38	0.400882	0.916130	0.437582	2.285285	1.091549	2.494499	0.0251028	1.1269678	22
39	0.401149	0.916013	0.437929	2.283476	1.091688	2.492842	0.0251585	1.1254498	21
40	0.401415	0.915896	0.438276	2.281669	1.091827	2.491187	0.0252143	1.1239342	20
41	0.401681	0.915779	0.438622	2.279865	1.091966	2.489535	0.0252703	1.1224211	19
42	0.401948	0.915663	0.438969	2.278064	1.092105	2.487885	0.0253263	1.1209103	18
43	0.402214	0.915546	0.439316	2.276264	1.092245	2.486238	0.0253824	1.1194019	17
44	0.402480	0.915429	0.439663	2.274467	1.092384	2.484593	0.0254386	1.1178959	16
45	0.402747	0.915311	0.440011	2.272673	1.092524	2.482950	0.0254948	1.1163922	15
46	0.403013	0.915194	0.440358	2.270881	1.092664	2.481310	0.0255512	1.1148910	14
47	0.403279	0.915077	0.440705	2.269091	1.092804	2.479672	0.0256076	1.1133921	13
48	0.403545	0.914960	0.441053	2.267304	1.092944	2.478037	0.0256642	1.1118955	12
49	0.403811	0.914842	0.441400	2.265518	1.093085	2.476403	0.0257208	1.1104014	11
50	0.404078	0.914725	0.441748	2.263736	1.093225	2.474773	0.0257775	1.1089095	10
51	0.404344	0.914607	0.442095	2.261955	1.093366	2.473144	0.0258343	1.1074201	9
52	0.404610	0.914490	0.442443	2.260177	1.093506	2.471518	0.0258912	1.1059329	8
53	0.404876	0.914372	0.442791	2.258402	1.093647	2.469894	0.0259482	1.1044481	7
54	0.405142	0.914254	0.443139	2.256628	1.093788	2.468273	0.0260053	1.1029656	6
55	0.405408	0.914136	0.443487	2.254857	1.093929	2.466654	0.0260625	1.1014855	5
56	0.405673	0.914018	0.443835	2.253089	1.094070	2.465037	0.0261197	1.1000077	4
57	0.405939	0.913900	0.444183	2.251322	1.094212	2.463423	0.0261771	1.0985321	3
58	0.406205	0.913782	0.444532	2.249558	1.094353	2.461811	0.0262345	1.0970589	2
59	0.406471	0.913664	0.444880	2.247796	1.094495	2.460201	0.0262920	1.0955881	1
60	0.406737	0.913545	0.445229	2.246037	1.094636	2.458593	0.0263497	1.0941195	0
Minutes	Cosine	Sine	Cotangent	Tangent	Cosecant	Secant	Read Down	66°–67° Involute	Minutes

↑ 113° or 293° 66° or 246° ↑

↓ 24° or 204° Trigonometric and Involute Functions 155° or 335° ↓

Minutes	Sine	Cosine	Tangent	Cotangent	Secant	Cosecant	Involute 24°–25°	Read Up	Minutes
0	0.406737	0.913545	0.445229	2.246037	1.094636	2.458593	0.0263497	1.0941195	60
1	0.407002	0.913427	0.445577	2.244280	1.094778	2.456988	0.0264074	1.0926532	59
2	0.407268	0.913309	0.445926	2.242525	1.094920	2.455385	0.0264652	1.0911892	58
3	0.407534	0.913190	0.446275	2.240772	1.095062	2.453785	0.0265231	1.0897275	57
4	0.407799	0.913072	0.446624	2.239022	1.095204	2.452186	0.0265810	1.0882680	56
5	0.408065	0.912953	0.446973	2.237274	1.095347	2.450591	0.0266391	1.0868109	55
6	0.408330	0.912834	0.447322	2.235528	1.095489	2.448997	0.0266973	1.0853560	54
7	0.408596	0.912715	0.447671	2.233785	1.095632	2.447405	0.0267555	1.0839034	53
8	0.408861	0.912596	0.448020	2.232043	1.095775	2.445816	0.0268139	1.0824531	52
9	0.409127	0.912477	0.448369	2.230304	1.095917	2.444229	0.0268723	1.0810050	51
10	0.409392	0.912358	0.448719	2.228568	1.096060	2.442645	0.0269308	1.0795592	50
11	0.409658	0.912239	0.449068	2.226833	1.096204	2.441062	0.0269894	1.0781156	49
12	0.409923	0.912120	0.449418	2.225101	1.096347	2.439482	0.0270481	1.0766743	48
13	0.410188	0.912001	0.449768	2.223371	1.096490	2.437904	0.0271069	1.0752352	47
14	0.410454	0.911881	0.450117	2.221643	1.096634	2.436329	0.0271658	1.0737983	46
15	0.410719	0.911762	0.450467	2.219918	1.096777	2.434756	0.0272248	1.0723637	45
16	0.410984	0.911643	0.450817	2.218194	1.096921	2.433184	0.0272839	1.0709313	44
17	0.411249	0.911523	0.451167	2.216473	1.097065	2.431616	0.0273430	1.0695011	43
18	0.411514	0.911403	0.451517	2.214754	1.097209	2.430049	0.0274023	1.0680732	42
19	0.411779	0.911284	0.451868	2.213038	1.097353	2.428484	0.0274617	1.0666474	41
20	0.412045	0.911164	0.452218	2.211323	1.097498	2.426922	0.0275211	1.0652239	40
21	0.412310	0.911044	0.452568	2.209611	1.097642	2.425362	0.0275806	1.0638026	39
22	0.412575	0.910924	0.452919	2.207901	1.097787	2.423804	0.0276403	1.0623835	38
23	0.412840	0.910804	0.453269	2.206193	1.097931	2.422249	0.0277000	1.0609665	37
24	0.413104	0.910684	0.453620	2.204488	1.098076	2.420695	0.0277598	1.0595518	36
25	0.413369	0.910563	0.453971	2.202784	1.098221	2.419144	0.0278197	1.0581392	35
26	0.413634	0.910443	0.454322	2.201083	1.098366	2.417595	0.0278797	1.0567288	34
27	0.413899	0.910323	0.454673	2.199384	1.098511	2.416048	0.0279398	1.0553206	33
28	0.414164	0.910202	0.455024	2.197687	1.098657	2.414504	0.0279999	1.0539146	32
29	0.414429	0.910082	0.455375	2.195992	1.098802	2.412961	0.0280602	1.0525108	31
30	0.414693	0.909961	0.455726	2.194300	1.098948	2.411421	0.0281206	1.0511091	30
31	0.414958	0.909841	0.456078	2.192609	1.099094	2.409883	0.0281810	1.0497095	29
32	0.415223	0.909720	0.456429	2.190921	1.099239	2.408347	0.0282416	1.0483122	28
33	0.415487	0.909599	0.456781	2.189235	1.099386	2.406813	0.0283022	1.0469169	27
34	0.415752	0.909478	0.457132	2.187551	1.099532	2.405282	0.0283630	1.0455238	26
35	0.416016	0.909357	0.457484	2.185869	1.099678	2.403752	0.0284238	1.0441329	25
36	0.416281	0.909236	0.457836	2.184189	1.099824	2.402225	0.0284847	1.0427441	24
37	0.416545	0.909115	0.458188	2.182512	1.099971	2.400700	0.0285458	1.0413574	23
38	0.416810	0.908994	0.458540	2.180836	1.100118	2.399176	0.0286069	1.0399729	22
39	0.417074	0.908872	0.458892	2.179163	1.100264	2.397656	0.0286681	1.0385905	21
40	0.417338	0.908751	0.459244	2.177492	1.100411	2.396137	0.0287294	1.0372102	20
41	0.417603	0.908630	0.459596	2.175823	1.100558	2.394620	0.0287908	1.0358320	19
42	0.417867	0.908508	0.459949	2.174156	1.100706	2.393106	0.0288523	1.0344559	18
43	0.418131	0.908387	0.460301	2.172491	1.100853	2.391593	0.0289139	1.0330820	17
44	0.418396	0.908265	0.460654	2.170828	1.101000	2.390083	0.0289756	1.0317101	16
45	0.418660	0.908143	0.461006	2.169168	1.101148	2.388575	0.0290373	1.0303403	15
46	0.418924	0.908021	0.461359	2.167509	1.101296	2.387068	0.0290992	1.0289727	14
47	0.419188	0.907899	0.461712	2.165853	1.101444	2.385564	0.0291612	1.0276071	13
48	0.419452	0.907777	0.462065	2.164198	1.101592	2.384063	0.0292232	1.0262436	12
49	0.419716	0.907655	0.462418	2.162546	1.101740	2.382563	0.0292854	1.0248822	11
50	0.419980	0.907533	0.462771	2.160896	1.101888	2.381065	0.0293476	1.0235229	10
51	0.420244	0.907411	0.463124	2.159248	1.102036	2.379569	0.0294100	1.0221656	9
52	0.420508	0.907289	0.463478	2.157602	1.102185	2.378076	0.0294724	1.0208104	8
53	0.420772	0.907166	0.463831	2.155958	1.102334	2.376584	0.0295349	1.0194573	7
54	0.421036	0.907044	0.464185	2.154316	1.102482	2.375095	0.0295976	1.0181062	6
55	0.421300	0.906922	0.464538	2.152676	1.102631	2.373608	0.0296603	1.0167572	5
56	0.421563	0.906799	0.464892	2.151038	1.102780	2.372122	0.0297231	1.0154103	4
57	0.421827	0.906676	0.465246	2.149402	1.102930	2.370639	0.0297860	1.0140654	3
58	0.422091	0.906554	0.465600	2.147768	1.103079	2.369158	0.0298490	1.0127225	2
59	0.422355	0.906431	0.465954	2.146137	1.103228	2.367679	0.0299121	1.0113817	1
60	0.422618	0.906308	0.466308	2.144507	1.103378	2.366202	0.0299753	1.0100429	0
Minutes	Cosine	Sine	Cotangent	Tangent	Cosecant	Secant	Read Down	65°–66° Involute	Minutes

↑ 114° or 294° 65° or 245° ↑

↓ 25° or 205° Trigonometric and Involute Functions 154° or 334° ↓

Minutes	Sine	Cosine	Tangent	Cotangent	Secant	Cosecant	Involute 25°–26°	Read Up	Minutes
0	0.422618	0.906308	0.466308	2.144507	1.103378	2.366202	0.0299753	1.0100429	60
1	0.422882	0.906185	0.466662	2.142879	1.103528	2.364727	0.0300386	1.0087062	59
2	0.423145	0.906062	0.467016	2.141254	1.103678	2.363254	0.0301020	1.0073714	58
3	0.423409	0.905939	0.467371	2.139630	1.103828	2.361783	0.0301655	1.0060387	57
4	0.423673	0.905815	0.467725	2.138009	1.103978	2.360314	0.0302291	1.0047080	56
5	0.423936	0.905692	0.468080	2.136389	1.104128	2.358847	0.0302928	1.0033794	55
6	0.424199	0.905569	0.468434	2.134771	1.104278	2.357382	0.0303566	1.0020527	54
7	0.424463	0.905445	0.468789	2.133156	1.104429	2.355919	0.0304205	1.0007281	53
8	0.424726	0.905322	0.469144	2.131542	1.104580	2.354458	0.0304844	0.9994054	52
9	0.424990	0.905198	0.469499	2.129931	1.104730	2.352999	0.0305485	0.9980848	51
10	0.425253	0.905075	0.469854	2.128321	1.104881	2.351542	0.0306127	0.9967661	50
11	0.425516	0.904951	0.470209	2.126714	1.105032	2.350088	0.0306769	0.9954495	49
12	0.425779	0.904827	0.470564	2.125108	1.105184	2.348635	0.0307413	0.9941348	48
13	0.426042	0.904703	0.470920	2.123505	1.105335	2.347184	0.0308058	0.9928221	47
14	0.426306	0.904579	0.471275	2.121903	1.105486	2.345735	0.0308703	0.9915114	46
15	0.426569	0.904455	0.471631	2.120303	1.105638	2.344288	0.0309350	0.9902027	45
16	0.426832	0.904331	0.471986	2.118706	1.105790	2.342843	0.0309997	0.9888959	44
17	0.427095	0.904207	0.472342	2.117110	1.105942	2.341400	0.0310646	0.9875912	43
18	0.427358	0.904083	0.472698	2.115516	1.106094	2.339959	0.0311295	0.9862883	42
19	0.427621	0.903958	0.473054	2.113925	1.106246	2.338520	0.0311946	0.9849875	41
20	0.427884	0.903834	0.473410	2.112335	1.106398	2.337083	0.0312597	0.9836886	40
21	0.428147	0.903709	0.473766	2.110747	1.106551	2.335648	0.0313250	0.9823916	39
22	0.428410	0.903585	0.474122	2.109161	1.106703	2.334215	0.0313903	0.9810966	38
23	0.428672	0.903460	0.474478	2.107577	1.106856	2.332784	0.0314557	0.9798035	37
24	0.428935	0.903335	0.474835	2.105995	1.107009	2.331355	0.0315213	0.9785124	36
25	0.429198	0.903210	0.475191	2.104415	1.107162	2.329928	0.0315869	0.9772232	35
26	0.429461	0.903086	0.475548	2.102837	1.107315	2.328502	0.0316527	0.9759360	34
27	0.429723	0.902961	0.475905	2.101261	1.107468	2.327079	0.0317185	0.9746507	33
28	0.429986	0.902836	0.476262	2.099686	1.107621	2.325658	0.0317844	0.9733673	32
29	0.430249	0.902710	0.476619	2.098114	1.107775	2.324238	0.0318504	0.9720858	31
30	0.430511	0.902585	0.476976	2.096544	1.107929	2.322820	0.0319166	0.9708062	30
31	0.430774	0.902460	0.477333	2.094975	1.108082	2.321405	0.0319828	0.9695286	29
32	0.431036	0.902335	0.477690	2.093408	1.108236	2.319991	0.0320491	0.9682529	28
33	0.431299	0.902209	0.478047	2.091844	1.108390	2.318579	0.0321156	0.9669790	27
34	0.431561	0.902084	0.478405	2.090281	1.108545	2.317169	0.0321821	0.9657071	26
35	0.431823	0.901958	0.478762	2.088720	1.108699	2.315761	0.0322487	0.9644371	25
36	0.432086	0.901833	0.479120	2.087161	1.108853	2.314355	0.0323154	0.9631690	24
37	0.432348	0.901707	0.479477	2.085604	1.109008	2.312951	0.0323823	0.9619027	23
38	0.432610	0.901581	0.479835	2.084049	1.109163	2.311549	0.0324492	0.9606384	22
39	0.432873	0.901455	0.480193	2.082495	1.109318	2.310149	0.0325162	0.9593759	21
40	0.433135	0.901329	0.480551	2.080944	1.109473	2.308750	0.0325833	0.9581153	20
41	0.433397	0.901203	0.480909	2.079394	1.109628	2.307354	0.0326506	0.9568566	19
42	0.433659	0.901077	0.481267	2.077847	1.109783	2.305959	0.0327179	0.9555998	18
43	0.433921	0.900951	0.481626	2.076301	1.109938	2.304566	0.0327853	0.9543449	17
44	0.434183	0.900825	0.481984	2.074757	1.110094	2.303175	0.0328528	0.9530918	16
45	0.434445	0.900698	0.482343	2.073215	1.110250	2.301786	0.0329205	0.9518405	15
46	0.434707	0.900572	0.482701	2.071674	1.110406	2.300399	0.0329882	0.9505912	14
47	0.434969	0.900445	0.483060	2.070136	1.110562	2.299013	0.0330560	0.9493436	13
48	0.435231	0.900319	0.483419	2.068599	1.110718	2.297630	0.0331239	0.9480980	12
49	0.435493	0.900192	0.483778	2.067065	1.110874	2.296248	0.0331920	0.9468542	11
50	0.435755	0.900065	0.484137	2.065532	1.111030	2.294869	0.0332601	0.9456122	10
51	0.436017	0.899939	0.484496	2.064001	1.111187	2.293491	0.0333283	0.9443721	9
52	0.436278	0.899812	0.484855	2.062472	1.111344	2.292115	0.0333967	0.9431338	8
53	0.436540	0.899685	0.485214	2.060944	1.111500	2.290740	0.0334651	0.9418973	7
54	0.436802	0.899558	0.485574	2.059419	1.111657	2.289368	0.0335336	0.9406627	6
55	0.437063	0.899431	0.485933	2.057895	1.111814	2.287997	0.0336023	0.9394299	5
56	0.437325	0.899304	0.486293	2.056373	1.111972	2.286629	0.0336710	0.9381989	4
57	0.437587	0.899176	0.486653	2.054853	1.112129	2.285262	0.0337399	0.9369697	3
58	0.437848	0.899049	0.487013	2.053335	1.112287	2.283897	0.0338088	0.9357424	2
59	0.438110	0.898922	0.487373	2.051818	1.112444	2.282533	0.0338778	0.9345168	1
60	0.438371	0.898794	0.487733	2.050304	1.112602	2.281172	0.0339470	0.9332931	0
Minutes	Cosine	Sine	Cotangent	Tangent	Cosecant	Secant	Read Down	64°–65° Involute	Minutes

↑ 115° or 295° 64° or 244° ↑

↓ **26° or 206°** **Trigonometric and Involute Functions** **153° or 333°** ↓

Minutes	Sine	Cosine	Tangent	Cotangent	Secant	Cosecant	Involute 26°–27°	Read Up	Minutes
0	0.438371	0.898794	0.487733	2.050304	1.112602	2.281172	0.0339470	0.9332931	60
1	0.438633	0.898666	0.488093	2.048791	1.112760	2.279812	0.0340162	0.9320712	59
2	0.438894	0.898539	0.488453	2.047280	1.112918	2.278455	0.0340856	0.9308511	58
3	0.439155	0.898411	0.488813	2.045771	1.113076	2.277099	0.0341550	0.9296328	57
4	0.439417	0.898283	0.489174	2.044263	1.113234	2.275744	0.0342246	0.9284162	56
5	0.439678	0.898156	0.489534	2.042758	1.113393	2.274392	0.0342942	0.9272015	55
6	0.439939	0.898028	0.489895	2.041254	1.113552	2.273042	0.0343640	0.9259886	54
7	0.440200	0.897900	0.490256	2.039752	1.113710	2.271693	0.0344339	0.9247774	53
8	0.440462	0.897771	0.490617	2.038252	1.113869	2.270346	0.0345038	0.9235680	52
9	0.440723	0.897643	0.490978	2.036753	1.114028	2.269001	0.0345739	0.9223604	51
10	0.440984	0.897515	0.491339	2.035256	1.114187	2.267657	0.0346441	0.9211546	50
11	0.441245	0.897387	0.491700	2.033762	1.114347	2.266315	0.0347144	0.9199506	49
12	0.441506	0.897258	0.492061	2.032268	1.114506	2.264976	0.0347847	0.9187483	48
13	0.441767	0.897130	0.492422	2.030777	1.114666	2.263638	0.0348552	0.9175478	47
14	0.442028	0.897001	0.492784	2.029287	1.114826	2.262301	0.0349258	0.9163490	46
15	0.442289	0.896873	0.493145	2.027799	1.114985	2.260967	0.0349965	0.9151520	45
16	0.442550	0.896744	0.493507	2.026313	1.115145	2.259634	0.0350673	0.9139568	44
17	0.442810	0.896615	0.493869	2.024829	1.115306	2.258303	0.0351382	0.9127633	43
18	0.443071	0.896486	0.494231	2.023346	1.115466	2.256974	0.0352092	0.9115715	42
19	0.443332	0.896358	0.494593	2.021865	1.115626	2.255646	0.0352803	0.9103815	41
20	0.443593	0.896229	0.494955	2.020386	1.115787	2.254320	0.0353515	0.9091932	40
21	0.443853	0.896099	0.495317	2.018909	1.115948	2.252996	0.0354228	0.9080067	39
22	0.444114	0.895970	0.495679	2.017433	1.116108	2.251674	0.0354942	0.9068219	38
23	0.444375	0.895841	0.496042	2.015959	1.116269	2.250354	0.0355658	0.9056389	37
24	0.444635	0.895712	0.496404	2.014487	1.116431	2.249035	0.0356374	0.9044575	36
25	0.444896	0.895582	0.496767	2.013016	1.116592	2.247718	0.0357091	0.9032779	35
26	0.445156	0.895453	0.497130	2.011548	1.116753	2.246402	0.0357810	0.9021000	34
27	0.445417	0.895323	0.497492	2.010081	1.116915	2.245089	0.0358529	0.9009239	33
28	0.445677	0.895194	0.497855	2.008615	1.117077	2.243777	0.0359249	0.8997494	32
29	0.445937	0.895064	0.498218	2.007152	1.117238	2.242467	0.0359971	0.8985767	31
30	0.446198	0.894934	0.498582	2.005690	1.117400	2.241158	0.0360694	0.8974056	30
31	0.446458	0.894805	0.498945	2.004229	1.117563	2.239852	0.0361417	0.8962363	29
32	0.446718	0.894675	0.499308	2.002771	1.117725	2.238547	0.0362142	0.8950687	28
33	0.446979	0.894545	0.499672	2.001314	1.117887	2.237243	0.0362868	0.8939027	27
34	0.447239	0.894415	0.500035	1.999859	1.118050	2.235942	0.0363594	0.8927385	26
35	0.447499	0.894284	0.500399	1.998406	1.118212	2.234642	0.0364322	0.8915760	25
36	0.447759	0.894154	0.500763	1.996954	1.118375	2.233344	0.0365051	0.8904151	24
37	0.448019	0.894024	0.501127	1.995504	1.118538	2.232047	0.0365781	0.8892559	23
38	0.448279	0.893894	0.501491	1.994055	1.118701	2.230753	0.0366512	0.8880985	22
39	0.448539	0.893763	0.501855	1.992609	1.118865	2.229459	0.0367244	0.8869426	21
40	0.448799	0.893633	0.502219	1.991164	1.119028	2.228168	0.0367977	0.8857885	20
41	0.449059	0.893502	0.502583	1.989720	1.119192	2.226878	0.0368712	0.8846361	19
42	0.449319	0.893371	0.502948	1.988279	1.119355	2.225590	0.0369447	0.8834853	18
43	0.449579	0.893241	0.503312	1.986839	1.119519	2.224304	0.0370183	0.8823361	17
44	0.449839	0.893110	0.503677	1.985400	1.119683	2.223019	0.0370921	0.8811887	16
45	0.450098	0.892979	0.504041	1.983964	1.119847	2.221736	0.0371659	0.8800429	15
46	0.450358	0.892848	0.504406	1.982529	1.120011	2.220455	0.0372399	0.8788988	14
47	0.450618	0.892717	0.504771	1.981095	1.120176	2.219175	0.0373139	0.8777563	13
48	0.450878	0.892586	0.505136	1.979664	1.120340	2.217897	0.0373881	0.8766154	12
49	0.451137	0.892455	0.505502	1.978233	1.120505	2.216621	0.0374624	0.8754762	11
50	0.451397	0.892323	0.505867	1.976805	1.120670	2.215346	0.0375368	0.8743387	10
51	0.451656	0.892192	0.506232	1.975378	1.120835	2.214073	0.0376113	0.8732028	9
52	0.451916	0.892061	0.506598	1.973953	1.121000	2.212802	0.0376859	0.8720685	8
53	0.452175	0.891929	0.506963	1.972530	1.121165	2.211532	0.0377606	0.8709359	7
54	0.452435	0.891798	0.507329	1.971108	1.121331	2.210264	0.0378354	0.8698049	6
55	0.452694	0.891666	0.507695	1.969687	1.121496	2.208997	0.0379103	0.8686756	5
56	0.452953	0.891534	0.508061	1.968269	1.121662	2.207732	0.0379853	0.8675478	4
57	0.453213	0.891402	0.508427	1.966852	1.121828	2.206469	0.0380605	0.8664217	3
58	0.453472	0.891270	0.508793	1.965436	1.121994	2.205208	0.0381357	0.8652972	2
59	0.453731	0.891139	0.509159	1.964023	1.122160	2.203948	0.0382111	0.8641743	1
60	0.453990	0.891007	0.509525	1.962611	1.122326	2.202689	0.0382866	0.8630531	0
Minutes	Cosine	Sine	Cotangent	Tangent	Cosecant	Secant	Read Down	63°–64° Involute	Minutes

↑ **116° or 296°** **63° or 243°** ↑

↓ 27° or 207° Trigonometric and Involute Functions 152° or 332° ↓

Minutes	Sine	Cosine	Tangent	Cotangent	Secant	Cosecant	Involute 27°–28°	Read Up	Minutes
0	0.453990	0.891007	0.509525	1.962611	1.122326	2.202689	0.0382866	0.8630531	60
1	0.454250	0.890874	0.509892	1.961200	1.122493	2.201433	0.0383621	0.8619334	59
2	0.454509	0.890742	0.510258	1.959791	1.122659	2.200177	0.0384378	0.8608154	58
3	0.454768	0.890610	0.510625	1.958384	1.122826	2.198924	0.0385136	0.8596990	57
4	0.455027	0.890478	0.510992	1.956978	1.122993	2.197672	0.0385895	0.8585841	56
5	0.455286	0.890345	0.511359	1.955574	1.123160	2.196422	0.0386655	0.8574709	55
6	0.455545	0.890213	0.511726	1.954171	1.123327	2.195173	0.0387416	0.8563592	54
7	0.455804	0.890080	0.512093	1.952770	1.123494	2.193926	0.0388179	0.8552492	53
8	0.456063	0.889948	0.512460	1.951371	1.123662	2.192681	0.0388942	0.8541408	52
9	0.456322	0.889815	0.512828	1.949973	1.123829	2.191437	0.0389706	0.8530339	51
10	0.456580	0.889682	0.513195	1.948577	1.123997	2.190195	0.0390472	0.8519286	50
11	0.456839	0.889549	0.513563	1.947183	1.124165	2.188954	0.0391239	0.8508249	49
12	0.457098	0.889416	0.513930	1.945790	1.124333	2.187715	0.0392006	0.8497228	48
13	0.457357	0.889283	0.514298	1.944398	1.124501	2.186478	0.0392775	0.8486222	47
14	0.457615	0.889150	0.514666	1.943008	1.124669	2.185242	0.0393545	0.8475233	46
15	0.457874	0.889017	0.515034	1.941620	1.124838	2.184007	0.0394316	0.8464259	45
16	0.458133	0.888884	0.515402	1.940233	1.125006	2.182775	0.0395088	0.8453300	44
17	0.458391	0.888751	0.515770	1.938848	1.125175	2.181543	0.0395862	0.8442358	43
18	0.458650	0.888617	0.516138	1.937465	1.125344	2.180314	0.0396636	0.8431431	42
19	0.458908	0.888484	0.516507	1.936082	1.125513	2.179086	0.0397411	0.8420519	41
20	0.459166	0.888350	0.516875	1.934702	1.125682	2.177859	0.0398188	0.8409623	40
21	0.459425	0.888217	0.517244	1.933323	1.125851	2.176635	0.0398966	0.8398743	39
22	0.459683	0.888083	0.517613	1.931946	1.126021	2.175411	0.0399745	0.8387878	38
23	0.459942	0.887949	0.517982	1.930570	1.126191	2.174189	0.0400524	0.8377029	37
24	0.460200	0.887815	0.518351	1.929196	1.126360	2.172969	0.0401306	0.8366195	36
25	0.460458	0.887681	0.518720	1.927823	1.126530	2.171751	0.0402088	0.8355376	35
26	0.460716	0.887548	0.519089	1.926452	1.126700	2.170534	0.0402871	0.8344573	34
27	0.460974	0.887413	0.519458	1.925082	1.126870	2.169318	0.0403655	0.8333785	33
28	0.461232	0.887279	0.519828	1.923714	1.127041	2.168104	0.0404441	0.8323013	32
29	0.461491	0.887145	0.520197	1.922347	1.127211	2.166892	0.0405227	0.8312255	31
30	0.461749	0.887011	0.520567	1.920982	1.127382	2.165681	0.0406015	0.8301513	30
31	0.462007	0.886876	0.520937	1.919619	1.127553	2.164471	0.0406804	0.8290787	29
32	0.462265	0.886742	0.521307	1.918257	1.127724	2.163263	0.0407594	0.8280075	28
33	0.462523	0.886608	0.521677	1.916896	1.127895	2.162057	0.0408385	0.8269379	27
34	0.462780	0.886473	0.522047	1.915537	1.128066	2.160852	0.0409177	0.8258698	26
35	0.463038	0.886338	0.522417	1.914180	1.128237	2.159649	0.0409970	0.8248032	25
36	0.463296	0.886204	0.522787	1.912824	1.128409	2.158447	0.0410765	0.8237381	24
37	0.463554	0.886069	0.523158	1.911469	1.128581	2.157247	0.0411561	0.8226745	23
38	0.463812	0.885934	0.523528	1.910116	1.128752	2.156048	0.0412357	0.8216125	22
39	0.464069	0.885799	0.523899	1.908765	1.128924	2.154851	0.0413155	0.8205519	21
40	0.464327	0.885664	0.524270	1.907415	1.129096	2.153655	0.0413954	0.8194928	20
41	0.464584	0.885529	0.524641	1.906066	1.129269	2.152461	0.0414754	0.8184353	19
42	0.464842	0.885394	0.525012	1.904719	1.129441	2.151268	0.0415555	0.8173792	18
43	0.465100	0.885258	0.525383	1.903374	1.129614	2.150077	0.0416358	0.8163246	17
44	0.465357	0.885123	0.525754	1.902030	1.129786	2.148888	0.0417161	0.8152715	16
45	0.465615	0.884988	0.526125	1.900687	1.129959	2.147699	0.0417966	0.8142199	15
46	0.465872	0.884852	0.526497	1.899346	1.130132	2.146513	0.0418772	0.8131698	14
47	0.466129	0.884717	0.526868	1.898007	1.130305	2.145327	0.0419579	0.8121211	13
48	0.466387	0.884581	0.527240	1.896669	1.130479	2.144144	0.0420387	0.8110740	12
49	0.466644	0.884445	0.527612	1.895332	1.130652	2.142962	0.0421196	0.8100283	11
50	0.466901	0.884309	0.527984	1.893997	1.130826	2.141781	0.0422006	0.8089841	10
51	0.467158	0.884174	0.528356	1.892663	1.131000	2.140602	0.0422818	0.8079413	9
52	0.467416	0.884038	0.528728	1.891331	1.131173	2.139424	0.0423630	0.8069000	8
53	0.467673	0.883902	0.529100	1.890001	1.131348	2.138247	0.0424444	0.8058602	7
54	0.467930	0.883766	0.529473	1.888671	1.131522	2.137073	0.0425259	0.8048219	6
55	0.468187	0.883629	0.529845	1.887344	1.131696	2.135899	0.0426075	0.8037850	5
56	0.468444	0.883493	0.530218	1.886017	1.131871	2.134727	0.0426892	0.8027495	4
57	0.468701	0.883357	0.530591	1.884692	1.132045	2.133557	0.0427710	0.8017156	3
58	0.468958	0.883221	0.530963	1.883369	1.132220	2.132388	0.0428530	0.8006830	2
59	0.469215	0.883084	0.531336	1.882047	1.132395	2.131221	0.0429351	0.7996520	1
60	0.469472	0.882948	0.531709	1.880726	1.132570	2.130054	0.0430172	0.7986223	0
Minutes	Cosine	Sine	Cotangent	Tangent	Cosecant	Secant	Read Down	62°–63° Involute	Minutes

↑ 117° or 297° 62° or 242° ↑

↓ 28° or 208° Trigonometric and Involute Functions 151° or 331° ↓

Minutes	Sine	Cosine	Tangent	Cotangent	Secant	Cosecant	Involute 28°–29°	Read Up	Minutes
0	0.469472	0.882948	0.531709	1.880726	1.132570	2.130054	0.0430172	0.7986223	60
1	0.469728	0.882811	0.532083	1.879407	1.132745	2.128890	0.0430995	0.7975941	59
2	0.469985	0.882674	0.532456	1.878090	1.132921	2.127727	0.0431819	0.7965674	58
3	0.470242	0.882538	0.532829	1.876774	1.133096	2.126565	0.0432645	0.7955421	57
4	0.470499	0.882401	0.533203	1.875459	1.133272	2.125405	0.0433471	0.7945182	56
5	0.470755	0.882264	0.533577	1.874145	1.133448	2.124246	0.0434299	0.7934958	55
6	0.471012	0.882127	0.533950	1.872834	1.133624	2.123089	0.0435128	0.7924748	54
7	0.471268	0.881990	0.534324	1.871523	1.133800	2.121933	0.0435957	0.7914552	53
8	0.471525	0.881853	0.534698	1.870214	1.133976	2.120778	0.0436789	0.7904370	52
9	0.471782	0.881715	0.535072	1.868906	1.134153	2.119625	0.0437621	0.7894203	51
10	0.472038	0.881578	0.535446	1.867600	1.134329	2.118474	0.0438454	0.7884050	50
11	0.472294	0.881441	0.535821	1.866295	1.134506	2.117324	0.0439289	0.7873911	49
12	0.472551	0.881303	0.536195	1.864992	1.134683	2.116175	0.0440124	0.7863786	48
13	0.472807	0.881166	0.536570	1.863690	1.134860	2.115027	0.0440961	0.7853676	47
14	0.473063	0.881028	0.536945	1.862390	1.135037	2.113882	0.0441799	0.7843579	46
15	0.473320	0.880891	0.537319	1.861091	1.135215	2.112737	0.0442639	0.7833497	45
16	0.473576	0.880753	0.537694	1.859793	1.135392	2.111594	0.0443479	0.7823429	44
17	0.473832	0.880615	0.538069	1.858496	1.135570	2.110452	0.0444321	0.7813374	43
18	0.474088	0.880477	0.538445	1.857202	1.135748	2.109312	0.0445163	0.7803334	42
19	0.474344	0.880339	0.538820	1.855908	1.135926	2.108173	0.0446007	0.7793308	41
20	0.474600	0.880201	0.539195	1.854616	1.136104	2.107036	0.0446853	0.7783295	40
21	0.474856	0.880063	0.539571	1.853325	1.136282	2.105900	0.0447699	0.7773297	39
22	0.475112	0.879925	0.539946	1.852036	1.136460	2.104765	0.0448546	0.7763312	38
23	0.475368	0.879787	0.540322	1.850748	1.136639	2.103632	0.0449395	0.7753342	37
24	0.475624	0.879649	0.540698	1.849461	1.136818	2.102500	0.0450245	0.7743385	36
25	0.475880	0.879510	0.541074	1.848176	1.136997	2.101370	0.0451096	0.7733442	35
26	0.476136	0.879372	0.541450	1.846892	1.137176	2.100241	0.0451948	0.7723513	34
27	0.476392	0.879233	0.541826	1.845610	1.137355	2.099113	0.0452801	0.7713598	33
28	0.476647	0.879095	0.542203	1.844329	1.137534	2.097987	0.0453656	0.7703696	32
29	0.476903	0.878956	0.542579	1.843049	1.137714	2.096862	0.0454512	0.7693808	31
30	0.477159	0.878817	0.542956	1.841771	1.137893	2.095739	0.0455369	0.7683934	30
31	0.477414	0.878678	0.543332	1.840494	1.138073	2.094616	0.0456227	0.7674074	29
32	0.477670	0.878539	0.543709	1.839218	1.138253	2.093496	0.0457086	0.7664227	28
33	0.477925	0.878400	0.544086	1.837944	1.138433	2.092376	0.0457947	0.7654394	27
34	0.478181	0.878261	0.544463	1.836671	1.138613	2.091258	0.0458808	0.7644574	26
35	0.478436	0.878122	0.544840	1.835400	1.138794	2.090142	0.0459671	0.7634768	25
36	0.478692	0.877983	0.545218	1.834130	1.138974	2.089027	0.0460535	0.7624976	24
37	0.478947	0.877844	0.545595	1.832861	1.139155	2.087913	0.0461401	0.7615197	23
38	0.479203	0.877704	0.545973	1.831594	1.139336	2.086800	0.0462267	0.7605432	22
39	0.479458	0.877565	0.546350	1.830327	1.139517	2.085689	0.0463135	0.7595680	21
40	0.479713	0.877425	0.546728	1.829063	1.139698	2.084579	0.0464004	0.7585942	20
41	0.479968	0.877286	0.547106	1.827799	1.139879	2.083471	0.0464874	0.7576217	19
42	0.480223	0.877146	0.547484	1.826537	1.140061	2.082364	0.0465745	0.7566505	18
43	0.480479	0.877006	0.547862	1.825277	1.140242	2.081258	0.0466618	0.7556807	17
44	0.480734	0.876867	0.548240	1.824017	1.140424	2.080154	0.0467491	0.7547123	16
45	0.480989	0.876727	0.548619	1.822759	1.140606	2.079051	0.0468366	0.7537451	15
46	0.481244	0.876587	0.548997	1.821503	1.140788	2.077949	0.0469242	0.7527793	14
47	0.481499	0.876447	0.549376	1.820247	1.140971	2.076849	0.0470120	0.7518149	13
48	0.481754	0.876307	0.549755	1.818993	1.141153	2.075750	0.0470998	0.7508517	12
49	0.482009	0.876167	0.550134	1.817741	1.141336	2.074652	0.0471878	0.7498899	11
50	0.482263	0.876026	0.550513	1.816489	1.141518	2.073556	0.0472759	0.7489294	10
51	0.482518	0.875886	0.550892	1.815239	1.141701	2.072461	0.0473641	0.7479703	9
52	0.482773	0.875746	0.551271	1.813990	1.141884	2.071367	0.0474525	0.7470124	8
53	0.483028	0.875605	0.551650	1.812743	1.142067	2.070275	0.0475409	0.7460559	7
54	0.483282	0.875465	0.552030	1.811497	1.142251	2.069184	0.0476295	0.7451007	6
55	0.483537	0.875324	0.552409	1.810252	1.142434	2.068094	0.0477182	0.7441468	5
56	0.483792	0.875183	0.552789	1.809009	1.142618	2.067006	0.0478070	0.7431942	4
57	0.484046	0.875042	0.553169	1.807766	1.142802	2.065919	0.0478960	0.7422429	3
58	0.484301	0.874902	0.553549	1.806526	1.142986	2.064833	0.0479851	0.7412930	2
59	0.484555	0.874761	0.553929	1.805286	1.143170	2.063748	0.0480743	0.7403443	1
60	0.484810	0.874620	0.554309	1.804048	1.143354	2.062665	0.0481636	0.7393969	0
Minutes	Cosine	Sine	Cotangent	Tangent	Cosecant	Secant	Read Down	61°–62° Involute	Minutes

↑ 118° or 298° 61° or 241° ↑

↓ **29° or 209°** **Trigonometric and Involute Functions** **150° or 330°** ↓

Minutes	Sine	Cosine	Tangent	Cotangent	Secant	Cosecant	Involute 29°–30°	Read Up	Minutes
0	0.484810	0.874620	0.554309	1.804048	1.143354	2.062665	0.0481636	0.7393969	60
1	0.485064	0.874479	0.554689	1.802811	1.143539	2.061584	0.0482530	0.7384508	59
2	0.485318	0.874338	0.555070	1.801575	1.143723	2.060503	0.0483426	0.7375061	58
3	0.485573	0.874196	0.555450	1.800341	1.143908	2.059424	0.0484323	0.7365626	57
4	0.485827	0.874055	0.555831	1.799108	1.144093	2.058346	0.0485221	0.7356204	56
5	0.486081	0.873914	0.556212	1.797876	1.144278	2.057269	0.0486120	0.7346795	55
6	0.486335	0.873772	0.556593	1.796645	1.144463	2.056194	0.0487020	0.7337399	54
7	0.486590	0.873631	0.556974	1.795416	1.144648	2.055120	0.0487922	0.7328016	53
8	0.486844	0.873489	0.557355	1.794188	1.144834	2.054048	0.0488825	0.7318645	52
9	0.487098	0.873347	0.557736	1.792962	1.145020	2.052976	0.0489730	0.7309288	51
10	0.487352	0.873206	0.558118	1.791736	1.145205	2.051906	0.0490635	0.7299943	50
11	0.487606	0.873064	0.558499	1.790512	1.145391	2.050837	0.0491542	0.7290611	49
12	0.487860	0.872922	0.558881	1.789289	1.145578	2.049770	0.0492450	0.7281291	48
13	0.488114	0.872780	0.559263	1.788068	1.145764	2.048704	0.0493359	0.7271985	47
14	0.488367	0.872638	0.559645	1.786847	1.145950	2.047639	0.0494269	0.7262691	46
15	0.488621	0.872496	0.560027	1.785628	1.146137	2.046575	0.0495181	0.7253410	45
16	0.488875	0.872354	0.560409	1.784411	1.146324	2.045513	0.0496094	0.7244141	44
17	0.489129	0.872212	0.560791	1.783194	1.146511	2.044451	0.0497008	0.7234885	43
18	0.489382	0.872069	0.561174	1.781979	1.146698	2.043392	0.0497924	0.7225642	42
19	0.489636	0.871927	0.561556	1.780765	1.146885	2.042333	0.0498840	0.7216411	41
20	0.489890	0.871784	0.561939	1.779552	1.147073	2.041276	0.0499758	0.7207193	40
21	0.490143	0.871642	0.562322	1.778341	1.147260	2.040220	0.0500677	0.7197987	39
22	0.490397	0.871499	0.562705	1.777131	1.147448	2.039165	0.0501598	0.7188794	38
23	0.490650	0.871357	0.563088	1.775922	1.147636	2.038111	0.0502519	0.7179614	37
24	0.490904	0.871214	0.563471	1.774714	1.147824	2.037059	0.0503442	0.7170446	36
25	0.491157	0.871071	0.563854	1.773508	1.148012	2.036008	0.0504367	0.7161290	35
26	0.491411	0.870928	0.564238	1.772302	1.148200	2.034958	0.0505292	0.7152147	34
27	0.491664	0.870785	0.564621	1.771098	1.148389	2.033910	0.0506219	0.7143016	33
28	0.491917	0.870642	0.565005	1.769896	1.148578	2.032863	0.0507147	0.7133898	32
29	0.492170	0.870499	0.565389	1.768694	1.148767	2.031817	0.0508076	0.7124792	31
30	0.492424	0.870356	0.565773	1.767494	1.148956	2.030772	0.0509006	0.7115698	30
31	0.492677	0.870212	0.566157	1.766295	1.149145	2.029729	0.0509938	0.7106617	29
32	0.492930	0.870069	0.566541	1.765097	1.149334	2.028686	0.0510871	0.7097548	28
33	0.493183	0.869926	0.566925	1.763901	1.149524	2.027645	0.0511806	0.7088491	27
34	0.493436	0.869782	0.567310	1.762705	1.149713	2.026606	0.0512741	0.7079447	26
35	0.493689	0.869639	0.567694	1.761511	1.149903	2.025567	0.0513678	0.7070415	25
36	0.493942	0.869495	0.568079	1.760318	1.150093	2.024530	0.0514616	0.7061395	24
37	0.494195	0.869351	0.568464	1.759127	1.150283	2.023494	0.0515555	0.7052387	23
38	0.494448	0.869207	0.568849	1.757936	1.150473	2.022459	0.0516496	0.7043392	22
39	0.494700	0.869064	0.569234	1.756747	1.150664	2.021425	0.0517438	0.7034408	21
40	0.494953	0.868920	0.569619	1.755559	1.150854	2.020393	0.0518381	0.7025437	20
41	0.495206	0.868776	0.570004	1.754372	1.151045	2.019362	0.0519326	0.7016478	19
42	0.495459	0.868632	0.570390	1.753187	1.151236	2.018332	0.0520271	0.7007531	18
43	0.495711	0.868487	0.570776	1.752002	1.151427	2.017303	0.0521218	0.6998596	17
44	0.495964	0.868343	0.571161	1.750819	1.151618	2.016276	0.0522167	0.6989673	16
45	0.496217	0.868199	0.571547	1.749637	1.151810	2.015249	0.0523116	0.6980762	15
46	0.496469	0.868054	0.571933	1.748456	1.152001	2.014224	0.0524067	0.6971864	14
47	0.496722	0.867910	0.572319	1.747277	1.152193	2.013200	0.0525019	0.6962977	13
48	0.496974	0.867765	0.572705	1.746098	1.152385	2.012178	0.0525973	0.6954102	12
49	0.497226	0.867621	0.573092	1.744921	1.152577	2.011156	0.0526928	0.6945239	11
50	0.497479	0.867476	0.573478	1.743745	1.152769	2.010136	0.0527884	0.6936389	10
51	0.497731	0.867331	0.573865	1.742571	1.152962	2.009117	0.0528841	0.6927550	9
52	0.497983	0.867187	0.574252	1.741397	1.153154	2.008099	0.0529799	0.6918723	8
53	0.498236	0.867042	0.574638	1.740225	1.153347	2.007083	0.0530759	0.6909907	7
54	0.498488	0.866897	0.575026	1.739053	1.153540	2.006067	0.0531721	0.6901104	6
55	0.498740	0.866752	0.575413	1.737883	1.153733	2.005053	0.0532683	0.6892313	5
56	0.498992	0.866607	0.575800	1.736714	1.153926	2.004040	0.0533647	0.6883533	4
57	0.499244	0.866461	0.576187	1.735547	1.154119	2.003028	0.0534612	0.6874765	3
58	0.499496	0.866316	0.576575	1.734380	1.154313	2.002018	0.0535578	0.6866009	2
59	0.499748	0.866171	0.576962	1.733215	1.154507	2.001008	0.0536546	0.6857265	1
60	0.500000	0.866025	0.577350	1.732051	1.154701	2.000000	0.0537515	0.6848533	0
Minutes	Cosine	Sine	Cotangent	Tangent	Cosecant	Secant	Read Down	60°–61° Involute	Minutes

↑ **119° or 299°** **60° or 240°** ↑

↓ 30° or 210° **Trigonometric and Involute Functions** 149° or 329° ↓

Minutes	Sine	Cosine	Tangent	Cotangent	Secant	Cosecant	Involute 30°–31°	Read Up	Minutes
0	0.500000	0.866025	0.577350	1.732051	1.154701	2.000000	0.0537515	0.6848533	60
1	0.500252	0.865880	0.577738	1.730888	1.154895	1.998993	0.0538485	0.6839812	59
2	0.500504	0.865734	0.578126	1.729726	1.155089	1.997987	0.0539457	0.6831103	58
3	0.500756	0.865589	0.578514	1.728565	1.155283	1.996982	0.0540430	0.6822405	57
4	0.501007	0.865443	0.578903	1.727406	1.155478	1.995979	0.0541404	0.6813720	56
5	0.501259	0.865297	0.579291	1.726248	1.155672	1.994976	0.0542379	0.6805045	55
6	0.501511	0.865151	0.579680	1.725091	1.155867	1.993975	0.0543356	0.6796383	54
7	0.501762	0.865006	0.580068	1.723935	1.156062	1.992975	0.0544334	0.6787732	53
8	0.502014	0.864860	0.580457	1.722780	1.156257	1.991976	0.0545314	0.6779093	52
9	0.502266	0.864713	0.580846	1.721626	1.156452	1.990979	0.0546295	0.6770465	51
10	0.502517	0.864567	0.581235	1.720474	1.156648	1.989982	0.0547277	0.6761849	50
11	0.502769	0.864421	0.581625	1.719322	1.156844	1.988987	0.0548260	0.6753244	49
12	0.503020	0.864275	0.582014	1.718172	1.157039	1.987993	0.0549245	0.6744651	48
13	0.503271	0.864128	0.582403	1.717023	1.157235	1.987000	0.0550231	0.6736070	47
14	0.503523	0.863982	0.582793	1.715875	1.157432	1.986008	0.0551218	0.6727500	46
15	0.503774	0.863836	0.583183	1.714728	1.157628	1.985017	0.0552207	0.6718941	45
16	0.504025	0.863689	0.583573	1.713583	1.157824	1.984028	0.0553197	0.6710394	44
17	0.504276	0.863542	0.583963	1.712438	1.158021	1.983039	0.0554188	0.6701858	43
18	0.504528	0.863396	0.584353	1.711295	1.158218	1.982052	0.0555181	0.6693333	42
19	0.504779	0.863249	0.584743	1.710153	1.158415	1.981066	0.0556175	0.6684820	41
20	0.505030	0.863102	0.585134	1.709012	1.158612	1.980081	0.0557170	0.6676319	40
21	0.505281	0.862955	0.585524	1.707872	1.158809	1.979097	0.0558166	0.6667828	39
22	0.505532	0.862808	0.585915	1.706733	1.159007	1.978115	0.0559164	0.6659349	38
23	0.505783	0.862661	0.586306	1.705595	1.159204	1.977133	0.0560164	0.6650881	37
24	0.506034	0.862514	0.586697	1.704459	1.159402	1.976153	0.0561164	0.6642425	36
25	0.506285	0.862366	0.587088	1.703323	1.159600	1.975174	0.0562166	0.6633980	35
26	0.506535	0.862219	0.587479	1.702189	1.159798	1.974195	0.0563169	0.6625546	34
27	0.506786	0.862072	0.587870	1.701056	1.159996	1.973218	0.0564174	0.6617123	33
28	0.507037	0.861924	0.588262	1.699924	1.160195	1.972243	0.0565180	0.6608712	32
29	0.507288	0.861777	0.588653	1.698793	1.160393	1.971268	0.0566187	0.6600311	31
30	0.507538	0.861629	0.589045	1.697663	1.160592	1.970294	0.0567196	0.6591922	30
31	0.507789	0.861481	0.589437	1.696534	1.160791	1.969322	0.0568206	0.6583544	29
32	0.508040	0.861334	0.589829	1.695407	1.160990	1.968351	0.0569217	0.6575177	28
33	0.508290	0.861186	0.590221	1.694280	1.161189	1.967381	0.0570230	0.6566822	27
34	0.508541	0.861038	0.590613	1.693155	1.161389	1.966411	0.0571244	0.6558477	26
35	0.508791	0.860890	0.591006	1.692031	1.161589	1.965444	0.0572259	0.6550143	25
36	0.509041	0.860742	0.591398	1.690908	1.161788	1.964477	0.0573276	0.6541821	24
37	0.509292	0.860594	0.591791	1.689786	1.161988	1.963511	0.0574294	0.6533509	23
38	0.509542	0.860446	0.592184	1.688665	1.162188	1.962546	0.0575313	0.6525209	22
39	0.509792	0.860297	0.592577	1.687545	1.162389	1.961583	0.0576334	0.6516919	21
40	0.510043	0.860149	0.592970	1.686426	1.162589	1.960621	0.0577356	0.6508641	20
41	0.510293	0.860001	0.593363	1.685308	1.162790	1.959659	0.0578380	0.6500374	19
42	0.510543	0.859852	0.593757	1.684192	1.162990	1.958699	0.0579405	0.6492117	18
43	0.510793	0.859704	0.594150	1.683077	1.163191	1.957740	0.0580431	0.6483871	17
44	0.511043	0.859555	0.594544	1.681962	1.163393	1.956782	0.0581458	0.6475637	16
45	0.511293	0.859406	0.594937	1.680849	1.163594	1.955825	0.0582487	0.6467413	15
46	0.511543	0.859258	0.595331	1.679737	1.163795	1.954870	0.0583518	0.6459200	14
47	0.511793	0.859109	0.595725	1.678626	1.163997	1.953915	0.0584549	0.6450998	13
48	0.512043	0.858960	0.596120	1.677516	1.164199	1.952961	0.0585582	0.6442807	12
49	0.512293	0.858811	0.596514	1.676407	1.164401	1.952009	0.0586617	0.6434627	11
50	0.512543	0.858662	0.596908	1.675299	1.164603	1.951058	0.0587652	0.6426457	10
51	0.512792	0.858513	0.597303	1.674192	1.164805	1.950107	0.0588690	0.6418298	9
52	0.513042	0.858364	0.597698	1.673086	1.165008	1.949158	0.0589728	0.6410150	8
53	0.513292	0.858214	0.598093	1.671982	1.165210	1.948210	0.0590768	0.6402013	7
54	0.513541	0.858065	0.598488	1.670878	1.165413	1.947263	0.0591809	0.6393887	6
55	0.513791	0.857915	0.598883	1.669776	1.165616	1.946317	0.0592852	0.6385771	5
56	0.514040	0.857766	0.599278	1.668674	1.165819	1.945373	0.0593896	0.6377666	4
57	0.514290	0.857616	0.599674	1.667574	1.166022	1.944429	0.0594941	0.6369571	3
58	0.514539	0.857467	0.600069	1.666475	1.166226	1.943486	0.0595988	0.6361488	2
59	0.514789	0.857317	0.600465	1.665377	1.166430	1.942545	0.0597036	0.6353415	1
60	0.515038	0.857167	0.600861	1.664279	1.166633	1.941604	0.0598086	0.6345352	0
Minutes	Cosine	Sine	Cotangent	Tangent	Cosecant	Secant	Read Down	59°–60° Involute	Minutes

↑ 120° or 300° 59° or 239° ↑

↓ 31° or 211° Trigonometric and Involute Functions 148° or 328° ↓

Minutes	Sine	Cosine	Tangent	Cotangent	Secant	Cosecant	Involute 31°–32°	Read Up	Minutes
0	0.515038	0.857167	0.600861	1.664279	1.166633	1.941604	0.0598086	0.6345352	60
1	0.515287	0.857017	0.601257	1.663183	1.166837	1.940665	0.0599136	0.6337300	59
2	0.515537	0.856868	0.601653	1.662088	1.167042	1.939726	0.0600189	0.6329259	58
3	0.515786	0.856718	0.602049	1.660994	1.167246	1.938789	0.0601242	0.6321229	57
4	0.516035	0.856567	0.602445	1.659902	1.167450	1.937853	0.0602297	0.6313209	56
5	0.516284	0.856417	0.602842	1.658810	1.167655	1.936918	0.0603354	0.6305199	55
6	0.516533	0.856267	0.603239	1.657719	1.167860	1.935983	0.0604412	0.6297200	54
7	0.516782	0.856117	0.603635	1.656629	1.168065	1.935050	0.0605471	0.6289212	53
8	0.517031	0.855966	0.604032	1.655541	1.168270	1.934119	0.0606532	0.6281234	52
9	0.517280	0.855816	0.604429	1.654453	1.168475	1.933188	0.0607594	0.6273266	51
10	0.517529	0.855665	0.604827	1.653366	1.168681	1.932258	0.0608657	0.6265309	50
11	0.517778	0.855515	0.605224	1.652281	1.168887	1.931329	0.0609722	0.6257363	49
12	0.518027	0.855364	0.605622	1.651196	1.169093	1.930401	0.0610788	0.6249427	48
13	0.518276	0.855214	0.606019	1.650113	1.169299	1.929475	0.0611856	0.6241501	47
14	0.518525	0.855063	0.606417	1.649030	1.169505	1.928549	0.0612925	0.6233586	46
15	0.518773	0.854912	0.606815	1.647949	1.169711	1.927624	0.0613995	0.6225681	45
16	0.519022	0.854761	0.607213	1.646869	1.169918	1.926701	0.0615067	0.6217786	44
17	0.519271	0.854610	0.607611	1.645789	1.170124	1.925778	0.0616140	0.6209902	43
18	0.519519	0.854459	0.608010	1.644711	1.170331	1.924857	0.0617215	0.6202028	42
19	0.519768	0.854308	0.608408	1.643634	1.170538	1.923937	0.0618291	0.6194164	41
20	0.520016	0.854156	0.608807	1.642558	1.170746	1.923017	0.0619368	0.6186311	40
21	0.520265	0.854005	0.609205	1.641482	1.170953	1.922099	0.0620447	0.6178468	39
22	0.520513	0.853854	0.609604	1.640408	1.171161	1.921182	0.0621527	0.6170635	38
23	0.520761	0.853702	0.610003	1.639335	1.171368	1.920265	0.0622609	0.6162813	37
24	0.521010	0.853551	0.610403	1.638263	1.171576	1.919350	0.0623692	0.6155000	36
25	0.521258	0.853399	0.610802	1.637192	1.171785	1.918436	0.0624777	0.6147198	35
26	0.521506	0.853248	0.611201	1.636122	1.171993	1.917523	0.0625863	0.6139407	34
27	0.521754	0.853096	0.611601	1.635053	1.172201	1.916611	0.0626950	0.6131625	33
28	0.522002	0.852944	0.612001	1.633985	1.172410	1.915700	0.0628039	0.6123853	32
29	0.522251	0.852792	0.612401	1.632918	1.172619	1.914790	0.0629129	0.6116092	31
30	0.522499	0.852640	0.612801	1.631852	1.172828	1.913881	0.0630221	0.6108341	30
31	0.522747	0.852488	0.613201	1.630787	1.173037	1.912973	0.0631314	0.6100600	29
32	0.522995	0.852336	0.613601	1.629723	1.173246	1.912066	0.0632408	0.6092869	28
33	0.523242	0.852184	0.614002	1.628660	1.173456	1.911160	0.0633504	0.6085148	27
34	0.523490	0.852032	0.614402	1.627598	1.173665	1.910255	0.0634602	0.6077437	26
35	0.523738	0.851879	0.614803	1.626537	1.173875	1.909351	0.0635700	0.6069736	25
36	0.523986	0.851727	0.615204	1.625477	1.174085	1.908448	0.0636801	0.6062045	24
37	0.524234	0.851574	0.615605	1.624418	1.174295	1.907546	0.0637902	0.6054364	23
38	0.524481	0.851422	0.616006	1.623360	1.174506	1.906646	0.0639005	0.6046694	22
39	0.524729	0.851269	0.616408	1.622303	1.174716	1.905746	0.0640110	0.6039033	21
40	0.524977	0.851117	0.616809	1.621247	1.174927	1.904847	0.0641216	0.6031382	20
41	0.525224	0.850964	0.617211	1.620192	1.175138	1.903949	0.0642323	0.6023741	19
42	0.525472	0.850811	0.617613	1.619138	1.175349	1.903052	0.0643432	0.6016110	18
43	0.525719	0.850658	0.618015	1.618085	1.175560	1.902156	0.0644542	0.6008489	17
44	0.525967	0.850505	0.618417	1.617033	1.175772	1.901262	0.0645654	0.6000878	16
45	0.526214	0.850352	0.618819	1.615982	1.175983	1.900368	0.0646767	0.5993277	15
46	0.526461	0.850199	0.619221	1.614932	1.176195	1.899475	0.0647882	0.5985686	14
47	0.526709	0.850046	0.619624	1.613883	1.176407	1.898583	0.0648998	0.5978104	13
48	0.526956	0.849893	0.620026	1.612835	1.176619	1.897692	0.0650116	0.5970533	12
49	0.527203	0.849739	0.620429	1.611788	1.176831	1.896803	0.0651235	0.5962971	11
50	0.527450	0.849586	0.620832	1.610742	1.177044	1.895914	0.0652355	0.5955419	10
51	0.527697	0.849433	0.621235	1.609697	1.177257	1.895026	0.0653477	0.5947877	9
52	0.527944	0.849279	0.621638	1.608653	1.177469	1.894139	0.0654600	0.5940344	8
53	0.528191	0.849125	0.622042	1.607609	1.177682	1.893253	0.0655725	0.5932822	7
54	0.528438	0.848972	0.622445	1.606567	1.177896	1.892368	0.0656851	0.5925309	6
55	0.528685	0.848818	0.622849	1.605526	1.178109	1.891485	0.0657979	0.5917806	5
56	0.528932	0.848664	0.623253	1.604486	1.178322	1.890602	0.0659108	0.5910312	4
57	0.529179	0.848510	0.623657	1.603446	1.178536	1.889720	0.0660239	0.5902829	3
58	0.529426	0.848356	0.624061	1.602408	1.178750	1.888839	0.0661371	0.5895355	2
59	0.529673	0.848202	0.624465	1.601371	1.178964	1.887959	0.0662505	0.5887890	1
60	0.529919	0.848048	0.624869	1.600335	1.179178	1.887080	0.0663640	0.5880436	0
Minutes	Cosine	Sine	Cotangent	Tangent	Cosecant	Secant	Read Down	58°–59° Involute	Minutes

↑ 121° or 301° 58° or 238° ↑

↓ 32° or 212° Trigonometric and Involute Functions 147° or 327° ↓

Minutes	Sine	Cosine	Tangent	Cotangent	Secant	Cosecant	Involute 32°–33°	Read Up	Minutes
0	0.529919	0.848048	0.624869	1.600335	1.179178	1.887080	0.0663640	0.5880436	60
1	0.530166	0.847894	0.625274	1.599299	1.179393	1.886202	0.0664776	0.5872991	59
2	0.530413	0.847740	0.625679	1.598265	1.179607	1.885325	0.0665914	0.5865555	58
3	0.530659	0.847585	0.626083	1.597231	1.179822	1.884449	0.0667054	0.5858129	57
4	0.530906	0.847431	0.626488	1.596199	1.180037	1.883574	0.0668195	0.5850713	56
5	0.531152	0.847276	0.626894	1.595167	1.180252	1.882700	0.0669337	0.5843307	55
6	0.531399	0.847122	0.627299	1.594137	1.180468	1.881827	0.0670481	0.5835910	54
7	0.531645	0.846967	0.627704	1.593107	1.180683	1.880954	0.0671627	0.5828522	53
8	0.531891	0.846813	0.628110	1.592078	1.180899	1.880083	0.0672774	0.5821144	52
9	0.532138	0.846658	0.628516	1.591051	1.181115	1.879213	0.0673922	0.5813776	51
10	0.532384	0.846503	0.628921	1.590024	1.181331	1.878344	0.0675072	0.5806417	50
11	0.532630	0.846348	0.629327	1.588998	1.181547	1.877476	0.0676223	0.5799067	49
12	0.532876	0.846193	0.629734	1.587973	1.181763	1.876608	0.0677376	0.5791727	48
13	0.533122	0.846038	0.630140	1.586949	1.181980	1.875742	0.0678530	0.5784397	47
14	0.533368	0.845883	0.630546	1.585926	1.182197	1.874876	0.0679686	0.5777076	46
15	0.533615	0.845728	0.630953	1.584904	1.182414	1.874012	0.0680843	0.5769764	45
16	0.533861	0.845573	0.631360	1.583883	1.182631	1.873148	0.0682002	0.5762462	44
17	0.534106	0.845417	0.631767	1.582863	1.182848	1.872286	0.0683162	0.5755169	43
18	0.534352	0.845262	0.632174	1.581844	1.183065	1.871424	0.0684324	0.5747886	42
19	0.534598	0.845106	0.632581	1.580825	1.183283	1.870564	0.0685487	0.5740612	41
20	0.534844	0.844951	0.632988	1.579808	1.183501	1.869704	0.0686652	0.5733347	40
21	0.535090	0.844795	0.633396	1.578792	1.183719	1.868845	0.0687818	0.5726092	39
22	0.535335	0.844640	0.633804	1.577776	1.183937	1.867987	0.0688986	0.5718846	38
23	0.535581	0.844484	0.634211	1.576761	1.184155	1.867131	0.0690155	0.5711609	37
24	0.535827	0.844328	0.634619	1.575748	1.184374	1.866275	0.0691326	0.5704382	36
25	0.536072	0.844172	0.635027	1.574735	1.184593	1.865420	0.0692498	0.5697164	35
26	0.536318	0.844016	0.635436	1.573723	1.184812	1.864566	0.0693672	0.5689955	34
27	0.536563	0.843860	0.635844	1.572713	1.185031	1.863713	0.0694848	0.5682756	33
28	0.536809	0.843704	0.636253	1.571703	1.185250	1.862860	0.0696024	0.5675565	32
29	0.537054	0.843548	0.636661	1.570694	1.185469	1.862009	0.0697203	0.5668384	31
30	0.537300	0.843391	0.637070	1.569686	1.185689	1.861159	0.0698383	0.5661213	30
31	0.537545	0.843235	0.637479	1.568678	1.185909	1.860310	0.0699564	0.5654050	29
32	0.537790	0.843079	0.637888	1.567672	1.186129	1.859461	0.0700747	0.5646896	28
33	0.538035	0.842922	0.638298	1.566667	1.186349	1.858614	0.0701931	0.5639752	27
34	0.538281	0.842766	0.638707	1.565662	1.186569	1.857767	0.0703117	0.5632617	26
35	0.538526	0.842609	0.639117	1.564659	1.186790	1.856922	0.0704304	0.5625491	25
36	0.538771	0.842452	0.639527	1.563656	1.187011	1.856077	0.0705493	0.5618374	24
37	0.539016	0.842296	0.639937	1.562655	1.187232	1.855233	0.0706684	0.5611267	23
38	0.539261	0.842139	0.640347	1.561654	1.187453	1.854390	0.0707876	0.5604168	22
39	0.539506	0.841982	0.640757	1.560654	1.187674	1.853548	0.0709069	0.5597078	21
40	0.539751	0.841825	0.641167	1.559655	1.187895	1.852707	0.0710265	0.5589998	20
41	0.539996	0.841668	0.641578	1.558657	1.188117	1.851867	0.0711461	0.5582927	19
42	0.540240	0.841511	0.641989	1.557660	1.188339	1.851028	0.0712659	0.5575864	18
43	0.540485	0.841354	0.642399	1.556664	1.188561	1.850190	0.0713859	0.5568811	17
44	0.540730	0.841196	0.642810	1.555669	1.188783	1.849352	0.0715060	0.5561767	16
45	0.540974	0.841039	0.643222	1.554674	1.189005	1.848516	0.0716263	0.5554731	15
46	0.541219	0.840882	0.643633	1.553681	1.189228	1.847681	0.0717467	0.5547705	14
47	0.541464	0.840724	0.644044	1.552688	1.189451	1.846846	0.0718673	0.5540688	13
48	0.541708	0.840567	0.644456	1.551696	1.189674	1.846012	0.0719880	0.5533679	12
49	0.541953	0.840409	0.644868	1.550705	1.189897	1.845179	0.0721089	0.5526680	11
50	0.542197	0.840251	0.645280	1.549715	1.190120	1.844348	0.0722300	0.5519689	10
51	0.542442	0.840094	0.645692	1.548726	1.190344	1.843517	0.0723512	0.5512708	9
52	0.542686	0.839936	0.646104	1.547738	1.190567	1.842687	0.0724725	0.5505735	8
53	0.542930	0.839778	0.646516	1.546751	1.190791	1.841857	0.0725940	0.5498771	7
54	0.543174	0.839620	0.646929	1.545765	1.191015	1.841029	0.0727157	0.5491816	6
55	0.543419	0.839462	0.647342	1.544779	1.191239	1.840202	0.0728375	0.5484870	5
56	0.543663	0.839304	0.647755	1.543795	1.191464	1.839375	0.0729595	0.5477933	4
57	0.543907	0.839146	0.648168	1.542811	1.191688	1.838550	0.0730816	0.5471005	3
58	0.544151	0.838987	0.648581	1.541828	1.191913	1.837725	0.0732039	0.5464085	2
59	0.544395	0.838829	0.648994	1.540846	1.192138	1.836901	0.0733263	0.5457175	1
60	0.544639	0.838671	0.649408	1.539865	1.192363	1.836078	0.0734489	0.5450273	0
Minutes	Cosine	Sine	Cotangent	Tangent	Cosecant	Secant	Read Down	57°–58° Involute	Minutes

↑ 122° or 302° 57° or 237° ↑

↓ 33° or 213° Trigonometric and Involute Functions 146° or 326° ↓

Minutes	Sine	Cosine	Tangent	Cotangent	Secant	Cosecant	Involute 33°–34°	Read Up	Minutes
0	0.544639	0.838671	0.649408	1.539865	1.192363	1.836078	0.0734489	0.5450273	60
1	0.544883	0.838512	0.649821	1.538885	1.192589	1.835256	0.0735717	0.5443380	59
2	0.545127	0.838354	0.650235	1.537905	1.192814	1.834435	0.0736946	0.5436495	58
3	0.545371	0.838195	0.650649	1.536927	1.193040	1.833615	0.0738177	0.5429620	57
4	0.545615	0.838036	0.651063	1.535949	1.193266	1.832796	0.0739409	0.5422753	56
5	0.545858	0.837878	0.651477	1.534973	1.193492	1.831977	0.0740643	0.5415895	55
6	0.546102	0.837719	0.651892	1.533997	1.193718	1.831160	0.0741878	0.5409046	54
7	0.546346	0.837560	0.652306	1.533022	1.193945	1.830343	0.0743115	0.5402205	53
8	0.546589	0.837401	0.652721	1.532048	1.194171	1.829527	0.0744354	0.5395373	52
9	0.546833	0.837242	0.653136	1.531075	1.194398	1.828713	0.0745594	0.5388550	51
10	0.547076	0.837083	0.653551	1.530102	1.194625	1.827899	0.0746835	0.5381735	50
11	0.547320	0.836924	0.653966	1.529131	1.194852	1.827085	0.0748079	0.5374929	49
12	0.547563	0.836764	0.654382	1.528160	1.195080	1.826273	0.0749324	0.5368132	48
13	0.547807	0.836605	0.654797	1.527190	1.195307	1.825462	0.0750570	0.5361343	47
14	0.548050	0.836446	0.655213	1.526222	1.195535	1.824651	0.0751818	0.5354563	46
15	0.548293	0.836286	0.655629	1.525253	1.195763	1.823842	0.0753068	0.5347791	45
16	0.548536	0.836127	0.656045	1.524286	1.195991	1.823033	0.0754319	0.5341028	44
17	0.548780	0.835967	0.656461	1.523320	1.196219	1.822225	0.0755571	0.5334274	43
18	0.549023	0.835807	0.656877	1.522355	1.196448	1.821418	0.0756826	0.5327528	42
19	0.549266	0.835648	0.657294	1.521390	1.196677	1.820612	0.0758082	0.5320791	41
20	0.549509	0.835488	0.657710	1.520426	1.196906	1.819806	0.0759339	0.5314062	40
21	0.549752	0.835328	0.658127	1.519463	1.197135	1.819002	0.0760598	0.5307342	39
22	0.549995	0.835168	0.658544	1.518501	1.197364	1.818199	0.0761859	0.5300630	38
23	0.550238	0.835008	0.658961	1.517540	1.197593	1.817396	0.0763121	0.5293927	37
24	0.550481	0.834848	0.659379	1.516580	1.197823	1.816594	0.0764385	0.5287232	36
25	0.550724	0.834688	0.659796	1.515620	1.198053	1.815793	0.0765651	0.5280546	35
26	0.550966	0.834527	0.660214	1.514661	1.198283	1.814993	0.0766918	0.5273868	34
27	0.551209	0.834367	0.660631	1.513704	1.198513	1.814194	0.0768187	0.5267199	33
28	0.551452	0.834207	0.661049	1.512747	1.198744	1.813395	0.0769457	0.5260538	32
29	0.551694	0.834046	0.661467	1.511790	1.198974	1.812598	0.0770729	0.5253886	31
30	0.551937	0.833886	0.661886	1.510835	1.199205	1.811801	0.0772003	0.5247242	30
31	0.552180	0.833725	0.662304	1.509881	1.199436	1.811005	0.0773278	0.5240606	29
32	0.552422	0.833565	0.662723	1.508927	1.199667	1.810210	0.0774555	0.5233979	28
33	0.552664	0.833404	0.663141	1.507974	1.199898	1.809416	0.0775833	0.5227360	27
34	0.552907	0.833243	0.663560	1.507022	1.200130	1.808623	0.0777113	0.5220749	26
35	0.553149	0.833082	0.663979	1.506071	1.200362	1.807830	0.0778395	0.5214147	25
36	0.553392	0.832921	0.664398	1.505121	1.200594	1.807039	0.0779678	0.5207553	24
37	0.553634	0.832760	0.664818	1.504172	1.200826	1.806248	0.0780963	0.5200967	23
38	0.553876	0.832599	0.665237	1.503223	1.201058	1.805458	0.0782249	0.5194390	22
39	0.554118	0.832438	0.665657	1.502275	1.201291	1.804669	0.0783537	0.5187821	21
40	0.554360	0.832277	0.666077	1.501328	1.201523	1.803881	0.0784827	0.5181260	20
41	0.554602	0.832115	0.666497	1.500382	1.201756	1.803094	0.0786118	0.5174708	19
42	0.554844	0.831954	0.666917	1.499437	1.201989	1.802307	0.0787411	0.5168164	18
43	0.555086	0.831793	0.667337	1.498492	1.202223	1.801521	0.0788706	0.5161628	17
44	0.555328	0.831631	0.667758	1.497549	1.202456	1.800736	0.0790002	0.5155100	16
45	0.555570	0.831470	0.668179	1.496606	1.202690	1.799952	0.0791300	0.5148581	15
46	0.555812	0.831308	0.668599	1.495664	1.202924	1.799169	0.0792600	0.5142069	14
47	0.556054	0.831146	0.669020	1.494723	1.203158	1.798387	0.0793901	0.5135566	13
48	0.556296	0.830984	0.669442	1.493782	1.203392	1.797605	0.0795204	0.5129071	12
49	0.556537	0.830823	0.669863	1.492843	1.203626	1.796825	0.0796508	0.5122585	11
50	0.556779	0.830661	0.670284	1.491904	1.203861	1.796045	0.0797814	0.5116106	10
51	0.557021	0.830499	0.670706	1.490966	1.204096	1.795266	0.0799122	0.5109635	9
52	0.557262	0.830337	0.671128	1.490029	1.204331	1.794488	0.0800431	0.5103173	8
53	0.557504	0.830174	0.671550	1.489092	1.204566	1.793710	0.0801742	0.5096719	7
54	0.557745	0.830012	0.671972	1.488157	1.204801	1.792934	0.0803055	0.5090273	6
55	0.557987	0.829850	0.672394	1.487222	1.205037	1.792158	0.0804369	0.5083835	5
56	0.558228	0.829688	0.672817	1.486288	1.205273	1.791383	0.0805685	0.5077405	4
57	0.558469	0.829525	0.673240	1.485355	1.205509	1.790609	0.0807003	0.5070983	3
58	0.558710	0.829363	0.673662	1.484423	1.205745	1.789836	0.0808322	0.5064569	2
59	0.558952	0.829200	0.674085	1.483492	1.205981	1.789063	0.0809643	0.5058164	1
60	0.559193	0.829038	0.674509	1.482561	1.206218	1.788292	0.0810966	0.5051766	0
Minutes	Cosine	Sine	Cotangent	Tangent	Cosecant	Secant	Read Down	56°–57° Involute	Minutes

↑ 123° or 303° 56° or 236° ↑

↓ **34° or 214°** **Trigonometric and Involute Functions** **145° or 325°** ↓

Minutes	Sine	Cosine	Tangent	Cotangent	Secant	Cosecant	Involute 34°–35°	Read Up	Minutes
0	0.559193	0.829038	0.674509	1.482561	1.206218	1.788292	0.0810966	0.5051766	60
1	0.559434	0.828875	0.674932	1.481631	1.206455	1.787521	0.0812290	0.5045376	59
2	0.559675	0.828712	0.675355	1.480702	1.206692	1.786751	0.0813616	0.5038995	58
3	0.559916	0.828549	0.675779	1.479774	1.206929	1.785982	0.0814943	0.5032621	57
4	0.560157	0.828386	0.676203	1.478846	1.207166	1.785213	0.0816273	0.5026255	56
5	0.560398	0.828223	0.676627	1.477920	1.207404	1.784446	0.0817604	0.5019897	55
6	0.560639	0.828060	0.677051	1.476994	1.207641	1.783679	0.0818936	0.5013548	54
7	0.560880	0.827897	0.677475	1.476069	1.207879	1.782913	0.0820271	0.5007206	53
8	0.561121	0.827734	0.677900	1.475144	1.208118	1.782148	0.0821606	0.5000872	52
9	0.561361	0.827571	0.678324	1.474221	1.208356	1.781384	0.0822944	0.4994546	51
10	0.561602	0.827407	0.678749	1.473298	1.208594	1.780620	0.0824283	0.4988228	50
11	0.561843	0.827244	0.679174	1.472376	1.208833	1.779857	0.0825624	0.4981918	49
12	0.562083	0.827081	0.679599	1.471455	1.209072	1.779095	0.0826967	0.4975616	48
13	0.562324	0.826917	0.680025	1.470535	1.209311	1.778334	0.0828311	0.4969322	47
14	0.562564	0.826753	0.680450	1.469615	1.209550	1.777574	0.0829657	0.4963035	46
15	0.562805	0.826590	0.680876	1.468697	1.209790	1.776815	0.0831005	0.4956757	45
16	0.563045	0.826426	0.681302	1.467779	1.210030	1.776056	0.0832354	0.4950486	44
17	0.563286	0.826262	0.681728	1.466862	1.210270	1.775298	0.0833705	0.4944223	43
18	0.563526	0.826098	0.682154	1.465945	1.210510	1.774541	0.0835058	0.4937968	42
19	0.563766	0.825934	0.682580	1.465030	1.210750	1.773785	0.0836413	0.4931721	41
20	0.564007	0.825770	0.683007	1.464115	1.210991	1.773029	0.0837769	0.4925481	40
21	0.564247	0.825606	0.683433	1.463201	1.211231	1.772274	0.0839127	0.4919249	39
22	0.564487	0.825442	0.683860	1.462287	1.211472	1.771520	0.0840486	0.4913026	38
23	0.564727	0.825278	0.684287	1.461375	1.211713	1.770767	0.0841847	0.4906809	37
24	0.564967	0.825113	0.684714	1.460463	1.211954	1.770015	0.0843210	0.4900601	36
25	0.565207	0.824949	0.685142	1.459552	1.212196	1.769263	0.0844575	0.4894400	35
26	0.565447	0.824785	0.685569	1.458642	1.212438	1.768513	0.0845941	0.4888207	34
27	0.565687	0.824620	0.685997	1.457733	1.212680	1.767763	0.0847309	0.4882022	33
28	0.565927	0.824456	0.686425	1.456824	1.212922	1.767013	0.0848679	0.4875845	32
29	0.566166	0.824291	0.686853	1.455916	1.213164	1.766265	0.0850050	0.4869675	31
30	0.566406	0.824126	0.687281	1.455009	1.213406	1.765517	0.0851424	0.4863513	30
31	0.566646	0.823961	0.687709	1.454103	1.213649	1.764770	0.0852799	0.4857359	29
32	0.566886	0.823797	0.688138	1.453197	1.213892	1.764024	0.0854175	0.4851212	28
33	0.567125	0.823632	0.688567	1.452292	1.214135	1.763279	0.0855553	0.4845073	27
34	0.567365	0.823467	0.688995	1.451388	1.214378	1.762535	0.0856933	0.4838941	26
35	0.567604	0.823302	0.689425	1.450485	1.214622	1.761791	0.0858315	0.4832817	25
36	0.567844	0.823136	0.689854	1.449583	1.214866	1.761048	0.0859699	0.4826701	24
37	0.568083	0.822971	0.690283	1.448681	1.215109	1.760306	0.0861084	0.4820593	23
38	0.568323	0.822806	0.690713	1.447780	1.215354	1.759564	0.0862471	0.4814492	22
39	0.568562	0.822641	0.691143	1.446880	1.215598	1.758824	0.0863859	0.4808398	21
40	0.568801	0.822475	0.691572	1.445980	1.215842	1.758084	0.0865250	0.4802312	20
41	0.569040	0.822310	0.692003	1.445081	1.216087	1.757345	0.0866642	0.4796234	19
42	0.569280	0.822144	0.692433	1.444183	1.216332	1.756606	0.0868036	0.4790163	18
43	0.569519	0.821978	0.692863	1.443286	1.216577	1.755869	0.0869431	0.4784100	17
44	0.569758	0.821813	0.693294	1.442390	1.216822	1.755132	0.0870829	0.4778044	16
45	0.569997	0.821647	0.693725	1.441494	1.217068	1.754396	0.0872228	0.4771996	15
46	0.570236	0.821481	0.694156	1.440599	1.217313	1.753661	0.0873628	0.4765956	14
47	0.570475	0.821315	0.694587	1.439705	1.217559	1.752926	0.0875031	0.4759923	13
48	0.570714	0.821149	0.695018	1.438811	1.217805	1.752192	0.0876435	0.4753897	12
49	0.570952	0.820983	0.695450	1.437919	1.218052	1.751459	0.0877841	0.4747879	11
50	0.571191	0.820817	0.695881	1.437027	1.218298	1.750727	0.0879249	0.4741868	10
51	0.571430	0.820651	0.696313	1.436136	1.218545	1.749996	0.0880659	0.4735865	9
52	0.571669	0.820485	0.696745	1.435245	1.218792	1.749265	0.0882070	0.4729869	8
53	0.571907	0.820318	0.697177	1.434355	1.219039	1.748535	0.0883483	0.4723881	7
54	0.572146	0.820152	0.697610	1.433466	1.219286	1.747806	0.0884898	0.4717900	6
55	0.572384	0.819985	0.698042	1.432578	1.219534	1.747078	0.0886314	0.4711926	5
56	0.572623	0.819819	0.698475	1.431691	1.219782	1.746350	0.0887732	0.4705960	4
57	0.572861	0.819652	0.698908	1.430804	1.220030	1.745623	0.0889152	0.4700001	3
58	0.573100	0.819486	0.699341	1.429918	1.220278	1.744897	0.0890574	0.4694050	2
59	0.573338	0.819319	0.699774	1.429033	1.220526	1.744171	0.0891998	0.4688106	1
60	0.573576	0.819152	0.700208	1.428148	1.220775	1.743447	0.0893423	0.4682169	0
Minutes	Cosine	Sine	Cotangent	Tangent	Cosecant	Secant	Read Down	55°–56° Involute	Minutes

↑ **124° or 304°** **55° or 235°** ↑

↓ 35° or 215° **Trigonometric and Involute Functions** 144° or 324° ↓

Minutes	Sine	Cosine	Tangent	Cotangent	Secant	Cosecant	Involute 35°–36°	Read Up	Minutes
0	0.573576	0.819152	0.700208	1.428148	1.220775	1.743447	0.0893423	0.4682169	60
1	0.573815	0.818985	0.700641	1.427264	1.221023	1.742723	0.0894850	0.4676240	59
2	0.574053	0.818818	0.701075	1.426381	1.221272	1.742000	0.0896279	0.4670318	58
3	0.574291	0.818651	0.701509	1.425499	1.221521	1.741277	0.0897710	0.4664403	57
4	0.574529	0.818484	0.701943	1.424617	1.221771	1.740556	0.0899142	0.4658496	56
5	0.574767	0.818317	0.702377	1.423736	1.222020	1.739835	0.0900576	0.4652596	55
6	0.575005	0.818150	0.702812	1.422856	1.222270	1.739115	0.0902012	0.4646703	54
7	0.575243	0.817982	0.703246	1.421977	1.222520	1.738395	0.0903450	0.4640818	53
8	0.575481	0.817815	0.703681	1.421098	1.222770	1.737676	0.0904889	0.4634940	52
9	0.575719	0.817648	0.704116	1.420220	1.223021	1.736958	0.0906331	0.4629069	51
10	0.575957	0.817480	0.704551	1.419343	1.223271	1.736241	0.0907774	0.4623205	50
11	0.576195	0.817313	0.704987	1.418466	1.223522	1.735525	0.0909218	0.4617349	49
12	0.576432	0.817145	0.705422	1.417590	1.223773	1.734809	0.0910665	0.4611499	48
13	0.576670	0.816977	0.705858	1.416715	1.224024	1.734094	0.0912113	0.4605657	47
14	0.576908	0.816809	0.706294	1.415841	1.224276	1.733380	0.0913564	0.4599823	46
15	0.577145	0.816642	0.706730	1.414967	1.224527	1.732666	0.0915016	0.4593995	45
16	0.577383	0.816474	0.707166	1.414094	1.224779	1.731953	0.0916469	0.4588175	44
17	0.577620	0.816306	0.707603	1.413222	1.225031	1.731241	0.0917925	0.4582361	43
18	0.577858	0.816138	0.708039	1.412351	1.225284	1.730530	0.0919382	0.4576555	42
19	0.578095	0.815969	0.708476	1.411480	1.225536	1.729819	0.0920842	0.4570757	41
20	0.578332	0.815801	0.708913	1.410610	1.225789	1.729110	0.0922303	0.4564965	40
21	0.578570	0.815633	0.709350	1.409740	1.226042	1.728400	0.0923765	0.4559180	39
22	0.578807	0.815465	0.709788	1.408872	1.226295	1.727692	0.0925230	0.4553403	38
23	0.579044	0.815296	0.710225	1.408004	1.226548	1.726984	0.0926696	0.4547632	37
24	0.579281	0.815128	0.710663	1.407137	1.226801	1.726277	0.0928165	0.4541869	36
25	0.579518	0.814959	0.711101	1.406270	1.227055	1.725571	0.0929635	0.4536113	35
26	0.579755	0.814791	0.711539	1.405404	1.227309	1.724866	0.0931106	0.4530364	34
27	0.579992	0.814622	0.711977	1.404539	1.227563	1.724161	0.0932580	0.4524622	33
28	0.580229	0.814453	0.712416	1.403675	1.227818	1.723457	0.0934055	0.4518887	32
29	0.580466	0.814284	0.712854	1.402811	1.228072	1.722753	0.0935533	0.4513159	31
30	0.580703	0.814116	0.713293	1.401948	1.228327	1.722051	0.0937012	0.4507439	30
31	0.580940	0.813947	0.713732	1.401086	1.228582	1.721349	0.0938493	0.4501725	29
32	0.581176	0.813778	0.714171	1.400224	1.228837	1.720648	0.0939975	0.4496018	28
33	0.581413	0.813608	0.714611	1.399364	1.229092	1.719947	0.0941460	0.4490318	27
34	0.581650	0.813439	0.715050	1.398503	1.229348	1.719247	0.0942946	0.4484626	26
35	0.581886	0.813270	0.715490	1.397644	1.229604	1.718548	0.0944435	0.4478940	25
36	0.582123	0.813101	0.715930	1.396785	1.229860	1.717850	0.0945925	0.4473261	24
37	0.582359	0.812931	0.716370	1.395927	1.230116	1.717152	0.0947417	0.4467589	23
38	0.582596	0.812762	0.716810	1.395070	1.230373	1.716456	0.0948910	0.4461924	22
39	0.582832	0.812592	0.717250	1.394213	1.230629	1.715759	0.0950406	0.4456267	21
40	0.583069	0.812423	0.717691	1.393357	1.230886	1.715064	0.0951903	0.4450616	20
41	0.583305	0.812253	0.718132	1.392502	1.231143	1.714369	0.0953402	0.4444972	19
42	0.583541	0.812084	0.718573	1.391647	1.231400	1.713675	0.0954904	0.4439335	18
43	0.583777	0.811914	0.719014	1.390793	1.231658	1.712982	0.0956406	0.4433705	17
44	0.584014	0.811744	0.719455	1.389940	1.231916	1.712289	0.0957911	0.4428081	16
45	0.584250	0.811574	0.719897	1.389088	1.232174	1.711597	0.0959418	0.4422465	15
46	0.584486	0.811404	0.720339	1.388236	1.232432	1.710906	0.0960926	0.4416856	14
47	0.584722	0.811234	0.720781	1.387385	1.232690	1.710215	0.0962437	0.4411253	13
48	0.584958	0.811064	0.721223	1.386534	1.232949	1.709525	0.0963949	0.4405657	12
49	0.585194	0.810894	0.721665	1.385684	1.233207	1.708836	0.0965463	0.4400069	11
50	0.585429	0.810723	0.722108	1.384835	1.233466	1.708148	0.0966979	0.4394487	10
51	0.585665	0.810553	0.722550	1.383987	1.233726	1.707460	0.0968496	0.4388911	9
52	0.585901	0.810383	0.722993	1.383139	1.233985	1.706773	0.0970016	0.4383343	8
53	0.586137	0.810212	0.723436	1.382292	1.234245	1.706087	0.0971537	0.4377782	7
54	0.586372	0.810042	0.723879	1.381446	1.234504	1.705401	0.0973061	0.4372227	6
55	0.586608	0.809871	0.724323	1.380600	1.234764	1.704716	0.0974586	0.4366679	5
56	0.586844	0.809700	0.724766	1.379755	1.235025	1.704032	0.0976113	0.4361138	4
57	0.587079	0.809530	0.725210	1.378911	1.235285	1.703348	0.0977642	0.4355604	3
58	0.587314	0.809359	0.725654	1.378067	1.235546	1.702665	0.0979173	0.4350076	2
59	0.587550	0.809188	0.726098	1.377224	1.235807	1.701983	0.0980705	0.4344555	1
60	0.587785	0.809017	0.726543	1.376382	1.236068	1.701302	0.0982240	0.4339041	0
Minutes	Cosine	Sine	Cotangent	Tangent	Cosecant	Secant	Read Down	54°–55° Involute	Minutes

↑ 125° or 305° 54° or 234° ↑

↓ 36° or 216° Trigonometric and Involute Functions 143° or 323° ↓

Minutes	Sine	Cosine	Tangent	Cotangent	Secant	Cosecant	Involute 36°–37°	Read Up	Minutes
0	0.587785	0.809017	0.726543	1.376382	1.236068	1.701302	0.0982240	0.4339041	60
1	0.588021	0.808846	0.726987	1.375540	1.236329	1.700621	0.0983776	0.4333534	59
2	0.588256	0.808675	0.727432	1.374699	1.236591	1.699941	0.0985315	0.4328033	58
3	0.588491	0.808504	0.727877	1.373859	1.236853	1.699261	0.0986855	0.4322540	57
4	0.588726	0.808333	0.728322	1.373019	1.237115	1.698582	0.0988397	0.4317052	56
5	0.588961	0.808161	0.728767	1.372181	1.237377	1.697904	0.0989941	0.4311572	55
6	0.589196	0.807990	0.729213	1.371342	1.237639	1.697227	0.0991487	0.4306098	54
7	0.589431	0.807818	0.729658	1.370505	1.237902	1.696550	0.0993035	0.4300631	53
8	0.589666	0.807647	0.730104	1.369668	1.238165	1.695874	0.0994584	0.4295171	52
9	0.589901	0.807475	0.730550	1.368832	1.238428	1.695199	0.0996136	0.4289717	51
10	0.590136	0.807304	0.730996	1.367996	1.238691	1.694524	0.0997689	0.4284270	50
11	0.590371	0.807132	0.731443	1.367161	1.238955	1.693850	0.0999244	0.4278830	49
12	0.590606	0.806960	0.731889	1.366327	1.239218	1.693177	0.1000802	0.4273396	48
13	0.590840	0.806788	0.732336	1.365493	1.239482	1.692505	0.1002361	0.4267969	47
14	0.591075	0.806617	0.732783	1.364660	1.239746	1.691833	0.1003922	0.4262548	46
15	0.591310	0.806445	0.733230	1.363828	1.240011	1.691161	0.1005485	0.4257134	45
16	0.591544	0.806273	0.733678	1.362996	1.240275	1.690491	0.1007050	0.4251727	44
17	0.591779	0.806100	0.734125	1.362165	1.240540	1.689821	0.1008616	0.4246326	43
18	0.592013	0.805928	0.734573	1.361335	1.240805	1.689152	0.1010185	0.4240932	42
19	0.592248	0.805756	0.735021	1.360505	1.241070	1.688483	0.1011756	0.4235545	41
20	0.592482	0.805584	0.735469	1.359676	1.241336	1.687815	0.1013328	0.4230164	40
21	0.592716	0.805411	0.735917	1.358848	1.241602	1.687148	0.1014903	0.4224789	39
22	0.592951	0.805239	0.736366	1.358020	1.241867	1.686481	0.1016479	0.4219421	38
23	0.593185	0.805066	0.736815	1.357193	1.242134	1.685815	0.1018057	0.4214060	37
24	0.593419	0.804894	0.737264	1.356367	1.242400	1.685150	0.1019637	0.4208705	36
25	0.593653	0.804721	0.737713	1.355541	1.242666	1.684486	0.1021219	0.4203357	35
26	0.593887	0.804548	0.738162	1.354716	1.242933	1.683822	0.1022804	0.4198015	34
27	0.594121	0.804376	0.738611	1.353892	1.243200	1.683159	0.1024389	0.4192680	33
28	0.594355	0.804203	0.739061	1.353068	1.243467	1.682496	0.1025977	0.4187351	32
29	0.594589	0.804030	0.739511	1.352245	1.243735	1.681834	0.1027567	0.4182029	31
30	0.594823	0.803857	0.739961	1.351422	1.244003	1.681173	0.1029159	0.4176713	30
31	0.595057	0.803684	0.740411	1.350601	1.244270	1.680512	0.1030753	0.4171403	29
32	0.595290	0.803511	0.740862	1.349779	1.244539	1.679853	0.1032348	0.4166101	28
33	0.595524	0.803337	0.741312	1.348959	1.244807	1.679193	0.1033946	0.4160804	27
34	0.595758	0.803164	0.741763	1.348139	1.245075	1.678535	0.1035545	0.4155514	26
35	0.595991	0.802991	0.742214	1.347320	1.245344	1.677877	0.1037147	0.4150230	25
36	0.596225	0.802817	0.742666	1.346501	1.245613	1.677220	0.1038750	0.4144953	24
37	0.596458	0.802644	0.743117	1.345683	1.245882	1.676563	0.1040356	0.4139682	23
38	0.596692	0.802470	0.743569	1.344866	1.246152	1.675907	0.1041963	0.4134418	22
39	0.596925	0.802297	0.744020	1.344049	1.246421	1.675252	0.1043572	0.4129160	21
40	0.597159	0.802123	0.744472	1.343233	1.246691	1.674597	0.1045184	0.4123908	20
41	0.597392	0.801949	0.744925	1.342418	1.246961	1.673943	0.1046797	0.4118663	19
42	0.597625	0.801776	0.745377	1.341603	1.247232	1.673290	0.1048412	0.4113424	18
43	0.597858	0.801602	0.745830	1.340789	1.247502	1.672637	0.1050029	0.4108192	17
44	0.598091	0.801428	0.746282	1.339975	1.247773	1.671985	0.1051648	0.4102966	16
45	0.598325	0.801254	0.746735	1.339162	1.248044	1.671334	0.1053269	0.4097746	15
46	0.598558	0.801080	0.747189	1.338350	1.248315	1.670683	0.1054892	0.4092532	14
47	0.598791	0.800906	0.747642	1.337539	1.248587	1.670033	0.1056517	0.4087325	13
48	0.599024	0.800731	0.748096	1.336728	1.248858	1.669383	0.1058144	0.4082124	12
49	0.599256	0.800557	0.748549	1.335917	1.249130	1.668735	0.1059773	0.4076930	11
50	0.599489	0.800383	0.749003	1.335108	1.249402	1.668086	0.1061404	0.4071741	10
51	0.599722	0.800208	0.749458	1.334298	1.249675	1.667439	0.1063037	0.4066559	9
52	0.599955	0.800034	0.749912	1.333490	1.249947	1.666792	0.1064672	0.4061384	8
53	0.600188	0.799859	0.750366	1.332682	1.250220	1.666146	0.1066309	0.4056214	7
54	0.600420	0.799685	0.750821	1.331875	1.250493	1.665500	0.1067947	0.4051051	6
55	0.600653	0.799510	0.751276	1.331068	1.250766	1.664855	0.1069588	0.4045894	5
56	0.600885	0.799335	0.751731	1.330262	1.251040	1.664211	0.1071231	0.4040744	4
57	0.601118	0.799160	0.752187	1.329457	1.251313	1.663567	0.1072876	0.4035599	3
58	0.601350	0.798985	0.752642	1.328652	1.251587	1.662924	0.1074523	0.4030461	2
59	0.601583	0.798811	0.753098	1.327848	1.251861	1.662282	0.1076171	0.4025329	1
60	0.601815	0.798636	0.753554	1.327045	1.252136	1.661640	0.1077822	0.4020203	0
Minutes	Cosine	Sine	Cotangent	Tangent	Cosecant	Secant	Read Down	53°–54° Involute	Minutes

↑ 126° or 306° 53° or 233° ↑

↓ 37° or 217° Trigonometric and Involute Functions 142° or 322° ↓

Minutes	Sine	Cosine	Tangent	Cotangent	Secant	Cosecant	Involute 37°–38°	Read Up	Minutes
0	0.601815	0.798636	0.753554	1.327045	1.252136	1.661640	0.1077822	0.4020203	60
1	0.602047	0.798460	0.754010	1.326242	1.252410	1.660999	0.1079475	0.4015084	59
2	0.602280	0.798285	0.754467	1.325440	1.252685	1.660359	0.1081130	0.4009970	58
3	0.602512	0.798110	0.754923	1.324638	1.252960	1.659719	0.1082787	0.4004863	57
4	0.602744	0.797935	0.755380	1.323837	1.253235	1.659080	0.1084445	0.3999762	56
5	0.602976	0.797759	0.755837	1.323037	1.253511	1.658441	0.1086106	0.3994667	55
6	0.603208	0.797584	0.756294	1.322237	1.253787	1.657803	0.1087769	0.3989578	54
7	0.603440	0.797408	0.756751	1.321438	1.254062	1.657166	0.1089434	0.3984496	53
8	0.603672	0.797233	0.757209	1.320639	1.254339	1.656529	0.1091101	0.3979419	52
9	0.603904	0.797057	0.757667	1.319841	1.254615	1.655893	0.1092770	0.3974349	51
10	0.604136	0.796882	0.758125	1.319044	1.254892	1.655258	0.1094440	0.3969285	50
11	0.604367	0.796706	0.758583	1.318247	1.255169	1.654623	0.1096113	0.3964227	49
12	0.604599	0.796530	0.759041	1.317451	1.255446	1.653989	0.1097788	0.3959175	48
13	0.604831	0.796354	0.759500	1.316656	1.255723	1.653355	0.1099465	0.3954129	47
14	0.605062	0.796178	0.759959	1.315861	1.256000	1.652722	0.1101144	0.3949089	46
15	0.605294	0.796002	0.760418	1.315067	1.256278	1.652090	0.1102825	0.3944056	45
16	0.605526	0.795826	0.760877	1.314273	1.256556	1.651458	0.1104508	0.3939028	44
17	0.605757	0.795650	0.761336	1.313480	1.256834	1.650827	0.1106193	0.3934007	43
18	0.605988	0.795473	0.761796	1.312688	1.257113	1.650197	0.1107880	0.3928991	42
19	0.606220	0.795297	0.762256	1.311896	1.257392	1.649567	0.1109570	0.3923982	41
20	0.606451	0.795121	0.762716	1.311105	1.257671	1.648938	0.1111261	0.3918978	40
21	0.606682	0.794944	0.763176	1.310314	1.257950	1.648309	0.1112954	0.3913981	39
22	0.606914	0.794768	0.763636	1.309524	1.258229	1.647681	0.1114649	0.3908990	38
23	0.607145	0.794591	0.764097	1.308735	1.258509	1.647054	0.1116347	0.3904004	37
24	0.607376	0.794415	0.764558	1.307946	1.258789	1.646427	0.1118046	0.3899025	36
25	0.607607	0.794238	0.765019	1.307157	1.259069	1.645801	0.1119747	0.3894052	35
26	0.607838	0.794061	0.765480	1.306370	1.259349	1.645175	0.1121451	0.3889085	34
27	0.608069	0.793884	0.765941	1.305583	1.259629	1.644551	0.1123156	0.3884123	33
28	0.608300	0.793707	0.766403	1.304796	1.259910	1.643926	0.1124864	0.3879168	32
29	0.608531	0.793530	0.766865	1.304011	1.260191	1.643303	0.1126573	0.3874219	31
30	0.608761	0.793353	0.767327	1.303225	1.260472	1.642680	0.1128285	0.3869275	30
31	0.608992	0.793176	0.767789	1.302441	1.260754	1.642057	0.1129999	0.3864338	29
32	0.609223	0.792999	0.768252	1.301657	1.261036	1.641435	0.1131715	0.3859406	28
33	0.609454	0.792822	0.768714	1.300873	1.261317	1.640814	0.1133433	0.3854481	27
34	0.609684	0.792644	0.769177	1.300090	1.261600	1.640194	0.1135153	0.3849561	26
35	0.609915	0.792467	0.769640	1.299308	1.261882	1.639574	0.1136875	0.3844647	25
36	0.610145	0.792290	0.770104	1.298526	1.262165	1.638954	0.1138599	0.3839739	24
37	0.610376	0.792112	0.770567	1.297745	1.262448	1.638335	0.1140325	0.3834837	23
38	0.610606	0.791935	0.771031	1.296965	1.262731	1.637717	0.1142053	0.3829941	22
39	0.610836	0.791757	0.771495	1.296185	1.263014	1.637100	0.1143784	0.3825051	21
40	0.611067	0.791579	0.771959	1.295406	1.263298	1.636483	0.1145516	0.3820167	20
41	0.611297	0.791401	0.772423	1.294627	1.263581	1.635866	0.1147250	0.3815289	19
42	0.611527	0.791224	0.772888	1.293849	1.263865	1.635251	0.1148987	0.3810416	18
43	0.611757	0.791046	0.773353	1.293071	1.264150	1.634636	0.1150726	0.3805549	17
44	0.611987	0.790868	0.773818	1.292294	1.264434	1.634021	0.1152466	0.3800689	16
45	0.612217	0.790690	0.774283	1.291518	1.264719	1.633407	0.1154209	0.3795834	15
46	0.612447	0.790511	0.774748	1.290742	1.265004	1.632794	0.1155954	0.3790984	14
47	0.612677	0.790333	0.775214	1.289967	1.265289	1.632181	0.1157701	0.3786141	13
48	0.612907	0.790155	0.775680	1.289192	1.265574	1.631569	0.1159451	0.3781304	12
49	0.613137	0.789977	0.776146	1.288418	1.265860	1.630957	0.1161202	0.3776472	11
50	0.613367	0.789798	0.776612	1.287645	1.266146	1.630346	0.1162955	0.3771646	10
51	0.613596	0.789620	0.777078	1.286872	1.266432	1.629736	0.1164711	0.3766826	9
52	0.613826	0.789441	0.777545	1.286099	1.266719	1.629126	0.1166468	0.3762012	8
53	0.614056	0.789263	0.778012	1.285328	1.267005	1.628517	0.1168228	0.3757203	7
54	0.614285	0.789084	0.778479	1.284557	1.267292	1.627908	0.1169990	0.3752400	6
55	0.614515	0.788905	0.778946	1.283786	1.267579	1.627300	0.1171754	0.3747603	5
56	0.614744	0.788727	0.779414	1.283016	1.267866	1.626693	0.1173520	0.3742812	4
57	0.614974	0.788548	0.779881	1.282247	1.268154	1.626086	0.1175288	0.3738026	3
58	0.615203	0.788369	0.780349	1.281478	1.268442	1.625480	0.1177058	0.3733247	2
59	0.615432	0.788190	0.780817	1.280709	1.268730	1.624874	0.1178831	0.3728473	1
60	0.615661	0.788011	0.781286	1.279942	1.269018	1.624269	0.1180605	0.3723704	0
Minutes	Cosine	Sine	Cotangent	Tangent	Cosecant	Secant	Read Down	52°–53° Involute	Minutes

↑ 127° or 307° 52° or 232° ↑

↓ 38° or 218° Trigonometric and Involute Functions 141° or 321° ↓

Minutes	Sine	Cosine	Tangent	Cotangent	Secant	Cosecant	Involute 38°–39°	Read Up	Minutes
0	0.615661	0.788011	0.781286	1.279942	1.269018	1.624269	0.1180605	0.3723704	60
1	0.615891	0.787832	0.781754	1.279174	1.269307	1.623665	0.1182382	0.3718942	59
2	0.616120	0.787652	0.782223	1.278408	1.269596	1.623061	0.1184161	0.3714185	58
3	0.616349	0.787473	0.782692	1.277642	1.269885	1.622458	0.1185942	0.3709433	57
4	0.616578	0.787294	0.783161	1.276876	1.270174	1.621855	0.1187725	0.3704688	56
5	0.616807	0.787114	0.783631	1.276112	1.270463	1.621253	0.1189510	0.3699948	55
6	0.617036	0.786935	0.784100	1.275347	1.270753	1.620651	0.1191297	0.3695214	54
7	0.617265	0.786756	0.784570	1.274584	1.271043	1.620050	0.1193087	0.3690485	53
8	0.617494	0.786576	0.785040	1.273820	1.271333	1.619450	0.1194878	0.3685763	52
9	0.617722	0.786396	0.785510	1.273058	1.271624	1.618850	0.1196672	0.3681045	51
10	0.617951	0.786217	0.785981	1.272296	1.271914	1.618251	0.1198468	0.3676334	50
11	0.618180	0.786037	0.786451	1.271534	1.272205	1.617652	0.1200266	0.3671628	49
12	0.618408	0.785857	0.786922	1.270773	1.272496	1.617054	0.1202066	0.3666928	48
13	0.618637	0.785677	0.787394	1.270013	1.272788	1.616457	0.1203869	0.3662233	47
14	0.618865	0.785497	0.787865	1.269253	1.273079	1.615860	0.1205673	0.3657544	46
15	0.619094	0.785317	0.788336	1.268494	1.273371	1.615264	0.1207480	0.3652861	45
16	0.619322	0.785137	0.788808	1.267735	1.273663	1.614668	0.1209289	0.3648183	44
17	0.619551	0.784957	0.789280	1.266977	1.273956	1.614073	0.1211100	0.3643511	43
18	0.619779	0.784776	0.789752	1.266220	1.274248	1.613478	0.1212913	0.3638844	42
19	0.620007	0.784596	0.790225	1.265463	1.274541	1.612884	0.1214728	0.3634183	41
20	0.620235	0.784416	0.790697	1.264706	1.274834	1.612291	0.1216546	0.3629527	40
21	0.620464	0.784235	0.791170	1.263950	1.275128	1.611698	0.1218366	0.3624878	39
22	0.620692	0.784055	0.791643	1.263195	1.275421	1.611106	0.1220188	0.3620233	38
23	0.620920	0.783874	0.792117	1.262440	1.275715	1.610514	0.1222012	0.3615594	37
24	0.621148	0.783693	0.792590	1.261686	1.276009	1.609923	0.1223838	0.3610961	36
25	0.621376	0.783513	0.793064	1.260932	1.276303	1.609332	0.1225666	0.3606333	35
26	0.621604	0.783332	0.793538	1.260179	1.276598	1.608742	0.1227497	0.3601711	34
27	0.621831	0.783151	0.794012	1.259427	1.276893	1.608153	0.1229330	0.3597094	33
28	0.622059	0.782970	0.794486	1.258675	1.277188	1.607564	0.1231165	0.3592483	32
29	0.622287	0.782789	0.794961	1.257923	1.277483	1.606976	0.1233002	0.3587878	31
30	0.622515	0.782608	0.795436	1.257172	1.277779	1.606388	0.1234842	0.3583277	30
31	0.622742	0.782427	0.795911	1.256422	1.278074	1.605801	0.1236683	0.3578683	29
32	0.622970	0.782246	0.796386	1.255672	1.278370	1.605214	0.1238527	0.3574093	28
33	0.623197	0.782065	0.796862	1.254923	1.278667	1.604628	0.1240373	0.3569510	27
34	0.623425	0.781883	0.797337	1.254174	1.278963	1.604043	0.1242221	0.3564931	26
35	0.623652	0.781702	0.797813	1.253426	1.279260	1.603458	0.1244072	0.3560359	25
36	0.623880	0.781520	0.798290	1.252678	1.279557	1.602873	0.1245924	0.3555791	24
37	0.624107	0.781339	0.798766	1.251931	1.279854	1.602290	0.1247779	0.3551229	23
38	0.624334	0.781157	0.799242	1.251185	1.280152	1.601706	0.1249636	0.3546673	22
39	0.624561	0.780976	0.799719	1.250439	1.280450	1.601124	0.1251495	0.3542122	21
40	0.624789	0.780794	0.800196	1.249693	1.280748	1.600542	0.1253357	0.3537576	20
41	0.625016	0.780612	0.800674	1.248948	1.281046	1.599960	0.1255221	0.3533036	19
42	0.625243	0.780430	0.801151	1.248204	1.281344	1.599379	0.1257087	0.3528501	18
43	0.625470	0.780248	0.801629	1.247460	1.281643	1.598799	0.1258955	0.3523972	17
44	0.625697	0.780067	0.802107	1.246717	1.281942	1.598219	0.1260825	0.3519448	16
45	0.625923	0.779884	0.802585	1.245974	1.282241	1.597639	0.1262698	0.3514929	15
46	0.626150	0.779702	0.803063	1.245232	1.282541	1.597061	0.1264573	0.3510416	14
47	0.626377	0.779520	0.803542	1.244490	1.282840	1.596482	0.1266450	0.3505908	13
48	0.626604	0.779338	0.804021	1.243749	1.283140	1.595905	0.1268329	0.3501406	12
49	0.626830	0.779156	0.804500	1.243009	1.283441	1.595328	0.1270210	0.3496909	11
50	0.627057	0.778973	0.804979	1.242268	1.283741	1.594751	0.1272094	0.3492417	10
51	0.627284	0.778791	0.805458	1.241529	1.284042	1.594175	0.1273980	0.3487931	9
52	0.627510	0.778608	0.805938	1.240790	1.284343	1.593600	0.1275869	0.3483450	8
53	0.627737	0.778426	0.806418	1.240052	1.284644	1.593025	0.1277759	0.3478974	7
54	0.627963	0.778243	0.806898	1.239314	1.284945	1.592450	0.1279652	0.3474503	6
55	0.628189	0.778060	0.807379	1.238576	1.285247	1.591877	0.1281547	0.3470038	5
56	0.628416	0.777878	0.807859	1.237839	1.285549	1.591303	0.1283444	0.3465579	4
57	0.628642	0.777695	0.808340	1.237103	1.285851	1.590731	0.1285344	0.3461124	3
58	0.628868	0.777512	0.808821	1.236367	1.286154	1.590158	0.1287246	0.3456675	2
59	0.629094	0.777329	0.809303	1.235632	1.286457	1.589587	0.1289150	0.3452231	1
60	0.629320	0.777146	0.809784	1.234897	1.286760	1.589016	0.1291056	0.3447792	0
Minutes	Cosine	Sine	Cotangent	Tangent	Cosecant	Secant	Read Down	51°–52° Involute	Minutes

↑ 128° or 308° 51° or 231° ↑

↓ 39° or 219° Trigonometric and Involute Functions 140° or 320° ↓

Minutes	Sine	Cosine	Tangent	Cotangent	Secant	Cosecant	Involute 39°–40°	Read Up	Minutes
0	0.629320	0.777146	0.809784	1.234897	1.286760	1.589016	0.1291056	0.3447792	60
1	0.629546	0.776963	0.810266	1.234163	1.287063	1.588445	0.1292965	0.3443359	59
2	0.629772	0.776780	0.810748	1.233429	1.287366	1.587875	0.1294876	0.3438931	58
3	0.629998	0.776596	0.811230	1.232696	1.287670	1.587306	0.1296789	0.3434508	57
4	0.630224	0.776413	0.811712	1.231963	1.287974	1.586737	0.1298704	0.3430091	56
5	0.630450	0.776230	0.812195	1.231231	1.288278	1.586169	0.1300622	0.3425678	55
6	0.630676	0.776046	0.812678	1.230500	1.288583	1.585601	0.1302542	0.3421271	54
7	0.630902	0.775863	0.813161	1.229769	1.288887	1.585033	0.1304464	0.3416870	53
8	0.631127	0.775679	0.813644	1.229038	1.289192	1.584467	0.1306389	0.3412473	52
9	0.631353	0.775496	0.814128	1.228308	1.289498	1.583900	0.1308316	0.3408082	51
10	0.631578	0.775312	0.814612	1.227579	1.289803	1.583335	0.1310245	0.3403695	50
11	0.631804	0.775128	0.815096	1.226850	1.290109	1.582770	0.1312177	0.3399315	49
12	0.632029	0.774944	0.815580	1.226121	1.290415	1.582205	0.1314110	0.3394939	48
13	0.632255	0.774761	0.816065	1.225393	1.290721	1.581641	0.1316046	0.3390568	47
14	0.632480	0.774577	0.816549	1.224666	1.291028	1.581078	0.1317985	0.3386203	46
15	0.632705	0.774393	0.817034	1.223939	1.291335	1.580515	0.1319925	0.3381843	45
16	0.632931	0.774209	0.817519	1.223212	1.291642	1.579952	0.1321868	0.3377488	44
17	0.633156	0.774024	0.818005	1.222487	1.291949	1.579390	0.1323814	0.3373138	43
18	0.633381	0.773840	0.818491	1.221761	1.292256	1.578829	0.1325761	0.3368793	42
19	0.633606	0.773656	0.818976	1.221036	1.292564	1.578268	0.1327711	0.3364454	41
20	0.633831	0.773472	0.819463	1.220312	1.292872	1.577708	0.1329663	0.3360119	40
21	0.634056	0.773287	0.819949	1.219588	1.293181	1.577148	0.1331618	0.3355790	39
22	0.634281	0.773103	0.820435	1.218865	1.293489	1.576589	0.1333575	0.3351466	38
23	0.634506	0.772918	0.820922	1.218142	1.293798	1.576030	0.1335534	0.3347147	37
24	0.634731	0.772734	0.821409	1.217420	1.294107	1.575472	0.1337495	0.3342833	36
25	0.634955	0.772549	0.821897	1.216698	1.294416	1.574914	0.1339459	0.3338524	35
26	0.635180	0.772364	0.822384	1.215977	1.294726	1.574357	0.1341425	0.3334221	34
27	0.635405	0.772179	0.822872	1.215256	1.295036	1.573800	0.1343394	0.3329922	33
28	0.635629	0.771995	0.823360	1.214536	1.295346	1.573244	0.1345365	0.3325629	32
29	0.635854	0.771810	0.823848	1.213816	1.295656	1.572689	0.1347338	0.3321341	31
30	0.636078	0.771625	0.824336	1.213097	1.295967	1.572134	0.1349313	0.3317057	30
31	0.636303	0.771440	0.824825	1.212378	1.296278	1.571579	0.1351291	0.3312779	29
32	0.636527	0.771254	0.825314	1.211660	1.296589	1.571025	0.1353271	0.3308506	28
33	0.636751	0.771069	0.825803	1.210942	1.296900	1.570472	0.1355254	0.3304238	27
34	0.636976	0.770884	0.826292	1.210225	1.297212	1.569919	0.1357239	0.3299975	26
35	0.637200	0.770699	0.826782	1.209509	1.297524	1.569366	0.1359226	0.3295717	25
36	0.637424	0.770513	0.827272	1.208792	1.297836	1.568815	0.1361216	0.3291464	24
37	0.637648	0.770328	0.827762	1.208077	1.298149	1.568263	0.1363208	0.3287216	23
38	0.637872	0.770142	0.828252	1.207362	1.298461	1.567712	0.1365202	0.3282973	22
39	0.638096	0.769957	0.828743	1.206647	1.298774	1.567162	0.1367199	0.3278736	21
40	0.638320	0.769771	0.829234	1.205933	1.299088	1.566612	0.1369198	0.3274503	20
41	0.638544	0.769585	0.829725	1.205219	1.299401	1.566063	0.1371199	0.3270275	19
42	0.638768	0.769400	0.830216	1.204506	1.299715	1.565514	0.1373203	0.3266052	18
43	0.638992	0.769214	0.830707	1.203793	1.300029	1.564966	0.1375209	0.3261834	17
44	0.639215	0.769028	0.831199	1.203081	1.300343	1.564418	0.1377218	0.3257621	16
45	0.639439	0.768842	0.831691	1.202369	1.300658	1.563871	0.1379228	0.3253414	15
46	0.639663	0.768656	0.832183	1.201658	1.300972	1.563324	0.1381242	0.3249211	14
47	0.639886	0.768470	0.832676	1.200947	1.301287	1.562778	0.1383257	0.3245013	13
48	0.640110	0.768284	0.833169	1.200237	1.301603	1.562232	0.1385275	0.3240820	12
49	0.640333	0.768097	0.833662	1.199528	1.301918	1.561687	0.1387296	0.3236632	11
50	0.640557	0.767911	0.834155	1.198818	1.302234	1.561142	0.1389319	0.3232449	10
51	0.640780	0.767725	0.834648	1.198110	1.302550	1.560598	0.1391344	0.3228271	9
52	0.641003	0.767538	0.835142	1.197402	1.302867	1.560055	0.1393372	0.3224098	8
53	0.641226	0.767352	0.835636	1.196694	1.303183	1.559511	0.1395402	0.3219930	7
54	0.641450	0.767165	0.836130	1.195987	1.303500	1.558969	0.1397434	0.3215766	6
55	0.641673	0.766979	0.836624	1.195280	1.303817	1.558427	0.1399469	0.3211608	5
56	0.641896	0.766792	0.837119	1.194574	1.304135	1.557885	0.1401506	0.3207454	4
57	0.642119	0.766605	0.837614	1.193868	1.304453	1.557344	0.1403546	0.3203306	3
58	0.642342	0.766418	0.838109	1.193163	1.304771	1.556803	0.1405588	0.3199162	2
59	0.642565	0.766231	0.838604	1.192458	1.305089	1.556263	0.1407632	0.3195024	1
60	0.642788	0.766044	0.839100	1.191754	1.305407	1.555724	0.1409679	0.3190890	0
Minutes	Cosine	Sine	Cotangent	Tangent	Cosecant	Secant	Read Down	50°–51° Involute	Minutes

↑ 129° or 309° 50° or 230° ↑

↓ 40° or 220° Trigonometric and Involute Functions 139° or 319° ↓

Minutes	Sine	Cosine	Tangent	Cotangent	Secant	Cosecant	Involute 40°–41°	Read Up	Minutes
0	0.642788	0.766044	0.839100	1.191754	1.305407	1.555724	0.1409679	0.3190890	60
1	0.643010	0.765857	0.839595	1.191050	1.305726	1.555185	0.1411729	0.3186761	59
2	0.643233	0.765670	0.840092	1.190347	1.306045	1.554646	0.1413780	0.3182637	58
3	0.643456	0.765483	0.840588	1.189644	1.306364	1.554108	0.1415835	0.3178517	57
4	0.643679	0.765296	0.841084	1.188941	1.306684	1.553571	0.1417891	0.3174403	56
5	0.643901	0.765109	0.841581	1.188240	1.307004	1.553034	0.1419950	0.3170293	55
6	0.644124	0.764921	0.842078	1.187538	1.307324	1.552497	0.1422012	0.3166189	54
7	0.644346	0.764734	0.842575	1.186837	1.307644	1.551961	0.1424076	0.3162089	53
8	0.644569	0.764547	0.843073	1.186137	1.307965	1.551425	0.1426142	0.3157994	52
9	0.644791	0.764359	0.843571	1.185437	1.308286	1.550890	0.1428211	0.3153904	51
10	0.645013	0.764171	0.844069	1.184738	1.308607	1.550356	0.1430282	0.3149819	50
11	0.645235	0.763984	0.844567	1.184039	1.308928	1.549822	0.1432355	0.3145738	49
12	0.645458	0.763796	0.845066	1.183340	1.309250	1.549288	0.1434432	0.3141662	48
13	0.645680	0.763608	0.845564	1.182642	1.309572	1.548755	0.1436510	0.3137591	47
14	0.645902	0.763420	0.846063	1.181945	1.309894	1.548223	0.1438591	0.3133525	46
15	0.646124	0.763232	0.846562	1.181248	1.310217	1.547691	0.1440675	0.3129464	45
16	0.646346	0.763044	0.847062	1.180551	1.310540	1.547159	0.1442761	0.3125408	44
17	0.646568	0.762856	0.847562	1.179855	1.310863	1.546628	0.1444849	0.3121356	43
18	0.646790	0.762668	0.848062	1.179160	1.311186	1.546097	0.1446940	0.3117309	42
19	0.647012	0.762480	0.848562	1.178464	1.311510	1.545567	0.1449033	0.3113267	41
20	0.647233	0.762292	0.849062	1.177770	1.311833	1.545038	0.1451129	0.3109229	40
21	0.647455	0.762104	0.849563	1.177076	1.312158	1.544509	0.1453227	0.3105197	39
22	0.647677	0.761915	0.850064	1.176382	1.312482	1.543980	0.1455328	0.3101169	38
23	0.647898	0.761727	0.850565	1.175689	1.312807	1.543452	0.1457431	0.3097146	37
24	0.648120	0.761538	0.851067	1.174996	1.313132	1.542924	0.1459537	0.3093127	36
25	0.648341	0.761350	0.851568	1.174304	1.313457	1.542397	0.1461645	0.3089113	35
26	0.648563	0.761161	0.852070	1.173612	1.313782	1.541871	0.1463756	0.3085105	34
27	0.648784	0.760972	0.852573	1.172921	1.314108	1.541345	0.1465869	0.3081100	33
28	0.649006	0.760784	0.853075	1.172230	1.314434	1.540819	0.1467985	0.3077101	32
29	0.649227	0.760595	0.853578	1.171539	1.314760	1.540294	0.1470103	0.3073106	31
30	0.649448	0.760406	0.854081	1.170850	1.315087	1.539769	0.1472223	0.3069116	30
31	0.649669	0.760217	0.854584	1.170160	1.315414	1.539245	0.1474347	0.3065130	29
32	0.649890	0.760028	0.855087	1.169471	1.315741	1.538721	0.1476472	0.3061150	28
33	0.650111	0.759839	0.855591	1.168783	1.316068	1.538198	0.1478600	0.3057174	27
34	0.650332	0.759650	0.856095	1.168095	1.316396	1.537675	0.1480731	0.3053202	26
35	0.650553	0.759461	0.856599	1.167407	1.316724	1.537153	0.1482864	0.3049236	25
36	0.650774	0.759271	0.857104	1.166720	1.317052	1.536631	0.1485000	0.3045274	24
37	0.650995	0.759082	0.857608	1.166033	1.317381	1.536110	0.1487138	0.3041316	23
38	0.651216	0.758893	0.858113	1.165347	1.317710	1.535589	0.1489279	0.3037364	22
39	0.651437	0.758703	0.858619	1.164662	1.318039	1.535069	0.1491422	0.3033416	21
40	0.651657	0.758514	0.859124	1.163976	1.318368	1.534549	0.1493568	0.3029472	20
41	0.651878	0.758324	0.859630	1.163292	1.318698	1.534030	0.1495716	0.3025533	19
42	0.652098	0.758134	0.860136	1.162607	1.319027	1.533511	0.1497867	0.3021599	18
43	0.652319	0.757945	0.860642	1.161923	1.319358	1.532993	0.1500020	0.3017670	17
44	0.652539	0.757755	0.861148	1.161240	1.319688	1.532475	0.1502176	0.3013745	16
45	0.652760	0.757565	0.861655	1.160557	1.320019	1.531957	0.1504335	0.3009825	15
46	0.652980	0.757375	0.862162	1.159875	1.320350	1.531440	0.1506496	0.3005909	14
47	0.653200	0.757185	0.862669	1.159193	1.320681	1.530924	0.1508659	0.3001998	13
48	0.653421	0.756995	0.863177	1.158511	1.321013	1.530408	0.1510825	0.2998092	12
49	0.653641	0.756805	0.863685	1.157830	1.321344	1.529892	0.1512994	0.2994190	11
50	0.653861	0.756615	0.864193	1.157149	1.321677	1.529377	0.1515165	0.2990292	10
51	0.654081	0.756425	0.864701	1.156469	1.322009	1.528863	0.1517339	0.2986400	9
52	0.654301	0.756234	0.865209	1.155790	1.322342	1.528349	0.1519515	0.2982512	8
53	0.654521	0.756044	0.865718	1.155110	1.322675	1.527835	0.1521694	0.2978628	7
54	0.654741	0.755853	0.866227	1.154432	1.323008	1.527322	0.1523875	0.2974749	6
55	0.654961	0.755663	0.866736	1.153753	1.323341	1.526809	0.1526059	0.2970875	5
56	0.655180	0.755472	0.867246	1.153075	1.323675	1.526297	0.1528246	0.2967005	4
57	0.655400	0.755282	0.867756	1.152398	1.324009	1.525785	0.1530435	0.2963140	3
58	0.655620	0.755091	0.868266	1.151721	1.324343	1.525274	0.1532626	0.2959279	2
59	0.655839	0.754900	0.868776	1.151044	1.324678	1.524763	0.1534821	0.2955422	1
60	0.656059	0.754710	0.869287	1.150368	1.325013	1.524253	0.1537017	0.2951571	0
Minutes	Cosine	Sine	Cotangent	Tangent	Cosecant	Secant	Read Down	49°–50° Involute	Minutes

↑ 130° or 310° 49° or 229° ↑

↓ 41° or 221° Trigonometric and Involute Functions 138° or 318° ↓

Minutes	Sine	Cosine	Tangent	Cotangent	Secant	Cosecant	Involute 41°–42°	Read Up	Minutes
0	0.656059	0.754710	0.869287	1.150368	1.325013	1.524253	0.1537017	0.2951571	60
1	0.656279	0.754519	0.869798	1.149693	1.325348	1.523743	0.1539217	0.2947724	59
2	0.656498	0.754328	0.870309	1.149018	1.325684	1.523234	0.1541419	0.2943881	58
3	0.656717	0.754137	0.870820	1.148343	1.326019	1.522725	0.1543623	0.2940043	57
4	0.656937	0.753946	0.871332	1.147669	1.326355	1.522217	0.1545831	0.2936209	56
5	0.657156	0.753755	0.871843	1.146995	1.326692	1.521709	0.1548040	0.2932380	55
6	0.657375	0.753563	0.872356	1.146322	1.327028	1.521201	0.1550253	0.2928555	54
7	0.657594	0.753372	0.872868	1.145649	1.327365	1.520694	0.1552468	0.2924735	53
8	0.657814	0.753181	0.873381	1.144976	1.327702	1.520188	0.1554685	0.2920919	52
9	0.658033	0.752989	0.873894	1.144304	1.328040	1.519682	0.1556906	0.2917108	51
10	0.658252	0.752798	0.874407	1.143633	1.328378	1.519176	0.1559128	0.2913301	50
11	0.658471	0.752606	0.874920	1.142961	1.328716	1.518671	0.1561354	0.2909499	49
12	0.658689	0.752415	0.875434	1.142291	1.329054	1.518166	0.1563582	0.2905701	48
13	0.658908	0.752223	0.875948	1.141621	1.329393	1.517662	0.1565812	0.2901908	47
14	0.659127	0.752032	0.876462	1.140951	1.329731	1.517158	0.1568046	0.2898119	46
15	0.659346	0.751840	0.876976	1.140281	1.330071	1.516655	0.1570281	0.2894334	45
16	0.659564	0.751648	0.877491	1.139613	1.330410	1.516152	0.1572520	0.2890554	44
17	0.659783	0.751456	0.878006	1.138944	1.330750	1.515650	0.1574761	0.2886779	43
18	0.660002	0.751264	0.878521	1.138276	1.331090	1.515148	0.1577005	0.2883008	42
19	0.660220	0.751072	0.879037	1.137609	1.331430	1.514646	0.1579251	0.2879241	41
20	0.660439	0.750880	0.879553	1.136941	1.331771	1.514145	0.1581500	0.2875479	40
21	0.660657	0.750688	0.880069	1.136275	1.332112	1.513645	0.1583752	0.2871721	39
22	0.660875	0.750496	0.880585	1.135609	1.332453	1.513145	0.1586006	0.2867967	38
23	0.661094	0.750303	0.881102	1.134943	1.332794	1.512645	0.1588263	0.2864218	37
24	0.661312	0.750111	0.881619	1.134277	1.333136	1.512146	0.1590523	0.2860473	36
25	0.661530	0.749919	0.882136	1.133612	1.333478	1.511647	0.1592785	0.2856733	35
26	0.661748	0.749726	0.882653	1.132948	1.333820	1.511149	0.1595050	0.2852997	34
27	0.661966	0.749534	0.883171	1.132284	1.334163	1.510651	0.1597318	0.2849265	33
28	0.662184	0.749341	0.883689	1.131620	1.334506	1.510154	0.1599588	0.2845538	32
29	0.662402	0.749148	0.884207	1.130957	1.334849	1.509657	0.1601861	0.2841815	31
30	0.662620	0.748956	0.884725	1.130294	1.335192	1.509160	0.1604136	0.2838097	30
31	0.662838	0.748763	0.885244	1.129632	1.335536	1.508665	0.1606414	0.2834383	29
32	0.663056	0.748570	0.885763	1.128970	1.335880	1.508169	0.1608695	0.2830673	28
33	0.663273	0.748377	0.886282	1.128309	1.336225	1.507674	0.1610979	0.2826968	27
34	0.663491	0.748184	0.886802	1.127648	1.336569	1.507179	0.1613265	0.2823267	26
35	0.663709	0.747991	0.887321	1.126987	1.336914	1.506685	0.1615554	0.2819570	25
36	0.663926	0.747798	0.887842	1.126327	1.337259	1.506191	0.1617846	0.2815877	24
37	0.664144	0.747605	0.888362	1.125667	1.337605	1.505698	0.1620140	0.2812189	23
38	0.664361	0.747412	0.888882	1.125008	1.337951	1.505205	0.1622437	0.2808506	22
39	0.664579	0.747218	0.889403	1.124349	1.338297	1.504713	0.1624737	0.2804826	21
40	0.664796	0.747025	0.889924	1.123691	1.338643	1.504221	0.1627039	0.2801151	20
41	0.665013	0.746832	0.890446	1.123033	1.338990	1.503730	0.1629344	0.2797480	19
42	0.665230	0.746638	0.890967	1.122375	1.339337	1.503239	0.1631652	0.2793814	18
43	0.665448	0.746445	0.891489	1.121718	1.339684	1.502748	0.1633963	0.2790151	17
44	0.665665	0.746251	0.892012	1.121062	1.340032	1.502258	0.1636276	0.2786493	16
45	0.665882	0.746057	0.892534	1.120405	1.340379	1.501768	0.1638592	0.2782840	15
46	0.666099	0.745864	0.893057	1.119750	1.340728	1.501279	0.1640910	0.2779190	14
47	0.666316	0.745670	0.893580	1.119094	1.341076	1.500790	0.1643232	0.2775545	13
48	0.666532	0.745476	0.894103	1.118439	1.341425	1.500302	0.1645556	0.2771904	12
49	0.666749	0.745282	0.894627	1.117785	1.341774	1.499814	0.1647882	0.2768268	11
50	0.666966	0.745088	0.895151	1.117130	1.342123	1.499327	0.1650212	0.2764635	10
51	0.667183	0.744894	0.895675	1.116477	1.342473	1.498840	0.1652544	0.2761007	9
52	0.667399	0.744700	0.896199	1.115823	1.342823	1.498353	0.1654879	0.2757383	8
53	0.667616	0.744506	0.896724	1.115171	1.343173	1.497867	0.1657217	0.2753764	7
54	0.667833	0.744312	0.897249	1.114518	1.343523	1.497381	0.1659557	0.2750148	6
55	0.668049	0.744117	0.897774	1.113866	1.343874	1.496896	0.1661900	0.2746537	5
56	0.668265	0.743923	0.898299	1.113215	1.344225	1.496411	0.1664246	0.2742930	4
57	0.668482	0.743728	0.898825	1.112563	1.344577	1.495927	0.1666595	0.2739328	3
58	0.668698	0.743534	0.899351	1.111913	1.344928	1.495443	0.1668946	0.2735729	2
59	0.668914	0.743339	0.899877	1.111262	1.345280	1.494960	0.1671301	0.2732135	1
60	0.669131	0.743145	0.900404	1.110613	1.345633	1.494477	0.1673658	0.2728545	0
Minutes	Cosine	Sine	Cotangent	Tangent	Cosecant	Secant	Read Down	48°–49° Involute	Minutes

↑ 131° or 311° 48° or 228° ↑

↓ **42° or 222°** **Trigonometric and Involute Functions** **137° or 317°** ↓

Minutes	Sine	Cosine	Tangent	Cotangent	Secant	Cosecant	Involute 42°–43°	Read Up	Minutes
0	0.669131	0.743145	0.900404	1.110613	1.345633	1.494477	0.1673658	0.2728545	60
1	0.669347	0.742950	0.900931	1.109963	1.345985	1.493994	0.1676017	0.2724959	59
2	0.669563	0.742755	0.901458	1.109314	1.346338	1.493512	0.1678380	0.2721377	58
3	0.669779	0.742561	0.901985	1.108665	1.346691	1.493030	0.1680745	0.2717800	57
4	0.669995	0.742366	0.902513	1.108017	1.347045	1.492549	0.1683113	0.2714226	56
5	0.670211	0.742171	0.903041	1.107369	1.347399	1.492068	0.1685484	0.2710657	55
6	0.670427	0.741976	0.903569	1.106722	1.347753	1.491588	0.1687857	0.2707092	54
7	0.670642	0.741781	0.904098	1.106075	1.348107	1.491108	0.1690234	0.2703531	53
8	0.670858	0.741586	0.904627	1.105428	1.348462	1.490628	0.1692613	0.2699975	52
9	0.671074	0.741391	0.905156	1.104782	1.348817	1.490149	0.1694994	0.2696422	51
10	0.671289	0.741195	0.905685	1.104137	1.349172	1.489670	0.1697379	0.2692874	50
11	0.671505	0.741000	0.906215	1.103491	1.349528	1.489192	0.1699767	0.2689330	49
12	0.671721	0.740805	0.906745	1.102846	1.349884	1.488714	0.1702157	0.2685790	48
13	0.671936	0.740609	0.907275	1.102202	1.350240	1.488237	0.1704550	0.2682254	47
14	0.672151	0.740414	0.907805	1.101558	1.350596	1.487760	0.1706946	0.2678722	46
15	0.672367	0.740218	0.908336	1.100914	1.350953	1.487283	0.1709344	0.2675194	45
16	0.672582	0.740023	0.908867	1.100271	1.351310	1.486807	0.1711746	0.2671671	44
17	0.672797	0.739827	0.909398	1.099628	1.351668	1.486332	0.1714150	0.2668151	43
18	0.673013	0.739631	0.909930	1.098986	1.352025	1.485856	0.1716557	0.2664636	42
19	0.673228	0.739435	0.910462	1.098344	1.352383	1.485382	0.1718967	0.2661125	41
20	0.673443	0.739239	0.910994	1.097702	1.352742	1.484907	0.1721380	0.2657618	40
21	0.673658	0.739043	0.911526	1.097061	1.353100	1.484433	0.1723795	0.2654115	39
22	0.673873	0.738848	0.912059	1.096420	1.353459	1.483960	0.1726214	0.2650616	38
23	0.674088	0.738651	0.912592	1.095780	1.353818	1.483487	0.1728635	0.2647121	37
24	0.674302	0.738455	0.913125	1.095140	1.354178	1.483014	0.1731059	0.2643630	36
25	0.674517	0.738259	0.913659	1.094500	1.354538	1.482542	0.1733486	0.2640143	35
26	0.674732	0.738063	0.914193	1.093861	1.354898	1.482070	0.1735915	0.2636661	34
27	0.674947	0.737867	0.914727	1.093222	1.355258	1.481599	0.1738348	0.2633182	33
28	0.675161	0.737670	0.915261	1.092584	1.355619	1.481128	0.1740783	0.2629708	32
29	0.675376	0.737474	0.915796	1.091946	1.355980	1.480657	0.1743221	0.2626237	31
30	0.675590	0.737277	0.916331	1.091309	1.356342	1.480187	0.1745662	0.2622771	30
31	0.675805	0.737081	0.916866	1.090671	1.356703	1.479718	0.1748106	0.2619309	29
32	0.676019	0.736884	0.917402	1.090035	1.357065	1.479248	0.1750553	0.2615850	28
33	0.676233	0.736687	0.917938	1.089398	1.357428	1.478779	0.1753003	0.2612396	27
34	0.676448	0.736491	0.918474	1.088762	1.357790	1.478311	0.1755455	0.2608946	26
35	0.676662	0.736294	0.919010	1.088127	1.358153	1.477843	0.1757911	0.2605500	25
36	0.676876	0.736097	0.919547	1.087492	1.358516	1.477376	0.1760369	0.2602058	24
37	0.677090	0.735900	0.920084	1.086857	1.358880	1.476908	0.1762830	0.2598619	23
38	0.677304	0.735703	0.920621	1.086223	1.359244	1.476442	0.1765294	0.2595185	22
39	0.677518	0.735506	0.921159	1.085589	1.359608	1.475975	0.1767761	0.2591755	21
40	0.677732	0.735309	0.921697	1.084955	1.359972	1.475509	0.1770230	0.2588329	20
41	0.677946	0.735112	0.922235	1.084322	1.360337	1.475044	0.1772703	0.2584907	19
42	0.678160	0.734915	0.922773	1.083690	1.360702	1.474579	0.1775179	0.2581489	18
43	0.678373	0.734717	0.923312	1.083057	1.361068	1.474114	0.1777657	0.2578075	17
44	0.678587	0.734520	0.923851	1.082425	1.361433	1.473650	0.1780138	0.2574665	16
45	0.678801	0.734323	0.924390	1.081794	1.361799	1.473186	0.1782622	0.2571258	15
46	0.679014	0.734125	0.924930	1.081163	1.362166	1.472723	0.1785109	0.2567856	14
47	0.679228	0.733927	0.925470	1.080532	1.362532	1.472260	0.1787599	0.2564458	13
48	0.679441	0.733730	0.926010	1.079902	1.362899	1.471797	0.1790092	0.2561064	12
49	0.679655	0.733532	0.926551	1.079272	1.363267	1.471335	0.1792588	0.2557673	11
50	0.679868	0.733334	0.927091	1.078642	1.363634	1.470874	0.1795087	0.2554287	10
51	0.680081	0.733137	0.927632	1.078013	1.364002	1.470412	0.1797589	0.2550904	9
52	0.680295	0.732939	0.928174	1.077384	1.364370	1.469951	0.1800093	0.2547526	8
53	0.680508	0.732741	0.928715	1.076756	1.364739	1.469491	0.1802601	0.2544151	7
54	0.680721	0.732543	0.929257	1.076128	1.365108	1.469031	0.1805111	0.2540781	6
55	0.680934	0.732345	0.929800	1.075501	1.365477	1.468571	0.1807624	0.2537414	5
56	0.681147	0.732147	0.930342	1.074873	1.365846	1.468112	0.1810141	0.2534051	4
57	0.681360	0.731949	0.930885	1.074247	1.366216	1.467653	0.1812660	0.2530693	3
58	0.681573	0.731750	0.931428	1.073620	1.366586	1.467195	0.1815182	0.2527338	2
59	0.681786	0.731552	0.931971	1.072994	1.366957	1.466737	0.1817707	0.2523987	1
60	0.681998	0.731354	0.932515	1.072369	1.367327	1.466279	0.1820235	0.2520640	0
Minutes	Cosine	Sine	Cotangent	Tangent	Cosecant	Secant	Read Down	47°–48° Involute	Minutes

↑ **132° or 312°** **47° or 227°** ↑

↓ 43° or 223° Trigonometric and Involute Functions 136° or 316° ↓

Minutes	Sine	Cosine	Tangent	Cotangent	Secant	Cosecant	Involute 43°–44°	Read Up	Minutes
0	0.681998	0.731354	0.932515	1.072369	1.367327	1.466279	0.1820235	0.2520640	60
1	0.682211	0.731155	0.933059	1.071744	1.367699	1.465822	0.1822766	0.2517296	59
2	0.682424	0.730957	0.933603	1.071119	1.368070	1.465365	0.1825300	0.2513957	58
3	0.682636	0.730758	0.934148	1.070494	1.368442	1.464909	0.1827837	0.2510622	57
4	0.682849	0.730560	0.934693	1.069870	1.368814	1.464453	0.1830377	0.2507290	56
5	0.683061	0.730361	0.935238	1.069247	1.369186	1.463997	0.1832920	0.2503963	55
6	0.683274	0.730162	0.935783	1.068623	1.369559	1.463542	0.1835465	0.2500639	54
7	0.683486	0.729963	0.936329	1.068000	1.369932	1.463087	0.1838014	0.2497319	53
8	0.683698	0.729765	0.936875	1.067378	1.370305	1.462633	0.1840566	0.2494003	52
9	0.683911	0.729566	0.937422	1.066756	1.370678	1.462179	0.1843121	0.2490691	51
10	0.684123	0.729367	0.937968	1.066134	1.371052	1.461726	0.1845678	0.2487383	50
11	0.684335	0.729168	0.938515	1.065513	1.371427	1.461273	0.1848239	0.2484078	49
12	0.684547	0.728969	0.939063	1.064892	1.371801	1.460820	0.1850803	0.2480778	48
13	0.684759	0.728769	0.939610	1.064271	1.372176	1.460368	0.1853369	0.2477481	47
14	0.684971	0.728570	0.940158	1.063651	1.372551	1.459916	0.1855939	0.2474188	46
15	0.685183	0.728371	0.940706	1.063031	1.372927	1.459464	0.1858512	0.2470899	45
16	0.685395	0.728172	0.941255	1.062412	1.373303	1.459013	0.1861087	0.2467614	44
17	0.685607	0.727972	0.941803	1.061793	1.373679	1.458562	0.1863666	0.2464332	43
18	0.685818	0.727773	0.942352	1.061174	1.374055	1.458112	0.1866248	0.2461055	42
19	0.686030	0.727573	0.942902	1.060556	1.374432	1.457662	0.1868832	0.2457781	41
20	0.686242	0.727374	0.943451	1.059938	1.374809	1.457213	0.1871420	0.2454511	40
21	0.686453	0.727174	0.944001	1.059321	1.375187	1.456764	0.1874011	0.2451245	39
22	0.686665	0.726974	0.944552	1.058703	1.375564	1.456315	0.1876604	0.2447982	38
23	0.686876	0.726775	0.945102	1.058087	1.375943	1.455867	0.1879201	0.2444724	37
24	0.687088	0.726575	0.945653	1.057470	1.376321	1.455419	0.1881801	0.2441469	36
25	0.687299	0.726375	0.946204	1.056854	1.376700	1.454971	0.1884404	0.2438218	35
26	0.687510	0.726175	0.946756	1.056239	1.377079	1.454524	0.1887010	0.2434971	34
27	0.687721	0.725975	0.947307	1.055624	1.377458	1.454077	0.1889619	0.2431728	33
28	0.687932	0.725775	0.947859	1.055009	1.377838	1.453631	0.1892230	0.2428488	32
29	0.688144	0.725575	0.948412	1.054394	1.378218	1.453185	0.1894845	0.2425252	31
30	0.688355	0.725374	0.948965	1.053780	1.378598	1.452740	0.1897463	0.2422020	30
31	0.688566	0.725174	0.949518	1.053166	1.378979	1.452295	0.1900084	0.2418792	29
32	0.688776	0.724974	0.950071	1.052553	1.379360	1.451850	0.1902709	0.2415567	28
33	0.688987	0.724773	0.950624	1.051940	1.379742	1.451406	0.1905336	0.2412347	27
34	0.689198	0.724573	0.951178	1.051328	1.380123	1.450962	0.1907966	0.2409130	26
35	0.689409	0.724372	0.951733	1.050715	1.380505	1.450518	0.1910599	0.2405916	25
36	0.689620	0.724172	0.952287	1.050103	1.380888	1.450075	0.1913236	0.2402707	24
37	0.689830	0.723971	0.952842	1.049492	1.381270	1.449632	0.1915875	0.2399501	23
38	0.690041	0.723771	0.953397	1.048881	1.381653	1.449190	0.1918518	0.2396299	22
39	0.690251	0.723570	0.953953	1.048270	1.382037	1.448748	0.1921163	0.2393101	21
40	0.690462	0.723369	0.954508	1.047660	1.382420	1.448306	0.1923812	0.2389906	20
41	0.690672	0.723168	0.955064	1.047050	1.382804	1.447865	0.1926464	0.2386715	19
42	0.690882	0.722967	0.955621	1.046440	1.383189	1.447424	0.1929119	0.2383528	18
43	0.691093	0.722766	0.956177	1.045831	1.383573	1.446984	0.1931777	0.2380344	17
44	0.691303	0.722565	0.956734	1.045222	1.383958	1.446544	0.1934438	0.2377165	16
45	0.691513	0.722364	0.957292	1.044614	1.384344	1.446104	0.1937102	0.2373988	15
46	0.691723	0.722163	0.957849	1.044006	1.384729	1.445665	0.1939769	0.2370816	14
47	0.691933	0.721962	0.958407	1.043398	1.385115	1.445226	0.1942440	0.2367647	13
48	0.692143	0.721760	0.958966	1.042790	1.385502	1.444788	0.1945113	0.2364482	12
49	0.692353	0.721559	0.959524	1.042183	1.385888	1.444350	0.1947790	0.2361321	11
50	0.692563	0.721357	0.960083	1.041577	1.386275	1.443912	0.1950469	0.2358163	10
51	0.692773	0.721156	0.960642	1.040970	1.386663	1.443475	0.1953152	0.2355010	9
52	0.692983	0.720954	0.961202	1.040364	1.387050	1.443038	0.1955838	0.2351859	8
53	0.693192	0.720753	0.961761	1.039759	1.387438	1.442601	0.1958527	0.2348713	7
54	0.693402	0.720551	0.962322	1.039154	1.387827	1.442165	0.1961220	0.2345570	6
55	0.693611	0.720349	0.962882	1.038549	1.388215	1.441729	0.1963915	0.2342430	5
56	0.693821	0.720148	0.963443	1.037944	1.388604	1.441294	0.1966613	0.2339295	4
57	0.694030	0.719946	0.964004	1.037340	1.388994	1.440859	0.1969315	0.2336163	3
58	0.694240	0.719744	0.964565	1.036737	1.389383	1.440425	0.1972020	0.2333034	2
59	0.694449	0.719542	0.965127	1.036133	1.389773	1.439990	0.1974728	0.2329910	1
60	0.694658	0.719340	0.965689	1.035530	1.390164	1.439557	0.1977439	0.2326789	0
Minutes	Cosine	Sine	Cotangent	Tangent	Cosecant	Secant	Read Down	46°–47° Involute	Minutes

↑ 133° or 313° 46° or 226° ↑

↓ 44° or 224° **Trigonometric and Involute Functions** 135° or 315° ↓

Minutes	Sine	Cosine	Tangent	Cotangent	Secant	Cosecant	Involute 44°–45°	Read Up	Minutes
0	0.694658	0.719340	0.965689	1.035530	1.390164	1.439557	0.1977439	0.2326789	60
1	0.694868	0.719138	0.966251	1.034928	1.390554	1.439123	0.1980153	0.2323671	59
2	0.695077	0.718936	0.966814	1.034325	1.390945	1.438690	0.1982871	0.2320557	58
3	0.695286	0.718733	0.967377	1.033724	1.391337	1.438257	0.1985591	0.2317447	57
4	0.695495	0.718531	0.967940	1.033122	1.391728	1.437825	0.1988315	0.2314341	56
5	0.695704	0.718329	0.968504	1.032521	1.392120	1.437393	0.1991042	0.2311238	55
6	0.695913	0.718126	0.969067	1.031920	1.392513	1.436962	0.1993772	0.2308138	54
7	0.696122	0.717924	0.969632	1.031319	1.392905	1.436531	0.1996505	0.2305042	53
8	0.696330	0.717721	0.970196	1.030719	1.393298	1.436100	0.1999242	0.2301950	52
9	0.696539	0.717519	0.970761	1.030120	1.393692	1.435669	0.2001982	0.2298862	51
10	0.696748	0.717316	0.971326	1.029520	1.394086	1.435239	0.2004724	0.2295777	50
11	0.696957	0.717113	0.971892	1.028921	1.394480	1.434810	0.2007471	0.2292695	49
12	0.697165	0.716911	0.972458	1.028323	1.394874	1.434380	0.2010220	0.2289618	48
13	0.697374	0.716708	0.973024	1.027724	1.395269	1.433952	0.2012972	0.2286543	47
14	0.697582	0.716505	0.973590	1.027126	1.395664	1.433523	0.2015728	0.2283473	46
15	0.697790	0.716302	0.974157	1.026529	1.396059	1.433095	0.2018487	0.2280406	45
16	0.697999	0.716099	0.974724	1.025931	1.396455	1.432667	0.2021249	0.2277342	44
17	0.698207	0.715896	0.975291	1.025335	1.396851	1.432240	0.2024014	0.2274282	43
18	0.698415	0.715693	0.975859	1.024738	1.397248	1.431813	0.2026783	0.2271226	42
19	0.698623	0.715490	0.976427	1.024142	1.397644	1.431386	0.2029554	0.2268173	41
20	0.698832	0.715286	0.976996	1.023546	1.398042	1.430960	0.2032329	0.2265124	40
21	0.699040	0.715083	0.977564	1.022951	1.398439	1.430534	0.2035108	0.2262078	39
22	0.699248	0.714880	0.978133	1.022356	1.398837	1.430109	0.2037889	0.2259036	38
23	0.699455	0.714676	0.978703	1.021761	1.399235	1.429684	0.2040674	0.2255997	37
24	0.699663	0.714473	0.979272	1.021166	1.399634	1.429259	0.2043462	0.2252962	36
25	0.699871	0.714269	0.979842	1.020572	1.400033	1.428834	0.2046253	0.2249931	35
26	0.700079	0.714066	0.980413	1.019979	1.400432	1.428410	0.2049047	0.2246903	34
27	0.700287	0.713862	0.980983	1.019385	1.400831	1.427987	0.2051845	0.2243878	33
28	0.700494	0.713658	0.981554	1.018792	1.401231	1.427564	0.2054646	0.2240857	32
29	0.700702	0.713454	0.982126	1.018200	1.401631	1.427141	0.2057450	0.2237840	31
30	0.700909	0.713250	0.982697	1.017607	1.402032	1.426718	0.2060257	0.2234826	30
31	0.701117	0.713047	0.983269	1.017015	1.402433	1.426296	0.2063068	0.2231815	29
32	0.701324	0.712843	0.983842	1.016424	1.402834	1.425874	0.2065882	0.2228808	28
33	0.701531	0.712639	0.984414	1.015833	1.403236	1.425453	0.2068699	0.2225805	27
34	0.701739	0.712434	0.984987	1.015242	1.403638	1.425032	0.2071520	0.2222805	26
35	0.701946	0.712230	0.985560	1.014651	1.404040	1.424611	0.2074344	0.2219808	25
36	0.702153	0.712026	0.986134	1.014061	1.404443	1.424191	0.2077171	0.2216815	24
37	0.702360	0.711822	0.986708	1.013471	1.404846	1.423771	0.2080001	0.2213826	23
38	0.702567	0.711617	0.987282	1.012882	1.405249	1.423351	0.2082835	0.2210840	22
39	0.702774	0.711413	0.987857	1.012293	1.405653	1.422932	0.2085672	0.2207857	21
40	0.702981	0.711209	0.988432	1.011704	1.406057	1.422513	0.2088512	0.2204878	20
41	0.703188	0.711004	0.989007	1.011115	1.406462	1.422095	0.2091356	0.2201903	19
42	0.703395	0.710799	0.989582	1.010527	1.406867	1.421677	0.2094203	0.2198930	18
43	0.703601	0.710595	0.990158	1.009939	1.407272	1.421259	0.2097053	0.2195962	17
44	0.703808	0.710390	0.990735	1.009352	1.407677	1.420842	0.2099907	0.2192996	16
45	0.704015	0.710185	0.991311	1.008765	1.408083	1.420425	0.2102764	0.2190035	15
46	0.704221	0.709981	0.991888	1.008178	1.408489	1.420008	0.2105624	0.2187076	14
47	0.704428	0.709776	0.992465	1.007592	1.408896	1.419592	0.2108487	0.2184121	13
48	0.704634	0.709571	0.993043	1.007006	1.409303	1.419176	0.2111354	0.2181170	12
49	0.704841	0.709366	0.993621	1.006420	1.409710	1.418761	0.2114225	0.2178222	11
50	0.705047	0.709161	0.994199	1.005835	1.410118	1.418345	0.2117098	0.2175277	10
51	0.705253	0.708956	0.994778	1.005250	1.410526	1.417931	0.2119975	0.2172336	9
52	0.705459	0.708750	0.995357	1.004665	1.410934	1.417516	0.2122855	0.2169398	8
53	0.705665	0.708545	0.995936	1.004081	1.411343	1.417102	0.2125739	0.2166464	7
54	0.705872	0.708340	0.996515	1.003497	1.411752	1.416688	0.2128626	0.2163533	6
55	0.706078	0.708134	0.997095	1.002913	1.412161	1.416275	0.2131516	0.2160605	5
56	0.706284	0.707929	0.997676	1.002330	1.412571	1.415862	0.2134410	0.2157681	4
57	0.706489	0.707724	0.998256	1.001747	1.412981	1.415449	0.2137307	0.2154760	3
58	0.706695	0.707518	0.998837	1.001164	1.413392	1.415037	0.2140207	0.2151843	2
59	0.706901	0.707312	0.999418	1.000582	1.413802	1.414625	0.2143111	0.2148929	1
60	0.707107	0.707107	1.000000	1.000000	1.414214	1.414214	0.2146018	0.2146018	0
Minutes	Cosine	Sine	Cotangent	Tangent	Cosecant	Secant	Read Down	45°–46° Involute	Minutes

↑ 134° or 314° 45° or 225° ↑

Table of Common Logarithms

	0	1	2	3	4	5	6	7	8	9
10	000000	004321	008600	012837	017033	021189	025306	029384	033424	037426
11	041393	045323	049218	053078	056905	060698	064458	068186	071882	075547
12	079181	082785	086360	089905	093422	096910	100371	103804	107210	110590
13	113943	117271	120574	123852	127105	130334	133539	136721	139879	143015
14	146128	149219	152288	155336	158362	161368	164353	167317	170262	173186
15	176091	178977	181844	184691	187521	190332	193125	195900	198657	201397
16	204120	206826	209515	212188	214844	217484	220108	222716	225309	227887
17	230449	232996	235528	238046	240549	243038	245513	247973	250420	252853
18	255273	257679	260071	262451	264818	267172	269513	271842	274158	276462
19	278754	281033	283301	285557	287802	290035	292256	294466	296665	298853
20	301030	303196	305351	307496	309630	311754	313867	315970	318063	320146
21	322219	324282	326336	328380	330414	332438	334454	336460	338456	340444
22	342423	344392	346353	348305	350248	352183	354108	356026	357935	359835
23	361728	363612	365488	367356	369216	371068	372912	374748	376577	378398
24	380211	382017	383815	385606	387390	389166	390935	392697	394452	396199
25	397940	399674	401401	403121	404834	406540	408240	409933	411620	413300
26	414973	416641	418301	419956	421604	423246	424882	426511	428135	429752
27	431364	432969	434569	436163	437751	439333	440909	442480	444045	445604
28	447158	448706	450249	451786	453318	454845	456366	457882	459392	460898
29	462398	463893	465383	466868	468347	469822	471292	472756	474216	475671
30	477121	478566	480007	481443	482874	484300	485721	487138	488551	489958
31	491362	492760	494155	495544	496930	498311	499687	501059	502427	503791
32	505150	506505	507856	509203	510545	511883	513218	514548	515874	517196
33	518514	519828	521138	522444	523746	525045	526339	527630	528917	530200
34	531479	532754	534026	535294	536558	537819	539076	540329	541579	542825
35	544068	545307	546543	547775	549003	550228	551450	552668	553883	555094
36	556303	557507	558709	559907	561101	562293	563481	564666	565848	567026
37	568202	569374	570543	571709	572872	574031	575188	576341	577492	578639
38	579784	580925	582063	583199	584331	585461	586587	587711	588832	589950
39	591065	592177	593286	594393	595496	596597	597695	598791	599883	600973
40	602060	603144	604226	605305	606381	607455	608526	609594	610660	611723
41	612784	613842	614897	615950	617000	618048	619093	620136	621176	622214
42	623249	624282	625312	626340	627366	628389	629410	630428	631444	632457
43	633468	634477	635484	636488	637490	638489	639486	640481	641474	642465
44	643453	644439	645422	646404	647383	648360	649335	650308	651278	652246
45	653213	654177	655138	656098	657056	658011	658965	659916	660865	661813
46	662758	663701	664642	665581	666518	667453	668386	669317	670246	671173
47	672098	673021	673942	674861	675778	676694	677607	678518	679428	680336
48	681241	682145	683047	683947	684845	685742	686636	687529	688420	689309
49	690196	691081	691965	692847	693727	694605	695482	696356	697229	698101
50	698970	699838	700704	701568	702431	703291	704151	705008	705864	706718
51	707570	708421	709270	710117	710963	711807	712650	713491	714330	715167
52	716003	716838	717671	718502	719331	720159	720986	721811	722634	723456
53	724276	725095	725912	726727	727541	728354	729165	729974	730782	731589
54	732394	733197	733999	734800	735599	736397	737193	737987	738781	739572
55	740363	741152	741939	742725	743510	744293	745075	745855	746634	747412
56	748188	748963	749736	750508	751279	752048	752816	753583	754348	755112
57	755875	756636	757396	758155	758912	759668	760422	761176	761928	762679
58	763428	764176	764923	765669	766413	767156	767898	768638	769377	770115
59	770852	771587	772322	773055	773786	774517	775246	775974	776701	777427

Table of Common Logarithms

	0	1	2	3	4	5	6	7	8	9
60	778151	778874	779596	780317	781037	781755	782473	783189	783904	784617
61	785330	786041	786751	787460	788168	788875	789581	790285	790988	791691
62	792392	793092	793790	794488	795185	795880	796574	797268	797960	798651
63	799341	800029	800717	801404	802089	802774	803457	804139	804821	805501
64	806180	806858	807535	808211	808886	809560	810233	810904	811575	812245
65	812913	813581	814248	814913	815578	816241	816904	817565	818226	818885
66	819544	820201	820858	821514	822168	822822	823474	824126	824776	825426
67	826075	826723	827369	828015	828660	829304	829947	830589	831230	831870
68	832509	833147	833784	834421	835056	835691	836324	836957	837588	838219
69	838849	839478	840106	840733	841359	841985	842609	843233	843855	844477
70	845098	845718	846337	846955	847573	848189	848805	849419	850033	850646
71	851258	851870	852480	853090	853698	854306	854913	855519	856124	856729
72	857332	857935	858537	859138	859739	860338	860937	861534	862131	862728
73	863323	863917	864511	865104	865696	866287	866878	867467	868056	868644
74	869232	869818	870404	870989	871573	872156	872739	873321	873902	874482
75	875061	875640	876218	876795	877371	877947	878522	879096	879669	880242
76	880814	881385	881955	882525	883093	883661	884229	884795	885361	885926
77	886491	887054	887617	888179	888741	889302	889862	890421	890980	891537
78	892095	892651	893207	893762	894316	894870	895423	895975	896526	897077
79	897627	898176	898725	899273	899821	900367	900913	901458	902003	902547
80	903090	903633	904174	904716	905256	905796	906335	906874	907411	907949
81	908485	909021	909556	910091	910624	911158	911690	912222	912753	913284
82	913814	914343	914872	915400	915927	916454	916980	917506	918030	918555
83	919078	919601	920123	920645	921166	921686	922206	922725	923244	923762
84	924279	924796	925312	925828	926342	926857	927370	927883	928396	928908
85	929419	929930	930440	930949	931458	931966	932474	932981	933487	933993
86	934498	935003	935507	936011	936514	937016	937518	938019	938520	939020
87	939519	940018	940516	941014	941511	942008	942504	943000	943495	943989
88	944483	944976	945469	945961	946452	946943	947434	947924	948413	948902
89	949390	949878	950365	950851	951338	951823	952308	952792	953276	953760
90	954243	954725	955207	955688	956168	956649	957128	957607	958086	958564
91	959041	959518	959995	960471	960946	961421	961895	962369	962843	963316
92	963788	964260	964731	965202	965672	966142	966611	967080	967548	968016
93	968483	968950	969416	969882	970347	970812	971276	971740	972203	972666
94	973128	973590	974051	974512	974972	975432	975891	976350	976808	977266
95	977724	978181	978637	979093	979548	980003	980458	980912	981366	981819
96	982271	982723	983175	983626	984077	984527	984977	985426	985875	986324
97	986772	987219	987666	988113	988559	989005	989450	989895	990339	990783
98	991226	991669	992111	992554	992995	993436	993877	994317	994757	995196
99	995635	996074	996512	996949	997386	997823	998259	998695	999131	999565
100	000000	000434	000868	001301	001734	002166	002598	003029	003461	003891
101	004321	004751	005181	005609	006038	006466	006894	007321	007748	008174
102	008600	009026	009451	009876	010300	010724	011147	011570	011993	012415
103	012837	013259	013680	014100	014521	014940	015360	015779	016197	016616
104	017033	017451	017868	018284	018700	019116	019532	019947	020361	020775
105	021189	021603	022016	022428	022841	023252	023664	024075	024486	024896
106	025306	025715	026125	026533	026942	027350	027757	028164	028571	028978
107	029384	029789	030195	030600	031004	031408	031812	032216	032619	033021
108	033424	033826	034227	034628	035029	035430	035830	036230	036629	037028
109	037426	037825	038223	038620	039017	039414	039811	040207	040602	040998

Index

G

U

V

W

Y

Z